# The biology of
# epithelial cell populations

## Volume 1

# The biology of epithelial cell populations

## Volume 1

**NICHOLAS WRIGHT**, M.A., M.D., Ph.D., M.R.C.Path.

*Professor of Histopathology,*
*Royal Postgraduate Medical School,*
*University of London*

and

**MALCOLM ALISON**, B.Sc., Ph.D.

*Lecturer in Experimental Pathology,*
*Royal Postgraduate Medical School,*
*University of London*

CLARENDON PRESS · OXFORD
1984

*Oxford University Press, Walton Street, Oxford OX2 6DP*

*London New York Toronto*
*Delhi Bombay Calcutta Madras Karachi*
*Kuala Lumpur Singapore Hong Kong Tokyo*
*Nairobi Dar es Salaam Cape Town*
*Melbourne Auckland*

*and associates in*
*Beirut Berlin Ibadan Mexico City Nicosia*

*Oxford is a trade mark of Oxford University Press*

*Published in the United States*
*by Oxford University Press, New York*



*British Library Cataloguing in Publication Data*
*Wright, Nicholas A.*
*The biology of epithelial cell populations.*
*Vol. 1*
*1. Epithelium    2. Mammals—Cytology*
*3. Exfoliative cytology*
*I. Title    II. Alison, Malcolm*
*599.08'7    QP88.4*
*ISBN 0-19-857558-0*

*Library of Congress Cataloging in Publication Data*
*Wright, Nicholas A.*
*The biology of epithelial cell populations.*
*Bibliography: v. 1, p.*
*Includes index.*
*1. Epithelium—Collected works.    2. Cytology—*
*Collected works.    I. Alison, Malcolm.    II. Title.*
*QP88.4.W75 1984    591.1'858    84-764*
*ISBN 0-19-857558-0 (v. 1)*

*Typeset by Hope Services, Abingdon, Oxon*
*Printed in Great Britain by*
*St Edmundsbury Press, Bury St Edmunds, Suffolk*

# Preface

Though by now it has become almost platitudinous to state, the purely histo-morphological avenue to the understanding of tissue organization is limited, not only by its essentially descriptive approach, but also by the intrinsically static nature of its observational structure. Into this breach stepped the study of *cell population kinetics*, which added the dimension of *time* to conventional histology, and by a logical intermarriage with the measurement of population *size*, allowed the measurement of *flux* parameters between tissue compartments. However, although such experiments have led, in many instances, to quantitative tissue descriptions of considerable detail, the approach remained descriptive. Latterly the field has expanded to encompass the more *operational* aspects of tissue organization, posing questions about the functional role of cells and cell compartments in the maintenance of tissue homeostasis, relating such consider-ations to the *control* of growth and population size.

Nowhere has this evolution been more apparent than in the epithelial tissue systems of the mammalian body; from the highly detailed morphological studies of the 1950s, quintessentially descriptive and replete with ultrastructural minutiae; through the quantitative definitions of the 1960s and early 1970s, made possible by the innovations of labelled DNA precursors, autoradiography, and flow cytometry; to the design of techniques for tracing cell lineages, for measuring clonogenic capacity, and to the search for the elusive stem cell, which have been the concerns of the past few years. Contemporaneous with these developments has been the application of the acquired knowledge and techniques to the understanding of tissue responses such as regeneration and hyperplasia, of the growth processes which accompany pathological states, and which, in some instances, can explain the histogenesis of some familiar histopathological patterns, and obviously to define the role of cell proliferation in carcinogenesis and the histogenesis of tumours; several of the major hypotheses of growth control, for example the negative feedback hypothesis and its offspring, the enigmatic chalone hypothesis, have emanated from studies in epithelia.

There are thus several reasons for presenting this book. Perhaps a major justification is our feeling that *epithelia are important*; there have been several attempts to review and summarize concepts in haemopoietic and tumour cell populations, but not in epithelia, though perusal of the necessarily selected bibliography of this volume will show that, while there has been no shortage of analysis, this is the first work of *synthesis* in epithelial cell populations. This work is directed primarily at those readers interested in the *organization* of proliferating and differentiating tissue systems, in terms of their contained cell lineages; we also believe that there are many lessons in epithelia for those interested in growth control, and consider that epithelia are, in many situations, ideal models upon which to test general ideas and concepts in growth control; in this field, there has been a major information explosion, but all experiments in growth control must perforce employ a measurement of proliferative rate, and much effort has, in our opinion, been negated or obscured by the choice of inappropriate methodology. This book has much to say on this point; methodology is perhaps nowhere so important as in the assessment of cell proliferation, and we make no apology for emphasizing this point. Conclusions in this field are heavily dependent upon techniques.

We also hope to attract those interested in the mechanisms of tissue response to induced change, be it physiological or pathological, mechanisms of hyperplasia and atrophy, regeneration and the responses that follow cell death induced by cytotoxic insult, circadian variation and hormone action, and the interaction of epithelial cell populations with carcinogens.

So much for the reasons for writing this book. We hope that it will promote the study of epithelial cell populations, provide an easily-assimilated, up-to-date synthesis of current thinking, point the direction of the most important new research areas, and also serve as a source of advice in the practical approach to problems in epithelial cell proliferation and its control, for beginners and *aficionados* alike, for we have not hesitated to be critical and to say where we consider that techniques are non-robust or data overinterpreted.

Sir Gavin de Beer, in offering his volume on the vertebrate skull, hoped that its publication would not stifle research in that topic over the succeeding years; we have no such fear. We are beginning to understand how epithelial cell populations work, and we believe that the next ten years will see signal advances in this understanding. We would be very content if our efforts, in any way, engendered this.

<table>
<tr><td>Oxford and London</td><td align="right">N. A. W.</td></tr>
<tr><td>June 1983</td><td align="right">M. R. A.</td></tr>
</table>

# Acknowledgements

It is a pleasure to record our thanks to the following friends and colleagues who have been most helpful in making available illustrative material to us: Dr David Appleton, Professor John Azzopardi, Dr Peter Barnes, Dr W. W. L. Chang, Professor Rufus Clarke, Dr Z. Darzynkiewicz, Dr Paul Dyson, Dr T. Fujita, Dr Seymour Gelfant, Dr Robert Goodlad, Dr Niels Hartmann, Dr T. Hattori, Dr William Hume, Dr P. Jeffreys, Professor C. P. Leblond, Dr Michael Marsh, Dr David Menton, Dr Ulla Møller, Dr Adrian Morley, Dr Bruce Ponder, Dr G. W. Richter, Dr D. G. de Rooij, Professor P. Sengel, Dr W. Simpson, Susan Van Noorden, and Dr Alex Watson.

In particular we should like to thank Dr Julia Polak, Dr Christopher Potten, and Dr Ian Varndell. Mrs Jane Fallows drew many diagrams with great skill, and Mr Bill Hinks gave invaluable assistance with photomicrography.

We should also like to thank the following for permission to reproduce previously published illustrations: Academic Press (London), Fig. 7.5; Cambridge University Press, Figs. 5.1, 16.21, 16.23, 18.10, 18.11; Cancer Research, Figs. 2.8, 2.9, 2.10; Churchill Livingstone (Edinburgh), Figs. 12.1, 18.18, 19.14, 19.20, 21.10; Company of Biologists, Fig. 10.5; Pergamon Press (Oxford), Fig. 10.23; Springer Verlag (Heidelberg), Figs. 15.1C, 16.1, 19.25; Wistar Press, Figs. 8.2A, 8.8A,B, 8.9A,B.

There are also numerous individuals who have contributed in no small way, through numerous discussions, to the concepts and information presented in this book; among these we should mention David Appleton, Rufus Clarke, Niels Hartmann, Adrian Morley, and Christopher Potten; errors of omission and commission, however, are ours alone.

We must thank Elizabeth Madigan and Barbara Gill for typing a difficult text, the staff of Oxford University Press for friendly encouragement during a prolonged gestation period, and our respective families for very necessary forebearance.

Our personal research for many years has been supported by the Cancer Research Campaign.

# Contents

# Contents of Volume 2

# Abbreviations

| | |
|---|---|
| CCPR | The crypt cell production rate, measured in cells produced per crypt per hour. |
| FCM | Flow cytometry. |
| FLM | The fraction of labelled mitoses. |
| $G_0$ | A phase of post-mitotic proliferative quiescence. |
| $G_1$ | The phase between mitosis and the beginning of DNA synthesis. |
| $G_2$ | The phase between the completion of DNA synthesis and the beginning of mitosis. |
| $I_L$ | The continuous [$^3$H]-TdR labelling index; the proportion of cells labelled after prolonged exposure to [$^3$H]-TdR. |
| $I_M$ | The mitotic index; the proportion of cells in mitosis. |
| $I_{Meta}$ | The metaphase index; the proportion of cells in metaphase. |
| $I_P$ | The growth fraction; the ratio of proliferating to non-proliferating cells. |
| $I_S$ | The flash [$^3$H]-TdR labelling index; the proportion of cells labelled after a brief exposure (usually one hour) of the tissue to [$^3$H]-TdR. |
| $I_{S\,expt}$ | The experimental (observed) flash [$^3$H]-TdR labelling index. |
| $I_{S\,theor}$ | The theoretical flash [$^3$H]-TdR labelling index, equivalent to $I_{S\,expt}$ if all cells in the population are proliferating. |
| $k_B$ | The birth rate of new cells. |
| $k_G$ | The overall growth rate of the population (equivalent to the birth rate if no cell loss). |
| $k_L$ | The rate of cell loss, found by subtraction of $k_G$ from $k_B$. |
| $M$ | The phase of mitosis. |
| $P$ cells | Those cells born into the proliferative compartment. |
| $\Phi$ (phi) | The cell loss factor; the ratio of the cell loss rate to the cell birth rate. |
| $Q$ cells | Those cells born into the non-proliferative (quiescent) compartment. |
| $r_M$ | The rate at which cells enter mitosis. |
| $r_S$ | The rate at which cells enter DNA synthesis. |
| $S$ | The phase of DNA synthesis. |
| $T_C$ | The cell cycle time; the time between the completion of mitosis and the next mitosis in one or both of the daughter cells. |
| $T_{C(a)}$ | The apparent cell cycle time; the time taken to replace all the cells in the population (equals $T_C$ when the growth fraction is unity). |
| $t_D$ | The population doubling time, ideally related to a doubling of cell number, but more usually to a doubling of weight or volume. |

| | |
|---|---|
| $t_{G1}$ | The duration of the $G_1$ phase. |
| $t_{G2}$ | The duration of the $G_2$ phase. |
| $t_M$ | The duration of the mitotic phase. |
| $t_{PD}$ | The potential doubling time, the expected time taken for the cell population to double in number based upon the rate of cell production (used in the context of exponentially growing populations). Equals $t_D$ if there is no cell loss. |
| $t_S$ | The duration of the DNA synthesis phase. |
| $T_T$ | The transit time, usually within the context of a compartment. |
| $T$ | The turnover time; the time taken to replace all the cells in the population. Equivalent in duration to $t_{PD}$, but a more appropriate term in situations where the population size remains constant. |
| $t_2$ | $t_{G2} + \frac{1}{2} t_M$. |

PART 1

# The organization of epithelial cell populations

# 1 Introduction

## 1.1 Why epithelia?

Why do we want to enveigle the reader into reading a book devoted entirely to epithelia? Why do we think that epithelia are so important?

There are several answers to this question. From a purely practical viewpoint, most malignant tumours arise from epithelial cells; in 1973, out of the 181 000 cancer deaths registered in the United Kingdom, no less than 166 000 arose in epithelial tissues, mainly lung, gastrointestinal tract, breast, testis, ovary, liver, pancreas, endocrine organs, and skin. Putting the statistics another way, some 90 per cent of all cancers in Britain arise from epithelial cells (20 per cent from 'sex-associated' epithelia, i.e. breast, endometrium, prostate, etc., and 70 per cent from surface epithelia, the gut, bladder, and lung). Cairns (1975) has used the Danish statistics (Clemmesen 1965, 1969, 1974), which separate each organ site into tumours originating from epithelial and connective tissue elements, to emphasize the importance of epithelial tumours (see Table 1.1). In context, only

Table 1.1. Cancer incidence in Denmark, 1943–1967. (After Cairns (1975))

| Types | Common sites | Totals | % |
|---|---|---|---|
| Carcinomas | | | |
|     External epithelia | skin, colon, lung, stomach, cervix | 168 591 | 56 |
|     Internal epithelia | breast, prostate, ovary, bladder, pancreas | 110 182 | 36 |
| Sarcoma and leukaemias | | 23 801 | 8 |
| | Totals | 302 574 | 100 |

some 4000 deaths in the United Kingdom every year are due to leukaemia, and only about 4500 lymphoma deaths occur. Consequently the understanding of the control of cell proliferation in epithelial tissues becomes very important. Moreover, we shall argue quite strongly (see Chapter 2, section 9) that carcinogenesis is specifically a disease of *stem cells*; if this is true, the identity, life-history, and fate of epithelial stem cells is crucial to any theory of carcinogenesis we may formulate. In this respect, there appears to be general agreement that the induction of cell proliferation is essential, both for the initiation and promotion of carcinogenesis in many tissues, but in this field, cancer research workers have not availed themselves of the advances in cell proliferation methodology which have occurred in recent years, and many continue to use simplistic and potentially misleading techniques. The reason for this may lie in the fact that, as yet, no real synthesis of the methodology used in epithelia, its advantages and disadvantages, and of the results which are forthcoming, has been attempted.

Thus if we wish fully to appreciate the cellular events which accompany carcinogenesis, a thorough understanding, not only of the cell kinetic events which occur during carcinogenesis, but also of the kinetic organization of normal epithelia, becomes mandatory.

A second good reason for considering epithelial cell kinetics stems from a desire to know how tissues are *organized*. Descriptive histology can only go so far in the definition of tissues, and, although it has already become a cliché to say so, a dynamic approach to epithelial structure is only afforded through the study of epithelial cell kinetics. This is because all aspects of the life-history of all cells are theoretically quantifiable; we should be able to give stem cells, proliferating cells, differentiating, migrating, and dying cells dimensions, in terms of *identity*, and therefore of *size*, or numbers of cells involved, *flux*, the rate at which they perform life history functions, and *time*, or how long they take to perform these respective processes; similarly, the *origin* of individual cell types in epithelia, involving both the qualitative and quantitative aspects of *cell lineage*, is also encompassed by epithelial cell kinetics. A full understanding of kinetic organization also implies a knowledge of which cells are responsible for the origin of cytogenesis in normal epithelium, the *functional stem cells* (see Chapter 2, section 9), and the definition of the possible differentiation pathways. We are also concerned with those cells which are capable of repopulation after epithelial damage, the so-called *potential stem cells*. We should also be able to state the relationship between these several cell populations: whether a sharp distinction can be made between stem cells and proliferating cells, or whether a continuum of 'stemness' exists between them; the distinction between proliferating and the various types of non-proliferating cells should be manifest, and the mechanisms responsible for the transition defined.

If we can understand kinetic organization, we can progress almost immediately to its *control*; at its simplest level, this means control of cell population size, and hence the control of cell proliferation. We believe that epithelial cell systems are ideal for investigating the control of cell proliferation, and this is because, in many tissues, there is easy morphological distinction between *proliferative* and *functional* cells. For example, in the small intestine, the distinction between functional villus cells and proliferative crypt cells is readily made, and methods are available for the measurement of size, flux and time in each of these (see Chapter 17). We would maintain that, while the study of *in vitro* systems may point to potential control mechanisms (e.g. the *transition probability hypothesis*, see Chapter 2, section 11), in the last analysis, the only critical test of the physiological usefulness of a model is its applicability to *in vivo* populations, and again we would suggest that epithelia are excellent for the testing of such theories (see Chapters 14 and 23). It is no accident that the formalization of the *negative feedback hypothesis* (Weiss and Kavanau 1957) came from observations on the regeneration of liver epithelia, and that the first attempt, however imperfect, to provide a biochemical explanation, came with the advent of the *chalone hypothesis* (Bullough and Laurence 1960*a, b*), based on work with the squamous epithelium of the epidermis. It therefore comes as no surprise that the putative growth controlling role of the cyclic nucleotides, which produce such definite results *in vitro*, has led to a sharp controversy concerning their physiological and pathological role in the same squamous epithelium (see Chapter 14, section 2). Most, if not all of our present ideas concerning the trophic (or growth controlling) action of hormones has emanated from work on epithelial systems, cf. androgen action and the prostate (Chapter 31), oestrogens and the uterus etc.; the case for epithelia as excellent models for growth control is thus easily made.

Moreover, in addition to their importance as investigative models for the study of these rather large biological questions, an understanding of epithelial cell kinetics is mandatory for the adequate appreciation of the morphogenesis and indeed pathogenesis of several non-neoplastic disease states, such as the intestinal appearances in coeliac disease and parasitic infestations (Chapter 20), of the skin in psoriasis and other hyperproliferative skin diseases, and, while our knowledge of the kinetic changes are less extensive, it is evident that tissue reactions in prostatic and endometrial hyperplasia, non-neoplastic proliferations of the breast, and in thyroid and adrenal hyperplasias, all involve disturbances in cell proliferation.

From these remarks, we hope that the biological importance of epithelia are now apparent, but we are indeed a far cry from reaching the level of understanding demanded by the above statements. It would be a mistake to imply that measurements in cell proliferation studies are easy; because of their quantitative nature, observations are often laborious, and attempts to manufacture short cuts often lead to disaster. Furthermore, while the kinetic parameters which define any renewal system are quite easily defined, in many cases it is not even possible to obtain crude estimates of some parameters, in which case recourse to mathematical models, often computer based, becomes necessary.

It is the purpose of this volume to discuss, as critically as possible, the current concepts of the organization of epithelial tissues, our knowledge of the kinetic parameters which define the system, the factors which are thought to control cell proliferation in these tissues, and the kinetic responses therein to both physiological and pathological changes.

## 1.2  The general structure of epithelia

Epithelia can reasonably be defined as those tissues which cover the free surfaces of the body; this is self-evident where tissues such as the epidermis or gut epithelia are concerned, but we can also carry on the definition to include the glandular epithelia which line the smallest surfaces in internal organs, such as the glands of the endometrium, or the follicles of the thyroid. Thus our brief is wide indeed; as well as including the great epithelial renewal systems, a consideration of epithelial cell kinetics would also include the gut-associated glands such as the liver and pancreas, endocrine organs such as the thyroid and adrenal, and endocrine target tissues such as prostate, mammary gland, uterus, and gonadal epithelia.

The word *epithelium* is actually a good guide to its general structure; it was probably first used by the Dutch anatomist Ruysch, who is said to have coined it to portray the tissues growing (Gk *thelio*) upon (Gk *epi*) another tissue, although in fact the etymology of the word may be much more complex.* Nevertheless, the usage refers to the fact that all epithelia consist of contiguous cells which rest upon a basement membrane, which is an extracellular supporting layer composed of several fibrous proteins, particularly Type IV collagen. In all instances, epithelial cells have an *apex*, and a *base*, and in many cases show both apical and basal surface specializations. Such cells are said to be *polarized*, seen perhaps most strikingly in the mature absorptive columnar cells which cover the villus in the small intestine, with its microvillus brush border, supranuclear Golgi net, and apical distribution of mitochondria (Fig. 1.1), while in epithelia which are associated with water and electrolyte transport, such as the salivary gland ducts and renal tubular epithelium, the mitochondria are basally concentrated; the bronchial and Fallopian tube epithelium show surface *cilia*. Even in stratified epithelia, such as squamous epithelium, where there is of course limited opportunity for basal and apical specialization, there is distinct polarity in the sense that the nuclei are· orientated, above the basal layer, in a plane parallel to the basement membrane (Fig. 4.1 p. 206). Very little is known about the factors which control polarity in epithelial tissues, but, in neoplastic change, as occurs for example in squamous carcinoma *in situ*, loss of polarity is an important change.

Theoretically, at least, epithelia can be derived from all three of the primary embryonic germ layers, and their functions are, naturally, related to the fact that the tissues cover and line free surfaces. These include protection (cf. the epidermis), secretion and absorption of products into and from a lumen, as in the gut and renal tubules, while the exocrine and endocrine glands exclusively secrete products into duct systems and the bloodstream respectively. The presence of epithelia on exposed surfaces makes them ideally situated for both *sensory reception* and *antigen recognition*, and epithelia thus contain specialized structures such as taste buds on the tongue, while individual cells such as the Merkel cell and Langerhans cell in stratified squamous epithelia are now thought to serve these two respective functions.

*A literal translation gives $\dot{\epsilon}\pi\acute{\iota} + \theta\tilde{\eta}\lambda\tilde{\eta}$, the skin overlying the teat or nipple (Leach, J., personal communication), which would confine the field rather too much. The *OED* attributes the first usage to Hartley (1748) Observ. Man, 1 ii, 117 'The impression can easily penetrate the soft epithelium'.

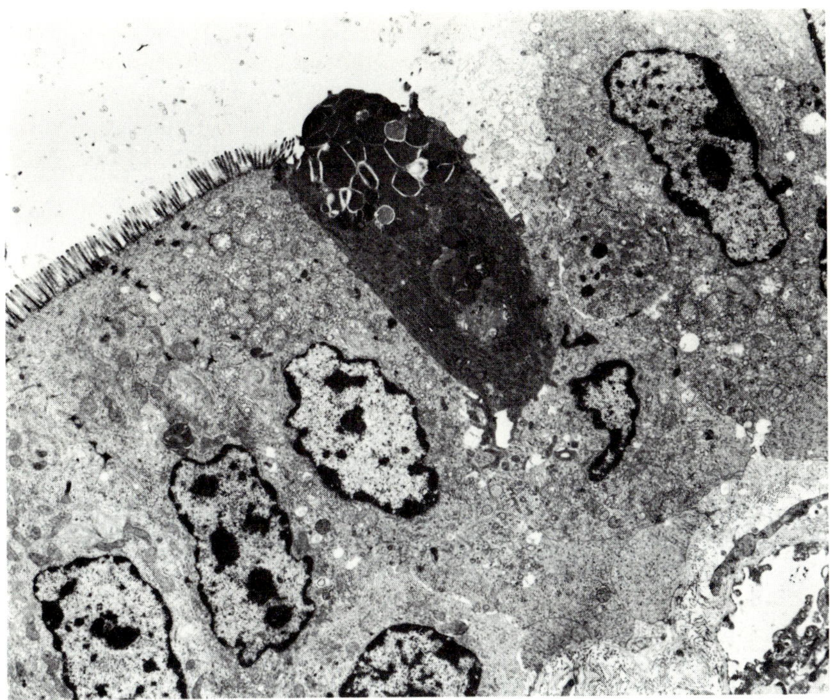

Fig. 1.1.   An electron micrograph of the villus epithelium of the rat. Several absorptive cells and a mucus-secreting goblet cell are shown. The *polarity* of epithelial cells is well depicted here: on the luminal surface of the intestine are closely packed and regularly arranged microvilli, beneath which is the endoplasmic reticulum and the Golgi complex, which is found immediately above the nucleus. Mitochondria are widely distributed in both infra- and supranuclear cytoplasm. (× 2100.)

There are at least eight different types of epithelia: an epithelium which consists of a single layer of cells is said to be *simple*, and in this group we can recognize (i) *simple squamous epithelium*, and here the word squamous comes from the L. *squama* to indicate that the cells are shaped like flat plates or scales. The cells interdigitate closely with neighbours, and are held in place by specialized attachment plates at the cell margins (see below). This arrangement is seen in the cells lining the alveoli in the lung, in the renal glomerulus, and in some renal collecting tubules (see Fig. 1.2A);  (ii) in *simple cuboidal epithelium*, the height of each component cell is approximately equivalent to its width, thus allowing more specialization and polarization, and cuboidal epithelium is extensively involved in secretion and absorption, as in much of the renal tubular system, and in the resting thyroid acinus (Fig. 1.2B); (iii) *simple columnar epithelial cells*

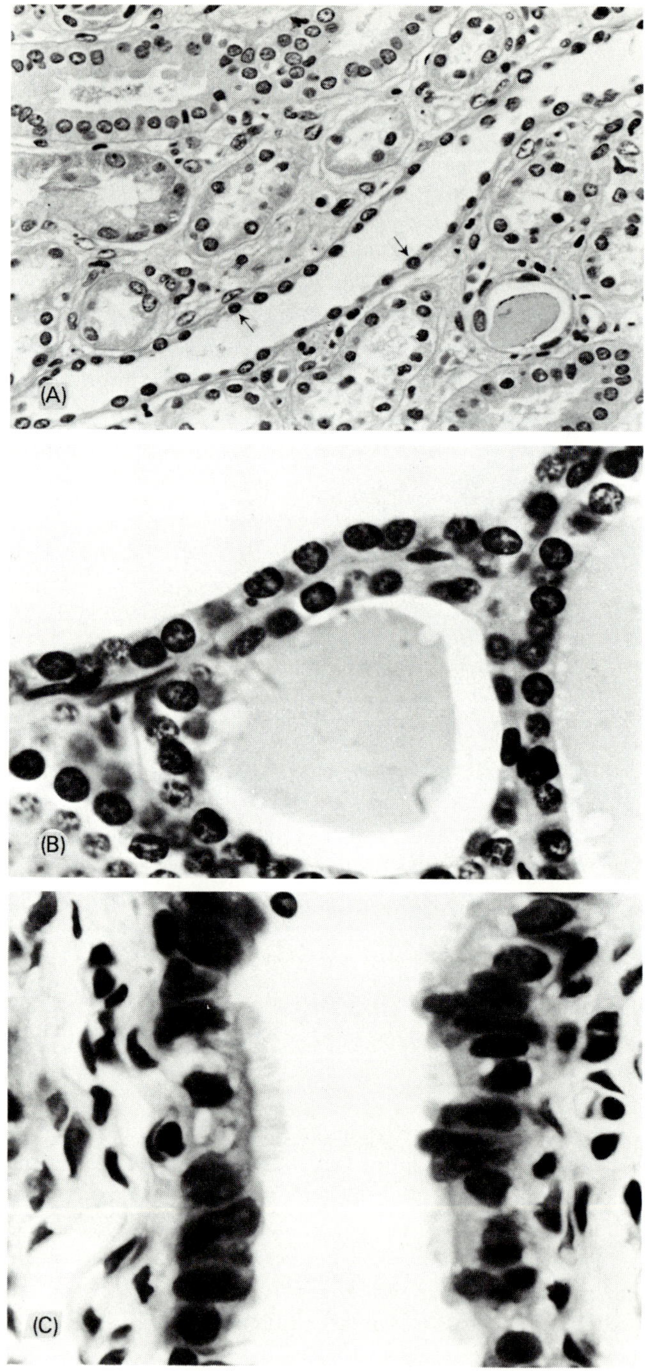

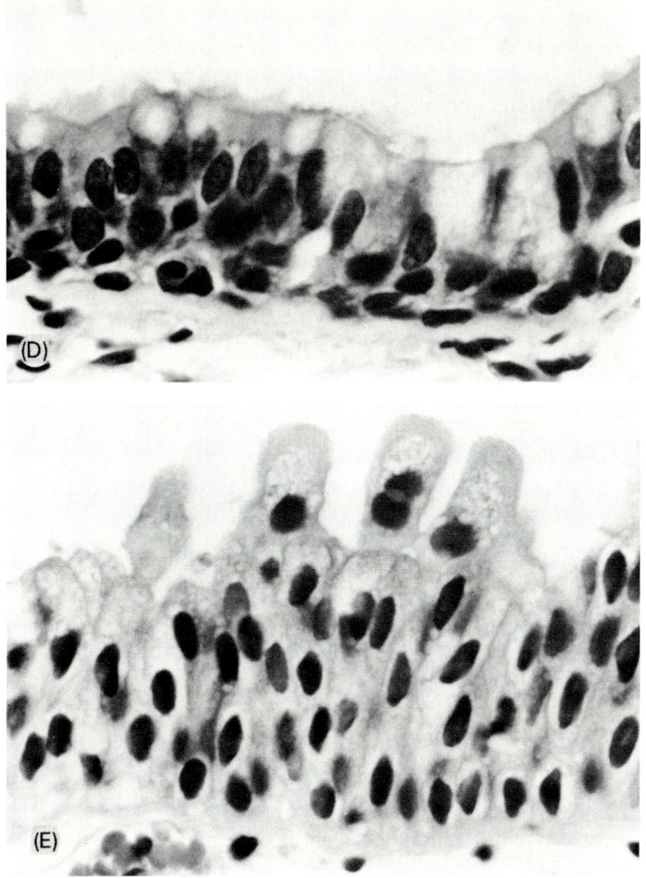

Fig. 1.2. (A) *Simple squamous epithelium* in the collecting tubules of the human kidney (arrowed). (Haematoxylin and eosin. ×318.) (B) Typical *simple cuboidal epithelium* as found in the human thyroid gland. (C) *Simple columnar epithelium* in the human Fallopian tube. Many of the cells show apical cilia in the luminal surface, while the non-ciliated cells are glandular-secretory in type. (D) *Pseudostratified epithelium* from the human trachea. All the cells in fact rest upon the basement membrane, but not all reach the surface. Ciliated cells predominate, but some mucous cells can be seen. (E) Transitional epithelium from the human bladder. The cells in the superficial layers are large and cuboidal, and the nuclei can be polyploid. In the intermediate and basal layers, the nuclei are diploid and much smaller. (All haematoxylin and eosin. ×954.)

are tall, and rectangular or pyramidal in section, and the height exceeds the width of the cell. These cells often show marked polarization, as in the villus epithelium in the small intestine and the ciliated lining epithelium of the Fallopian tube (Fig 1.2C). Often, epithelial cells are intermediate between cuboidal and columnar, exemplified by thyroid epithelium, where the height signifies the functional state of the cell, columnar cells indicating an active epithelium. Moving on to the more complex epithelia, where the cells are arranged in layers, we have (iv) *pseudostratified columnar epithelium*, which is usually found lining large ducts, such as the trachea, large bronchi and pharynx, vas deferens, and endocervical canal (Fig. 1.2D). Here an additional layer of cells is often present in the basal layer, but careful analysis reveals that all cells, including the more differentiated surface cells, maintain their attachment to the basement membrane, so that the epithelium is not really stratified. Truly stratified epithelium consists of two or more layers of cells, the superficial layers of which have lost contact with the basement membrane, and similarly, there are representatives of the above simple epithelia in the stratified group; by far the commonest is (v) the *stratified squamous epithelia*, where squamous cells are piled upon each other (Fig. 4.1, p. 206). Here there is some division of labour: in the simple epithelia the single layer of cells houses both the proliferative cells and the functional cells, but here the more cuboidal *basal cells* take over the proliferative function of the tissue, giving rise to differentiating cells which migrate through the epithelium, flattening and maturing as they go. These epithelia are obviously useful for protection, and indeed, in regions exposed to the air the superficial cells synthesize the protein keratin and become keratinized and anuclear, and the *stratum corneum* forms an effective surface barrier (Fig. 4.1A, p. 206). However, in stratified squamous epithelia which are usually covered with fluid, such as the cornea, mouth, oesophagus, and vagina, the epithelium contains much less keratin, surface cells retain their nuclei, and the epithelium is called 'non-keratinized' (cf. Figs. 4.1B and 4.1D, pp. 206 and 207). A frequent epithelial response to injury or altered conditions in these non-keratinized epithelia is a change to a keratinized format.

(vi) The *stratified cuboidal and columnar epithelia* are relatively sparsely distributed; in sweat gland ducts, and epithelia in the intermediate zones of the anal, conjunctival, and female urethral epithelium the superficial cells are more cuboidal than columnar, and are sometimes called *stratified cuboidal*, while stratified columnar epithelium is often seen in the zones intermediate between stratified squamous and pseudostratified epithelium, particularly in the pharynx and larynx, and is often seen lining large tubes such as the oesophagus in the embryo. Finally, we have (vii) *transitional epithelium*, which is found lining the urinary tract, in the bladder, ureters, and upper urethra (Fig. 1.2E). The surface cells are large and bulge into the lumen, while the basal cells are smaller, and interdigitate with the overlying cells. In the bladder, when the organ is distended, the epithelium becomes flattened, not due to the cells sliding past one another as was at one time thought, but probably due to flattening of the cells, with less interdigitation of cell processes.

Epithelial cells are also remarkably individually specialized in relationship to their function. The protective epithelia show lateral and basal surface specialization, which maintains cell contiguity and strengthens the epithelial sheet. Important among these are the areas of cell-to-cell contact, or intercellular junctions; the two main strengthening junctional complexes are (a) the *desmosome* or *macula adherens* (see Fig. 1.3A), which is particularly well developed in stratified squamous epithelium, particularly in the epidermis. The characteristic intracellular morphological feature is the dense attachment plaques, which occur in pairs *between* cells, but the single plaques which occur on the basal plasmalemma in some stratified squamous epithelia (cf. epidermis) are called *hemidesmosomes*. Into these are inserted the *tonofilaments*, which loop back into the cytoplasm. Microdissection studies show that the desmosome is a point of very strong attachment (Chambers and Renyi 1925), and more recent studies have shown that the narrow 20 nm intercellular space between the two halves of the desmosome is filled with low-density extracellular material containing sialic acid, acid mucopolysaccharide, and protein: secondly, (b) the *zonula adherens* resembles the desmosome, but does not possess a well-developed attachment plaque, and shows moderately dense material associated with the apical *terminal web*, a filamentous meshwork in the apical cytoplasm of cuboidal and columnar cells. In epithelial cells which line tubes or spaces the intercellular space is sealed from contact with the lumen by the *zonula occludens*, and here the plasmalemmae of the two adjacent cells are in contact with each other; these are also called *tight junctions*, because the outer leaflets of the two opposed membranes appear as a straight line, suggesting that the membranes really do fuse at this point; the zonula occludens network extends completely around the apical border of the cell and seals off the underlying intercellular space from the lumen. These several lateral specializations next to the lumen are collectively called the *junctional complex*, and can be made up of zonula occludens and adherens together with numerous desmosomes, as in the intestinal epithelium, or merely by a zonula occludens; the *terminal bar* is a light microscopical term referring to the belt-like zone around the lateral margin of the epithelial cells, and probably corresponds to the zonula adherens and occludens only.

The *gap junction*, or *nexus*, is a large plate-like junction which occurs on the deeper lateral aspects of epithelial cells (Fig. 1.3B); within these are extracellular hexagonal subunits arranged in a hexagonal fashion around other units, and the 2 nm gap is filled with lipid, carbohydrates, and proteins (Goodenough and Revel 1970). The gap junction may well be involved in small ion transport between cells.

Also contributing to the maintenance of tissue integrity is the more or less well-developed *cytoskeleton* which is found in epithelial cells; for example, stratified squamous epithelium contains thick bundles of filaments (tonofilaments) called *tonofibrils*, which insert into the desmosomes; most epithelial cells

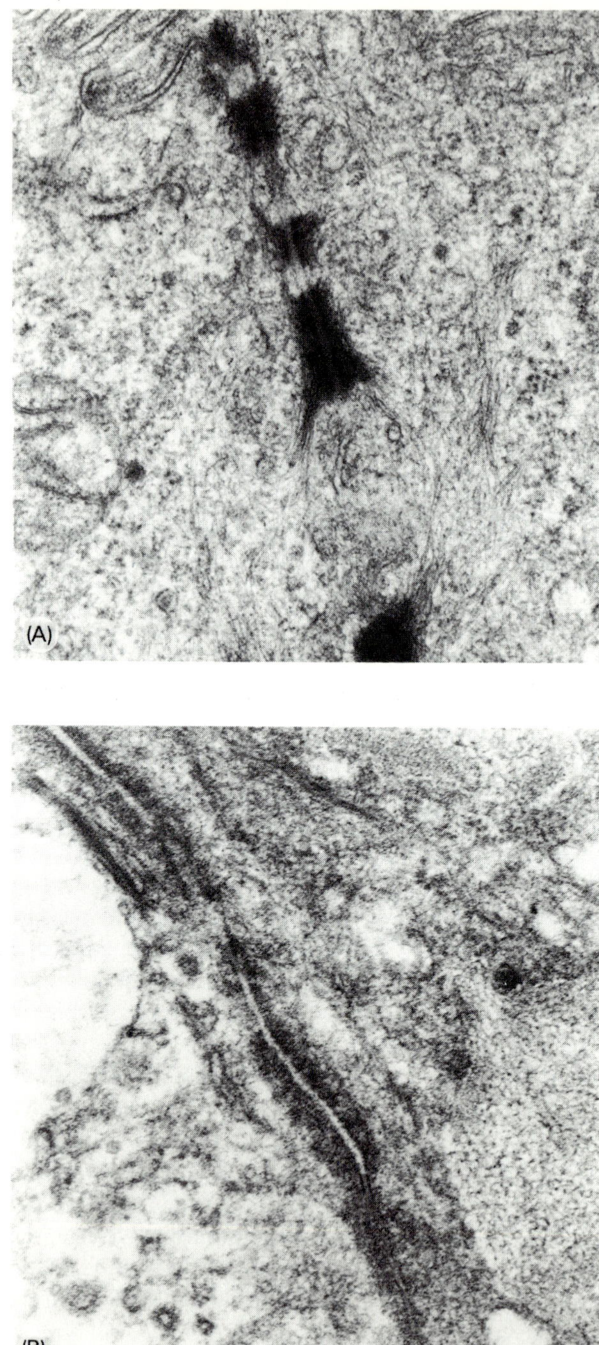
(A)
(B)

contain such microfilaments, which are especially prominent in the terminal web, and these, together with the microtubular system, make up the cytoskeleton of epithelial cells.

On the basal surface of epithelial cells the basement membrane is found: this is an extracellular condensation of mucopolysaccharide and proteins that consists of a dense filamentous sheet measuring 50–100 nm in thickness, called the *basal lamina*, which is attached to the plasmalemma of the epithelial cell, and also to the reticular tissue beneath. This underlying zone is sometimes called the *reticular lamina*, since it contains irregular collagen bundles called reticular fibres, and condensed ground substance. Under the light microscope, the reticular lamina is usually that portion resolved by silver stains (see Fig. 1.4), but the periodic acid–Schiff (PAS) method stains mucopolysaccharide components in both basal and reticular laminae (see also Fig. 1.4). The basal lamina is predominantly composed of Type IV collagen (Kefalides 1970). The epithelial cells probably do contribute to the synthesis of their own basement membrane, and while underlying fibroblasts contribute to the formation of the reticular lamina, corneal epithelium, isolated *in vitro*, can certainly produce the basal lamina in the absence of fibroblasts (Dodson and Hay 1971).

Secretory and absorptive cells also have specializations of the free surface, and comprise the light microscopic brush border of the proximal renal tubule and the striated border of the intestinal villus cell, and these undoubtedly increase the surface area and contribute to the absorptive, hydrolytic, and active transport mechanisms which are concentrated in this area (Fig. 1.5). *Cilia* are motile cell processes with a highly complex inner structure, and are found on the pseudo-stratified epithelium of the major bronchi and in the Fallopian tube, and, by the production of forward-spreading ciliary waves, effectively move mucus, for example, over the surface of the epithelium.

So much for the general structure of surface-covering epithelium. On the other hand, many epithelial tissues are arranged in *glands*, the *endocrine* and the *exocrine* glands. Endocrine glands secrete directly into the blood, while the secretory products of exocrine glands are usually transported directly, or by ducts, on to a free surface. Some glands, such as the liver, have an exocrine function in the production of bile, but also an endocrine function in that substances such as plasma proteins are secreted directly into the blood. Some endocrine glands retain their epithelial nature and line a space, for example the thyroid follicle or acinus (Fig. 1.2B), but, in most cases, endocrine glands such as the adrenal and the pituitary are composed of anastomosing sheets or cords of

---

Fig. 1.3.  (A) Desmosomes between two epithelial cells; the dense plaque of intracellular material on either side of the desmosome, subjacent to the cell membrane, are well shown (human pancreas (nesidioblastosis). (× 24 000.)  (B) A gap junction between two epithelial cells in the cat gastric mucosa. The 2 nm gap between the outer leaflets of the two apposed membranes can be seen. (Uranyl acetate/lead citrate. × 30 000.) (Both by courtesy of Dr J. Polak.)

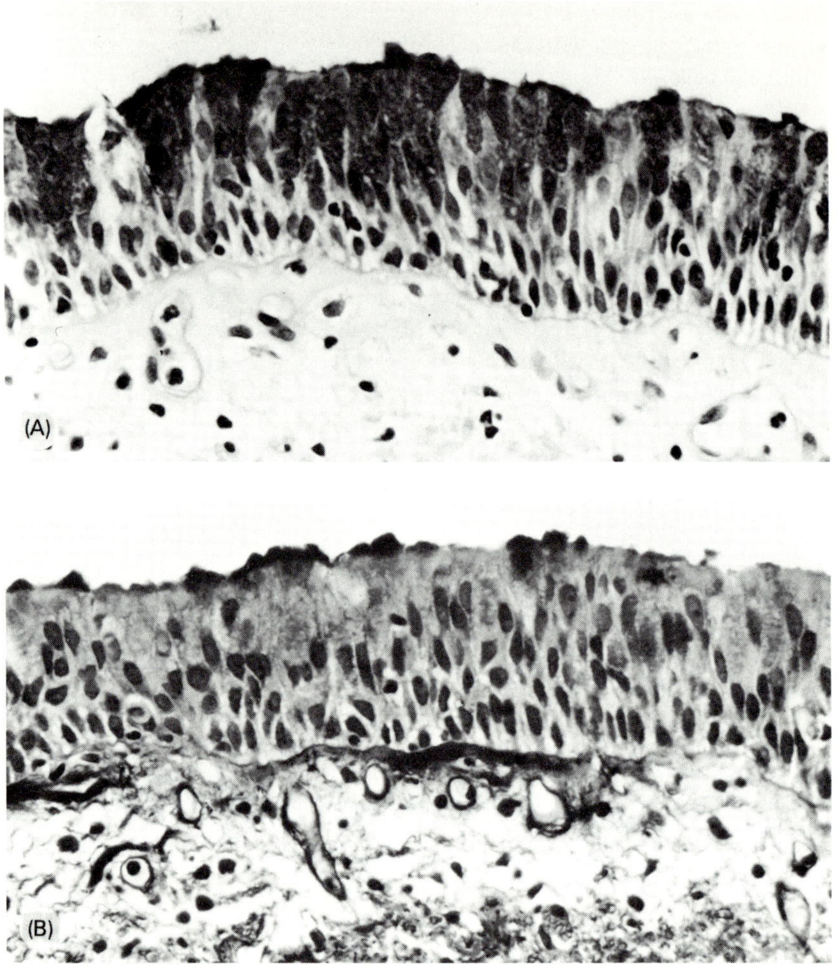

Fig. 1.4. The basal lamina of the human bronchial epithelium stained with (A) the diastase periodic acid–Schiff method; and (B) the silver methenamine method. (× 420.)

---

Fig. 1.5. Microvilli from the small intestine, shown by (A) scanning electron, and (B) transmission electron microscopy. Each microvillus is about 1.4 $\mu$m long and 0.08 $\mu$m, in diameter, and contains a core of longitudinal filaments which merge just beneath the microvillus border with the *terminal web*, a dense meshwork of filaments orientated parallel to the free surface of the cell, and insert laterally into junctional complexes at the cell border. (A, human jejunal villus cells, ×11 700, courtesy of Dr M. Marsh; B, rat ileal villus cells, ×7800, courtesy of Dr J. Polak.)

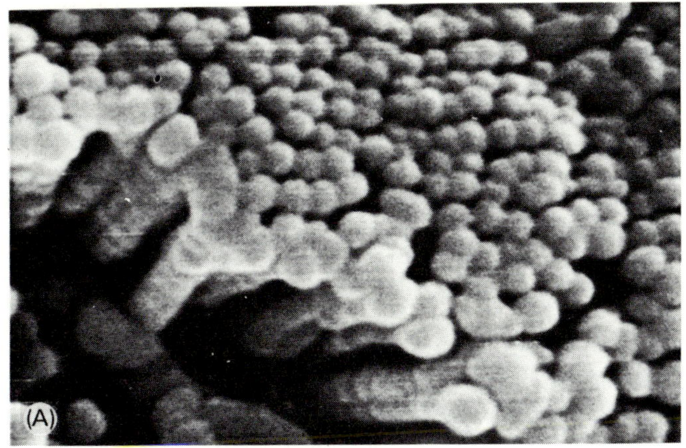
(A)

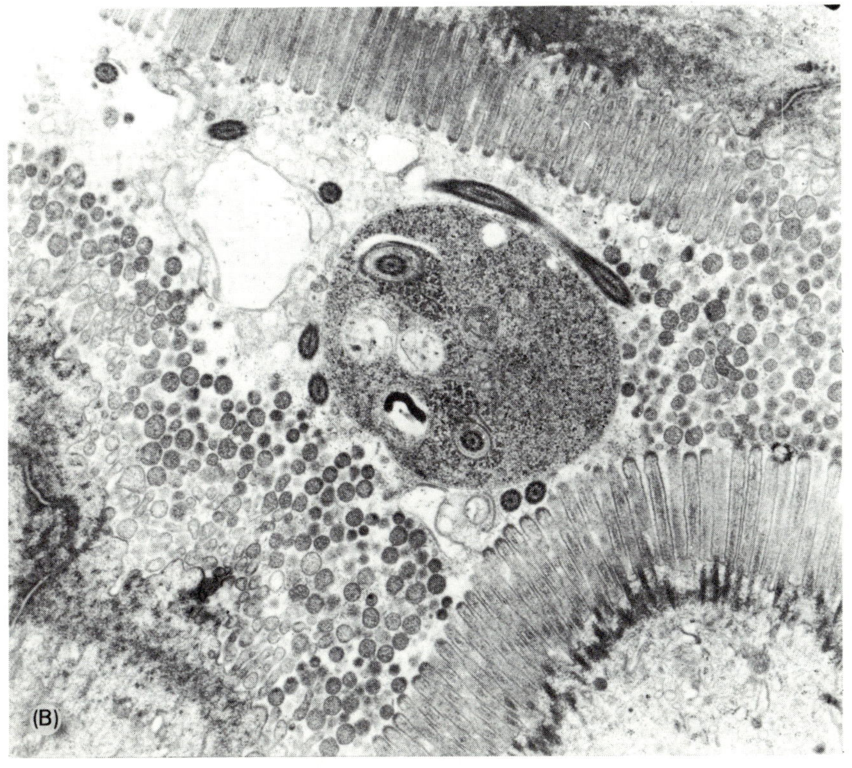
(B)

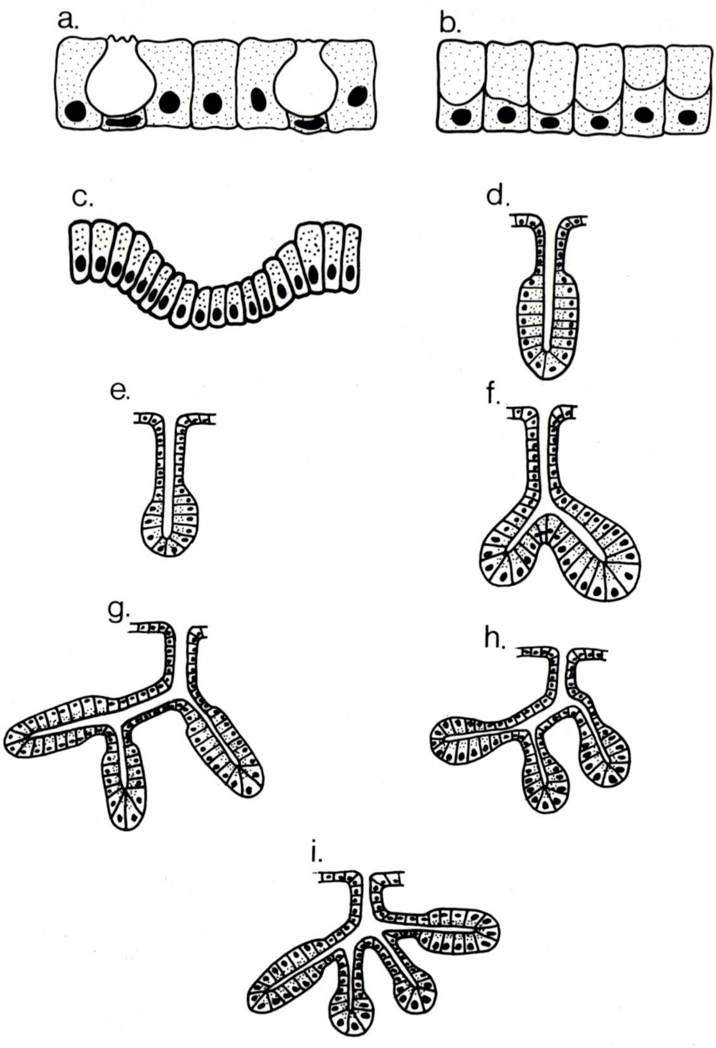

Fig. 1.6. Diagrammatic representation of the several types of glands. (a) Unicellular glands (goblet cells, gut endocrine cells). (b) Secretory sheet (surface of the gastric mucosa). (c) Intraepithelial gland (colon). (d) Simple tubular gland. (e) Simple acinar gland. (f) Simple branched gland (Meibomian gland of eyelid). (g) Compound tubular gland (cardiac glands of stomach). (h) Compound acinar gland. (i) Compound tubuloacinar gland (pancreas). (Redrawn after Krause and Cutts (1981).)

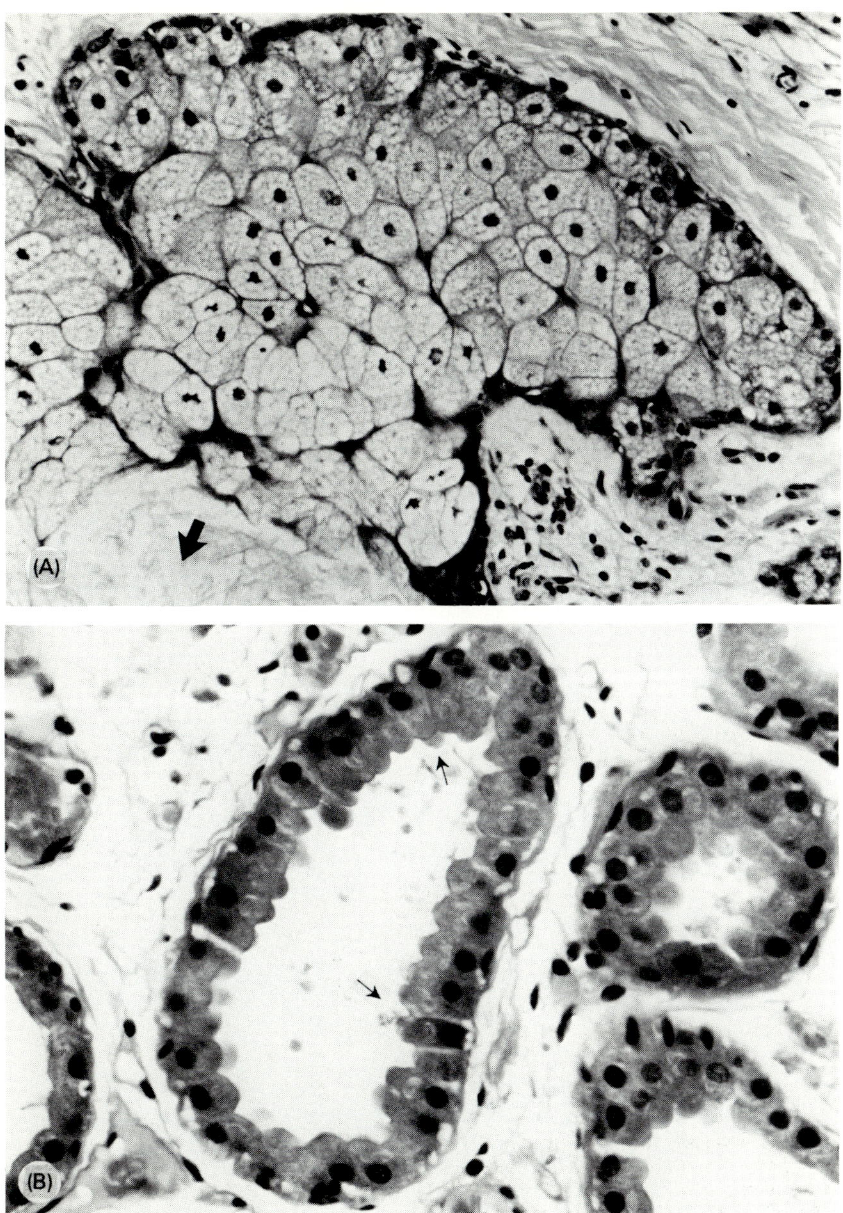

Fig. 1.7. (A) A typical *holocrine gland*, the human sebaceous gland. Cells are formed in the peripheral part, and migrate centrally to undergo complete degeneration and emerge as sebum as the sebaceous gland empties into the hair follicle (arrowed). (Human scalp, H and E. ×245.) (B) An *apocrine gland acinus* from the human axilla. In this sort of gland, the secretion is formed by a portion of the apex becoming 'pinched off' (arrowed). (H and E. ×420.)

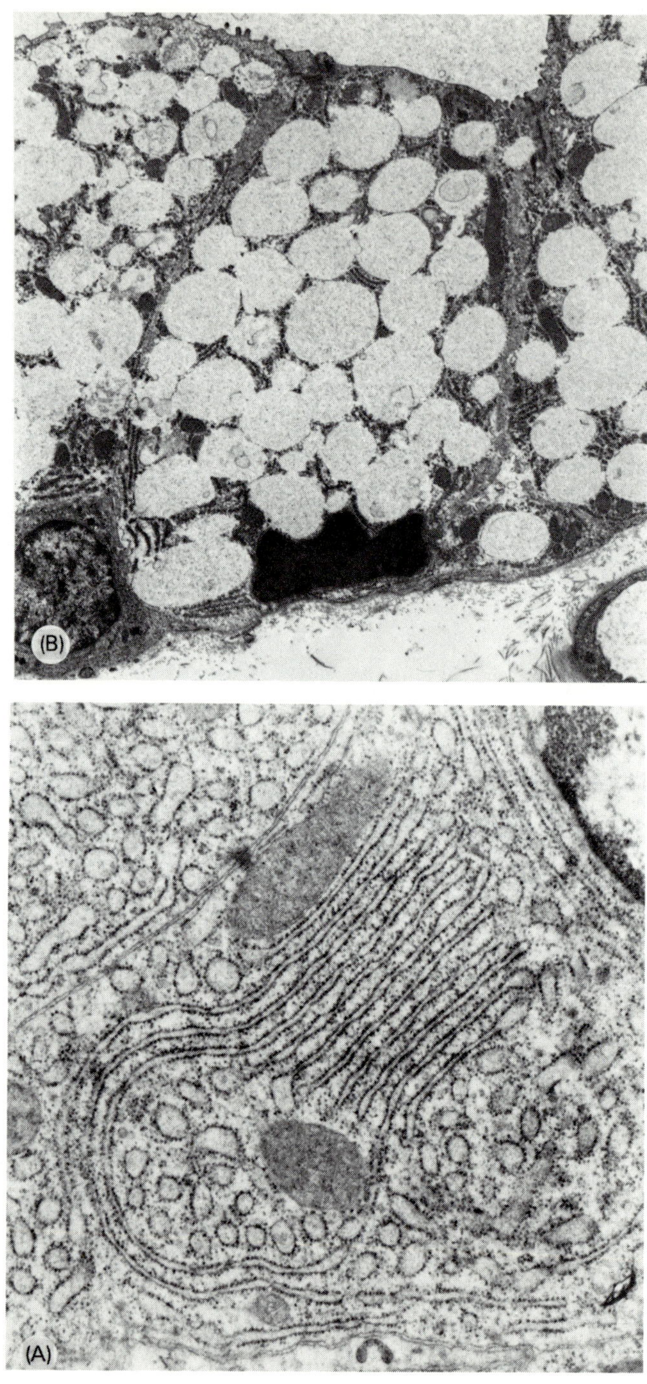

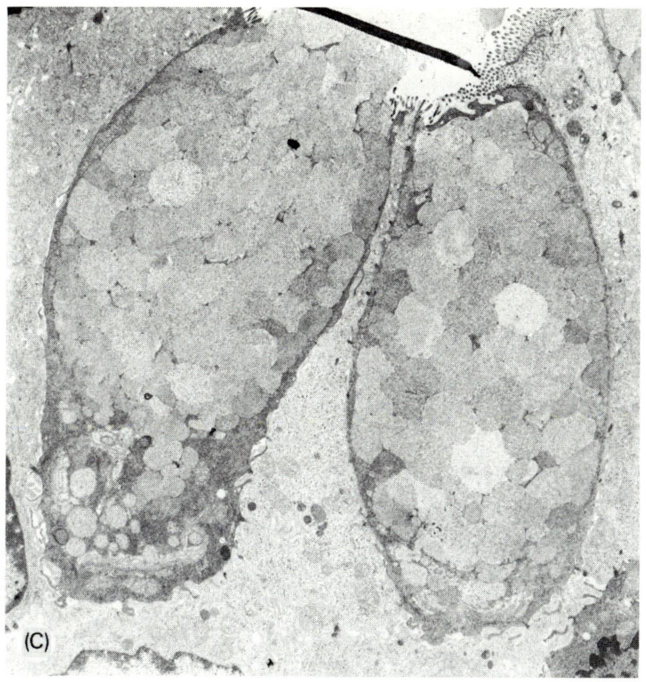

Fig. 1.8. Specialized features of glandular cells. (A) Abundant rough endoplasmic reticulum in the cytoplasm of a human pancreatic acinar cell. Note the numerous ribosomes attached to the cytoplasmic surface of the membrane. (×12 000.) (B) Zymogen granules in a peptic cell from the human stomach. The secretory granules contain *pepsinogen*, a precursor of the enzyme *pepsin*. (×3200.) (C) Goblet cells in the rat small intestine, one in the process of releasing its mucus into the intestinal lumen. Note the large number of mucus droplets in the supranuclear cytoplasm; in the apical portion, the droplets coalesce, and lose their surrounding membranes, and the secretion is released in an apocrine manner into the intestinal lumen. (×1600.) (Courtesy of Dr J. Polak.)

cells of epithelial character, and the liver, with both endocrine and exocrine functions, is of similar basic structure (see Chapter 25).

It is customary for exocrine glands to be classified according to the shape of the secretory units and the degree of branching of ducts; hence there are *simple* or *compound acinar* and *tubular* glands, or more complex forms such as the *compound tubulo-acinar* glands (see Fig. 1.6). Additionally, glands are sometimes classified according to their mode of secretion: where whole cells are incorporated into the secretion, as in the production of sebum by the sebaceous gland, the gland is called *holocrine* (Fig. 1.7A); where protrusions of the apical cytoplasm are lost, the term *apocrine* is used (Fig. 1.7B); this occurs in the apocrine sweat glands in the axilla and groin in man, and also in the breast.

Where no actual cytoplasm is lost, the gland is *merocrine*. Most glands, such as the sweat glands in the skin, and the salivary glands, are merocrine. The types of actual secretion can be *serous*, which is protein-rich and watery, and can be high in enzyme content, such as the pancreatic secretions. Serous cells contain abundant basal endoplasmic reticulum (Fig. 1.8A), giving the basal portion of the cell a strong affinity for basic dyes. Sometimes the protein precursor of the secretion accumulates in the apical cytoplasm in the form of zymogen granules (Fig. 1.8B); the nucleus is thus basal and rather small. *Mucous secretion* is viscous and rich in sugars, and the precursor forms masses in the apical cytoplasm of the mucous cell, compressing the nucleus to the base, and compressing it (Fig. 1.8C). Mucous glands can be *unicellular* in the form of goblet cells (see Fig. 15.5, p. 549), or *multicellular*, often forming a large part of the salivary glands or gastric tubules, where they are intimately intermingled with serous cells (see Fig. 15.1, p. 541).

Even such a brief survey of the general structure of epithelia indicates that the tissues have a distinct morphological organization, and it therefore comes as no surprise that the kinetic organization is similarly highly specialized, and in several instances, much more complex than mere histological observation would indicate (cf. epidermis; Chapter 8). In fact, most investigations suggest that epithelial renewal systems have a *highly ordered morphokinetic architecture*, with definable subpopulations, often with a defined *spatial* organization as well. This emphasizes that the kinetic organization of epithelia cannot be appreciated without a painstaking analysis of *three-dimensional* architecture. To regard such complex tissues as kinetically homogeneous is a perfect recipe for disaster.

# 2 The theory of renewing cell populations

Just as we have seen that epithelia have spatial architecture, so they also have kinetic architecture, and, with appropriate techniques, we are able to dissect this architecture into kinetic compartments, and assign each compartment its *size, flux,* and *time parameters.* This chapter examines the several component parameters within this framework, and the various kinetic compartments which are found in epithelial renewal systems.

## 2.1. Cell cycle events: the life-history of proliferating cells

Figure 2.1 shows the now familiar pie-diagram of the cell cycle, and denotes the phases in the life-cycle of all proliferating mammalian cells. The history of the discovery of the cell cycle and its component phases is now well known, and for this the reader is referred elsewhere (Aherne *et al.* 1977*a*). The period of DNA synthesis is called, not unreasonably, the $S$ phase, and in most epithelial renewal systems lasts between 7 and 19 hours; the interval between DNA synthesis and the next mitosis is referred to as the $G_2$ phase, and the interval between mitosis and the $S$ phase is the $G_1$ phase. In this nomenclature, $G_1$ and $G_2$ refer to 'gaps' in the cell cycle, but this should not be taken to indicate that these are mere staging posts in the cycle; far from it. During the $G_2$ phase there is biochemical preparation for the coming mitosis, including assembly of spindle protein, while much current thinking places a crucial control point, at which cells commit themselves to a further mitosis, firmly in the late $G_1$ phase; in late $G_1$ there is intense biosynthetic activity, with the assembly of enzymes necessary for DNA

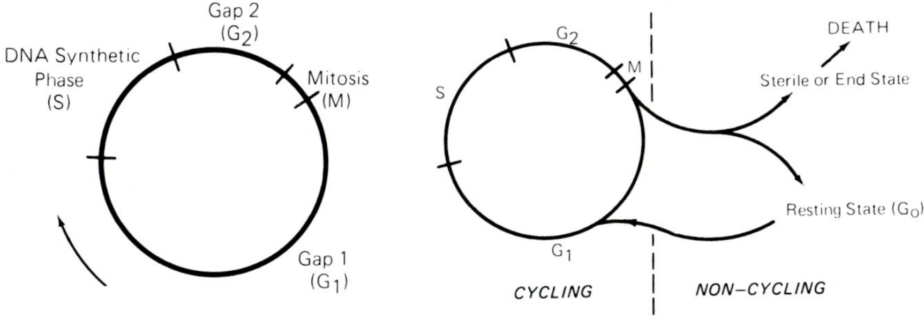


Fig. 2.1. A diagrammatic representation of the cell cycle. Proliferating cells traverse the cycle, but, mainly in the late $G_1$ phase, many cells leave the cell cycle, with the possibility of permanent decycling to differentiate, or temporary proliferative rest in the $G_0$ phase.

synthesis, which activity is presumably switched on after the cell has committed itself to DNA synthesis (Baserga 1976). This control point is indicated diagrammatically in Fig. 2.1; this point in the cell cycle is variously called a 'critical decision phase', 'restriction point', or 'transition point', and there are three potential avenues open to a cell reaching this point in the cell cycle: (i) to 'recycle' and embark on another round of DNA synthesis and mitosis; (ii) to 'decycle', and enter a resting phase, commonly called the $G_0$ phase, from which the cell can be ejected back into the cell cycle if conditions become appropriate, or (iii) the cell can decycle and progress upon a course that leads to differentiation, function, and eventual death. It is generally considered that the transition point in $G_1$ is the most important, and possibly the only control point in the cell cycle, although there is some evidence that cells can decycle in $G_2$, particularly in squamous epithelium (Gelfant 1976, 1977, and see Chapter 2, section 2, and Chapter 8); even if this is substantiated, only a small number of cells are involved, and the significance of the phenomenon is as yet unclear.

The $G_1$ phase is most subject to variation of all the cell cycle phases, and is largely responsible for the differences in the rate of transit around the cell cycle in different epithelial tissues; for example, in the mouse small intestine the duration of the $G_1$ phase ($t_{G_1}$) is between 2 and 3 hours, while in mouse dorsal epidermis, $t_{G_1}$ is thought to be about 100 hours; such is the variation in $t_{G_1}$ compared with $t_S$, $t_{G_2}$ and $t_M$ that Smith and Martin (1973), as the basis of their transition probability model, argued that $G_1$ was the only stochastic variable in the cell cycle, and that $t_S$, $t_{G_2}$, and $t_M$ were deterministic and did not vary; in point of fact, $t_S$, $t_{G_2}$, and $t_M$ do vary between cells in the same population and this has necessitated some modification of the hypothesis. The duration of $G_2$ was hitherto considered to be short and relatively constant, between 2 and 4 hours in most tissues, but the recent work of Clausen (Clausen *et al.* 1979, 1981) has shown that $t_{G_2}$ in mouse epidermis varies from about 1 hour to 7 hours, according to the time of day, and that there may be subpopulations with $t_{G_2}$ as long as 17 hours. The duration of mitosis is a very difficult parameter to measure (see Chapter 3, section 3), but in most epithelial tissues seems to last about 30 minutes to 1.5–2 hours; some measurements suggest that $t_M$ can be as long as 4 hours (Bullough and Laurence 1964*a*), but we shall see that here the methodology is suspect (see Chapter 3, section 3, and Chapter 7, section 3).

The time taken to go once around the cell cycle is called the *cell cycle time*, usually denoted by $T_C$. If all cells in the population are proliferating, and there is no cell loss, then the population will double in size in a period of time equal to $T_C$, and $T_C$ will equal the population doubling time, $t_D$. A concept that the reader must grasp, even this early, is that, in real cell populations, cells do not all traverse the cell cycle all at the same rate, as is suggested by the single line describing the cell cycle in Fig. 2.1; naturally there will be an *average* cell cycle time, but such a measurement is of strictly limited usefulness. More valuable by far is the *confidence interval* or *variance* of the cell cycle time, and some idea of the shape of the *frequency distribution*. It is a matter of some import to realize that all cell populations, including epithelia particularly, have a *distributed cell cycle* (Steel 1977), and do not have a strictly deterministic life cycle.

## 2.2. Cell cycle events: the life-history of non-proliferating cells

It is very rare to find all cells in the cell cycle at any time; when viewed in isolation, *portions* of an epithelial cell population may consist of proliferating cells only. For instance, there is good evidence that effectively all cells in the mid-crypt region of the rat jejunal crypt are cycling, but, when viewed as a whole, the crypt system is a *mixture* of proliferating and non-proliferating cells. If we follow the convention of Cairnie *et al.* (1965*b*), and designate those cells born into the proliferative category as $P$ cells, and those born into the non proliferative cateory as $Q$ cells (Fig. 2.1), it is axiomatic that a ratio of proliferating to non-proliferating cells must exist: this is readily defined, in index form, as the *growth fraction* ($I_P$), as

$$I_P = N_P/N_P + N_Q \qquad (2.1)$$

where $N_P$ and $N_Q$ are the fractions of cells in the $P$ and $Q$ compartments respectively; or more formally

$$I_P = N_C/N \qquad (2.2)$$

where $N_C$ is the number of proliferating cells, and $N$ the total number of cells in the population. Thus in the mouse jejunal crypt there are about 300 cells per crypt, and $N_C \sim 180$, giving an $I_P$ value of about 0.6 (see Chapter 18).

If some of the cells are non-proliferating, the rate of cell production for the population at large will be smaller that we would expect from inspection of the $T_C$ value, since the $Q$ cell compartment cannot, by definition, contribute to the cell production. Thus the proliferating cells must compensate, in terms of the cell production rate, for the non-proliferating cells, and the time taken for the population to double will be longer than the cell cycle time. This time period, i.e. the time taken for the whole population to double, or, perhaps more relevant for epithelia, the time taken to replace all cells in the population, has received various names: the *turnover time* is perhaps most apposite for normal epithelia, since in this time the whole population has doubled or 'turned over' (the indiscriminate and vague term 'cell turnover' should perhaps be eschewed, since the turnover time, and its reciprocal, the *turnover rate*, are specific kinetic parameters; see Table 2.2). The term *potential doubling time* ($t_{PD}$) was introduced by Steel (1968) with special reference to exponentially growing tumour populations, but, where epithelial populations are concerned, $t_{PD}$ is not appropriate, since in the steady state, the population size remains constant, due to balanced cell loss. Aherne *et al.* (1977*a*) have championed the use of the term 'apparent cell cycle time' as a useful synonym for the turnover time, but this appellation has met with less than general approval and usage. It is essential, at the outset, to distinguish carefully between the *cell cycle time* and the *turnover time*; they can equal each other, if all cells are proliferating, but in most cases, in epithelia at any rate, the turnover time exceeds the cell cycle time. Failure to appreciate the difference has led to the publication of a large number of bogus $T_C$ values, and to not a little confusion in some fields.

Aherne *et al.* (1977*a*) have drawn attention to the usage of the words 'cell proliferation' and 'cell production' in the cell kinetic context; here *cell proliferation* is a general term, used to denote the process by which cells traverse the cell cycle and produce new cells at mitosis, whereas *cell production* refers to the output of new cells in relation to the cell population at large. It is evident that the determinants of the cell production rate are the cell cycle time and the growth fraction, and that variation in either of these parameters will result in changes in the cell production rate; such changes are the basis of most kinetic responses. Naturally, limits are imposed on the increase in the cell production rate by the temporal nature of the cell cycle, and in epithelia there does appear to be a level below which $T_C$ cannot fall; in the rodent jejunal crypt $T_C$ only seems to be able to contract to about two-thirds of its control value (see Chapter 20), and in the human jejunum, to about one-half of normal (Wright *et al.* 1973*b*). Similarly the growth fraction cannot increase to more than 100 per cent. However, proliferative systems are not solely committed to changing only these parameters, since perhaps the most powerful method of increasing the cell production rate in epithelia is by an *absolute increase* in the number of proliferating cells, mediated by an expansion of the population size; in the jejunal crypt this would be brought about by a lengthening and widening of the crypt unit, and in the epidermis by an increase in the number of 'germinative' cell layers, in each case, giving an augmentation of the cell production rate, even though $T_C$ and $I_P$ may remain constant. On the other hand, there is theoretically no limit to the *decrease* in the cell production rate that can be caused by a lengthening of the cell cycle, although $I_P$ is of course limited to zero.

The fate of cells which decycle is shown in Fig. 2.1. The movement into the differentiation pathway may be accompanied by *cell migration*, as when intestinal crypt cells decycle and move onto the villus, or when basal epidermal cells ascend into the Malpighian layer, to die at the villus tip and in the stratum corneum respectively. Cells in the $G_0$ phase are considered to remain fertile and to retain the ability to re-enter the cell cycle if conditions become suitable. Perhaps the best example are the $G_0$ cells in the liver; most, if not all of the liver parenchymal cells are in $G_0$, but if part of the liver is removed, the remaining cells will be triggered into the cell cycle and the resulting cell division will replace the lost tissue mass; there are many examples of such resting cells which are in a putative $G_0$ phase, and these are discussed later (see Chapter 2, section 7, and Chapter 24).

## 2.3.  Age distribution of cells in a population

The concept of the *age distribution* is very important to the understanding of
the kinetic organization of epithelia. The age distribution merely means the
*frequency distribution of cells with respect to their age in the cell cycle*; here age
is measured from birth at mitosis. It is quite easy to identify the simple, well-
known age distributions; in the simplest case, we have the *rectangular age
distribution* (Fig. 2.2A), which is often mistakenly referred to as a 'steady state
age distribution', and generally assumed to apply indiscriminately to epithelial
cell populations which are in a kinetic steady state, i.e. where no parameter is
time-dependent, and population size remains constant, since influx and efflux
into and from the system are equal: certainly the rectangular age distribution is
consistent with a steady state, in that, on the average, at the end of the cycle at
mitosis one cell decycles and the other re-enters in the cell cycle at the beginning

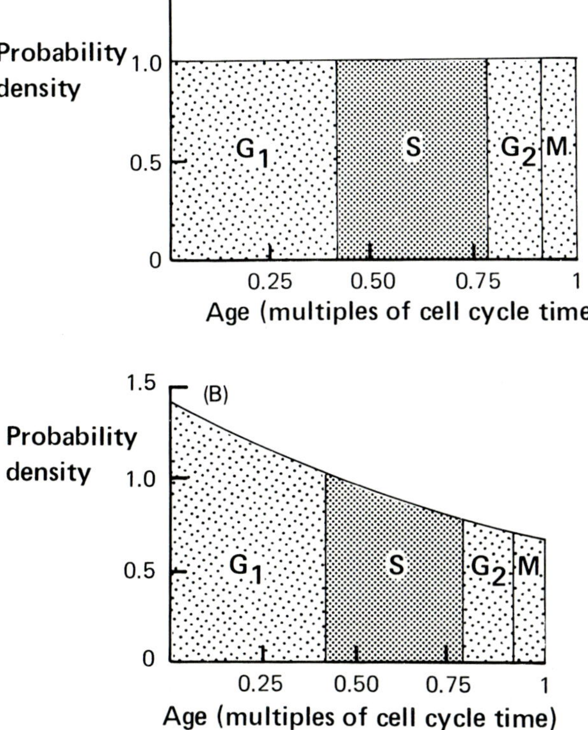


Fig. 2.2.  Two frequently used age distributions for cell populations.  (A) The rectangular
age distribution, and (B) the exponential age distribution.

of $G_1$; thus cell production rate equals cell loss, and there is no change in the population size. However, as we shall see, the proliferating portion of most renewal systems do not have a rectangular age distribution, even though the overall age distribution for the tissue compartment may be rectangular, as in the jejunal crypt (see Chapter 19, section 2). However, in some cases, for example the epidermis, even the overall age distribution for the tissue compartment that houses the proliferating cells, the basal layer, is non-rectangular.

In the case of Fig. 2.2A, the abscissa is cell age ($\tau$), while the ordinate gives the probability of finding a cell at age $\tau$; the number of cells in any phase is therefore proportional to the duration of that phase, and we have the following simple relationships

$$I_M = t_M/T_C \tag{2.3}$$

$$I_S = t_S/T_C \tag{2.4}$$

where $I_M$ is the *average mitotic index*, or fraction of cells in mitosis, and $I_S$ is the proportion of cells in the $S$ phase, the *average labelling index*.

The other limiting case is where, at the end of the cell cycle, all cells produced at mitosis re-enter the cycle; if we assume that there is no synchronization of cells in the cell cycle, then the population size (again in cells) will increase exponentially, and should follow a smooth exponential growth curve (Fig. 3.31B, p. 199). The growth curve of this population will then be

$$N_t = N_0\, e^{k_G t}$$

where $N_0$ and $N_t$ are the population sizes at times 0 and $t$, and $k_G$ is the growth constant. If there is no cell loss, and all cells are proliferating, the doubling time, $t_D = T_C$; it follows that,

$$k_G = \ln 2/t_D$$

or
$$k_G = \ln 2/T_C.$$

Since growth is constant, $k_G$ and hence $T_C$ remain constant. Then the rate of production of new cells in the population, the *birth rate*, is

$$(\mathrm{d}N/\mathrm{d}t)_t = \ln 2/T_C \cdot N_t \tag{2.5}$$

and the rate at which cells enter a new cell cycle is $2(\mathrm{d}N/\mathrm{d}t)$.

Let us now define cell age as a fraction $\tau$ of the whole cell cycle; here $T_C = 1$. Hence the number of cells at age $\tau$ is equal to the number of cells which were born at a time $\tau$ ago, or $2\ln 2(N_{t-\tau})/T_C$, which can be rewritten

$$\frac{2\ln 2}{T_C}\, N_t \exp\left[\frac{-\ln 2}{T_C}\,\tau\right]$$

and normalizing to give the probability of finding a cell at age $\tau$, $P_\tau$ (i.e. the height of the ordinate on the age distribution diagram at age $\tau$ in the cell cycle),

$$P_\tau = 2\frac{\ln 2}{T_C}\exp\left[-\frac{\ln 2}{T_C}\tau\right]$$
(2.6)

(Steel 1968). $P_\tau$ can then be calculated for different ages in the cell cycle, and the resulting age distribution is shown in Fig. 2.2B; the age distribution is bounded by an exponential curve, and the probability of finding a cell at zero age is twice the probability of finding a cell at age $T_C$; the cell flux from one phase to the next decreases with distance away from mitosis in the cell cycle. An important consequence is that the average rate of entry into DNA synthesis $(r_S)$ exceeds the average rate of entry into mitosis $(r_M)$. This age distribution is usually called *exponential*, and to find the equivalents of eqns 2.3 and 2.4 for this exponential age distribution, we could measure the ratio of the area under the exponential boundary curve of the appropriate phase to the total area under the curve, or, by integrating eqn 2.6 over the $M$ phase

$$I_M = \ln 2\, t_M/T_C$$
(2.7)

and similarly for cells in the $S$ phase

$$I_S = \lambda\, t_S/T_C$$
(2.8)

where $\lambda$ is a parameter which depends on the position of $S$ in the cell cycle; if $S$ is close to the end of the cycle, $\lambda \to \ln 2$, but if $S$ occurs at the beginning of the cell cycle, $\lambda \to 2\ln 2$ (Steel 1968) (see Chapter 2, section 7). Alternatively, Dendy and Cleaver (1965) have shown that

$$I_S = [\exp(t_S\ln 2/T_C) - 1]\,[\exp(t_2\ln 2/T_C)]$$
(2.9)

where $t_2 = t_{G_2} + \frac{1}{2}t_M$.

It is thus evident that calculation of the kinetic parameters, for example, $T_C$ from $I_S$ and $t_S$, is very much dependent on the choice of the age distribution. On the other hand, we are not often in a position to be able to produce the correct age distribution on demand, and frequently we are reduced to guessing the appropriate one; however, in some instances it is possible to interpolate the age distribution from measurements of $r_S$ and $r_M$; if these are equal, then we would have a rectangular age distribution, since, in Fig. 2.2A the ordinate at the beginning of the $S$ phase is the same as that at the end. Similarly, if $r_S > r_M$ by the appropriate amount, which can readily be measured from the ordinates on the age distribution diagram at the $G_1/S$ and $S/G_2$ boundaries for the particular set of cell cycle parameters, then an exponential age distribution is more appropriate; such measurements have been made by Schultze and her colleagues (1972) for the mouse jejunal crypt (see Chapter 19, section 2). Alternatively, it may be possible to *calculate* the age distribution if the mode of migration from the proliferative compartment can be defined, as in some squamous epithelia (see Chapter 6, section 7).

This far, we have proposed the simplest model: uniform cell cycle time, all cells proliferating, no cell loss, and, very important for epithelia, we have assumed *asynchrony*. We can remove some of these assumptions; we can introduce variation into $T_C$ and its component phases (Fig. 6.5, p. 245), and we can calculate age distributions for populations which contain non-proliferative cells and for those which are subject to cell loss (see sections 7 and 8). In epithelia the main cause of synchrony in the unperturbed state is *circadian variation*; all renewing epithelia, including the epidermis and gut, show prominent circadian variation in proliferative indices, denoting a major degree of synchrony, and any realistic age distribution diagram must take cognisance of this fact; such age distribution diagrams have been drawn (see Fig. 9.11, p. 366).

## 2.4. Cell birth and cell loss

The population size is jointly determined by the rate of addition of new cells, called the birth rate ($k_B$), and the rate of cell loss ($k_L$). The average birth rate is jointly determined by the growth fraction and the average cell cycle time. For a rectangular age distribution,

$$k_B = 1/T_C \tag{2.10}$$

or if $I_P < 1$

$$k_B = I_P/T_C \tag{2.11}$$

and for an exponential age distribution

$$k_B = \ln2/T_C \tag{2.12}$$

or if $I_P < 1$

$$k_B = \ln(1 + I_P)/T_C. \tag{2.13}$$

In each of these equations, $k_B = r_M$, and in eqn 2.10, $k_B = r_S = r_M$, but in eqn 2.12 $r_S > r_M$, since there is an exponential age distribution. Note also that the mitotic rate, or birth rate, can also be calculated from the mitotic index and the duration of mitosis

$$r_M = I_M/t_M. \tag{2.14}$$

The birth rate is expressed in cells produced per cell per hour, or as cells/1000 cells/hour. There is a convenient and simple relationship between the turnover time $T$ and the birth rate; from eqn 2.10

$$k_B = 1/T. \tag{2.15}$$

If we assume an exponential age distribution, and a steady state of exponential growth,

$$k_B = \ln2/t_{PD} \tag{2.16}$$

where $t_{PD}$ is the potential doubling time of the population, or the time in which the population would double in the absence of cell loss. Note that the use of the term turnover time implies *linear growth*, and that $t_{PD}$ is a fraction 0.693 of the turnover time, since some of the newly produced cells enlarge the $P$ cell population, increasing $k_B$ (Steel 1977). Hence we should use $t_{PD}$ whenever exponential growth seems more appropriate.

If no cells are lost from the population, then $k_B = k_G$, the growth constant; $k_G$, the *growth rate*, can be defined as the actual rate of increase in the number of cells in the population, and should be expressed in cells/cell/hour. However, in most kinetic situations there is at least some cell loss, and

$$k_G = k_B - k_L. \tag{2.17}$$

Perhaps more useful as an estimator of cell loss is the cell loss factor $\phi$, which is the cell loss rate expressed as a fraction or percentage of the total cell birth, and denotes the magnitude of the cell loss rate in relation to the birth rate. Since $k_B = \ln(1 + I_P)/T_C$ (eqn 2.13) and $k_G = \ln 2/t_D$,

$$k_L = \ln(1 + I_P)/T_C - \ln 2/t_D \tag{2.18}$$

and $\phi$, the cell loss factor, is the ratio of cell loss to cell birth, and is given by

$$\phi = k_L/k_B = 1 - t_{PD}/t_D. \tag{2.19}$$

In epithelial cell systems which are in a steady state, $k_B = k_L$, and $\phi = 1$, i.e. net cell birth = net cell loss and the population size is constant. In growing populations $k_B > k_L$, and so $k_G > 0$, and the population expands. In epithelial cell kinetics it is well to remember that net population growth can result from an increase in $k_B$ *or* a reduction in $k_L$.


## 2.5. Models for simple epithelial systems

It is immediately evident that all epithelial cell populations cannot obey the same rules; for example, the epidermis is continuously producing and losing cells, while in the mature liver and kidney, both cell loss and cell birth are negligible flux parameters, while in each case the population size remains constant. There are thus *different types of population within the range of epithelial tissues*.

Before the advent of radioactive DNA precursors, mitotic indices were, by and large, the only available kinetic measurements, and Cowdry (1950) divided cell populations into three types on the basis of the frequency of mitosis: (i) *vegetative intermitotics*, which showed frequent mitotic figures, as, for example, in the intestinal mucosa or the epidermis; (ii) *reverting intermitotics*, where mitoses are rare in physiological conditions, but rapidly appear in large numbers after an appropriate stimulus, as in the liver after partial hepatectomy, and (iii) *fixed postmitotics*, in which mitoses were never evident. There are no epithelial examples of this type, but neurones in the central nervous system in post-natal life are typical fixed post-mitotics.

Following the introduction of radioisotopes effective in labelling DNA synthesizing cells, Leblond and co-workers (Leblond and Walker 1956; Leblond *et. al.* 1959; Messier and Leblond 1960) introduced a similar classification based on the extent to which populations take up and retain injected tritiated thymidine ($[^3H]$-TdR). This 'uptake and retention' experiment was carried out by injecting rats of different ages, and preparing autoradiographs from the various tissues within a few hours of injection; young rats were also labelled and killed at six months later, and thus both cell formation and local retention of label could be studied. This work also led to the description of three types of population: (i) *renewing cell populations*, in which the proportion of labelled cells was relatively high a few hours after labelling, but fell rapidly to zero over the course of a few days, implying a continual need for cell replacement rather than a process of continuous tissue growth; these include of course, stratified squamous and gastrointestinal epithelia; (ii) *expanding cell populations*, in which only a small proportion of cells were labelled immediately after injection, and these persisted at the same low level for a period of several months. They included liver and kidney; since rapid proliferation in these organs is dependent upon an appropriate stimulus, these tissues are sometimes called *conditional renewal populations*. Thirdly, there were (iii) *stable, or static cell populations*, which take up and retain the label early in their life-span, but show no cell formation in adult animals; again these are exemplified by neurones in the central nervous system.

Having defined the various types of cell populations which occur in epithelia, we can proceed to analyse each in more detail in terms of their component subpopulations, and Gilbert and Lajtha (1965) were responsible for a most useful classification of cell populations *into compartments*, and these are shown in Fig. 2.3, and are, severally:

(i) *A closed static compartment*, in which there is no renewal of cells and

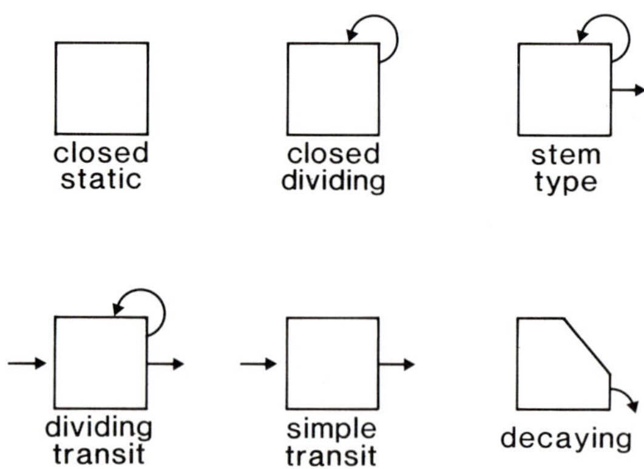


Fig. 2.3.  The various compartments which can be recognised in epithelial cell populations. (Redrawn from Gilbert and Lajtha (1965).)

which also has negligible cell loss, cf. the neurone population of the central nervous system—this compartment need not concern us more.

(ii) A *decaying compartment*, again with no cell renewal, and where the population size is declining; in normal tissues, this would be illustrated by the oogonial population in the adult ovary, where cells are regularly lost but not replaced. In abnormal circumstances, even compartments in epithelial renewal systems can become decaying compartments, as in the small intestinal villus cell population after irradiation or cytotoxic chemotherapy, where cell loss is proceeding at a normal rate, while input from the proliferative crypts is reduced or negligible.

(iii) A *stem cell compartment*; this has indigenous cell proliferation, but no input. Couched in these terms, a stem cell population or compartment is easy to define, but, as we shall see (section 2.9), the definition of an epithelial stem cell has evoked much argument. Nevertheless, this is a reasonable definition for normal tissues, and of course all renewal systems must have stem cells; older thinking suggested that all cells in the proliferating portion of a renewal population were stem cells, but, in recent years, the concept of small stem cell subpopulations has crystallized (see Chapters 8 and 19).

Thus stem cells must be capable of producing large numbers of descendants, and are responsible for feeding cells into succeeding compartments, which may be a *dividing transit compartment*, or a *simple transit compartment* (see below), and are also responsible for maintaining their own numbers; thus, on *average*, one daughter cell of a stem cell division will migrate and be lost from the stem cell compartment, while the other will remain to maintain the stem cell compartment size. Epithelial stem cells will obviously be housed within the proliferative compartment, and in a position appropriate for stem cell function; thus in the small intestinal crypt, we might expect the stem cells to be present at the base of the crypt (Chapter 19). Where there is no intrinsic spatial organization that would predict the site of the stem cell, as in the apparently proliferatively homogeneous basal layer of the epidermis, the recognition of the stem cell has been problematical.

(iv) A *dividing transit compartment*; here there is input (from a stem cell compartment or a previous dividing transit compartment), indigenous cell production, and output of cells into a succeeding compartment. With the proposal that stem cells are a minority of proliferating epithelial cells, it is evident that the greater part of the cell production in any renewal system is the province of the dividing transit compartment, which has thus been christened the 'amplification compartment' by the Manchester group, presumably because it amplifies the output of this stem cell compartment. Typical dividing transit compartments are found in the mid-crypt region of the intestinal mucosa, which receives cells from the stem cell compartment, and feeds cells into the maturation compartment, a simple transit compartment in the top of the crypt. In the epidermis, the dividing transit compartment is housed, with the stem cells, in the basal layer.

(v) A *simple transit compartment*; here cells have ceased proliferation, and although the compartment has input and output, there is no indigenous cell production. Typical examples include the functional villus epithelium in the small intestine, and the Malpighian layer in the epidermis.

(vi) A *closed dividing compartment*, with no input or output of cells, and retention of the cells produced. In epithelial systems, the liver regenerating after hepatectomy, or the castrate mouse prostate complex repopulating after androgen treatment, are examples. However, this would only be an ephemeral state, and the tissue would later revert to its resting low basal proliferative rate. Proliferating tissue cultures, developing organs and tumours in their early growth phase, with no or minimal cell loss, are also examples.

Using these compartments, it is possible to dissect all epithelial proliferative systems in terms of their compartmental structure; renewal populations such as the epidermis and the gastrointestinal epithelia consist of several of the above compartments linked together so that they feed sequentially one into another. Several different alternatives are depicted in Fig. 2.4, of increasing complexity. The addition of compartments in the more complex models is not through perversity, but because the simpler models do not comply with the available experimental evidence. For example, in the small bowel (see Fig. 15.4, p. 547) simple observation of the presence of mitoses in the crypt tells us that the crypts are the site of cell proliferation, while the villus epithelium does not display any mitoses, contains tall enzyme-replete cells (Fig. 1.1, p. 7) and may be regarded as the *functional compartment* of the epithelium. Hence we have at least two compartments, a *proliferative compartment* and a *functional compartment*. Similarly, in the epidermis, mitoses are largely confined to the basal layer, and the suprabasal layers are non-proliferative (see Fig. 4.1, p. 206). In these cases the simplest arrangement would be to regard all cells in the crypt or the basal layer as stem cells, and all cells in the villus and suprabasal layers as functional, simple transit cells (Fig. 2.4A). Now, careful analysis of the distribution of mitoses in the intestinal crypt shows that the uppermost crypt cells do not contain any mitoses (Fig. 18.5, p. 640) and are therefore also simple transit cells, probably in the process of differentiation, but still occupying the crypt; similarly, the Malpighian layer in the epidermis does not contain functional cells, but differentiating cells synthesizing keratin and other proteins in preparation for function in the stratum corneum. Hence we would have to place a further simple transit compartment into the scheme (Fig. 2.4B).

We have already alluded to the concept that stem cells in renewing epithelial cell populations form but a small subpopulation. The detailed evidence for this is discussed for the epidermis and gastrointestinal mucosa respectively in Chapters 8 and 19, but here we might temporise by saying that a combination of ultrastructural and radiobiological observations provide a strong argument for small stem cell subpopulations. In this case we would have to add a dividing transit compartment, as shown in Fig. 2.4C, where the input of the stem cell

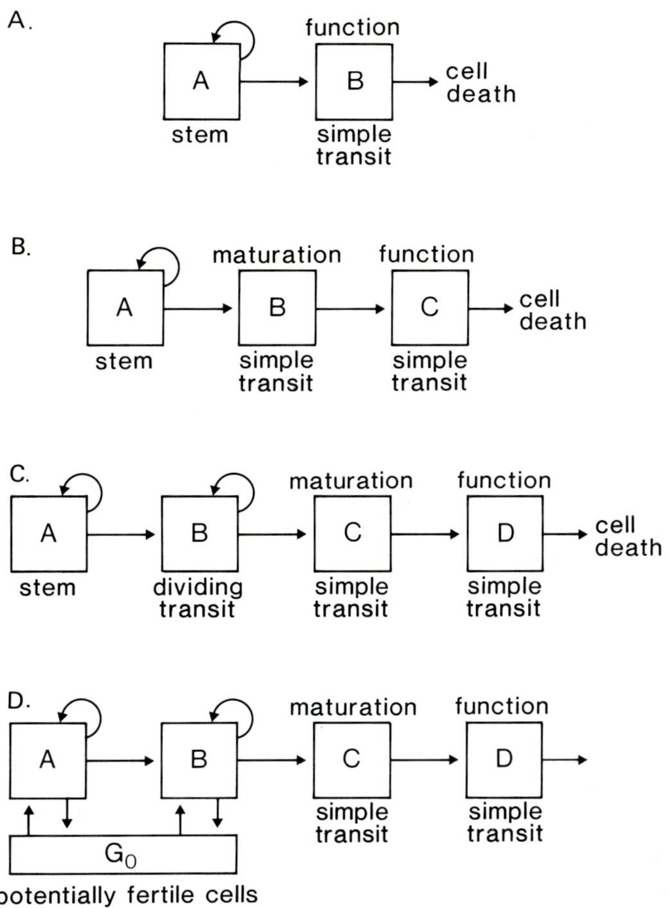



Fig. 2.4.  Sequences of compartments which make up epithelial renewal systems. (Redrawn from Cleaver (1967).)

compartment would be amplified. It will depend upon convention whether we add a separate dividing compartment for each cell position in the dividing transit compartment as Potten *et al.* (1982*b*) do, or whether we regard the compartment as a sequence of cells with an appropriate number of *transit divisions*. Of course, if there are morphological or other reasons for differentiating the component cells of a dividing transit compartment, we must naturally do so, but in many cases, cf. the intestinal crypt, there is no good reason for dividing up the compartment, and it is reasonable to regard it as homogenous.

Where do $G_0$ cells enter into this scheme? Well, those who advocate the presence of such a compartment in epithelial renewal populations will place it in the stem cell compartment, and possibly in the dividing transit compartment. Now there is good evidence that bone marrow stem cells have a $G_0$ phase (Lord 1976), and that they can be recruited into the cell cycle if conditions are appropriate (e.g. after haemorrhage). We shall however argue that there is little evidence for the presence of classical $G_0$ phases in epithelial *renewal* systems (Chapters 8 and 19), despite the extensive literature proposing this in the epidermis, for example (Chapter 7, section 4). However, in the conditional renewal tissues we firmly embrace the concept.

The schemes illustrated in Fig. 2.4 can be applied to all renewal systems, but, since many epithelial tissues cannot be classified as renewal populations, where would we put, for example, the conditional renewal systems as seen in the liver, kidney, and prostate? It is likely that some arrangement similar to that given in Fig. 2.4D would apply, but the centre of gravity of the population would be changed to the $G_0$ compartment. Even in the resting adult liver, we would see some cells in mitosis, so there are some proliferating stem or dividing transit cells. However, there is a subtle difference between such systems and the model shown in Fig. 2.4D; the $G_0$ cells are also the *functional* cells, and after stimulation, in the case of the liver by partial excision, all the remaining liver cells are capable of re-entering the cell cycle; similarly, in the mouse seminal vesicle, all cells re-enter the cell cycle after androgen stimulation. Hence, in the resting state, we would suggest the model shown in Fig. 2.12, p. 49, with most cells in $G_0$, but a few emerging to proliferate to compensate for the small amount of cell death due to 'wear and tear'. If there is a large amount of cell death, as occurs after partial hepatectomy, then the system would change so that the major flux would be from $G_0$ into the cell cycle. There are two unanswered questions here: what are the stem cells, and where does cell loss occur? Several experimental approaches have been used in an attempt to identify a stem cell subpopulation in the liver and prostate, without success (see Chapters 25 and 31), and it may be that all cells in conditional renewal populations have stem cell characteristics. Does cell loss occur from the $G_0$ compartment, or is there a separate simple transit compartment with its efflux towards cell death? We cannot answer this question at this time.


## 2.6. Kinetic parameters used to define epithelial renewal systems

One ultimate aim in epithelial cell kinetics is to be in a position to define any system in quantitative terms. This being so, how many parameters must we measure in order to define our system? This question was examined by Henry Quastler in his paper with Sherman in 1959. Consider a population containing $R$ different compartments, with *free transition* between each, and with cell birth and cell death occurring in each compartment. Such a system can be partially defined by the statistical parameters shown in Table 2.1.

We have already defined birth and death (loss) rates (see section 2.4); in

Table 2.1. Statistical parameters governing the kinetics of a system of $R$ compartments. (After Quastler and Sherman (1959))

| Parameter | Number of individual parameters |
| --- | --- |
| Size, $n_i$ | $R$ |
| Transit time, $t_i$ | $R$ |
| Birth and death rates, $k_{Bi}, k_{Li}$ | $2R$ |
| Flux from $i$th to $j$th compartment | $R(R-1)$ |
| Total | $R^2 + 3R$ |

kinetic terms, *absolute size* usually means the number of cells in a compartment, although sometimes the term is used to indicate the *fractional* or *relative* size of a compartment. The *transit time* is, or should be, the *average* time taken for a cell to traverse any compartment, be it a *spatial, kinetic* or *cell cycle phase* compartment, while the *flux* is the average rate of movement of cells from one compartment (here the $i$th) to another (here the $j$th compartment).

Epithelial renewal systems, in the unperturbed condition, are usually said to be in a *steady state*. This means that population size remains effectively constant, and, in the proliferative compartment, implies that on *average*, for every two cells produced at mitosis, one re-enters the cell cycle and one differentiates. The imposition of the steady state also implies that, on *average*, for each cell born at mitosis, one is also lost from the *functional compartment*. Such conditions are in fact found in the mouse small intestine, since Schultze *et al.* (1972) have shown that $r_S = r_M$ *for the whole crypt*, and Wright and Irwin (1982) have recently shown that the villus epithelium is in an overall steady state.

Our imposition of steady state conditions implies two further constraining factors. First, that the influx is equal to the efflux for each compartment, or using the notation shown in Table 2.1,

$$\sum_j k_{ij} = \sum_j k_{ji} \tag{2.20}$$

and secondly that the transit time is equal to the ratio of the compartment size to the total influx (or efflux), or

$$t_i = \frac{n_i}{\sum_j k_{ij}} \tag{2.21}$$

Thus for each compartment two constraining factors are postulated, equal to $2R$, reducing the number of parameters to $R^2 + R$.

Now consider the simple system shown in Fig. 2.4A, a two-compartment model. The total parameters (Table 2.1) are $R^2 + 3R$; where $R = 2$, this becomes $2^2 + (3 \times 2) = 10$ parameters. With the introduction of the steady state constraint,

$2 \times 2R$ or $4R$ parameters are removed (since here there are two compartments), leaving six individual parameters. In our two-compartment model, we have put the flux to cell death only from the functional compartment, and the arrow which indicates the direction of differentiation points only from the stem to the functional compartment; it is highly unlikely that functional cells will 'dedifferentiate' back to the stem cell compartment. Thus we can remove two further parameters. Up to this point, we have considered that birth can potentially occur into both compartments, but, as we have designated cell proliferation as occurring only in the stem cell compartment, with no cell production in the functional compartment, we can remove a further parameter.

These constraints thus reduce the number of overall independent parameters to *three*, $n_P$, $n_F$ and a single flux parameter, $k_{PF}$ (Fig. 2.5), where $P$ represents

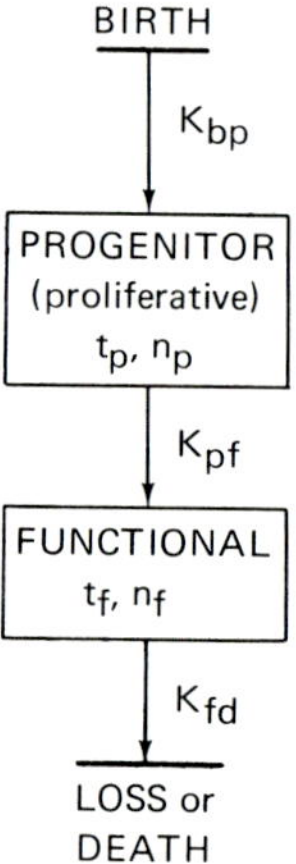


Fig. 2.5.   The parameters which define a simple two-compartment renewal system. (After Quastler and Sherman (1959).)

the progenitor or stem cell compartment and $F$ the functional compartment. Naturally, each compartment will have its own equivalent parameters of size, flux, and time, but, due to our steady state condition, $t_F = n_F/k_{PF}$. Since stem cells are, by definition, self-renewing and, on average, one cell remains at each stem cell division, the stem cell compartment does not have a transit time. An example of real-life analysis would be the villus epithelium of the small intestine (Fig. 2.6).

It should be appreciated that each individual average parameter may be the resultant of interaction between several individual parameters. Quastler and Sherman (1959) provide an important illustrative example of such interaction; the stem cell compartment could supposedly consist of two subpopulations of equal size, one of which had a cell cycle time of 24 hours, and the other of one month. The turnover time would be of the order of two days, but no single

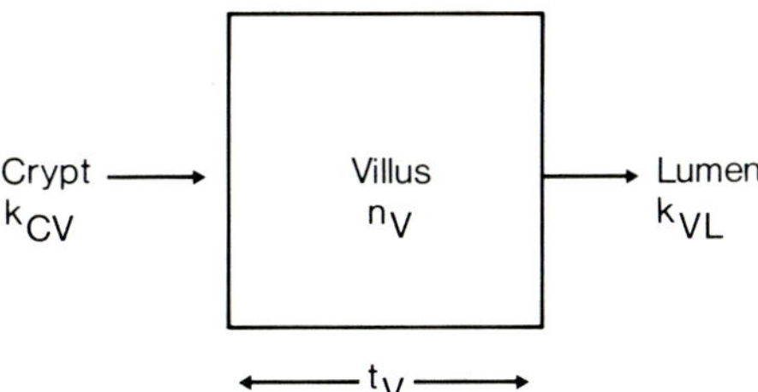



Fig. 2.6.   The parameters which define a kinetic compartment, here, for example, the functional villus epithelium of the small intestine: the influx ($k_{CV}$), the size or cell number ($n_V$), the transit time ($t_V$), and the efflux ($k_{VL}$). The villus epithelium is usually in a steady state, and hence $n_V$ remains constant, and $k_{CV} = k_{VL}$.

cell in the population would have a $T_C$ equal to this time; similarly, when we quote size, flux, and time parameters we should remember that each average parameter measured is the resultant of diverse parameters of the individual cells forming the compartment or population for which the parameter was measured.

Naturally, if we increase the number of compartments then we will increase the number of parameters which we have to define. In addition, there are one or two other parameters which it is convenient to use in the kinetic definition of renewal systems. Let us use an adapted version of the scheme shown in Fig. 2.4C, and displayed in Fig. 2.7; here we have merely lumped the maturation and functional compartments together as a single *non-dividing compartment*. The system is in a dynamic steady state; here a small stem cell compartment (actually one cell) feeds the dividing transit compartment, which in turn feeds a non-dividing simple transit compartment, which represents our maturation and functional compartments. While cells are in the dividing transit compartment

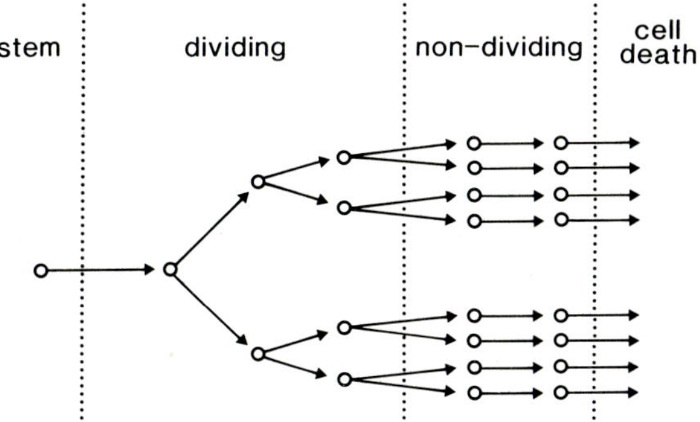



Fig. 2.7.   A simple model of a renewal system containing stem, dividing transit and non-dividing, simple transit compartments, the kinetic parameters of which are defined, in terms of the cell cycle time, in Table 2.2. (Redrawn from Cleaver (1967).)

they undergo three division cycles, or *transit divisions*. We can now proceed to dimensionalize the system in terms of a single parameter, the *cell cycle time*, assuming that duration of the cell cycle is constant throughout the renewal system, and that loss from the non-dividing compartment is the only flux to cell death (Table 2.2).

Table 2.2. Relationships between population compartments and their population kinetics

| Parameter | Stem | Dividing transit | Non-dividing (maturation or function) |
|---|---|---|---|
| Size | 1 | 7 | 16 |
| Cell cycle time $T_C$ (h) | $T_C$ | $T_C$ | – |
| Number of cell cycles ($m$) during transit | – | 3 | – |
| Cell birth rate, $k_B$ (cells/cell/h) | $1/T_C$ | $7/T_C$ | – |
| Input, $k_{in}$ (cells/cell/h) | 0 | $1/T_C$ | $8/T_C$ |
| Output, $k_{out}$ (cells/cell/h) | $1/T_C$ | $8/T_C$ | $8/T_C$ |
| Transit time, $T_T$ (h) | – | $3T_C$ | $2T_C$ |
| Turnover time, $T$ (h)* | $T_C$ | $7T_C/8$ | $2T_C$ |

*When a stem cell compartment cannot be distinguished from the dividing compartment, the turnover time in this combination of compartments is $T_C$. Turnover time is defined as the time taken to replace all cells in a given class.

We give the compartment sizes as the notional numbers of cells shown in the diagram in Fig. 2.7. If the number of transit divisions is $m$, this can be related to the fluxes into and out of the dividing transit compartment thus

$$k_{in} = 2^m k_{out} \qquad (2.22)$$

(where $k$ is the flux), and, since here $m = 3$, $k_{in} = 2^3 k_{out}$. We can solve eqn 2.22 in terms of common logarithms, and

$$m = 3.3 \log_{10} k_{out}/k_{in} \qquad (2.23)$$

(Cleaver 1967), which provides a useful way of computing $m$ if $k_{out}$ and $k_{in}$ are known (see Chapter 19, section 6). It is apparent from eqn 2.22 that the major determinant of the cell production rate (equal in effect to $k_{out}$) is the number of transit divisions in the dividing transit (proliferative) compartment; thus a major mechanism of augmenting the cell production rate is by *increasing the number of transit divisions*; the underlying reason why an increase in the growth fraction or an absolute increase in the proliferative compartment size elevates the cell production rate is because the number of transit divisions is increased, and thus $k_{out}$ increases. For the same reason that the stem cell compartment does not have a transit time, so the stem cell compartment, with no input, does not have transit divisions in the usual sense of the term. Naturally, the non-dividing compartment, with no cell production, does not have any transit divisions either.

The birth rate $k_B$ can be calculated from the reciprocal of the cell cycle time, assuming that all cells are in cycle (see eqn 2.10), and, in the stem cell compartment in Fig. 2.7, $k_B$ is $1/T_C$, while in the dividing compartment, which contains seven cells, $k_B$ will be $7(1/T_C)$, or $7/T_C$. The non-dividing compartment, of course, has no birth rate.

The next parameter to be considered is the influx into each compartment. By definition, the stem cell compartment has no influx, while the *influx* into the dividing compartment will equal the *efflux* from the stem cell compartment; the *efflux* from the stem cell compartment in turn equals the *birth rate* in that compartment, i.e. $1/T_C$. A useful point to remember is that *efflux from a compartment is the sum of the birth rates in all preceding compartments*; thus the *efflux* from the dividing compartment equals $1/T_C + 7/T_C$, or $8/T_C$. This will also equal the *influx* into the non-dividing compartment. It follows that, since the non-dividing compartment is a simple transit compartment, the influx will equal the efflux, and, with no birth rate, both the influx and the efflux of the non-dividing compartment will equal the efflux from the dividing transit compartment, or $8/T_C$.

There is no transit time in the stem cell compartment, but, in the dividing transit compartment, there are three transit divisions, so with a cell cycle time $T_C$, the transit time is $3T_C$. Now, since we have a steady state, eqns 2.20 and 2.21 apply, influx into the non-dividing compartment equals efflux, and from eqn 2.21, transit time = size/influx, or $16/8/T_C$, which simplifies to $2T_C$.

The *turnover time* is a useful concept, so long as we realize what it means, and calculation of the parameter for the different compartments shown in Fig. 2.7 exemplifies its application. Since we have imposed a growth fraction of unity in all proliferating compartments, the turnover time in the stem cell compartment is the cell cycle time $T_C$; in the dividing transit compartment, again since all cells are proliferating, the turnover time should equal the cell cycle time, but, since we have an input of $1/T_C$ from the stem cell compartment, and, since there are seven cells to be replaced, the turnover time is $7T_C/8$. Note that if we could not distinguish the stem cell compartment from the dividing transit compartment, the turnover time for both compartments combined would be $T_C$. In the non-dividing compartment, an important observation is that, since there is no cell division within the compartment, the turnover time will be equal to the time taken for any cell to pass through the compartment, or the transit time, $2T_C$.

A further parameter, the growth fraction, is in fact imposed by our constraints. We have made all cells in the stem and dividing transit compartments proliferative, by definition, so the growth fraction $I_P = 8/16$, or 0.5.

It might be appropriate to illustrate how far we can go to meet the demands made by the parameters in Table 2.2; Table 2.3 shows results for the mouse small intestinal renewal system (Fig. 15.4, p. 547). The first point at which we fall down is in the separation of the stem cell from the dividing compartment; we think we know where the stem cells are, but we do not know their number

Table 2.3. Kinetic parameters in the mouse jejunal crypt

| Parameter | Stem | Dividing transit (proliferative) | Non-dividing (maturation) | Non-dividing (functional) |
|---|---|---|---|---|
| Size* | 30 40 | 172 | 80 | 6150 |
| Cell cycle time (h) | 25 | 14 | – | – |
| Number of cell cycles in transit | – | 3 | – | – |
| Cell birth rate (cells/hour) | ? | 16 | – | – |
| Input (cells/h) | – | ? | 16 | 187[†] |
| Output (cells/h) | ? | 16 | 16 | 187[¶] |
| Transit time (h) | – | ~24 | 1.1[‖] | 37 |
| Turnover time (h) | ? | 24.8[‡] | 5[¶] | 37 |

*The figures given here are for 'functional stem cells' and 'clonogenic cells (see Chapter 19).

[†] Here the situation is complicated by the fact that the crypt:villus ratio, i.e. the number of crypts which feed one villus, is 11:1, therefore to find the influx of cells into the villus or functional compartment, the birth rate in the proliferative compartment is multiplied by the crypt:villus ratio.

[‡] All crypt cells.

[‖] For a totally non-proliferating cell.

[¶] Calculated from the influx and the compartment size.

accurately, or their detailed kinetics, so we are obliged to lump the stem cell and dividing transit compartments together; also, in this real renewal system, there is variability in cell cycle times in different parts of the proliferative compartment (see Chapter 18). We can, however, separate the dividing compartment from the non-dividing compartments, and indeed maturation from functional compartments; the maturation compartment occupies the upper part of the crypt, while the functional component is found on the villus; methods exist for independent measurement of parameters in each of these (see Chapter 17).

It is evident that we can, in fact, go a good way towards measuring all the parameters required, *so long as we can identify the compartments by spatial or morphological criteria*. In the intestine this is reasonably easy, but in other renewal systems, such as the epidermis, there is no such discrete spatial discrimination and thus measurement of the several parameters becomes very difficult, since we cannot by any means be sure which cells to place in which compartment (see Chapter 8).

In Chapter 1 we tried to emphasize the importance of population analysis, and in particular the three-dimensional aspects of morphology are critical in cell population kinetics; nowhere is this more important than in the small intestine. At first sight, it would appear as if the size of the functional compartment was much larger than the combined proliferative and maturation compartments in the crypt (see Fig. 15.4, p. 547). However, three-dimensional analysis has shown that, in the site selected (the jejunum) some 11 crypts feed one villus; this then accounts for the large influx into the villus epithelium, and, in point of fact, the crypt population (proliferative + maturation compartments), when calculated as the total crypt cells per villus, equals the villus cell population (Chapter 18).

## 2.7. More complex models: populations containing non-proliferative cells

### (i) The definition of a proliferating cell

Now might be the appropriate point to tackle the rather vexatious problem concerning the definition of a proliferative cell. How do we recognize proliferative cells? Well, the classical answer to this question would be those cells which are in DNA synthesis and mitosis at any time of observation. And how do we recognize non-proliferative cells? Again, these are cells which have acquired differentiation markers such as keratin in the epidermis or a goblet cell theca in the small intestine. However, if a cell has neither proliferative nor differentiation markers, then we are in some difficulty. Such a cell might be in $G_1$, $G_2$ or in some resting state which we might call '$G_0$'. Steel (1977) has suggested that it might be possible to define a proliferating cell as one which is *destined to divide*, but this is easier said than done. Even if a cell is in DNA synthesis, we cannot be absolutely certain that this cell is destined to divide: Claesson and Hartmann (1976) have shown that some 17 per cent of small thymocytes which enter DNA synthesis do not reach mitosis, but probably die instead. In psoriatic epidermis, Ralfs *et al.* (1981*a*) have shown that many more cells enter DNA synthesis than mitosis, and these could be lost by cell death, by decycling in $G_2$, or migration to a non-proliferative compartment, also with a $G_2$ DNA content; this phenomenon is also well recognized in organ culture (Simnett 1972). Thus despite the cliches which have appeared saying that $S$ phase cells are 'remarkably lacking in independence . . . without a life of its own . . . programmed and committed' — presumably to mitosis — there are now other reasons for doubting these and similar statements. Gelfant (1977) has proposed that there are $G_2$ blocked cells in many renewal and conditional renewal systems, and has supported this argument with much data (see below). Moreover, recent analysis of the data for both erythropoiesis and granulopoiesis indicate again that the flux into DNA synthesis exceeds the flux into mitosis by an amount which cannot be explained by age distribution considerations (Mackey and Dormer 1981; Mary 1981), and have led to the resurrection of a random proliferation–maturation scheme, whereby cells can leave the cell cycle and begin differentiation *at any point* in the cell cycle, rather than decycling at a specific point in the classical sequential models which we have hitherto proposed (Tarbutt and Blackett 1968).

Similarly with cells in $G_2$; we can detect these as tetraploid cells, and, since in renewal systems, at any rate, the maturing cells in the upper crypt in the intestine and Malpighian layer in the epidermis have a 2C complement of DNA, we can be reasonably sure that any such $G_2$ cell in a proliferative compartment will eventually divide; but if it is *blocked* in $G_2$ for some time, in a form of $G_2/G_0$ phase, we could be justified in calling this a non-proliferating cell. We encounter the same semantic problem when we consider the classical $G_0$ phase, in which cells with a $G_1$ DNA content are said to accumulate, in some tissues such as the liver. How do we distinguish these putatively non-proliferating cells from cells with a long cell cycle time (and hence a long $G_1$), but none the less orthodox proliferating cells; i.e. how long must a cell remain in $G_1$ before we call it a non-proliferating cell and place it in the $G_0$ phase?

There is no easy answer to this question, and the matter is not helped by the apparent differences between epithelial renewal and the so-called conditional renewal sysems, and by the fact that some authors argue, not for a true $G_0$ phase, but for phases of prolonged rest, which can occur in both the $G_1$ and the $G_2$ phases (Prescott 1976). Similarly, there are models of the cell cycle, such as that of Smith and Martin (1973, and see section 2.11), which eschew any reference to a $G_0$ phase, and state that, after mitosis, all cells enter an $A$ state, equivalent to the greater part of the $G_1$ phase; there they remain, but leave to enter DNA synthesis at random with a transition probability which depends on the tissue concerned or on its kinetic state. Hence tissues with a high proliferative rate, e.g. gut epithelia, would have a high transition probability, while tissues with a low proliferative rate, such as the liver, have a low transition probability, and most cells would remain in the $A$ state. On the other hand, there is the hypothesis which proposes that *all* cells enter a $G_0$ phase after mitosis, in which they remain for a period, and then leave to enter DNA synthesis and mitosis once more (Wolfsberg 1964; Burns and Tannock 1970; De Maertelaer and Galand 1977*a,b*); cells in renewal systems would remain for a short time, while those in conditional renewal populations would stay for a long period, unless stimulated out of the $G_0$ phase by a suitable stimulus. But, we would immediately say, how does the $A$ state differ from the $G_0$ phase, and the correct answer would be that there is, on this superficial level, no difference at all, providing that the transition probability could be held at a low enough level to account for the infrequency of mitosis in the normal liver, for example; a stimulus such as partial excision would result in an increase in the transition probability and hence the rate of cell cycle traverse. Smith and Martin's $A$ state would then be equivalent to the $G_0$ phase of Wolfsberg, Burns and Tannock, and De Maertelaer and Galand.

However, we believe that the problem is more than semantic, and this is because there appear to be distinct biochemical differences between putative $G_0$ and $G_1$ cells; there are discrete biochemical events which occur when classical $G_0$ populations, such as the liver or prostate, are ejected into the cell cycle, and these are discussed in Chapter 24. In addition to these differences, developments by Darzynkiewicz *et al.* (1979*a,b*) indicate that there are differences between $G_0$ and $G_1$ cells which can only be detected by multiparametric flow cytometric studies (see Chapter 3, section 8).

Thus it is possible to take the middle ground: because of the distinct biochemical differences between $G_0$ and $G_1$ cells, and the different biochemical sequence gone through by $G_0$ cells preparing for DNA synthesis compared with regular $G_1$ cells, we can propose *that a specific $G_0$ phase does exist*; it is probably largely confined, in epithelia, to conditional renewal systems, but that is not to deny its at least potential presence in certain compartments in renewal systems, although we shall personally argue against such a concept.

Is the $G_1/G_0$ phase the only possible resting compartment? Are there other resting phases in the cell cycle? This question has been considered in a series of salient publications, by Seymour Gelfant. Possibly because most cells appear to decycle with a $G_1$ DNA content (cf. epidermal suprabasal cells, Bullough 1965; small intestinal villus cells, Kember *et al.* 1961), most attention has focused on control points in $G_1$; viz. the transition point in the Smith and Martin (1973) model of the cell cycle is in late $G_1$, and Pardee (1974) has argued strongly for a unique restriction point '$R$' in mid-$G_1$ or at the $G_1$-$S$ boundary. However, in their classification of resting cell populations, Epifanova and Terskikh (1969) incorporated an $R_1$ population of $G_1$ arrested cells, and an $R_2$ population of cells arrested in $G_2$. Gelfant (1977) formalized this concept by recognizing four categories of cells: (i) *cycling cells*; (ii) *non-cycling $G_1$-blocked cells*; (iii) *non-cycling $G_2$-blocked cells*; (iv) *non-cycling $G_0$ cells*. These are shown schematically in Fig. 2.8. We should examine these concepts critically: Gelfant (1977) has summarized the methods which have been used to detect the presence of such subpopulations in several *in vivo* and *in vitro* situations; we hold no brief for the physiological importance of growth control experiments *in vitro*, so we will confine our comments largely to the *in vivo* situation. There are two main procedures: (i) some form of continuous [³H]-TdR labelling procedure, to label all cells in the population in the cell cycle, which may be followed by a stimulus such as wounding in the epidermis, which might be expected to stimulate any blocked cells; a search is made for *unlabelled cells* at appropriate times after the stimulation. These unlabelled cells are subjected to quantitative microdensitometry to distinguish $G_1$ from $G_2$ DNA contents (Gelfant 1976). Such unlabelled cells are deemed to be blocked in the appropriate phase; it is unfortunate that such a procedure involves the search for unlabelled cells, since we are aware of the problems of self-absorption and the difficulties that quantitative autoradiography can cause if such subpopulations are small (see Chapter 3, section 1). There are also the well known problems which beset such growth fraction studies with the continuous labelling method (see Chapter 3, section 5) and the additional hazard of non-labelling of cells in the $S$ phase: several authors (Clausen *et al.* 1980; Møller *et al.* 1982) have recognized this phenomenon, which appears to be widespread in the stratified squamous epithelium of the mouse epidermis and the hamster cheek pouch (see Chapter 6, section 1); since much of Gelfant's original work was done with the former tissue, without further studies on the part of the protagonists of such arrested subpopulations, this phenomenon could be held responsible for these apparently unlabelled cells.

The second procedure for the identification of arrested subpopulations is to

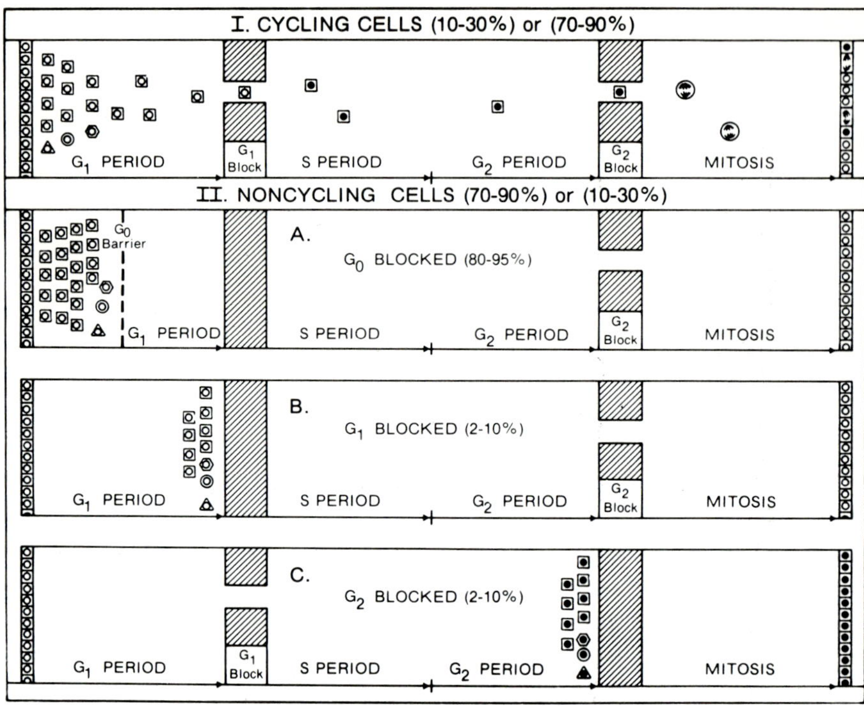



Fig. 2.8. A scheme representing the various compartments of cycling and non-cycling cells in a tissue system, as envisaged by Gelfant (1981); two possible situations are shown: a high growth fraction state and population percentages when only a few cells are cycling. (By courtesy of Dr S. Gelfant.)

experimentally open the blocks, as shown in Fig. 2.9; cells arrested in $G_0$ with a $G_1$ DNA content will show labelling only after the classical prereplicative period, followed at time $t_S + t_{G_2}$ later by a rise in the mitotic index (see Fig. 24.1); cells blocked in $G_1$ will rapidly enter $S$, and hence a more or less immediate rise in $I_S$ occurs without an intervening prereplicative phase. However, we must confess that, in renewing tissues, we do not see how one can operationally distinguish between a small $G_1$ blocked subpopulation triggered into $S$ and a general decrease in the cell cycle time produced by a contraction of $t_{G_1}$ in all cells, evoked by the stimulus; however, the *conceptual* difference is quite clear: we can either have cells that are blocked and released, or we increase the probability of cells entering $S$, perhaps by increasing the transition probability (see section 2.11); remember, on the Gelfant hypothesis, there are $G_1$ blocked cells and *normally cycling cells*, which are conceptually different populations. Thus the experimental evidence for $G_1$ blocked cells in renewing systems is equivocal; however, in populations where most cells are usually regarded as being in a

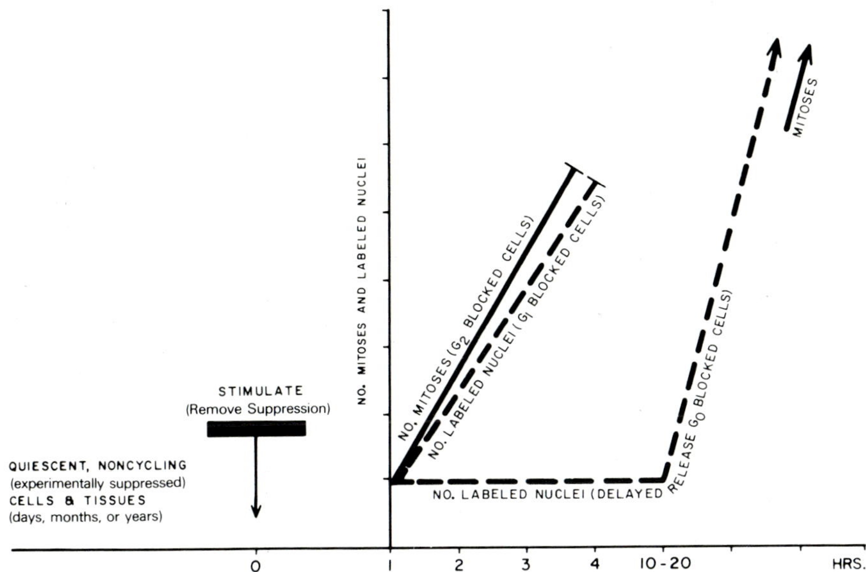



Fig. 2.9. The diagram shows the several methods used by Gelfant to distinguish the $G_0$ and blocked $G_1$ and $G_2$ populations postulated in both renewing and resting cell populations; the cell population is stimulated in some way, and the difference in timing of the rise in labelling and mitotic indices differentiates the various blocks. $G_2$ blocked cells lead to an immediate rise in the mitotic index, before the labelling index increase. $G_1$ blocked cells give rise to a rapid rise in labelling index, followed at a time $t = t_S + t_{G_2}$ later by a rise in the mitotic index as the stimulated cells reach mitosis. Cells in a putative $G_0$ phase take a finite time (the prereplicative period) before entering DNA synthesis and showing increasing labelling indices; the mitotic index will then increase at a later time point, also equal to $t_S + t_{G_2}$. (By courtesy of Dr S. Gelfant.)

classical $G_0$ phase, if stimulation evokes an immediate increase in the labelling index, how are we to explain it? Such a situation apparently occurs in hepatocytes stimulated with lead acetate (Fig. 2.10). Apart from incriminating such unlikely proposals as a marked contraction of the usually protracted biochemical preparation for DNA synthesis, we must postulate a $G_1$-blocked population.

Upon stimulation, $G_2$ blocked cells should enter mitosis before increased labelling indices become apparent; hence there should be a rise in $I_M$ before the increase in $I_S$. Figure 2.10 shows such a sequence. However conclusive this appears for the existence of $G_2$ cells, we should note the conclusions of Clausen and Thorud (1980); in stimulated mouse epidermis, treated with a variety of agents, there is an early increase in mitotic index. A detailed classical cell kinetic and flow cytometric analysis showed that this early increase in mitotic activity was due to an increased transit time through $G_2$ and mitosis; we shall also see that the early increase in mitotic index which follows carcinogen treatment is also due to a prolonged mitotic phase (Chapter 13). Clausen considers that, in

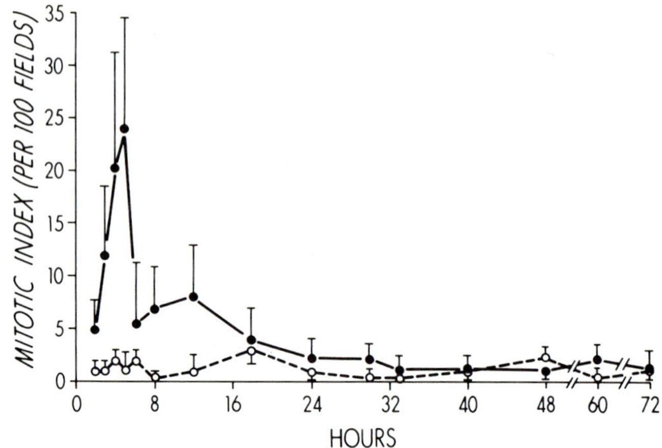

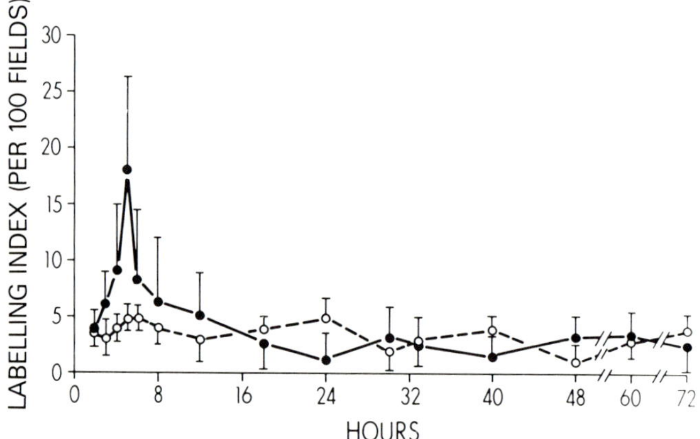

Fig. 2.10. The effect of lead acetate injection on proliferative indices in the mouse liver; note the rapid increase in the mitotic index, which rises and peaks at the same point as the labelling index curve. This indicates that a considerable number of hepatocytes are blocked in the $G_2$ phase, and are released by the lead acetate. The rapid rise in labelling index also occurs in a shorter time than would be expected if cells were being ejected from a $G_0$ phase by the stimulation (compare the prereplicative period after partial hepatectomy, cf. Chapter 25), and could represent the release of $G_1$-blocked cells in addition; on this interpretation, the secondary rise in the mitotic index between 8 and 12 hours represents the released $G_1$ cohort reaching mitosis (● experimental; ○ control). (By courtesy of Dr G. W. Richter.)

the epidermis at least, increased transit times were the cause of increases in $G_2 + M$ subpopulations after stimulation, in which case, perhaps we should beware interpreting mere proliferative index studies in terms of $G_2$ subpopulations (see Table 3, Gelfant 1977), until they are confirmed by rate measurements and flow cytometric studies. Work by Sauerborn and his colleagues (1978) has also negated $G_2$ subpopulations in epidermis.

We take up the discussion of the kinetic and proliferative status of arrested cells again in our consideration of conditional renewal systems, where Gelfant's concepts may have more applicability. In this respect, we should note that Darzynkiewicz *et al.* (1979*b*), on the basis of multiparametric FCM studies, and the denaturation properties of DNA, has proposed that not only can cells in classical $G_1/G_0$ be detected in lymphocyte populations, but that cells are in fact arrested in a resting phase in $G_1, S,$ and $G_2$ (see Fig. 3.28). It is generally true to say that $G_0$ populations do have a discrete prereplicative period, which all cells usually go through (cf. Fig. 24.1). On the other hand, there does appear to be a distribution of times for cells to leave $G_0$ after a proliferative stimulus, and we wonder whether the cells in $S$, for example, do not prepare for DNA synthesis faster than $G_1$ cells; however, most $G_0$ cell populations certainly behave as though most, if not all cells, have decycled in $G_1$, and we would question the size of these putative subpopulations. In any case, we shall see that, as yet, there is little in the way of definitive experimental evidence for $G_1$ and $G_2$ blocked cells in renewing epithelial cell populations; moreover, we shall argue that there is little in the way of *experimental* evidence for $G_0$ populations in renewing epithelial populations either.

We should also consider the relationship of the differentiated state to the resting $G_0$ phase. In renewing epithelial systems, for whatever reason, functional cells are terminally differentiated, and will not, short of nuclear transplantation with an $S$ phase nucleus, ever enter the cell cycle again. However, in conditional renewal systems, such as the liver, kidney or castrate seminal vesicle (a component of the mouse prostate complex), this relationship is by no means as clear cut. Partial hepatectomy, or unilateral nephrectomy, induces, after a short latent period, compensatory cell proliferation, in which many cells in the population participate; we might therefore say that most, if not all, hepatocytes are in the $G_0$ phase, but are also capable of performing their respective metabolic functions. However, resting, functional seminal vesicle cells in the adult animal are non-proliferating, fully differentiated protein-secreting cells (Fig. 2.11 B); upon castration, with the removal of testicular androgen, there is much cell loss, but the residual cells, which are derived from the previously functional cells (Alison *et al.* 1976), are again non-proliferative, but are small and metabolically inactive (Fig. 2.11 C). Stimulation with androgen injection produces a typical sequence of induced DNA synthesis from a $G_0$ compartment, and, with continuous androgen, the cells differentiate again and become secreting functional cells. Now, which are in $G_0$? It cannot be the differentiated cells, since these do not respond to androgen (Morley 1970), so it must be the small inactive cells which do respond. But the differentiated cells are *potentially* in a $G_0$ phase, to which they revert with the removal of androgen, and with later androgen treatment, enter the differentiated state again. In this case it is necessary to set up a model with a $G_0$ compartment and a functional compartment, with androgen-induced fluxes between them (Fig. 2.12). In fact this type of model can be quite successful in

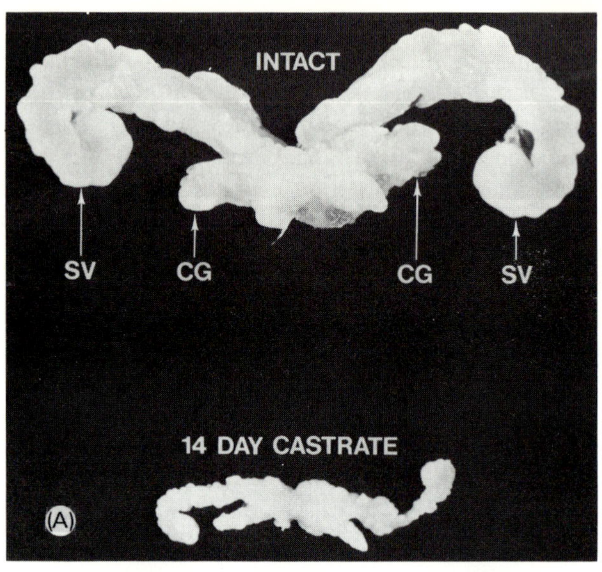
INTACT
SV
CG
CG
SV
14 DAY CASTRATE
(A)

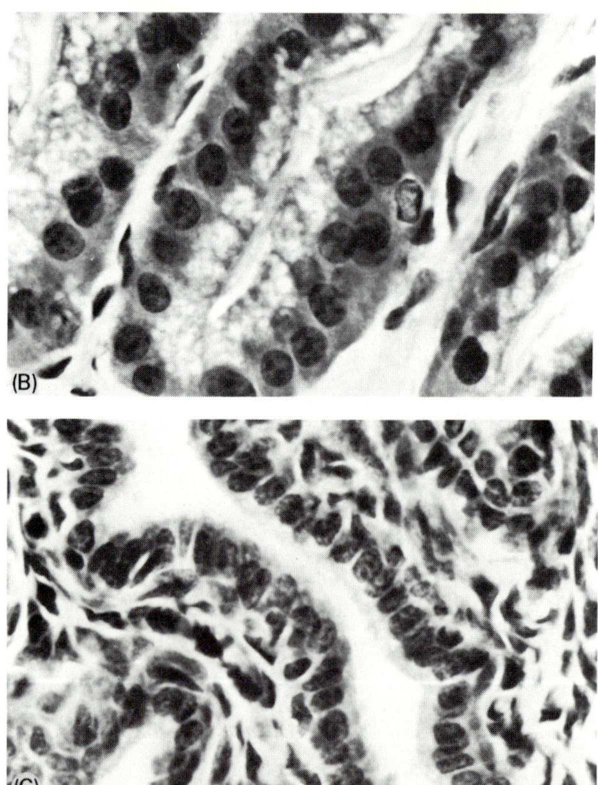
(B)
(C)

simulating experimental data derived from this system (Appleton *et al.* 1973; and see Chapter 31).

In the context of a model, it is of course possible to know which cells are proliferative, and which are non-proliferative, since we can follow Cairnie *et al.* (1965*b*) and Steel (1977) and call them *P* and *Q* cells for cells born into the respective categories. This indeed removes the problems of definition, and we can follow the fate of the respective *P* and *Q* compartments, but *only in the context of the model*.

### (ii) *The age structure of populations containing non-proliferating cells*

In our simple model of an exponential age distribution we assumed that, at

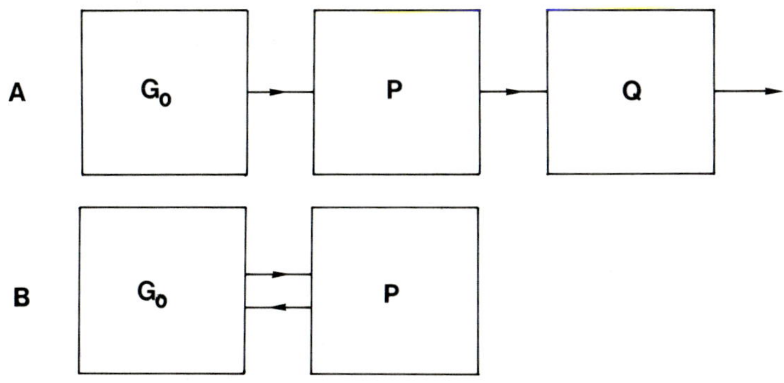


Fig. 2.12. To show the difference between $G_0$ populations as seen in the mouse or rat prostate complex (A) or the rat liver (B). The liver is the simplest case, and is seen in most parenchymal conditional renewal populations: the $G_0$ cells are also the *functional* cells, and on stimulation by partial hepatectomy these cells all become proliferative and replace the lost tissue mass. Upon completion of the response, the erstwhile proliferative cells revert to a functional $G_0$ condition again, i.e. in this model, the functional cells themselves are released from $G_0$. In the androgen-dependent prostate, the situation is quite different: the functional cells, although in a resting, i.e. non-proliferative phase, do not respond to androgen challenge. After castration, the cells lose functional characteristics (see Fig. 2.11C), but enter a true $G_0$ phase, from which they can be stimulated into proliferation by androgen injection. They then revert to the functional state again. Thus in this model, functional cells are not the $G_0$ cells themselves, but have to go through a sequence of changes before they become androgen-sensitive.

---

Fig. 2.11. (A) The mouse prostate complex, before and after castration. Note the marked reduction in size, caused by both a reduction in cytoplasmic mass, and also by a considerable depletion in the cellularity of the glands (SV = seminal vesicle; CG = coagulating gland). (B) The glandular epithelium of the intact mouse seminal vesicle. The epithelial cells are columnar, with abundant contained secretion. (Haematoxylin and eosin. ×900.) (C) The epithelium of the mouse seminal vesicle at 14 days after castration; the cells are almost cuboidal, with little evidence of apical secretion. There is no luminal secretion, and the amount of intervening connective tissue stroma between the glands is increased. (Haematoxylin and eosin. ×300).

cell division, both cells re-entered the cell cycle, Let us propose instead that, at mitosis, a proportion of cells opt out of the proliferative cycle (Steel 1968); let us not pursue the fate of these non-proliferative cells for the moment, other than making them non-proliferative; they could be moving into a maturation compartment in preparation for differentiation, or into a $G_0$ compartment from which they might return; again we impose a uniform cell cycle time. The convention is to denote the number of proliferating cells ($P$) produced at mitosis as $\alpha$; in an exponential age distribution, $\alpha = 2$, but if cells are decycling at mitosis, a number $2-\alpha$ are lost to the non-proliferating cell ($Q$) compartment

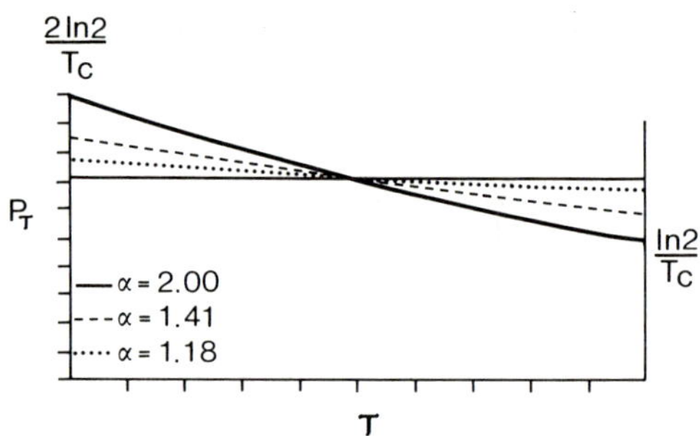


Fig. 2.13. The age distribution of cell populations which contain different numbers of $P$ cells produced ($\alpha$) at mitosis. With decreasing values of $\alpha$ the initial ordinate on the age distribution is reduced, until, when $\alpha = 1$, the rectangular age distribution is attained (cf. Fig. 2.2A). (Redrawn from Steel (1968).)

(Fig. 2.13). The reduction in the ordinate of the age distribution curves would then be expressed in terms of the growth fraction, $I_P$, as

$$\alpha = (I_P + 1) < 2$$

and the probability of decycling would be

$$2 - \alpha = 1 - I_P.$$

Now, since we have not imposed any cell loss flux, the general form of the exponential growth equation, with $t = T_C$, is

$$N_t = N_0 e^{k_G t}$$

where, with all cells proliferating, $k_G = \ln 2/T_C$. Now adding our recycling portion $\alpha$, $k_G = \ln \alpha/T_C$, and the population of $P$ cells grows according to the equation

$$N_t = N_0 \exp \left[ \frac{\ln \alpha}{T_C} t \right].$$

If we put $N_t = t_D$, the doubling time (i.e. $2N_0$) of the $P$ cells is

$$2N_0 = N_0 \exp\left[\frac{\ln 2}{t_D} t_D\right]$$

and hence the doubling time is given by

$$t_D = \frac{\ln 2}{\ln\alpha} T_C \quad \text{(Steel 1968).} \tag{2.24}$$

Now, the rate of production of new cells in the population (eqn 2.5) is $(dN/dt)_t = \ln 2 (N_t)/T_C$: at every mitosis, $\alpha$ new cells are born, and $\alpha - 1$ new $P$ cells are added to the population. Note that $2 - \alpha$ $Q$ cells are also added at this time. Thus the rate at which new cells begin a new cell cycle is given by

$$\frac{\alpha}{\alpha - 1}\left[\frac{\ln\alpha}{T_C} N_t\right].$$

As before (eqn 2.6), and following Steel (1968, 1977), the number of cells at age $\tau$ is

$$\frac{\alpha}{\alpha - 1}\frac{\ln\alpha}{T_C} N_{t-\tau} = \frac{\alpha}{\alpha - 1}\frac{\ln\alpha}{T_C} N_t \exp\left[\frac{-\ln\alpha}{T_C}\tau\right]$$

and normalizing to find the probability of finding a cell at age $\tau$ (Steel 1968) gives

$$P_\tau = \frac{\alpha}{\alpha - 1}\frac{\ln\alpha}{T_C} \exp\left[\frac{-\ln\alpha}{T_C}\tau\right] \tag{2.25}$$

The similarity between this equation and that derived by Appleton *et al.* (1977; see eqn 6.7) should be noted (see Chapter 6, section 7). The form of this distribution is shown in Fig. 2.13; again note the similarity of this diagram to Fig. 6.3; in the present case $\alpha$ is the general case for cells recycling, whereas in Fig. 6.3 $p$ is the specific fraction of cells remaining in the basal layer of the epidermis after mitosis, because of the horizontal orientation of the mitotic axis of the dividing cells in the basal layer. From eqn 2.25, the probability of finding a cell at zero age is 2, of which by definition $\alpha$ will be $P$ cells and $2 - \alpha$ will be $Q$ cells. Steel (1968) has calculated age distributions for both $P$ and $Q$ cells in this system (Fig. 2.14A), normalizing the distributions in order to make the probability of finding a cell about to enter mitosis unity. Note that the format of the curves is dependent on the value of $\alpha$; if $\alpha = 2$ the age distribution is exponential, but where $\alpha \to 1$ the age distribution tends towards the rectangular limit. Since here we have assumed that $Q$ cells are retained within the population, the age distribution for $Q$ cells has the same exponential form as for $P$ cells.

Just as we were able to integrate eqn 2.6 to find the mitotic index, we can

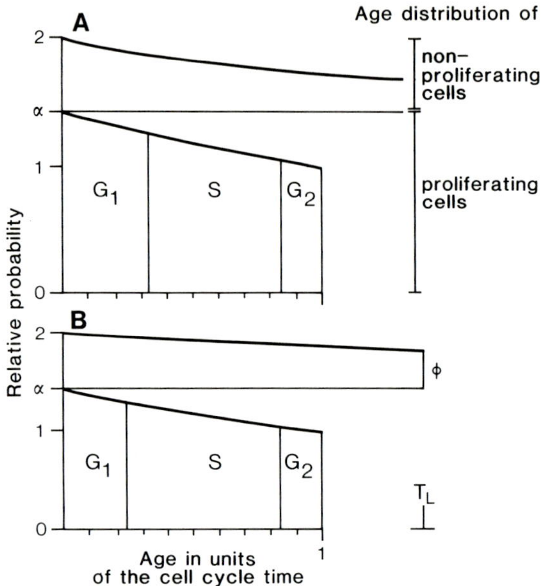



Fig. 2.14.  The age distribution for cell populations containing non-proliferating cells: (A) a no cell loss population in which there is a fixed probability for the production of non-proliferating cells at each mitosis; again $\alpha$ is the average number of cells at each mitosis that continue to proliferate. In this population, the age distribution of the non-proliferating cells stretches to infinite age.  (B) The age distribution of a cell population where the non-proliferating cells are lost at a fixed age: for example, this fixed age might represent the time taken for the non-proliferating cells found in the upper part of the intestinal crypt to migrate onto the villus, or for the post-mitotic cells in the basal layer of the epidermis to migrate into the Malpighian layer. In real tissue systems of course, there will not be a rigidly fixed maximum age for the non-proliferating cells, but some form of distribution of times to loss from the appropriate compartment. (Redrawn from Steel (1977).)

integrate eqn 2.25 over any desired portion of the cell cycle to derive the labelling or mitotic index in this case (Steel 1977). Following eqn 2.8, the fraction of cells in DNA synthesis, commonly called the *tritiated thymidine labelling index*, for $P$ cells ($I_{S(P)}$) can be written

$$I_{S(P)} = \mu t_S/T_C.  \tag{2.26}$$

Now the labelling index for the whole population ($I_S$) is

$$I_S = I_{S(P)} \cdot I_P  \tag{2.27}$$

(from which we derive the familiar relationship $I_P = I_{S\text{expt}}/I_{S\text{theor}}$, see Chapter 3, section 9). Combining eqns 2.26 and 2.27, we obtain

$$I_S = \mu t_S/T_C.(\alpha - 1)  \tag{2.28}$$

and, since $t_D = \ln 2 T_C/\ln\alpha$ (eqn 2.24), we can substitute for $T_C$ thus

$$t_D = \mu\frac{\ln 2}{\ln\alpha}(\alpha - 1)\frac{t_S}{I_S}$$

and putting $\lambda = \mu\ln 2/\ln\alpha \cdot (\alpha - 1)$

$$t_D = \lambda t_S/I_S. \tag{2.29}$$

Thus when $\alpha \to 2$, i.e. towards an exponential age distribution, $\ln 2(\alpha - 1)/\ln\alpha \to 1$, and $\mu$ takes values between $\ln 2$ or $0.693$ and rather greater than unity, and when $\alpha \to 1$, $\ln 2(\alpha - 1)/\ln\alpha \to 0.693$, and $\mu \to 1$. Steel (1968) has derived equations for calculating $\lambda$ when $\alpha$ and $\mu$ are unknown, but where experimental measurements of $t_D$, $t_S$, and $t_{G_2}$ are available; this has found most usage in tumour cell kinetics.

However, $\mu$ has been determined for the basal cells in mouse epidermis (Hegazy and Fowler 1973$a$). From eqn 2.28, since $(\alpha - 1) = I_P$, we have

$$I_S = \mu I_P t_S/T_C. \tag{2.30}$$

From this equation, Hegazy and Fowler calculated $\mu = 1.00 \pm 0.14$ in normal mouse epidermis, and $\mu = 0.92$ in epidermis where the hair had been plucked (i.e. where cell proliferation had been stimulated). In normal epidermis $\mu$ can equal 1 if $t_{G_1}$ is very short so that $t_S$ occupies the whole of the cell cycle, or if $t_{G_1} = t_{G_2}$, so that $t_S$ is in the middle of the cell cycle; neither of these conditions exists even in stimulated mouse epidermis (Chapters 7 and 10); moreover, since a random removal process for basal cells was proposed by Hegazy and Fowler, this precludes a rectangular age distribution, and an exponential age distribution was expected (see Chapter 6, section 7). For the cell cycle durations that they had obtained, Hegazy and Fowler considered that $\mu$ should equal 0.78 and 0.73 for normal and plucked epidermis respectively, and could not suggest an explanation for the disparity in the values; Duffill *et al.* (1977) suspected that a proportion of the mitotic axes in the mouse epidermis studied were vertically orientated, so that some cells were lost immediately after mitosis. This would have the effect of lowering $\alpha$ on the age distribution diagram, and of changing the value for $\mu$ (see Chapter 6, section 7).

Sometimes a more useful concept of decycling cells is provided by looking exclusively at the proportions which are recycling and decycling respectively; Bresciani (1968) represented the proportion of cells which decycle as $\overrightarrow{\eta}$, and the fraction which recycle as $\overleftarrow{\eta}$; on this notation the ratio $\overleftarrow{\eta}/\overrightarrow{\eta}$ is called the *distribution ratio;* on the Steel (1968) notation this would be $\alpha/2 - \alpha$. With application to our two compartment model (Fig. 2.4A), we could say that, for steady state, $\overleftarrow{\eta} = 0.5$ and $\overleftarrow{\eta}/\overrightarrow{\eta} = 1$. In many instances it is easier to work (and think) in terms of the *decycling probability* $\overrightarrow{\eta}$, since this is amenable to experimental measurement by the grain count halving technique (see Chapter 3, section 6); in fact the decycling probability can be used as the central component of a useful model for the analysis of induced DNA synthesis in conditional renewal systems (Appleton *et al.* 1973; and see Chapter 24).

## 2.8. More complex systems: the age structure of populations with cell loss

It is time we removed the final constraints on our models, and looked at the situation where (i) cell loss occurs, and (ii) where the cell cycle time varies. Steel (1968) calculated the age distribution for a variety of populations with different varieties of cell loss, i.e. loss from *different phases of the cell cycle*, and from *different compartments*. There are two reasons for pursuing such considerations here: first, in epithelial systems the age distribution is intimately connected with the mode of migration from the proliferative compartment; by *mode of migration*, here we mean from which cell cycle phase cells migrate into further compartments, and in particular into the differentiating compartment. For example, in our three-compartment model (Fig. 2.7), cells must leave the proliferating compartment to enter the maturation compartment in some phase of the cell cycle; this can be *random* with respect to phase, or migration can be a *cell cycle phase specific event* (in $G_1$, for example). Whichever mode of migration we select will have important consequences for the age distribution of the proliferating and maturing cells. The second reason is that we can attempt to model continuous labelling curves, putting forward certain modes of cell loss (see Chapter 3, section 5).

While most of Steel's models are directly applicable mainly to tumour cell populations, there are in fact important lessons within them for epithelial cell populations. It is usual to propose that, in epithelial populations, cell loss takes place as a consequence of cell maturation, in which process cell loss occurs from the oldest non-proliferating cells in the population, and this may be the arrangement in cell populations where we propose that cell migration from compartment to compartment is a strictly sequential process, for example as Fig. 2.7; the cells decycle at the dividing transit:maturation boundary, and cells feed into the maturation compartment. With 'first in–first out' kinetics, the cells that are lost to the functional compartment and hence move on to cell death are the oldest cells in the population (Fig. 2.14B). In real cell systems, this may apply in the small intestinal crypt (see Chapter 19) and in the EPU model of cell proliferation in the epidermis (Chapter 8), where cells move from a proliferative compartment in the crypt or in the basal layer, into a maturation compartment, still in the crypt and basal layer, before being lost onto the villus or into the Malpighian layer respectively. However, Appleton *et al.* (1977) have considered the situation where cells are lost from the basal layer of the epidermis directly into the suprabasal layers at *various different ages in the cell cycle*, i.e. we have loss of cells directly from the proliferative compartment without a maturation stage in the basal layer, which situation may be particularly applicable to non-stacked, rapidly proliferating epidermis. We can generalize

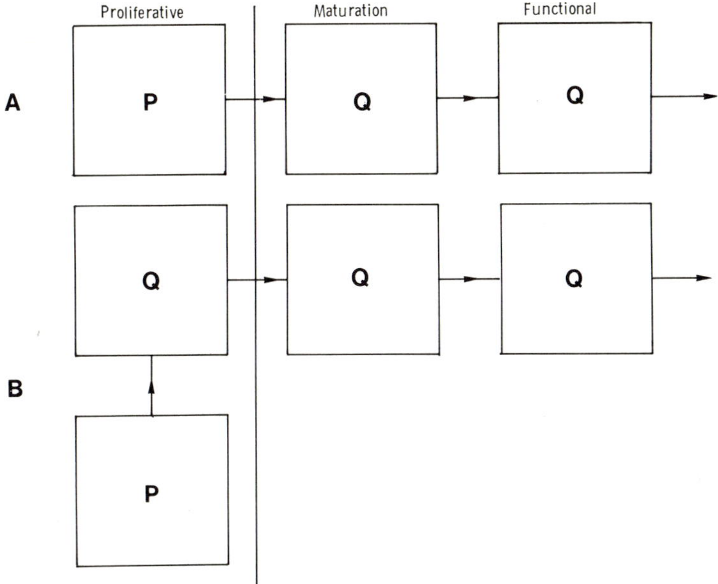


Fig. 2.15.  To illustrate the conceptual difference in the age distribution between the proliferative compartment containing wholly proliferative cells (A) and a 'proliferative' compartment which also houses non-proliferating $Q$ cells (B).

these alternatives by drawing Fig. 2.15.

It is considered a general rule that, in the absence of cell loss directly from the proliferative compartment, a steady state situation demands a rectangular age distribution in the proliferating cells; however, if cell loss is occurring directly from the $P$ cells, then we may not have a rectangular age distribution. We are particularly interested in the age distribution of this anatomical proliferative compartment; in model A (Fig. 2.15) the proliferative compartment contains $P$ cells only, i.e. $I_P = 1$, and this feeds cells into the maturation compartment of of $Q$ cells. This would be the case where the basal layer of Fig. 4.1 consists entirely of proliferating cells, and cells are lost directly into the $Q$ cell Malpighian layer. In model B (Fig. 2.15), the proliferative compartment contains both $P$ and $Q$ cells, and also feeds into the maturation $Q$ cell compartment; this would be the situation in the EPU model of Potten (1976; and see Chapter 8, section 2), and also in the sequential model of the intestinal crypt, where we regard the whole crypt as the proliferative compartment. Hence, if we need the age distribution for the proliferative compartment, in model A the age distribution would lie between a rectangular format and an exponential one, but in model B the age distribution must be rectangular. Lest this distinction appear artificial to the reader, we should perhaps emphasize that the models shown in Fig. 2.15 are in fact very much different if we are considering alternative kinetic organization schemes for the basal layer in stratified squamous epithelium, where we are testing different modes of migration for the basal layer cells (see Chapter 6, section 7).

The simplest case is perhaps where the cells that are lost are the oldest in the $Q$ cell compartment, and that they are lost at a well-defined age $t_{mat}$; $t_{mat}$ may be the time required to migrate through the maturation compartment in the intestinal crypt (Chapter 18) or through post-mitotic cell positions in the EPU model (see Chapter 8) (Fig. 2.14B). We have two limiting situations: where the age distribution is *exponential*, and where it is *rectangular*. As Steel (1977) points out, with an exponential age distribution, the imposition of cell loss at a certain age, if such loss continues, will lead to an increase in the growth fraction, but the growth rate will remain constant. But *we* need to consider populations which are in an overall steady state. As mentioned above, such systems generally have a rectangular age distribution. An important point to realize is that, notwithstanding this proposition (for which there is experimental evidence in the small intestine, see Burholt *et al.* 1976, and Chapter 19), the age distribution of component parts of the population *may very well be non-rectangular*. To illustrate, let us consider the dividing transit compartment of a proliferative compartment in a model such as that shown in Fig. 2.7; if all cells in the compartment were proliferative, and decycling was confined to the last division in the compartment, the age distribution of the population when the decycling divisions are excluded is exponential. This is, in fact, the kinetic organization as seen in the intestinal crypt, where the age distribution of the lowermost 10 cell positions is exponential, but the age distribution of succeeding cell positions, where decycling divisions are occurring, is so organized as to give an overall rectangular age distribution (Cairnie *et al.* 1965*b*; Schultze *et al.* 1972; Burholt *et al.* 1976; and see Chapter 19); the age distribution of the $Q$ cells is also rectangular, but only if the aging sequence is strictly ordered. In fact, experimental evidence suggests that there is a distribution of $t_{mat}$ in the small intestine (Chapter 18) so that the age distribution of $Q$ cells would not be strictly rectangular. The age distribution of such sequential populations is discussed, with reference to the intestinal crypt, in Chapter 19.

In the alternative model, we have loss of proliferative cells at some point in the cell cycle; remember we have to maintain a steady state, so that, on average, for every cell which is added to the population, one should leave. Let us introduce cell loss at mitosis: to fulfill the steady state requirement, at mitosis, on average, one cell would decycle, and one would recycle; we would have, by definition, a rectangular age distribution. We have just this situation in the epidermis of the embryonic mouse: here all cells divide so that the mitotic axis is orientated in a vertical plane (Smart 1970*a*; Fig. 2.16A): this means that, at each mitosis, one cell remains in the basal layer, to proliferate, and one

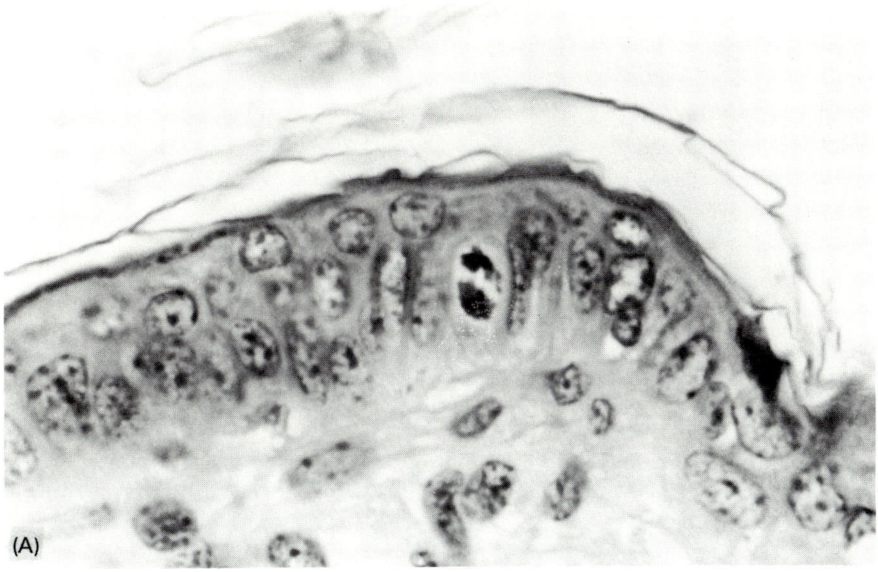

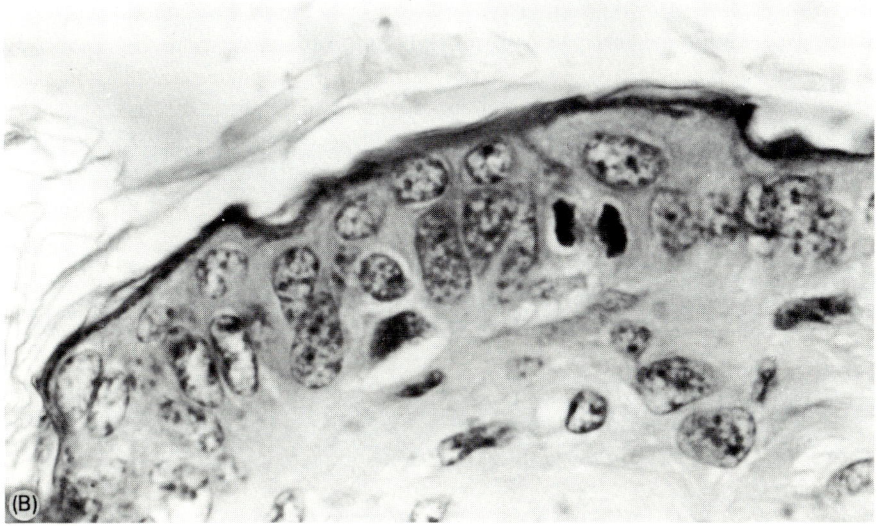

Fig. 2.16. The direction of the mitotic axis in stratified squamous epithelium: (A) a mitosis dividing with a vertical mitotic axis, i.e. with the result that one daughter cell is immediately removed into the Malpighian layer, and one remains. (B) a mitosis with a horizontal axis. Here both daughter cells remain in the basal layer, and an adjacent cell is ejected from the basal layer. (Mouse dorsal epidermis. Haematoxylin and eosin. × 1350.)

decycles into the Malpighian layer. This means that the age distribution in the basal layer must be exactly rectangular (see Chapter 6, section 7).

The other limiting case, which would lead to an exponential age distribution, is where cell loss from the proliferative compartment is random with respect to age. This situation would occur in our basal layer model if cells divide in a horizontal direction (Fig. 2.16B) and, by population pressure, eject a cell at *random* with respect to age. The horizontal mitoses mean that, at mitosis, two new cells are added to the population, so that $\alpha = 2$, as for an exponential age distribution; cells then leave the cell cycle at random, which will lead to an exponential decline in population density as cells age in the cell cycle: we thus have our promised age distribution. Note that we need not have either of the two limits: if there is a distribution of mitotic axes, i.e. some cells divide horizontally and some vertically, then we would have an intermediate but non-exponential age distribution similar to that shown in Fig. 2.13, where $\alpha$ takes different values. This is, in fact, a simplistic summary of quite a complex kinetic situation, where experimental evidence is conflicting, and this is considered further in Chapter 6, section 7, and Chapter 8.

There remains the problem of variation in the cell cycle time. Ignoring time-lapse cinematography, which is not applicable to *in vivo* epithelial systems, we cannot directly measure the distribution of cell cycle times or of the component phases; hence any variation must be imposed by a chosen distribution in terms of a model. Such information can be derived from a fraction of labelled mitoses curve (see Chapter 3, section 4) by making appropriate assumptions about the cell cycle time distribution (e.g. lognormal, Steel 1977; gamma, Appleton *et al.* 1977) and assuming an exponential age distribution. Figure 6.5 (p. 245) shows the age distribution of an exponentially distributed population (a hypothetical basal layer where all mitotic axes are in the place of the basal layer, i.e. horizontal) for values of the coefficient of variation of the cell cycle time set at 20, 30, 40, and 50 per cent; here the age distribution is drawn in terms of the distribution of times to mitosis, but can readily be transformed to the more familiar format by reversing the picture.

From these and later considerations (see Chapter 6, section 7, and Chapters 8 and 19) it will be clear to the reader that, far from being the simpler alternative to exponentially *growing* populations such as experimental tumours, the age structure of steady state cell populations, as exemplified by epithelia, can be extremely complex. Moreover, there is little doubt that the age distributions we are considering are in fact too simple: for example, in Chapter 9 we shall see how circadian variation in phase fractions can modify the age distribution, and, in many kinetic models for real systems (e.g. the EPU argument, see Chapter 8, section 2) there is quite extreme variation in the duration of the cell cycle between the several portions of the proliferative compartment. But, as Aherne *et al.* (1977*a*) remark, inconsistencies between model and reality should not scandalize the reader; better experimental data will lead to better models.

## 2.9.  Stem cells and clonogenic cells in epithelia

*Definition*

It is probable that the term 'stem cell' means different things to different people, and nowhere is this more evident than in the consideration of epithelial stem cells. Because steady state populations continue to turnover in the unperturbed state, there is no doubt that, somewhere in the population, is a compartment which contains cells capable of giving rise to large numbers of descendants. We could then regard stem cells as cells with 'unlimited growth potential', which, in the context of *in vivo* systems, we would have to qualify 'unlimited' to apply only to the life of the organism (Lajtha 1967). Perhaps the most reasonable definition is given by Steel (1977), that *stem cells have the capacity to produce a large family of descendants within their own natural environment*, and here 'large' would have to be sufficient to cope with the lifetime needs of the organism; for example, if we assume that all cells are proliferating in the mouse intestinal crypt, with a (maximum) cell cycle time of 24 hours, over 700 divisions would be necessary to maintain a mouse over two years, and, with the same assumptions in mouse epidermis, with a maximum cell cycle time of five days, nearly 150 divisions would be required over the same period, and that is to ignore the number of cell divisions needed to form the neonatal animal from the zygote. Moreover, haemopoietic stem cells from the mouse can repopulate the bone marrow of an irradiated younger mouse, and mouse skin has been maintained by serial transplantation for about five years in syngeneic animals. There is thus little evidence that stem cells have a limit of around 70 divisions imposed upon them, as the *in vitro* arguments of Hayflick (1965) would suggest.

In epithelial systems, the term *stem cell*, largely as a consequence of our sequential models of the organization of cell populations, has come to mean those cells which are most unspecialized or primitive, i.e. most lacking in differentiation features, but which give rise to a large number of cells which do differentiate, and also to new stem cells, i.e. *they maintain their own numbers*; for the stem cell population to remain constant, *on average* one cell would remain, and one leave to differentiate. How can they do this? Again this debate has generated more heat than light. Since the steady state is a statistical phenomenon, we should consider the population at large rather than single cells, and the various possibilities are shown in Fig. 2.17; in (i) each mitosis would give rise to a stem cell and to what we might call, for the moment, a non-stem cell. This division is called the *differential* or *asymmetric mitosis*, implying that some mechanism exists for the endowment of the daughter cells with different genetic material leading them to different fates; (ii) here one stem cell would give rise to a pair of new stem cells, while another produces a pair of cells destined to differentiate, a *symmetric cell division*. This would need some mechanism which ensures that stem cells which divide produce equal numbers of paired stem and non-stem cells. In (iii) cell (a) gives rise to two new stem cells,

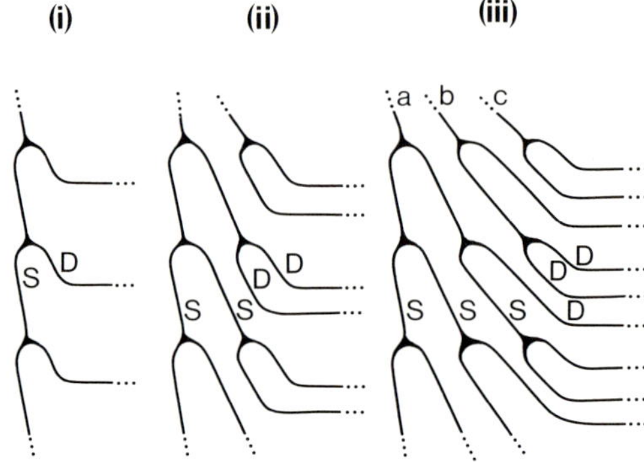



Fig. 2.17.   Three schemes describing the alternatives for stem cell divisions, to satisfy the condition that equal numbers of stem (S) and non-stem (D) cells are produced. (Redrawn from Leblond *et al.* (1967).)

or a stem and a non-stem cell, while stem cell (b) provides two cells which will differentiate. The pattern of these divisions may occur at random, or there may be some organized sequence (Leblond *et al.* 1967).

This far, so good. However, until now we have only considered the normal, unperturbed state, in which stem cells operate in a 'functional' (Cairnie 1976) or 'effective' (Steel 1977) role, giving rise to their descendants in *normal growth conditions*. However, should we disturb the system by, say, irradiation, there may be other proliferating cells, destined later to differentiate, that can be stimulated to act as stem cells, i.e. to produce large numbers of descendants, and to contribute, perhaps largely, to repopulation; these are called *potential stem cells*, to denote that they do not normally have stem cell functions, but can act as such in a pinch; it is unfortunately true that the morphological identification of epithelial stem cells is not reliable, and that we are obliged to rely for their recognition and their enumeration, largely on the use of *functional assays*, namely their performance in tests of *clonogenic capacity*; however, we should distinguish each of the previous definitions from the term 'clonogenic cell'; etymologically this means a cell which is capable of forming *a clone*. Although normal tissues do contain notional clones, in the sense that there are families of descendants of single cells present, here the term indicates an experimentally recognizable clone of descendants, it is to be hoped from a single cell. While there is no problem in defining a clone operationally in terms of cell number for work in cell culture (Nias *et al.* 1965), a clonogenic assay in epithelial tissues implies studying the behaviour of the cells in a perturbed, test environment, where the cell population has usually been reduced to but a few clonogenic cells, which then repopulate to form an identifiable clone. We must therefore follow Steel (1977) and *distinguish* between clonogenic cells and stem cells;

clonogenicity refers to the cells' capacity to repopulate in the test environment. It is generally hoped that clonogenic cells and stem cells are one and the same population: but there is no guarantee of this, and indeed, little experimental proof (see below, and Chapters 8 and 19).

*Models for stem cell behaviour*

In our introduction to the sequential models of renewal systems, we noted that it is customary to separate proliferative compartments into 'stem' and 'dividing transit' subcompartments: this signifies that we generally do not think that all proliferative cells are stem cells; why is this? Such a distinction is, in fact, relatively recent. Largely emanating from the work of Leblond and colleagues in the sixties, it was assumed that the cells arising from stem cells comprise equal numbers of stem and differentiating cells (Fig. 2.17; Leblond *et al.* 1967). While based on observation, this proposal remains an assumption; Leblond *et al.* (1967) noted that for dividing basal cells in the oesophagus (a typical non-keratinizing stratified squamous epithelium), infusion of [³H]-TdR for a sufficient period labelled all basal cells, which were then *all* deemed capable of cell division. A single injection of [³H]-TdR labelled a fraction of basal cells, and, of course, in the fullness of time, practically all these labelled cells divided with mitotic axes effectively wholly in the plane of the basal layer; each gave rise to a pair of daughter cells with similar grain density. The distribution of pairs of cells which will be (a) wholly in the basal layer; (b) wholly in the suprabasal layer; and (c) mixed pairs, i.e. one daughter in the basal layer and one above in the suprabasal layer (Fig. 2.18), was found to be random; it was concluded that *all* basal layer

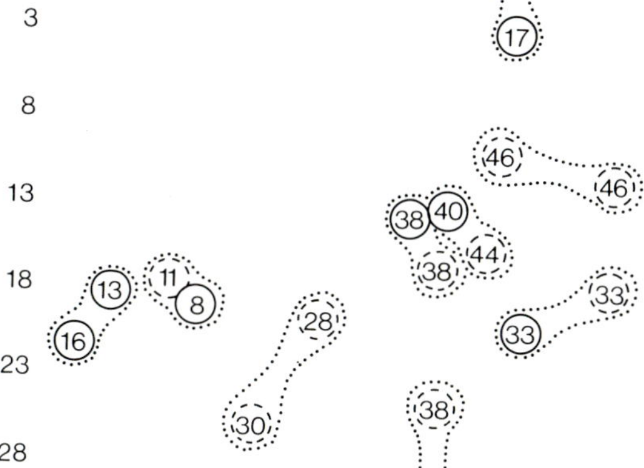

Fig. 2.18. A map of the distribution of pairs of labelled cells at 48 hours after [³H]-TdR injection in the mouse oesophageal stratified squamous epithelium, showing the various types of paired cells recognized in serial sections: basal cells – solid line circles; differentiating suprabasal cells – broken line circles, and mixed pairs. Numbers on the left represent serial section numbers, and those in the cells indicate the grain count. (Redrawn from Leblond *et al.* (1967).)

cells were stem cells, and that a random removal mechanism ensured that all basal cells had the same chance of being ejected into the suprabasal layer. Hence either one, both or neither of a pair of daughter cells will migrate (see Fig. 2.18). Thus the migration pattern would appear to fit scheme (iii) in Fig. 2.17, with a random distribution of migration patterns. Here the division sequence is *determined* by the migration pattern, and the motive force for migration was assumed to be simple population pressure induced by cell division, with (nearly) all mitotic axes in a horizontal direction. Note that this hypothesis fits all the observations, but also note that the proposal that all basal cells are stem cells *remains an assumption*.

Similarly in the intestinal crypt (Fig. 2.19): all proliferating cells were assumed to be stem cells, and again scheme (iii) in Fig. 2.17 was considered to apply; cells in the lower part of the crypt gave rise to two new stem cells (a division sequence shown by (a) type stem cells in Fig. 2.17iii), while cells in the intermediate zone give rise to mixed pairs of stem and differentiating cells, or to two differentiating cells. However, this scheme was seen to differ from that in the oesophagus in that an ordered migration sequence operated, rather than a random removal mechanism. In the rat and human testis, despite the fact that there are several different types of spermatogonia (variously named, see Chapter 31), and that three or four transit divisions were thought to occur in transit from the $A_d$ spermatogonia in man and the $A_1$ spermatogonia in the rat to the first non-proliferating, differentiating cell, all spermatogonia were considered to be stem cells (Leblond *et al.* 1967). The type (ii) division mechanism in Fig. 2.17 was considered appropriate.

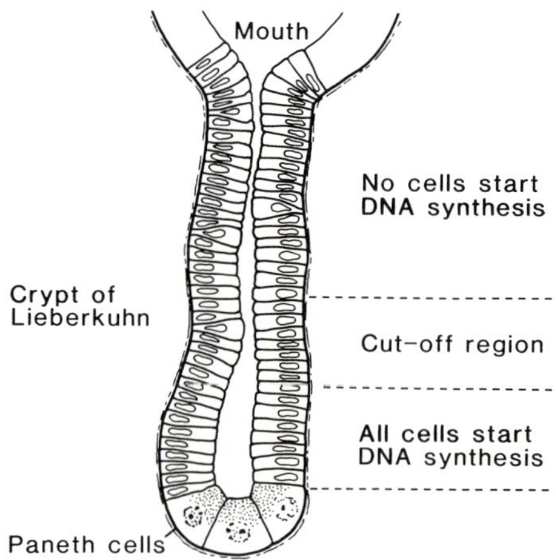


Fig. 2.19. The organization of stem cell divisions in a more complex system such as the small intestine, as envisaged by Leblond *et al.* (1967), from a consideration of the division sequences shown in Fig. 2.17. (Redrawn from Leblond *et al.* (1967).)

The alternative viewpoint was put forward by Lajtha and coworkers (Lajtha 1963, 1967; Potten *et al.* 1979), who considered that normal renewing populations in fact consisted of cells which could be called stem cells under our above definition, and others which were indeed proliferating, but which only had a limited number of transit divisions (see Fig. 2.7), before finally giving rise to non-proliferative, differentiating cells. Thus the essence of the difference between these models is the addition of a non-stem dividing transit compartment, i.e. the model shown in Fig. 2.4C instead of Fig. 2.4A.

This suggestion is founded upon two main items of experimental observation:

(a) information from irradiation-induced clonal regeneration studies indicate that clonogenic cells are but a small fraction of the total proliferative cellularity;

(b) that some epithelia have a distinct ordered structure: this implies that there may be a distinct role for each cell within this structure, and that the cell at the origin of the ordered arrangement, i.e. where flux through the system begins, is the stem cell for the system concerned.

*Clonogenic cells in epithelia*

The first evidence that clonogenic cells were a small proliferative subpopulation of a renewal system came from work on the bone marrow: when mice are lethally irradiated, injected syngeneic bone marrow cells become localized in the spleen where they form macroscopically evident colonies, in which all haemopoietic cell lines can be recognized (Till and McCullough 1961). Of the injected bone marrow cells, only some 1 per cent have this spleen colony forming (CFU-S) ability, and thus these cells form only a small fraction of the bone marrow cellularity; the remaining (i.e. non-CFU-S) proliferative cells have considerable ability to divide, and indeed, can form colonies in agar, but unlike the CFU-S cells, cannot be transferred, i.e. are not self-maintaining; moreover, $[^3H]$-TdR suicide techniques indicate that less than 10 per cent of the CFU-S are in DNA synthesis (Becker *et al.* 1965), and many are considered to be in the $G_0$ state (Lajtha 1963). Thus the appropriate sequential model for haemopoiesis might be that shown in Fig. 2.4D, with the $G_0$ phase constrained to the stem cell compartment. The stem cells form a small subpopulation of the proliferating cells, while the remainder are amplifying divisions of cells committed to a particular differentiation pathway, be it granulocytic or erythrocytic.

However, the enumeration of clonogenic cells in epithelial tissues is not nearly so easy; since epithelial cells cannot generally be transplanted in the same way, and no adequate *in vitro* assay system is yet forthcoming, all epithelial clonogenic assays must, of necessity, be made *in situ*. This introduces many logistic problems: it in effect means that the actual population under consideration must itself be exposed to graded doses of (usually) radiation (but see Moore 1979 and Chapter 21), and, in some way, clonogenic cell survival must be measured. Immediately the definition of a clone becomes a problem, and this was imaginatively approached by Withers (1967*a,b*; Withers and Elkind 1969,

1970), who developed a *macrocolony* (i.e. clones which can be seen by the naked eye) technique for both epidermis and intestine, and a *microcolony assay* (a microscopic assay) for the intestine, while Al-Barwari and Potten (1976) have similarly designed a microcolony method for use in the epidermis (for details, see Chapters 8, 12, 19, and 21).

In these assays, when an appropriate denominator is chosen, such as crypts/ circumference in the small intestine, the results can be plotted as a survival curve, which, on a semi-logarithmic plot, has a general shape shown in Fig. 2.20.

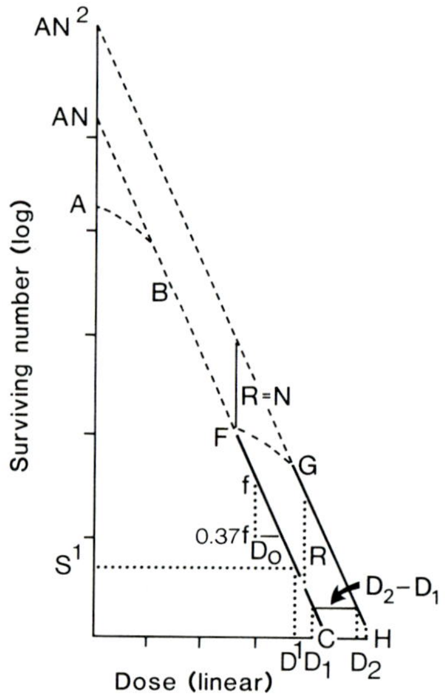


Fig. 2.20. A general survival curve (ABC) for clonogenic cells after graded doses of irradiation, together with a split-dose curve (FGH). $N$ is the extrapolation number, and the initial number of cells, $A$ is deduced from single dose extrapolation to zero dose ($AN$). $N$ is equal to or slightly greater than $R$ the recovery factor, which is deduced from the two-dose curve and calculated from eqn 8.1. $D_2-D_1$ equals the additional dose required to reach the same survival level when the dose is given in two fractions (eqn 8.2). (Redrawn from Potten and Hendry (1973).)

The resulting curve shows an exponential decrease in survival, but only after a considerable shoulder region is present because of the multiple numbers of clonogenic cells which are present in the intestinal crypt, for example, and also because of the capacity of cells to repair irradiation-induced damage. The parameters of this curve include the $D_0$ dose, formally defined as the dose required to reduce the surviving fraction to $e^{-1}$, but more usually thought of as the slope of the curve; the $D_Q$, or 'quasithreshold dose', which defines the length of the shoulder region, and the intercept on the ordinate, which is the product of the clonogenic cell multiplicity, i.e. clonogenic cells per crypt, and the cell survival curve shoulder, or the extrapolation number of the *individual cells* in the crypt. This sort of curve has been used extensively to measure the effects of irradiation on, for example, crypt survival in the intestine particularly (cf. Withers *et al.* 1974), but here we are particularly concerned with its use as an indicator of the identity and number of clonogenic cells: still considering the intestine, we have to relate crypt survival to cell survival, by a series of assumptions. These are, severally:

(a) that one surviving crypt cell is sufficient for the regeneration of a micro-colony, which is in fact implicit in the definition of a clonogenic cell. This would appear to be reasonable on the straight line portion of the graph, where unit increase in radiation dose causes unit decrease in crypt survival;

(b) that clonogenic cells survive independently of each other; the shoulder of a survival curve for such an organized population as the intestinal crypt or the epidermis may not only depend upon cellularity and repair potential alone; cells may be capable of 'helping each other to survive' (Potten *et al.* 1979);

(c) we also in effect assume that the survival curve, at the asymptote, is an exponential characterized by a slope $D_0$ and an intercept on the ordinate.

To get at the actual number of clonogenic cells we must separate multiplicity from the cell survival curve shoulder, and this can be done by using a two-dose survival curve, where the initial shoulder size is assumed to be equal to or greater than the maximum split-dose recovery factor (Hendry and Potten 1974; Potten and Hendry 1975). The details of this method are described in Chapters 8 and 19, for epidermis and intestinal crypts respectively, but there are several general problems: the errors are large; for example, in the mouse small intestinal crypt, values for clonogen number of $86 \pm 48$ (SD) cells per crypt are obtained (Potten and Hendry 1975), while, in the mouse epidermis, analysis of the available survival curves (Potten 1978*b*; Potten *et al.* 1979) indicate that there could be anything from 300 to 1100 clonogenic cells per $mm^2$. However, these may be maximal values, since, if the interval between the two doses is insufficient for full recovery, or if the true asymptote is not reached by the first or second dose, the technique would overestimate the clonogenic cell number (Potten *et al.* 1979). The possibility also exists that some clones may be transitory clones derived from non-clonogenic cells, but this is generally not considered likely, since survival curves produced after an initial priming dose of irradiation have identical slopes.

A further major problem is that the equations used to analyse the survival curves (see Chapter 8) are developed on the assumption that the population is homogeneous in its response. This is most unlikely to be true for the epidermis or the intestinal crypt. Potten (1977) and Barnes (1983) have shown that, in the intestinal crypt, some cells in the crypt base, the so-called *presumptive stem cells*, are extremely radiosensitive, with $D_0$ values as small as 0.1 Gy, while the cells whose radiation survival characteristics are measured by the crypt survival curve have a $D_0$ of about 1.3 to 1.6 Gy; a similar situation may occur in the epidermis (Potten 1978; Potten *et al.* 1979). While the reasons for this variation in radiosensitivity are unclear (and there is some evidence that radioresistance increases with decreasing proliferative capability, i.e. that the most radioresistant cells are those with the least number of transit divisions remaining), what it does mean is that, if there is a distribution of radiosensitivity among clonogenic cells, then the survival curve can be very complex, depending on the type of curve proper to each subpopulation; even if we assume that the survival curves of each subpopulation are shouldered and have a terminal exponential portion, for the extrapolation number of the composite curve to be even a weighted average, we need to assume that the slope, i.e. the $D_0$ of the subpopulations is the same (Alper 1979). Naturally, especially in the intestinal crypt, we cannot assume this; even so, since we are ignorant of the sizes of the component subpopulations, we could not weight the curve anyway. To add to these confounding factors, we should add that the proliferative populations show varying degrees of asynchrony, and have a marked radiation age response function in the cell cycle; moreover, split-dose irradiation could well synchronize the population (Hall 1975).

These considerations leave us with the conclusion that epithelial stem cell assays have no intrinsic degree of precision, and are based upon numerous, not yet founded assumptions. Nevertheless, since these are, as yet, our only lead to clonogen number in epithelia, and because the argument for the two compartment proliferative system is largely based on the clonogen values the methods give, we should examine them: the 86 clonogenic cells in the intestinal crypt are out of a total proliferative cellularity of $164 \pm 5$, while the epidermis has 300 to 1100 clonogenic cells/mm$^2$ compared with a proliferative cellularity of 9000 cells/mm$^2$ (using our best estimate of 0.6 for the growth fraction in the mouse epidermis, see Chapter 7).

Though the results in the epidermis are perhaps more convincing than those in the intestinal crypt (see Chapters 8 and 19), Potten *et al.* (1979) nevertheless consider that these results indicate that, as in the bone marrow, clonogenic cells are but a fraction of the larger dividing transit population, and making the clonogenic cell synonymous with the functional stem cell (a large step, see below), have challenged the original proposals of Leblond *et al.* (1967).

A second observation which might suggest that the fraction of stem cells is small in epithelial renewal systems, and that most cell production is achieved by the amplifying dividing transit compartment, is that most epithelial systems have a highly ordered structure. For example, in some types of slowly renewing epidermis the cells are arranged into hexagonally shaped columns, or stacks, of epidermal cells, each with about 10 basal cells; these columns have been called 'epidermal proliferative units' (EPU) (see Fig. 8.4, p. 295). Similarly, detailed study of the tongue filiform papilla (Hume and Potten 1976) has revealed a highly intricate ordered arrangement, with cell flux beginning at the bottom of a series of columns of cells (see Fig. 8.16, p. 316). The highly ordered structure of the intestinal crypt has already been referred to. In such a structured tissue the cells (or their progeny) at the origin of the flux can be considered to be the only permanent cells present in the tissue; these could be said to be 'fixed' or 'anchored' (Potten *et al.* 1979), while the remaining proliferative cells have strictly limited proliferative capacity, with distinctly limited life expectancy. On this viewpoint, since both migration into the superficial portions of the EPU, and also mitotic activity, are both concentrated in the periphery of the EPU, the maturer cells, and the later dividing transit cells, are likely to be peripherally positioned, and thus the stem cells for the epidermis would be found in the basal layer near the centre of the EPU: in the filiform papilla the presumptive stem cell lies at the origin of the flux at the base of the cell columns, and in the intestinal crypt, at or near the crypt base.

Here we are concerned with generalities, and the specifics will follow in later sections, but even this brief look suggests that stem cell populations in epithelia could be quite complex in their makeup. In our opinion, there has not yet been a critical experiment which distinguishes between the two stem cell models in epithelia, viz. the proliferative cell = stem cell hypothesis of Leblond *et al.* (1967), and the stem cell subpopulation hypothesis of Lajtha and colleagues; if we proceed by the Popperian method (which might have been propounded with cell kinetics in mind), there is not yet any single experiment which disproves either hypothesis. However, there are many instances, as we shall see, where the multicompartmental hypotheses of Potten and colleagues provide more convincing explanations for the observations.

*The relationship between the various classes of stem cells*

We should look critically at the relationships between the several classes of cells which we have defined above, i.e. functional stem cells, potential stem cells, and clonogenic cells. On the multicompartmental stem cell hypothesis, the functional stem cells are those at the origin of the cell flux. Indeed, in the intestinal crypt there is substantial evidence that the basally sited crypt cells, by definition at the origin of the flux, are the precursors of all four lineages (columnar, goblet, Paneth, and endocrine) of the crypt; the cells are thus considered to be pluripotent, and respond to factors which initiate different avenues of differentiation (Bjerknes and Cheng 1981*a,b,c*; and Chapter 19). It is highly likely that these cells are responsible for the repopulation of the intestinal crypt in normal circumstances, and are obvious candidates for functional stem cells. In the EPU, the central cell is the candidate functional stem cell. Before the immunological role of the Langerhans cell was established, because it was found centrally in the EPU, at least in the mouse, suspicion existed that it might fulfil a functional stem cell role (Potten and Allen 1976), but since this concept has been abandoned, some physically closely related centrally placed cell must operate as the functional stem cell. In the filiform papilla, cells at the column base are presumptive functional stem cells, while in the testis the identity of the functional stem cell is in doubt, but topographical studies suggest that an early class of spermatogonia may be the one.

What is the relationship of these functional stem cells to the *clonogenic* cell population? This question is especially difficult to answer. Originally, Hendry and Potten (1974), using what we might call the 'principle of numerical equivalence', believed that the clonogenic cell population of the small intestinal crypt was equivalent to the slowly proliferating subpopulation of functional stem cells in the crypt base, since initial numerical estimates suggested that putative functional stem cell and clonogenic subpopulations each contained some 40 cells. However, later, adjusted values indicated a mean of about 80 clonogenic cells per crypt, and thus the clonogenic cell population, if it did indeed encompass the basal crypt functional stem cells, would also overlap with the rapidly proliferating cells which occupy the dividing transit compartment, and are, in normal conditions, discarded into the differentiation pathway. These remaining cells are by definition, *potential* stem cells, which are stimulated, possibly by extensive death of functional stem cells, to take on a stem cell function.

The relationship between functional and potential stem cells is conjectural: of course it could be that the cells at the origin of the flux are *fully* effective stem cells, while the cells immediately below them in the hierarchical system possess different degrees, or 'shells' of 'stemness' (Potten *et al.* 1979); this 'stem cell continuum hypothesis' would then account for the potential stem cells (see below). What would then be the relationship of clonogenic cells to stem cells? We might expect them to equal functional stem cells + potential stem cells in number. However, in the intestine (and also the epidermis) some functional stem cells in the crypt base are very radiosensitive, with a $D_0$ of around 0.1 Gy, and these cells are also capable of being killed by only 2 $\mu$Ci/g body weight of [$^3$H]-TdR (Cheng and Leblond 1974c); this makes it possible that the clonogenic cell population, as measured by the current radiobiological methods, contains *few if any of the functional stem cells*. The possibility therefore exists that functional stem cells and clonogenic cells are different subpopulations.

In the epidermis the situation is less clear; the clonogenic cell population forms anything up to 10 per cent of the basal cell population, the best estimate being about 900 clonogenic cells/mm². Since there are about 1100 EPU/mm² in the mouse, again by the principle of numerical equivalence there should be one clonogenic cell per EPU; whether this cell is identical to the stimulus responsive, slowly cycling central cell in the EPU is again conjectural (see Potten 1976; Potten *et al.* 1979), but the situation in epidermis may be more complex than this (see Chapters 8 and 12). Since there is, as yet, no clonogenic assay for the filiform papilla, we cannot comment on population equivalences.

Thus, while we suspect that clonogenic and functional stem cell populations are but a fraction of the proliferative cellularity of epithelial renewal systems, we are only able to speculate on the identity of such cells. There is a pressing need for an assay (?*in vitro*) which might allow recognition of the component cells.

*The general properties of epithelial stem cells*

Since we do not know for certain which cells constitute the stem cell population, how can we discuss their properties? A good question. Perhaps we should take refuge in the terms 'presumptive' or 'putative' stem cells for those cells we consider candidates for a stem cell role. Potten (1978*a*; Potten *et al.* 1979) has listed the more important attributes of epithelial stem cells, and because of their importance, we should examine these critically; the details of the experimental data upon which these conclusions are based are considered in the relevant section later on; here we merely draw some general threads together.

*(a)  Epithelial stem cells are slowly cycling*
There is little doubt that the experimental evidence indicates that presumptive stem cell regions contain cells which have long cell cycle times compared with the bulk of the proliferative cells in that tissue, both in the intestinal crypt base (Al-Dewachi *et al.* 1974; Wright *et al.* 1975; Wright 1978), and in the central area of the EPU (Potten 1974). Potten *et al.* (1979) have, largely by analogy with the bone marrow, where the CFU-S are considered to spend some time in the $G_0$ phase, proposed that epithelial stem cells also enter $G_0$, and, by implication, that the apparently long cell cycle times in presumptive stem cell compartments are a reflection of the time spent in $G_0$. This long non-cycling state is envisaged as being necessary for the process of 'genetic housekeeping', whereby the integrity of the stem cell genome is maintained (see below). However, there are experimental methods available for the identification of cells in the $G_0$ phase, and as yet, there is no firm *experimental* evidence for a $G_0$ population in epithelial renewal systems in normal circumstances (see Chapters 8 and 19). However, there is little doubt that presumptive stem cells can be triggered into rapid cycle; after hydroxyurea-induced cell death in the rat intestinal crypt, basal crypt cells initiate crypt repopulation (Al-Dewachi *et al.* 1977; Wright 1978), and EPU central cells respond to wounding (Potten 1974), changes which we would regard as due to a decrease in cell cycle time rather than ejection from a $G_0$ compartment. Lest this should appear a semantic argument in isolation, we should refer the reader to section 2.7 where the distinction between a cycling $G_1$ and a resting $G_0$ period is discussed. The proposal that epithelial stem cells have a $G_0$ phase is essentially non-robust (see Chapters 8 and 19).

*(b)  Epithelial stem cells are pluripotent*
There is evidence that the several different types of bone marrow cells are all derived from a pluripotent precursor stem cell (Lajtha 1979). In the intestinal crypt there is accumulating evidence from experiments with labelled phagolyso-somes (Cheng and Leblond 1974*c*), the clonal analysis of tumours (Isaacson 1981) and allophenic mice (Ponder 1983), that the various lines of differentiation (into Paneth, goblet cells, etc.) originate from a common precursor cell. In the

skin, the evidence is much more tenuous, but Pinkus (Pinkus and Mehregan 1981) have developed a hypothesis, based largely on the differentiation patterns seen in basal cell and adnexal tumours in the skin, that an 'adult adnexal matrix cell' can give rise to both sebaceous and hair elements. In this scheme, however, the 'adult epidermal matrix cell' produces exclusively epidermal cells (Fig. 2.21).

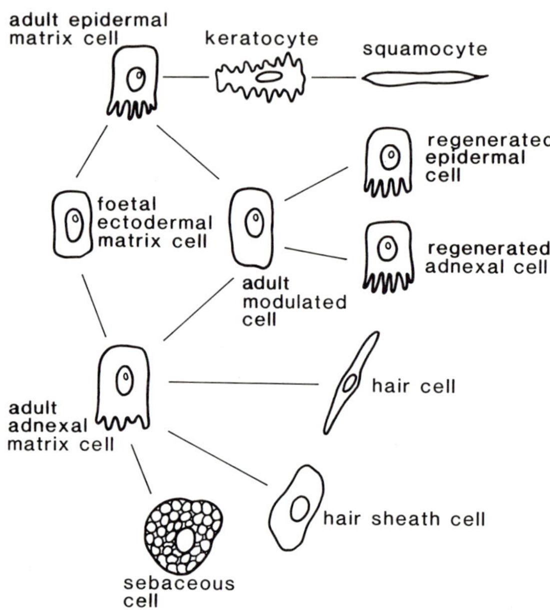


Fig. 2.21. A possible schema for the lineage of epidermal derivatives as envisaged by Pinkus and Mehregan (1981). The foetal ectodermal cell has epidermal and adnexal lineage potential; the adult ectodermal cell, whether of epidermal or adnexal type, preserves its potential to develop into the modulated cells seen during the process of epidermal wound healing. Such modulated cells can differentiate in an epidermal or adnexal direction. (Redrawn from Pinkus and Mehregan (1981).)

On this view, the pluripotential nature of the central EPU cell is thus in question; the anatomical location of the adult adnexal matrix cell is obscure. However, since hair follicle cells do contribute in some degree to wound healing (Argyris 1968), a location in the hair follicle is possible. In this respect, there have been certain speculations on the relationship of follicular stem cells to interfollicular stem cells (Potten 1978*b*; Potten *et al.* 1979), but as yet, the relationship is unclear (Chapter 8). Certainly different types of keratinization can occur in oral and tongue epithelium, but this is scarcely evidence for pluripotency of the parent stem cells; similarly, there is, as yet, no evidence that pluripotent *stem cells* occur in the testis; presumably the frequency with which trophoblastic elements occur in seminoma is a reflection of a teratomatous nature.

On the other hand, against these statements we must match the views of Laurence (1980), who does not agree that there is any evidence for a pluripotent cell in the epidermis, while Holtzer (1979) believes that no cell is pluripotent; Holtzer argues that, for pluripotency to be demonstrated, it must be shown that a cell A in a given set must transfer directly, without an intervening cell cycle, into a cell B or C of a given set; in fact Holtzer propounds that cells in set A can generate cells in set B or C only as the result of a *quantal cell cycle*, and are, at the most, bipotential; differentiating signals would decide whether A → 2A, 2B, 2C or B + C cells. These results were largely based upon the study of myogenic and chondrogenic cell lines (Holtzer 1978).

On this hypothesis, the fact that Paneth, endocrine, mucous, and columnar cells are derived from a single stem cell would be explained by the stem cell giving rise to two daughters, each of a different type; the four sets of terminally differentiated cells are derived from the daughters only after a further cell division. Holtzer also makes the point that claims for pluripotency are usually based upon the observation of *groups of cells* rather than single cells, cells which have probably passed through several cell cycles, and that no experiments have demonstrated the production of more than two sets from one set; this is certainly so for the Cheng and Leblond *unitarian hypothesis for cytogenesis in the small intestine*, where no attempt is made to show that one cell, without division, gives rise to the four different cell lines. Lajtha (1979) counters this argument by asking what we mean by pluripotentiality; if a cell can, even several cell cycles later, give rise to four cell types, then is that cell not, in a sense, pluripotential, notwithstanding the fact that several divisions are required? The term would, if used in this sense, distinguish such cells from those which can only yield one or two different lineages. This may be special pleading, but, in the final analysis, *operationally* Lajtha is correct; it is in the consideration of mechanisms of differentiation control that the number of steps between a stem cell and its differentiated progeny become important.

Even assuming (as seems distinctly possible) that stem cells are (ultimately) pluripotent, the actual relationship between stem cells and differentiating dividing transit and maturing progeny is unclear; the alternatives are that each stem cell may be capable of following any pathway of differentiation, or the situation may exist whereby the early divisions that occur within the stem cell and dividing transit cell populations are in bipotential cells: in effect the quantal cell cycle concept of Holtzer (1978), in which a differentiation signal induces a cell cycle in a stem cell, and this cell cycle ends with an asymmetric division. This is the *asymmetric division* of Osgood (1957), which of course is a basic concept of the Cairns (1975) hypothesis of selective segregation of stem cell DNA at mitosis; however, in this scheme, stem cells would be also capable of a *symmetric or logarithmic division* (Osgood 1957) to produce two new stem cells, possibly in regenerative situations. Because, on this hypothesis, there are two types of stem cell division, there must be *two control mechanisms*, one for the asymmetric, and one for the symmetric division (Lajtha 1979). If most cell divisions are symmetric, stem cells should not be available for differentiation, but, in point of fact, even in maximum cycling conditions in regenerating bone marrow cells, some 40 per cent of stem cell daughters enter the differentiation pathway (Lajtha 1979). In order to account for this phenomenon, Lajtha (1979) proposed that signals for differentiation and proliferation were *dissociated*; assuming that, in steady state conditions, most stem cells are in $G_0$, differentiation stimuli affect cells in $G_0$, which are thus removed from the stem cell population irrespective of whether they proliferate or not, and the two random events, triggering into cycle and into differentiation, are independent mechanisms. Differentiation signals could be generated from the functional cell mass, and in conditions of increased demand, there would be a conflict between increasing the depleted stem cell numbers and a demand for increased functional cells, which, in the bone marrow, is met by ensuring that only 40 per cent of stem cell progeny differentiate; this mechanism apparently will allow maintenance of stem cell numbers. We have yet to see if such a model would be appropriate for epithelia; we cannot yet measure stem cell number with any precision, let alone the number of cells committed to differentiation.

A major problem lies in defining how stem cells 'decide' when to give rise to the differentiated progeny of the several varieties. Potten *et al.* (1979) have speculated that (presumably) chemical initiators of differentiation may exist, and that stem cells may have differential susceptibility to these initiators, either intrinsically or because the stem cells are of different maturities and therefore abilities to respond to initiators, or because they are, possibly through their position in the tissue (see below), shielded from the effects of the initiators.

(c) *Epithelial stem cells are the source of circadian rhythm in phase fractions*
It has been suggested that the cells in the presumptive stem cell regions are the origin of the circadian variation which is such a prominent feature of epithelial renewal systems, in intestine and epidermis (Potten *et al.* 1977) and in the tongue filiform papilla (Hume and Potten 1979), and possibly passed on to other proliferating cells by a process of entrainment. However, a detailed analysis of circadian variation in the small intestinal crypt, on a cell positional basis, could not confirm this (Al-Dewachi *et al.* 1976; Wright 1978; and see Chapter 20), although the data in the tongue appear most convincing (Potten *et al.* 1978; and see Chapter 8, section 4).

(d) *Epithelial stem cells have long-lived thymidine pools*
Again presumptive stem cell regions of epidermis (Potten 1974; Hume and Potten 1982) apparently have the ability to store [$^3$H]-TdR in a (?cytoplasmic) pool, and only incorporate it at times up to 17 hours later. The functional significance of this phenomenon is obscure, but more recent evidence (Møller and Keiding 1982) indicates that this may be a feature of all proliferating cells in the squamous epithelium of the hamster cheek pouch.

(e) *Stem cells are radiosensitive*
We have seen above that current studies indicate that at least some cells in stem cell regions are very radiosensitive, particularly in the intestinal crypt; however, because these studies are basically non-quantitative, it is difficult to know whether such cells constitute the whole or only part of the population. Potten *et al.* (1979) have speculated that they may be cells with a very short $G_0$ period, and thus with little time to repair DNA damage. In the testis, it is the $A_1$–$A_4$ and $B$ spermatogonia which are more sensitive to irradiation than the putative $A_0/A_S$ stem cell (Clow and Gillette 1970; Withers *et al.* 1974), so an evident difference exists between the testis and the gut.

The biological role of such extreme radiosensitivity is obscure, unless it is associated with an extremely low tolerance for irradiation-induced errors (see (g) below).

(f) *Epithelial stem cells are clonogenic*
While it is probable that stem cells can be equated with clonogenic cells in the bone marrow, we have seen that there is, as yet, little evidence that presumptive stem cells are clonogenic; in fact, if some presumptive stem cells are as radio-sensitive as current experiments indicate, then these at least cannot form part of the clonogenic population indicated by clonal regeneration experiments.

(g) *Epithelial stem cells have a mechanism for protecting their genome*
Since stem cells (or their progeny) have to provide cells throughout the lifetime of the organism, it would be extremely advantageous if genetic errors in stem cells could be kept to a minimum (Cairns 1975); this would be especially so if there is a small number of such cells (cf. one in the EPU, see Chapter 8), where effectively, an assymetrical stem cell division is mandatory. Cairns suggested that one way in which such errors could be kept to a minimum is where the remaining stem cells retain the older DNA, and the newly synthesized strands, which would presumably be more prone to error, are discarded into the cell destined to differentiate. Thus this selective DNA segregation is an *effectively asymmetrical division*. Potten *et al.* (1978) have produced some early experimental evidence suggesting that such a process occurs in the presumptive stem cell region in the filiform papilla of the mouse (see Chapter 8, section 4), but in the small intestine, the experiments were difficult to interpret (Chapter 19).

This is an interesting hypothesis, which will bear much more substantiation; however, we might note two corollaries which are interesting; first that any extensive sister chromatid exchange would render the selective segregation ineffective. This could lead to an 'intolerance' of the type of damage repaired by such a process, which might account for the extreme radiosensitivity of stem cells (Potten *et al.* 1979). Secondly, it is critical that the older, or 'immortal' strands should be efficiently repaired, since, on this hypothesis, these DNA strands would be present throughout the lifetime of the organism. Lajtha (1979) and Potten *et al.* (1979) have suggested that this is the reason for the presence of the $G_0$ period in bone marrow stem cells, where genetic housekeeping would be carried out before each division occurred. Whether this hypothesis is also applicable to epithelia is a moot point.

*Models for stem cell organization*

Most current models of stem cell proliferation have been designed with the bone marrow in mind, and, by the principle of homology, and also possibly a feeling that there must be a certain economy of design features in renewal systems, models for epithelial tissues have been adapted to fit the general haemo-poietic model. So it is that perhaps the most powerful model put forward for epithelial stem cell systems was designed primarily for the bone marrow by Potten *et al.* (1979) and hereafter called the Potten model. Haemopoietic stem cells, the CFU-S population, are probably formed of cells with differing proliferative capabilities; Schofield (1978) has suggested that this is organized so that CFU-S progressively lose proliferative capacity as stem cells as they mature; however, the earliest stem cells are envisaged as occupying a 'niche', or focal point, which prevents loss of the cell to the environment outside the niche (Fig. 2.22), where they would be subject to the factors inducing transit divisions

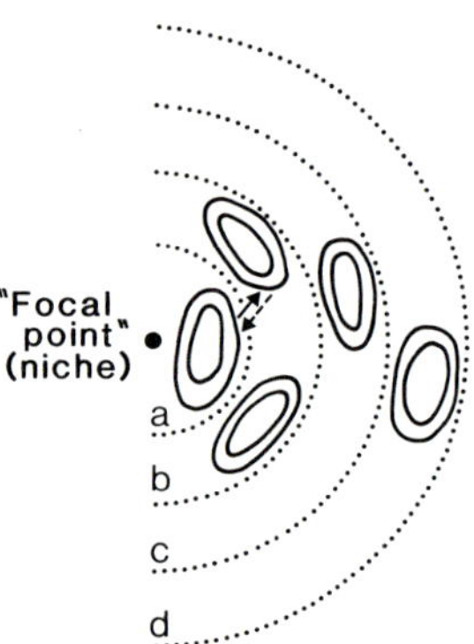


Fig. 2.22.   The concept of the stem cell *niche*. The pluripotent stem cell occupies the putative focal point, and is surrounded by (notional) shells indicating diminishing stem cell quality (a–d). Niche cells have high replicative probability but are protected from differenti-ation signals. Extraniche cells have higher differentiation probability but can replace niche cells in times of need. (Redrawn from Potten *et al.* (1979).)

and maturation in their progeny. Thus the stem cell line is ensured, and, as they migrate from the focal point, they lose stem cell characteristics until they become mere transit cells; thus the stem cell population has a distinct age structure (Hellman and Botnick 1977).

The location of the niche can be thought of as a discrete anatomical location. In the bone marrow it is thought that the niches may be lacunae on the endosteal surface of the bone; irradiated bone marrow can be repopulated by grafted stem cells capable of entering and repopulating the niches vacated by the irreversibly damaged, irradiated stem cells. Alternatively, the niche concept can be viewed as a more general model, as notional and non-spatial in the sense that it only implies a spatial relationship to the other cell types, and a more general statement of the model would be Fig. 2.23. The possession of the position in the niche would

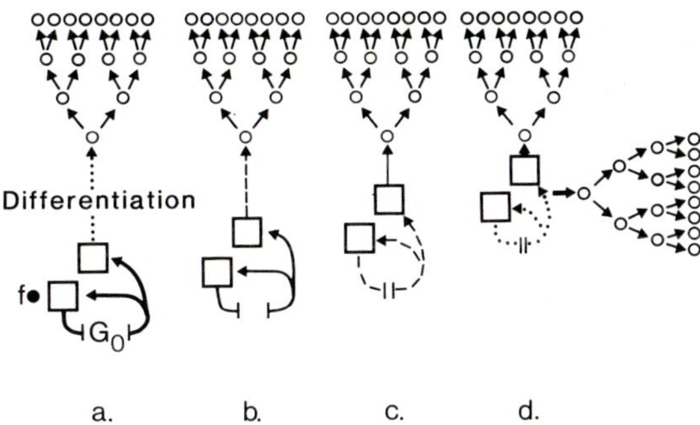



Fig. 2.23. A general scheme for the organisation of stem cell proliferation. The stem cells occupying the niche position have polarity imposed by the niche (f), and can segregate old and new DNA strands. Stem cells need (long) periods in $G_0$ for genetic housekeeping, and are shielded from differentiation initiation. At mitosis, one daughter remains in the niche and the other moves through a series of shells (b–d) of declining stem cell quality and increasing differentiation probability. (Boxes = stem cells; circles = transit cells.) (Redrawn from Potten *et al.* (1979).)

exert certain constraints on the contained stem cell: (i) all stem cell divisions would be, in effect, asymmetrical divisions, not imposed by any genetic condition, but purely because of the position in the niche, since, by definition, only one daughter can remain. However, since displaced CFU-S can repopulate the niche after irradiation, some of the younger cells (where age is measured by distance, in terms of divisions, from the niche position), can evidently revert to a 'true' stem cell role; hence we have a *stem cell continuum type* of model; (ii) niche cells have a low probability of entering a differentiation pathway, and hence a high self-renewal capacity, while extra-niche cells have a high probability of differentiating and a low self-renewal capacity; in the context of the stem cell continuum model, it could be considered that, as the cells migrate away from the focal point, they enter and leave shells of differing degrees of stemness (Fig. 2.23), with a decreasing probability of self-renewal, and increasing probability of differentiation; (iii) the *polarity* of the stem cell division could well be very important; obviously, if the niche is only a microenvironment and not an actual lacuna, one daughter must in some way remain anchored and be prevented from migrating, and one way of doing this is by an imposed polarity of the stem cell division. Moreover, if, as Cairns (1975) and Potten *et al.* (1979) suggest, the immortal DNA strands have nuclear membrane attachments, at one pole of the nucleus, ensuring their preservation, again the polarity of the division would be critical. A further constraint, not imposed by the niche hypothesis, but by the foreseen need for genetic housekeeping, is the imposition of a high probability for niche stem cells to enter a $G_0$ phase, i.e. the probability of entering $G_0$ decreases with increasing 'distance' from the focal point; the life history of a cell displaced from the niche can then be traced (Fig. 2.23).

In epithelial tissues, with their discrete spatial arrangement, the presumptive stem cell zones will be the location of the niche or anchorage point, somewhere in the crypt base or in the centre of the EPU, at the base of the column in the filiform papilla (Fig. 8.16, p. 316), and on the basal lamina in the seminiferous tubule. Whether this anchorage point is physical, i.e. in some way the cell is bound to the basal lamina, or whether the polarity of the stem cell division is maintained by factors intrinsic to the cell itself, is not known. It is certainly not simple: Cairn's (1975) original suggestion, that all cell divisions in the basal layer of the epidermis are horizontal (see Chapter 6, section 7), hence allowing

Table 2.4. The properties of stem and non-stem cell populations on the Potten model, with particular reference to epithelial cell renewal systems. (Adapted from Potten *et al.* (1979))

| Characteristic | Niche cells | Extra-niche cells | Comments |
|---|---|---|---|
| Probability of triggering into cycle | low | high | little firm evidence of a $G_0$ phase in epithelial renewal systems |
| Probability of triggering into differentiation | low | high | model imposed |
| Turnover rate | low | high | presumptive stem cells in intestine and EPU have long cell cycle times |
| Mean $G_0$ time | long | short | no direct experimental evidence in epithelia; said to be required for genetic housekeeping |
| Self-renewal capacity | high | low | model imposed; little direct evidence that clonogenic cells and stem cells are the same population in epithelial systems |
| Probability of mobilization | low | high | model imposed |
| Probability of anchorage | high | low | model imposed |
| Probability of asymmetrical division | high | low | model imposed; because of niche hypothesis, divisions are perforce asymmetrical, because of physical/polarity constraints. Evidence for genetically ordered asymmetrical division exists in the filiform papilla |

stem cells to enter mitosis with 'duplicated chromosomes correctly aligned' is unlikely to be true for many squamous epithelia, since many divisions are in fact vertical; if it is true for stem cell divisions only, then some mechanism would have to be advanced for ensuring that stem cell divisions are selectively polarized, while the others are not. This is of course conceivable: the mechanism would be interesting.

The Potten hypothesis is a useful one, and Table 2.4 places most of the points raised by Potten *et al.* (1979) in experimental context, with special reference to epithelial renewal systems; some of the points amenable to experimental check do agree, while points of disagreement could be explained by a surfeit of homology, with respect to the bone marrow.

*Stem cells, carcinogenesis, and cancer*

It is sometimes stated, largely as a result of rationalistic reasoning, that carcinogenesis and cancer are processes which are exclusively the property of stem cells (Wright 1978; Potten *et al.* 1979); in view of the relatively short transit times of transit populations compared with the time scale of multistep carcinogenic mechanisms, the only cells whose DNA remains in the target tissue for long enough to interact with the carcinogen are the stem cells; thus the main target cells for malignant transformation should be the stem cell. This would of course account for the long latent period between carcinogen treatment and the appearance of tumours (cf. epidermal carcinogenesis, Chapter 13), and also the phenomenon of the age-dependent incidence of human neoplasms; since carcinogenesis, *inter alia*, may be taken as indicating that some genotypic change has occurred, and that such change probably occurs during normal or induced cell reproduction, carcinogenesis probably occurs only in proliferating cells, of which stem cells, of course, form a component part. In renewing populations such as the gut and the epidermis, there is no problem in producing enough proliferating cells, although even here, application of carcinogens evokes a prominent proliferative response (see Chapters 13 and 22). On the other hand, in conditional renewal systems such as the liver, experimental carcinogenesis is associated with induced cell proliferation, but in normal human beings, liver cell carcinomas are very rare (0.4 per cent of all cancers in the United Kingdom, compared with 90 per cent for all epithelial neoplasms); that this may be related to the incidence of mitotic events in the liver is suggested by the fact that the majority of these tumours appear in people with cirrhosis of the liver, in which cell proliferation as part of a continuing regenerative process is a prominent feature. Though we can think of numerous exceptions to the rule, we might argue that carcinogenesis is the exclusive province of proliferative cells and, possibly, of stem cells.

Why this should be is interesting; the probability of an event leading to malignant transformation affecting any proliferative cell is small unless the cell is in, or will shortly enter, the cell cycle. Peto (1977) considered that this could be due to damage occurring exclusively at mitosis, or perhaps more likely, by a mutation becoming 'fixed' in the population by daughter cell inheritance. If this genetic change occurred long before mitosis, it could be repaired by genetic housekeeping, but if the event occurred shortly before mitosis, then there would be less chance of correcting it.

In either case, the *incidence rate* of an event leading to cancer is a function of the number of cell divisions in the target population; in many multistage models of carcinogenesis, $I(t)$, the incidence rate among individuals at age $t$, is proportional to $t^k$, i.e. plotting log $I$ against log age for the tumour gives a straight line with slope $k$, and $I(t) \propto t^k$ (Peto 1977). If we denote the number of cell divisions in the target population as $N$, then $I(t)$ is not proportional to $t^k$, but $I(t) \propto N^k$. Note that this assumes that $N \propto t$: this may be the case if cell divisions *in utero* do not count, but if they do, or there are other factors in the organization of the proliferative population which makes $N$ not proportional to $t$ (e.g. non-random removal to a $G_0$ phase), then $I(t) \propto t^k$ is not equivalent to $I(t) \propto N^k$; to know how these expressions differ, it would become critical to analyse the kinetic architecture of the population, with special reference to carcinogen target cells.

These concepts effectively exclude non-proliferating cells from a role in carcinogenesis, but if proliferative cells are important, we need to know which proliferative cells are involved. Stem cells we say; it stands to reason. But possibly one of the reasons that we have belaboured the concepts of stem cells and their potential hierarchical structure is that our concept of a stem cell target will in turn depend on our concept of what stem cells are. And we have already conceded that we do not really know. We know conceptually what they are, and what they are supposed to do, but which, in our breakdown of stem cells into functional, potential and clonogenic, is our carcinogen target?

First let us exclude clonogenic cells, since they must be included in either or both of the functional and potential stem cell populations. Continuing on a rationalistic basis, we should say that effective or functional stem cells, i.e. those cells that repopulate in normal conditions, are the most appropriate target cell; they are small in number, and their progeny, and hence the contained DNA, last throughout the life of the organism. Consequently, stem cell DNA, and its preservation, is of the utmost importance, and to vouchsafe its purity, the Cairns and Potten hypothesis enables 'immortal strands' to be maintained by 'selective segregation' at mitosis, and what damage is tolerated by the stem cells is repaired by genetic housekeeping during the putative $G_0$ period, while, if the damage is too great, stem cells do not tolerate it and die.

But this argument makes several assumptions: first, that we accept the tenets of a stem cell hierarchical system, which, though we might say that it seems the most reasonable hypothesis, in epithelia largely stands and falls on a suspect numerical assay, and our principle of numerical equivalence; if we did accept the uniform stem cell hypothesis of Leblond and his colleagues, this would radically change our thinking about stem cells and carcinogenesis. A second assumption is that, during carcinogen treatment, functional and only functional stem cells respond. We shall see that carcinogen treatment is uniformly associated with cell damage and cell proliferation; which cells repopulate during this process? Could potential stem cells proliferate, transform, and become a dormant tumour cell in the general sense of the multistage process of carcinogenesis? In other words, we need to assume that only functional stem cells are involved in the carcinogenic process, and negate any role for potential stem cells. In the gut, the situation is probably clearer: there is a substantial body of opinion, not yet disproven, that cells other than functional stem cells can repopulate in abnormal circumstances. Such circumstances could also apply to carcinogenesis, in which case the potential stem cell population would constitute a target. We cannot really answer this question, unless we can evolve a marker for different stem cell populations, and moreover, a marker which will withstand a carcinogenic event, to allow the source of the lesion to be traced.

It has to be said, therefore, that purely cell kinetic approaches to the origin of epithelial tumours have not answered our question about target cells, even though we are beginning to dissect stem cell hierarchies by these methods. But there are related questions which cell kinetic approaches might be expected to answer in the not too distant future. For example, in the stem cell-target cell hypothesis, even if we incriminate a multistage mechanism for carcinogenesis, why are epithelial tumours comparatively rare (rare that is, compared with the number of stem cell mitoses)? And why do they tend to occur in later life? Peto (1977), comparing mouse and man, has calculated that humans should have between $30^4$ and $30^6$ times the risk of cancer induction per epithelial cell than mice, yet, in the natural state, cancer induction probabilities are not greatly different, given that there is no evidence that resistance to chemical trans-formation *in vitro* is not dissimilar. We have seen that selective segregation and/or genetic housekeeping have been suggested as a means of protecting the stem cell genome from potentially harmful changes; Peto (1977) has speculated that there may be some sort of protection from cancer induction offered by the stem cell/proliferative cell hierarchy, and disturbance of this relationship, perhaps by increasing the proliferative activity of the stem cells (we have seen that cells in presumptive stem cell compartments are slowly cycling) could possibly be dangerous in the sense that more stem cell targets are proffered. Certainly cell proliferation, presumably including stem cell proliferation, is induced during chemical carcinogenesis in many *in vivo* epithelial models. Furthermore, it is becoming increasingly recognized that hyperplasia, induced before or during the phase of treatment with carcinogens, can increase the tumour yield in models of intestinal carcinogenesis (see Chapter 22), and individuals with chronic ulcerative colitis, who show elevated proliferative rates in the colonic mucosa, have a greatly increased risk of developing intestinal cancer. It is possible that the increased number of stem cell divisions could be an important factor in these phenomena.

Increased stem cell production could lead to increased stem cell number (cf.

the increase in stem cell number which follows partial small intestinal resection (Bjerknes and Cheng 1981*d*; see Chapter 20, section 4); thus an initial event in carcinogenesis could be an increase in the number of target cells, and hence an increase in the rate at which even protected immortal DNA strands have to be made, with resulting increases in the chances of a mutation (Cairns 1975; Peto 1977). Alternatively, and possibly associated with an increase in the proliferative rate, a first stage in the multi-step process could be an increase in the survival of an altered (stem) cell, a primary mutation (Cairns 1975), which, with its increased survival advantage, would displace other cells, increasing its own number. The chances of progressing to the next step are thus increased (Cairns 1975). In fact, Cairns (1975) used this argument to support the concept that epithelial cells should be organized into stem cell territories which could discourage colonization of the tissue by aberrant stem cells, citing the intestinal crypt as an example. If variant stem cells do colonize a whole crypt, as Chang (1978, 1980) has suggested occurs in the dimethylhydrazine-induced colonic cancer model in mice, colonization of other crypt units would mean growth into adjacent crypts against the prevalent cell flux. Presumably the EPU hypothesis would also come in here, although there are no physical barriers in the basal layer (Fig. 8.3, p. 293). Possibly the existence of a stem cell, with some kind of spatial arrangement, and its own territory (and its niche?), is a form of protection against colonization by variant cells. Certainly squamous carcinoma, as seen for example in the mouse epidermis, hamster cheek pouch or human cervix uteri, begins as a clonal proliferation of rapidly proliferating squamous epithelial cells (see Chapter 13, section 3), probably indicating that a priming mutation has increased the rate at which the mutant cells accumulate.

If carcinogenesis is a process affecting effective or functional stem cells, we might reasonably expect the derived tumours to be monoclonal in origin, i.e. take origin from a single cell, particularly if we embrace a multiple step mutation model of carcinogenesis; this is because, on the hierarchical hypothesis, stem cells are small subpopulations (cf. one per EPU; Potten 1976). Evidence accruing from the study of epithelial tumours in individuals who display polymorphic variants of the enzyme glucose-6-phosphate dehydrogenase have shown that, by and large, carcinomas show a single enzyme phenotype (Fialkow 1976), including squamous carcinoma of the cervix (Smith *et al.* 1971), nasopharynx, palate, and thyroid (Fialkow *et al.* 1972), although the data in other epithelia (cf. colon) is not as clear. In experimental animals, Iannaconne *et al.* (1978), using allophenic mice chimaeric for different electrophoretic variants of the enzyme glucose phosphate isomerase, concluded that epidermal tumours induced by a variety of carcinogens could not have arisen from more than eight cells, and the most probable interpretation of this data is that the tumours were monoclonal. Thus origin from a single cell seems most probable for most carcinomas yet studied: the problem remains, which cell?

## 2.10. Some general remarks on growth control in epithelial tissues

It is, in many ways, an indictment of modern biological research that we do not, at present, have anything in the way of a widely accepted general biological model of growth control. True, there are several hypotheses, which have been generated to fit certain specific instances, such as density dependent or contact inhibition for cells growing *in vitro*, and some of these have been compounded to make impressive edifices, such as the *chalone hypothesis*; as we shall see, there is by no means any form of general agreement about this.

Such is also the case in epithelial tissues; although we believe that we are beginning to unravel the kinetic architecture of epithelial cell populations, we are far from identifying the controls which act upon cell proliferation and differentiation.

In the genesis of growth control hypotheses, which phenomena do we have to account for? There are several (Allen and Smith 1979): (i) *compensatory growth*, a term usually applied to the process by which organs undergo hyperplasia after unilateral ablation of a pair (the kidney) or partial excision (the liver); (ii) *target organ growth*, which occurs without loss of tissue, for example the stimulation of adrenocortical growth which follows ACTH treatment (Bransome 1971); (iii) *reparative growth*, when a wounded or damaged tissue is replaced, as seen in epidermal wounding, and (iv) *malignant growth*, where growth controlling mechanisms have apparently broken down.

There are several hypotheses which do deserve our attention in epithelia; here we shall confine ourselves to some general remarks about each. In our discussion of each tissue system we shall encounter the (sometimes) meagre information available. Growth control in epithelia essentially means the way in which the population size is maintained at an appropriate level; in normal steady state conditions this would be orientated towards the control of the functional compartment size at a level which presumably allows adequate function, whether we assume a hypothesis based upon functional demand, or propose that the population size must be, in some way, genetically predetermined (Goss 1972). However, any global theory must account for the phenomena mentioned above, those of physiological adaptation to an induced environmental change, wound healing and compensatory growth, and for the changes in population size which accompany hyperplasia, however caused.

A further major ingredient of a comprehensive growth hypothesis should include the explanation for compensatory growth being *tissue specific*; partial hepatectomy does not induce lung growth, and vice versa. At the same time the hypothesis must account for the observation that, whereas some types of regeneration are local (e.g. an epidermal abrasion on one knee, which does not induce cell proliferation on the other knee and is an essentially local response), some growth phenomena, such as regeneration after unilateral nephrectomy, are associated with epithelial regeneration at a distance. In still other organs which are also paired, such as the seminal vesicle and the thyroid, unilateral resection does not induce cell proliferation in the remaining gland or lobe (Alison *et al.* 1979; Wynford-Thomas *et al.* 1982*a*).

These are the basic largely unarguable tenets upon which comprehensive growth theories must be based, and we should note, at the outset, that it is extremely difficult to encompass all these findings in one general theory, avoiding, if possible, the prospect of specific growth control mechanisms for specific tissues and situations. However, it is distinctly possible that there is more than one growth control mechanism which acts on the same tissue; in fact, this is probable.

Before proceeding further, and being teleological for a moment, we should perhaps ask how the cell population 'senses' its cell number, and how it realises that, if appropriate, a proliferative response is indicated. Further, after a proliferative response of any sort, how does the population recognize that the proliferative rate should be curtailed, and how does the population revert to what we might call a *stable focus*, i.e. to normal cell numbers again? Well, the sensor could be sited at the cellular level entirely; this would suggest that the normal population size is genetically preprogrammed. Such control could only operate at the level of the stem cells, which give rise to differentiation-committed proliferating cells which are programmed for a certain number of divisions, and which will differentiate at a certain point. The mean lifetime of the functional cells would also be genetically programmed into them from the stem cell stage. After removal of tissue, the stem cells must 'sense' in some way the deficit, and increase the number of differentiating proliferating cells with the same number of predetermined divisions, or change the predetermined differentiation point; the population then expands until the predetermined population size is reached. It is difficult, on this view, to account for hyperplasia, induced, for example, by chronic epidermal abrasion (MacKenzie 1974*a,b*), or increased food intake (Al-Mukhtar *et al.* 1982*b*), in which the actual size of the population increases and remains enlarged.

This theory needs not only a sensing mechanism, to assess cell loss, but also to know when the normal population size has been reached; while the latter could conceivably be genetic, it is difficult to see how cell loss can be assessed by genetic mechanisms. How such genetic mechanisms could work is, at this time, difficult to say.

Alternatively a population size could be determined by the functional demands made upon it (Goss 1964). Normal function would impose a certain cell number; if functional demands are increased, either by the removal of some functional cell mass, as in partial hepatectomy, or by increasing demand artificially, as with increased food intake, increased growth would ensue until the functional demand has been met; in these examples, by regrowth back to approximately normal levels, in the case of the liver, or by hyperplasia of the intestine; decreased demand (cf. starvation in the gastrointestinal tract), leads to decreased tissue mass, or *atrophy*. This hypothesis may or may not need a sensor, since, in its simplest form, functional demand itself could be a sensor, an increase in which directly stimulates cell production, and, as the demand decreases, so the stimulus for cell production would decrease. However, functional demand could act through an intermediary, i.e. by stimulating the production of some factor which itself stimulates cell proliferation, for example, the *enteroglucagon hypothesis* in intestinal adaptation (see Chapter 23). Goss (1972) has discussed the evidence for the functional demand hypothesis in some detail, and has largely concluded that it is difficult to separate the hyperplasia which follows unilateral extirpation of a paired organ, or indeed induced local damage, from a functional demand on the tissue. However, unlike the protagonists of the tissue mass hypothesis, adherents of the functional demand school have yet to elaborate a tenable hypothesis at the biochemical or molecular level, which translates functional demand into cell proliferation.

A third hypothesis would imply that it is the actual *cell number* that is important: the steady state cell number in some way acts upon the proliferative compartment to regulate cell production; again, there must be some sensing mechanism and some effector mechanism. This is usually accommodated either by proposing that the normal cell number is responsible for maintaining the concentration of some chemical signal which *inhibits* cell proliferation, effectively a hypothesis of growth control by *self-inhibition*, or by invoking a stimulator of cell proliferation, produced either by damaged cells in wounding, or by the remnant after procedures such as partial hepatectomy. This might be appropriate to circumstances where there is actual tissue damage, but to explain the cell proliferation which follows unilateral nephrectomy we would have to postulate a separate sensing mechanism.

Several of these hypotheses need a sensor to detect when a decrease in tissue mass becomes evident, and an effector to increase the proliferative rate; obviously the most attractive hypothesis (but not *necessarily* the best) would be the simple case whereby the sensor and the effector operated by the same mechanism. Perhaps the best of the early attempts to produce a global model was due to Weiss and Kavanau (1957) who proposed that cell population size was controlled by two factors, 'templates' which stimulated synthetic processes and growth, and 'antitemplates' which specifically inhibit the action of the templates (Fig. 2.24, and see Chapter 24). A simplification, implicit in this hypothesis, is that only a cell proliferation inhibitor specific for that tissue is produced. If this inhibitor is produced by the differentiated or functional cells and the *concentration* of the inhibitor is made proportional to the *number* of differentiated cells, then we have both a sensor and an effector. A decreased functional cell mass, induced for example by partial hepatectomy, would produce a proportional decrease in the concentration of the inhibitor; cells would then be released from the inhibition and begin to proliferate, and the differentiation of increased numbers of progeny would increase the functional

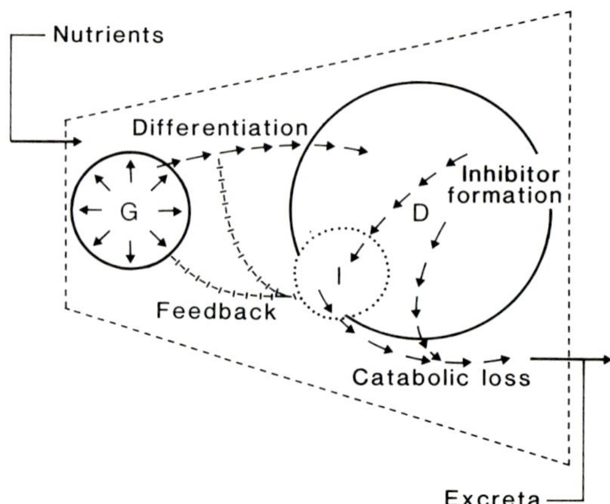



Fig. 2.24. The Weiss–Kavanau model of negative feedback from a differentiated compartment on to the proliferative or generative compartment. (G = generative cell mass; D = differentiated cell mass; I = inhibiting principle.) (Redrawn from Weiss and Kavanau (1957).)

cell mass and therefore increase the concentration of the inhibitor; cell proliferation would then be curtailed. Since the curtailment of any proliferative response needs increased inhibitor concentration and therefore increased numbers of differentiated cells, control of the differentiation rate is also important. In essence, these concepts crystallize the concept of *growth control by self-inhibition*. The proposal that an inhibitory signal is produced by the differentiated cell mass would then make this a *negative feedback hypothesis*, in cybernetic terms (Iversen 1961).

Since its proposal by Weiss and Kavanau (1957), this hypothesis has found considerable favour. The usual way of approaching the hypothesis on an experimental level is to appropriately perturb the system and analyse the ensuing proliferative response; the experiments of Bullough and Laurence (1960), on the pattern of cell proliferation in the mouse ear epidermis after wounding (see Fig. 14.1, p. 446) were said to support the hypothesis of growth control by some hypothetical inhibitor. However, such experiments are always open to the criticism that an exogenous stimulator (perhaps from the dermis; Allen and Smith 1979), might be involved. Consequently, we believe that to search for evidence of negative feedback using studies on cell proliferation only are inappropriate; since we are discussing the control of population size, we should consider measurements of *cell population size* after perturbation (not, of course, neglecting to measure the proliferative rate). While this has been done in conditional renewal systems induced to proliferate (e.g. the mouse prostate complex after androgen treatment; Alison *et al.* 1974), there has been very little work on variation in population size after perturbation in epithelial *renewal* systems, where the self-inhibition hypothesis should be no less important. A second point to make is that, unlike some investigators, we would not expect there to be anything in the way of a critical experiment which would establish or deny the hypothesis, which in the end will probably have to be failed to have been disproved.

Having said this, what sort of experimental evidence would favour a self-inhibition or negative-feedback hypothesis? There are perhaps three main pointers: (a) if the differentiated cell number were proportional to the inhibitor concentration, then perturbation of the system would lead to a series of damped oscillations about the mean value (see Figs. 23.3–5, pp. 846, 847, and 848). Only recently have workers in renewal systems considered cell number to be a parameter worth measuring, and damped oscillations about the mean are now described in the epidermis after wounding (Argyris 1976; and see Chapter 10) and in the small intestine after irradiation (Sato *et al.* 1972) and cytotoxic injury (Wright and Al-Nafussi 1982). Secondly (b) there should be a distinct relationship between the number of differentiated cells and the number of proliferating cells (which will reflect the cell production rate), or the cell production rate itself. The only tissues where this could be done satisfactorily are those where the functional cells can be distinguished from the proliferating cells, as in the small intestine, where functional villus cells can be distinguished from proliferating crypt cells; Fig. 23.4 (p. 847) shows that changes in the villus cell population would appear to effect crypt cell repopulation after cytotoxic damage, in that crypt repopulation does not begin until the differentiated villus cell population is depleted, but once the functional compartment does become depleted, crypt repopulation ensues (see Chapter 23). Studies of this nature are, however, unusual and it would be of great interest to see if similar relationships apply to other renewing epithelia, or if the same phenomena are seen in the intestine in other, more physiological conditions. Thirdly (c) it should be possible to fit the curves of the various parameters which describe the proliferative response after perturbation, including the population size measurements, by models which incorporate feedback terms in them. This has proved possible in the epidermis (Bjerknes and Iversen 1974), small intestine (Britton *et al.* 1982; and see Chapter 23), and in the castrate mouse seminal vesicle (Appleton *et al.* 1973). Thus, in these various tissues, experimental analysis has, as yet, failed to disprove the concept of growth control by a negative feedback effect of functional tissue mass upon cell proliferation.

Having taken the step of proposing that growth control by self-inhibition is a reasonable hypothesis, and not disproved by experiment, we must ask how such a general hypothesis works. Potential mechanisms of self-inhibition are several: the inhibition that follows cell contact during growth *in vitro* could be viewed as one mechanism of self-inhibition, although the importance of this in *in vivo* tissue systems is difficult to assess; while such extremely local inhibition may play a part in the curtailment of, say, epidermal growth after wounding (see

Chapter 10), its role after partial hepatectomy or unilateral nephrectomy is problematical. There is also the consideration that tissue mitoses, in both normal and regenerating tissues, are not confined to those cells out of contact with each other.

It is true to say that most interest in recent years has devolved upon a chemical mechanism of self-inhibition; such a concept is very attractive in renewal systems generally, and theoretically, at any rate, can explain the observed phenomena. This is the *chalone hypothesis*, originally formulated by Bullough and Laurence in the 1960s, and since then embroidered, propounded, castigated, and ridiculed by innumerable authors. This hypothesis states that the rate of cell production in all tissues is under the control of a tissue-specific (but not species-specific) freely diffusible chemical agent, the chalone, produced locally by the functional cells of a tissue, which acts to inhibit cell proliferation. This concept then explains several growth phenomena: removal of functional tissue decreases the local chalone concentration, allowing cells to divide, and, with the production of new functional cells which synthesize chalone, the local chalone concentration increases, and inhibits cell proliferation again; thus the initiation and curtailment of the proliferative response to wounding and partial extirpation is accounted for, and slight modification allows compensatory growth after unilateral removal of an anatomically distant paired organ to be brought in. The original concept also accounts for the tissue-specific and purely local phenomena associated with wound healing; proposing that malignant cells are insensitive to normal chalone concentrations could also account for malignant changes. To fulfil these roles, chalones must be water-soluble and freely diffusible, and indeed, most chalones are prepared, at least initially, by water extraction.

Although earlier ideas in epithelial chalones suggested an action in $G_2$, current biological thought and indeed, chalone concepts, argue for an action in late $G_1$ (Iversen 1981), which would suggest a more important biological action; indeed, the functional role of the epidermal $G_2$ chalone, which inhibits $G_2$ to $M$ or $S$ to $G_2$ flux, must be in doubt (see Chapter 14, section 1). Many chalone preparations now exist, for a whole range of tissues (Allen and Smith 1979; Laurence 1980; Iversen 1981), which appear to be active in inhibiting cell proliferation at several points in the cell cycle, in both *in vitro* and *in vivo* experimental systems, although the fact remains that no confirmed report of a pure chalone preparation has yet appeared.

It is a truism that the chalone hypothesis (theory for its adherents) has petrified into an elaborate belief system for the protagonists, a chopping block for its critics, so much so that dispassionate comment for the purposes of journal refereeing becomes difficult to obtain; however, the non-aligned assessor, with the best will in the world, must confess to several large difficulties with this hypothesis as it currently stands.

Some of these problems are conceptual: what is a tissue? Thornley's (1979) epidermal chalone preparation apparently affects trachea and bladder epithelia, and the small bowel chalone (Sassier and Bergeron 1977, 1980) has action in the colon. While the putative changes in chalone concentrations after wound healing are clearly postulated (see Fig. 14.1), how is the concentration gradient organized in normal tissues, and how does this gradient change after partial organ extirpation or paired organ removal (Allen and Smith 1979)? Also in normal tissues, what controls the rate of differentiation, since most observations indicate that, in most tissues, proliferative activity and the degree of differentiation are inversely correlated. Do chalones act directly on differentiation, or are differentiation changes a direct result of the effects on cell proliferation?

There is no experimental evidence that chalones affect the differentiation rate in epithelial tissues, despite a wholly theoretical proposal by Bullough and Mitrani (1976) that the basic action of the epidermal chalone is to divert proliferating cells into a keratinization and aging pathway. Iversen (1981) riposted that if chalone *production* is associated with maturation, with mature cells producing more chalone per cell per unit time, then it becomes unnecessary to postulate an action of the chalone on the differentiation rate. Moreover, the chalone hypothesis maintains that the target cells for the antiproliferative effect are the proliferating cells in the system, and that the source of the chalone is the maturing cell population; it is therefore surprising to note that the fetal calf thymus is a convenient source of granulocyte chalone (Maurer *et al*. 1976), and that epidermal basal cells synthesize the $G_2$ chalone (Elgjo 1972, 1974*b*; and see Chapter 14, section 1). While post-mitotic maturing cells do, in all probability, occur in the basal layer (see Chapter 8), there would have to be a difference between post-mitotic maturing cells in the basal layer and those in the suprabasal layers. Thus proliferative compartments do appear to produce chalones.

Other problems are more prosaic, and refer to experimental method: the first of these is the extraction procedure. In epithelial tissues (e.g. epidermis) most workers have used tissue homogenates, rather than eluates; this is itself a hazardous step, since it is distinctly possible that water-soluble factors other than growth control substances are extracted, whereas an eluate of intact cells is unlikely to give rise to artefacts of this type. The use of distilled water was criticized by Allen and Smith (1979), since it does lead to aggregation, insolubility, and binding of small molecules, such as polyamines, to proteins. In this respect, Dewey's (1973) melanocyte chalone turned out to be spermidine (Dewey 1978).

A further consideration is the amount of chalone extracted. Allen and Smith (1979) have calculated that, for the best current molecular weight estimate of 2000, and if chalones were active at 'hormonal concentrations' of say 10 nM, then 50 kg of tissue should be extracted to obtain 1 mg of pure chalone. The corollary of this is that activities of mg/ml *in vitro* (Kiger *et al*. 1972) or injectates of mg quantities *in vivo* should be looked at askance. Allen and Smith (1979) have also complained that preparative methods detect $\mu$g quantities, whereas a chalone could well be active in ng quantities; thus either tissues contain large amounts of chalones (in which case pure preparatons should be simple), or else most chalone preparations are highly impure. In fact, the high molecular weight of many chalone preparations is probably due to non-specific aggregation of the active component to a protein by non-covalent binding (see Chapter 14, section 1).

There are also major problems to contend with in the chalone assay systems: should we favour *in vitro* systems, on primary explants, with their well-known disruption of proliferative rates and of cell cycle phase distributions (see Chapter 6, section 5), or cell lines, with disruption of normal tissue architecture, or should we prefer *in vivo* assays with the problems attendant upon injecting quantities of foreign protein? There is no easy answer. Since most chalones appear to act in late $G_1$, the favoured measurement has been the incorporation of [$^3$H]-TdR into DNA, a technique replete with problems (see Maurer 1981, and Chapter 3, section 2), or where $G_2$ inhibitors are being assessed, by (usually) inappropriate application of the metaphase arrest method (see Chapter 14, section 1). There is also the choice of controls: first, particlarly for *in vivo* work, if we inject a quantity of foreign protein, often in hypertonic solution, along with the chalone preparation, are saline-injected controls sufficient? Secondly, in the search for tissue or cell line specificity, which tissues do we select for controls? As indicated by Laurence *et al.* (1979), it is fatuous to compare, say, epidermal chalone action on epidermis with its action on liver or kidney cells, where the cell kinetics of inhibition may be completely different; therefore the control tissue must have similar cell kinetics to the target. The work of Taylor (1977) has emphasized that the time points at which measurements are made after chalone application are critical, in that the time relationships of inhibition differ considerably in different tissues.

A third problem of controls is one which has been largely ignored, and that is the possibility of a *cytotoxic effect*: just because an action is reversible, this does not automatically exclude a transient cytotoxic effect. This can be so particularly in *in vitro* studies (Iversen 1981), and few if any investigators have looked at, for example, the oxidative metabolism of the target cell population, or assessed cell viability with the vigour with which Drewinko *et al.* (1978) have approached the topic, and most workers rely on the non-discriminating dye exclusion test (cf. Houck *et al.* 1971), which only distinguishes live from dead cells, and not from lethally damaged ones.

A corollary of the impure state of most chalone preparations is the distinct possibility of a cytotoxic effect; these need not be chemical; bacterial contamination, namely with Clostridial spores, was responsible for the oncolytic effect of a tissue extract on experimental tumours (cf. Mohr *et al.* 1972*a*; and see Chapter 5, section 2). The potential chemical artefacts are legion; the presence of thymidine in the extract can dilute intracellular pools and block TdR uptake, a noteworthy pitfall of [$^3$H]-TdR incorporation (see Chapter 3, section 2). Though usually regarded as growth stimulators, the presence of polyamines in extracts can lead to the inhibition of cell proliferation; thus Allen *et al.* (1977)

showed that the inhibitory action of the lymphocyte chalone prepared by Kiger *et al.* (1973) was due to spermine, and we have already mentioned that Dewey's (1973) melanocyte chalone is in fact spermidine (Dewey 1978). It is evident that definitive chemical analysis would need much chemical work.

There are obviously major conceptual and experimental difficulties in accepting the chalone hypothesis as a *general* model for growth control. The main current problem is the lack of a pure preparation, and though there are of course special problems with chalones, the field does compare rather badly with the many success stories in recent years, where new hormones and peptides have been speedily isolated and purified; as Allen and Smith (1979) remark, the preparative quality of current preparations is quite inadequate to support the many claims of biological functions made for chalones. Even the acceptance of inhibitors in the tissue extracts does not confirm their physiological role; again, this will have to await the isolation of the active component, exploration of its antigenic qualities, and, with the preparation of a specific antibody, its cellular localization, and the development of sensitive radioimmunoassay.

Moving on, can we find any role in the field of growth control for endogenous stimulators of cell proliferation, the antitemplates of Weiss and Kavanau (1957)? There are an impressive array of such stimulators in the haemopoietic field (Lord 1979), but there are few such claims in epithelia. Leaving aside the undoubted trophic effects of established hormones (e.g. ACTH, TSH, FSH, etc.) on adrenal, thyroid and gonads, and androgens and oestrogens on their respective target tissues (see Chapter 31), there are no clear indications for an action of endogenous stimulators in the maintenance of normal homeostasis or in adaptation to new, imposed conditions. However, we might note that recent work might indicate a trophic role for a gut hormone, active in intestinal adaptation to partial resection (see Chapter 23), and the list of putative hepatotrophic factors is endless (see Chapter 25). There is also a whole host of recently isolated growth factors, of which the potentially most important to our present field is epidermal growth factor (EGF), which appears to have stimulatory effects on cell proliferation in both epidermis and gut epithelia (Chapter 14, section 3, and Chapter 23). In addition, there have been several theoretical attempts to incriminate an 'antichalone', interacting with a chalone to modulate proliferative rates in the appropriate direction; one such was the 'mesenchymal factor' of Bullough (1975*a,b*), elaborated by the dermis, and stimulating epidermal cell proliferation; if these factors exist, they have yet to be characterized.

It is difficult to be sure of the physiological role of any of these stimulators; if they are indeed active, how do they fit into a global picture where negative feedback systems also operate?

Finally, we might anticipate Chapter 14, section 2 by noting that the intracellular levels of the cyclic nucleotides cAMP and cGMP appear, in some way, to be related to the proliferative rate, and particularly in the epidermis, there has been a great deal of interest in the levels of these cyclic nucleotides and their relationship to cell proliferation, especially in the skin disease psoriasis. The evidence is complex, and often confusing, and is considered in Chapter 14, section 2. There are suggestions that the chalones and hormones active in cell proliferation, for example adrenaline, act via the cyclic nucleotide cascade (see Chapter 14, section 2).

## 2.11. Working models of the cell cycle

In several epithelia, kinetic models have been formulated with a specific kinetic situation in mind; for example, the mouse prostate model after androgen stimulation (Appleton *et al.* 1973; see Chapter 31) and the small bowel stem cell feedback model (Britton *et al.* 1982, and see Chapter 23, section 1). But there is a species of general model which provides a viable alternative to the orthodox model of the cell cycle, in which cells pass sequentially through the various phases until mitosis is reached, after which some sort of decision is made to decycle and differentiate, or to enter a resting phase, or to recycle and complete another cell cycle. In many instances the orthodox model can cope with cell proliferation phenomena, but in recent years there has been a move to subdivide the cell cycle in different ways, a view which has been largely based on the observation that the intrinsic variation in the cell cycle is confined mainly to the $G_1$ phase, and the $S$, $G_2$, and $M$ phases are largely constant for any one proliferating cell system. That is to say, while the $S$, $G_2$, and $M$ phases are *deterministic*, and not subject to a great deal of variation (and in some models, none at all), the $G_1$ phase is *stochastic* and introduces most of the variation seen in the cell cycle time distribution. It is therefore appropriate to ask why there is so much variation in $t_{G_1}$.

This alternative concept of the cell cycle was introduced by Burns and Tannock (1970), specifically for slowly cycling cells in the epidermis, but the general case was made by Smith and Martin (1973). These models divide the cell cycle into two parts, the one of relatively constant duration, corresponding to the $S$, $G_2$, $M$ phases and the last part of $G_1$, and called the $C$ phase (Burns and Tannock 1970) or the $B$ phase (Smith and Martin 1973) and a second phase of indeterminant length, called the $G$ or $A$ states respectively (Fig. 2.25); cells remain in this state for any length of time, but the rate of transit into the $B$ phase is controlled by a transition probability $(P)$, which under steady state conditions is *constant*. This means that the cells leave the $A$ state *exponentially* with respect to the time spent in that phase. Support for this proposal is provided by the so-called $\alpha$ curves; these are generated by plotting the distribution of frequencies of the intermitotic times for various *in vitro* cell lines, obtained by time lapse cinematography. The term $\alpha$ is simply $(N - $ (the fraction of cells already divided)$/N)$, and $\log \alpha$ is usually plotted against time. The resulting curves (Fig. 2.26) show *exponential decay*, indicating that cell division is initiated by a random process, rather than at the end of a regular sequence of events.

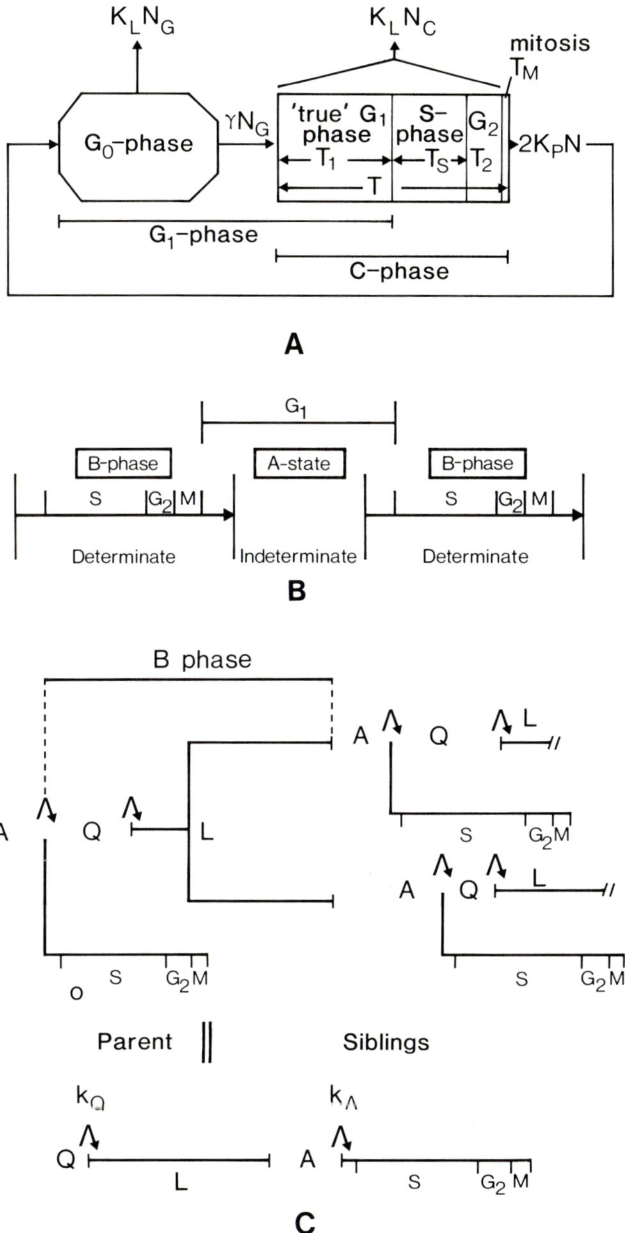



Fig. 2.25. Alternative models of the cell cycle: (A) due to Burns and Tannock (1970); (B) of Smith and Martin (1973); and (C) the modification introduced by Brooks *et al.* (1980), which proposes that there are two controlling points in the cell cycle. The symbols and the identity of the compartments and controlling parameters are described in the text.

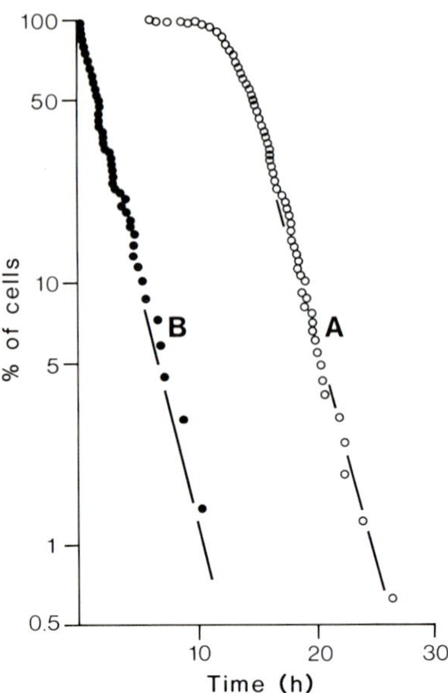



Fig. 2.26. The $\alpha$ curve (A) and the $\beta$ curve (B). (Redrawn from Shields and Smith (1977).) The $\alpha$ curve is the fraction of cells with intermitotic time $\geqslant t$, and the $\beta$ curve the fraction of sibling cells with intersibling time $\geqslant t$. Results are for SV3T3 cells grown in Dulbeccos modified Eagles medium supplemented with 10 per cent fetal calf serum. ($\alpha$ curve, $N = 280$; $\beta$ curve, $N = 107$.)

The shoulder seen on the curves in some cases was ascribed to the small but detectable stochastic component of the $B$ phase, which may be caused by small differences in cell size among the population (Shields 1977). The transition probability hypothesis also predicts that the differences in the intermitotic time of sibling cells (the intersibling time) is also exponentially distributed; this is the $\beta$ curve (Fig. 2.26), where $\beta$ represents the fraction of cells whose intersibling time is equal to or greater than the age difference shown on the abscissa. Again $\log \beta$ is plotted against time, and this also is exponential (Minor and Smith 1974), implying that siblings have $B$ phases of almost identical duration. Moreover, the transition probability hypothesis would predict that the slope of the curve is identical to the exponential portion of the $\alpha$ curve, if the duration of the $B$ phase $(T_B)$ is similar for the sibling cells, and Fig. 2.26 shows that this is so for SV3T3 cells in culture (Shields and Smith 1977). It is important that the $\beta$ curve be exponential over its whole length, since this indicates the presence of an exponentially distributed phase of the cell cycle, and Shields and Smith (1977) maintained that while a variety of models could be devised to give satisfactory fits to the $\alpha$ curve, only models with such an exponentially distributed phase would realize $\beta$ curves (but see Nelson and Green 1981; and below). Even so, it is interesting that if a $G_0$ population existed in the cells shown in the $\alpha$ plot in Fig. 2.26, it would not exceed 0.6 per cent of the population, since the $\alpha$ plot is linear right down to this level.

Theoretically, there are several ways, not mutually exclusive, in which cells might increase their proliferation rate: there could be a reduction in $T_B$, a decrease in the fraction of cells in the non-cycling $G_0$ compartment, or an increase in the transition probability, i.e. an increase in $P$. When the proliferative rate of 3T3 cells is reduced by serum starvation and by cell crowding, the resulting decrease in rate was due solely to a reduction in $P$, and $T_B$ remained largely constant; there was no increase in the fraction of cells in $G_0$. Thus change in the value of $P$ was proposed as the mechanism of changes in the inter-mitotic time, both *in vitro* and *in vivo*, over a wide range of intermitotic times. It was further noted that mitotically selected cells entered $S$ with exponential kinetics (Smith and Martin 1973), and that serum-stimulated quiescent fibro-blasts also entered the $S$ phase exponentially (Brooks 1976). Shields (1977) maintained that this series of observations could only be explained by the random transition probability hypothesis.

As a direct result of such experiments, Shields and Smith (1977) could find no room for a $G_0$ compartment in defining proliferative rates; quiescent or $G_0$ populations were those where the transition probability was so low that no proliferative activity was discernable. The differences between cycling cells and putative $G_0$ cells (viz. membrane transport (Sander and Pardee 1972); polyribosome content (Becker *et al.* 1971); chromatin template activity (Rovera and Baserga 1975)) were seen as occupying different ends of a continuum of cell proliferative rates, or possibly related to the means by which quiescence was achieved, e.g. serum starvation could induce membrane changes (Kohn 1975). These conclusions were also considered to apply to *in vivo* populations, since free-living organisms (*Euglena*) behave according to the model (Smith and Martin 1974), and the characteristic shape of many FLM curves from *in vivo* populations can be accurately predicted from the model (Burns and Tannock 1970; Smith and Martin 1973; De Maertelaer and Galand 1975, 1977*a*)—along, of course, with a whole host of other models (see Chapter 3, section 4).

There have been serious criticisms of these concepts: the first of these is mainly statistical and concerns the fitting of the $\alpha$ and $\beta$ curves (Nelson and Green 1981). Central to this criticism is the proposal that goodness of fit does not imply that the model is correct; it is possible to fit $\alpha$ (Morris *et al.* 1979; Castor 1980) and $\beta$ (Castor 1980; Green 1980) curves, and although Shields (1977) claims that the $\beta$ curve can only be fitted with the random transition probability model, in fact a large class of other models give an exponential distribution of intersibling times (Nelson and Green 1981); the applicability of the model based solely on its ability to fit $\alpha$ and $\beta$ curves was therefore questioned.

On the experimental side, it has been claimed that, far from being deterministic, the $B$ phase is as relatively variable as $t_{G_1}$ (Guignet *et al.* 1980); aging cells in culture do not show the characteristics predicted by the model (Grove and Cristalafo 1976); the $\alpha$ plot showed a lengthening of $T_B$, and did not possess the conventional straight line portion (but see Smith 1977).

The other main criticism has been conceptual. Baserga (1978) considered that the model was basically descriptive, had little informational content, and that the term $G_0$ was as good, or as bad, as the $A$ state, since they were both used to clothe ignorance about the mechanism that caused some cells to have short, and others long, times between mitosis and DNA synthesis. More emphatically, however, the many biochemical differences between rapidly cycling cells and quiescent populations were considered to be real (Baserga 1976); the distinct differences between resting lymphocytes and mitogen stimulated cells, demonstrated by flow cytometric analysis by Darzynkiewicz *et al.* (1978); and see Chapter 3, section 8) were free of the criticisms levelled at serum depleted cells, and supported a discrete, unique $G_0$ phase. Pardee (1978) also maintained that, if the $A$ state was an obligatory phase of the cell cycle, the transition probability differed between rapidly cycling and quiescent cells, and thus their biochemical properties differed, and they therefore could not be in the same state.

There were also several points which made the original proposers dissatisfied with the original, simple model: although siblings had the same $B$ phase duration, all cells in the population did not, since the distribution of intermitotic times was not exponential, and it is widely agreed that the $B$ phase is variable in general but identical in siblings; it is clear that the duration of the $B$ phase is not fixed. Quiescent cells, produced by serum starvation, only commence $S$ after a significant lag phase, usually longer than $t_{G_1}$ for that cell, and often similar in duration to the *minimum intermitotic time* (Brooks *et al.* 1980). Under the original model, quiescent cells were held in the $A$ state. There is good evidence that the initiation of DNA synthesis follows shortly after a change in random transition, and therefore the lag must represent the time taken to change the transition probability. This is problematical, since in many situations, a rapid influx into $S$ can be stimulated within a very short time (cf. refeeding after starvation; Chapter 20, section 3). Moreover, when extra serum was added to cells previously stimulated with a suboptimal dose, a secondary increase in the rate of entry into $S$ occurred only after another lag phase, lasting nearly as long as the first one (Brooks 1976). To account for these phenomena, Brooks *et al.* (1980) introduced a second model, which contains two transition points (Fig. 2.25) (there are other models with two transition points, see Appleton *et al.* 1973 and Chapter 31). Quiescent cells are trapped in some state $Q$, which they leave, with rate constant $K_Q$; they complete a process $L$ to pass from $Q$ to the $A$ state, and leave $Q$ to initiate $L$ at random. The indeterminate stage $Q$ and process $L$ are integral parts of the cell cycle and not peculiar to quiescent cells. After completing $S$, $G_2$, and $M$ cells re-enter $Q$ (Fig. 2.25, p. 91), to leave at random and begin $L$ again. $L$ can be initiated at any stage of the cell cycle (but not of course in the $A$ state). With $K_Q$ sufficiently high, cells initiate $L$ before dividing, but do not complete it, and siblings are at the same stage of $L$ at birth, and enter the $A$ state together. The phases $S + G_2$ are *relatively* constant, $A$ is the only state where sibling differences are correlated, and $\beta$ curves are thus exponential. Unrelated cells initiate $L$ at different times, and so $B$ phase varies in the population, but sibling intermitotic times correlate. Quiescent cells are in both $Q$ and $A$, and both transition probabilities respond to serum.

This new model accounts for some of the anomalies described above, and also allowed precise calculation of the $\alpha$ and $\beta$ curves, and the sibling correlation coefficient could be calculated from experimentally determined values of $K_Q$ and $K_A$. $L$ was seen as the time required to assemble some structure, and it was noted that centriole duplication showed striking similarities with the hypothetical process $L$. Brooks *et al.* (1980) saw no reason why this model should not apply *in vivo*.

It is always difficult to make the jump from *in vitro* to *in vivo* conditions, and, in our own opinion, studies on *in vitro* populations can be misleading where the *in vivo* organization of cell populations is concerned. Nevertheless, there have been some interesting attempts to apply the transition probability group of models to *in vivo* problems. The original model was formulated with the stratified squamous epithelium of the hamster cheek pouch in mind (Burns and Tannock 1970). The model predicted FLM and continuous labelling curves which agreed reasonably well with the experimental curves of Brown (1968; Brown and Berry 1968; Brown and Oliver 1968; and see Chapter 7); experiments on wounded rat epidermis also supported a constant probability per unit time of a cell being released from the $A$ state (see Chapter 10). The parameters of the model were investigated by De Maertelaer (De Maertelaer and Galand 1975, 1977*a*; De Maertelaer 1978), and it was possible to fit a selection of FLM curves for steady state and exponential populations, culled from the *in vivo* and *in vitro* literature, and to predict the flash labelling index function for basal cells in mouse vaginal epithelium. However, it must be stressed that such fitting is perfectly possible using the orthodox model with appropriately distributed transit times through the phases, and also that these studies assumed a constant $T_B$ throughout the population, an assumption which is certainly not justified. Domingo *et al.* (1978) claimed that calculations of the proportion of cells showing spatial mitotic coincidence in the chick embryo liver, after a metaphase arrest period, supported the transition probability hypothesis: here the fraction of sibling pairs in synchronous mitosis after a metaphase block of duration $t$, was equated with the probability that the difference in their time of progression through the cell cycle is less than $t$, the duration of the block. However, this does assume that if the time between mitosis in the two sister cells is less than $t$, then both mitoses will occur before time $t$, when in some pairs one cell may enter mitosis just before $t$, and the other just after it; therefore the time between mitoses will be $< t$, but they will not be in synchronous mitosis at time $t$

(Curnow and Livingstone 1984).

Do we gain anything by applying the transition probability hypothesis to *in vivo* kinetic problems? At this stage, probably not. On the *in vitro* side, because of the ability to fit the experimental data with several different types of model, it is possible that studies in the kinetics of cell cycle traverse will not allow a decision between $G_0$ and $A/B$ state models, and it may well be that specific markers for the various states of $G_1$ are required (Pardee 1978). In our opinion, there is insufficient discriminatory evidence for a definitive decision at present.

# 3 Methodology in epithelial cell kinetics

It has been said, with some justification, that cell kinetics is methodology. However, it is certainly true that if techniques are inappropriate, or wrongly applied, the chances of robust conclusions being obtained are small. The gathering of data in cell proliferation studies is often a tedious and painstaking business, but unfortunately, errors and misconceptions have abounded in this field, so that many otherwise elegantly constructed studies remain inconclusive.

In this section we survey the numerous methods which can be mustered to solve general problems in cell proliferation; later, as individual tissues are encountered, more specialized techniques will be introduced. To the beginner in the field, the theoretical basis of the technology often appears rabbinical in its complexity, but its importance cannot be overemphasized; a short introduction is already available, which sets out the main methods in more elementary form (Aherne *et al.* 1977*a*).

However, it is to be hoped that even the seasoned cell kineticist, when perusing this section, will not ignore it to get at the kinetics of the various tissue systems, but will find something of interest in this part of the book, which exemplifies methodology as it applies especially to *epithelia*. We believe that careful attention to methodology is of primary importance. Many of the controversies we will encounter are pure methodological disagreements.

## 3.1. Proliferative indices

These are numerical values of the fraction of the cell population occupying any phase of the cell cycle; formerly, only two of these were exploitable, the mitotic index ($I_M$), and the flash labelling index ($I_S$), the latter usually measured by high-resolution autoradiography, after between 40 and 60 minutes exposure of the tissue to tritiated thymidine. However, with the advent of flow cytometry (see section 3.8), it is now possible to derive the $G_1$ population density, while the $G_2$ fraction can be measured by subtraction of the mitotic index from the $G_2 + M$ population density.

### (i) *The mitotic index*

This is simply the proportion of the cell population in mitosis at any time, and is derived by simple observation. Because epithelial tissues, at least in the *in vivo* situation, are usually viewed in sections, there are several resulting important technical factors:

(a) *The definition of the target cell population* can readily modify the magnitude of the count, and must be defined at the outset. For example, in stratified squamous epithelium, proliferating cells are confined to the basal layers, and a mitotic index for that layer must be higher than a count which includes all the nucleated cells in the epithelium; similarly, a count confined to the lower part of the intestinal crypt (Tutton 1973*a*), where higher mitotic indices are found (see Chapter 18), will of perforce be higher than if the whole crypt epithelium were included in the count. Therefore we must define the cell population considered. This is a simple problem when the spatial or anatomical boundaries are succinct, but there are occasions when there are considerable difficulties in defining the population in morphological terms, which, after all, is the only criterion which usually exists; this occurs, for example, in the human epidermis in the hyper-proliferative skin disease psoriasis, where the definition of the germinative compartment has been problematical, and has led to consequent large disparities in the mitotic indices from investigator to investigator.

In the intestine, however, where spatial differentiation between proliferating and non-proliferating cells is possible, the concept of refined proliferative indices has been introduced, where mitotic indices have been corrected for the presence of non-proliferating cells by excluding them, from a knowledge of the growth fraction. This allows an index to be quoted which refers to the proliferating cells only, and allows direct comparison between kinetic situations where the growth fraction (which after all does directly modify the mitotic index; see Chapter 2, section 3, and Thrasher and Greulich 1965*a*; Wright *et al.* 1973*a*) varies. However it is distinctly unusual to be in a position where mitotic indices can be so corrected; similar considerations apply for the labelling index, and this is especially important, since an accurately measured labelling index, for a defined cell population, is mandatory for a precise measurement of the growth fraction (see section 3.9). This is particularly relevant where non-cycling cells are suspected in a compartment which usually contains exclusively cycling cells, as is seen in the small intestinal crypt during starvation (see Chapter 20).

These concepts are especially applicable to tissues which have a specific kinetic structural organization, such as the intestinal crypt, gastric gland or epidermis, but where parenchymal organs such as the liver are concerned, where as yet no proliferative subpopulations have been defined, and each and every cell is regarded as being of equal proliferative status, the measurement of the mitotic index becomes simpler. On the other hand, it must not be assumed that the study of parenchymal organs cannot benefit from the analysis of anatomically defined subpopulations, otherwise the now well-recognized, spatially-orientated wave of proliferative activity, moving centripetally from portal area to central vein, would not be appreciated (Grisham 1972; and see Chapter 25), nor the prominent spatial distribution of cell proliferation in the adrenal cortex (Wright 1971*a*; Wright and Voncina 1977; and see Chapter 29). Indeed, the logical extension of these studies in epithelial subpopulations is to measure proliferative indices for each individual cell position in a tissue, viz. the intestinal crypt (Cairnie *et al.* 1965*a*; Wright *et al.* 1972*a*; and see Chapter 18, section 2), the mouse filiform papilla (Hume and Potten 1976; and see Chapter 8), and for accurately defined subpopulations in the mouse epidermis (Hamilton and Potten 1972; and see Chapter 8), always with sharp insight into the structural organization of these tissues.

In this context, it must not be considered that *tumours* constitute a homogeneous population; there is every indication that profound regional variation in proliferative indices in tumours exists, both from the viewpoint of the relationship of the field selected to the tumour periphery (Hermens and Barendsen 1969; Aherne *et al.* 1977*a*) and in relationship to the tumour blood vessels (Tannock 1968, 1970), and hence to $O_2$ concentration; this applies especially to human tumours. Table 3.1 shows the *mitotic index*, the *raw data* from which

Table 3.1. Mitotic indices in blocks taken from five sites in five nephroblastomas. Figures in brackets are the number of mitoses counted. (After Aherne *et al.* (1977*b*))

| Block | 1 | 2 | 3 | 4 | 5 | $I_M \pm$ SEM |
|---|---|---|---|---|---|---|
| Tumour | | | | | | |
| $N_1$ | 0.50 (225) | 0.63 (291) | 0.68 (244) | 0.71 (305) | 1.03 (365) | $0.71 \pm 0.009$ |
| $N_2$ | 0.61 (312) | 0.66 (305) | 0.69 (292) | 0.72 (328) | 0.79 (290) | $0.69 \pm 0.13$ |
| $N_3$ | 0.72 (301) | 0.86 (407) | 0.97 (424) | 1.05 (494) | 1.43 (662) | $1.01 \pm 0.12$ |
| $N_4$ | 0.78 (425) | 1.00 (418) | 1.03 (444) | 1.09 (494) | 1.14 (494) | $1.01 \pm 0.06$ |
| $N_5$ | 1.06 | 1.08 | 1.10 | 1.36 | 1.50 | $1.22 \pm 0.009$ |
| | | | | | Mean | $0.93 \pm 0.11$ |

Within tumour, $\chi^2 = 311$, d.f. $= 20$, $p < 0.001$.
Between tumours, $\chi^2 = 782$, d.f. $= 24$, $p < 0.001$.

the mitotic index was calculated, and a $\chi^2$ *analysis* on five blocks taken from five advanced cases of nephroblastoma; between blocks the $\chi^2$ value is 311 with 20 degrees of freedom ($p < 0.001$), between individual tumours $\chi^2 = 782$ with 24 degrees of freedom (again $p < 0.001$) and although there are larger differences between individual tumours than from block to block, it is clear that there are highly significant differences between individual blocks from the same tumour, and thus we cannot regard neoplastic tissues as homogeneous (Aherne *et al.* 1977*b*); similar results were derived for mitotic indices in mammary and colorectal carcinomas, and for the *in vitro* labelling of mammary carcinoma (Aherne *et al.* 1977*b*).

We may conclude that the definition of the population under study is a critical step which must be countenanced at the beginning of any kinetic investigation.

(b) *The relationship of the count to the interphase population* is critical in the identification of any induced change in the mitotic index; this is because the question arises as to how many cells we should count in order to obtain a reliable estimate of the mitotic index. Before proceeding, we had better ask the meaning of the word *reliable*; in this context, it is important to distinguish between the *accuracy* of a measurement, which indicates the proximity of the estimated mean to the population mean, and the *precision* of the count, which implies its interval estimate or standard error.

If we can assume that there is no great kinetic heterogeneity, the section examined can be regarded as being representative of the whole tissue; in the case of the measurement of the mitotic index for a kinetically homogeneous tissue, since the nuclear count $n$ will be large, we can use the *normal approximation* of the *binomial distribution* (which is the appropriate theory for a dichotomous population of mitotic and non-mitotic cells, see below). The 95 per cent confidence limits are then $I_M + 1.96\sqrt{(\hat{p}\hat{q}/n)}$ and $I_M - 1.96\sqrt{(\hat{p}\hat{q}/n)}$, where $\hat{p}$ estimates $I_M$ and $\hat{q} = 1 - \hat{p}$.

However, in the present context we are rarely interested in a proliferative index and its confidence interval for its own sake, and usually wish to discriminate between two sets of mitotic or labelling indices. Here Aherne *et al.* (1977*a*) have introduced a useful practice: mitotic and interphase cells are binomially distributed, but, if the number of counts is large, it is possible to use the normal approximation; say we wished to distinguish between $I_{M(1)} = 0.01$ and $I_{M(2)} = 0.005$ in two different tissues; the variance of $I_{M(1)} - I_{M(2)}$ is $[(\hat{p}\hat{q}/n)_1 + (\hat{p}\hat{q}/n)_2]^{1/2}$, and $p_1 = 0.01$ and $p_2 = 0.005$. If we have counted the same number of cells in each sample, i.e. $n_1 = n_2$, this variance is equal to $13.9 \times 10^{-3}/n$, which gives a standard error of $s_M = 0.1182/\sqrt{n}$. Assuming the normal distribution, $n$ must satisfy $(0.01 - 0.005)/(0.1182/\sqrt{n}) > 2$, and $n = >2235$ cells. Hence, if we count in excess of 2300 cells and their contained mitoses, we should be able to discriminate between these two mitotic index values.

Now this analysis is only reasonable if we have complete kinetic uniformity, which is, in real tissue systems, extremely unlikely. Since the section examined will not then be representative of the tissue, then any confidence interval we may calculate relates to the accuracy of counting only, i.e. the *counting error*, rather than the *sampling error*. In this instance, we can take as many tissue samples as are practically possible, which here will be our number $(n)$, and assume that this is normally distributed. Then the upper and lower confidence limits are $I_M + t(s/\sqrt{n})$ and $I_M - t(s/\sqrt{n})$.

Now this argument may appear self-evident, but there have been numerous instances where the number of cells counted endow the measurement with very little statistical discrimination; a case in point is the human epidermis in the skin disease psoriasis, which occurs in the form of raised plaques on the skin. There is known to be marked kinetic heterogeneity, with highest proliferative indices at the plaque margins, and the lowest values found at the centre. Nevertheless, investigators tend to make their measurements at the plaque margins, assuming kinetic homogeneity between successive biopsies (see Ralfs *et al.* 1981*a*). Whilst this may be understandable in the clinical situation, some workers (cf. Weinstein and Frost 1968) have included as few as 600 cells in the interphase population sample for the determination of proliferative indices; the preceding argument would counsel strongly against such a practice.

Again relating to the interphase population, there is the vexatious question of errors incurred from geometric artefact, accruing from the use of histological sections, which of course is most pertinent to studies on epithelia, since most counts therein are done on sections, and it is worthwhile being aware of the pitfalls which await the unwary. These errors are usually implicit in the histological method; for example, in a section of infinite thickness (Rogers 1973), say between 5 and 10 $\mu$m, there will be a tendency for cells with a large nuclear diameter to be counted more frequently than those with a small diameter, which would have repercussions if selective indices are sought in a histologically complex tissue. Similarly, we might note that Michael Abercrombie (1946) wondered 'whether the size of mitotic figures is not sometimes relevant in the very numerous investigations of mitotic rate' (!), and indeed, since the size of mitoses often differs from that of interphase nuclei, it becomes necessary to correct for this in accurate work (see Fig. 3.1). A further, related problem is the size of the nuclei in the interphase population; in sections, some or all of the cells are cut into fragments, with the result that the number of fragments in a series of sections is greater than the number of whole cells present in the intact tissue; this is also true of nuclei or indeed of any other cellular component; we should correct for such fragmentation.

One of the earliest approaches to these problems was that of Agduhr (1941), who proposed that the amount of fragmentation of a nuclear population was a function of the section thickness, and of the nuclear diameter. Assuming that the population consists of spherical nuclei, randomly distributed in space, Agduhr derived

$$N = \frac{2Ta - D}{aT(a + 1)}\, n \qquad (3.1)$$

where $N$ is the total number of whole nuclei with their midpoints in the section, $n$ is the number of apparent nuclei counted (whole and fragmented), $T$ is the section thickness (in $\mu$m), $D$ is the mean nuclear diameter at right angles to the plane of section, and $a$ is the number of sections within which the representative nucleus falls. This relationship was investigated graphically by Floderus (1944), who found it valid only if the nuclear diameter was an integer multiple of the section thickness; Floderus then argued that, for the same spherical nuclear

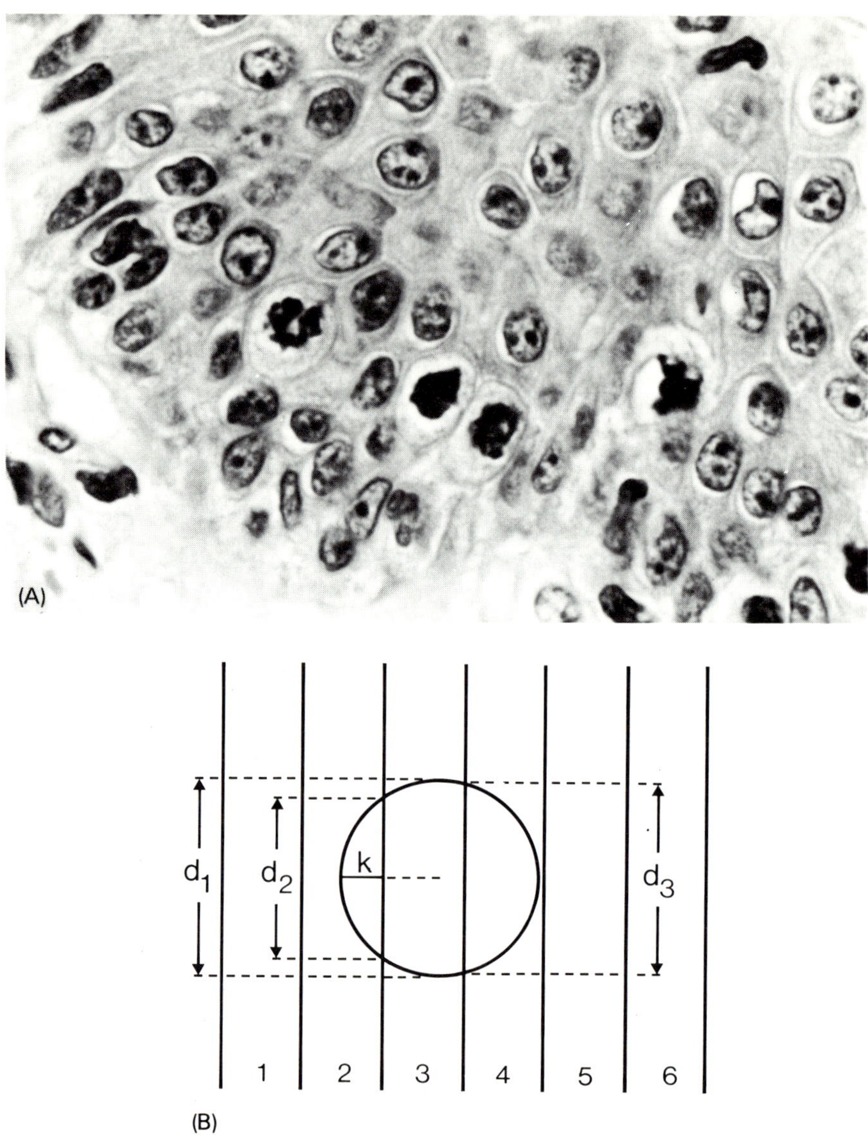

Fig. 3.1. (A) A photomicrograph of metaphases arrested with vincristine in human epidermis in psoriasis. Note that several of the arrested metaphases are larger than the adjacent interphase nuclei, and that many interphase nuclei are not sectioned in their maximum diameter, and appear as nuclear fragments in the section. (B) How a nucleus can be fragmented in serial sections: 1–6 are consecutive serial sections in lateral view. A nucleus is cut into three fragments: a major zone with an orthogonally 'projected' diameter $d_1$, equal to the diameter of the sphere, and two minor segments of diameter $d_2$ and $d_3$. The thickness $k$ of the smallest visible fragment is measured along a plane perpendicular to the plane of the section. (Redrawn from Marrable (1962).)

population, identical in diameter,

$$N = \frac{T}{T + (D - 2k)^n} \qquad (3.2)$$

where $k$ is the thickness of the smallest fragment that can be resolved, and is inserted because, in practice, some of the smallest fragments would not be seen, and therefore would be missed in the count. The variable $a$ in Agduhr's formula was considered unimportant and was omitted.

Abercrombie (1946) also emphasized the presence of nuclear fragments in sections and their importance in the determination of nuclear populations (and naturally, similar considerations will apply when counting mitoses). Abercrombie's correction factor, widely used for correcting the apparent nuclear number to the actual value for whole nuclei, is

$$N = \frac{T}{T + D} \, n \qquad (3.3)$$

and a similar relationship was derived by Gray and Scholes (1951); it should be noticed that, in this formula, $k$ is assumed equal to 0. Marrable (1962) showed that eqn 3.3 holds for situations where the nuclear diameter is less than the section thickness, i.e. $T > D$, where $T < D < 2T$, and also for the more general case where $fT < D < (f + 1)T$, where $f$ is an integer $< D/T$. Note should also be taken of a further method, proposed by Bok and Kip (1939) and advocated by Ebbeson and Tang (1965). In this, sections are cut at two thicknesses, $T_1$ and $T_2$ (say 15 and 10 $\mu$m respectively), and the number of fragments in both sections $(n_1 + n_2)$ are counted. Then

$$N = \frac{T_1}{T_1 - T_2} \, (n_1 - n_2). \qquad (3.4)$$

These formulae are not often used, but commonly the assumptions upon which they are based are assaulted; for example, many nuclear populations contain oval nuclei, and these will have major and minor axes; orientation then becomes important, for if the major axis is perpendicular to the cutting plane, as for example in the bladder epithelium, the chances of a nucleus being sectioned are greater than if the section was in the opposite plane (cf. also the pseudo-stratified bronchial epithelium). If, however, the distribution of axes is random, then the nuclear dimensions will vary between the minor and major axes. But the degree of fragmentation of an individual nucleus is independent of its shape, and dependent only upon its linear dimensions at right angles to the cutting plane where large numbers of nuclei are concerned, and this can be investigated by the use of a frequency distribution (Marrable 1962). Even if nuclei are in fact approximately spherical in shape, the nuclear size may vary in any one nuclear population; thus measurements perpendicular to the cutting plane will vary, and these can then again be arranged as a frequency distribution; the arithmetic mean can be substituted into the appropriate correction factor (Marrable 1962). It is also most unlikely that the nuclei will be randomly arranged in any tissue, and thus this assumption is difficult to sustain.

In practical terms, we must measure the diameter of our stated object, be it nucleus or mitotic figure. In the case of nuclei, in the light microscope, the lateral resolution sets an ultimate limit to the precision of linear measurements, and class intervals are rarely smaller than 0.5 $\mu$m, and more usually nearer 1 $\mu$m. The arithmetic mean of the measurements is used as $D$ in the formula. In this context, Simnett (1968) has described a useful way of determining the nuclear diameter of a cell population, which certainly deserves to be better known: measurements are made of a suitable number (say 400) of nuclear fragments, and cumulative count (or frequency distribution) is calculated (Table 3.2).

Table 3.2.   The measured and calculated distribution of nuclei and nuclear fragments in the zona glomerulosa of the AGrb mouse

| Nuclear diameter ($\mu$m) | Number of nuclei | Cumulative count | Calculated distribution |
|---|---|---|---|
| 10 | 2 | 2 | 133 |
| 9 | 6 | 8 | 143 |
| 8 | 35 | 43 | 154 |
| 7 | 98 | 141 | 167 |
| 6 | 87 | 228 | 182 |
| 5 | 81 | 309 | 200 |
| 4 | 51 | 360 | 222 |
| 3 | 28 | 388 | 250 |
| 2 | 12 | 400 | 286 |

The closest correspondence between the measured distribution and the calculated distribution is therefore 7 $\mu$m, and on the graphical analysis (Fig. 3.2A), 6.8 $\mu$m.

We then need to know the level, in the cumulative count column, at which nuclei are sectioned to project their whole diameter—the level at which nuclear fragments have their midpoints in the section. Then, substituting in Abercrombie's formula (eqn 3.3) for the nuclear fragment number and the section thickness, a calculated distribution of fragments can be made. Graphing these distributions (Fig. 3.2A) gives the value for the nuclear fragment diameter giving closest correspondence between the measured and calculated distributions, and is the nearest value to the true nuclear diameter; for mouse adrenocortical cells from the zona glomerulosa this value is 6.8 $\mu$m.

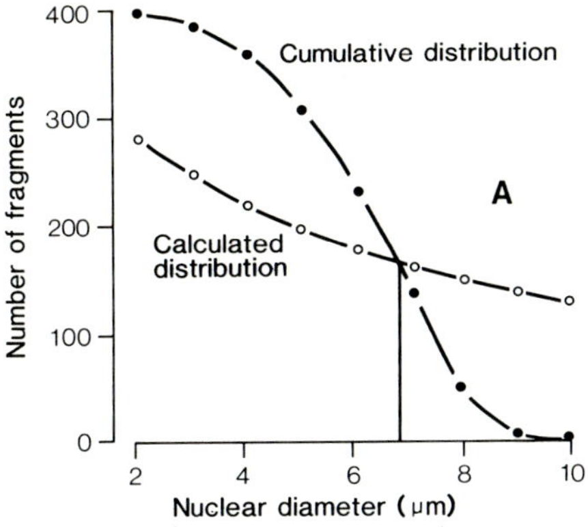

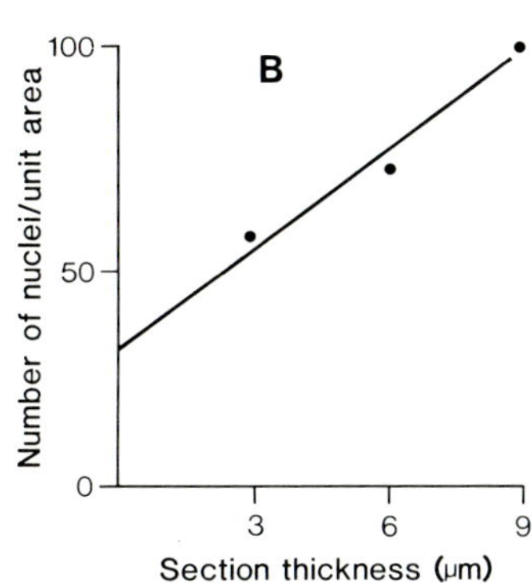


Fig. 3.2. (A) A graphical comparison of the measured and calculated cumulative distribution of nuclear and nuclear fragment diameters in the zona glomerulosa of the AGrb mouse. The nuclear diameter (mean) is here 6.8 μm. (B) The number of nuclei per unit area at different section thicknesses in the striated muscle of the mouse thigh. (Redrawn from Simnett (1968).)

This argument applies for spherical nuclei. If elongated nuclei are present, randomly orientated, then a similar argument can be used; if there is regular orientation, however, then considerable problems are encountered, since we cannot ensure that nuclei are sectioned in a major or minor axis. But Simnett (1968) counted the number of nuclear fragments in sections of different thickness, and graphed them against section thickness (Fig. 3.2B); the intercept on the ordinate gives the number of nuclear fragments of which the midpoint is not in the section (i.e. 33 minor fragments per field), while the slope gives the number of nuclear fragments per section thickness (here equal to 7.4). This method is independent of orientation, but does encounter problems where paraffin sections are required at less than 3 μm.

The measurement of the section thickness can be difficult, and, depending on the individual microtome, it may be unwise to rely upon the feed setting, particularly when using serial sections; in this case, it is probably better to count the fragments in a large series of sections and apply the correction factor to their sum, using the feed value as the mean section thickness. It is well known that, after the first few strokes of a microtome at one setting, the section thickness seems to settle down to a steady value (Marrable 1962), and the careful studies of Clarke (1968), using the Marengo (1964) method of re-embedding individual sections, re-sectioning at right angles to the original plane of section, and then measuring the section thickness, indicated that the first section cut after a change in nominal thickness should be excluded, but its successors are more reliable. But Clarke (1968) showed that three simple methods of measuring section thickness gave reliable results: weighing individual sections, the Marengo method (see also Simnett 1968), and measuring the regression of the block after cutting a known number of sections. For more accurate work, interference microscopy is available.

If the Floderus formula is to be used, then we should correct for fragments not resolved: one can actually measure the thickness of the smallest fragment by focusing on its upper and lower limits, or, by assuming that nuclei are spherical, calculating $k$ from the smallest fragments recorded. Alternatively, and more reasonably, if a frequency distribution is drawn, the size of the smallest fragment seen is readily discernible (Fig. 3.3). The percentage of non-resolvable fragments falling below this value can be read off.

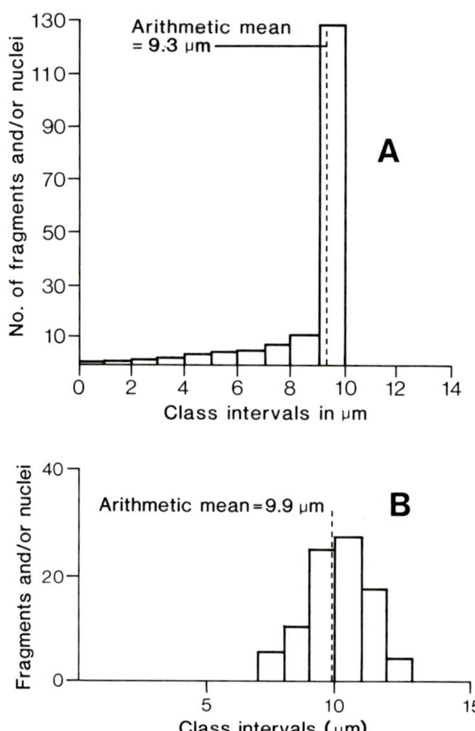


Fig. 3.3.   (A) Theoretical frequency distributions of 'projected' diameters of nuclei and nuclear fragments in histological sections. Diameter of spherical model nuclei = 10 µm. Thickness of section = 15 µm. Arithmetic mean = 9.3 µm. (B) Actual frequency distribution of nuclei and nuclear fragments in sections of teleost blastula. Section thickness = 7 µm. Arithmetic mean = 9.9 µm. (Redrawn from Marrable (1962).)

The above considerations may be all very well for the determination of the interphase nuclear count, and also for the number of prophases and telophases, where the shape of the nucleus is more regular, but for metaphase and anaphase the situation is more complex. Marrable (1962) regarded these phases as cylinders:

metaphases as short, wide cylinders, and anaphases as taller, slimmer ones, in which case eqn 3.3 will also apply for a randomly distributed population of cylinders. However, experience shows that it is possible to resolve quite small fragments of mitotic figures (see Fig. 3.1), and this probably also varies with the mitotic phase; this compounds the difficulty.

Despite the problems involved in this approach, if *the* mitotic index must be measured, as opposed to a relative measurement for comparison with controls, the correction is probably worth making, as it does affect the accuracy of the estimate. For example, Marrable (1962) measured the uncorrected mitotic index in a teleost blastula at 21.1 per cent, but when corrected for fragmentation and size differentials of both mitotic figures and interphase nuclei, this was reduced to 18.4 per cent, a correction worth making.

Abercrombie's correction can be modified to give a correction factor by which to multiply the mitotic index to correct for differences between mitotic and interphase nuclear diameters without the need for independent calculation (Simnett 1968): suppose we need to correct a metaphase index $I_{\text{meta}}$ in a stathmokinetic experiment (see section 3.3). In this

$$I_{\text{meta}} = n_{\text{meta}}/n \tag{3.5}$$

using the usual notation (see eqn 3.1), but the true metaphase index is $N_{\text{meta}}/N$. From Abercrombie's formula

$$N_{\text{meta}} = \frac{n_{\text{meta}}T}{D_{\text{meta}} + T}. \tag{3.6}$$

Since the number of metaphases is small with respect to the total number of nuclei, the mean diameter of the whole population will be the same as the mean interphase diameter, and

$$N = \frac{n \cdot T}{D + T} \tag{3.7}$$

and combining eqns 3.6 and 3.7

$$\frac{N_{\text{meta}}}{N} = \frac{n_{\text{meta}}}{n} \times \frac{D + T}{D_{\text{meta}} + T}. \tag{3.8}$$

Therefore the correction factor $f$, by which the crude metaphase index should be multiplied, is $D + T/D_{\text{meta}} + T$. For 5 $\mu$m sections of mouse lung, Simnett (1968) measured $D$ at 7.75 $\mu$m, and $D_{\text{meta}}$ at *11* $\mu$m; $f$ is then 0.797. Further comment is made on the role of correction factors in metaphase indices in section 3.3.

The reader may ask two questions: which is the best method to use? Well, Clarke (1968) investigated several methods of measuring the interphase nuclear population in the duodenum of the chick embryo, comparing the 'subtraction' method of Ebbeson and Tang (1965 see above) with the 'calculation' method of Floderus (1944), and using a painstaking photographic reconstruction method on serial sections as an independent assessor. Clarke concluded that the subtraction and the calculation method gave reliable results, but that the reconstruction method overestimated the population. Marrable (1962) also noted this in teleost blastulae. The second question is, are such corrections really necessary? If purely comparative data for the mitotic index are required, probably not, unless there is good reason to suppose that experimental treatment, or natural interspecimen variation, might modify the nuclear characteristics; if *the* mitotic index is needed, for a kinetic calculation (viz. the turnover time) or for comparison with a different tissue, then *yes*. If the nuclear population is needed, to be expressed per unit area of a surface epithelium or per anatomical unit (cf. a colonic crypt), then such corrections are essential.

Before leaving this survey of geometrical artefacts, of particular interest to epithelia is the well known tendency for mitoses to be found near the centre of crypt lumina in the intestine; if crypts are viewed in longitudinal section, then, because of the singular arrangement, more mitoses are seen than interphase nuclei (Tannock 1967; and see Chapter 17, section 1).

(c) *Other factors affecting the mitotic index*: although the mitotic index is generally a reliable estimator of proliferative rate, like many similar *state parameters*, it is dependent, not only on the rate of transit of cells into mitosis, but also upon how long they remain in the phase, i.e.

$$I_M = r_M \cdot t_M \tag{3.9}$$

and thus an elevation of the mitotic index can be caused by an increase in the duration of mitosis; there is some evidence that circadian changes in $I_M$ are due mainly to variations in $t_M$ (Evensen 1965; Møller *et al.* 1979; and see Chapter 9) and we should beware such influences in crude $I_M$ estimations. In certain circumstances the metaphase arrest method can compensate (see section 3.3). Because the mitotic index is so time dependent, its value will largely be dependent on which phases are included in the estimate. Prophase and telophase give rise to some difficulties in recognition in sections, and, if these are not all included in the count, this will affect the magnitude of the index. Because of this, some workers, studying tissues with a high proliferative rate, prefer to rely upon metaphase and anaphase counts only (for data of comparative worth, that is).

There are some indications that delay in fixation of the tissue might lead to a decrease in the proportion of mitoses, presumably due to existing mitoses continuing to completion (Bullough 1950; Rajewsky 1965; Denekamp and Kallman 1973; Graem 1979). One possible way around this, and one which does aid in the recognition of mitoses, is to 'explode' mitoses in ice cold *hypotonic* saline; it is hoped that the temperature of the saline inhibits mitosis (Simnett and Heppleston 1966), but one should always be on guard against the possible anomalies of delayed fixation, particularly in studies on human tissues.

### (ii) *The labelling index*

This is the other useful and easily measured proliferative index, and, all other things being equal, gives the proportion of cells which are in the $S$ phase; we will note, only in passing, that the procedure necessitates the preparation of *auto-radiographs*. We will assume that the reader is cognisant of the principles of this now elementary technique, and technical aspects will only be mentioned insofar as they impinge on cell kinetic measurements; the reader lacking autoradiographic technical skill and knowledge is referred to Rogers (1973) or Gahan (1972), or to the care of a competent autoradiographic technologist.

The metabolism of thymidine is discussed in section 3.2, but here we should note that the nucleotide thymidine is not a component of the *de novo* DNA synthetic pathway, but is incorporated by the enzyme thymidine kinase to thymidylic acid, which is then further phosphorylated to the triphosphate form before polymerization (see Fig. 3.4). This has become known as the 'salvage'

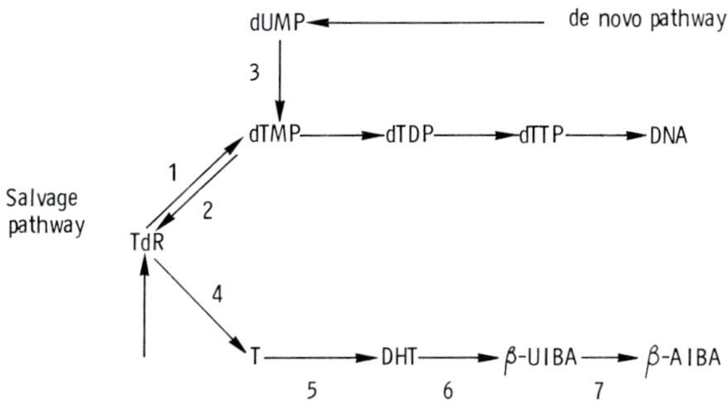

Fig. 3.4.   Metabolic pathways of thymidine. (Redrawn from Maurer (1981).) Substrates: dUMP = deoxyuridine monophosphate; dTMP, dTDP, dTTP = deoxythymidine mono-, di-, triphosphate; T = thymine; DHT = dihydrothymine; $\beta$-UIBA = $\beta$-ureidoisobutyric acid; $\beta$-AIBA = $\beta$-aminobutyric acid. Enzymes: 1 = thymidine kinase; 2 = thymidine phosphorylase; 3 = thymidylate synthetase; 4 = thymidine hydrolase; 5 = thymine dehydrogenase; 6 = di-hydrothymine hydrolase; 7 = decarbamylase.

pathway, and, as we shall see, there are numerous factors which affect [³H]-TdR or [¹⁴C]-TdR incorporation other than changes in the rate of DNA synthesis. These problems are more germane to measurements of the absolute amount of [³H]-TdR incorporated into DNA using the method of scintillation counting, and are discussed in section 3.2. Autoradiographic identification of [³H]-TdR . uptake is always to be preferred to crude estimations of the DNA specific activity, and one of the many reasons for this is that we do have independent visual confirmation of [³H]-TdR uptake in the form of silver grains located in the emulsion over a cell, which tell us that cell was in DNA synthesis at the time of exposure to [³H]-TdR. Because of this, it becomes important to decide how to recognize a labelled cell, and this means that we must consider the number of grains over a nucleus which heralds an $S$ phase cell.

All too often, one reads of authors apparently arbitrarily selecting a grain count threshold of five grains (though some choose three and others eight), and then choose to believe that above this level, all nuclei are labelled and are thus in DNA synthesis; many of these authors are also blessed with perfect autoradiographic technique, and it is interesting to note how often one reads that 'background was generally negligible', whatever that means. For critical work, and we would hope that most work is critical, we should seriously consider our reasons for regarding a cell as labelled, because, in the last analysis, this is a major factor which will determine the magnitude, not only of the labelling index, but also the format of the FLM curve (Shackney *et al.* 1973; and see section 3.4); this is a serious problem, and is worth some attention by all who perform such measurements.

On what does the grain count over a nucleus depend? There are several points to consider.

(a) *The DNA synthetic rate*: it might be thought, *a priori*, that the DNA synthetic rate would determine the amount of [³H]-TdR taken up, and therefore largely be responsible for the grain count. However, there has been a considerable controversy as to the advisability of equating the mean grain count with the rate of DNA synthesis. Dendy and Cleaver (1964) compared the ratio of biochemical measurements of DNA synthesized, in control and UV irradiated mouse L cells in culture, with the same ratio calculated from mean grain counts in the same cells, and concluded that, in the absence of systematic changes in *self-absorption*, the mean grain count was a valid measurement of the mean rate of DNA synthesis (even though, in fact, the point estimates of the ratios differed by nearly 40 per cent, which variation was put down to experimental error).

On the other hand, Baserga and Malamud (1969) have argued strongly that grain counts do not measure DNA synthetic rate. The rate at which DNA becomes labelled is a function of both the rate of DNA synthesis and the specific activity of the precursor pool; because the rate at which the injected [³H]-TdR is incorporated into thymidine triphosphate competes with cold thymidine triphosphate synthesis from the *de novo* pathway (see Fig. 3.4, and section 3.2), to measure the DNA synthetic rate we would have to know many things: the size of the endogenous thymidine pool, the specific activity of the precursor pool, the activity of the TdR kinase and TdR catabolic enzymes, and the degree of competition from the *de novo* pathway. These arguments are very difficult to refute, and perhaps more experimental evidence than that presented by Dendy and Cleaver (1964) is required before we can equate the mean grain count with DNA synthetic rate. Perhaps it is best to take a nihilistic viewpoint, and agree that interphase grain count is a reasonable guide to the *rate of [³H]-TdR incorporation* into DNA.

Several authors have claimed that the interphase grain count follows Poisson statistics (Koberg 1963; Hilscher and Maurer 1962), in tissues such as the mouse intestinal crypt and rat spermatogonia. In fact the distribution of grain counts in these tissues was somewhat wider than a Poisson, but this was attributed to the influence of a finite availability time for the injected $[^3H]$-TdR. Adductive evidence for this viewpoint was proffered by following the *mean metaphase grain count* over the $S$ period; Koberg (1963) found that labelled metaphases in crypt cells showed a constant value throughout the first peak of an FLM curve, and claimed that $[^3H]$-TdR uptake and DNA synthesis in mouse tissues was constant throughout the $S$ phase. Koberg proceeded to measure $t_S$ of one cell type by comparison of the mean interphase grain count of the cell in question with that of another cell type for which the duration of $S$ had been independently determined.

However, this has not been the experience of many workers, who found that the interphase grain count distribution was usually very much wider than a Poisson (Lajtha *et al.* 1960; Painter *et al.* 1960; Cleaver 1967). Wright (1971*b*) also showed a broad grain count distribution in adrenocortical cells (Fig. 3.5A), where it is clear that the spread is far wider than would be expected from a Poisson distribution, and this is further shown by the value of the ratio $\sigma^2/\bar{x}$ (variance/mean), which for a Poisson distribution, should equal 1; in fact, for the adrenocortical cells shown, $\sigma^2/\bar{x}$ was 9.4. This variation in $[^3H]$-TdR uptake through the $S$ phase is also evidenced by the variation in the metaphase grain count with time after $[^3H]$-TdR injection; practically, the mean grain count over labelled metaphases is counted and graphed against time. Cells taking up $[^3H]$-TdR at the end of DNA synthesis will be the first to appear as labelled metaphases after a time equal to $t_{G_2} + \frac{1}{2}t_M$, while cells at the onset of the $S$ phase at the time of $[^3H]$-TdR injection will appear at approximately time $(t_{G_2} + \frac{1}{2}t_M) + t_S$ later (Fig. 3.5B). There is first a slow rise in early $S$ to reach a peak near the middle of the $S$ phase, followed by a more precipitous fall at the end of $S$; similar results have been recorded for other rodent cell types by Cleaver (1967) (also combining Feulgen microdensitometry and mean grain counts), Odell *et al.* (1968), Schaier *et al.* (1971), and Sawicki and Rowinski (1968), and it is clear that variation in $[^3H]$-TdR uptake during the $S$ phase is an important factor in producing the large variance in interphase grain counts, and therefore the grain count of individual nuclei is largely dependent on the position of the cell in $S$ at the time of labelling.

Few authors are aware of this variation in $[^3H]$-TdR uptake through $S$, even though it has indirectly provoked an entertaining but bitter controversy (Shackney 1975*b*; Nicolini 1975) concerning the discreteness of the phases of the cell cycle, which has, however, attracted little comment from non-aligned parties, when one considers its importance to cell proliferation studies; this defect is rectified below.

It is now becoming apparent that DNA synthesis is a remarkably asynchronous

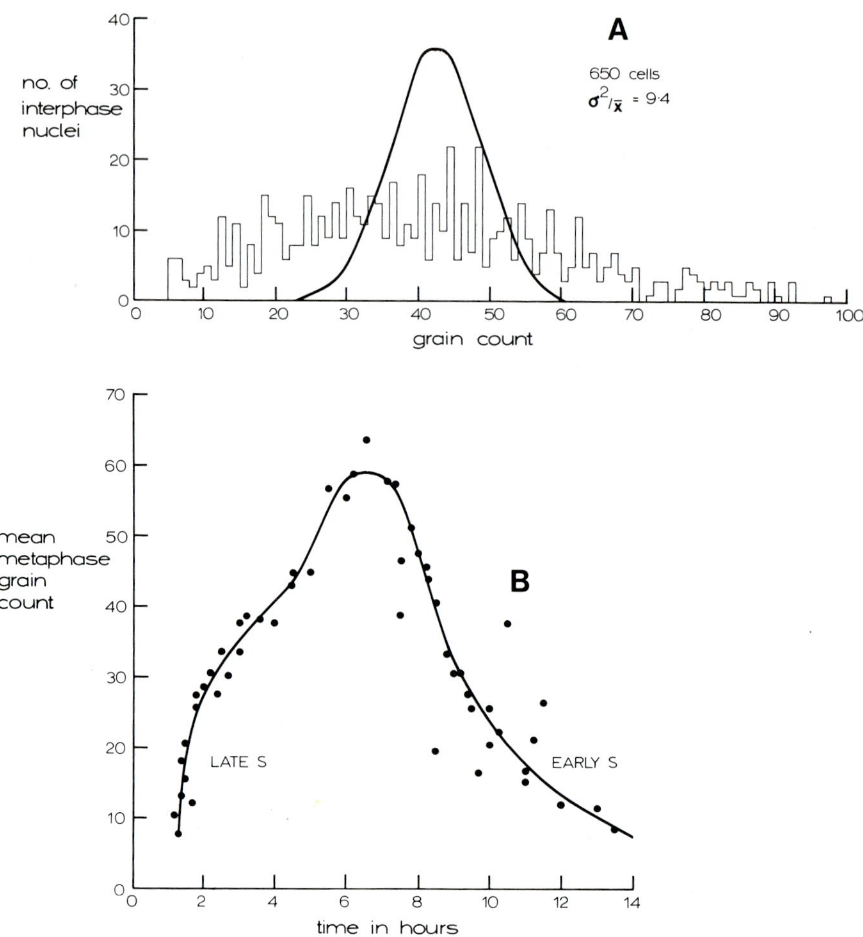


Fig. 3.5.   (A) The distribution of grain counts in interphase nuclei of the 14-day-old rat adrenal cortex (zona glomerulosa). [³H]-TdR (5 Ci/mM) was given in dosage of 1 μCi/g body weight, and the exposure period was 28 days using Kodak A.R. 10. The distribution shown is a Poisson drawn for the mean grain count; it is clear that the distribution is much wider than that expected from a random radioactive disintegration process (i.e. ratio of the variance:mean, $\sigma^2/\bar{x}$, is >1).   (B) The variation in metaphase grain count with time after injection in the cells of the 14-day-old rat adrenal cortex (zona glomerulosa). The cells which appear early in the graph were those in late $S$ at the time of thymidine injection, and the early $S$ phase cells appear towards the end of the experiment.

event: in apical root meristems Howard and Dewey (1961) have even demonstrated, again using metaphase grain counts, that [³H]-TdR uptake apparently ceases *in the middle of the S phase*, resuming shortly afterwards, and it is interesting to note that the DNA synthesis rate function is becoming important in the theoretical analysis of FCM data (see section 3.8); it thus becomes an important concept. Cleaver (1967) has proposed a simple model based on the asynchronous initiation of individual replicons, which does explain such variations in DNA synthetic rate.

(b) *The 'background'*: this can lead to a uniform, or localized, distribution of grains which are unrelated to DNA synthetic processes, and is possibly due to inherent background in the stock emulsion, environmental radioactivity emanating from the glass slides, or careless storage of the emulsion. However, Nicolini (1975) has found considerable increases in background counts in mouse intestinal crypt cells when both $[^3H]$-TdR dosage and exposure period themselves are increased, and there is some evidence that $[^3H]$-TdR can be non-specifically bound to proteins, withstand acid treatment, and so show up as cytoplasmic radioactivity in autoradiographs and in scintillation counting (see section 3.2).

Now these considerations will seriously affect the labelling index values if uncorrected, so what can we do about it? There are in fact several alternatives: (i) counting grains over nuclear areas of standard sections submitted to the autoradiographic procedure, but not exposed to radioactivity, has been suggested (Sillstrom 1963), but it could be argued that this would not be suitable if there were significant spatial variations in the background count. Indeed, Shackney (1975*a*) found it necessary to perform background corrections for each cell in an *in vitro study*; this would be difficult *in vivo*! Moreover, on occasion the test background distribution is incompatible with the experimental distribution; in this case simple computer methods are available to obtain the background correction (Benassi *et al.* 1973); (ii) sometimes, plots of the interphase grain count distribution reveal a clear peak associated with labelling, separated from the distribution of nuclei with low grain counts, in which case the choice of grain count threshold is simple; (iii) practically speaking, a useful method in flash labelling, where one is sure that the time elapsed after injection is less than the minimum duration of $t_{G_2}$, is to perform a grain count distribution over metaphases (Baserga and Malamud 1969; Wright 1971*b*; Nicolini 1975). These metaphases should not be labelled, and hence a measurement of the grain count threshold actually on the section used can be obtained. Baserga and Malamud (1969) recommend ignoring the highest 5 per cent of counts.

(c) *Self-absorption*: efficiency in autoradiographs is measured by the number of radioactive silver grains produced in the emulsion by the radioactive disintegrations in the nucleus. Because tissue sections are, by necessity, of finite thickness, there will always be some loss of efficiency through self-absorption by the section. For example, Maurer and Primbusch (1963) calculated that some 90 per cent of grains produced from labelled nuclei came from the upper 2.8 $\mu$m of the section. Routine paraffin sections are usually cut at a thickness of about 5 $\mu$m, so there will inevitably be some loss of grains; for $^3$H, Baserga and Malamud (1969) have calculated that less than 10 per cent of the irradiation reaches the emulsion if the source is 2 $\mu$m away from it.

This difficult and important problem of self-absorption in autoradiographs has been tackled by Shackney (1975$a$). In the absence of intercell variations in DNA synthetic rate throughout the $S$ phase, the grain count distribution would be a function of a random radioactive disintegration process (Cleaver 1967; Wright 1971$b$), i.e.

$$a(m;n) = n^m \, e^{-n}/m! \tag{3.10}$$

where $a(m;n)$ is the probability that $m$ disintegrations actually do occur over the exposure period in a nucleus which will have an average $n$ disintegrations over the same period. For commonly used emulsions such as Kodak A.R.10 and Ilford K5, which are of intermediate sensitivity, where one disintegration produces at most one grain, if $p_{gr}$ is the probability that a radioactive disintegration produces a visible grain, $P_{em}$ is the probability that a particle is emitted towards the emulsion without self-absorption, and $P_{eff}$ is the probability that a particle reaching the emulsion produces a visible grain, then

$$P_{gr} = P_{em} \cdot P_{eff}. \tag{3.11}$$

Shackney calculated that $P_{gr}$ varied from 0.048 for a section of 2 $\mu$m in thickness, to 0.014 for a section 7 $\mu$m thick; he proceeded to show that, if $b(l;m)$ is the probability that a nucleus which has $m$ disintegrations will produce $l$ visible grains, then $c(l;n)$, the probability that a nucleus with a mean number of $n$ disintegrations over the exposure period will give rise to $l$ visible grains, is

$$c(l;n) = \sum_{m=l}^{m \to \infty} b(l;m)n \cdot a(m;n) \tag{3.12}$$

Assuming that *sectioned* nuclei are cylinders rather than spheroids, Shackney then defined an *autoradiographic transfer function*, which is 'the probability that a cell whose tritium content gives rise to a mean number of $n$ disintegrations during exposure will exceed the background grain count threshold'; this can be computed for various backgrounds, and is shown for sections of 2 and 5 $\mu$m in Fig. 3.6. It is evident that, in the presence of heavily labelled cells, some cells which have incorporated significant amounts of [$^3$H]-TdR would appear unlabelled.

Now the practical consequences of this are several. If we assume that the variation in [$^3$H]-TdR uptake through the $S$ phase shown in Fig. 3.5B represents changes in DNA synthetic rate, as labelling intensity increases, then the duration

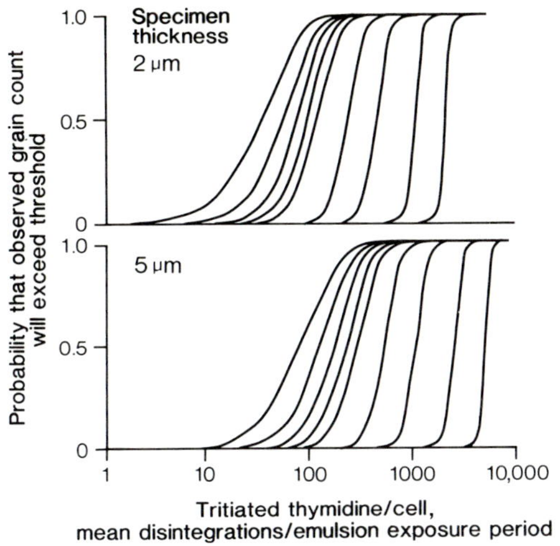


Fig. 3.6.   Illustrating the *autoradiographic transfer function*, defined as the probability of tritium detection as a function of uptake, for different background thresholds in 2 µm and 5 µm sections. (Redrawn from Shackney (1975a).) The curves are drawn for background thresholds (from left to right) of greater than 1, 2, 3, 4, 5, 10, 20, 50, and 100 grains respectively.

of the detectable $S$ phase would also increase. In populations, or subpopulations, with long cell cycle times, which may therefore have long $t_S$ values, the overall rate of DNA synthesis should fall, since the total amount of DNA synthesized remains the same; so that, on this argument, if we were considering a heterogeneous population (say the whole basal layer in the epidermis or the whole intestinal crypt population), grain count *per se* would not be a reliable guide to the localization of cells in the cycle. In practical terms, this concept was held to affect the FLM curve particularly; Fig. 3.7 shows some *in vitro* FLM curves for sarcoma 180, redrawn from Shackney *et al.* (1973) and Shackney (1975a), which were constructed using a variety of grain count thresholds. The counts were made on metaphase spreads, where cell membranes were hypotonically ruptured to minimize self-absorption, emulsion thickness was held to be uniform (though dipping was used), and stringent background corrections were applied. It is evident that the shape of the FLM curve depends on both grain count and exposure time. At low thresholds, $T_C$ would appear to be about 10 to 16 hours, but at higher thresholds, a more rapidly proliferating subpopulation appears, and 5–15 per cent of the cells have $T_C < 8$ hours; note also that $t_S$ increases and $t_{G_1}$ decreases with decreasing grain count thresholds.

We could, however, argue that these limits are within the sort of distribution we would expect from a typical distribution of cell cycle times (see section 3.4), and would probably be given by a $T_C$ distribution calculated from the FLM

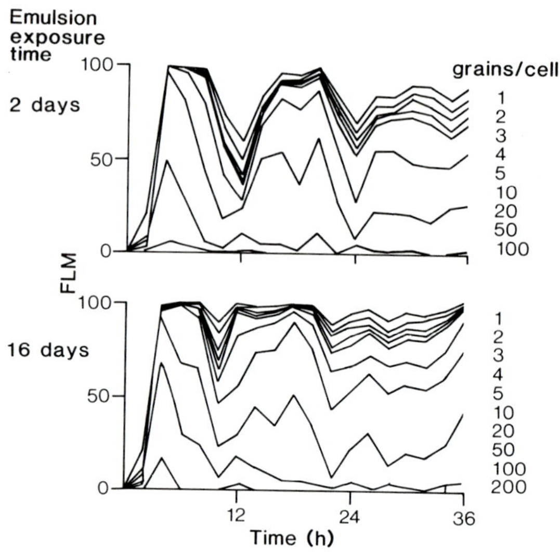



Fig. 3.7. The influence of the grain count threshold on the format of the FLM curve for Sarcoma 180; multithreshold curves obtained at emulsion exposure times of 2 and 16 days, at day 2 of population growth. (Redrawn from Shackney (1975*a*).) The numbers in the grains/cell column refer to the grain count threshold used to construct the respective curve.

curve for a notional grain count, i.e. five grains. But it would appear that the distribution of the $t_S$ value indicated may be wider than we would expect from such an analysis (see section 3.4). This brings to mind some recent data of Potten *et al.* (1982*b*), who found a grain count threshold-dependent peak in FLM curves of mouse epidermis (see Chapter 7, section 2), of Clarkson *et al.* (1967) and Harris and Holzer (1971), Friedmann *et al.* (1973) and Sheehy *et al.* (1974), all of which have described differing FLM formats based upon the grain count threshold. Consequently, grain count threshold can define $t_S$ and therefore $t_{G_1}$, and the duration of the cell cycle can only be defined if the autoradiographic parameters are themselves standardized; in fact, on this hypothesis, the possibility that nuclear DNA synthesis occurs continuously throughout the entire cell cycle, albeit at different rates, is raised by the observation that, with relatively prolonged exposure periods and low grain count thresholds, cells remain labelled throughout the first trough of the curve, i.e. in $G_1$ and in $G_2$ (see Fig. 3.7); the so-called *continuous cell cycle model*.

However, Nicolini (1975) reported that the mean mitotic grain count, at six hours after [³H]-TdR, was constant in the mouse small intestine, colon, and oesophagus, where $T_C$ varies by up to three to four times; since only a single time point was used, we should properly reserve judgement, particularly since the rate of [³H]-TdR uptake through the $S$ phase does vary (see Fig. 3.5B).

However, there is scant evidence that, as $T_C$ increases, $t_S$ also increases; in different epithelial tissues $t_S$ appears to remain largely constant as $T_C$ increases (see Tables 7.2 and 18.3); in the *same* tissue, where spatial variations in $T_C$ are held to occur, the evidence is conflicting; in the small bowel crypt, where $T_C$ decreases with increasing cell position in the crypt (counting from the bottom upwards), $t_S$ also decreases (Al-Dewachi *et al.* 1974), while in the epidermis, where a hierarchical arrangement may also exist, the long cell cycle time claimed for the stem cell is said to be associated with only a very short $t_S$ (see Potten *et al.* 1982*b*; and Chapter 8, section 4). In this respect, Shackney's proposal should be treated as not proven.

These concepts have been vigorously and caustically attacked by Nicolini (1975) on several counts; (i) Nicolini wondered why there was no change in the background correction with increasing exposure period, and also about the role of background in the evolution of the variable grain count threshold FLM; (ii) the equivalence of the grain count with the DNA synthetic rate was queried; (iii) Nicolini pointed to the now numerous FLM studies on homogeneous *in vitro* cell populations which indicate that discrete phases of the cell cycle do exist; (iv) the anomalous FLM curves can be explained by stochastic variation in residence times with a log-normal distribution, and finally (v) Nicolini considered that the more recently acquired data using quantitative cytochemistry, image analysis, biochemical analysis, and FCM studies indicated that DNA synthesis was a discrete event in a non-continuous cell cycle.

These points were taken sequentially by Shackney (1975*b*) with some effect, and he did emphasize that all populations did not possess this continuous DNA synthesis rate function, although for the population he himself considered, he remained adamant. We think that judgement in this dispute is difficult: there are numerous unsubstantiated assumptions on both sides, and we should like to see some epithelial cell populations explored in the same detailed manner in which Shackney *et al.* (1973) have studied tumour cells. More data on the elusive DNA synthesis rate, possibly with the newer FCM methods, are needed; having said this, while we are not convinced that the present evidence, insofar as it goes, disproves the concept of a discrete model of the cell cycle, we do believe that Shackney has a serious point. The definition of the $S$ phase does depend on the stated autoradiographic parameters, and when we compare $t_S$ values from different tissues or populations, one with another, we must remember Shackney's caveat. We are also sure that the last word has not been written in the controversy, and look forward to a resumption of hostilities at an early date.

One factor which has emerged from this debate is the dependence of the mean interphase grain count and the labelling index on the emulsion exposure time, and this message was reinforced by Simpson Herren *et al.* (1976), who found large increases in the $I_S$ values from a range of transplantable tumours over exposure times up to 40 days. Thus we should also beware when comparing $I_S$ values in different situations.

Finally, can we, and should we, correct for self-absorption in autoradiographs? Some authors have found it necessary to apply a correction factor to compensate for the limited range of $\beta$-particles in tissue sections; Evensen (1961) recorded that, in sections of mouse epidermis measuring 5 to 7 $\mu$m in thickness, the true number of labelled nuclei was 40 per cent higher than that detected by the analysis of the autoradiographic image, and thus a correction factor of 1.4 was indicated. This was fastened upon by Weinstein and Frost (1968) to provoke the running controversy that exists in epidermis. However, experimental experience has been mixed; we shall see (Chapter 6, section 1) that Gelfant (1962, 1976) has found no effect of section thickness on $I_S$ in mouse epidermis over a range of section thickness from 2 to 7 $\mu$m, while Simnett (1968) in lung tissue, found $I_S$ to be *higher* in sections of 5 $\mu$m than 2 and 3 $\mu$m, ascribed to the low flux from the thinner sections; above 5 $\mu$m, however, $I_S$ fell. Simnett concluded that all nuclei in his 5 $\mu$m sections were able to produce an image. It is difficult to generalize: such experiments should ideally be done for each tissue and experimental condition (Budtz 1982).

In the context of autoradiographic technique, there are several factors implicit in the procedure which are usually assumed to be constant when comparing controls and experimental groups of observations, but which naturally can effect the grain count; these are (i) *specific activity and dose of [$^3$H]-TdR*; (ii) *the exposure period*: the mean grain count usually increases linearly with exposure period for times up to about a year, apart from some latent image fading with long exposure periods (i.e. the number of disintegrations required to activate silver grains increases as exposure time increases; Baserga and Nemeroff 1962), although Nicolini (1975) showed a considerable increase in *background* count with increasing [$^3$H]-TdR dosage and exposure periods. For example, Clarkson *et al.* (1967) increased the exposure period from two to four weeks, and $I_S$, measured at a grain count threshold of eight, rose to the same value as that found using a five grain threshold after only two weeks exposure; (iii) *histological procedures*: it is now widely appreciated that the Feulgen reaction to stain nuclei, involving, as it does, acid hydrolysis in $N$HCl, can lead to significant loss of label. Cleaver (1967) showed that hydrolysis times in excess of 12 minutes can lead to a rapid diminution in the grain count. Fortunately, adequate staining can usually be achieved with shorter hydrolysis periods; (iv) *the availability of [$^3$H]-TdR*: we have seen above that the level of detection of a labelled cell is influenced by exposure time and tritium content; a further consideration is the period of availability of the [$^3$H]-TdR, depending somewhat on the route of injection. The rapid metabolism of TdR *in vivo* generally confines its availability to DNA synthesizing cells to about 30 minutes or less. However, in some epithelia there is evidence that cells have the ability to store [$^3$H]-TdR in the intracellular pool, and to incorporate it at a later time; this can be up to 17 hours later for putative epidermal stem cells (Hume and Potten 1982) but Møller and Keiding (1982) concluded that this was a general phenomenon for all basal cells in the hamster cheek pouch (see Chapter 6, section 1). Because of this, it may be a reasonable precaution, in studies on epithelia, to survey the dependence of the mean interphase grain count on the sampling time after [$^3$H]-TdR.

We should also note that, in the autoradiographic method, we have an effective way of monitoring [$^3$H]-TdR availability. In comparison of experimental groups with controls, we should beware claims of differences in the rate of cell proliferation based on labelling index studies alone, unless the data is accompanied by interphase grain counts; particularly in the field of growth factor research (cf. Chapter 14), experimental treatments can affect TdR availability, by affecting pool sizes or enzyme activity (see section 3.2); thus the apparent action of gastrin on the small intestinal mucosa, acting to stimulate DNA synthesis (Johnson 1976), is probably due to nothing more than a stimulation of precursor uptake (Oscarson *et al.* 1977); comparison of grain counts in autoradiographs is a useful method of screening for availability effects; (v) *[$^3$H]-TdR reutilization*: if long-term experiments with [$^3$H]-TdR are entertained, it is always wise to beware the potential anomalies introduced by the reutilization of [$^3$H]-TdR. This takes place when nuclei containing [$^3$H]-TdR are degraded. Naturally this is a major worry when working with irradiation or some other form of cytotoxic insult, but even in normal tissues, this can be a problem; nuclei are degraded in the stratum corneum of the epidermis, and their incorporated [$^3$H]-TdR is potentially available for re-incorporation (Cutright and Bauer 1967); labelled crypt cells are lost into the bowel lumen after about 40 hours in the mouse and the [$^3$H]-TdR is again available. Thus skin grafted to the host at 48 hours after the host had received [$^3$H]-TdR became labelled, and Dethlefsen (1969), measuring [$^3$H]-TdR reutilization by comparing the half-time for the loss of the isotope with the generation time of the crypt cells, found significant reutilization in mouse duodenum. If this is likely to be a hazard, then one should consider the possibility of using an isotope such as [$^{131}$I]-iododeoxyuridine ([$^{131}$I]-UdR), or better for autoradiography, [$^3$H]-IUdR, if one can withstand the expense; these are reutilized to a significantly less degree than [$^3$H]-TdR (Dethlefsen 1969).

Before leaving this discussion, we should consider the potential anomalies introduced by the use of [$^3$H]-TdR and cold TdR themselves; despite the copious literature on the use of TdR as an inhibitor of DNA synthesis and cell proliferation (Cleaver 1967), it has been claimed that 10 $\mu$g of cold TdR (or 12 $\mu$Ci of [$^3$H]-TdR, specific activity 3.6 Ci/mM) *stimulated* mitotic activity in mouse duodenum (Greulich *et al.* 1961), but Møller *et al.* (1974) could find no effects of cold TdR, in similar dosage, on the squamous epithelium of the hamster cheek pouch, and provided that a reasonably high specific activity [$^3$H]-TdR is used, *in vivo* effects due to TdR can probably be ignored. However, there is more evidence for a *radiation* effect of [$^3$H]-TdR on cell proliferation; 5 $\mu$Ci/g effectively kills mouse spermatogonia (Johnson and Cronkite 1959) and 2 $\mu$Ci/g also leads to cell death in basally sited jejunal crypt cells (Cheng and Leblond 1974c). In squamous epithelium there are a whole host of largely conflicting reports on the effects of [³H]-TdR on cell proliferation (Gibbs and Cassarret 1969; Møller *et al.* 1974; Olsson 1976; and see Chapter 9), few of which agree on the action or timing of the effect, and in the mouse ileum, Post and Hoffman (1968) have claimed that [³H]-TdR causes a dosage-dependent anomaly in the second peak of the FLM curve at doses over 2 $\mu$Ci/g, but this could not be confirmed by Steel *et al.* (1966) in a transplantable rat tumour at doses between 0.5 and 5 $\mu$Ci/g. Moreover, Shackney (1975*a*) has questioned whether Post and Hoffman's data might not be due to an effect similar to the exposure period difficulties Shackney has described in FLM curves. We might conclude by agreeing with Steel (1977) that the popular 0.5 to 1 $\mu$Ci/g body weight is probably all right, provided the specific activity is relatively high (and therefore small amounts of TdR are given). This is so particularly in flash labelling: if we are worried about ³H or TdR effects in longer-term experiments, then independent checks, such as the mitotic index (Wright *et al.* 1973*c*) can be instituted throughout the experiment.

We conclude that, with suitable care, $I_S$ values are reliable estimators of the proliferative rate; because of the longer duration of $S$, $I_S$ is certainly easier to work with than the mitotic index, and, to our knowledge, there is only one report where the labelling index has changed because of an increased duration of $S$, rather than an increase in $r_S$ (Lewis 1978), despite the many fears to the contrary (Pearse and Marks 1977; Wright 1980*c*). All that we would suggest is that some thought be given to the standardization of the method, and where worries exist about the procedure steps, a more rigorous analysis of the methodology.

## 3.2. Measurements of the incorporation of [³H]-TdR into DNA

It is true to say that measurement of [³H]-TdR incorporation into DNA has proved a very popular measurement, and it is not difficult to see why this is so: the ease of radioactivity assay with the use of liquid scintillation counters, the ready availability of high specific activity [³H]-TdR, the belief that [³H]-TdR is incorporated selectively into DNA alone, and the general impression that something useful is being measured. Because of the rapidity and convenience of the procedure, many workers have relied upon it alone, as the final (and only) criterion in the assessment of proliferative rate, particularly in growth factor research (cf. the epidermal chalone, Chapter 14, section 1), the investigation of hormone action and role in intestinal adaptation (Chapter 23), and in the analysis of ephemeral kinetic responses (Chapter 24). Even from relatively earlier days, there were problems in this court (Cleaver 1967), and despite numerous warnings (for example Maurer and Laerum 1976), many investigators persist in basing their conclusions exclusively upon [³H]-TdR incorporation measurements. Whole theoretical edifices have been built up upon this technique alone (cf. the supposed role of gastrin on cell proliferation in the intestine, Johnson 1976; and see Chapter 23), but before we accept such claims we must seriously enquire after the basis of the methodology.

We have already seen that such factors as radioactivity effects, chemical effects and problems related to the DNA synthetic rate can be important (section 3.1), but there are many other potential anomalies arising from metabolic considerations, factors associated with the extranuclear binding of [$^3$H]-TdR, and biological and experimental factors which are very important to consider.

### (i) *Potential artefacts due to [$^3$H]-TdR metabolism*

(a)  *Uptake and phosphorylation*:  as we have already seen, [$^3$H]-TdR is incorporated into DNA exclusively via the salvage pathway, into deoxythymidine monophosphate, and then by further phosphorylation into DNA (see Fig. 3.4), and this pathway is subsidiary to, and competes with, thymidylate monophosphate synthesis via the *de novo* pathway, which is fed by thymidylate synthetase action. Consequently, incorporation of [$^3$H]-TdR is heavily dependent on one enzyme, *thymidine kinase*; TdR kinase activity is a rate-limiting step in the entry of TdR into the cell (Barlow and Ord 1975) and increased [$^3$H]-TdR uptake correlates mainly with TdR kinase activity. In this respect, TdR kinase can be completely inactivated by the absence of TdR (Kit *et al*. 1965), an important consideration when one remembers that TdR incorporation via the salvage pathway is usually in abeyance in normal cells, due to the usual absence of intracellular and extracellular TdR. Møller and Keiding (1982) have also suggested that TdR kinase activity can vary throughout the day, and Duell *et al*. (1975) have reported an important and unfortunately poorly publicized result: an epidermal chalone preparation appeared to inhibit the growth of dermal fibroblasts in culture, as assessed by [$^3$H]-TdR incorporation, an anomalous result when we consider that a cornerstone of the chalone hypothesis is the tissue specificity of the chalone. However, further analysis indicated that this effect was due to depression of TdR kinase activity in the fibroblasts by the chalone. Moreover, Potter *et al*. (1974) have emphasized that enhanced [$^3$H]-TdR uptake does not necessarily mean net increase in DNA synthesis, based on experience with the effect of AMP on TdR transport.

The essentially practical nature of these caveats is indicated by the report of the effects of an alkylating agent triethyleneiminobenzoquinone, which causes a rapid and profound decrease in [$^3$H]-TdR uptake into Ehrlich and Yoshida ascites cells (Grunicke *et al*. 1975). This was shown to be due to inhibition of TdR transport, and the cytostatic effect of the agent was expressed, not in DNA synthesis or late $G_1$, but in $G_2$; DNA synthesis itself was not inhibited. The recent report that colonic uptake of [$^3$H]-TdR is stimulated by pentagastrin, whereas the cell production rate remained unaffected (Stevens *et al*. 1981), could well be explained by a stimulation of precursor uptake by the pentagastrin.

(b)  *Pool sizes*:  the amounts of TdR in the intracellular pool is normally very small (Rubini 1966), and consequently changes in pool size, produced by any experimental manoeuvre, could lead to artefactual changes in the pool size entered by the labelled precursor, and hence in the amount of $[^3H]$-TdR taken up into DNA. Thus Smets (1969) made the anomalous observation that *in vivo* irradiation of T lymphocytes decreased $[^3H]$-TdR incorporation but increased $[^3H]$-CdR incorporation, probably due to an increase in TdR pool size, and a suppression of the biosynthesis of CdR derivatives. This point is particularly germane to studies in which material containing putative growth substances, particularly chalones, are introduced to an experimental system, especially *in vitro*. Thus bovine and human endometrial extracts contain enough TdR to inhibit $[^3H]$-TdR uptake in HeLa, Ehrlich, and Novikoff cells (Chevalier and Verly 1975).

This is a rather difficult problem to solve experimentally; more recently, perhaps due to the caution expressed by Maurer (Maurer and Laerum 1976, 1981), some workers have measured the radioactivity in the TCA-soluble washings in an attempt to distinguish chalone-induced pool effects from proliferative effects (Laurence *et al.* 1979); unfortunately this is an isolated instance, and most other authors (e.g. Burns 1981; Johnson and Guthrie 1980) gaily continue to use uncontrolled $[^3H]$-TdR uptake as the sole criterion of proliferative rate.

While some, if not all, of these comments also apply to $[^3H]$-TdR detection by autoradiography, there have been some reports which suggest a lack of correlation between $I_S$ values and $[^3H]$-TdR incorporation as measured by scintillation counting; thus Youdale and McManus (1975) failed to demonstrate the calcium-induced proliferative stimulation of thymic lymphocytes using $[^3H]$-TdR incorporation measurements, even though this was readily appreciated by autoradiographic detection. In this model, DNA synthesis could well be stimulated by the *de novo* pathway, with an increase in intracellular thymidylate, and a resulting decrease in $[^3H]$-TdR incorporation, and a similar mechanism could be responsible for the failure of Simnett and Fisher (1973*a*) to correlate incorporation and autoradiographic measurements in newborn mouse lung in organ culture. If this is the reason, the fact that proliferative cells are detected by autoradiography but not $[^3H]$-TdR incorporation may be caused by sufficient $[^3H]$-TdR being incorporated to label nuclei in autoradiographs, but the actual *amount* is small and does not therefore correlate with the magnitude of the proliferative response, notwithstanding the fact that the $I_S$ is also unlikely to reflect the magnitude of the proliferative rate.

Of special relevance to epidermis are the studies of Davidson *et al.* (1979); studying the growth of human keratinocytes *in vitro*, they found that $[^3H]$-TdR incorporation totally failed to correlate with measurements of cell number and DNA content during the growth phase of these cells; moreover, TdR kinase activity actually *decreased* during the growth phase. On the other hand, studies on incorporation of a pyrimidine nucleotide precursor, $[^{14}C]$-aspartate, showed that the *de novo* pathway was stimulated, indicating an inverse relationship between the salvage pathway and the *de novo* pathway, or that *de novo* thymidylate was inhibiting TdR uptake.

Before leaving pool size considerations, we should take special note of the practice of Johnson and co-workers (e.g. Johnson and Guthrie 1980), who scrape gastrointestinal mucosa into medium containing [³H]-TdR and measure its incorporation after one hour. Cells and nuclei will be damaged and release nucleosides, with undoubted effect on the local TdR pool. What these measurements mean is anybody's guess.

### (ii) *Potential artefacts related to DNA synthesis*

These have been mentioned above, since they apply equally well to autoradiography, and, in relation to epithelial cells, the phenomenon of late labelling in squamous epithelium (Hume and Potten 1982 Møller and Keiding 1982) is especially relevant. We should also, at this point, note the lack of kinetic discrimination in the [³H]-TdR incorporation procedure; decreased [³H]-TdR incorporation can be due to a decreased number of cells synthesizing DNA, or to a reduction in the net incorporation per cell, i.e. a reduction in the DNA synthetic rate itself.

### (iii) *Other potential pitfalls*

These include incorporation of [³H]-TdR into macromolecules other than DNA (Bryant 1966; Houck and Dougherty 1974); degradation of [³H]-TdR prior to its incorporation into DNA (Maurer 1981) and the problem of distinguishing semiconservative DNA synthesis from mitochondrial and repair synthesis, especially in situations, for example after UV-irradiation, where repair synthesis is stimulated.

### (iv) *The choice of a denominator*

In all correctness, the dpm measured on the scintillation counter should be related to the net amount of DNA, i.e. dpm/µg of DNA, the specific activity of DNA. However, it is quickly seen that many authors do not do this; they prefer, for some reason, to relate their measurements to per mg/protein, or even to wet or dry weight. This practice has been particularly rife in the gut epithelium, in the search for trophic factors (e.g. Johnson 1976; Johnson and Guthrie 1980; see Al-Mukhtar *et al.* 1982a for review, and Chapter 17, section 1, and Chapter 23). A moment's thought will indicate that this is a dangerous thing to do; any change in cytoplasmic mass in the absence of changes in cell number, or, if using the wet weight, changes in the cell volume or extracellular fluid, will affect the results even though the proliferative rate may be unchanged. Given the inherent problems with the method, it is difficult to defend such a stance.

### (v) *The target population*

To our minds, a major difficulty with this technique is the problem of defining the population under study. Say we were trying to measure [³H]-TdR in epithelial crypt cells of the small intestine; should we follow previous usage

(Johnson 1976; Williamson *et al.* 1978) of whole intestine, including non-proliferative villus cells, lymphoid, smooth muscle, stromal and serosal cells, or of mucosal scrapes (epithelial, lymphoid, and stromal cells)? We should remember that the lamina propria of the small intestine contains lymphoid cells fully capable of proliferating. In fact, Clarke (1971) found that 10 per cent of all mucosal mitoses were non-epithelial, and of course the peri-cryptal fibroblast sheath readily labels with [$^3$H]-TdR (Marsh and Trier 1974).

Consequently, if we do this, how do we know that what we are measuring has anything to do with epithelial cells, or that any experimental manoeuvre we make is not affecting non-epithelial cells? The answer is that we don't and can't know, unless we obtain a denominator related to epithelial cells; thus Hagemann *et al.* (1970*a,b*) measured dpm/*crypt* by counting [$^3$H]-TdR in microdissected crypts, thus confining the measurement to the epithelial compartment only. Unless this is done, we must distrust any measurement done on whole tissue, which is thus masquerading as an index of cell proliferation.

*Conclusions*

After this discussion, it is difficult to see how anyone can rely upon this measurement as any discriminator of cell proliferation, let alone the only one. Authors often promise to bolster any such 'preliminary measurements' with more definitive kinetic parameters, a promise which, as Rufus Clarke has noted, is seldom kept. Just think of the work one would have to do to make this measurement definitive: isolate the target cells, measure TdR kinase activity, endogenous pool size, *de novo* thymidylate synthesis, and finally [$^3$H]-TdR incorporation. Having done this, we obtain a mere state parameter which might correlate with proliferative rate. Is it worth it? The supposed advantages of the method are speed and convenience; if these disappear, we should perhaps forget it, and move to some measurements which, although more demanding, actually mean something. We find it difficult to comprehend why so many workers persist with the use of this hopelessly non-robust method, in the absence of appropriate and careful control conditions. Could it be an eagerness for results, or lack of willingness to spend the time necessary for meaningful measurements in cell proliferation studies? Whatever the cause, they should appreciate that others in the field, with any critical faculty whatever, will regard their results with extreme suspicion.

*Thymidine analogues*

Because of the relative lack (compared with [$^3$H]-TdR) of reutilization, the loss of radioactivity due to [$^{131}$I]- or [$^{125}$I]-UdR from a labelled population has been used as an index of cell loss (Begg 1977; Wright 1981). Most of this work has been in tumours, and very little use has been made of this technique in epithelia (but see Dethlefsen 1969). We should note one recent result of relevance in this field: Hamilton and Dobbin (1982) have recently shown that [$^3$H]-UdR labels many more nuclei in an experimental tumour than [³H]-TdR; the [³H]-TdR was found to label < 50 per cent of nuclei thought to be in the $S$ phase; the reasons for this were not clear, but it is comforting for those of us who work with epithelia to note that there was no difference between $I_S$ values in mouse colon whether measured with [³H]-TdR or [³H]-UdR!


## 3.3. The metaphase arrest method

In the analysis of epithelial cell kinetics, many induced responses are ephemeral, and occur far too quickly to permit application of, say, the fraction of labelled mitoses (FLM) method (see section 3.4) for the calculation of cell cycle phase durations and proliferative rates. In these latter situations we have to rely on the measurement of the proliferative indices, the mitotic and labelling indices. We have seen that these indices reflect, not only changes in the proliferative rate, but also changes in the phase duration, the mitotic duration, and the duration of DNA synthesis respectively.

For these several reasons, namely the time expenditure of the autoradiographic method, and the inherent vagaries of proliferative index measurements, investigators have turned to a time-honoured method, the metaphase arrest or stathmokinetic technique. Correctly applied, this method provides a firm kinetic parameter, the *rate of entry into mitosis, or the birth rate*, sometimes called the *mitotic rate*. From the birth rate, we can calculate the cell cycle time if all cells are proliferating and the growth fraction is unity; if there is a growth fraction of less than 100 per cent, a further useful parameter emerges, which might be called the *turnover time* in normal tissues, or the *potential doubling time* in tumours (Aherne *et al.* 1977*a*). This method can give results over as short a period as 2.5 hours, and involves the study of paraffin sections only.

The earlier literature in this field was summarized by Eigsti and Dustin (1955), but, in a timely review, Tannock (1967) laid down some of the requirements for suitable stathmokinetic agents, requirements that were most adequately met by the *Vinca* alkaloid *vincristine*. However, in addition to using an agent with adequate metaphase arresting properties, there are several critical experimental points which must be fulfilled. Having carried out the experiment in as optimal a manner as possible, serious problems are encountered in the analysis of the data.

In this chapter we shall identify these problems in experimental technique and in data analysis; furthermore, there are numerous instances in the recent literature where the method has been incorrectly applied, and inappropriate measurements obtained, which are often euphemistically dignified by the term 'mitotic rate'; it is clearly important for the continued use of this method that these abuses be identified. There is increasing use of the method in clinical situations, notably for measuring the birth rate in human tumours (Camplejohn *et al.* 1973; Wright *et al.* 1977; Taylor *et al.* 1977; Tabuchi *et al.* 1980), and if maximum yield in terms of data is to be gained from these procedures, then it is essential that the scope and limitations of the method be appreciated. It is its incorrect application that has brought the metaphase arrest method into something like disrepute (Clarke 1971); when correctly applied, the method has discrimination and measurable precision: when incorrectly used, it can be potentially misleading.

*Metaphase arrest agents*

There are many compounds which have been shown to have metaphase arresting properties; of these, the earliest used was colchicine, and latterly its less toxic derivative, colcemid (*N*-desacetyl-*N*-methylcolchicine). More recently, the *Vinca* alkaloids vincristine and vinblastine have found use as stathmokinetic agents. The arresting action of these compounds depends on an interaction with the mitotic spindle or its subunits, which results in the failure of the assembly or disruption of the mitotic spindle, with the production of the characteristically abnormal mitotic figures. These arrested metaphases are usually readily distinguished from normal metaphases, and have been called, not surprisingly, 'colchicine metaphases' (Clarke 1971) – see Fig. 3.1.

Investigators have varied in their choice of specific agent; of those who have compared the metaphase arresting properties of the several stathmokinetic agents, Tannock (1967) and Smith *et al.* (1974) agreed that vincristine was superior, while colcemid proved the worst agent of all. However, Clarke (1971) and Nome (1975*a*) have produced evidence that colcemid can be useful in several tissues (see below).

*Metaphase arrest techniques*

Basically, the method hinges upon the measurement of *the slope of the line* which describes the rate of accumulation of arrested metaphases. The results of a typical experiment is shown in Fig. 3.8, which depicts the accumulation of arrested metaphases in the crypts of the normal human jejunum; vincristine sulphate was injected intravenously at time 0, and jejunal biopsies were taken at the times indicated. The measurement of the slope of the line gives the estimate of the birth rate, mitotic rate, or rate of entry into mitosis. It cannot be emphasized too strongly that the slope of the line is the *only measurement which can be called the mitotic rate*; with this method, no other measurement will suffice.

It is easiest to approach the correct use of the technique in terms of a series of questions:

(i)    *What dosage to use?*    It is of considerable importance to determine the optimal dosage for both the stathmokinetic agent and the tissue to be studied; too few authors, when describing stathmokinetic experiments, have shown dose-response curves for the agent and the tissue discussed. This latter point is especially relevant, since Tannock (1967) and Nome (1975*a*) have shown that optimal doses for the various agents may differ in the same tissue, and moreover,

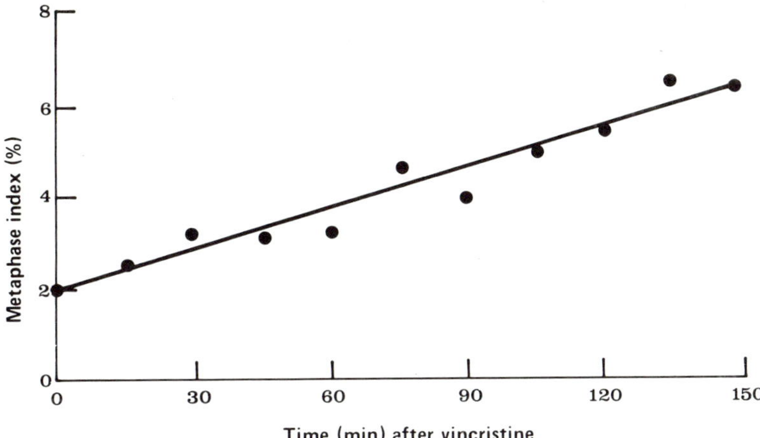


Fig. 3.8.   Mitotic index accumulation data for the normal human jejunal mucosa after an intravenous injection of vincristine (0.045 mg/kg body weight), followed by serial biopsy using the Quinton multiple biopsy capsule. (Redrawn from Wright *et al*. (1973*b*).)

different tissues may have different optimal doses for the same agent. Tannock (1967) has analysed the characteristics of dose-response curves of several metaphase arrest agents; Fig. 3.9A shows a dose-response curve for vincristine in the epidermis of patients with psoriasis; vincristine was injected, in the doses shown, intradermally into stable psoriatic plaques, and biopsies were taken at 2.5 hours after injection (Ralfs *et al*. 1981*a*). Also shown in Fig. 3.9C is Tannock's analysis of such a curve. With increasing dose the metaphase index increases (curve A), attributed by Tannock to an increasing proportion of metaphases arrested, but also probably to the disappearance of any delay period before metaphase arrest is complete, since, at least *in vitro*, this delay period is dose-dependent (Taylor 1965; and see below). With increasing dose, Tannock found that degenerative, pyknotic metaphases predominated, leading to loss of arrested metaphases and thus a decrease in the number of metaphases, represented by curve B. Increasing doses of stathmokinetic agents appear to be toxic, and Fitzgerald and Brehaut (1970) also showed depression of metaphase indices with increasing doses of colcemid in lymphocytes *in vitro*. In the dose-response curve shown for vincristine in psoriatic epidermis, we cannot be sure that the increase in metaphase arrest effect and the metaphase degeneration effect are not overlapping, with consequent lowering of the peak value and possible underestimation of the birth rate.

A different type of dose-response curve, this time for colcemid, is shown in Fig. 3.9B; the results are for primary explants of the mouse coagulating gland in organ culture, and here the readings were taken at four hours after the

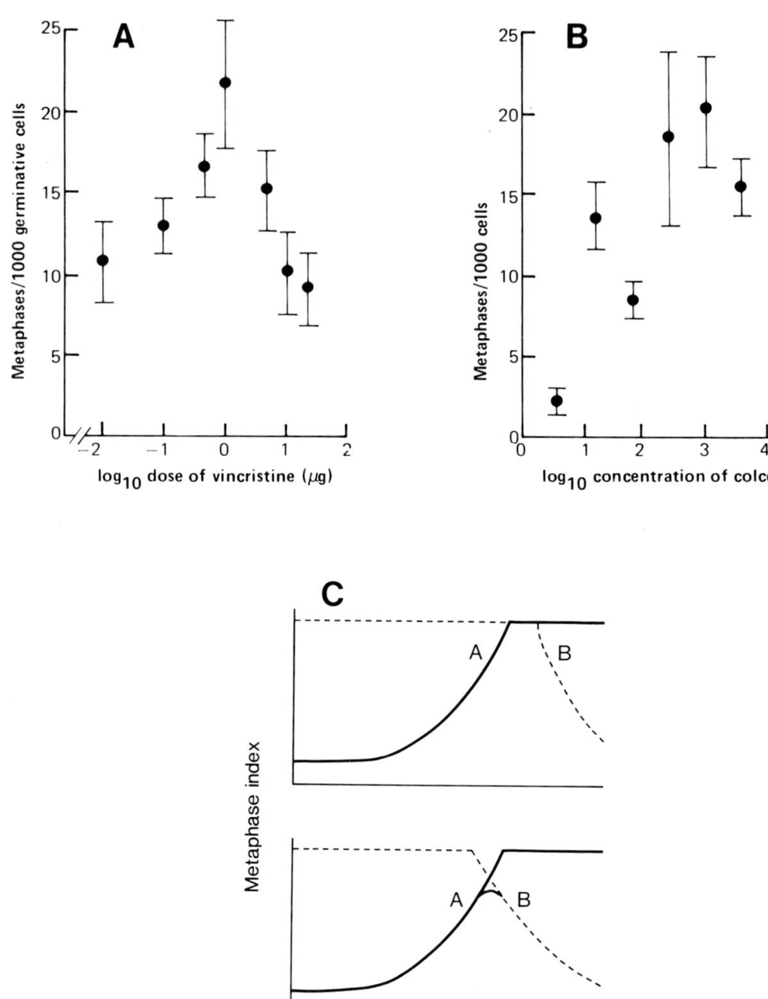



Fig. 3.9. (A) A dose-response curve for vincristine in human psoriatic epidermis *in vivo*. Vincristine was injected in the doses indicated into the periphery of stable psoriatic plaques, and biopsies taken 2.5 hours later under local anaesthesia. Vertical bars indicate standard errors for the points. (Redrawn from Duffill *et al.* (1977).) (B) A dose-response curve for colcemid in the mouse coagulating gland *in vitro*. Explants were incubated with colcemid in the doses shown, and readings were made at four hours later. Vertical bars represent standard errors for the points. (Courtesy of Dr A. R. Morley). (C) An analysis of the competing factors affecting a dose-response curve with a stathmokinetic agent: the metaphase index increases with increasing dose, as more cells are arrested (curve A). Increasing dose leads to metaphase degeneration and a reduction in the metaphase index (curve B).

introduction of colcemid in the concentrations shown. The curve rises to a sustained plateau which persists over the dose range used. On the Tannock model (Fig. 3.9C), the increase in the metaphase index is again due to the increase in the proportion of metaphases arrested, and this is supported by measurements of the proportions of mitotic phases with increasing doses of colcemid; after the plateau is attained anaphases are no longer present. In this instance a broadly peaked dose-response curve is evident, and consequent decreases in metaphase index due to metaphase degeneration (curve B) have not yet decreased the peak index attained. Assuming that there is a critical concentration below which no metaphase degeneration occurs, then this concentration is higher than the minimum concentration required for maximum metaphase arrest.

In rat small bowel, Tannock (1967) obtained dose-response curves of the type shown in Fig. 3.9B for colchicine, vinblastine, and vincristine, and in tumour isografts for colchicine and vincristine only; in tumour isografts colcemid and vinblastine showed curves similar to that of Fig. 3.9A. Tannock's results for colcemid have been criticized by Clarke (1971), on the basis of small sample size in terms of animals, and also that the within-animal sample or variability was not given, and also by Nome (1975a), who considered that Tannock's collection period of four hours was too long to be optimal for colcemid. Comparing the effects of dose ranges of 1–5 mg/kg of colcemid and vinblastine on the rat small bowel crypt, Clarke (1971) showed a decreasing number of arrested metaphases with increasing dose with both agents, over a period of 2.5 hours; the larger doses increased the number of overtly degenerate metaphases. With colcemid, however, variation in metaphase collection, and degeneration, were less apparent. Furthermore, Nome (1975a) also demonstrated a broadly peaked dose-response curve for colcemid in several tissues. Smith *et al.* (1974) concluded that, over the dose range of 1–4 mg/kg in a transplantable mouse carcinoma, colcemid showed no variation in metaphase collection over a two hour period, nor vincristine over a four hour period.

Thus, although Tannock (1967) considered that colcemid showed poor dose-response relationships, the results given in Fig. 3.9B for the mouse coagulating gland, with those of Clarke (1971), Smith *et al.* (1974) and Nome (1975a), indicate that, if the experimental conditions are analysed critically, colcemid may give reliable dose-response characteristics.

A further criterion of an optimal dose of concentration is completeness of metaphase arrest, i.e. an absence of *anaphase escape*. If metaphase arrest is incomplete, then of course the birth rate will be underestimated; therefore it is critical that anaphase escape be excluded when entertaining an adequate concentration of the agent. For example, Clarke (1971) reported anaphase escape in small bowel crypts with doses of colcemid below 2.5 mg/kg in the rat, while Riches *et al.* (1979) have shown that doses of vincristine, which produce complete arrest in maternal tissues, show anaphase escape in fetal tissues; whether or not this is due to some differential sensitivity of fetal tissues or to difficulties with transplacental passage, is not yet clear. In this respect, Williams (1973) reported that although maternal injection gave complete metaphase arrest in fetal gut and brain, increasing doses of colcemid or vinblastine gave diminishing metaphase indices; this was considered to reflect an embryo-specific inhibitory effect on cells in DNA synthesis.

This leads us to a further consideration: an important requirement for an optimal dose is that it should not affect the flux of cells through the other phases of the cell cycle; however, Fitzgerald and Brehaut (1970), studying cultured lymphocytes, reported that large doses of colchicine depressed DNA synthetic rate and flux through $G_2$, although Hell and Cox (1963) showed that this effect was less marked with the same dose of colcemid in epidermis. On the other hand, Clarke (1971) concluded, from his comparison of [$^3$H]-TdR and stathomokinetic agents in the small bowel, that metaphase-blocking agents in suitable doses do not affect the rate of entry of cells into mitosis, and, also in the small intestine, Jellinghaus *et al.* (1977) maintain that, in the usual doses, vincristine does not alter the flux of cells into or out of DNA synthesis, as indicated by a detailed double-labelling analysis. Moreover, similar methodology in ascites tumour cells show no effect on the kinetics of $S$ influx and efflux (Camplejohn *et al.* 1980). Ideally such effects should be monitored for each system, though this has seldom been done.

In conclusion, we should determine the dose-response characteristics for the selected agent in the tissues to be studied; a suitable curve would have a plateau value, and the optimal dose or concentration is then the lowest that will arrest all metaphases, and prevent anaphase escape over the experimental period chosen *for that tissue*; once this has been decided, then other factors, such as the delay period, linearity of metaphase collection and metaphase degeneration, can be assessed for that dose.

(ii) *What cells to count, and how many?* Having ensured that metaphase arrest is complete in all available readings, there remains the problem of what to count. There will be prophases and arrested metaphases in the samples, and some workers have counted both of these (see, for example, Wright *et al.* 1972*b*), whereas others have counted metaphases only (Smith *et al.* 1974). One disadvantage of the first procedure is the problem of recognizing prophases, and since the criteria for their recognition do vary, and problems are also experienced in recognizing early forms, perhaps the best advice is to count metaphases only: this should make reproducibility between observers possible, but it should be noted that a separate count of prophases is always advisable to allow a check to be made on any changes in mitotic influx or in the duration of prophase during the experiment. Piscitelli (1956) observed a considerable decrease in bone marrow prophase incidence shortly after colcemid injection, but such changes do not appear to occur in epithelial tissues such as the epidermis (Thorud *et al.* 1980) or intestine (Wright 1973). A point of practical importance arises in calculations of the mitotic duration. It is shown below that the mitotic duration is usually calculated from the ratio of the native mitotic index to the mitotic rate; if metaphases only are included in the native mitotic index, then the duration of metaphase will be estimated. If, however, all phases of mitosis are included, a value for the total mitotic duration may be derived.

The number of arrested metaphases should of course be related to the total interphase nuclear population counted *en passant*, and not to some anatomical denominator such as a unit length of the basement membrane in the epidermis; this would necessitate expression of the birth rate as cells produced per unit length of epidermis, and in comparison of birth rates in experiments, involves the important assumption that nuclear population density remains constant (see section 3.1). If certain experimental conditions produce an increase in nuclear density, then the cell production rate/unit length of epidermis would apparently rise, although the birth rate in cells/cell/hour may have remained constant.

Some investigators have increased the interphase nuclear population by counting all metaphases in a large measured area, and then computing the total nuclear population in the area on the basis of sample counts in smaller areas; this may be helpful in instances where metaphase indices are particularly low, as in adult mouse lung (Simnett and Heppleston 1966). Smith *et al*. (1974) used a modified technique employing a point counting system to measure the tumour cell component of tumour isografts and computed the number of tumour cells and metaphases per unit volume; this allows calculation of the cell production rate for tumour cells only.

In the intestine, crypts are closed systems from which arrested metaphases are unable to escape during the experiment. Microdissection and counting of arrested metaphases allows the calculation of the cell production rate per crypt (Wimber and Lamerton 1963, Clarke 1971; and see Chapter 17, section 3). If the turnover time is to be calculated, then these metaphases must be expressed as a fraction of the interphase population. However, this method is also sensitive to changes in the crypt population, and an increased population size may give an elevated crypt cell production rate even though the birth rate in cells/cell/hour is unchanged (Kovacs *et al*. 1982). A particular advantage of this method is the avoidance of any problems which might accrue from differences in the diameter of arrested metaphases and interphase nuclei (Floderus 1944; and see section 3.1).

Indeed, correction factors have been used to compensate for the observation that arrested metaphases differ from associated interphase nuclei in size, at least in tissue sections; for example Simnett (1968) considered that colcemid-arrested metaphases, with chromosomes spread by exposing the tissue to hypotonic saline, were larger than interphase nuclei, and calculated a correction factor of 0.797 by which the metaphase index should be multiplied. In tumour isografts, Smith *et al*. (1974) showed that vincristine-arrested metaphases (measured at an unspecified time after vincristine) showed no effective difference in diameter compared with interphase nuclei. Clarke (1968, 1970*a*) concluded that, in chick duodenum, colcemid-arrested metaphases were *smaller* than interphase nuclei, and suggested a correction factor of 1.54. Now these correction factors were assumed to be constant, but Tannock (1967) proposed that although a newly arrested metaphase is larger than an interphase nucleus, with advancing age they decrease in size, and that these cell size considerations tend to cancel themselves out. While correction factors should definitely be investigated in the tissue in question, they may well change with time, which would complicate the analysis. In Tutton and Barkla's (1976) analysis, there was a tendency for arrested metaphases in normal colon to decrease in size with time (see Table 3.3),

Table 3.3.  Diameter of metaphase and non-metaphase nuclei in epithelial cells of the rat descending colon. (After Tutton and Barkla (1976))

| | Diameter of metaphases | | Diameter of non-metaphases | |
|---|---|---|---|---|
| | 1 h post-VB | 4 h post-VB | 1 h post-VB | 4 h post-VB |
| Descending colon | 4.4 ± 0.1 | 3.6 ± 0.1‡ | 5.4 ± 0.1† | 6.2 ± 0.2†‡ |
| DMH-treated colon | 4.3 ± 0.2 | 4.2 ± 0.1* | 4.6 ± 0.2* | 5.5 ± 0.2†‡ |
| Adenocarcinoma | 4.7 ± 0.2 | 4.6 ± 0.1* | 5.3 ± 0.2 | 5.3 ± 0.2* |

    * $p < 0.05$ for same figure at same time in normal colon.
    † $p < 0.05$ for metaphase in same tissue at same time.
    ‡ $p < 0.05$ for same tissue, at same time, same figure one hour post-VB.
VB = vinblastine;  DMH = dimethylhydrazine.

although this was not evident in hyperplastic or neoplastic tissue (Table 3.3); in this study arrested metaphases were uniformly smaller than interphase nuclei. Some idea of the importance of geometrical correction factors can be assessed from Table 3.4, also from Tutton and Barkla (1976), which compares the differences in the metaphase index when corrections for nuclear size are made,

Table 3.4.  The influence of applying correction factors to observed metaphase indices in tissues treated with vinblastine for four hours. (After Tutton and Barkla (1976))

| | Normal descending colon | DMH-treated colon | Adenocarcinoma |
|---|---|---|---|
| Observed $I_{meta}$ | 0.137 | 0.117 | 0.113 |
| Percentage increase in $I_{meta}$ due to size correction (Floderus 1944) | +21.3 | +6.5 | −2.5 |
| Percentage decrease due to Tannock's factor | −41 | −41.9 | N/A |
| Corrected $I_{meta}$ | 0.098 | 0.073 | 0.110 |

$I_{meta}$ = metaphase index at four hours after vinblastine.

and also when corrections are made for the fact that mitoses in intestinal crypts are found out of alignment with the interphase nuclear column (Tannock 1967; and see Chapter 17), and hence tend to be overcounted in axial crypt sections: in normal rat colon the correction due to nuclear size certainly appears worth making, although in the other tissues this difference is not so great. In each case, the correction due to the non-alignment of nuclei is considerable.

Care should therefore be taken not to bias the count; similarly, if the tissue is heterogeneous with respect to its kinetic architecture, the constituent parts of the tissue should be sampled in proportion to their volume if an unbiased overall birth rate is to be calculated. It may indeed be unrealistic to attempt to describe the activity of a heterogeneous tissue by a single birth rate, but if this is to be done, it should be noticed that the numbers of cells to be counted, which will depend on the error we are prepared to accept, cannot be determined purely on the basis of binomial counting error; variability between sections may well be as important as variability within sections. This is exemplified by the observations of Aherne *et al.* (1977*b*), who have considered this important question, which is particularly relevant to measurements carried out on human material, where samples are limited in number and are generally hard to come by (see Table 3.1); some workers have felt obliged to pool metaphase arrest results from different patients (cf. Camplejohn *et al.* 1973); if the experiment is carried out on $M$ patients with one sample being taken at each time point, $t_i$ ($i = 1, 2, 3, \ldots M$), we need to assume that the procedure is equivalent to obtaining $M$ samples from a single host, even though we do not know how much kinetic variation exists in any specimen we take; we have seen above that, in human tissues (in this case tumours, Aherne *et al.* 1977*b*), there is at least as much within-tumour variation as between-tumour variation (cf. section 3.1). In these conditions our metaphase index $I_{M(i)}$ is

$$I_{M(i)} = a + bt_i + e_i + c_i \qquad (i = 1, 2, 3, \ldots M)$$

where $a$ and $b$ are the usual regression constants; $e_i$ is the counting error (the error of a proportion), and $c_i$ is the between tissue or patient variation. The binomial error can be minimized by large numbers of cells, and, if $c_i$ is normally distributed,

$$I_{M(i)} = a + bt_i + c_i; \qquad \text{where } c \sim N(0, \sigma_c^2)$$

the usual regression equation; hence, in this instance, $c$ becomes the more important error (Aherne *et al.* 1977*a*).

There remains the problem of how many cells to count. This will again depend on the relative standard error we are prepared to accept, and, since the metaphase index, in the statistical context, is equivalent to the mitotic index, the statistical remarks made in section 3.1 on proliferative indices apply equally well here.

(iii) *How many readings to take, and when?* As mentioned above, the stathmo-kinetic method depends on the measurement of the slope of the metaphase accumulation line. Therefore it is critically important to establish linearity of metaphase collection. In recent years this has all too often been disregarded. A very common error is to take only one reading, at three or four hours after injection or application of the stathmokinetic agent, and to call this the mitotic rate (Bertalanffy 1960; Bullough and Laurence 1968*b*, 1970; Elgjo and Hennings 1971; Laurence and Randers Hansen 1972; Clausen 1975; Argyris 1977) or mitotic index (Simnett and Heppleston 1966; Chopra 1973). Others have divided this hourly reading by the collection period, and calculated an hourly 'mitotic rate' (Elgjo 1968*a*; Hennings and Elgjo 1970; Simnett 1972; Tvermyr 1972; Iversen *et al*. 1976; Bertsch *et al*. 1976). Inspection of Fig. 3.10 will show

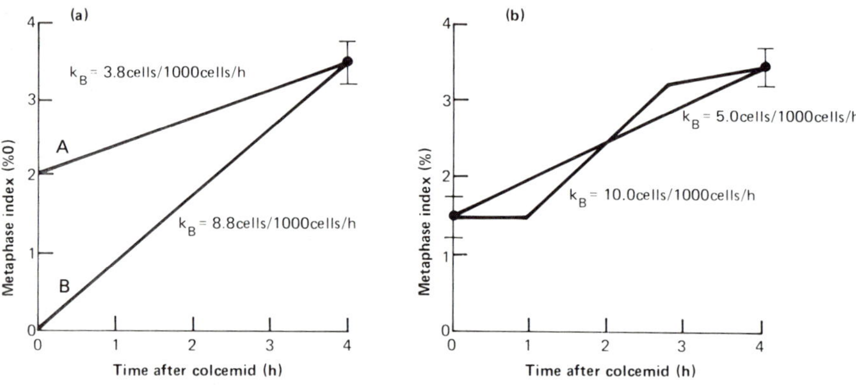


Fig. 3.10. (a) A possible error which can ensue from using a single reading of the mitotic index after injection of a stathmokinetic agent. The single point is consistent with both lines which give very different estimates of the birth rate, $k_B$. (b) When two points are available, large errors can still be made because of the possibility of a delay period or of metaphase degeneration. Both the straight line and the curve, which give quite different estimates of $k_B$, are possible.

that this is a highly dangerous thing to do. Suppose we have taken a single reading at, say, four hours after injection of a stathmokinetic agent, in two tissues in which the birth rate must be compared. Figure 3.10a shows the single reading in one of the tissues; obviously, such a single reading would be compatible with both line A and line B in which the *birth rate differs significantly*. Consequently, any comparison we wish to make of this reading with another tissue would be meaningless, since we have not measured the birth rate; it is clearly possible to obtain the same single four hour reading in those kinetic situations where the birth rate (or mitotic rate) differs significantly. Implicit in the single reading method is an assumption that there is linearity of metaphase collection, and that the line extrapolates to the origin. Detailed analysis of metaphase collection in the intestine (Wright *et al*. 1972a; Al-Dewachi *et al*. 1975b) and in squamous epithelium (Nome 1975a) shows that very frequently the line does not extrapolate to the origin. This particular method has often been used to monitor ephemeral responses, in situations where the native mitotic index itself is changing rapidly (e.g. after various perturbations); this will compound the error, and can be misleading (e.g. see Clausen 1975; Bertsch *et al.* 1976).

An example of the possible errors which can be incurred by this procedure is shown by Simnett (1972), who claimed that the rate of entry into DNA synthesis was some three times greater than the rate of entry into mitosis, in mouse lung in organ culture; here the flux into mitosis was measured by a single reading after four hours metaphase collection with colcemid, and dividing this value by the collection period. Since the native mitotic index was not taken into consideration, and linearity of metaphase collection was not shown, we cannot draw any conclusion regarding the true mitotic rate to compare with the flux into $S$. Furthermore, any delay period (see below) before colcemid became effective would compound the error.

Another common practice is to take two readings; the native mitotic index at time zero and another reading at 3-4 hours later. This is often the case in the human situation, where, because of ethical considerations, there is a limitation in the number of biopsies which can be taken (Iversen 1967; Taylor *et al.* 1977; Wright *et al.* 1977). There is a further tacit assumption here; that the agent becomes active immediately after injection, and that metaphase collection occurs in a linear manner thereafter (see Fig. 3.10b). Figure 6.6 (p. 248) shows such an example where this is not the case; the instance is the mouse oesophagus, and the agent is vincristine in a dose of 1 mg/kg. If only the readings at 0 hours and 4 hours were used, then the birth rate would be underestimated, because of the presence of a delay period *before metaphase arrest becomes evident*, in this case some 75 minutes after injection.

This phenomenon of a delay period before metaphase arrest becomes complete is well recognized *in vitro*, where the delay period has been shown to be concentration related, decreasing in duration with increasing dose; Taylor (1965) showed that when the time of appearance of the first arrested metaphase is plotted against the reciprocal of the colchicine concentration, the duration of the lag phase follows a line of decreasing slope extrapolating to 0.9 hours, the duration of metaphase and anaphase; at higher concentrations the curve approaches a limiting value of zero. This indicates that different phases of mitosis are differentially sensitive to inhibition, and also that, at higher concentrations, cells in metaphase at the start of the arrest period can be arrested very shortly after addition of colchicine. Using [$^3$H]-colchicine, Taylor was able to show that the rate of binding to microtubular protein was proportional to concentration; similarly, Inoué (1952) in *Chaetopterus*, and Gaulden and Carlson (1951) in grasshopper neuroblast showed that increasing concentration of colchicine reduced the time for the spindle to disappear. However, Puck and Steffen (1963) could not confirm this in their *in vitro* work with colcemid, where only cells entering mitosis after colcemid application were arrested, and recently Pritchett *et al.* (1982) could not conclude that the delay period was dose-responsive in human colorectal mucosa and carcinomas growing in organ culture, and measured with vincristine.

It is also clear that the kinetics of metaphase arrest may differ in different tissues; for example there is a delay period of 20 minutes in mouse jejunal mucosa with vincristine in a dose of 1 mg/kg (Al-Dewachi *et al.* 1975*b*), whereas no such delay occurs in rat jejunum with the same experimental conditions. Moreover, Nome (1975*a*) showed that optimal doses of colcemid produced delay periods in mouse epidermis and colon, but not in jejunum and squamous fore-stomach. However, Elgjo (quoted by Taylor *et al.* 1977) reported that the metaphase arrest line extrapolated back to the origin in mouse epidermis; this indicated that a delay period equal to the duration of mitosis was evident, and a similar situation occurred in the studies of Brown and Oliver (1968) in their analysis of hamster cheek pouch epithelium. It cannot be assumed that this is always the case: none of Nome's (1975*a*) metaphase collection lines extrapolate to the origin in several different mouse tissues, and neither does the line for mouse oesophagus shown in Fig. 6.6 (p. 248). Of course, if metaphase arrest became immediate, then the line would extrapolate, not to the origin or to the native mitotic index, but to the native metaphase index. The data in Fig. 6.6 (p. 248) would be consistent with an immediate onset of metaphase arrest with a metaphase index of 1 per cent. However, the *in vitro* work of Taylor (1965), referred to above, indicates that the delay period is dose-dependent, and that the decreasing delay period is due to more rapid arrest of metaphases with increasing dose; as yet, there is no evidence that the delay period is dose-dependent *in vivo*, and the reason for the delay period must await further work.

A further claim made by Taylor *et al.* (1977) was that the native mitotic index was almost always higher than initial counts after colcemid. In this context, Thorud *et al.* (1980) found a transient increase in the fraction of metaphases at one hour after colcemid injection, an effect also noted by Piscitelli (1956), in the bone marrow, the cause of which is not clear. On the other hand, the detailed studies of Wright *et al.* (1972*b*) and of Al-Dewachi *et al.* (1975*b*) indicate that no such phenomenon is evident in the rodent or human (see Fig. 3.8) intestine. Moreover, Taylor (1965) has analysed in detail the kinetics of metaphase arrest during the first hour of colchicine action *in vitro*; at concentrations where no delay was evident, there was of course no initial fall in mitotic index, but *neither was there when a delay period was evident*. It is clear that it is not possible to generalize; the behaviour of the mitotic index at early time periods is a reflection of the individual agent, the concentration or dose, and the tissue in question. The route of administration can also be important, since metaphase arrest after intravenous injection of colchicine is established before arrest after subcutaneous or intraperitoneal injection (Stevens Hooper 1961).

Consequently it is crucial to establish the presence of a delay period and to exclude it from the measurement of the birth rate; in fact, this very point has been the subject of a detailed study by Thorud *et al.* (1980), who reported that there was no difference in the mitotic rate when calculated from a single reading taken at four hours, and that measured by serial readings after colcemid injection; the extrapolated lines also went through the origin. While this may be true for the dorsal epidermis of the hairless mouse, Fig. 6.6 (p. 248) shows that we again cannot generalize, and, if the single point method is to be used, then stringent preliminary investigation is required. If experimental interference, for example treatment with cytotoxic drugs, is anticipated, it would be really necessary to ensure that the experiment has not interfered with the linearity of arrest; such precautions would be negated by the performance of a full experiment as shown in Fig. 3.8. In human work, where the frequency of observations may prohibit the measurement of the delay period, it is essential to appreciate the assumptions made with limited numbers of readings. A further drawback of using only one or two readings is that it is not possible to calculate the confidence interval of the birth rate.

(iv) *What is the optimal collection period?* Tannock (1967) considered that the birth rate should be measured over the linear portion of the collection line; he demonstrated linearity of collection for differing periods with differing agents, which in the rat small bowel were up to a maximum of four hours after injection, with vinblastine. However, Frei *et al.* (1964) claimed that vincristine showed linear accumulation for 12 hours, Smith *et al.* (1974) reported a 10 hour linear collection, also with vincristine, while Taylor (1965) obtained 11 hours linearity, with colchicine *in vitro*. However, it is necessary to view these results with some caution, because of the phenomenon of *metaphase degeneration*, which will probably occur to some extent in all experiments lasting more than three hours; for example Clarke (1971) notes a considerable amount of metaphase degeneration at only 3.5 hours after injection of colcemid in the rat small intestine (2.5 mg/kg), while Morris (1967) reported degeneration occurring after three hours in mouse duodenum, and after 2.5 hours in the thymus. However, Shackney *et al.* (1976) considered that degeneration occurred only after eight hours in mouse bone marrow.

In the mathematical analysis of the stathmokinetic technique, Trucco (1965), while appreciating that metaphases degenerate over 'relatively short periods of time', assumed no loss of metaphases, but Puck *et al.* (1964) corrected their counts of arrested metaphases in culture. The *in vivo* problem has been attacked by Aherne and Camplejohn (1972) in mice, and in human tissues by Ralfs *et al.* (1981$a$) in the epidermis of subjects with psoriasis. Figure 3.11 summarizes their results. By giving an injection of tritiated thymidine at the same time as the stathmokinetic agent, the fate of unlabelled arrested metaphases (which were in $G_2$ at the time of the injections) can be followed. If degeneration did not occur, then the unlabelled $G_2$ cohort of metaphases would plateau and remain constant over the duration of the experimental period, as suggested by line B in Fig. 3.11.

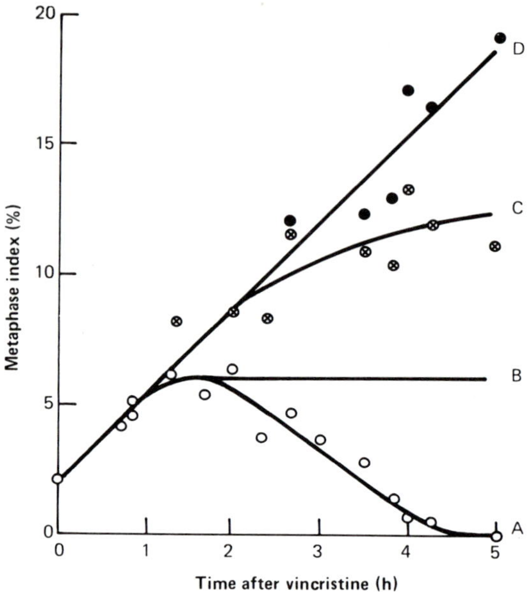



Fig. 3.11. Metaphase degeneration in the mouse jejunal crypt epithelium. Vincristine (1 mg/kg) and [³H]-TdR (1 μCi/g) were injected at time zero. Line A (○—○) shows the proportion of all cells which are unlabelled metaphases; line B is the plateau to be expected if these did not degenerate; line C (⊕—⊕) is the proportion consisting of all metaphases; and line D (●—●) is formed by correcting line C for the degeneration of line A. (Redrawn from Aherne and Camplejohn (1972).)

In fact, after two hours in intestine and four hours in psoriatic epidermis the $G_2$ cohort of arrested metaphases decreased in number, and had largely disappeared by the end of the experiment, indicating that all metaphases which were in $G_2$ at the time of injection had degenerated (curve A). Hence it is necessary to correct the observed metaphase index (curve C) for metaphase loss after two hours in the intestine and after four hours in the epidermis. Assuming that metaphase loss was random, Aherne and Camplejohn derived a function $\exp(\lambda - \beta)(t - m)$ by which the observed metaphase index must be multiplied, where $m$ is the minimum life span of an arrested metaphase, $t$ is the time from the beginning of the experiment, and $\beta$ and $\lambda$ are the rates at which metaphases are formed and degenerate, respectively.

Thus although we might be suspicious of the results of Fisher (1968a), who used arrest periods as long as 12 hours in human epidermis without proof of linearity, it is also reasonable to ask if the 10 hours linear collection claimed by Smith *et al.* (1974) could not be subject to a systematic error after a relatively short period, which would decrease the real birth rate.

There is also some evidence that the rate of metaphase degeneration is dose-dependent; Tannock (1967) and Clarke (1971) both noticed increased numbers of degenerate forms with higher doses of stathmokinetic agents, and particularly of vinblastine.

Ideally, one should investigate, in the same way, the kinetics of metaphase degeneration in each tissue to be studied, and only use that period before metaphases begin to degenerate. Because of this consideration, several workers have confined their observations to relatively short periods, for example 2.5 hours in the small intestine (Clarke 1970*a*; Al-Dewachi *et al.* 1977). This may be reasonable when the cell cycle time is short in relation to the arrest period, but minimizing metaphase loss in this way will increase the sampling error, and if cell cycle times are to be calculated in tissues of long cycle time, will increase the distance of the extrapolation required. If longer arrest periods are contemplated, then it is necessary to explore metaphase degeneration in some detail.

(v) *Having collected the data, how is the line fitted?* When the problems of adequate dosage (or concentration), linearity, delay period and metaphase degeneration are solved, or avoided, there remains the problem of the best way to analyse the data. Many authors have merely applied linear regression analysis to the untransformed data, and duly calculated the slope of the line (for example, Camplejohn *et al.* 1973; Smith *et al.* 1974). It is well to consider the several factors which form the components of any line we may fit. One important consideration is the *age distribution* of the population under study.

First a word about nomenclature: some confusion has occurred concerning the use of the term 'cell cycle time' as measured by stathmokinetic experiments: it is a cell cycle time only if all cells are proliferating and the growth fraction is unity. Consequently the cell cycle time can be calculated if the growth fraction is known, or more properly included in the equation of the line, see eqn 3.20 below. However, if there is no data about the growth fraction, then the parameter measured has variously been called the *potential doubling time, apparent cell cycle time* or *turnover time*. A reasonable compromise is to call the parameter turnover time or apparent cell cycle time in renewing cell populations such as the small intestine, and potential doubling time in tumours (Aherne *et al.* 1977*a*).

If the age distribution is rectangular, then the correct line to fit to the mitotic or metaphase index data is

$$M_{(t)} = \frac{t_M + t}{T_C}.$$  (3.13)

If, however, the age distribution is known to be exponential, we should use the collection function of Puck and Steffen (1963) and fit

$$\log_{10}(1 + M_{(t)}) = \frac{\log_{10} 2}{T_C}(t_M + t)$$  (3.14)

or more simply

$$\log_2 (1 + M_{(t)}) = \frac{t_M + t}{T_C}. \tag{3.15}$$

We can also treat populations with intermediate age distributions; consider a population of cells maintaining a constant size, by, for example, cell birth and migration. At each mitosis two new cells are produced, of which $1 + p$ remain in the population ($0 \leqslant p \leqslant 1$); the remainder $(2 - (1 + p))$ leaving the population (migrating) at each mitosis. If the probability $P_{(\tau)}$ that a cell will be at age $\tau$ for an exponential age distribution is

$$P_{(\tau)} = 2 \frac{\log_e 2}{T_C} \exp \left[ -\frac{\log_e 2}{T_C} \tau \right] \tag{2.6}$$

(Steel 1968; Appleton *et al.* 1977), then, for a population of cells with $(1 + p)$ cells remaining in the population at mitosis

$$P_{(\tau)} = \frac{1 + p}{p} \frac{\log_e(1 + p)}{T_C} \exp \left[ -\frac{\log_e(1 + p)}{T_C} \tau \right] \tag{3.16}$$

(Appleton *et al.* 1977).

In this case we would fit

$$\log_{(1 + p)}(1 + pM_{(t)}) = \frac{t_M + t}{T_C}. \tag{3.17}$$

It may be that we wish to use information about the growth fraction; in the same population, with $1 + p$ cells remaining at mitosis, there is also a probability that cells remain in the proliferative cycle. If cells migrate at random with respect to age and proliferative status, the growth fraction $\rho$ is

$$\rho = \frac{(1 + p)\eta - 1}{p} \tag{3.18}$$

where $\eta$ is the probability of remaining in the proliferative cycle. The age distribution is then

$$P_{(\tau)} = \frac{1 + p\rho}{p} \frac{\log_e(1 + p\rho)}{T_C} \exp \left[ -\frac{\log_e(1 + p\rho)}{T_C} \tau \right]. \tag{3.19}$$

In this case the correct line to fit to the stathmokinetic data is

$$\log_{(1 + p\rho)}(1 + pM_{(t)}) = \frac{t_M + t}{T_C}. \tag{3.20}$$

While it is important to realize that the curve-fitting procedure depends in principle on the format of the age distribution, it is not necessarily of importance when estimating the birth rate (Duffill *et al.* 1977). However, the age distribution should be remembered in one practical situation: when the age distribution can be measured or calculated, it is worthwhile incorporating it into the curve fitting, because derived values such as the apparent cell cycle time or potential doubling time depend on it. Instances where this is possible include the mouse jejunal crypt, where Schultze *et al.* (1972) have shown that fluxes into and out of $S$ are equal for the whole crypt, and thus a rectangular age distribution seems appropriate. In crypt bases, however, the flux into $S$ exceeds the $S$ efflux by a factor of 1.4 (Burholt *et al.* 1976), which would be expected on the basis of an exponential age distribution, with the cell cycle time as measured; indeed, such an age distribution was used by Tutton (1973*a*) in the analysis of his stathmo-kinetic data for the rat jejunal crypt, which included only the basal 10 cell positions. It is also possible to calculate the age distribution in layered tissues such as stratified squamous epithelium from a consideration of the direction of the mitotic axis (Appleton *et al.* 1977; and see Chapter 6, section 7).

In rapidly growing transplantable tumours of experimental animals, in the logarithmic growth phase, a reasonable assumption is an exponential age distribution. However, in older tumours, where extreme degrees of cell loss may decrease the initial ordinate on the age distribution curve, a rectangular age distribution has been proposed (Camplejohn *et al.* 1973); indeed, if cell death is occurring principally at mitosis and is leading to tumour regression, even a rectangular age distribution may be a poor approximation. This may be specially relevant for studies of human tumours.

Whittal *et al.* (1975) and Duffill *et al.* (1977) considered that the choice of the correct age distribution in stathmokinetic experiments was critically important for estimates of the turnover time; for example, an experiment with vincristine in human psoriatic epidermis, using a rectangular age distribution, gave a mean value for the birth rate of 0.013 cells/cell/hour, while application of the (correct) exponential age distribution gave 0.012 cells/cell/hour, providing mean values for the turnover time of 79 and 56 hours respectively (Duffill *et al.* 1977). Similarly, in the mouse oesophageal basal cells depicted in Fig. 6.6 (p. 248), mean birth rate values of 0.020 and 0.019 cells/cell/hour were derived with rectangular and exponential age distributions respectively, equivalent to turnover times of 49 and 36 hours. The correct age distribution, as calculated from the direction of the mitotic axes, has $p = 0.37$, and, using eqn 3.17 to fit the line, gives a turnover time of 43 hours (see Chapter 6, section 7).

It is never possible to tell from the data which age distribution should be used, as suggested by Steel (1977), however good the data. If the mitotic index $M_{(t)}$ appears to be linearly related to time, then so do all functions of the form $\log(1 + pM_{(t)})$. In summary, we should stress that age distribution considerations are only important when a derived value such as the turnover time is needed from stathmokinetic data; the birth rate is not nearly so dependent on the age distribution, and is therefore much more useful.

A further problem met with in analysing the data is the statistical method used to fit the line; most workers have used an unweighted least squares technique (Clarke 1971; Tutton 1973*a*; Camplejohn *et al.* 1973; Smith *et al.* 1974; Thorud *et al.* 1980). Since the variance of the mitotic (or metaphase) index after arrest is usually proportional to the mitotic index itself, Whittal *et al.* (1975) considered it appropriate to minimize a weighted sum of squared residuals, and found that the procedure reduced the variance of the cell cycle time estimate obtained.

One problem which has received insufficient attention is the measurement of the birth rate in situations where the native mitotic index is itself increasing. This is very common in cell kinetic studies and occurs for example during the recovery of tissues from perturbation, for example in renewal systems after irradiation, and in instances of induced DNA synthesis, such as liver regeneration. If a metaphase arrest experiment is carried out during such a proliferative response, the rate of entry into mitosis will not be constant over the experimental period, as is assumed in a linear model, but will increase with time; this can be appreciated by inspection of Fig. 3.12a, which shows the changes in the

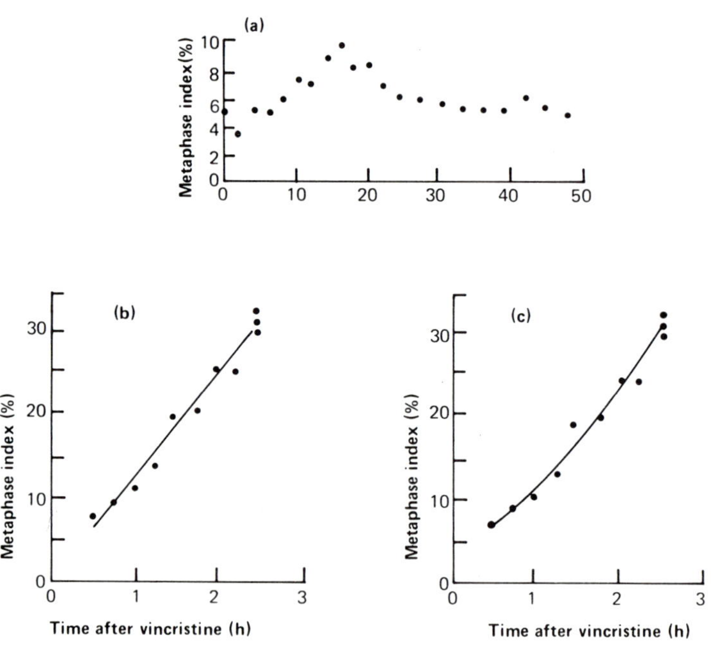

Fig. 3.12. Metaphase accumulation when the native mitotic index is itself increasing. (a) Shows the mitotic index in the jejunal crypts of rats refed after four days starvation. (b) Shows the results of a metaphase accumulation experiment with vincristine (1 mg/kg) at 16 hours after refeeding, the line being fitted assuming a rectangular age distribution. (c) Shows a quadratic equation fitted to the same data; since the native mitotic index is changing with time the birth rate must be too, and it can be calculated approximately from the tangent to the curve. (Redrawn from Al-Dewachi *et al.* (1975*a*).)

native mitotic index in the jejunal crypts of rats refed after four days starvation (Al-Dewachi *et al.* 1975*a*). Figure 3.12b shows the results of a metaphase arrest experiment with vincristine (1 mg/kg) carried out at 16 hours after refeeding, fitted with a linear model. Since the mitotic index is itself increasing, then theoretically the line fitted should be a curve, and Fig. 3.12c shows the results of fitting a quadratic expression to the data. The mitotic index at any time could be found by the tangent to the curve, but this does not mean that the curve describes the data any better than the linear model; in fact the fit is no better than would be expected on the basis of adding an extra parameter.

In some tissues, the density function of cells in the mitotic phase is not stationary, but changes diurnally. Møller *et al.* (1979) have used the metaphase arrest method to monitor the non-stationary probability density function for mitotic cells in the hamster cheek pouch by hourly metaphase collections with colcemid (see Chapter 9).

(vi) *What sort of precision is obtained with the method?* Having fitted the line using the appropriate equation and obtained a value for the birth rate from the slope of the line, the turnover time or apparent cell cycle time is readily calculated from the reciprocal of the slope of the line. As with any kinetic parameter, it is essential to quote a confidence interval for the birth rate and the turnover time; this is easily calculated from the regression line in the usual way.

For the basal cells of the mouse oesophagus (Fig. 6.6, p. 248) the birth rate was 0.020 cells/cell/hour with 95 per cent confidence limits of 0.013 and 0.027 cells/cell/hour, and thus the point estimate of the turnover time is 43 hours with a 95 per cent confidence interval of 32 to 67 hours. This does not indicate a high degree of precision, and yet Table 3.5, which gives a selection of cell production

Table 3.5. Cell production rates in the gastrointestinal tracts of mice and rats, measured by the metaphase arrest method with vincristine using microdissection. Values are mean ±SEM and are expressed as cells produced/crypt or gland/hour.

| Anatomical site* | Rats | Mice |
|---|---|---|
| Stomach (antrum) | 2.2 ± 0.4 | — |
| Small intestine | | |
| 10% | 29.7 ± 2.3 | 12.4 ± 1.0 |
| 25% | 16.9 ± 1.6 | 16.4 ± 1.1 |
| 50% | 24.7 ± 3.9 | 11.6 ± 0.7 |
| 75% | 18.7 ± 2.7 | 10.6 ± 0.8 |
| 95% | 28.6 ± 2.1 | 13.0 ± 0.8 |
| Colon | | |
| 50% | 5.7 ± 1.8 | 10.3 ± 1.2 |
| 90% | 9.8 ± 1.3 | 11.3 ± 1.8 |

*Unless specified, these refer to fractions of the length along the respective organ, i.e. 25 per cent in the small intestine refers to a portion of tissue taken from this point, allowing reasonable comparison between studies.

rates in the rat and mouse gastrointestinal tract shows that the relative standard errors of the estimates are quite small, of the order of 10 per cent or less in most cases, and this gives the measurement a considerable degree of discrimination in statistical tests; comparison of respective birth rate values need only a '$t$' test on the slopes, and consequently an important factor is the number of individual points on the metaphase accumulation line, which will determine the number of degrees of freedom in the test. This emphasizes the importance of measuring the slope of the line; if only one or two points are used it is not even possible to calculate approximate confidence limits.

If there is more than a trivial amount of variability in cell cycle times, a correction factor is required to compensate for the fact that any derived turnover time, or cell cycle time, is biased towards the faster cycling cells, and since these provide the most metaphases, the estimate of the turnover time obtained is not the mean turnover time. Appleton *et al.* (1977) give a factor by which to multiply the estimate to reduce the bias; this amounts to dividing by $1 - CV^2$, where $CV$ is the coefficient of variation of the cell cycle time. Unfortunately this factor will seldom be known with any accuracy, but its relevance should be remembered.

(vii) *How do stathmokinetic estimates compare with standard autoradiographic techniques?*    Table 3.6 shows some instances where the birth rates have been measured in the same tissue, and kinetic situation, with both stathmokinetic and autoradiographic methods: in this case the FLM method has been used to calculate the birth rate, from values for the cell cycle time and growth fraction.

Table 3.6.  A comparison of estimates of cell birth rates ($k_B$) in various tissues, found from metaphase arrest and fraction labelled mitoses experiments: $k_B$ is expressed in cells/1000 cells/hour

| Tissue | Kinetic state | Metaphase arrest $k_B$ | FLM $k_B$ |
|---|---|---|---|
| Mouse jejunum | Normal | 52 | 49 |
| Mouse jejunum | 15 hours after ara-C | 73 | 83 |
| Rat jejunum | Normal | 72 | 55 |
| Rat jejunum | After 96 hours starvation | 38 | 38 |
| Rat jejunum | 15 hours after hydroxyurea | 63 | 62 |
| Mouse descending colon | Normal | 26 | 19 |
| Mouse transverse colon | Normal | 13 | 14 |
| Mouse ascending colon | Normal | 16 | 16 |
| Mouse caecum | Normal | 18 | 20 |
| Mouse sarcoma | 7 days after implantation | 37 | 30 |
| Mouse sarcoma | 14 days after implantation | 36 | 29 |
| Mouse sarcoma | 21 days after implantation | 33 | 32 |

Note: (i) a rectangular age distribution was assumed for each tissue except the sarcoma for which an exponential age distribution was used; (ii) the growth fraction was estimated in every case from the ratio of observed to theoretical labelling indices, the latter being calculated from phase durations derived from the FLM curve which was analysed by the method of Gilbert (1972).

In such comparison it is critical that these criteria be met: a comparison such as that of Bertalanffy (1964), who compared tritiated thymidine and stathmo-kinetic results in different strains and ages of the same species, is not realistic, remembering the considerable strain and age differences which have been described. Similarly, Tannock (1967), in a comparison of the cell cycle time as measured by FLM and metaphase arrest techniques, compared measurements of the jejunal mucosa in two *different* strains of rats; further, in a transplantable tumour, Tannock compared doubling time measurements from the growth curves with potential doubling times from the metaphase arrest data; such a comparison makes the absolute assumption of no cell loss (Steel *et al.* 1966), an unusual situation even in tumours growing exponentially.

It can be seen from Table 3.6 that, generally speaking, agreement between mean values from the two techniques is good, and we can be reasonably certain that we are measuring the same variable using the two techniques. At the same time we should recall that, although the FLM method is widely regarded as the best available method for the estimation of the cell cycle time, and its constituent phases, the technique also has its problems (see section 3.4).

(viii) *Can we measure the mitotic duration using the metaphase arrest method?* The mitotic duration can only be roughly estimated from FLM curves, and is usually measured from $2(t_2 - t_{G_2})$, where $t_2 = t_{G_2} + \frac{1}{2}t_M$ (see section 3.4). The value for $t_{G_2}$ is usually the minimum value, taken as soon as labelled mitoses appear, and $t_2$ is measured at the 50 per cent point on the incline of the first peak. Now this is, in many cases, difficult, since in systems with a short cell cycle time (e.g. the small intestine), the incline of the first peak is very steep. Because of this, investigators have looked to the stathmokinetic method for accurate measurements of the mitotic duration (Evensen 1963), and in the small intestine, for instance, have quoted precise estimates (Clarke 1970*a*; Wright *et al.* 1972*b*; Pelc 1971; Tutton 1973*a*; Taylor *et al.* 1977), often to two decimal places, and usually without confidence limits. Now how accurate are these estimates?

The mitotic duration is usually calculated from the relation

$$t_M = \frac{M_{(0)}}{k_B} \tag{3.21}$$

where $M_{(0)}$ is the native mitotic index, and $k_B$ the birth rate as measured from the metaphase accumulation line. Now this is derived from the line shown in Fig. 3.8; for an exponential age distribution,

$$\log_2 (1 + M_{(t)}) = \frac{t_M + t}{T_C} \tag{3.15}$$

and when $\log_2 (1 + M_{(t)}) = 0$, $t_M = -t$, and is equal to the negative intercept on

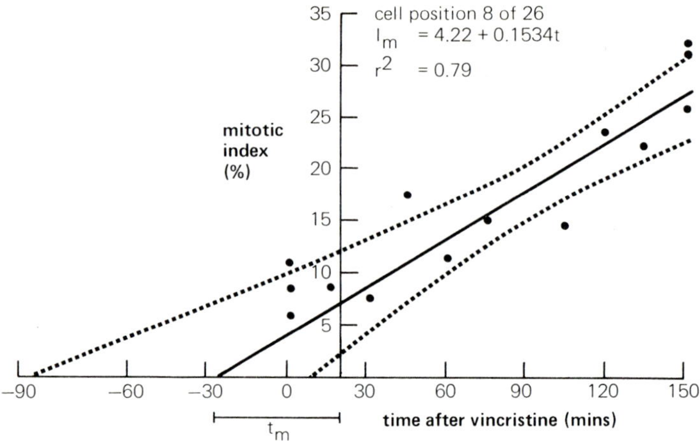


Fig. 3.13. Metaphase accumulation in the mouse jejunal crypt epithelium after injection of vincristine (1 mg/kg). The data are shown for cell position 8 of a crypt column of length 26 cells (mean). There is a delay period of at least 20 minutes before the line begins to rise; back-extrapolation to the abscissa gives the mitotic duration, but the 95 per cent confidence limits for the line (dotted lines) must also be extrapolated: the lag period before metaphase arrest is complete must also be added to the mitotic duration. This illustrates that the measurement of the mitotic duration, particularly in circumstances where there may be delay in achieving complete metaphase arrest, is an imprecise procedure.

the abscissa. Figure 3.13 shows a metaphase accumulation line for the mouse small intestine (Al-Dewachi *et al.* 1975*b*). If we decide that there is no delay period after vincristine, then $t_M = -t$. However, in this instance we cannot exclude the presence of a delay period before metaphase arrest becomes complete; any delay period would have to be added to the negative intercept to obtain the mitotic duration. The limiting case is when all mitoses present at the time of injection divide without arrest; the equation of the line is then

$$\log_2 (1 + M_{(t)}) = \frac{t}{T_C} \tag{3.22}$$

and when $\log_2 (1 + M_{(t)}) = 0$, $t_M = +t$, where $t$ here is the delay period.

Now there is no doubt that these methods will give values for the mitotic duration, but how precise are they? Eqn 3.21 is a quotient, and since the native mitotic index and the mitotic rate have standard errors, the standard error of the quotient will be compounded. The standard error of a quotient can be determined approximately from the standard error of the numerator and denominator, i.e.

$$\text{S.E.M.} (t_M) = t_M \left\{ \left[ \frac{\text{s.e.m.} M_{(0)}}{M_{(0)}} \right]^2 + \left[ \frac{\text{s.e.m.} k_B}{k_B} \right]^2 \right\}^{\frac{1}{2}}.$$

For the data in Fig. 6.6 (p. 248), the mouse oesophagus, $M_{(0)} = 0.036 \pm 0.0030$, and $k_B = 0.020 \pm 0.0035$, so that $t_M$ is estimated at 1.8 hours. In this case the standard error of $t_M$ is 0.35 hours, so that the 95 per cent confidence interval for $t_M$ is about 1.1 to 2.5 hours, which is too wide to be very useful. Similarly, in Fig. 3.13, $t_M$ is 0.9 hours with 95 per cent confidence interval from 0.5 to 1.7 hours; in this case, even if the assumption about the delay period is correct, the confidence interval is again wide, and this should be remembered when comparison of mitotic durations are made using this method (cf. Bullough and Laurence 1964*a*).

(ix) *How useful is the method for measuring proliferative rates in human tissues?* Since metaphase arrest methods do not have the radioactive drawbacks which accompany work with $[^3\text{H}]$-TdR, many workers have applied the stathmokinetic method to the measurement of proliferative rates in human tissues and tumours. For example, Refsum and Berdal (1967) reported a mean potential doubling time of 2.6 days in a group of human tumours, mainly of the mouth and throat, while Fulker *et al.* (1971) found $t_{PD}$ values in vesical carcinomas which varied from 22.2 days in well differentiated tumours to 2.2 days in undifferentiated carcinomas, and considered that there was a negative correlation between the degree of differentiation and the $t_{PD}$. Aherne and Buck (1971) observed a mean $t_{PD}$ of 4.2 days in a group of five cases of neuroblastoma and 7.8 days for a similar group of nephroblastomas. In Burkitt's lymphoma, $t_{PD}$ values ranging from 25 to 64 hours are apparent (Iversen *et al.* 1974*b*), while, in the Kaposi sarcoma, $t_{PD}$ estimates vary between 88 and 462 hours.

Now these results have depended upon one or two readings after injection of the arresting agent, and, as discussed above, there are numerous important assumptions made when using so few readings, and moreover, it is not possible to obtain any idea of the worth of the data. On the other hand, Meyer and Donaldson (1969) measured $t_{PD}$ in oral squamous carcinomas, using serial biopsy after vinblastine injection, with values varying between six and nine days; Camplejohn *et al.* (1973) used small numbers of biopsies in 19 patients with carcinoma of the rectum to produce a composite metaphase accumulation line giving a $t_{PD}$ of 192 hours, compared with 82 hours in the adjacent rectal mucosa. Wright *et al.* (1977) studied metaphase accumulation in human gastric carcinomas for varying periods after vincristine injection, taking up to nine biopsies at four time intervals, and compared the cell production rate in the carcinomas with those in the adjacent mucosa; in the carcinomas the cell production rate ranged from 3 to 24 cells/1000 cells/hour, and in the pyloric mucosa from 9 to 15 cells/1000 cells/hour. There was great uncertainty in each estimate, with 95 per cent confidence limits of zero for the birth rate, not from any statistical considerations, but from the assumption that cells are indeed being born.

A complicating factor which has been largely ignored in human tumour studies is that of tumour heterogeneity in proliferative rate; we have already referred to the analysis of Aherne *et al.* (1977*b*) which shows that human rectal carcinomas and neuroblastomas show differences in mitotic index between different areas of the same tumour which are of the same order as differences between individual tumours. This consideration will affect stathmokinetic studies, and could possibly be minimized by taking large numbers of biopsies from different parts of the tumour, preferably in proportion to the volume of the various parts, as discussed above.

In view of these findings, it is legitimate to ask if any useful information is being obtained, using the method within the constraints of human experimentation. Results in non-tumorous tissues have been better. Figure 3.8 shows data obtained from the jejunum of a patient with a morphologically normal mucosa, using serial biopsies with the Quinton multiple biopsy capsule *in situ*, after vincristine in a dose of 0.045 mg/kg injected intravenously. There is no apparent delay period, and linearity is evident over the whole experimental period; there was no evidence of anaphase escape. The data is fitted assuming a rectangular age distribution, and gives a birth rate of 25 cells/1000 cells/hour, with 95 per cent confidence limits of 20.4 and 29.4 cells/1000 cells/hour. This compares with the 30.1 and 36.4 cells/1000 cells/hour reported for two patients with normal jejunal mucosa, obtained using the FLM technique by Weinstein *et al.* (1973).

Similarly, in the common skin disease psoriasis, there is known to be an increase in epidermal proliferative rates (Weinstein and Frost 1968). A composite metaphase arrest accumulation line was derived from eight patients with psoriasis, after intradermal injection of 1 $\mu$g of vincristine, followed by serial biopsies (Duffill *et al.* 1976, 1977; Ralfs *et al.* 1981*a*). The line was fitted using a growth fraction of unity and an exponential age distribution was assumed (Appleton *et al.* 1977). This gave a mean birth rate of 12.4 cells/1000 cells/hour, with a 95 per cent confidence interval of 7.1 to 18.7. Results from an FLM curve from a patient with psoriasis gave a value of 12.3 cells/1000 cells/hour, while calculating the birth rate from the flash labelling index and the duration of $S$ gave a value of 12.7 cells/1000 cells/hour. It is not entirely clear whether it is better to find the mean of individual $k_B$ estimates, or to pool the data and find a single estimate, but the first method is normally to be preferred, if there are enough observations for each subject.

Camplejohn *et al.* (1973) reported a value of 12.2 cells/1000 cells/hour for the birth rate in human rectal mucosa, as measured with vincristine; using the growth fraction of 0.5 and the cell cycle time of 42 hours derived from autoradiographic studies by Lipkin *et al.* (1963), we obtain 11.9 cells/1000 cells/ hour. Furthermore, in human pyloric mucosa, Wright *et al.* (1977) found a cell production rate of 13 cells/1000 cells/hour, again with vincristine, which also agrees reasonably with the 10 cells/1000 cells/hour which can be calculated from the data of Shorter *et al.* (1964), and given by Lipkin (1965).

It thus appears that, in studies of non-tumorous human tissues, stathmokinetic results do compare with those given by the more sophisticated [$^3$H]-TdR techniques; however, we should be guarded in our acceptance of such results, unless optimal conditions for the stathmokinetic experiment are realized, enabling confidence intervals for the estimates to be quoted. It is also worth remembering that the above comparisons are made on different individuals in possibly different metabolic states, and, as in the experimental studies, we should really compare the two techniques carried out in the same individuals at the same time; an impossible objective?

## Conclusions

It is clear, from the numerous instances which have been quoted in this review, that the metaphase arrest technique has been subject to a good deal of ill-treatment in recent years; such abuse led Clarke (1971) to observe that the method 'enjoys a certain disrepute to this day'. It has been the purpose of this chapter to highlight the point that much of this disrepute stems from the ill-conceived manner in which many workers have applied the technique, and particularly in the several short cuts which have been employed.

The metaphase arrest technique gives a 'hard' kinetic parameter, the birth rate: when analysed correctly, appreciation of the worth of the data can be provided in the form of a confidence interval for the birth rate, and it allows the calculation of a turnover time for the population, again with its confidence interval. It is our opinion that a confidence interval for the birth rate must always be quoted, as there are certain situations, especially in human work, where the wide confidence intervals are often more relevant than the point estimate of the birth rate.

## 3.4.  The fraction labelled mitoses (FLM) method

It is customary to describe the FLM method as the 'key' or the 'cornerstone' of experimental methods for investigating the kinetics of cell populations. In the sense that, if successful, the technique allows (a) a description of the mean or median transit time through the cell cycle and of its component phases; (b) an approach to the calculation of the growth fraction; and (c) in some instances, some idea of the distribution of cell cycle times, then one can understand the reasons why. Indeed, the method was largely conceived for the analysis of cell proliferation in the small intestinal mucosa (Quastler and Sherman 1959), and since then many epithelial cell populations have been subjected to this type of analysis, which is also very popular with experimental tumour cell kineticists (Steel 1977).

However, from a time when no self-respecting kinetic study was complete without its FLM curve, there has been a growing tendency for workers, particularly in epithelial studies, to seriously question whether this is the most appropriate technique for their needs; many of its implicit assumptions are assailed with *in vivo* populations (see below), but perhaps the most serious defect is the time dependent nature of the FLM method, which makes it unsuitable for the monitoring of the many ephemeral tissue responses which accompany cytotoxic perturbation or hormonal action upon epithelia. Nevertheless, properly applied, it remains a powerful method, and will repay some attention. Steel (1977) has analysed the method and its output with particular reference to tumours, but there are many facets which are peculiar to epithelia. Within this brief, the special problems which apply to individual tissues are discussed in Chapters 6, 17, and 31.

## (i) *The principle of the FLM method*

This is outlined in Fig. 3.14. An important point to realize is that $t_M$ is much shorter than $t_S$, and this principle is exploited in this method. Basically we are

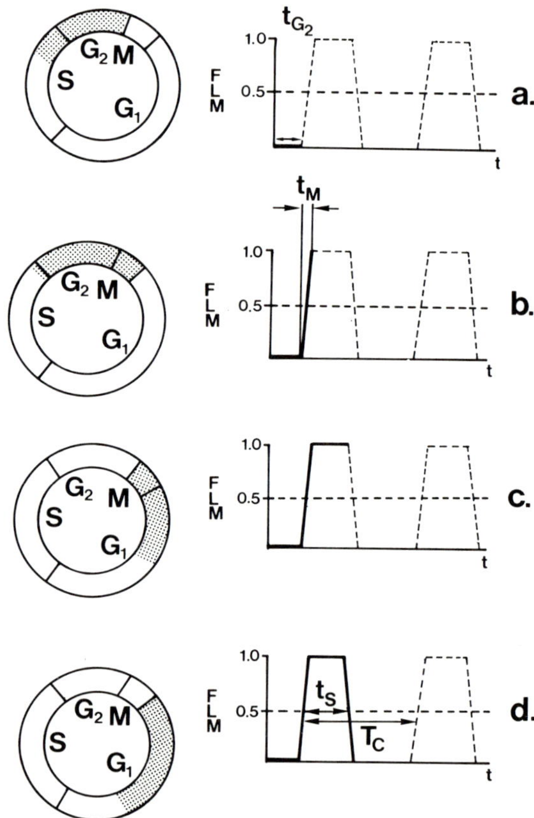

Fig. 3.14. A diagrammatic representation of the events in the construction of a fraction labelled mitoses (FLM) curve in a cell population without variation in phase transit times between cells. See text for explanation.

going to label a cohort of cells in the $S$ phase with a flash exposure to [$^3$H]-TdR; then we are going to follow the progress of this cohort around the cell cycle. In doing this, we take advantage of the convenient fact that we can all recognize mitosis as a stage in the cell cycle. Consequently, by using mitosis as a convenient window, we can peer at the labelled cohort as it passes by the mitotic window. This visual monitoring is carried out simply by seeing how many mitoses are labelled with [$^3$H]-TdR (the 'fraction of labelled mitoses') as our labelled cohort marches past. Naturally, the FLM will vary with time, since the labelled cohort only occupies (initially anyway) the fraction $t_S/T_C$ of the cell cycle. Figure 3.14 shows how the FLM varies with time, in an idealized cell population, all with the same cell cycle time. In the initial period following labelling, the first labelled cells emerge from the $S$ phase and enter $G_2$; the FLM will remain at zero for a time equal to the minimum duration of $G_2$ (Fig. 3.14a), after which the first labelled mitoses are seen. Then, because mitosis is so short, the mitotic window is flooded with labelled cells, and the FLM rapidly rises from 0 to 1.0, and this will occur in a time equal to the duration of mitosis. The mitotic window remains filled with labelled cells, and the FLM at 1.0, while the entire labelled cohort goes past; then, as the unlabelled cells (originally in $G_1$ at the time of labelling) come into view, the idealized FLM falls to zero, again within the duration of mitosis (Fig. 3.14c). The FLM remains at zero while the labelled cohort passes through $G_1$, $S$, and $G_2$, and, as they come around into the mitotic window again, the sequence would be repeated.

So much for an idealized cell population: let us now inspect a typical curve obtained for a fast cycling epithelial cell population, small intestinal crypt cells (Fig. 3.15). Now, even though the cells have a cell cycle time as short as 14 hours, there are considerable differences between the experimental and the theoretical curves: the first peak is well-defined, but the first trough does not

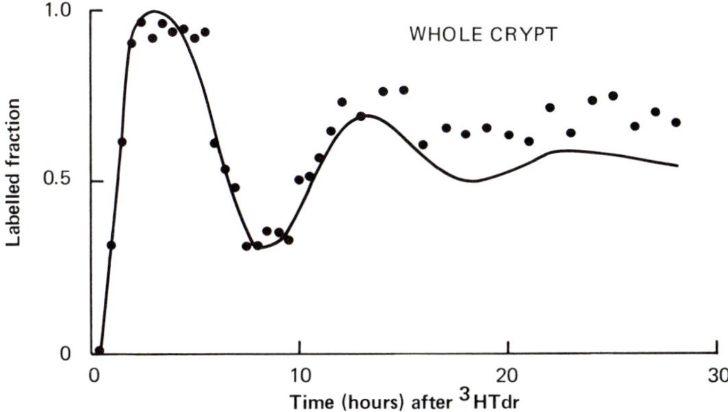


Fig. 3.15. A labelled mitoses curve from a real cell population – the jejunal crypt cells of the rat. The line is fitted using the Gilbert (1972) method.

reach zero, while the second peak by no means reaches 1.0, as the theoretical analysis would indicate. This phenomenon is called *damping*, and, when we come to examine curves from populations with longer cell cycle times, we will see that, generally speaking, the curves become progressively more damped with increasing cell cycle time, and that some curves are so heavily damped that no second peak can be discerned at all. This occurs, for example, in most curves from squamous epithelia, where cell cycle times vary from about 40 to 120 hours. This gives us a clue as to the cause of the damping effect, since, with increasing cell cycle time, we would postulate that there would be increasing opportunity for variation in phase transit times to occur within cells. Remember we have labelled a cohort of cells which are initially all synchronized in the $S$ phase; if they remained synchronized, and progressed around the cell cycle at exactly the same rate, with no interphase variation, then our idealized curve shown in Fig. 3.14 would materialize. However, cell cycle traverse is a *stochastic process*; each phase, and therefore the cell cycle time, has a distribution of residence times, and consequently, after a period which will vary according to the degree of intercell variation, the labelled cohort will become desynchronized, and the curve becomes damped and flattened as labelled cells become randomized throughout the phases of the cell cycle. This is an important point to appreciate.

### (ii) *Experimental aspects of the FLM curve*

There are broadly two ways in which we can approach the performance of an FLM experiment in, say, a rodent tissue; [3H]-TdR can be given to all animals at the same time, and readings are taken over the anticipated cell cycle duration (see below). Alternatively we can inject animals at different times, and sample simultaneously. The first of these has been generally followed in epithelial tissues; the only difference between them is that the former method gives information about the cohort in $S$ at the time of labelling, while the latter describes the characteristics of the cohort in $M$ at the sampling point. Practically speaking, the first alternative is the most expedient, since animals can be injected *en masse* far quicker than tissues can be obtained, gently treated, and placed in appropriate fixative.

In experimental work on epithelia it is usual for one animal to be used for each point, and it is unusual for investigators to take serial biopsies, even for the accessible epidermis; in larger animals, for example, the hamster, Møller *et al.* (1979) have removed one of the two cheek pouches to obtain more readings per animal, without evident effect on cell proliferation in the other pouch in the short term. In human studies, two experimental procedures have been used; for tissues such as the epidermis or surface tumours, [3H]-TdR can be locally injected and serial biopsies obtained (see Chapter 6, section 1). The other possibility is systemic exposure to [3H]-TdR, usually by intravenous bolus injection, and naturally, because of the radiobiological hazards of [3H]-TdR, this can only be done in individuals of strictly limited life expectancy. Much of the kinetic information obtained for gastrointestinal epithelia in man has been derived from FLM studies in individuals with advanced cancer (Lipkin 1965; Weinstein *et al.* 1973), and, since food intake and nutritional status can seriously modify gut proliferative rates (see Chapter 20), to say nothing of the effect of a tumour burden (Kovacs *et al.* 1982), the data in these cases must be interpreted with care.

A purely practical detail is the actual timing of the readings. Many authors have grouped data points at the same time, and thus obtained a mean and a standard error of the mean for all time points. The value of this practice is dubious. Increasing the precision of individual points is not necessarily going to increase the precision of the derived kinetic parameters, since, in fitting the curve, each reading is regarded as a single point. Every experiment is, to a greater or lesser extent, a compromise, and animal or biopsy numbers are thus limited. Since we are seeking the course of a time-dependent function, we consider that it is advisable to spread the available readings singly over the projected experimental period, rather than group them and seek the interval estimate for each point, which is not then used analytically. This problem is well illustrated by the present authors being consulted about a projected FLM curve in the rat colon, where there were some 48 rats per curve available: it was proposed to obtain three readings per hour for 16 hours, and this in a system where $T_C$ is about 48 hours (see Chapter 18, section 3). A little preparation will prevent this happening.

As to the actual timing of the readings, experience with the tissue concerned will point to the time over which the experiment should last, and once this is known, the detailed timing will depend on what is required. Inspection of Fig. 3.14 will show that, if the median duration of $t_S$ and/or $T_C$ are needed, the steepest part of the ascending and descending limbs of the first wave are the most important to define accurately; it is fruitless to waste readings on the sometimes prolonged apex of the first peak. If the curve is to be prolonged to compute the variances of the cell cycle parameters (see below), then the inflexion points of the curve become important. It is perhaps wise to perform a pilot experiment in any situation.

There remains the problem of how many mitoses to count. This should not be a consideration in rapidly dividing epithelial renewal systems such as the small intestine, but in normal epidermis, such as that of the mouse, there are epithelial subpopulations which have a low mitotic index, and require many hours of microscopy to obtain a suitable number of mitoses. What is a suitable number of mitoses? There is no correct number: the sampling error is binomially distributed (see section 3.1), and because $n$ is small, we cannot apply our usual normal approximation. The number counted will depend on the number of labelled mitoses, and also on the standard deviation of the count which we are prepared to accept;* even so, for the CV to always be less than 5 per cent, we must count

---

*The SD would then be $p(1-p)/n$, where $p$ is the proportion of labelled mitoses at any time.

at least 75 mitoses per count, no matter what the proportion of mitoses which is labelled. Most workers regard 100 as an appropriate number, which indeed gives an acceptable SD at all FLM levels.

We should, at this stage, note the innovation introduced by Schultze *et al.* (1972), in which the FLM method was combined with the double-labelling method with [$^3$H]-TdR and [$^{14}$C]-TdR; the isotopes are given three hours apart, and isolate a cohort of cells labelled only with [$^3$H]-TdR, which can be followed through successive mitoses. This was applied to the mouse jejunal crypt, and gave rise to FLM curves whose first peak had a width of exactly three hours — the duration of the interval between the injections, and a peak-to-peak $T_C$ of 14 hours. Later, Schultze *et al.* (1979) were able to use this modification to measure the means and variances of the cell cycle phases in the same tissue (see Chapter 18, section 2).

(iii) *How do we extract information from the curve?*

There are several points to consider about the format of the FLM curve: (a) the time from any point on the first peak to any point on the second peak will give a value for the cell cycle time. Because of the damping effect, it is generally difficult to measure $T_C$ from the peak regions of the curve, as is possible in the idealized curve (Fig. 3.14), and because of this, Quastler and Sherman (1959) recommended that the 0.5 labelling level on the peaks would be a practically useful alternative: for a well-defined curve such as that shown in Fig. 3.15, the value given by the 0.5 labelling level might be expected to be close to the median value, notwithstanding the fact that, even in the best curves, the second peak does not reach 100 per cent, owing to the desynchronization effect.

Similarly, since the width of the first peak is largely dependent on the time taken for the labelled cohort to pass the mitotic window, the width of the first peak will give the width of the labelled cohort, which is, of course, $t_S$. This is also usually measured at the 0.5 labelling level, and, in a well-defined curve, would be equal to the median $t_S$ value.

Sometimes, however, the first peak does not reach 1.0, because of large intercell variation in $t_{G_2}$; in this case, it is usual to take $t_S$ (and possibly $T_C$) measurements at the half-peak level, although Quastler (1963), concerned that the 0.5 labelling level might overestimate $t_S$, advised a measurement of the first peak at the 0.37 level. It has been advocated (Mendelsohn 1965; Bresciani 1965) that the area under the first peak is a more accurate $t_S$ measurement than its width, but this is often compromised by the failure of the descending limb to descend far enough to make the area measurement meaningful. These empirical measurements are still in frequent use, notwithstanding the evolution of automatic computer-based methods of FLM curve analysis (see (vi) below). As we shall see, there is good reason to advocate their use, so long as we have two (or more) well-defined peaks.

It is when we have marked damping of the curve that problems arise: there are many instances in epithelial cell kinetics where the second peak is either very damped, or non-existent, and the problem is then to obtain a cell cycle time from the data. Some authors have complained that, in the absence of a second peak, the cell cycle time cannot be measured (Hell and Hodgson 1978; Grove 1979): this is of course not so. Indeed, in many circumstances we would not expect a second peak. The reason for this lies in the *coefficient of variation of the cell cycle time*, and is illustrated in Fig. 3.16. This shows a family of FLM

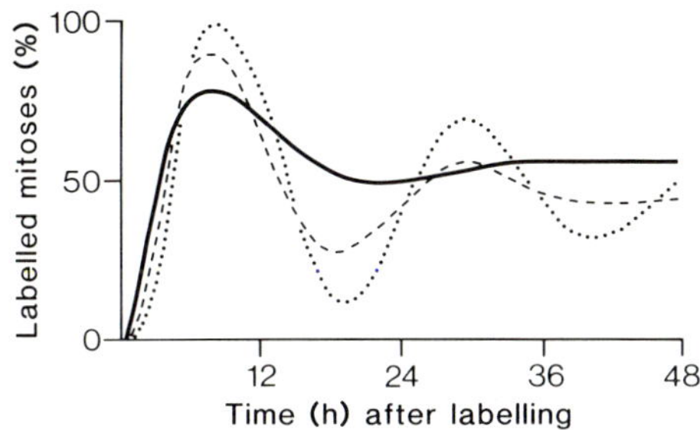


Fig. 3.16. A family of simulated FLM curves, using the same transit times through the cell cycle, but where the only difference between the curves is the coefficient of variation of the cell cycle time. The Gilbert method is again used to simulate the curves. (Redrawn from Appleton *et al.* (1977).) (—— $CV_{T_C}$ 59%;  - - - - $CV_{T_C}$ 26%;  . . . . . $CV_{T_C}$ 19%.)

curves which have been simulated on the computer: the mean cell cycle time in each is some 21 hours and the only difference between the curves lies in the coefficient of variation of the cell cycle time. When $CV_{T_C}$ is small, at around 20 per cent, there are two well-defined peaks. This is the level of variation which we find in the small intestine. However, as the $CV_{T_C}$ gets higher, and reaches 60 per cent, we see a reasonable first peak, but then there is effective flattening of the curve, with no second peak evident. This is the sort of variation seen in some stratified squamous epithelia (see Chapter 7, section 2), the adrenal cortex (Chapter 29) and, to a lesser extent, in the colonic crypts of rats and mice (see Chapter 18, section 3).

If we need to attempt to analyse such extensively damped curves using empirical methods, we would obviously be in some difficulty, and indeed, only able to extract $t_2$ and $t_S$. However, there are some more simple approaches to the analysis of damped FLM curves which are worth looking at. Mendelsohn (1963, 1965) proposed that, when the FLM achieves a plateau value, there is total randomization of cells in the cell cycle; that is to say, that labelled cells are dispersed uniformly throughout the phases of the cell cycle. Thus the proportion of labelled mitoses at the plateau value will reflect the value of $t_S/T_C$. This is because cells in mitosis are, by definition, in the cell cycle, and since we consider only mitotic cells in the FLM curve, non-cycling cells are excluded. When total randomization has occurred, the proportion of cells in each phase, including mitosis, and in the cell cycle as a whole will be the same; this means that if 30 per cent of mitoses are labelled, it follows that 30 per cent of all cycling cells are labelled. This is the labelling index for cycling cells only, and will be equal to $t_S/T_C$ for those cycling cells, for a rectangular age distribution, or $\exp(t_S\ln2/T_C-1)$ . $\exp(t_2\ln2/T_C)$ for an exponential age distribution. This will then allow the calculation of $T_C$, since $t_S$ can be read from the first peak of the FLM curve. The fact that the plateau value equals $t_S/T_C$ also allows the calculation of a value for the growth fraction from the expression

$$I_P = I_{S\,\text{expt}}/I_{S\,\text{theor}} \tag{3.23}$$

since the plateau value will also give a value for $I_{S\,\text{theor}}$ (see section 3.9). There are obviously problems involved in this approach; we must be sure that total randomization has occurred, and this may entail following the curve for a long time. There is also the problem of drawing the line at the correct level when there is a large amount of scatter about any line that we might draw. There are also problems that might accrue from dilution of the label due to division of labelled cells, and of reutilization of $[^3\text{H}]$-TdR (see section 3.1).

A further approach to the analysis of damped FLM curves was made by Barlow and MacDonald (1973); they derived an expression relating $T_C$ to $t_S$, and $p$, the number of daughter cells which continue to proliferate after mitosis (i.e. $p \leqslant 2$)

$$T_C = [-t_S\ln p]/[\ln\{1-(p-1)I_S p^{-0.5a-b/T_C}\}]$$

where $a = \{\ln 1 + (p-1)I_M\}(\ln p)^{-1}$, and $b = t_2+t_S-\omega$, where $\omega$ is the time for which $[^3\text{H}]$-TdR is available. However, we are often not in possession of such information as a value for $p$ or $\omega$, and have to make assumptions, about both the decycling probability, which is very difficult to measure, and usually involves some grain counting manoeuvre (Morley *et al.* 1973; and see Chapter 31), and also the period of $[^3\text{H}]$-TdR availability, which might be simple in such tissues as the small intestinal mucosa, but in epidermis there is now substantial evidence that $[^3\text{H}]$-TdR is available for incorporation for periods of up to four hours. Thus this method has little general applicability.

Thus far, when considering FLM curve analysis by empirical methods, the curve has been hand-fitted, and any such heuristic procedure will imply a good deal of subjective bias; naturally different workers would fit different curves to the same data, with differing results. The use of hand-fitting procedures have also led to the publication of many extremely optimistic curves (see Burns and Scheving 1975; and Chapter 9). Moreover, it is not possible to use the degree of damping of the curve as a guide to the likely coefficient of variation of $T_C$ in other than broad qualitative terms. For these several reasons, and stimulated mainly by mathematical colleagues, cell kineticists have looked for automatic optimization methods for FLM curve analysis. This was first proposed by Barrett (1966), and since then numerous methods have become available (Trucco and

Brockwell 1970; Hartmann and Pedersen 1970; Takahashi *et al.* 1971; Steel and Hanes 1971; Gilbert 1972; Valleron and Frindel 1973; MacDonald 1970; and many more (see Steel 1977, for review).

These methods are basically similar; they involve the building of a model of the cell cycle, in which the transit times or residence times through the phases of the cell cycle conform to a chosen statistical distribution; while some models are restricted to a single distribution, for example normal or log-normal, others are more adaptable in that they do allow the selection of an appropriate distribution. Several distributions have been used, the most popular being the log-normal distribution (Barrett 1966), or some other analytically tractable function such as the gamma distribution (Gilbert 1972). Having chosen the distribution of residence times, the computer model is used to fit the experimental data using an appropriate fitting procedure, such as the maximum likelihood (Hartmann and Pedersen 1970) or the least squares (Gilbert 1972) methods. A representative FLM curve, from the rat small intestine, and analysed by the Gilbert (1972) method, is shown in Fig. 3.15, and it is evident that the fit is good. Similarly the method will automatically fit FLM data with heavily damped second peaks, such as that shown in Fig. 3.17, taken from the rat forestomach, where no second peak is evident.

It is obvious that such analysis means going to some trouble, so should we perhaps examine the pros and cons of the procedure: the automatic methods generate *mean* values, whereas the use of the 0.5 peak method gives median values. Naturally, if the distribution is fairly symmetrical about the mean, the

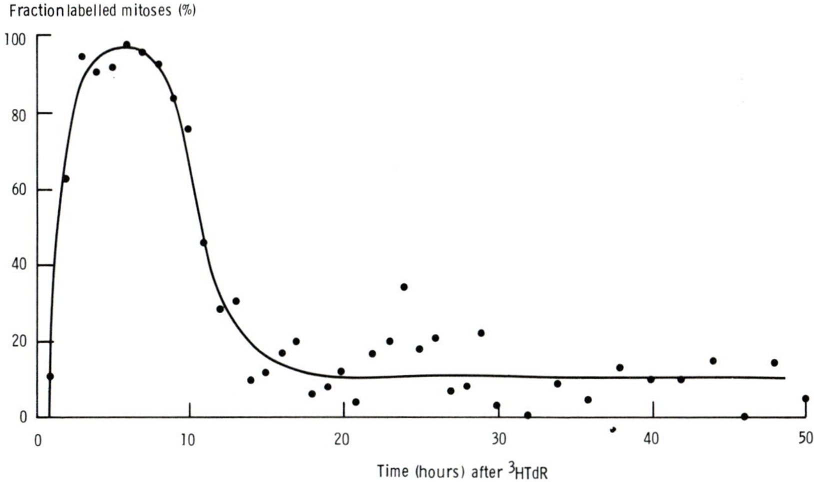


Fig. 3.17. An FLM curve for the basal cells of the squamous epithelium of the rat fore-stomach. Here the variation in transit times (particularly through the $G_1$ phase) is so great that no second peak is obtained (fitted by the Gilbert method).

mean and median should be close together; however, if there is a skewed distribution of cycle times, then the mean and median could be quite different, and it could be argued that the median is the more meaningful measurement. In the small intestine, we have built up a series of FLM curves, where the median value for $T_C$ and $t_S$ are evident from inspection, and the mean values are available from the Gilbert (1972) method, and it appears that, in this system at least, there is reasonable agreement between the median and the mean, perhaps indicating that we have a symmetrical distribution of cell cycle times about the mean, and also that, if two well-defined peaks are present, then the empirically derived median value is a good guide to the mean cell cycle time; this latter conclusion may not, of course, be applicable in all cases, and should perhaps be checked for individual tissues. While Steel and Hanes (1971) concluded that so long as the second peak rises above 0.6, the 0.5 peak measurement is a reasonable estimator of the median cell cycle time, Mendelsohn and Takahashi (1971) considered that empirical rules have large errors in heavily damped curves. Here too, Mendelsohn and Takahashi compared mean and median values, and, evidently unlike the data obtained in the intestine, cell cycle time distributions in these heavily damped curves could be more skewed than data from the gut, and hence explain the differences *vis à vis* empirical and automatic methods. But where the modelling methods are particularly valuable is in the analysis of these damped curves, where $T_C$ values cannot be extracted by empirical methods.

The automatic methods do obviate the subjective practice of hand-fitting, and should therefore eliminate bias, ensuring that a line of best fit will be drawn; however, this line will depend on the applicability (or otherwise) of the model to the actual kinetic situation. It is also valuable to obtain an estimate of the $CV_{T_C}$ (Gilbert 1972), the SD (Hartmann and Pedersen 1970; MacDonald 1970), or the distribution of cell cycle times (Steel and Hanes 1971). Obviously the shape of the output distribution will depend on the input distribution, but, in many instances, it is extremely useful to get some indication of the cell cycle time distribution. For example, it is unreasonable to attempt to optimize chemotherapy for the hyperproliferative skin disease psoriasis by attempting to cover a mean cell cycle time of, say, 37 hours, with a cytotoxic concentration of methotrexate (Weinstein 1971), if psoriatic cells have a $CV_{T_C}$ of around 75 per cent (Duffill *et al.* 1976).

The problem of the cell cycle time distribution and its confidence limits leads us onto a vexatious problem, which, in view of its importance to epithelial cell kinetics, has received little airing; this is how we can tell if sets of FLM data differ from each other. Presumably such considerations are not of great importance to tumour cell kineticists, since Steel (1977), in his masterly analysis of the FLM method and its contribution to the study of tumours, barely considers the topic, but in epithelial cell kinetics it looms large indeed. There remain numerous instances where investigators wish to test the effect of some experimental manoeuvre on cell cycle transit times, as measured with the FLM method.

Indeed, since there are no special rules to rely upon, this has forced many authors to use their own intuition. For example Baserga and Malamud (1969) were prepared to consider that median phase durations or cell cycle times were significantly different from one another if the difference between them exceeded 20 per cent, but of course such a rule of thumb cannot be generally applicable without knowing the variance of the individual values; even so, many have claimed differences in residence times based on very minor differences in median values, in the absence of any information on the variance of the cell cycle time. The archetypal example was the bitter controversy between Weinstein (1975) and Goodwin *et al.* (1975) on the question of the cell cycle time in the human epidermis in psoriasis, where Weinstein's value of 37 hours clashed with the estimate of 91 hours of Goodwin *et al.*; neither group calculated the variance of the cell cycle time, and when this is done, there is little doubt that the values do not differ (Wright 1980*a*).

If we do wish to know if cell cycle parameters differ, then we do need variance estimates, and this means automatic analysis. It is true that many programs include SEM or SD values in the output, but in view of the doubt cast on the applicability of these estimates (Aherne *et al.* 1977*a*) can we merely transpose these values into significance tests between groups of observations? The answer is probably not. Al-Dewachi *et al.* (1974), for example, considered that the SEM values generated by the Gilbert method could be about four or five times too small in reality, and in no case has the validity of an FLM curve standard deviation been compared with the standard deviation of cell cycle transit times measured directly, by say time-lapse cinematography, which would be one way of establishing the validity of the calculated values. It is true that the method of MacDonald (1970) may be more valuable, since it gives some weight to the number of mitoses counted per point, but the fact remains that the generated parameters are only as good as the fit, and even if the fit is good, on the apposition of the chosen distribution to the real one. Of course, if large differences are shown in curves where two well-defined peaks are seen, as between plucked and unplucked mouse epidermis, where $T_C$ was shown to be 45 and 110 hours respectively (Hegazy and Fowler 1973*a,b*), then we are probably justified in concluding that a difference does exist, but, in most situations, the differences are not as large as this. Lewis (1978), in a painstaking analysis, has measured $t_S$ in various cell populations in the central nervous system in a considerable number of individual FLM curves in control populations, and has calculated a mean and SEM from these; a $t_S$ value, which fell outside the 'normal range' would then be considered to be different. This is probably the most certain method of knowing if there are differences in experimental populations, but the instances where this is possible are indeed few. We would advocate that, if you wish to show that one population is proliferating at a different rate from another, then the FLM method is not the appropriate one to use, because of the difficulty in defining the interval estimate of the FLM derived parameters.

There are other problems in fitting the data; given the constraints of the model, there are curves which it is impossible to fit. Theoretically, the area under each wave of labelled mitoses should be the same, but Steel (1977), from a large experience of FLM curve simulation, has drawn attention to the several types of curves which tend to show differences from the fitted line. There are basically three examples of this phenomenon: (a) *limiting*, (b) *enhancement*, and (c) *fade*. *Limiting* is where the data points lie uniformly below the fitted line at the apex of the first peak, or above the line at the first trough (Fig. 3.18A). If this is seen, the first thing to think of is a potential artefact; scoring of labelled mitoses as unlabelled will lead to a low first peak, while confusing labelled nuclei with labelled mitoses will lead to a high first trough and should be looked for. Failure of $[^3H]$-TdR to reach all cells (owing to diffusion difficulties, and probably only seen in tumours), could result in first peak limiting, while $[^3H]$-TdR reutilization would have the same effect on the first trough. It is obvious that the definition of the grain count threshold (Shackney 1975*a*) will affect the height of the first trough. However, if these artefacts can be excluded, then the phenomenon must mean that there are subpopulations of cells with differing transit times through cell cycle phases (Steel 1977).

*Fade* is when the data in the second peak falls progressively below the fitted line (Fig. 3.18B). An obvious candidate cause is dilution of label by mitosis, causing labelled nuclei to fall below the grain count threshold; alternatively, $^3H$ radiotoxicity could prevent the cycle progression of labelled cells. Exclude these, then biological possibilities include the rapid influx of a large number of unlabelled cells into mitosis; this indeed would be expected in epithelial cell populations if an FLM curve was initiated at the beginning of an induced proliferative response, in say, the androgen-treated castrate prostate, or the partially excised liver (see Morley *et al.* 1973, and Chapters 25 and 31); as cells enter the cell cycle from the proliferative stimulus, the system would indeed be flooded with unlabelled cells, which would swamp the second peak, even though the first peak might be well fitted. Steel (1977) also makes the theoretical argument that a slowly proliferating stem cell subpopulation with a smaller $t_S/T_C$ ratio than the prevailing proliferative cells would lead to the later, progressive appearance of more cells in mitosis, which, at the time of labelling, were very primitive, with a low labelling index. Experimentally, in a stem cell hierarchical system such as the small intestinal crypt, although the basally sited 'stem cells' do have relatively long $T_C$ values, the $t_S/T_C$ ratio does not increase proportionally to $T_C$. However, in the epidermis, where modern theory also advocates a stem cell hierarchy, we do not have values for $t_S/T_C$ in the hierarchy. In this respect, Potten *et al.* (1982*b*) have successfully fitted a series of kinetic experiments in the epidermis, including prolonged FLM curves, by assuming that $T_C$ in the stem cells was very long, while $t_S$ was very short, both compared with the proliferative, differentiating cells. Although Potten *et al.* did not fit their FLM data with a conventional program, the very low FLM indices, after a

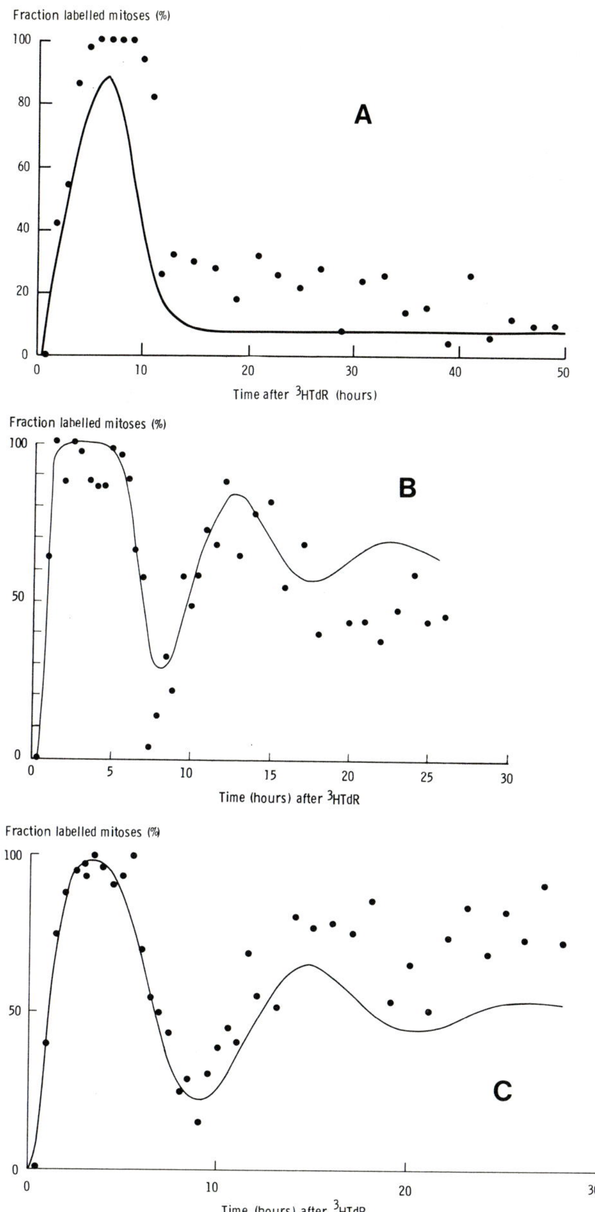

Fig. 3.18.  FLM curves from different epithelial tissues, illustrating the phenomenon of (A) 'trough limiting' in the zona reticularis of the adrenal cortex of the 14-day-old rat; (B) 'fade', and (C) 'enhancement' in different cell positions of the rat jejunal crypt column. The FLM curve from Fig. 3.18C also shows evidence of 'non-independence of phase durations'. The lines are fitted by the Gilbert (1972) method.

comparatively well-defined first peak, would indicate that fade was at least a possibility (see Chapter 8).

In *enhancement*, data points on the second peak lie above the best line (Fig. 3.18C). Again $^3$H reutilization remains the best bet, but if the stem cell population has a high $I_S$ value compared with the proliferating differentiating cells, then this phenomenon could occur (Tannock 1968; Steel 1977).

In the automatic analysis of FLM curves, it is generally assumed that there is no correlation between the phases of the cell cycle; the method of MacDonald (1970) allowed the simulation of populations in which negative correlation between the phases is allowed, and the degree of correlation could also be estimated. However, how relevant this is for epithelial cell populations is not clear, since Schultze *et al.* (1979) used their double-labelling FLM technique to estimate the variances of the individual cell cycle phases in the mouse jejunal crypt, and showed that it was the variances themselves which were additive, rather than the standard deviations, in which case there is no evidence for correlation between cell cycle phase durations, at least in the mouse small intestine; this is discussed in Chapter 18, section 2). Hence the caveats delineated by Schotz and Zelen (1971), and due mainly to correlation between phase durations, may not be applicable to epithelia.

Having said all this, and given the fact that there are many models available, if we decide to use an automatic method, then which one do we choose? A comparison of several of the available methods, carried out by the simple expedient of sending the same sets of FLM data, which included damped curves, to the several participants, showed that all the methods tested gave more or less similar results; most methods were able to cope more or less successfully with even heavily damped curves (Hartmann *et al.* 1975). We could conclude that, should we choose to use such methods, it does not really matter which one we choose, so long as we remember the limitations (and strengths) of the whole approach.

### (iv) *Limitations of the FLM method in epithelial cell kinetics*

It is probably true to say that the FLM method has been the lynch-pin of experimental tumour cell kinetics (Steel 1972, 1977), and in many respects, has also given us a good deal of knowledge about the various epithelia which we are about to describe in this volume. However, over the past few years, it has become increasingly evident that the FLM method is not the universal panacea it once appeared; in epithelia there are several major drawbacks to its general usage, and there are many situations where its use can be misleading.

(a) *The effect of synchronization*: one assumption implicit in the method is that cells are completely asynchronous with respect to phase in the cell cycle; there are numerous instances where this premise is challenged. We shall see that, in squamous epithelium (and also in the gut), there are large circadian swings in proliferative indices due to synchronization in one of several phases of the cell cycle. If cells become synchronized in say, the $S$ phase, then this can have profound influences on the FLM curve and its interpretation — these effects are discussed in Chapter 9, section 3.

In some circumstances synchrony is induced by a particular experimental manoeuvre. For example, after partial hepatectomy, liver cells move synchronously into $S$ and subsequently $M$, giving rise to waves of $I_S$ and $I_M$ respectively, and similar events follow androgen or oestrogen challenge of the prostate or uterus, folate treatment of the kidney, and numerous other examples. In these instances of induced DNA synthesis, synchrony is a constant accompaniment, and will modify an FLM curve initiated at any part of the response (see Chapter 24). Because of this, attempts to analyse such systems have been, of necessity, model based (Appleton *et al.* 1973; Aroesty *et al.* 1977) in an attempt to extract data from classical FLM experiments in these responses. Similarly, treatment with a cytotoxic or cytostatic agents, which inhibits cell cycle traverse, can effectively synchronize cells in a particular phase; hydroxyurea (Al-Dewachi *et al.* 1977) and cytosine arabinoside (Al-Dewachi *et al.* 1980) both inhibit $G_1$ to $S$ transition and kill cells in the $S$ phase, resulting in an increased $G_1$ to $S$ flux at a later time. Attempts to analyse subsequent proliferation with the FLM method could be affected by this synchronization.

(b) *The time base of the FLM curve*: perhaps the major drawback of the FLM method for epithelial cell populations is its long time base. Even in the rapidly proliferating rodent gastrointestinal mucosa, it takes as long as 15 hours to derive a cell cycle time value. Therefore its use in ephemeral responses cannot be recommended, since most of the action could be over before the end of the procedure.

(c) *The FLM method and kinetic microarchitecture*: in some epithelial cell systems, such as the intestinal crypt, the population can effectively be visualized as an array of proliferating cells arranged in layers (see Chapter 17), and cells migrate through the array to leave the crypt at the top. It is thought that cells at the base of the crypt, i.e. at the bottom of the array, are slowly cycling, and the cell cycle time gets progressively shorter as we move to the top of the crypt (Cairnie *et al.* 1965a; Al-Dewachi *et al.* 1974, 1979). However, if we were to carry out FLM analysis of each layer in this array, we would meet the problem that a labelled cell in, say, layer 3, which contributed therein to the first peak of labelled mitoses, and at that point was cycling at a particular rate, would, because of the finite time period between the first and second waves, have time to migrate into a higher layer in the array, and contribute to the second wave of labelled mitoses in a layer where the cell cycle time was much shorter. The effect that this *migration artefact* would have on the FLM curves for different layers would depend on the kinetics of the migration process. On migration from layer $i$ to layer $i+1$, in say, $G_1$ or $S$, cells could retain the cell cycle time of the layer in which they were born, until the next mitosis in layer $i + 1$. Alternatively, they could take on the new cell cycle parameters of the new layer $i + 1$ immediately on migration into it, or of course, they could gradually speed up over the remainder of the cell cycle (the restriction point for cell cycle traverse changes may not of course be mitosis, but could be the transition between the $A$ and the $B$ phase, on the Smith and Martin (1973) or Burns and Tannock (1970) models, but the principle of the argument remains the same).

Whichever of these alternatives obtains will have considerable effects on the interpretation of the FLM curve for each layer; in Chapter 6, section 7 and Chapter 19, section 2 we shall consider the age distribution of layered tissues, and, even without introducing the complications of circadian variation, the age distributions in higher layers are complex. The only way of critically analysing FLM curve families in layered tissues would be to combine an age distribution model for each layer with the possible alternatives for cell cycle traverse changes listed above. This problem is also met with in other systems, cf. plant meristems (Barlow and MacDonald 1973), and is a limiting factor in our appreciation of kinetic microarchitecture. It is a difficult problem; we have been unable to persuade a whole series of mathematical colleagues to even attempt it.

(d) *The FLM curve as a measurement of proliferative rate*: we have already mentioned this. If we need to assess proliferative rate, and the effect of an experimental change upon it, this would not be the method of choice. Although we get a complete breakdown of phase durations, etc., we would indeed have problems in testing these measurements against each other (see above).

(e) *Factors due to [$^3$H]-TdR*: in our discussion of the labelling index, we mentioned such problems as the reutilization of [$^3$H]-TdR, which, in a prolonged FLM curve, could materially affect the result. The question of radiotoxicity has also been mentioned, and recently, Beck (1982) produced evidence from flow cytometric studies that mouse fibroblasts exposed to 0.3 $\mu$Ci/ml of [$^3$H]-TdR in culture showed slowing of $S$ and $G_2$-$M$ transit, and that this had quite serious effects on the FLM curve; mention has already been made of the work of Post and Hoffman (1968) on the putative effect of [$^3$H]-TdR radiotoxicity on the FLM curve.

*Conclusions*

This discussion of the FLM curve has been general, and directed exclusively at epithelia; more specific points in specific epithelia are taken up when these tissues are examined. We might conclude, as is usual with any kinetic technique, by asking 'what do we want to measure?' If we need to know the duration of the cell cycle, and of its component phases, and to assess their variance, then we would be well advised to use the FLM method. However, for other purposes, such as the assessment of proliferative rate, and in situations where ephemeral responses or synchronization is expected, the method is certainly not that of choice.

## 3.5. The continuous labelling method

This technique is also based on autoradiography: it depends upon making [³H]-TdR available to the population for a time which is usually not less than $t_{G_2} + t_M + t_{G_1}$. The principle is shown in Fig. 3.19; the [³H]-TdR is present

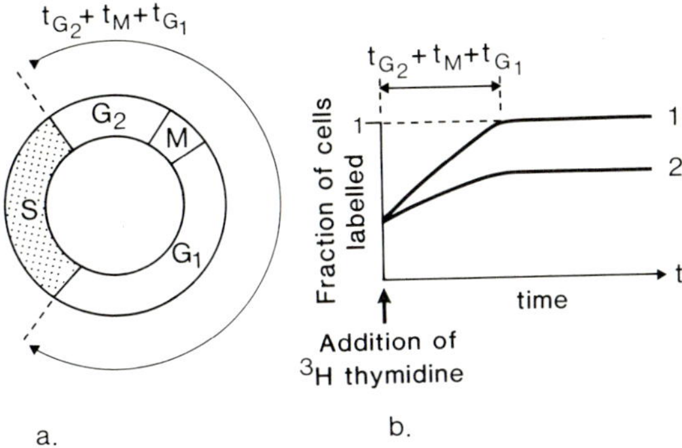



Fig. 3.19. The principles of the continuous labelling method: [³H]-TdR is made continuously available to the cell population under study, and the continuous labelling index ($I_L$) should climb to the plateau value in a time equal to $t_{G_2} + t_M + t_{G_1}$. If the growth fraction is less than unity, there is a widespread hope that this will be reflected by the plateau value; this is seldom realized in renewal populations for the reasons discussed in the text.

continuously, the initial exposure labels all cells in the $S$ phase, and, as cells move into $S$, they should all become labelled. The labelling index (here symbolized by $I_L$), will rise from the level of the flash labelling index, increasing at a rate equal to the flux into the $S$ phase, for a time equal to the minimum duration of $t_{G_2}$; theoretically there should be a doubling of the slope of the line at this point, which should approximate to the duration of $G_2$. However, the variation in $t_{G_2}$ will usually be sufficient to 'iron out' any such inflexion point, and inspection of continuous labelling curves from *in vivo* populations rarely if ever shows such an appearance. Labelled cells then divide, and the rate of increase in $I_L$ thereafter will be a function of the rate of entry into $S$ and the later division of the cells thus labelled. The rate of increase in $I_L$ is also dependent on the distribution of transit times through $G_2$, $M$, and $G_1$, and very sensitive to the age distribution of non-proliferating cells in the population (such as the post-mitotic maturing cells in the epidermis, or the decycled maturing cells at the top of intestinal crypts). This age distribution in turn depends on the age at which the non-proliferating cells migrate into the Malpighian layer or villus population respectively; in tumour cell populations cells can be lost from the cell cycle at various points, for example in mitosis, and this will affect the shape of the continuous labelling curve.

However, in the absence of cell loss or migration, the $I_L$ rises to reach a plateau value; the point of inflexion (Fig. 3.19) should be reached in a time equal to $T_C - t_S$, since the time taken to label all proliferating cells should take the remainder of the cell cycle after the initial $S$ phase cohort has been labelled. However, if a decycling event occurs after mitosis in *labelled* cells, non-proliferating cells will also become labelled; the obvious conclusion is that if the non-proliferating cells in the population are subject to turnover, the curve will reach 100 per cent.

### (i) *Experimental details*

The incorporation of [³H]-TdR throughout the cell cycle is of course no problem in *in vitro* work, where the label can be added to the medium, and cultures taken down and fixed at appropriate times. For *in vivo* use, there are naturally difficulties, and three main methods have been employed. The commonest practice for experimental animals is to give a series of injections, usually intraperitoneal, over the expected duration of $t_{G_2} + t_M + t_{G_1}$. If this is done, it is critical to ensure that the time period between injections is less than the minimum duration of the $S$ phase; if this is not rigidly adhered to, there is the distinct possibility that cells can enter and leave $S$ and not become labelled. This means that some information about $t_S$ must be available beforehand, and the choice of the interval will obviously be influenced by $t_S$. Steel (1977) maintains that, so long as the injections are given at less than the median duration of $t_S$, the fraction of cells missed will be within the limits of experimental error.

In epithelia the method has been popular with workers in squamous epithelia, and here intervals ranging from three hours (Potten and Major 1980; Potten 1981) to six hours (Iversen *et al.* 1968; Denekamp *et al.* 1976) have been used, while the few investigators using this method in gut epithelia have generally used six hour intervals (Wright 1978; Sunter 1981). In other epithelia, such as prostate and adrenal, authors have felt safe with six-hourly injections (Alison *et al.* 1976; Wright *et al.* 1973c), and these times are probably a reasonable compromise, in view of what we know about the duration of the $S$ phase in these tissues. It is, however, important to realize that such information is useful before setting out.

It is usual, when using the serial injection method, to kill one animal at one hour after each injection, and to prepare autoradiographs in the usual way. It goes almost without saying that, because of the large amount of handling and the trauma of repeated injections, there is a great deal of disturbance and hence potential perturbation of renewing populations. Stress evokes the production of catecholamines and corticosteroids, both powerful mitotic inhibitors; circadian rhythms may also be upset by the procedure, to say nothing of the protracted skin handling if the epidermis is being studied.

Alternatively, [³H]-TdR can be administered in the drinking water, a method used extensively by Cameron (1966, 1972a,b); it would, however, be unwise to rely upon this technique for critical work, since it is highly likely that rodents, being mainly nocturnal animals, consume most fluid at night, and may go for long periods during the day without drinking, with the result that cells could traverse $S$ without becoming labelled.

The third method, which is also applicable to humans, is by continuous intravenous infusion; this is usually carried out by an indwelling intravenous (or intraperitoneal) catheter (Mendelsohn 1962), although the advent of osmotic infusion pumps which can be implanted subcutaneously might be expected to minimize the (relatively speaking) minor amount of disturbance involved in this procedure. This is probably the method of choice, and has found use in human subjects, usually for the study of tumours in individuals of limited life expectancy by systemic infusion, or sometimes by isolated limb perfusion. However, Gelfant (1976) has been able to obtain epithelial tissues from these procedures, and derive useful information.

### (ii) *The extraction of information from continuous labelling curves*

Theoretically speaking, the continuous labelling method should provide us with a good deal of information; however, there are several important caveats which decrease its power; this has not prevented its advocates from overinterpreting and misinterpreting the data. The first parameter which is allegedly available from the technique is *the duration of the period* $t_{G_2} + t_M + t_{G_1}$. The problem here is the accurate definition of the inflexion point on the graph; this is usually poorly defined, because the increase in $I_L$ is gradual, and the achievement of the plateau value is not a sudden event. Nevertheless, Hegazy and Fowler (1973*a*) have obtained results with the continuous labelling method in mouse epidermis which compare well with FLM studies in the same tissue; secondly, the *growth fraction*; many workers have looked hopefully at the continuous labelling method in respect of its use as a measure of the growth fraction, and, as a result, there are now numerous continuous labelling curves for such tissues as the epidermis (see Chapter 7, section 4), and such measurements of growth fraction are also frequent in conditional renewal systems (see Chapter 24). There are several reasons why we should be wary of growth fraction values derived from continuous labelling curves. It is usual to regard the plateau value as the equivalent of the growth fraction, which is generally inaccurate, for the following reasons. In a *renewing system*, the cells continue to divide, and thus the plateau value is continually being augmented by proliferating cells which have labelled more than once. Moreover, the plateau value is sensitive to the age distribution of cell loss or cell migration (see below); the plateau value in tumour cell populations depends on the prevailing site of deletion of cells in the cell cycle, and in renewing populations on the mechanism of migration (see, for example, squamous epithelia, Chapter 8, sections 4 and 6). The time $(t_{G_2} + t_M + t_{G_1})$ taken to achieve the plateau value is also affected by the cell loss mechanism; non-proliferating cells which migrate from the population (cf. post-mitotic maturing cells in squamous epithelia; see Chapter 8, section 4) will not be recorded as labelled cells nor will they be recorded as non-labelled cells, even though they form part of the non-growth fraction. They will be replaced in a time which is equivalent to the turnover time for these non-proliferating cells; the continuous labelling curve will then tend to unity. We should also mention the possibility that the transit time through a $G_0$ phase might be less than the duration of the experiment, in which case even $G_0$ phase cells will become labelled; however this does take us into the semantic $G_1/G_0$ area, which is explored elsewhere.

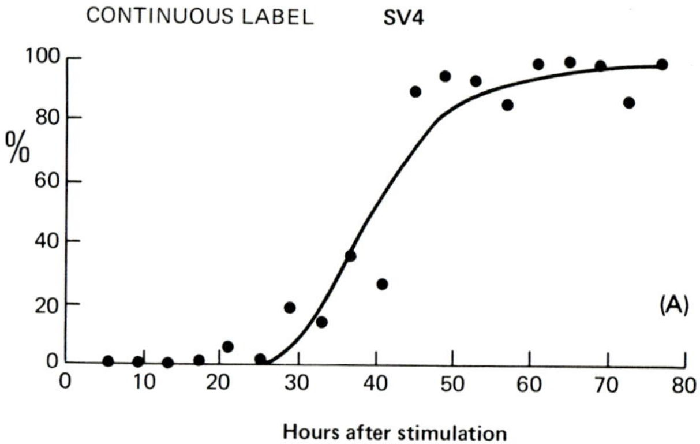


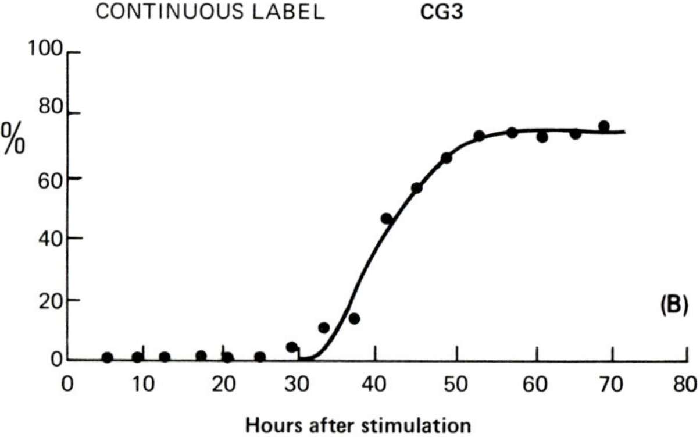


Fig. 3.20. Continuous [³H]-TdR experiments in the seminal vesicle (SV) and coagulating gland (CG) of the mouse prostate complex stimulated with testosterone at 12 and 9 days after castration respectively. The lines are fitted by the computer model of Appleton *et al.* (1973). The plateau value for the seminal vesicle reaches 100 per cent (A), and the coagulating gland plateaus at about 80 per cent (B). Since labelling was continued through the temporally limited proliferative response, all cells which entered $S$ were labelled, and providing allowance is made for those cells which have divided more than once in the response (readily found from grain count halving procedures; see section 3.6), the proliferating population size can be measured.

For these several reasons, it is not advisable to use the continuous labelling method to express the growth fraction in renewing epithelia or tumours which are subject to cell loss or migration. If there is only limited proliferation in a system, and no evident cell loss, then a continuous labelling experiment may be a rough guide to the relative size of the proliferative population, particularly if correction can be made for the proportion of cells which divide more than once and inflate the final $I_L$. This is encountered in conditional renewal systems subjected to a stimulus which evokes induced DNA synthesis; androgen stimulation of the castrate prostate is a typical example (see Chapter 31). Continuous labelling over the whole proliferative response labels every cell which enters the cell cycle (Fig. 3.20); the proportion of cells recycling at each mitosis can be estimated by interphase grain counts at frequent intervals (see section 3.6 and Chapter 31). The problem of cell loss still arises, however, but if the continuous labelling index reaches 100 per cent, we can make the reasonable conclusion that all cells in the population have responded to the proliferative stimulus.

### (iii) *The flux into DNA synthesis and the duration of DNA synthesis*

Some workers (e.g. Fukada *et al.* 1978*a* in epidermis, and Pozharisski *et al.* 1980*a, b* in DMH-induced colonic tumours in the rat) have proposed that $S_{\text{influx}}$ and $t_S$ can be measured from continuous labelling curves. The theory is shown in Fig. 3.21. The line of continuous labelling is extrapolated backwards until it cuts

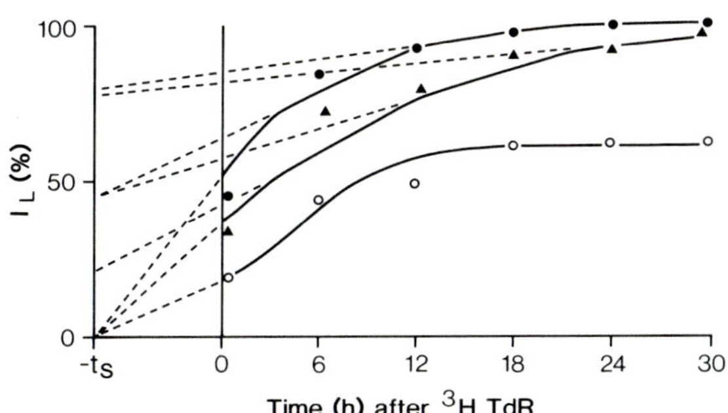


Fig. 3.21. Continuous labelling curves in the crypts of the large intestine of the rat, illustrating the problems involved in backwards-extrapolation and in identifying plateau values on the basis of few readings. The inflexion points in these curves are largely notions). (Redrawn from Pozharisski *et al.* (1980*a, b*).)

the abscissa; the negative intercept then indicates $t_S$ while the slope of the line reflects $S_{\text{influx}}$. Pozharisski *et al.* (1980*b*) have even been able to descry subpopulations in DMH-induced colonic tumours, with different $t_S$ and $S_{\text{influx}}$ values, by identifying several different points of inflexion on the curve, and extrapolating a family of lines of different slopes backwards to the abscissa.

This is a horrendous manoeuvre, when we remember that the $S_{\text{influx}}$ is only given by the slope of the line before labelled cells have divided, and the division of labelled cells complicates the picture; the slope of the line is highly dependent on the cell loss mechanism, particularly in tumour cell populations; if the initial labelling index is high, the extrapolation will be long, with resulting wide confidence limits (see measurement of the mitotic duration by the same procedure, in metaphase arrest curves in section 3.3); the identification of a single inflexion point is difficult enough, but the recognition of several such points is giving the method a precision which it does not inherently possess, even if the preceding caveats are ignored. This practice cannot be recommended.

There are other methods which can be used to derive $t_S$ from the continuous labelling curve; $S_{influx}$ could be found from the slope of the line before labelled cells have divided, and $t_S$ is then found from $t_S = I_S/S_{influx}$. Alternatively, the mean interphase grain count should reach a plateau under continuous labelling conditions in a time equal to $t_S$. Neither of these methods possesses any intrinsic degree of precision; the first suffers from the drawbacks listed for $t_M$ measurement by the same method (see section 3.3), while very frequent measurement of $I_L$ would be needed to define the grain count plateau by the second method; neither has been used much in epithelia.

### (iv) *The assessment of cell loss and migration*

As mentioned above, the continuous labelling curve is sensitive to the age distribution of cell loss from the population. This feature cannot be read directly from the curve, and must be approached by computer simulation. This was first carried out by Steel *et al.* (1966), and since then several methods have appeared for the computer analysis of continuous labelling curves (e.g. Dombernowsky and Hartmann 1972). These have usually been applied to tumours, but Iversen *et al.* (1968) and Potten and Major (1980) have assessed the mode of migration from the basal layer of epidermis by this method (see Chapter 8, section 6).

The principle is shown in Fig. 3.22. We formulate a suitable model, and introduce such cell loss mechanisms as we consider biologically appropriate. We also need values for cell cycle parameters and usually their variances; the theoretical lines are simulated by the computer according to the several cell loss mechanisms we have chosen, and the line which most reasonably fits the data points may indicate the appropriate cell loss pathway. In some models, it is possible to fit the data points by calculating the contribution of each cell loss mechanism to the line (Fig. 3.22). It is usually possible to quote a value for the growth fraction from the model, since $I_P$ is needed in the simulation.

There are obvious drawbacks to this approach. The cell loss mechanisms are model defined, and since we can rarely say which avenues are being used with any degree of certainty, we have problems. Moreover, the model usually requires the assumption of a steady state of exponential growth, and thus cannot be used for analysing perturbed systems. However, it is instructive to look at the results of such an analysis in a transplantable mouse tumour: Fig. 3.23 (p. 172)

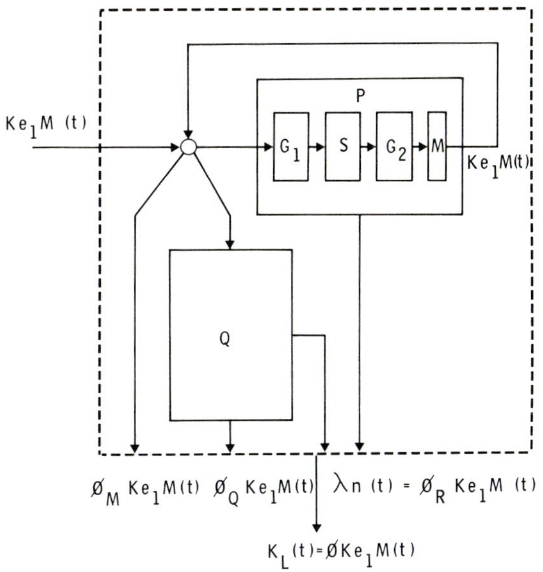



Fig. 3.22. A model suitable for the analysis of cell loss patterns, here for an experimental tumour. There are three avenues of cell loss: $\phi_M$ is the component of the cell loss factor (defined as the ratio of cell loss rate to cell birth rate) due to cell loss at mitosis, $\phi_Q$ is the component due to cell loss in the $Q$ or resting cell compartment (so-called 'pipe-line cell loss') and $\phi_R$ is the component of random loss in the cell cycle. This model was conceived by Dr N. R. Hartmann.

shows representative results, and Table 3.7 (below) compares values for $I_P$ and $\phi$ (see section 3.9) as calculated from the FLM curve, the metaphase arrest

Table 3.7. A summary of kinetic parameters generated by computer analysis of continuous labelling curves in a mouse sarcoma. (Analysis by courtesy of Dr N. R. Hartmann)

| Time after transplantation | $\phi_M$ | $\phi_Q$ | $\phi_R$ | $\phi$ | $I_P$ | $I_P^*$ | $I_P\dagger$ |
|---|---|---|---|---|---|---|---|
| 7 days | 0.000 | 0.192 | 0.000 | 0.192 (0.43)‡ | 0.52 | 0.45 | 0.50 |
| 14 days | 0.509 | 0.000 | 0.297 | 0.806 (0.78)‡ | 0.49 | 0.38 | 0.38 |
| 21 days | 0.549 | 0.000 | 0.377 | 0.926 (0.93)‡ | 0.46 | 0.31 | 0.35 |

*Growth fraction calculated from $I_{Sexpt}/I_{Stheor}$.

†Growth fraction calculated from $T_C/t_{PD}$, where $t_{PD}$ was calculated by metaphase arrest measurements with vincristine.

‡The cell loss factor calculated by the usual method $(k_L/k_B)$.

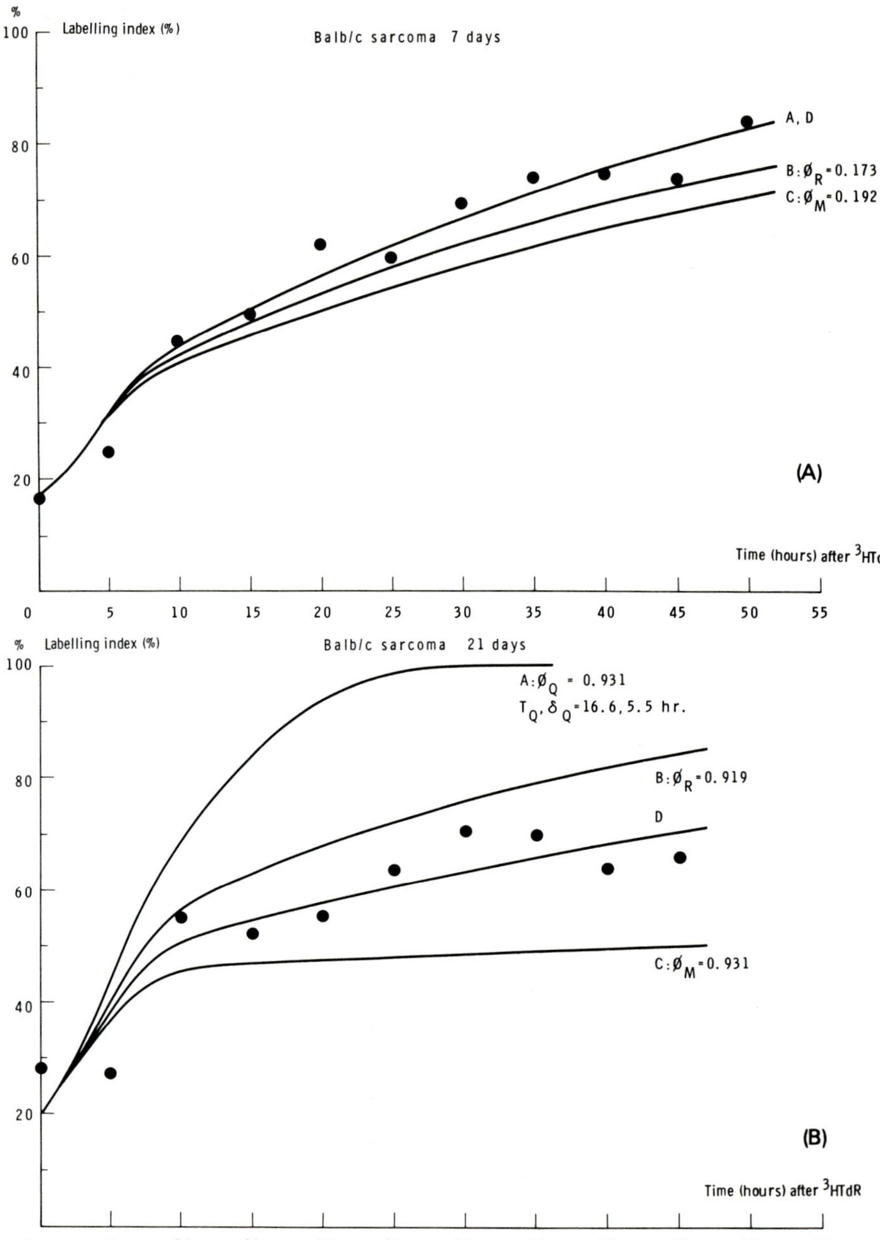



Fig. 3.23. Continuous [$^3$H]-TdR labelling curves in a rapidly growing mouse sarcoma. The experimental points are indicated by the solid circles, and the lines show the best fits to the data assuming the given avenues of cell loss. The best fitted line (D) is in fact achieved by proposing that the total $\phi$ is made up of several contributing cell loss mechanisms (Table 3.7). The program also generates values for the growth fraction (Table 3.7).

method, and the continuous labelling method; avenues of cell loss are also given. The model values for the growth fraction do not compare well with other methods, but for $\phi$, the results are more reasonable, and do at least show the same trend. Moreover, mechanisms of cell loss appear to vary with time after transplantation (see section 3.9).

In conclusion, the modelling of CL curves, despite its limitations, may give some insight into cell loss mechanisms, and does provide an approach to the measurement of the growth fraction from the curves. However, as we shall see (Chapter 8, section 5), this is a very difficult procedure in renewing populations.

*Conclusions*

The continuous labelling technique has been popular with workers in the field of epithelial cell population kinetics, so much so that some have relied on it exclusively (see, for example, Fukuda *et al.* 1978*a*). We think that this is a mistake; we would regard the method as a component part, but only a component part, of any comprehensive investigation of a steady state problem. Many of the parameters which are supposedly measurable by the method are largely dependent on information from other methods; essentially it remains one of the corroborative cell kinetic methods, especially in epithelia.


## 3.6. Grain count halving

This method is based on the apparently simple concept that labelled cells halve their content of label, and therefore their grain count, at each division. If this is true, then the time to the first grain count halving will be approximately $t_S + t_{G_2}$, and subsequent halvings should conform to the cell cycle time. This technique is usually only used for tissues of apparently long generation time, and, practically speaking, it appears a simple matter to take readings over a suitable period of time, and to explore the changes in the interphase grain count.

In epithelial tissues, however, there are numerous problems with the method: the number which actually halves is the *average grain count*, and since we are usually obliged to use a flash label *in vivo*, asynchrony of individual replicons (see section 3.1) and the short labelling period will mean that not all portions of the newly synthesized DNA will be labelled; this will result in chromatids which are labelled in a certain area, and those which are not (Taylor *et al.* 1957). However, since all the labelled chromatids do not move to the same pole at mitosis, the average grain count will probably halve. Much will obviously depend on the degree of asynchrony of DNA synthesis; if this is extreme, then whole portions of the chromatin of some cells may be unlabelled, with the production of non-labelled progeny at a time less than the simple model will predict.

A further factor that could seriously modify the grain count halving distribution is *non-random segregation of chromosomes at mitosis*, so that most, or all, of the newly synthesized DNA is passed on to only one of the descendants.

Indeed, Cairns (1975) has proposed that this would be a useful way in which stem cells might maintain a complement of intact DNA, while the DNA which may have been subject to genetic damage is discarded into the daughter cell which migrates with the doomed proliferative progeny. Experimental evidence for this was produced by Potten *et al.* (1978), who showed that the presumptive stem cell position in the mouse filiform papilla, which labels totally with [³H]-TdR at 0300 hours (due to strong circadian rhythm), contains *unlabelled cells* after what appears to be the first division after labelling (see Chapter 8, section 5). This may also occur in the basal intestinal crypt (see Chapter 19), and the process would obviously cause serious anomalies in a grain count halving experiment carried out on these cells.

A further important consideration is the probability of *decycling divisions*. There is always a probability that cells may leave the division cycle after a mitosis, and if this occurs in a labelled cell, say in the first division after labelling, then the daughter cells will remain labelled but will not halve their grain count at the next mitosis; this will ensure that the grain count halving period will be increased. Baserga *et al.* (1963) proposed that, unless all cells which had incorporated [³H]-TdR divided, the CV of the grain count would progressively increase, if cells with high grain counts refused to divide at the appropriate time during the experiment; in pulmonary metastases of the Ehrlich ascites tumour, there was no effective change in the CV of the grain count as the average grain count fell to low levels, indicating that *no* decycling divisions had occurred, and that $I_P$ was 100 per cent throughout the growth period. It is of course possible to confine the grain count distribution to proliferative cells only by counting grains exclusively over metaphases (Mendelsohn 1963).

On the other hand, it is possible to make use of the fact that the method is sensitive to decycling divisions to make estimates of the proportion of cells which are decycling in a population, the *decycling probability* (Morley and Wright 1972). Figure 3.24 shows the initial interphase grain count distribution in the seminal vesicle of the castrate mouse prostate complex stimulated with testosterone at one hour after [³H]-TdR injection; FLM analysis indicates that the cell cycle time is 17 hours in this system, and Fig. 3.24 shows the observed grain count distribution at 17 hours after [³H]-TdR, compared with the theoretical grain count distribution, calculated assuming that all cells halve their grain count at mitosis. Comparison of the two distributions indicates that about 65 per cent of the cells have decycled after the first division, giving a decycling probability of 0.65. The maintenance of such a decycling probability throughout the androgen-induced proliferative response means that the cell proliferation will eventually be curtailed, and this indeed is an example of *induced DNA synthesis* where the proliferative response can be simulated assuming a decycling probability model (Appleton *et al.* 1973).

Finally, one mundane practical point which must be remembered is that, with each cell division, the grain count will become progressively less as the label

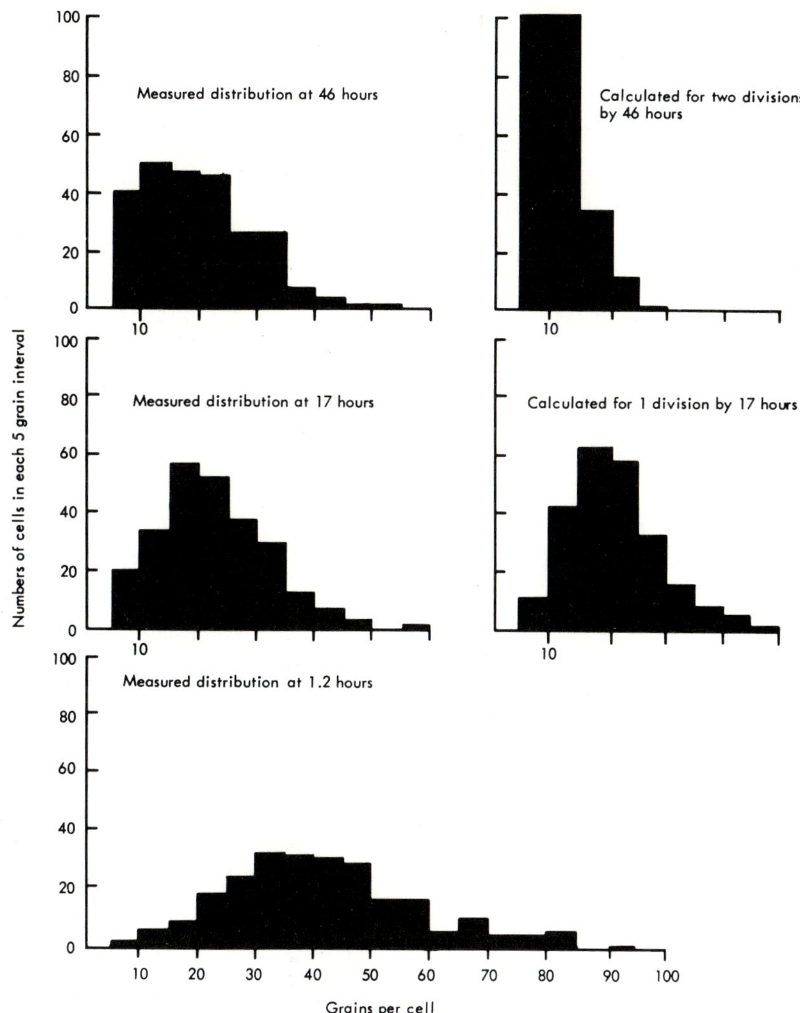



Fig. 3.24.  Grain count distributions in the castrate mouse seminal vesicle after testosterone stimulation. Animals were left for 14 days after castration before androgen challenge; the measured distribution is shown at 1.2 hours after [³H]-TdR injection. Thereafter, the calculated distribution is compared with the measured distribution at 17 and 46 hours after [³H]-Tdr. The cell cycle time was measured at 17 hours, so the time between successive divisions could be calculated. The difference between the two distributions gives the proportion of cells which have decycled at each division, from which the *decycling probability* can be calculated. (Redrawn from Morley and Wright (1972).)

becomes diluted with successive divisions, and the point will be reached, sooner or later, where labelled cells cannot be distinguished from background. Naturally, if the initial grain count is reasonably high, this might not happen to any great extent until after several divisions have occurred; but, even when the grain count is high, there will be some cells, at the beginning or the end of the *S* phase, which will be lightly labelled, and at division may dilute below the grain count threshold; if this occurs to any great extent, the grain count halving time could be prolonged by it. For this reason, some workers have confined their counts to the heavily labelled cells only (Clarkson *et al*. 1967). In these long-term experiments, we should also remember the possibility of [$^3$H]-TdR reutilization.

In summary, we could say that this method has not had much application in normal epithelia, but has been used, with some effect, in tumours, particularly human tumours (Shirakawa *et al*. 1970), where limitations to the number of biopsies available exist. The potential of the measurement of the decycling probability should be remembered.


### 3.7. The double labelling method

This method has come to mean a cell kinetic analysis carried out using two applications of injections of isotope, either [$^3$H]-TdR and [$^{14}$C]-TdR, or two injections of [$^3$H]-TdR. Obviously, there are no problems in detecting labelled cells when [$^3$H]-TdR only is used, but if both $^3$H- and $^{14}$C-labelled thymidine are employed, the two varieties of labelled cells must be identified and differentially counted. Since $^{14}$C has a much greater penetrating power than $^3$H, there needs to be a thicker emulsion to demonstrate the higher energy particles. In earlier work with one thick layer of emulsion (Wimber and Quastler 1963), it was only possible to detect $^3$H-only labelled cells from all $^{14}$C-labelled cells, because $^3$H labelling could not be distinguished from the widely dispersed $^{14}$C grains. Originally introduced by Baserga (1962), and championed in epithelia by Schultze *et al*. (1976), is the practice of *double emulsion autoradiography*, where two layers of emulsion are applied, separated by a layer of celloidin. The first layer is applied separately, and developed at a time optimal for [$^3$H]-TdR, then the second layer of emulsion is added, and exposed separately to catch the $^{14}$C grains. The microscope objective is focused on the first layer, and then on the second, and it should be possible to distinguish nuclei which have incorporated $^{14}$C and $^3$H only, or those with both.

In the hands of careful investigators, the method has given interesting and reliable results (see Schultze *et al*. 1976), and was used by them to compare fluxes into and out of the *S* phase, so that, while the overall age distribution in the intestinal crypt proved to be rectangular, the lower portion of the crypt shows an exponential age distribution.

Practically speaking, [$^3$H]-TdR is usually given first, followed at some time later by the injection of [$^{14}$C]-TdR. The usual procedure is for the interval $t$ between the injections to be $< t_{G_2}$, so that the system is not contaminated with $^3$H-labelled cells which have divided (Fig. 3.25). In this instance there will be (i) $^3$H-only labelled cells occupying $G_2$; (ii) $^{14}$C-only labelled cells in early *S*;

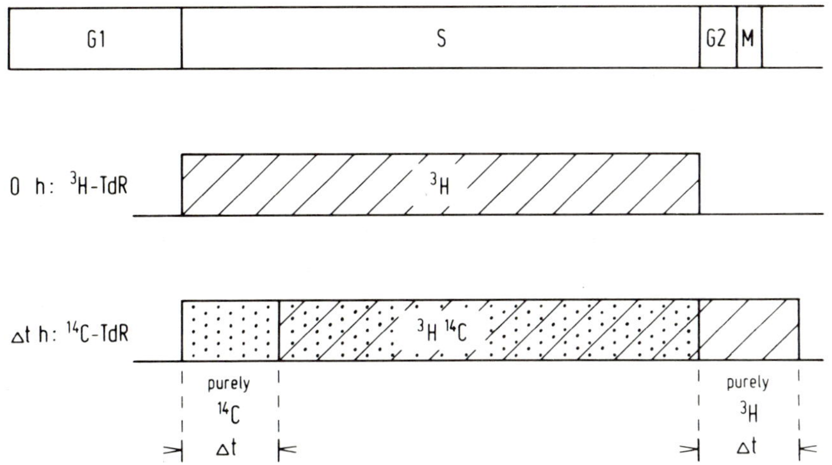


Fig. 3.25. A diagrammatic representation of the double-labelling method. At the top is shown the respective phases of the cell cycle. When an injection of [³H]-TdR is given at 0 time, it labels all the cells in the $S$ phase; at $t$ time later (usually one hour), [¹⁴C]-TdR is given, which creates the respective labelled populations shown at the bottom of the figure, which can be distinguished using double-emulsion autoradiography. (Courtesy of Professor B. Schultze.)

and (iii) double-labelled cells in the later part of the $S$ phase. Accurate discrimination between the different types of labelling is only possible if the activity ratio of [³H]- to [¹⁴C]-TdR is sufficiently high; this is because of the necessity of distinguishing spurious grains produced in the first emulsion by the $\beta$-particles from the ¹⁴C, and those which are due to genuine [³H]-TdR labelling. Schultze *et al.* (1976) have described an experimental approach to this problem.

If double-emulsion autoradiography is used successfully, and ¹⁴C- and ³H-labelled cells can be distinguished, assuming a rectangular age distribution

$$t_S = \frac{I_S[^{14}C] \cdot t}{I_S[^3H]}$$

enabling $t_S$ to be calculated. If double-emulsion autoradiography is not available, ³H-only cells can be distinguished from all ¹⁴C-labelled cells, and the ³H-only labelled fraction represents $S_{\text{efflux}}/t$, enabling the $S_{\text{efflux}}$ to be calculated. If we are prepared to assume that $S_{\text{influx}} = S_{\text{efflux}}$, or if we reverse the injection schedule to measure $S_{\text{influx}}$ itself, this equals $r_S$ and

$$t_S = \frac{I_S}{r_S} . \tag{3.24}$$

In the original injection schedule, the ³H-only labelled cells represent the cells which have left $S$ in the time $t$, and hence the rate of exit ($S_{\text{efflux}}$) from the $S$

phase can be determined. The comparison of $S_{\text{influx}}$ and $S_{\text{efflux}}$ is important in the experimental approach to the age distribution in the intestinal crypt (see Chapter 19) and also in the investigation of the mechanisms underlying circadian variation in proliferative indices (see Chapter 9).

If the age distribution is non-rectangular, then it may be necessary to correct for the overestimate in $t_S$ which is likely to occur. This error will only be large if the cell cycle time is short or $t_S$ is long compared with the cell cycle time, so that there is a sharp fall in the ordinate on the age distribution diagram over the interval $t$ between injections. In epidermis, for example, where $t_S$ is about 8 hours and $T_C$ about 110 hours, any correction factor will be negligibly small.

The interval $t$ is itself amenable to manipulation; Wimber and Quastler (1963) explored the possibilities of allowing $t$ to increase until all $^3$H-labelled cells had moved out of the $S$ phase before the $[^{14}$C$]$-TdR is given, and theoretically speaking, a complete cell cycle analysis is possible, but unfortunately the randomization which occurs after the injections does not make this procedure practically feasible. However, in slowly proliferating tissues such as the epidermis, putting $t$ at $<t_{G_2}$ does not make for a large number of $^3$H-only labelled cells, if $t_{G_2}$ is about 2-3 hours and $T_C$ around 110 hours (see Table 7.2, p. 258). A useful way of increasing $t$ without accruing excess $^3$H-only cells due to division is to give a metaphase arrest agent at an appropriate point to accumulate such cells in mitosis, in which case $t$ can safely be increased. This method has been used, not with $[^{14}$C$]$- and $[^3$H$]$-TdR, but with two injections of $[^3$H$]$-TdR (see below), in squamous epithelia (Tvermyr 1972).

A simpler method of approaching the method, without using double isotopes or double-emulsion autoradiography, is to use two injections of $[^3$H$]$-TdR, separated by a time interval $t < t_{G_2}$, or $>t_{G_2}$ if a metaphase arrest agent is used as well (Tvermyr 1972; Hartmann and Møller 1978). There are also two alternatives here; it is possible to give two differing doses of $[^3$H$]$-TdR—say 1 $\mu$Ci/g body weight, and 10 $\mu$Ci/g body weight, and to attempt to distinguish the two labelled cohorts on the basis of the differing grain counts. If the radiotoxicity is a worry *in vivo*, as in human studies, then the second procedure can be carried out *in vitro*, after explantation. How wise this method is, is a different matter: as we have seen (section 3.1), the grain count is not only a function of the specific activity or dosage of the $[^3$H$]$-TdR given, but the position of the cell in the $S$ phase is also critical; those cells at the beginning and the end of the $S$ phase show a low grain count. To avoid any confusion between the two cohorts it would be necessary to demonstrate that the grain count distributions of the 1 $\mu$Ci and 10 $\mu$Ci doses are completely distinguishable, with little or no overlap in the distributions; this is possible (Barlow 1984), but is rarely done.

The second alternative is a simplistic procedure which has usually been applied to human tissues, and notably to epidermis (Ralfs *et al.* 1981*a,b*); two sites are selected, and both are injected (intradermally in this example) with $[^3$H$]$-TdR; one hour later, one site is biopsied, and the other is injected with

[$^3$H]-TdR, and is itself biopsied one hour later. The difference between the labelling indices gives the rate of entry into the $S$ phase, providing $t_{G_2}$ is greater than two hours, as it appears to be in human epidermis, even in proliferative skin disease. There are no special precautions to be taken, but there is the problem that only a small increase in $I_S$ is being sought upon what might be a high initial reading (cf. psoriasis, where $I_S$ = 10 per cent). It is also necessary to make the usual assumptions about intradermal injection (see Chapter 6, section 1) and [$^3$H]-TdR not affecting cell cycle traverse.

The double-labelling method has given excellent agreement with results for $t_S$ from FLM curves (Schultze *et al.* 1972), but it must be stressed that in many reports the values obtained for $t_S$ are short (cf. Galand *et al.* 1967*b*; Harriss and Hoelzer 1971; Dyson and Heppleston 1976), and the reason for this is not at present evident, unless we entertain faulty technique regarding the differentiation of the two labelled cohorts. We should also note that a double [$^3$H]-TdR method applied to human epidermis gives $r_S$ values which are much higher than $r_M$ values measured in the same tissue, and also in the same individual (Ralfs *et al.* 1981*a*). Few of the published papers on this topic contain any internal check on the methodology (Steel 1977), such as plotting the incidence of $^3$H-only cells against the duration of the separating time interval $t$; Lala *et al.* (1965) could demonstrate a straight line passing through the origin, which however was not present in the results of Harriss and Hoelzer (1971), whom Schultze *et al.* (1976) criticized on the basis of inadequate attention to the grain count threshold and activity ratio. In particular, Harriss and Hoelzer (1971) found 4 per cent of $^3$H-only labelled cells present in double emulsion autoradiography of leukaemic cells after *only [$^{14}$C]-TdR* had been given, but apparently did not consider the grain count threshold, particularly in the first emulsion.

On the other hand, some results obtained in the human epidermis (Pullman *et al.* 1974) give longer values for $t_S$ than those derived from the FLM curve; again the reason is not clear, but it should be noted that the work of Pullman *et al.* does involve an *in vitro* step, which may be relevant.

In conclusion, we might say that this method has distinct advantages, in that only one sample is taken, and a good deal of information can be extracted, namely the $S_{\text{influx}}$, $S_{\text{efflux}}$, and $t_S$ itself. However, it must be conceded that great attention to autoradiographic technical detail, the control of emulsion thickness and the $^3$H:$^{14}$C activity ratio are mandatory if misleading results are not to be obtained.


## 3.8. Integrating microdensitometry and flow cytometry

The recognition that certain stains react stoichiometrically with DNA has opened many possibilities in cell population kinetics. In earlier days the stain used was Feulgen, and it was then necessary to laboriously measure the amount of light absorbed by each stained nucleus, at a specific wavelength, using an integrating microdensitometer, and carefully masking each nucleus; some cells with a known 2C DNA content, such as a resting lymphocyte, must also be measured. In this way, a form of distribution of DNA content of cells in any population can be built up (see Fig. 3.26), from which it is theoretically possible, at least, to derive phase fractions of cells in the cell cycle.

Even so, there is usually some difficulty in analysing such a distribution, because of the indigenous overlap between cells in the several phases, introduced by the presence of $S$ phase cells with intermediate DNA contents. It is usual to exclude such cells by an autoradiographic step, recording the site of each labelled cell on the slide, and then removing the silver grains by bleaching out with potassium ferricyanide (Cooper and Wickramsinghe 1969). In a population where all cells are proliferating, the number of cells in all phases of the cell cycle can be identified; $S$ phase cells by autoradiography, $M$ cells in mitosis, and $G_1$ and $G_2$ cells can be distinguished by microdensitometry. Therefore the age distribution of the population can be constructed, but only the relative durations of the cell cycle phases can be discerned, unless it is possible, using some other method, to measure, say the duration of the $S$ phase, in which case the absolute durations of the other phases can be calculated.

Now this is the ideal case, but where there are non-proliferating cells, there is no real way of dissecting the $G_1$ peak, which, in any tissue, may contain proliferating $G_1$ cells, $G_0$ cells, and also terminally differentiated cells with a $G_1$ DNA content. The only way to minimize this error is to confine the measurement to the proliferative portions of any cell population, such as the basal layer(s) of squamous epithelium or the lower crypt area in the intestine, but this introduces a further complication: for optimal results, the densitometry should be done on whole nuclei, either in touch preparations (which is only really possible in a parenchymatous organ such as the liver), or in squash preparations. To delineate a proliferative compartment accurately, it would be necessary to use Feulgen stained sections, in which case, even if so-called optimally sectioned nuclei are used, the problems of nuclear sectioning and section thickness variation obtrude; even so, Grove (1979) has been forced to this procedure to obtain meaningful results in human epidermis. The application of the technique to tumours, particularly human tumours, is limited by the degree of aneuploidy which might be found in the cell population.

This method has found little application in epithelial cell kinetics (apart from the liver), mainly because of its extremely laborious nature when it is combined with autoradiography, which it must be to get reasonable results. There is the added problem that, because of the work involved, it is not possible to count more than a few hundred nuclei at most. This difficulty, at least, was solved by the introduction of the flow cytometer; individual cells, stained with a fluorescent dye such as ethidium bromide or propidium iodide, are passed in suspension through an exciting laser beam. The resulting fluorescence is collected through a lens, and observed by a photomultiplier; a multi-channel pulse height

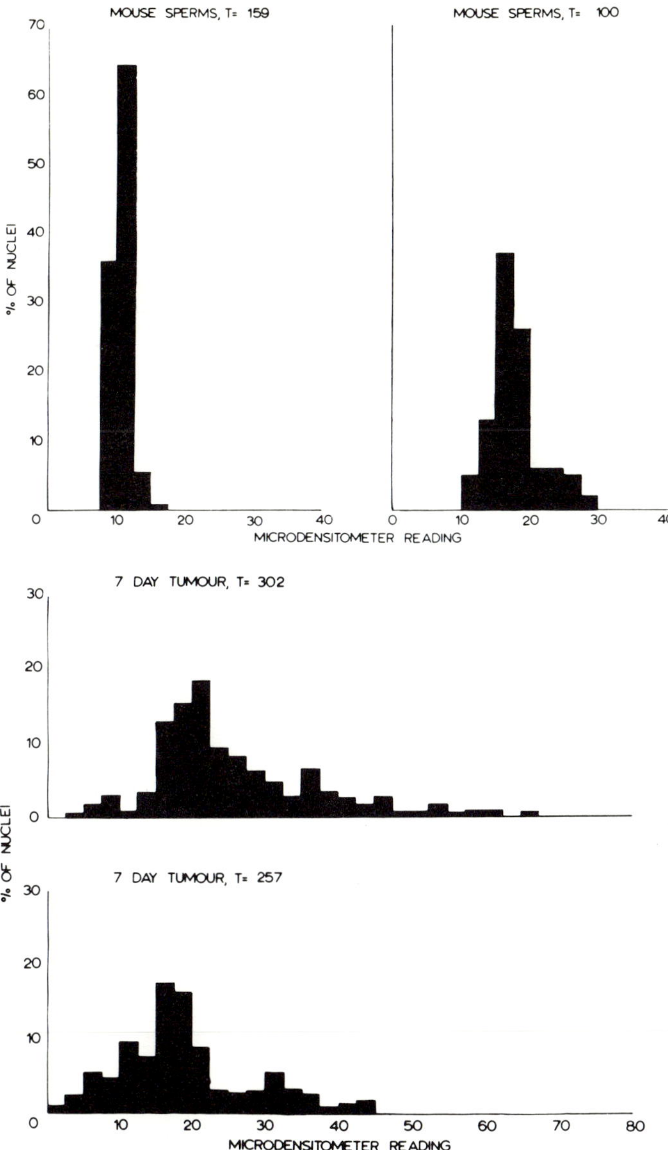



Fig. 3.26. DNA content histograms prepared by the use of an integrating microdensitometer (Vickers). The histogram for mouse spermatazoa is necessary for the calculation of the modal DNA value for the 2N mouse tumour cells, since the microdensitometric measurement is arbitrary. There is a prominent $G_1/G_0$ peak, and in both tumour histograms a $G_2$ peak can also be discerned. Even at seven days after transplantation there is a considerable degree of aneuploidy and polyploidy, which increases with tumour age. The nuclei on the left-hand side of the histogram are presumably measurements on nuclei fragmented by the smear procedure (Feulgen nuclear staining).

analyser then records the height of the pulses, and the data is expressed as a distribution of cellular DNA contents (Fig. 3.27). The attractiveness of this methodology is based on (i) a remarkable *precision* of measurement, with CVs of fluorescence or scatter as low as 1 per cent; (ii) *speed*, in that between 500 and

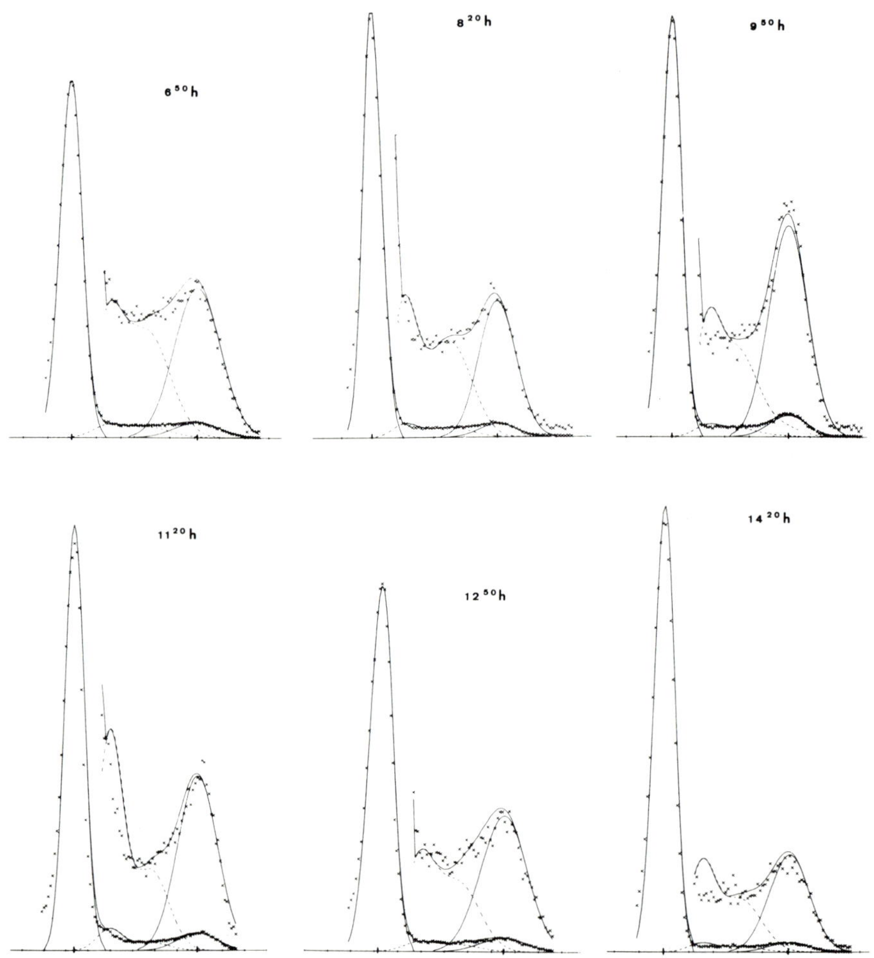



Fig. 3.27. Typical DNA histograms prepared by flow cytometry of the stratified squamous epithelium of the hamster cheek pouch. The cells are disaggregated by an appropriate dispersion procedure, stained with propidium iodide, and their individual DNA content measured. The ordinate is the cell number and the respective channel number is plotted on the abscissa. There is a prominent $G_1/G_0$ peak, and a less well-defined $G_2$ peak, while the cells with intermediate DNA content are in the $S$ phase. Also shown are magnifications of the $S$ and $G_2$ portions of the curve, while the lines represent attempts to fit the data and to calculate the numbers of cells in the respective phases. The histograms are for the indicated times of the day. (From Møller and Larsen (1978), by courtesy of Dr U. Møller.)

1000 cells can be analysed per second; (iii) the ability to measure *more than one parameter* in the same cell suspension, by exciting at more than one wavelength when cells are stained with a dye, such as acridine orange, which fluoresces with DNA and RNA at different wavelengths (Fig. 3.28), or by using two different stains showing different properties in a dual laser instrument; (iv) the ability to *sort* cells, for example in different parts of the $S$ phase, from the remaining population, for further study of their properties; (v) the method, in combination with a metaphase arrest agent, does allow the *measurement of phase durations* for cells in culture (Barfod 1980).

Thus, even in its simplest form, the method should allow the rapid and precise determination of the DNA distribution in a monodispersed cell population, which might be thought to be a powerful method for the study of epithelia. However, there are several problems which have tended to decrease the applicability of the technique; the first of these is the achievement of *monodispersion*; this is a major difficulty in epithelia. By and large, the only groups to use the method in experimental epidermal studies are Clausen and his colleagues in Oslo, who have applied the Laerum and Boyum (1969) method of differential trypsinization to separate basal and suprabasal differentiating cells with minimal contamination, and Møller and her colleagues in Copenhagen, who have used ultrasonication to prepare single-cell suspensions in the hamster cheek pouch. However, several methods are now available in human epidermis, which have given some insight into the proliferative status of the epidermis in psoriasis and the effects of therapy. In the intestine, however, the achievement of a single-cell suspension of crypt cells dissociated from villus cells is a more difficult matter, and to date only one or two workers (Pallavacini 1982) have been able to separate the two populations, by a combination of DNA content and sizing out, and have applied the method to analyse the effects of ara-C on the mouse small intestine.

The analysis of the DNA histogram usually involves some form of computer simulation of an assumed mathematical model; the $G_1$ and $G_2$ peaks are usually well defined, allowing their relative proportions to be accurately measured, but the assessment of the number of cells with intermediate DNA contents, which overlap the $G_1$ and $G_2$ peaks, is a more difficult matter.

As with the automatic analysis of FLM curves, there are now several computing methods for the analysis of flow cytometric (FCM) data. These have ranged from simple planimetric methods of measuring the areas under the various portions of the curve, to computer-based parametric models where the distribution of cells in the $S$ phase is described by second-degree polynomials (Dean and Jett 1974), or by the exponential function of a polynomial (Christensen *et al.* 1978). There are also non-parametric methods based upon dividing the $S$ phase into subcompartments, in which the numbers of cells is estimated; the distribution in each subcompartment is Gaussian (Fried 1976), or lognormal. The lines are fitted either by least squares or by maximum likelihood. An alternative approach has been used by Gray (1976) and Watson (1977), who

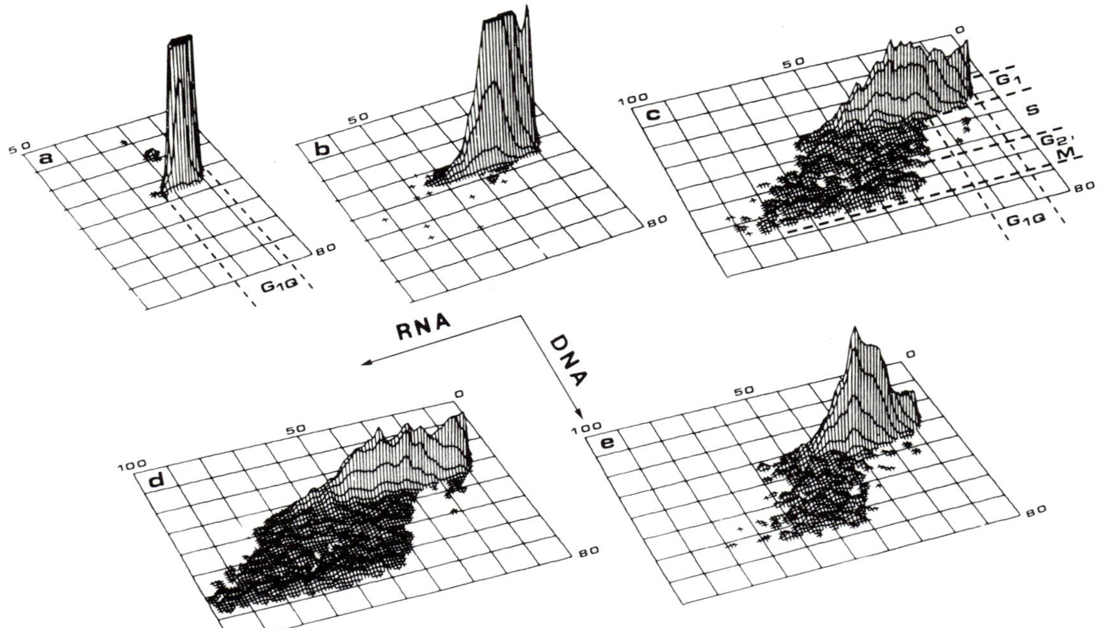


Fig. 3.28.  Three-dimensional histograms showing changes in lymphocyte properties after stimulation with phytohaemagglutinin (PHA). The vertical dimension is the number of cells. Cells were stained with acridine orange and examined in a FC 200 Cytofluorograf (Ortho). Simultaneous measurements of cellular DNA ($F_{530}$, measured in a band at 515–757 nm) and RNA ($F_{>600}$), allows the analysis of the procession of cells from the quiescent compartment ($G_{1Q}$) into the cell cycle (indicated by an increase in cellular RNA) and their further progression through the cell cycle, through $S$, $G_2$, and $M$ (from changes in $F_{530}$). In (a) quiescent cells ($G_{1Q}$) are characterized by a $G_1$ DNA content, and a low RNA content; in (b), at 24 hours after PHA stimulation, about 50 per cent of cells have RNA contents above the $G_{1Q}$ level while a few lymphocytes have entered the $S$ phase; at 48 hours (c), 80 per cent of cells have increased RNA content, and numerous cells are in the $S$, $G_2$, and $M$ phases of the cell cycle; at 72 hours after PHA, in (d), stimulated cells have maximal RNA content, and there is the largest proportion of cycling cells; after 96 hours (e) cells are losing RNA content and re-entering the quiescent $G_{1Q}$ compartment. (From Darzynkiewicz *et al.* (1981), by courtesy of Dr Z. Darzynkiewicz.)

have analysed the DNA histogram by simulations based on the age distribution of asynchronously growing or distorted cell populations. There are also computer models for the estimation of changes in the fractions of cells in $G_1$, $S$, and $G_2 + M$ phases in the sequential analysis of DNA histograms (Baisch *et al.* 1982). Baisch *et al.* (1982) have carried out a survey of the various methods of analysing FCM histograms, employing several modifications of the basic types of computer methods mentioned above, by comparing the performance of the several models against simulated histograms with known phase populations in both perturbed and unperturbed situations, and also by comparing $S$ phase estimates with autoradiographic $S$ phase measurement. In general, there was a tendency for most methods to give $G_1$ fractions that were too high, and $S$ phase fractions that were low, and the errors ranged from 10 to 30 per cent. There was an interesting tendency for the deviation to increase as the coefficient of variation of the peaks increased. Moreover, when the $S$ phase fraction in the simulated histograms was high, and the $G_1$ fraction was low, all methods behaved badly. There was also quite prominent deviation from the $S$ phase index as measured by autoradiography; one reason for this might be the non-labelling of some $S$ phase cells (see Chapter 7, section 1).

There remains the problem, referred to above in connection with integrating microdensitometry, of dissecting the $G_1$ peak. For example, in epidermis, a suspension of basal cells would contain cells in $G_1$, post-mitotic maturing cells, dendritic cells with a $G_1$ DNA content, and possibly $G_0$ cells, which are potentially proliferative. A singular advance has been made by Darzynkiewicz *et al.* (1979*a*), using one stain: two-colour fluorescence acridine orange. RNA is removed, the DNA is partially denatured by dilute acid treatment, and, after staining, the two colours are used to assess changes in chromatin denaturability; this method allowed separation of PHA-stimulated lymphocytes, not only into $S$, $G_2$, and $M$, but also into $G_1$ and $G_0$ compartments, based on plotting the DNA quantity (the sums of the red and the green fluorescence), and the chromatin stability to acid denaturation (the ratio of the colours). In Chinese hamster cells Darzynkiewicz *et al.* (1979*a, b*) have produced evidence indicating, not only that the RNA content (and probably the number of ribosomes per cell) correlates with the duration of $G_1$ and the $S$ phase, but also, through studies of PHA-stimulated lymphocytes, that the RNA content correlates with the position of the cell in the cell cycle, with $G_0$ cells containing small amounts of RNA (Fig. 3.28); dual parameter FCM therefore offers a further approach to the detection of $G_0$ cells. It can be seen that this development augurs well for the future analysis of non-proliferating cells in epithelia.

We should also add the proviso that, if we do prepare cell suspensions from epithelial tissues which have quite a sophisticated kinetic microarchitecture, evident in both epidermis and gut epithelia, it is difficult to know what sort of weight to put on the overall results, particularly as differential responses can occur in the several parts of the kinetic system (cf. Al-Dewachi *et al.* 1977;

and see Chapters 19 and 22), which would not be detected in the resultant given by FCM methods.

We could summarize by saying that this method has yet to make its impact on epithelia: where will the eventual (and inevitable) impact be? As Iversen has said, it is always difficult to prophesy, especially about the future, but we would suspect that it is with the analysis of perturbed epithelial cell populations, such as circadian variation and recovery from cytotoxic damage, that the method will come into its own. We can also expect advances in the biology of non-proliferating cells from the work with multiparameter histograms. It is a growing point to watch.


## 3.9. The measurement of derived kinetic parameters

### (i) *Techniques for measuring the growth fraction*

We have already discussed the theoretical and operational definition of the term *growth fraction* (see Chapter 2), and agreed that such a parameter can only have meaning if the proliferative cells are carefully defined within the context of the population under consideration. In epithelia we are in fact on stronger ground in considering non-proliferating cells, since, although we may argue about the semantic and experimental distinction between $G_0$ and $G_1$ cells, there is little doubt that proliferative compartments such as the intestinal crypt and the basal layer of the epidermis contain cells which have irreversibly left the cell cycle and are waiting to migrate; these cells are not in the growth fraction, and their size becomes important in the understanding of the system. Moreover, if $G_0$ cells are a reality, then cell populations such as the castrate prostate or the salivary gland acinar cells are prime candidates.

There are basically three practical methods for the estimation of the growth fraction, two of which derive from the FLM curve. The first of these requires a good two peak FLM curve, so that both $t_S$ and $T_C$ can be measured accurately. Then, since

$$I_P = N_C/N$$

(where $N_C$ and $N$ are the fraction of cycling cells and the total cell population, repectively), in terms of the labelling index.

$$I_P = (N_S/N)/(N_S/N_C)$$

where $N_S$ is the number of cells in the $S$ phase, $N_S/N$ is thus the actual counted labelling index, and $N_S/N_C$ is the *theoretical labelling index*, the labelling index we would expect to find if all cells in the population were proliferating. In more familiar form this is

$$I_P = I_{S\,\mathrm{expt}}/I_{S\,\mathrm{theor}} \tag{3.23}$$

and, when all cells are cycling, $I_{S\,\mathrm{expt}}$ is equal to $I_{S\,\mathrm{theor}}$. $I_{S\,\mathrm{expt}}$ is found from observation, while $I_{S\,\mathrm{theor}}$ is calculated from $t_S$ and $T_C$. For a rectangular age distribution, $I_{S\,\text{theor}} = t_S/T_C$, but if an exponential age distribution is suspected and we have some information about its form, we can correct for this as well (see Chapter 6, section 7).

In many ways this is probably the best method we have for a direct calculation of the growth fraction, yet it does present several problems in application, and it does share the limitations of the FLM method itself, which have already been listed (section 3.4). In the context of the growth fraction the most important limitation of the FLM curve is its insensitivity to the presence of non-proliferating cells, induced because of the inherent bias towards those cells which display high mitotic indices. Because of this, most FLM derived $I_P$ estimates may well be measurements of the fraction of the population which have relatively short cycle times, while the non-growth fraction is that portion at the extreme tail of the distribution of cell cycle times; in other words, we are defining 'proliferative cells' as those whose growth pattern is defined by the FLM curve.

Furthermore, eqn 3.23 gives $I_P$ as a quotient value, which will affect its precision, since

$$(\text{CV } I_{S\,\text{expt}}/I_{S\,\text{theor}})^2 = (\text{CV } I_{S\text{expt}})^2 + (\text{CV } I_{S\text{theor}})^2$$

approximately, and in the few instances where the standard deviation of the growth fraction has been calculated, this is usually large (Wright 1980c).

Appleton and Sunter (1979) have pointed out an important and serious correction which should be made in calculating the growth fraction from eqn 3.23: this stems from the assumption, implicit in eqn 3.23, that the variability in the cell cycle time and its component phase durations is negligible. This point has not been generally appreciated, and is worth an examination. We have seen that the age distribution $y(x)$ is given by

$$y(x) = \frac{1+p}{p} \frac{\ln(1+p)}{T_C} \exp\left[-\frac{\ln(1+p)}{T_C} x\right] \tag{3.16}$$

and when $p \to 1$ we have an exponential age distribution, and $p \to 0$ gives a rectangular age distribution (Chapter 2, section 3). We of course know that $T_C$ is subject to variability, and Appleton *et al.* (1977) showed that, if the distribution of cell cycle times follows a gamma distribution, with mean $n/\lambda$ and CV of $n^{-\frac{1}{2}}$, $g(x)$, the distribution of times to mitosis, is given by

$$g(x) = \frac{1+p}{p} \ln(1+p) \frac{\lambda^n}{\Gamma(n)} \int_0^\infty (t+x)^{n-2} \exp\left\{\frac{-t\ln(1+p)}{t+x}\right\} \exp\{-\lambda(t+x)\}\, dt$$

(see Chapter 6, section 7). If the age distribution is rectangular, then

$$g(x) = \frac{n\,e^{-nx}}{\Gamma(n)} \int_0^\infty (t+nx)^{n-2}\, e^{-t}\, dt \tag{3.25}$$

or, for an exponential age distribution,

$$g(x) \;=\; 2\ln 2\,\frac{n\,e^{-nx}}{\Gamma(n)}\int_0^{\infty}(t+nx)^{n-2}\exp\left\{\frac{-t\ln 2}{t+nx}\right\}\,e^{-t}\,dt. \qquad (3.26)$$

The theoretical labelling index is given by

$$I_{S\,\mathrm{theor}} \;=\; \int_{(t_{G_2}+t_M)/T_C}^{(t_S+t_{G_2}+t_M)/T_C} g(x)\,dx. \qquad (3.27)$$

Eqns 3.25, 3.26, and 3.27 can be evaluated for a range of values of $I_{S\,\mathrm{theor}}$ and of $\mathrm{CV}_{T_C}$. Hence it is assumed that $t_S$, $t_{G_2}$, and $t_M$ remain constant (a deterministic $B$ phase) and that variability in $T_C$ is produced only by the variability in $t_{G_1}$.* A correction factor can then be calculated by dividing the value given for $I_{S\,\mathrm{theor}}$ by eqn 3.27 by the value calculated for $I_{S\,\mathrm{theor}}$ from $t_S/T_C$, or from eqn 2.9 (p. 27). These values are shown, for both rectangular and exponential age distributions, in Table 3.8. The tabulated values can be used to multiply uncorrected values of $I_{S\,\mathrm{theor}}$ to obtain corrected values.

Table 3.8.    Approximate correction factors by which to multiply the $I_{S\mathrm{theor}}$ when the coefficient of variation of the cell cycle time is known. (After Appleton and Sunter (1979))

| | | Coefficient of variation of $T_C$ (%) | | | | | | |
|---|---|---|---|---|---|---|---|---|
| | | 20 | 30 | 40 | 50 | 60 | 70 | 80 |
| Initial estimate | 10 | 1.04 | 1.10 | 1.19 | 1.32 | 1.49 | 1.66 | 1.80 |
| of $I_{S\mathrm{theor}}$* | 20 | 1.04 | 1.10 | 1.18 | 1.28 | 1.36 | 1.40 | 1.42 |
| | 30 | 1.04 | 1.09 | 1.15 | 1.19 | 1.20 | 1.19 | 1.15 |
| | 40 | 1.04 | 1.08 | 1.09 | 1.08 | 1.05 | 1.00 | 0.95 |
| | 50 | 1.03 | 1.03 | 1.01 | 0.97 | 0.91 | 0.85 | 0.79 |
| | 60 | 0.99 | 0.95 | 0.91 | 0.85 | 0.79 | 0.72 | 0.67 |
| Initial estimate | 10 | 1.05 | 1.12 | 1.24 | 1.41 | 1.62 | 1.82 | 1.94 |
| of $I_{S\mathrm{theor}}$† | 20 | 1.06 | 1.14 | 1.25 | 1.37 | 1.46 | 1.50 | 1.51 |
| | 30 | 1.06 | 1.14 | 1.22 | 1.26 | 1.27 | 1.24 | 1.20 |
| | 40 | 1.06 | 1.12 | 1.14 | 1.13 | 1.09 | 1.04 | 0.98 |
| | 50 | 1.05 | 1.06 | 1.04 | 1.00 | 0.94 | 0.88 | 0.82 |
| | 60 | 1.00 | 0.99 | 0.94 | 0.88 | 0.82 | 0.75 | 0.70 |

*Where the $I_{S\mathrm{theor}}$ is calculated from $t_S/T_C$, i.e. for a rectangular age distribution.
†Where $I_{S\mathrm{theor}}$ is calculated as $\exp(t_S\ln 2/T_C-1)\cdot\exp(t_2\ln 2/T_C)$.

*This assumption is inconsistent with a gamma distribution for $T_C$, but not with the general shape of the distribution (Appleton and Sunter 1979).

Since the correction factors may vary from around 0.7 to almost 2, it is clear that if there is considerable variation in $T_C$, then considerable errors may be incurred, which may underestimate or overestimate $I_P$, depending on the magnitude of $I_{S\,\text{theor}}$. Few authors have yet taken cognisance of this correction factor, which will be more important in some epithelia than others; FLM data for the whole jejunal crypt of the rat indicates a $CV_{T_C}$ of about 20 per cent, so here the error will be small. However, in the rat colonic crypt, $CV_{T_C}$ values of up to 50 per cent are seen (Sunter *et al.* 1978*a,b*), and this therefore should be taken into consideration in calculations of the growth fraction. While some epidermal FLM curves suggest a small $CV_{T_C}$ (Hegazy and Fowler 1973*a*), these are probably the exception, and, certainly in human epidermis, albeit in the hyperproliferative skin disease psoriasis, $CV_{T_C}$ approaches 75 per cent and, with a $T_C$ of 55 hours and $t_S$ at around eight hours (Duffill *et al.* 1976), $I_{S\,\text{theor}}$ will have to be multiplied by 1.8 to correct for the large $CV_{T_C}$ encountered in this situation.

The second method of calculating $I_P$ from the FLM curve has already been mentioned, and is mainly used when a second peak is not obtainable. We have seen (section 3.4) that when successive waves of labelled mitoses damp out, the FLM tends to an equilibrium value, equal to $N_S/N_C$ when the labelled cohort is completely desynchronized; the proportion of cells labelled in each of the cycle phases and in the cell cycle itself is the same; in this instance we are looking at mitoses, and if the FLM curve stabilizes at 20 per cent, $I_{S\,\text{theor}}$ is then 20 per cent and $I_P$ is calculated from eqn 3.23. How accurate this method can be is another matter; successive divisions mean dilution of label, long-term experiments mean [$^3$H]-TdR reutilization, and each experiment must be judged upon its own merits.

Steel (1977), in the context of his definition of the growth fraction only in the framework of a model, has proposed that the optimal way to approach the calculation of $I_P$ from the FLM curve is to simulate the curve using such a model (see section 3.4), obtain the distribution of cell cycle times and of the component phase durations, and construct the age distribution diagram. Integration of this age distribution would then give a value for $I_P$. This method has the advantage that a real (rather than a relative) value for the labelling index does not have to be measured (a measurement more difficult than meets the eye in some epithelia, see section 3.1), but does introduce any worry that we may have had about non-applicability of the model, and indeed raises the uncomfortable fact that many different models can fit the same data.

A third method of approaching the growth fraction is via the continuous labelling curve; unfortunately, as we have seen, this is by no means simple, and we shall encounter numerous instances where investigators have assumed that the plateau value achieved on the continuous labelling curve is equal to the growth fraction (see section 3.5, and, for examples of the practice, especially Chapter 7, section 4). The problems with this method have already been discussed, but here we will introduce an approach which might be of use in the analysis of such curves.

We have seen that the continuous labelling curve is especially sensitive to the mode of cell loss in the population under consideration; theoretically, cells might be lost during any phase; in epithelia such as the epidermis and intestinal mucosa, cells appear to be lost from the population some time in $G_1$, after completing a final mitosis in the proliferative compartment. Cells then remain in the basal layer, or in the crypt, begin differentiation and emerge into the Malpighian layer or the villus as young maturing cells. We shall see that changing the assumption about the phase at which cells migrate makes considerable differences to the conformation of the continuous labelling curve. In tumours the process is more complex, as here, particularly in older tumours, there are likely to be large numbers of hypoxic cells, probably in the $G_0$ phase, from which cell loss by ischaemic necrosis is likely to be a major cell loss flux parameter; malignant cells accumulate genetic abnormalities with increasing time after transplantation, and cell death at mitosis becomes common; there is also the possibility of random cell loss in the cell cycle.

It is obvious that, in order to derive $I_P$ values from a continuous labelling curve, we must take some cognizance of mechanisms of cell loss. We have already given some consideration to the age distribution of populations which are subject to cell loss (see Chapter 2), and it is through these age distributions that it is possible to simulate continuous labelling curves. Here we are concerned with $I_P$, and the mechanisms of cell loss are further discussed below, but suppose we are faced with the analysis of a family of CL curves shown in Fig. 3.23, which were obtained for a transplantable mouse sarcoma at the given times after transplantation. For the several reasons given in section 3.5, we cannot proceed to read off the growth fraction from the plateau values, even if we were able to identify an inflexion point. Such curves need the formulation of an appropriate mathematical model; the principle of this model is shown in Fig. 3.22, and is due to N. R. Hartmann, and is fully described by Dombernowsky *et al.* (1974). There are two compartments, $P$ or proliferating cells, and $Q$, the non-proliferating cells. In the unperturbed tumour it is unlikely that there will be any $Q \rightarrow P$ recycling, and we need to know the proportion of $P$ cells at the stated times. There are three cell loss mechanisms: $Q$ cell (so-called 'pipe-line' cell loss), $\phi_Q$, cell loss at mitosis, $\phi_M$, and random cell loss in the cell cycle, $\phi_R$. The model is based on a set of integral equations, assumes exponential growth, and is parameterized by cell cycle parameters derived from FLM curves carried out at the appropriate times. Figure 3.23 also shows the fits to the data points, along with the fractional contributions from the three cell-loss pathways.

The program obviously needs a value for $I_P$, which will satisfy the continuous labelling curve and the cell cycle time values; these are given in Table 3.7, along with the values for $I_P$ calculated from the FLM curve using the methods given in eqn 3.23 above, and, since we also have measurements of $k_B$, and thus $t_{PD}$ from metaphase arrest data (see section 3.3), we can calculate $I_P$ from $T_C/t_{PD}$; this is because $k_B = I_P/T_C$, $t_{PD} = 1/(I_P/T_C)$, and hence $I_P/T_C = 1/t_{PD}$, and

$I_P = T_C/t_{PD}$; alternatively, the method can be regarded as being equivalent to eqn 3.23, but here we are dividing the experimental $T_C$ by the potential doubling time, or the 'apparent cell cycle time' (Aherne *et al.* 1977*a*), which should equal $T_C$ if all cells are proliferating. If $I_P < 1$, $t_{PD} > T_C$. This is not a method of choice, since, if we have a measurement of $T_C$ and its variance, we should probably use eqn 3.23 above. But if we also have reliable stathmokinetic measurements of the birth rate, this method provides an independent check on the growth fraction value calculated from the labelling index.

Table 3.7 shows that there is good agreement between the $I_P$ value calculated from the FLM curve, and the metaphase arrest method at all times after transplantation, but, as compared with these values, the continuous labelling method systematically overestimates the growth fraction. The assumption of exponential growth should not cause this disparity, since an exponential age distribution was also assumed in calculating $I_P$ from the FLM curve, and in fitting the metaphase arrest line; the fit to the continuous labelling curve is generally good, at all times, and if the $I_P$ calculated from the first two methods represents the actual position, then the errors incurred by the continuous labelling method lie either in the experimental method (animal stress, failure to label, etc.), or to defects in the analysis, particularly in the assumptions made in the formulation of the model.

Before concluding this section, we should note that there are several methods designed to measure the growth fraction in special situations, e.g. the Brown and Oliver (1968) method in epidermis (see Chapter 6, section 2), and the analysis of the labelling index distribution curve in the intestinal mucosa (see Chapter 17), which are considered when these particular epithelia are encountered. In tumours, the Pittsburgh group (Schiffer *et al.* 1976) has made considerable use of the PDP assay (the primer-dependent DNA polymerase assay), which can be done on touch preparations of tumour cells, and does appear to correlate well with tumour growth fraction measured by more conventional techniques. However, attempts to measure the growth fraction in epithelia using this method have not yet been successful, though few investigators have tried as yet (Weinstein *et al.* 1983).

In conclusion, we might say that the growth fraction in epithelial renewal systems has a definite meaning, and is not a mere semantic convention, introduced to account for those cells which make up the tail of the cell cycle time distribution curve; there are epithelial cells which are non-proliferative, and these do occupy the putative proliferative compartment, such as the epidermal basal layer and the intestinal crypt. These cells can be excluded by an appropriate calculation, as detailed above. The other part of the question, on the possibility of genuine $G_0$ cells in epithelia, is quite a different matter; we are resigned to the classification of cells in renewal systems as $G_0$ (see Chapter 2, section 7), but, as we shall see, it is most difficult to confirm their presence experimentally in renewing epithelia; we shall in fact argue that, in normal circumstances, there is no evidence for $G_0$ to denote temporarily decycled cells which are potentially proliferative, with either constant or stochastic transit times in the $G_0$ compartment.


(ii) *The measurement of cell loss*

Cell loss may have to be measured in three different situations: (a) *physiological cell loss*, say from the stratum corneum on the surface of the epidermis, or from the top of the intestinal villus; (b) *cell loss from a tumour*, either as part of the usual ageing process, and caused presumably by progressive lack of nutrients and oxygen, or as part of an experimental or therapeutic manoeuvre; and (c) the assessment of *clonogenic cell loss* after irradiation or cytotoxic chemotherapy, and this, of course, may be either in normal tissues or in tumours.

In the case of physiological cell loss from normal cell populations, it is true to say that this aspect has been seriously under-represented in comparison to measurements related to the production of cells. In the two major epithelial renewal systems, physiological cell loss is measured in a handful of studies only; in epidermis, Marks *et al.* (1980) have described an ingenious instrument which measures the number of cells desquamated from the surface after a standard period of 'scrubbing'; in the gut, measurements of DNA content in luminal perfusates are the usual techniques, but these are associated with a whole host of technical problems (see Chapter 17, section 3), and attempts to circumvent these have not been wholly successful (Clarke 1973).

Much more attention has been paid to the measurement of cell loss in tumours, and these should engage our attention, since there is also room for their applicability to normal growing epithelia (cf. Wright 1981). The basis of such measurements is the determination of the difference between the growth rate which would be expected from observations of the rate of cell production and the actual growth rate. The foundation of such a measurement is the construction of an accurate growth curve, and the measurement of the growth rate by an appropriate method (see section 3.11); the specific growth rate is then expressed in cells/cell/hour, either by blatantly assuming that mg/mg/hour can be equated with cells/cell/hour, or preferably by making a suitable correction. It then becomes necessary to measure the birth rate; this can be calculated from the FLM curve, making the usual assumptions (see Chapter 2, section 4), from $k_B = I_S/t_S$, or $\lambda I_S/t_S$ if $\lambda$ is known (see Chapter 2, section 4); or if we know $I_P$ and $T_C$, from $I_P/T_C$, or $\ln(1 + I_P)/T_C$. The assumptions and unknowns in these methods can be obviated by measuring $k_B$ directly and accurately from a metaphase arrest experiment (section 3.3).

Having measured $k_G$ and $k_B$, $k_L$, in cells/cell/hour, is readily found from

$$k_L = k_B - k_G$$

In Chapter 2, section 4, we introduced $\phi$, the cell loss factor, equal to $k_L/k_B$. Since $t_{PD}$ and $t_D$ are simply related to $k_B$ and $k_L$, we can calculate $\phi$ from

$$\phi = 1 - (t_{PD}/t_D) \tag{2.11}$$

$t_D$ is readily found from the growth rate (see section 3.11), while $t_{PD} = \ln2/k_B$.

We have already discussed the approach to the measurement of cell loss which is possible through computer analysis of continuous labelling curves. In fact, the model described in Fig. 3.22 is an adaptation of the earlier SAB models of Steel *et al.* (1966), and of Steel and Hanes (1971). We have already noted that the shape of the continuous labelling curve is very much dependent on the cell loss mechanism, and advantage can be taken of this to measure $\phi$ from such curves, using an appropriate model. However, as previously noted, since the pathway of cell loss determines the shape of the curve, we can make deductions about possible cell loss avenues. In Fig. 3.22 we have introduced $\phi_M$, cell loss at mitosis, $\phi_Q$, pipeline cell loss from decycled $Q$ cells, and $\phi_R$, random cell loss in the cell cycle. Table 3.7 summarizes the $\phi$ values calculated from a Gompertz analysis of the growth curve and a metaphase arrest experiment, and where $\phi$ is calculated from the modelling of the continuous labelling curves shown in Fig. 3.23.

As would be anticipated, $\phi$ values increase markedly from seven days onwards, until, at 21 days, we have $\phi$ values approaching 0.90; this means that, for every 100 cells that are born in the tumour, no less than 90 are lost, probably, in this particular tumour, by necrosis. While all methods agree that $\phi$ increases with time, it is interesting to note that the FLM and vincristine metaphase-arrest experiments agree very well (as was noted in the case of the growth fraction, see also Table 3.7), but the continuous labelling method underestimates $\phi$ compared with the first two methods. These results again suggest that stathmo-kinetic experiments give very comparable results, when done alongside [³H]-TdR techniques, for far less effort, and, in our opinion, is the method of choice whenever a $k_B$ value is needed. We note again that the continuous labelling method does not compare well; nevertheless, the same trend in $\phi$ values is evident.

It is perhaps instructive to look at the changing cell loss patterns; at seven days, cell loss appears to occur exclusively via the pipeline $Q$ cell mechanism. However, as time after transplantation proceeds, the avenues of cell loss apparently change, with increasing proportions of cell loss at $M$ and randomly in the cell cycle; this would reflect the increasing proportions of cells with aneuploid and polyploid nuclei that are known to occur in this tumour with age, and which therefore may not be able to complete mitosis successfully (Aherne *et al.* 1977a; Wright 1981). An increasing proportion of random cell loss could mean more rapid onset of necrosis with increasing age, possibly in the form of local vascular occlusion and loss of cycling cells. Before accepting such an analysis as more than an indicator, we must however remember that the values derived are largely dependent on the mechanisms of cell loss which are chosen; naturally there are several alternative cell loss avenues which could conceivably be incorporated into the model.

Thymidine analogues, which are not reutilized nearly so much as [³H]-TdR, have been used, mainly in tumours, to monitor cell loss. This is carried out either by the extraction of the DNA and measurement of its specific activity, or, because $^{125}$I or $^{131}$I are gamma emitters, measurement can be made serially in the same tumour by counting (usually mice) in a well counter; a good deal of accuracy has been claimed for this method, although reutilization does become a problem when there is extensive cell loss (Dethlefsen 1971). Investigators have not been very interested in this method as far as epithelia are concerned, although Dethlefsen (1971) did use the duodenum as a reference tissue, and Riches *et al.* (1976) have championed the use of $[^{131}$I]-UdR, mainly as an index of cell proliferation, in human prostatic explants.

## 3.10. The measurement of population size

Practically the entire content of the preceding sections on methodology has been devoted to the analysis of flux and time, but now we come to *size*. This is not to decry its importance, but it does reflect perhaps the relative amount of time and effort which has been devoted to population size measurement. It is difficult to understand this neglect, since, in the definition of any population, size is at least as important as flux and time, and, in many instances, for example in the investigation of hypotheses of growth control, it is absolutely essential to be in a position to measure the functional compartment size with both accuracy and precision. Moreover, such measurements are critical for the correct appreciation of many kinetic parameters: a transit rate, measured in a compartment such as the villus epithelium in the small intestine, is dependent, not only on the influx into the compartment, but also upon its size. The rate of cell production in a proliferative unit such as the intestinal crypt is also dependent, not only on the cell cycle time and the growth fraction, but also on the size of the crypt population (see Chapter 17). It is thus a parameter to ignore at our peril.

In the first place it is necessary to identify the functional and proliferative units. In the gastrointestinal tract this is a relatively simple step, since these are housed in discrete anatomical units in the main; the small intestine contains predominantly proliferative crypts which feed into the wholly functional villi, while in stomach and colon, glands and crypts furnish cells for the functional surface epithelium. In epithelial glands, which are usually of small size, such as the prostate complex or the adrenal or thyroid, again definition of the population is relatively easy, since although there are no anatomical divisions into proliferative and functional units, the whole population can be regarded as a single unit. It is in the large surface epithelia, as are found in the epidermis, bladder, or tracheobronchoalveolar tract that problems are encountered, since these are (histologically speaking) simple covering or lining epithelia, with no anatomical landmarks; however, there is something that can be done.

There are three main approaches to population measurements: (i) tissue wet or dry weight; (ii) direct estimation of DNA content; and (iii) cell counting, and of course, each has its advantages and disadvantages. Weight measurements are the easiest and most convenient, but are only mentioned here to be condemned. Many workers persist in using measurements of wet or dry weight, particularly in tumours, but usage is also widespread in work on epithelia, and especially in studies of such tissues as the gastrointestinal mucosa (Robinson *et al.* 1982). It is hardly necessary to say that such crude weight measurements are sensitive, not only to increases in cell number, but also to changes in individual cell mass (hypertrophy or atrophy), in addition to which all contained cells, including connective tissue and lymphoid elements, will contribute to the weight. These measurements thus have little intrinsic accuracy, and should be eschewed in any work which has even pretensions to a critical basis.

DNA content is also relatively easy to determine, either by the diphenylamine or ethidium bromide methods, but there is no guarantee that the measurements made will reflect the size of the target population; in some instances the consequences of this will not be large; for example, in the seminal vesicle or coagulating gland in the mouse prostate complex, there is little connective tissue stroma, and it is likely that changes in DNA content can reasonably be related to changes in epithelial DNA content (Alison *et al.* 1974). However, if we were to measure the DNA content of whole small intestine, or even of mucosal scrapes, we would be including the not inconsiderable amount of DNA in the lamina proprial infiltrate or the smooth muscle of the muscularis propria, with the consequent blurring of accuracy in the face of selective increases in lymphoid populations in the lamina propria. Similarly, such estimates of the epithelial population of the endometrium in the uterus would be complicated by the stromal and muscular DNA. In such instances it would be mandatory to exclude the non-epithelial cell populations in some way, by multiplying the DNA content by the epithelial fraction of the whole tissue. This epithelial fraction can be found in several ways, but probably the simplest method is that employed by Altmann (1971) who measured it by point counting in sections (see below, section 3.11).

A further problem in DNA measurements is the need to assume that all nuclei have the same DNA content. In some tissues this is not tenable; for example, suppose we were hoping to measure cell population changes during liver regeneration. Many cells in the liver are tetraploid or octaploid, or have even more chromosomal complements, and, during the cell proliferation induced by the partial hepatectomy, there are extensive ploidy changes (see Chapter 25). Any attempt to measure cell number by DNA measurements would thus be complicated by such changes. Ploidy changes may also occur in the uterine epithelium after oestrogen treatment (Leroy *et al.* 1978), and, if this is a serious worry, ploidy change can be sought with integrating microdensitometry.

Cell counting techniques should be more reliable, providing certain precautions are taken. In the gastrointestinal mucosa, microdissection techniques after in-bulk Feulgen staining are particularly valuable, whereby the dissected structural units can be squashed and counted; epithelial cells can readily be distinguished from connective tissue and lymphoid cells, and a high degree of precision is possible (see Chapter 17); the method can also be adapted for colonic and gastric mucosa. Cell counting techniques can also be applied to epidermis; there are obvious problems in sections, but these can be overcome by the use of *en face* preparations (Potten 1978*b*), where values can be expressed as cells/unit area, usually cells/cm$^2$, and this method has been used to give a useful analysis of cell population sizes in normal and irradiated epidermis (Potten 1978*b*). In some types of epidermis, it is possible to identify structural units, called 'epidermal proliferative units', and the numbers of cells in these are readily measured in *en face* preparations (see Chapter 8).

In conclusion, we would encourage investigators, not only to look very carefully at any measurements of population size they might make, but also to always, where possible, include population measurements in their analysis of epithelial cell population kinetics.


## 3.11. The analysis of growth curves

There are several circumstances in which we may wish to analyse a growth curve: naturally, the most frequent occasion is when we wish to analyse the rate of growth of a tumour, and basic to any investigation of tumour growth kinetics is the establishment of an accurate growth curve, since several important kinetic parameters are derivative from it. There are numerous pitfalls which must be avoided, both in the collection and the analysis of the data, and the reader is referred to Gordon Steel's excellent review of the topic, as it relates to tumours. However, in normal tissues, there are instances when we wish to know the rate of growth, or *specific growth rate*; for example, any consideration of cytogenesis in the adrenal cortex must include a measurement of the growth rate to appreciate the cell loss rate from each zone (see Chapter 29), and Altmann and Enesco (1967) have produced important data for normal tissues from measurements based on growth curves.

In each study, important parameters are the specific growth rate ($k_G$) and the volume doubling time ($t_D$). Now ideally we would measure these by counting the actual number of cells in the organ or tumour, but in practice this is rarely done; we normally measure some variable which we hope will reflect the overall cell number, such as the organ weight (Wright and Voncina 1977), volume, diameter or DNA content (Altmann and Enesco 1967), and of course, such measurements not only assess the growth of the target epithelial cells, but are also subject to change through variation in other component cell populations, viz. connective, vascular, and immigrant cell populations. The problem is compounded in tumour cell populations, which are subject to haemorrhage, necrosis, and cyst formation. Thus for accurate work we have no alternative but to employ some form of simple stereologic manoeuvre to calculate the proportion of the weighed/measured tissue which is constituted by our target cell population (Smith *et al*. 1974; Wright and Voncina 1977; Davidson *et al*. 1981). Sometimes we need to measure the growth rate of a specific subpopulation within our target population (e.g. bile duct epithelium or Kupffer cells during liver regeneration), in which case a modification of the same procedure should suffice. For example, say we wished to measure the growth rate, or volume doubling time, of the zona reticularis (see Fig. 3.29) in the adrenal cortex of the growing rat. First we would collect a set of adrenals over the required period, say 1–60 days after birth, and carefully weigh them. This would give us a growth curve for the whole adrenal gland, in terms of its weight increase, medulla and all. Now, we would then prepare serial sections of each gland, and ideally, in all sections,

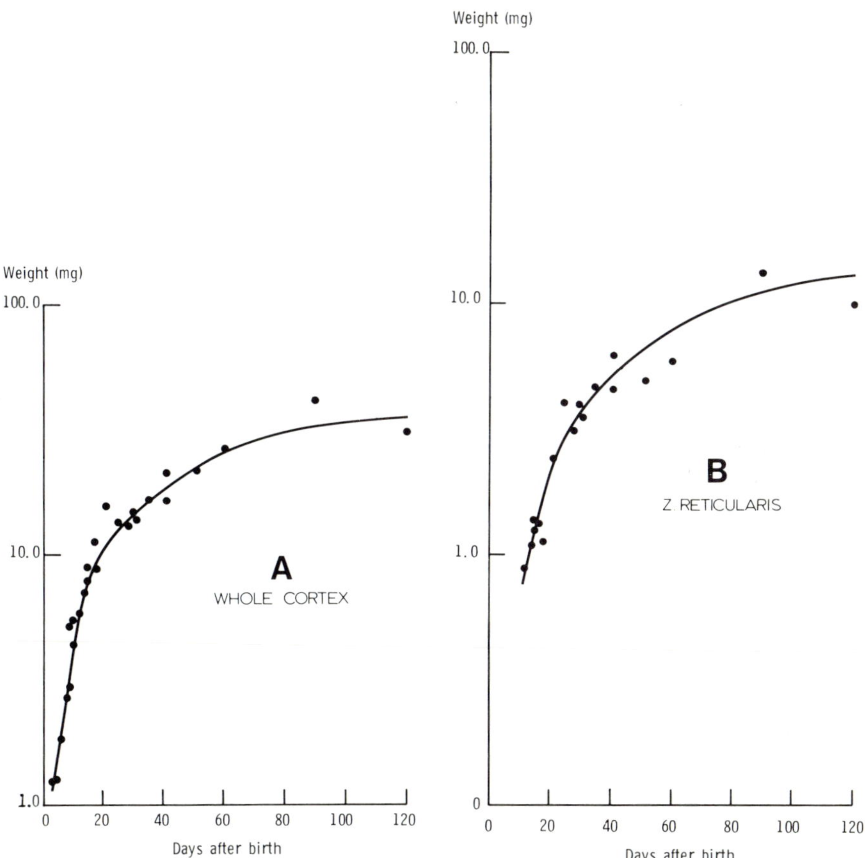


Fig. 3.29. Growth curves for (A) the whole adrenal cortex and (B) the zona reticularis of the rat: (A) shows the total weight of the cortex plotted against time after birth, while in (B) the fractional weight of the zona reticularis has been determined by the point counting method described in the text.

point count, using an appropriate graticule (see Aherne and Dunnill 1982), and record the number of points over the whole gland, and differentially over the zona reticularis. More practically, we would select a number of points which would give us an acceptable relative standard error, which, as Fig. 3.30 indicates,

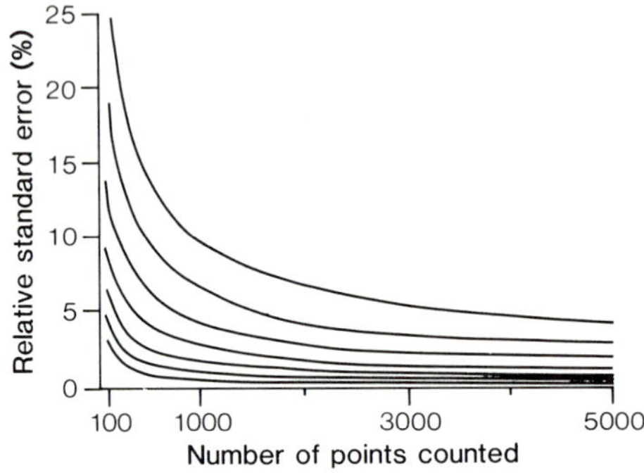



Fig. 3.30. The variation in the relative standard error $E_\rho$ of the estimate of the fractional volume $\rho$ with the number of points counted ($n$). (Redrawn from Weibel (1963).) The relative standard error is calculated from the equation $E_\rho = 0.6745\sqrt{(1-\rho/\rho.n)}$. The curves relate to fractional volumes of, from above downwards, 0.05, 0.1, 0.2, 0.4, 0.6, 0.8, and 0.9 respectively.

increases as the proportion of the desired component (of the whole) decreases (Hally 1964); we then have our area fraction for the zona reticularis $a/A$, where $a$ = the number of points counted over the zona reticularis, and $A$ the points over the whole gland. Then, by the Delisse principle (Aherne 1967),

$$\frac{a}{A} = \frac{w}{W}$$

where $W$ is the gland weight; we can then find $w$, the weight of the zona reticularis, assuming equality of tissue densities, and can then plot a zona reticularis growth curve.

It is here that we usually have to suspend disbelief, and equate weight or volume with cell number, because, in kinetic calculations, the growth rate is not expressed as mg/mg/day, but as cells/cell/day. To avoid this large assumption, we would have to measure the number of cells/mg, and, unless we are prepared to assume that there is no change in the cellular density as the organ or tissue ages, then we would have to measure the cellular density at each time point, obviously a most laborious task. Altmann and Enesco (1967) measured the amount of DNA per organ and graphed this against time to realize the growth curve; combined with a fractional point-counting method in sections of the organ to exclude non-epithelial DNA, this would allow assessment of the amount of DNA in the target population; values for DNA/nucleus are available for many tissues, and hence the cellularity of the organ can be calculated (assuming equality of DNA contents between cells).

In our opinion, accurate measurement of growth in normal tissues needs some such manoeuvre for reasonable results, and the practice of interpolating growth rate from weight alone is to be avoided if at all possible. Nevertheless, it has been the custom of tumour cell kineticists, over a long period of time, to plot weight or diameter measurements against time; corrections to measure the apparently viable tumour tissue mass are rare (but see Smith *et al.* 1974; Davidson *et al.* 1981), and measurements of cell number in solid tumours even rarer. The position has been critically reviewed by Steel (1977), but we should remember that most measurements of tumour $k_G$ we encounter in this volume are likely to be subject to this error.

Having made our measurements, how do we analyse the data? It is usual to plot our readings against time, on a semilogarithmic scale. Figure 3.31 shows

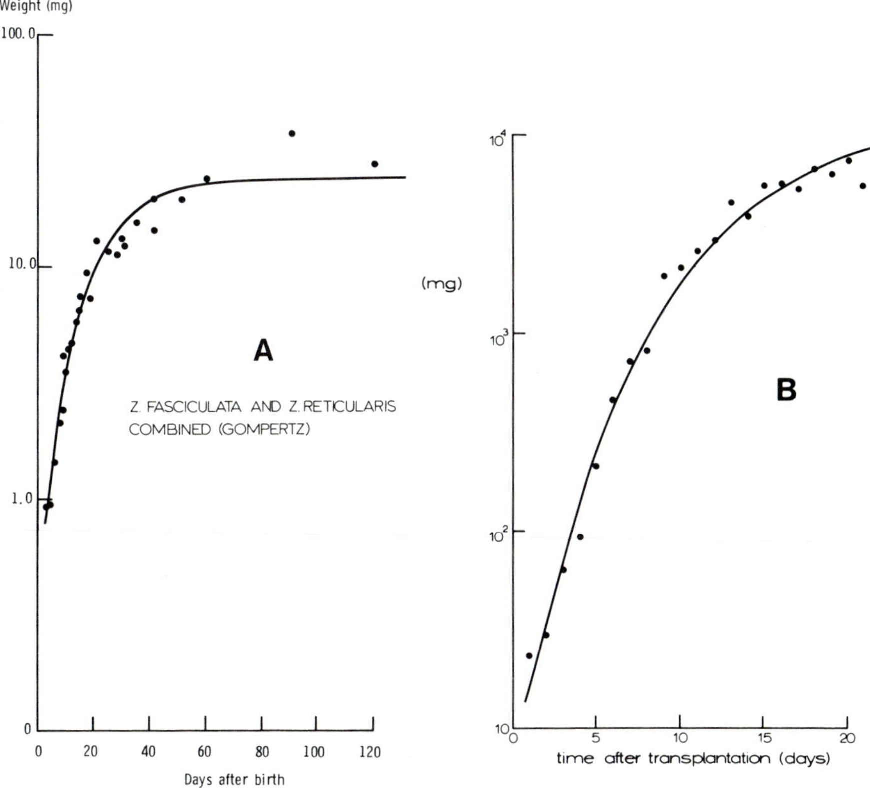


Fig. 3.31. (A) The growth curve for the zona fasciculata and reticularis of the rat adrenal, fitted by the Gompertz function; (B) shows the same function fitted to a growth curve for the transplantable mouse sarcoma described in the text.

growth curves for the adrenal cortex of the rat and for an experimental mouse tumour. Both show similarities; there is an initial period when the data shows a straight line. This means that during this period, growth is logarithmic, or *exponential*. After a time, in both tissues, the growth rate slows, the curve begins to plateau, and towards the end of the experimental period we have little in the way of net growth. How should we approach such a curve? Well, the initial exponential portion is easy to analyse: the growth rate, $k_G$, in mg/mg/day, or whatever, is given by the slope of the line, remembering that we are using a log scale. However, as growth slows, we cannot merely read off the slope, and unless we are prepared to fit a curve to the data (see below), we must recourse to the drawing of a tangent to the curve at the point where we desire a growth rate value, measuring the slope of the tangent, and assuming exponential growth at the point where the tangent touches the curve.

Now this is a difficult procedure: anyone who has tried to draw tangents to a curve will know how easy it is to vary the slope; moreover, it is transparently obvious that when the growth rate slows, growth is no longer exponential. This is a most unsatisfactory way of proceeding, so what can be done? We could propose some form of automatic curve-fitting procedure, based upon a model, which would give us access to the growth rate at any point in time; the usual way of doing this is to choose some form of growth equation, and, by an appropriate fitting procedure, obtain the best fit available. The simplest case is when growth is exponential, or if we are only interested in looking at the first part of a more complex curve such as those shown in Fig. 3.31. Here

$$V_t = V_0 e^{k_G(t)} \qquad (3.28)$$

where $V_t$ and $k_G(t)$ are the organ weight and growth rate at time $t$. Then

$$V_t/V_0 = 2 = e^{[k_G(t) \cdot t_D(t)]} \qquad (3.29)$$

and
$$t_D(t) = \ln 2/k_G(t). \qquad (3.30)$$

However, if we (as we usually do) wish to fit the curved part of the graph, we have perforce to use a more complex growth equation, what Steel (1977) has called a *comprehensive* growth equation. Inspection of the curves shown in Fig. 3.31 shows that the equation must allow exponential, or near exponential, growth initially, but then must be retarded as growth slows towards the asymptote. There have been several attempts to derive such equations, which include the Bertalanffy equation (Bertalanffy 1960), and the logistic equation (Verhagen 1960), but the most widely used model, certainly for tumour growth, but also for normal organs (Wright and Voncina 1977) is the Gompertz equation.

The Gompertz equation has the general form

$$V_t = V_0 \exp[A/\alpha(1 - e^{-\alpha t})] \qquad (3.31)$$

where $V_0$ and $V_t$ are the volumes at 0 and time $t$ respectively, $\alpha$ is a constant, and $A$ is the initial specific growth rate. This equation simplifies to

$$V_t = e^{a + b\rho t}$$

or
$$\log_e V_t = a + b\rho^t \qquad (3.32)$$

where $a$, $b$, and $\rho$ are constants; $a$ is related to the final height achieved by the curve (the asymptotic value of $V_t$ is $e^a$), $b$ is related to the height grown by the curve, and $\rho$ is related to the growth rate. In this simplification

$$V_0 = e^{a + b}$$

$$\alpha = -\log_e \rho$$

$$A = b \log_e \rho$$

The derivative of eqn 3.32 is the specific growth rate, which at time $t$ is

$$1/V \, \mathrm{d}V/\mathrm{d}t = b \log_e \rho \cdot \rho^t \qquad (3.33)$$

and this allows the growth rate to be calculated at any time.

The easiest way of fitting the curve is by an appropriate computer program, providing a least squares fit, and which generates values of $a$, $b$, and $\rho$ for the fitted line, together with standard errors (Wright and Voncina 1977). The growth rate at any time is found from eqn 3.33.

Although this is a convenient method, and allows easy calculation of $k_G$ at any time, the values obtained are only as good as the fit of the equation to the data points. Inspection of Fig. 3.31 shows that, particularly in the fit to the mouse tumour, the Gompertz function is less than optimal. This is because the Gompertz function is self-retarding from the outset, and so starts to bend at the beginning of the curve. However, in this early phase, growth is close to exponential, as indicated by the straight line format of the initial portion of the graph; it should also be noted that the line misses the tumour data points towards the end of the growth curve, since growth slows by more than the Gompertz function can cope with. In the event, therefore, the growth rate values generated by such an analysis are only as good as the fit of the curve to the data, and in some circumstances, we would be better advised to use the tangent method, for example to measure the growth rate at 21 days after transplantation of the tumour. This problem with the Gompertz function has provoked Wheldon (1983) to combine the Gompertz function with an exponential expression, a 'Gompex' equation, which allows exponential growth initially, but self-retardation also occurs towards the end of the growth process.

We might conclude by asking about the nature of growth curves and of their analysis; Steel (1977) has commented that the ability of a comprehensive growth equation to fit a growth curve does not necessarily provide any insight into the biological mechanisms of growth control. It is obvious that some controlling factor leads to growth retardation. In tumours this might be related to the provision of diffusible nutrients from the tumour blood supply, but in normal growing tissues, we usually assume that there is no growth limitation imposed by deficient nutrient supplies, and that all cells in these tissues are well oxygenated. Consequently, there should be some form of intrinsic growth control, possibly mediated by a negative feedback mechanism. We should not expect growth equations to give insight into these processes; they are essentially empirical, and merely provide an attractive approach to the non-heuristic analysis of growth data.

PART 2

# Organization and renewal
# in stratified squamous epithelia

# 4  Morphology

## 4.1. Structural organization

Stratified squamous epithelium is basically of two varieties—*non-keratinizing* and *keratinizing*. The former covers wet surfaces which are subjected to a degree of wear and tear; for example the oral cavity and oesophagus in man, and also the vagina and ectocervix; in the eye the cornea is covered by non-keratinizing epithelium. Keratin, a tough fibrillar protein, is formed whenever stratified squamous epithelium covers dry surfaces, such as the epidermal covering of the skin.

With only a few notable exceptions (see for example Wolfsberg 1964; Leblond *et al.* 1964) kinetic studies in stratified squamous epithelium have been largely carried out on epidermis, although it must be stressed that epidermis varies greatly in structure from site to site. Fortunately, however, the structural organization of both types of epithelium are basically similar, and can be appreciated by inspection of Fig. 4.1 A,B,C, which show vertical sections through the keratinizing squamous epithelium of the mouse epidermis; the organization of these different types will vary (see Chapter 8), but there are similar basic features.

Such an epithelium is divided into several zones: (i) the *basal layer*, sometimes referred to as the *stratum basale or germinativum*; (ii) the *prickle or spinous layer*, also called the *Malpighian layer*; (iii) the *granular layer*; and finally, (iv) the *corneal, horny or keratin layer*. The cells of the basal and spinous layers are often called keratinoblasts and keratinocytes respectively, or, etymologically more correct, keratoblasts and keratocytes (Hamilton 1972).

The basal cells are usually regarded as forming the proliferative compartment of the epidermis (Olson *et al.* 1969), from which cells migrate, differentiating as they progress, to be exfoliated or *desquamated* at the surface. In mouse ear epidermis, or mouse oesophagus, for example, there is good evidence that only one layer of proliferating cells exists in the basal layer (Leblond *et al.* 1964), but in normal human epidermis proliferating cells are known to be present above the basal layer (Penneys *et al.* 1970). Cells then migrate into the Malpighian layer, which varies a good deal in thickness. As they migrate they flatten, and, in keratinizing epithelia, begin the synthesis of keratin; however, staining of the epidermis with fluorescein isothiocyanate (FITC), said to localize keratin, shows that some cells begin keratin synthesis while still in the basal layer (Christophers 1971*b*). In epidermis, other differentiation changes which accompany migration, in addition to keratin synthesis, include production of other amorphous proteins

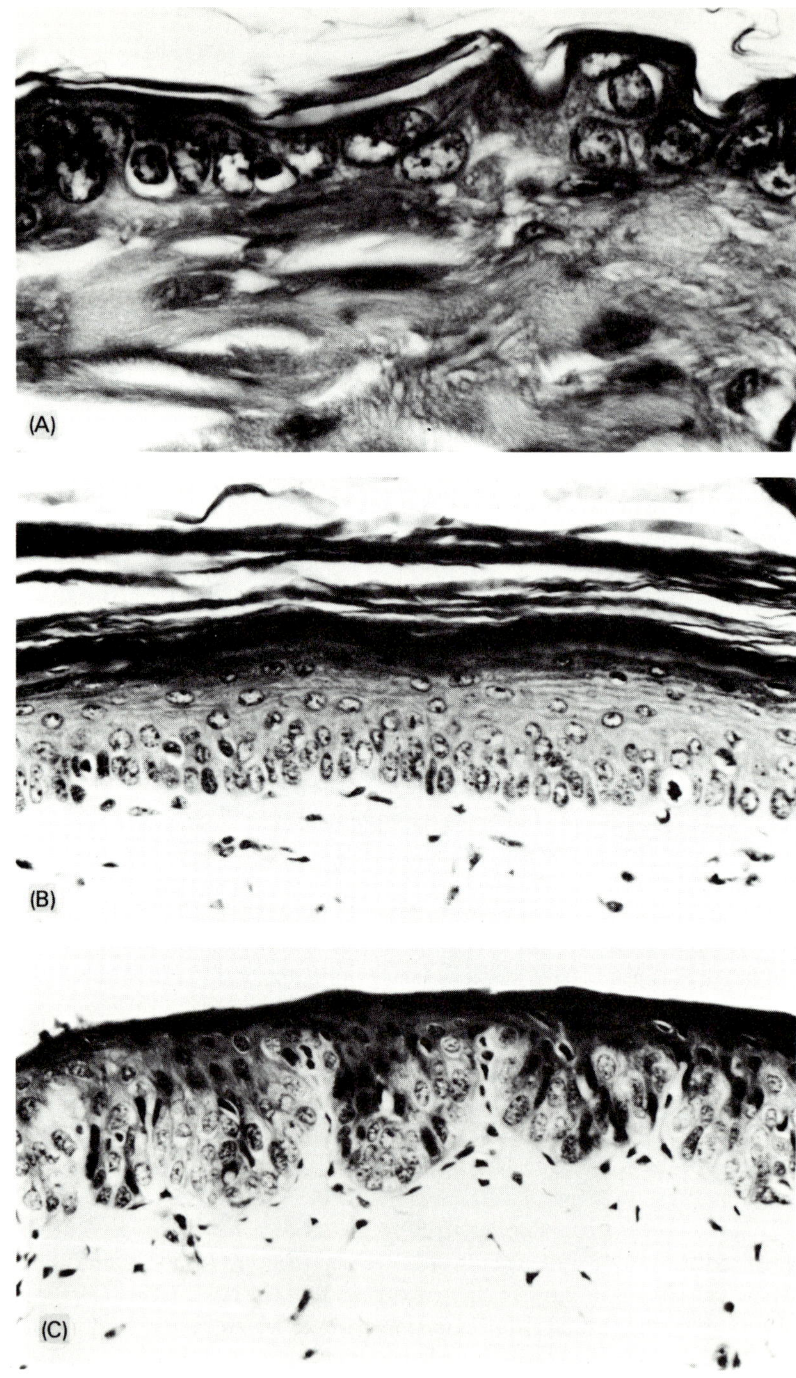

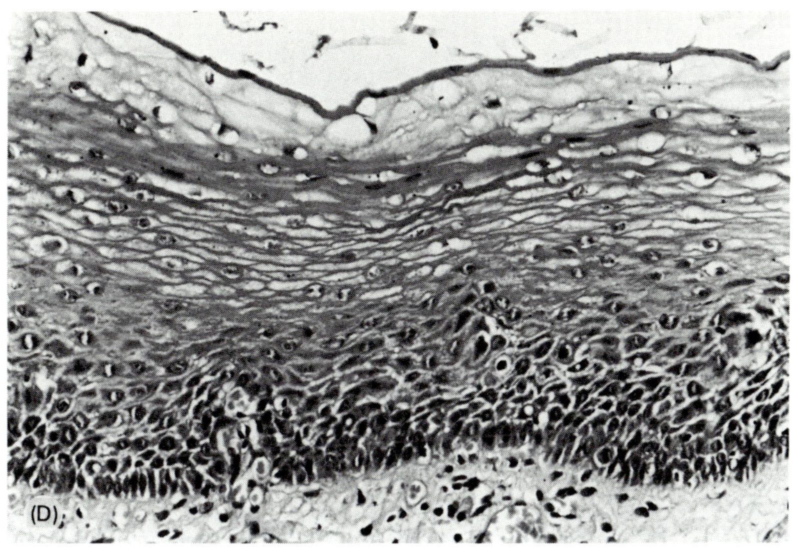

Fig. 4.1. (A) The keratinizing stratified squamous epithelium of the mouse dorsal epidermis; the basal layer is apparently homogeneous, and the Malpighian layer contains at most two cell layers before the keratinizing surface layers are reached. (×1105.) (B) Thicker keratinizing stratified squamous epithelium from the dorsal surface of the mouse foot. There are very many more Malpighian and corneocyte cell layers in this (relatively) rapidly proliferating epithelium. (×390.) (C) Stratified squamous epithelium where there are downward prolongations of the basal layers (the *rete pegs*) which are separated from each other by vascular connective tissue *dermal* papillae. Obviously, the organization of this epithelium, from the ventral surface of the mouse foot, is different from (A) and (B). (×390.) (D) So-called non-keratinizing stratified squamous epithelium from the human oesophagus. Although keratin is of course synthesized in the Malpighian layer in this tissue, the nuclei remain evident in the surface layers, and do not degenerate as in the epidermal tissues shown previously. (×260.)

and of membrane-coating granules (see Montagna and Parakkal 1974, for review).

The cells in the Malpighian layer are called prickle or spinous cells, because they are attached to each neighbour by numerous cytoplasmic processes called prickles, intercellular bridges, or more commonly *desmosomes*. These are found in all layers of the epidermis, including the basal layer, and also between the basal lamina and the deep surface of the basal layer, where they are called *hemidesmosomes*. The detailed structure of desmosomes has been reviewed by Campbell and Campbell (1971) and Allen and Potten (1974*b*) for the interested reader; here one might note that, since continued cell production is a function of the basal layer, desmosomal contacts are continuously broken down and reformed as cells migrate. Nafstad and Baker (1973) observed intracellular desmosomes inside dividing epidermal cells, and speculated that desmosomal contacts were withdrawn at mitosis. Allen and Potten (1974*b*) have made detailed ultrastructural studies on the progressive changes undergone by desmosomes during the process of differentiation, some of which are very relevant to the concept of ordered epidermal structure and are discussed in Chapter 8.

The granular layer is the transitional zone between the viable keratocytes below and the anuclear dead corneocytes on the surface. Granular cells contain basophilic *keratohyalin granules*; the nature of these structures is debatable. Early ideas suggested an early form of keratin, but Jarrett (1974) favours the view that they represent cell organelles which have undergone partial destruction by the hydrolytic enzymes demonstrable in the granular layer; it is at this level that nuclei are also resorbed and lost, also probably by lysosomal hydrolytic enzymes. The morphological changes which occur in the nucleus during the process of keratinization are well described by Weir *et al.* (1971) and DuBrule (1972). Iversen *et al.* (1971) also described degenerative changes in the nucleus during the keratinization process, and followed the changes in bound and acid-soluble $[^3H]$-TdR in corneocytes removed by tape stripping mouse epidermis; they concluded that nuclear DNA was broken down into small molecular parts which diffuse freely and rapidly.

The horny layer is formed of a variable number of flat, keratinized plates, or corneocytes, which in the mouse have a hexagonal shape (see Fig. 8.3, p. 293). Although in routine paraffin sections the stratum corneum may appear as a loose, basket-weave structure, haphazardly arranged, or a dense, amorphous mass, ultrastructural studies and special techniques have shown that, in certain species, there is a highly ordered organization, the understanding of which is critical to our appreciation of kinetic organization in epidermis (Raknerud 1974, 1977; and see Chapter 8). Cells are continuously desquamated from the surface of the epithelium: Spearman (1977) has fully described the ultrastructural appearances which occur during keratinization.

Although keratinizing squamous epithelium is found in the oesophagus and squamous stomach of some animals (for example the mouse), in man the oesophagus is lined with non-keratinizing epithelium and Fig. 4.1D shows a vertical section of human oesophagus, which can be regarded as an example of a non-keratinizing squamous epithelium, even though some keratohyalin granules may sometimes be evident. The basic structure in terms of cell layers is much the same as that described in Fig. 4.1A–C; but there is no granular or horny layer, and cells merely flatten as they migrate, while the nucleus becomes smaller; however, the nucleus is evident in the most superficial cells. It should be appreciated that, in certain circumstances, non-keratinizing epithelia can keratinize; for example the mouse or guinea pig vaginal epithelium at the time of oestrus.


## 4.2. Other cells in the epidermis

In addition to the keratocyte series of cells, there are other cells in the epidermis which in the mouse can number up to 15 per cent of the basal cell population (Potten 1975*b*); the fact that the basal layer is a heterogeneous population is often conveniently forgotten in kinetic studies. The most numerous of these are the pigmented *melanocytes*, usually found in a basal position sandwiched between the tightly packed basal cells. These cells are dendritic, and have no desmosomal attachments. Their unique characteristic, however, is the presence of *melanosomes*, which are distinctive organelles which represent stages in the formation of melanin granules. These melanin granules are then transferred to adjacent keratocytes by phagocytosis of the tips of the melanosome-filled dendrites of the melanocyte, which leads to pigmentation of the basal cells. Melanocytes are not distributed at random among the keratocytes, but there is an orderly structural and functional relationship between the two cell series, which is now dignified by the term *epidermal–melanin unit* (Fitzpatrick and Breathnach 1963); the ratio of melanocytes to basal cells in man is 1 : 10–13 in most parts of the body, but can reach 1 : 46 in the face (Szabo 1967). Epidermal melanin acts as a screen against solar irradiation.

A further cell, of debatable origin and function, is the *Langerhans cell*. These cells have a markedly indented nucleus, with clear cytoplasm, but without desmosomal contacts; they do, however, contain characteristic granules.

Histochemical recognition of these cells is a controversial topic (see Chapter 5, section 3) and the only sure method is by ultrastructural demonstration of the characteristic granules. The structural aspects of these cells are well reviewed by Breathnach and Wyllie (1966) and Wolff (1972). In guinea pig (Wolff 1972) and mouse (Potten and Allen 1976) epidermis they number about 900 cells/mm$^2$, in mouse basal layer in a ratio of about 1 : 10 with keratoblasts (Potten and Allen 1976). However, Langerhans cells are widely distributed, being found in cervix (Wolff 1972), oesophagus (Yassin and Toner 1977), palatal mucosa (Hill 1977), and dermis (Hashimoto and Tarnowski 1968). Moreover, cells containing granules similar to those of Langerhans cells are seen in the thymus gland and in the storage disease complex 'histiocytosis X' (Basset and Turiaf 1965; Tarnowski and Hashimoto 1967). There are differing views about the function of these cells; conflicting views include an antigen detecting function (Prunieras 1969), a role in the control of keratinization (Riley 1966), and even a central role in the control and maintenance of cell production in the epidermis (see Chapter 5, section 3, and Chapter 8, section 2). Wong and Buck (1971) made the interesting observation that Langerhans cells, although not usually present in trachea or bladder mucosa, appeared after induction of squamous metaplasia in vitamin A deficient animals; this would indicate a distinct relationship with epithelium of a squamous nature.

Although Langerhans cells do occur in a basal position (according to Lezzard *et al.* (1968), only about 20 per cent of the total complement) Langerhans cells are found in suprabasal layers, and evidently migrate with the keratocytes (see Breathnach 1964; Riley 1967).

In mammalian epidermis, in association with the tactile discs found adjacent to hair, especially in nasal skin follicles, are specialized cells called *Merkel cells* (see Winkelmann and Breathnach 1973, Winkelmann 1977). These cells are often associated with neural terminations, contain catecholamines, and are thought to be sensory receptors or transducers, and a recent theory is that they represent mechanoreceptors. They have irregular nuclei and prominent osmiophilic cytoplasmic granules. They are also especially common in anuran tadpoles, where recent studies point to a mechanoreceptor function (Tachibona 1978). Nafstad and Baker (1973) also regarded these cells as mechanoreceptors, but noted their association with numerous keratoblast mitoses, and proposed a relationship with keratocyte histodifferentiation.

In addition to these named cells, there are small numbers of dendritic cells of indeterminate nature scattered through the epidermis. Although some authors have attempted to classify these cells separately, Breathnach (1975) is violently against this, and insists that they are immature forms of the named dendritic cells. However, according to several other theories, the indeterminate cells are possibly transitional forms between melanocytes and Langerhans cells (see Chapter 5, section 3), but Breathnach (1975) has recorded their presence in embryonic epidermis transplanted before migration of melanoblasts from the neural crest has occurred, so they would appear to be unrelated to melanocytes.

# 5 Cell lineage in epidermis

## 5.1. The keratocyte series in the epidermis

While patterns of histo- and morphogenesis in the epidermis have been extensively studied and elegantly reviewed and summarized by Sengel (1976) and Garber (1976), there have been remarkably few studies of cell proliferation and kinetic organization in developing epidermis. The keratocyte series of cells is derived from embryonic ectoderm, and in the mouse, up to 8–12 days of gestation, is represented by a single layer of proliferating cells called the *stratum germinativum.* At about eight days a second layer of cells develops, the periderm, which is a transitory structure, disappearing at about 17 days. At 12–14 days, an intermediate layer develops, the *stratum intermedium*, which is also undifferentiated and able to proliferate (Sengel 1976). However, at day 15 the stratum intermedium begins to differentiate into the spinous layer, and begins to lose its proliferative ability, with all mitotic activity disappearing at about 16 days. It does appear that, during histogenesis, the earliest spinous cells are derived from the stratum intermedium, although at later times, spinous cells derive exclusively from the basal proliferating cells. At about 16 days desmosomal contacts are observed and the epithelium is now a fully stratified structure.

The granular layer appears at about 14 days in the mouse, followed by the stratum corneum at 16 days of gestation. The arrangement of 17-day mouse embryonic skin is shown in Fig. 5.1. The main features of early histogenesis of the epidermis in the avian embryo is not dissimilar (Sengel 1976), and the process in man has been reviewed by Hoyes (1968), although in this latter case, of course, the time relationships are much longer (Breathnach 1971; Holbrook and Odland 1975; see Fig. 5.2).

Kinetic information in development is sparse. In early chick embryonic epidermis, it is clear that both stratum germinativum and periderm show [$^3$H]-TdR labelling and mitotic activity (Wessels 1964). While cell production continues in the basal and intermediate layers, proliferation slowly ceases as the periderm begins to be shed, at about 16 days gestation, after which proliferative activity steadily decreases in the stratum intermedium and becomes confined to the basal layer (Rothberg and Ekel 1967, 1971; Sawyer 1972). At the time of periderm shedding, the minimum transit time from the basal layer to the surface is about 3–4 days (Sawyer 1972). Sawyer also did FLM studies in chick embryo epidermis; although there was no change in $t_S$ with stage of development, Sawyer claimed that there was a marked increase in the duration of $T_C$ as development proceeded, brought about by a prolongation of $G_1$; $T_C$ was 19 hours at 12 days

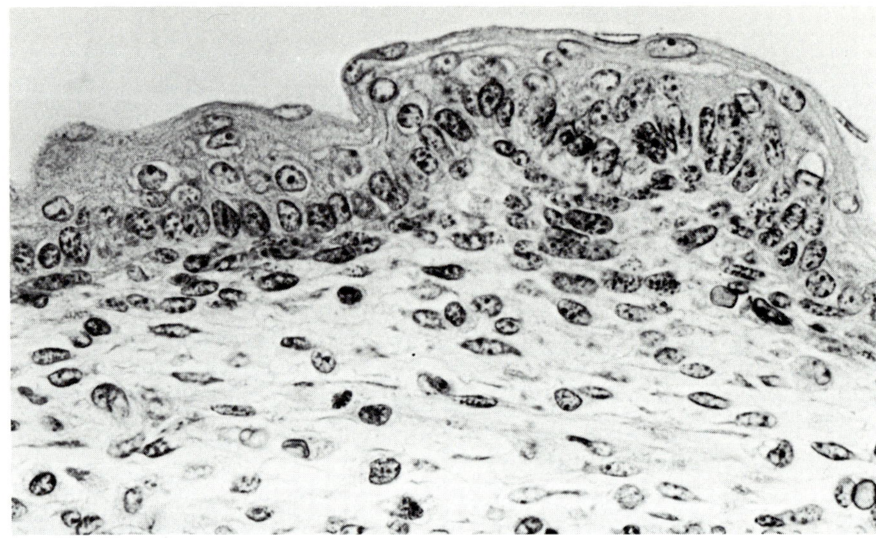

Fig. 5.1. A section of 17-day mouse embryo skin. Note that the stratum corneum is para-keratotic at this stage, and that viable nuclei are visible right up to the surface. (By courtesy of Professor P. Sengel.)

gestation, and 38 hours at 15 days. However, these values are really *turnover times* for the basal layer, obtained by a stage-duration calculation assuming a rectangular age distribution, and also assuming a growth fraction of unity; thus they cannot be regarded as cell cycle times, as claimed.

In the mammalian situation, Hanson (1947) showed a slow decline in mitotic activity of the stratum germinativum with time until birth in fetal rat epidermis, but Stern *et al.* (1971) have reinterpreted this data to indicate that peaking of cell production occurs shortly before birth. They then proceeded to study the changes in $I_S$ in the stratum germinativum of the fetal rat; values from 10–15 per cent were observed up to about day 15, when there was a sharp increase to around 30 per cent at day 18, followed by a fall back to about 10 per cent with birth at day 21. The $I_S$ had declined to 3 per cent by the fifth postnatal day. Stern *et al.* (1971) could not ascribe this increase in proliferation to growth alone, and suspected that it could be related to histodifferentiation control mechanisms. In human embryonic skin, the periderm also appears capable of proliferating, at least at six weeks gestation, where Breathnach and Robins (1969) reported seeing peridermal mitoses. Holbrook and Odland (1975) also observed mitotic activity in human periderm and stratum germinativum at 36–53 days, but reported that peridermal proliferative activity had ceased by 93 days; Hoyes (1968) considered that peridermal proliferative activity was lost at between 56–77 days gestation. However, by 126 days, Gerstein (1971) considered that [³H]-TdR labelling was confined to the basal layers, with no

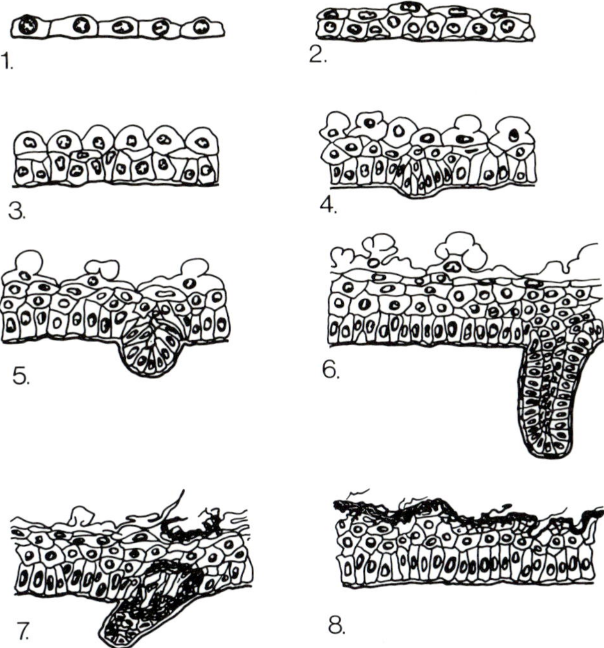

Fig. 5.2. The development of the human embryonic epidermis. There are eight stages recognized: (1) <36 days, *indifferent epithelium*; (2) 35–55 days, with a *flattened surface*; (3) 55–75 days, with an *elevated surface*; (4) 65–95 days, the stage of *incipient bleb formation* in the periderm; (5) 96–110 days, *single bleb formation*; (6) 95–120 days, *complex bleb formation*; (7) 110–160 days, phase of *periderm regression*; (8) 160 days, *cornification*. (Redrawn from Holbrook and Odland (1975.)

uptake by the periderm. Gerstein was able to inject [³H]-TdR intradermally into a 126-day fetus, and reported an $I_S$ value of 5.2 per cent; the stratum intermedium also appeared non-proliferative by this stage. Stern (1974), also studying human fetal epidermis, but this time *in vitro*, recorded a fall in $I_S$ from 8.1 per cent in the third month to 3.5 per cent in five-month old fetuses.

During the process of stratification of the developing epidermis there are, of course, likely to be marked changes in the kinetic organization of the epithelium of which, at the present time, few are appreciated. One approach to this problem is due to Smart (1970*a*, *b*) who studied the orientation of the mitotic axis during stratification in the histogenesis of mouse epidermis and oesophagus. These studies are of considerable importance in layered epithelia, and will be treated more fully when we discuss the age distribution in layered epithelia (Chapter 6, section 7), and the kinetic organization of the basal layer (Chapter 8), but we might note that, in fetal mouse epidermis, at the stage where a single layer of epithelium is present (i.e. the stratum germinativum), all mitoses are horizontal (for terminology see Chapter 6, section 7). Such a disposition could

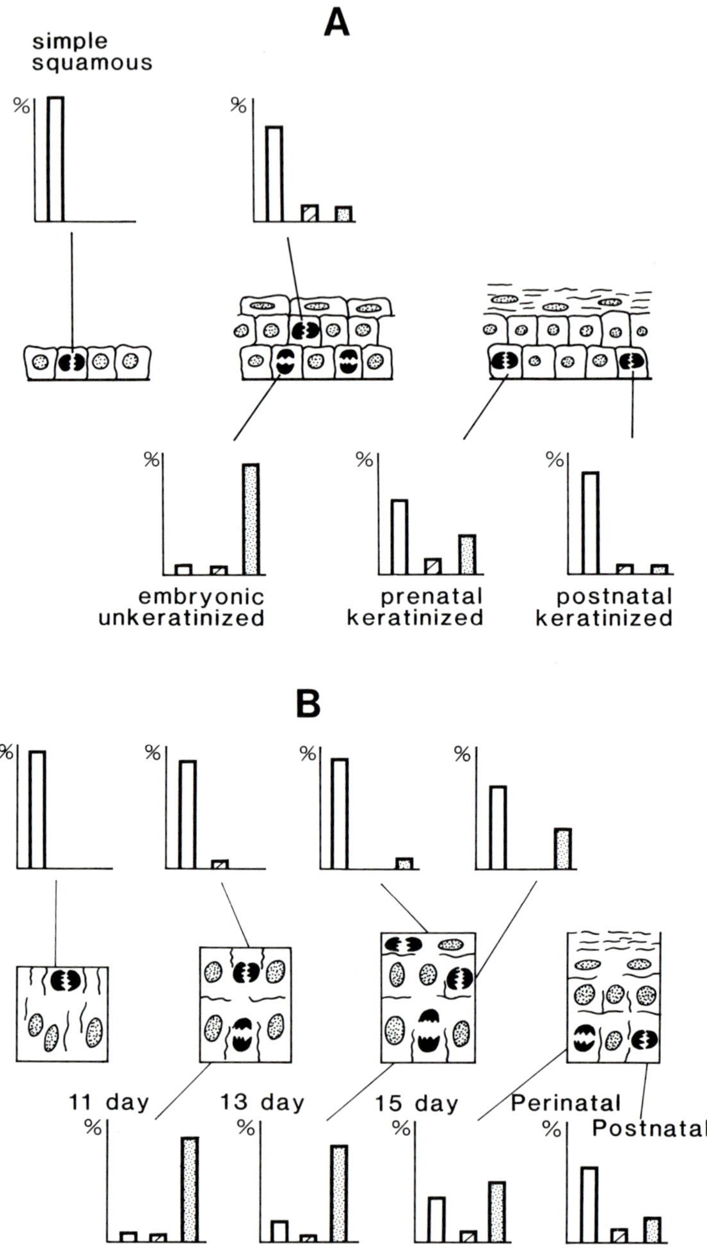



Fig. 5.3. Changes in the direction of the mitotic axis during the histogenesis of (A) the mouse epidermis, and (B) the mouse oesophageal squamous epithelium. (Key: unshaded, horizontal; cross-hatched, oblique; stippled, vertical.) (Redrawn from Smart (1970a,b).)

be contributory to the overall lateral growth of the surface. However, with the outset of stratification, the basal layer mitoses change to a vertical orientation, while those in the stratum intermedium (later to become the spinous layer) are mostly horizontal (see Fig. 5.3). Presumably basal mitoses are adding cells to the suprabasal layers, and contributing to stratification, while suprabasal mitoses are adding cells to the increasing area of the epidermis. When keratinization occurs, at about day 17, when mitoses are mainly confined to the basal layer, the orientation is again horizontal, which is said to be the normal adult situation. Thus interpolating, we might say that the stratification is due to the basal layer adding cells almost exclusively to the suprabasal layers, while the suprabasal layers then take responsibility for lateral growth. It is not clear whether this mechanism is also operative in the adult situation when stratification is being re-established, after say tape-stripping the epidermis or during an increase in epidermal thickness during the application of hyperplasiogenic agents; Pinkus and Hunter (1966) reported no change in the distribution of mitotic axes in regenerating, stripped epidermis, but Bullough and Mitrani (1976, 1978) have reported that the increased epidermal thickness which accompanies a hyper-plastic stimulus is associated with an increase in the number of vertical mitoses, indicating that vertically orientated mitoses may assist or indeed be responsible for the production of increased numbers of cell layers during regeneration; this point is returned to when kinetic organization is considered (see Chapter 8).

Some patterns of proliferation in development of the keratocyte series thus emerges; in early stages, all cell layers are proliferative, but, at times which vary with the species, cell production gradually becomes the province of the basal cells, which situation continues into neonatal and adult life. The kinetic events which are responsible for stratification are unclear, but variation in the orientation of the mitotic axis may be responsible. Which factors control mitotic axis direction are entirely in the realm of speculation; response to agents such as epidermal chalones do not appear to be established until some time after birth (see Chapter 14). There is of course, good evidence for extensive epithelial-mesenchymal interactions in epidermal morphogenesis, and some of this is discussed in Chapter 14.


## 5.2. The melanocyte series

There has been, and is, considerable controversy regarding the origin of the dendritic cell populations in the epidermis. As far as melanocytes are concerned, most opinion is agreed that melanocytes are derived from embryonic neural crest tissues and migrate early in development to their various sites in epidermis, hair follicles, retina, and central nervous system. This has been shown by Du Shane (1936) for amphibia, and for mammals (mice) by Rawles (1947) who demonstrated that melanocytes do not develop after transplantation of mouse somite tissue, with overlying ectoderm, into the coelomic cavity of chick embryos, so long as neural crest or cells originating from it were absent from the grafts at the time of transplantation. As is usual in dendritic cell biology, difficulties are experienced in identification; for example in early human fetuses, a DOPA reaction could not be obtained even though electron microscopy showed that pre-melanosomes were evident in epidermal melanocytes (Breathnach and Wyllie 1966), and ultrastructure remains the mainstay of these studies. In man it appears that melanocytes migrate from the neural crest in the sixth week of gestation and can be found in the skin from the tenth week onwards.

Many older studies have attempted to analyse the problem of dendritic cell replacement in mature tissues; unfortunately most of this work has relied mainly on morphological or histochemical demonstration of similarities, or of cells transitional between the putative precursor and the differentiated cell. For example Okun (1965) using just these criteria, considers that epidermal melanocytes may arise from connective tissue precursors, possibly mast cells. Many workers have relied upon the recognition of [$^3$H]-TdR-labelled cells in sections which are, autoradiographically speaking, of infinite thickness, inviting errors due to superposition. There have been few critical studies such as those of Leblond and his colleagues (see Chapter 16, section 2) on cell lineage in small intestinal cells, using autoradiography on thin and semithin sections.

Having said this, there is a considerable body of evidence which suggests that epidermal melanocytes are a self-renewing population. Giacometti *et al.* (1972) reviewed the early literature and noted that observers such as Bizzozero in 1906 had remarked on the presence of melanocytes in division, and moreover had noted that they retract their dendrites during mitosis; other workers have observed epidermal melanocytes in mitosis (see for example, Becker *et al.* 1952; Billingham and Medewar 1953; Billingham and Silvers 1960), and Giacometti *et al.* (1972) show some particularly convincing colour photographs of dividing melanocytes, interestingly enough, with dendrites retracted. Although Giacometti and Allegra (1967) could not show melanocyte labelling in unstimulated epidermis, they did consider that melanocytes were an (albeit slowly) renewing population. Shukla *et al* (1973) and Shukla (1977) were able to demonstrate, in isolated melanocyte preparations, flash labelling of guinea pig epidermal melanocytes in both stimulated (grafted) and resting epidermis; furthermore, Shukla proposed that human epidermal melanocytes had a defined life history, and graded them from 1 to 4, dependent upon their complement of dendrites; on reaching stage 4, with numerous dendrites, the melanocyte divided, giving rise to young melanocytes again. Shukla and Das (1976) also consider that the Langerhans cell derives from division of a stage 4 melanocyte (see below, section 5.3).

Thus melanocytes do occasionally show evidence of proliferative activity in resting skin and although Kitano and Hu (1970) showed that pigment cells divide *in vitro*, this may not reflect their proclivities *in vivo*. However, in stimulated epidermis, there is ample evidence of melanocyte cell production: in

UV-irradiated epidermis, it is well known that melanocyte counts per unit area increase (Snell 1962; Quevebo *et al.* 1965; Wolff and Winkelmann 1967*a*), but there remained the various possibilities that this increase was due to activation of originally DOPA-negative dendritic cells, or to migration from elsewhere (?dermis, ?surrounding epidermis), rather than to melanocyte proliferation. However, combining ARG with the DOPA reaction, Sato and Kowada (1972) and Giacometti *et al.* (1972) showed labelling of epidermal melanocytes in UV-irradiated hairless mouse and monkey epidermis respectively. In the monkey, melanocyte flash labelling indices of about 0.5 per cent were apparent 10 days after irradiation. Earlier observations also indicated that melanocytes migrated along with the migrating keratocytes during wound healing (Breathnach 1963; Snell 1963; Giacometti and Allegra 1967), albeit behind the leading edge of the keratocytes. Using this model in the guinea pig, Giacometti and Allegra (1967) showed that numerous melanocytes labelled with [³H]-TdR at six days after wounding. Similarly Shukla (1969*a*), arresting metaphases in wounded guinea pig epidermis with colchicine, reported mitoses in type 4 melanocytes, although his photographs are by no means convincing.

It appears that epidermal melanocytes are probably a slowly renewing population which, under appropriate stimuli, can increase their cell production rate, and may therefore qualify as a conditional renewal system (see Chapter 2, section 5). This is not to exclude the possibility of influx from another cell system; there is, however, nothing to suggest this at the moment.

Little or nothing is known about the control of melanocyte proliferation. There have been some attempts to show that water extracts of skin contain substances which inhibit proliferative activity in the malignant tumour of melanocytes, the malignant melanoma. Thus Bullough and Laurence (1968*c,e*) claimed that a 70–80 per cent ethanolic precipitate of pig skin extract inhibited $G_2$-$M$ flux *in vitro* in the Harding-Passey and Green Fortner melanomata, in mouse and hamster respectively. The substance did not act on epidermal cells, and was heat-labile, and this fulfils the criteria for a putative chalone; similar results were obtained with an 80 per cent ethanolic precipitate of a water extract of the melanoma itself. Furthermore, the action of both was only obtained in association with adrenalin and hydrocortisone. Working *in vivo*, Mohr *et al.* (1968) reported remission of the tumours, obtained with much the same extracts in the same experimental systems. However, as previously discussed (see Chapter 2, section 10), and as is again considered in the section on the epidermal chalone (Chapter 14), it is difficult to assess these results in the absence of suitable controls, and it is salutory to note that in later publications Mohr (Mohr *et al.* 1972*a,b*) reported contamination of the chalone extract with *Clostridium* spores, which were found to have a marked oncolytic effect; the action of the extract on the tumour was ascribed to these spores. Similarly Dewey (1973), observing that melanoma extract inhibited the growth of Harding-Passey melanoma cells in culture, and ascribing the effect to a dialysable polypeptide, considered that cells growing without serum were suitable controls. While we might question the physiological role of a factor acting on the $G_2$-$M$ transition, Seijii *et al.* (1974) considered that a 70–90 per cent ethanolic precipitate of Harding–Passey melanoma cells not only inhibited $G_2$-$M$ flux of melanoma cells *in vitro*, but also inhibited [³H]-TdR incorporation into DNA 12 h after injection of the extract *in vivo*. This was put forward as a $G_1$-$S$ effect even though no measurements of $t_S$ or $T_C$ were made, and the kinetics of the effect inadequately documented; again the $G_2$-$M$ action was in association with adrenaline and hydrocortisone. Seijii *et al.* recognized two fractions with inhibiting activity, only one of which was tissue specific. However, our previous observations on the significance of the $G_2$-$M$ action, and on the choice of controls apply, as do our remarks on the necessity of defining the system in kinetic terms, underlined recently by Laurence *et al.* (1979). Also operative is Maurer's (1981) extensive critique of the use of [³H]-TdR incorporation into DNA as a sole criterion of a $G_1$-$S$ action.

Even if these reports are substantiated, we would be unable to assess the physiological significance of the findings, since we are considering melanoma cells rather than epidermal melanocytes. It is also clear that factors other than 'chalones', present in extracts (cf. polyamines, Allen and Smith 1979) are inhibitory to melanoma cells. The situation remains confused.


## 5.3. The Langerhans cell

The origin of the Langerhans cell was for long a source of confusion, the main reason for which has been controversy about the best way of recognizing such cells. For example, the original gold impregnation techniques gave uncertain and unreliable results (Breathnach 1965), while such techniques as osmium iodide staining (Mishima and Miller-Milinska 1961) do not distinguish between melanocytes and Langerhans cells. Many workers, studying the potential renewal of Langerhans cells, have relied on the demonstration of ATPase in the cytoplasm (see for example Mackenzie 1975*b*), but this method has been attacked by Zelickson and Mottaz (1968), again on the grounds that it did not distinguish between melanocytes and Langerhans cells. However, Wolff and Winkelmann (1976*c*) and Wolff (1972) have claimed that, with the correct experimental conditions (namely correct fixation in 5 per cent phosphate buffered formal saline at pH 7.5 overnight), ATPase localization is specific for Langerhans cells, and confirmed the localization by ultrastructural studies. Reams and Tompkins (1973) also stress the importance of deionized water in the procedure. Breathnach (1965) characteristically regards the ultrastructural demonstration of specific Langerhans granules as the only definitive method of identification, and on this criterion, reported that Langerhans cells appear in human fetal skin at about 14 weeks gestation. More recently, with the finding that Ia antigens in mouse skin are predominantly expressed on Langerhans cells (Tamaki *et al.* 1979), more reliable recognition criteria are available.

With this background, we can consider the ontogeny of Langerhans cells; naturally, with the melanocyte origin from neural crest elements very much in mind, early workers were constrained to eliminate this possibility for the other major class of dendritic cells. Silvers (1957) reported that Langerhans cells still developed normally in mouse limb buds, after the neural crest had been removed, using the gold staining method to identify the cells. This study invited obvious criticism, and Breathnach *et al.* (1965), using the specific granules for localization, showed that Langerhans cells developed in neural crest-free embryonic skin implanted into host spleen. Unfortunately, it is now clear that cells with granules similar to those found in Langerhans cells are present in lymphoid tissue, for example in the storage complex of diseases called *histiocytosis X* (Tarnowsky and Hashimoto 1967), so the Langerhans cells could have taken origin from host spleen. This objection has apparently been resolved by Reams and Tompkins (1973), who implanted neural crest-free epidermis into chick embryo hosts; chick skin does not contain any Langerhans cells: ATPase-positive cells were found normally in the transplanted mouse epidermis. Since thymus and bone marrow develop at a later date, origin from these organs was apparently excluded (but see below). Reams and Tompkins (1973) concluded that an origin from keratocytes was likely, but Breathnach (1975) still unsatisfied, insisted on confirmatory ultrastructural studies, and pointed out that mesenchymal elements in the graft could well be the source of the Langerhans cells. It therefore does appear that Langerhans cells are not derived from neural crest tissues: if so, what is their nature?

There have been four contrasting theories: (a) that they represent neural elements; (b) they are effete derivatives or some other transformation product of melanocytes; (c) they represent an independent self-renewing population, and (d) that they are derived from macrophages or some other mesenchymal cell.

We can probably disregard (a), because of the lack of ultrastructural evidence, and because nervous communications have not been demonstrated. (b) is a hypothesis originally advanced by Masson (1948); Langerhans cells are melanocytes which can no longer elaborate melanin, and are swept away with epidermal migration to be lost at the surface. It was based upon (i) the similarity in shape between melanocytes and Langerhans cells; (ii) the allegation that although melanocytes are proliferative and do divide (see Chapter 5, section 2), the melanocyte population apparently remains constant in size (Snell and Bischitz 1963), without apparent proliferative activity in Langerhans cells (Billingham and Medewar 1953), and finally (iii) that there is a correlation in numbers and distribution between melanocytes and Langerhans cells (Billingham and Medewar 1953): after irradiating guinea pig skin with 2 Gy of X-rays, there is an increase in melanocytes and a reduction in Langerhans cells (Fau *et al.* 1959), and after application of the carcinogen dimethylbenzanthracene to mouse skin, an inverse relationship between melanocytes and Langerhans cells was noted (Tsuji *et al.* 1969). These data were considered consistent with a transformation pathway between melanocytes and Langerhans cells. Further evidence for the melanocyte origin of Langerhans cells came from Shukla (1969*b*, 1977); the earlier study was based upon morphological observations only, but later Shukla and Das (1976) studied melanocytes and Langerhans cells in isolated preparations, and melanocytes and indeterminate cells, but not Langerhans cells, were found to flash label with [$^3$H]-TdR; 12 hours after labelling, Langerhans cells with one dendrite appeared labelled, and by 14 days cells with many more dendrites became labelled. These findings indicate that only melanocytes divide (type 4, see above), and give rise to the Langerhans cell possibly via the indeterminate cell; the Langerhans cell migrates to be lost, while the melanocyte remains. A modified concept was considered by Breathnach (1963), who, discerning that the Langerhans cell was scarcely effete in appearance, suggested that each melanocyte division gave rise to two Langerhans cells, one of which migrated, while melanogenesis was induced in the other.

However, there are serious reasons for doubting this relationship: ultra-structural studies show little similarity between the two cells, and the appearance of the Langerhans cells, and its enzyme histochemical profile, are hardly characteristic of a worn-out cell (Riley 1967, 1974). The presence of melanin granules inside Langerhans cells is hardly proof of descent, since melanocytes are known to transfer granules to keratocytes (Breathnach and Wylie 1965). Moreover, Wolff and Winkelman (1967*a*), studying the effects of ultraviolet irradiation on red and albino guinea pig skin, found little increase in the numbers of ATPase-positive dendritic cells, but a large increase in DOPA-positive cells, indicating that the two cells form separate populations.

The third concept is that of a self-renewing, independent population of cells. There is now mounting evidence that Langerhans cells are fully capable of proliferation: Hashimoto and Tarnowski (1968) observed a mitosis in an unequivocal Langerhans cell, while Giacometti (Giacometti 1969; Giacometti and Montagna 1967) claimed that primate Langerhans cells flash labelled with [$^3$H]-TdR after epidermal wounding and UV-irradiation; labelling after DNCB application has also been seen (Gschait and Brenner 1979). Labelling of ATPase-positive cells has also been observed in resting epidermis, in the guinea pig at an incidence of 0.125 per cent, rising to 1.5 per cent after tape stripping (Schellander and Wolff 1967), and also in the mouse (Mackenzie 1975*b*), with a labelling index of 0.01 per cent; within one and five days of injection, labelled Langerhans cells were seen in pairs, separated by keratocytes, but up to 13 days after injection there was no increase in the labelling index, and cells possibly did not divide again. Thus Langerhans cells are self-maintained, but it appears that they only divide once after going through an $S$ phase as a recognizable (ATPase-positive) Langerhans cell. Because of the observed pairing of the progeny, transformation or recruitment, say from the dermis, was unlikely, even though Hashimoto and Tarnowski (1968) had previously observed Langerhans cells in the process of migrating from the dermis, and indeed the observations of

Mackenzie (1975*b*) do not exclude a source of Langerhans cells from elsewhere; a progenitor cell could migrate from the dermis and undergo a last $S$ phase as a recognizable Langerhans cell in the epidermis.

The location of Langerhans cells in the epidermis is also of some import: in resting epidermis, 17–20 per cent of Langerhans cells are in the basal layer in the guinea pig (Lezzard *et al.* 1966, 1968). The remainder are suprabasal. Tape stripping appears to remove all suprabasal Langerhans cells, but the basal Langerhans cells begin repopulation at 10 days, and by 21 days after damage the original Langerhans cell:keratocyte ratio is achieved; melanocytes remain constant. Since they did not see mitoses in Langerhans cells, Lezzard *et al.* considered that migration or transformation was the source of the Langerhans cells, although we now know that Langerhans cells can divide, at least at some stage of their life-cycle.

The problem lies in deciding whether this proliferation is sufficient to maintain the observed Langerhans cell numbers. The last proposal is that Langerhans cells originate from bone marrow elements. There is mounting evidence that Langerhans cells are concerned in some way with the afferent arm of the immune system: they express Ia antigens (Stingl *et al.* 1977; Rowden 1977; Klareskog *et al.* 1977; Forsum *et al.* 1978), which are of bone marrow origin (Frelinger *et al.* 1979); they have Fc and C3 complement receptors on their surface (Rowden 1977; Stingl *et al.* 1977), and are seen in close association with mononuclear cells in contact allergic dermatitis (Silberberg 1973). The studies on chimaeric mice by Katz *et al.* (1979) would support this mode of origin: donor transplanted skin becomes colonized by recipient Langerhans cells after three weeks, and irradiated recipients reconstituted with donor marrow showed donor Langerhans cells within three months. These experiments indicate that epidermal Langerhans cells are continuously repopulated by a mobile pool of precursor cells migrating from the bone marrow.

The turnover time of Langerhans cells is yet unknown, but 32 days after irradiation-induced chimaerization, 33 per cent of epidermal Langerhans cells are of donor origin (Tamaki and Katz 1980). In any computation, however, some notice would have to be taken of the slow, but detectable, *in situ* proliferation (MacKenzie 1975*b*).

The next question, of course, is which bone marrow cell? There is evidence for the existence of cells in lymph nodes and spleen with structural, antigenic and functional similarities to Langerhans cells—the so-called *interdigitating reticulum cells* (Rausch *et al.* 1977; Hoffman-Fezer *et al.* 1978), and Rowden (1980) has proposed a schema relating marrow monoblast, dermal histiocyte, and interdigitating cells (Fig. 5.4).

Thus the life-history of the Langerhans cells is becoming clearer: it takes origin in the bone marrow, and migrates to the epidermis where it has some *in situ* proliferative capacity; Langerhans cells then leave the epidermis. The relative contribution of migration and proliferation to the maintenance of Langerhans cell numbers is not yet clear; whether self-maintenance occurs,

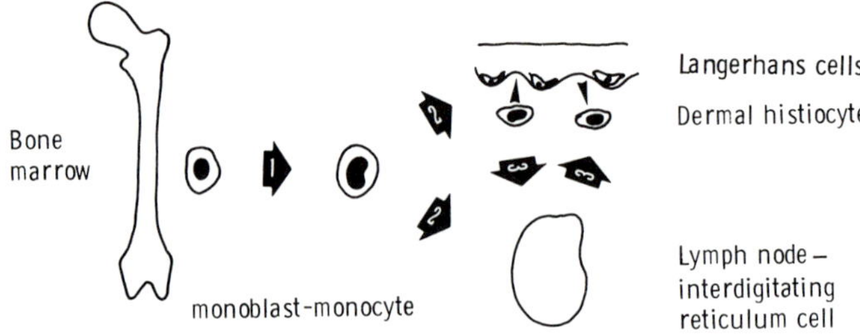



Fig. 5.4.  A possible flow diagram for the life-history of the Langerhans cell. Bone marrow monoblasts give rise to circulating monocytes (1); these cells migrate to the epidermis and become Langerhans cells, or to the lymph nodes where they appear as interdigitating reticulum cells (2). There is free exchange between the epidermis and the lymph node (3). (After Rowden (1980).)

at least in normal circumstances (Rowden 1980), or whether migration and emigration is the primary controlling mechanism in the renewal of Langerhans cells (Tamaki and Katz 1980).

## 5.4.  The Merkel cell

Comparatively little work has been done on the lineage of Merkel cells: Tachibona (1978) summarized the available evidence. Because Merkel cells bear desmosomes and hemidesmosomes, contain tonofilaments, and are phagocytic for melanin granules, authors such as Smith (1970) have proposed their origin from keratocytes. However, Breathnach and Robins (1970) and Hashimoto (1972) have noted their appearance in human fetal dermis at the 16th week, and later in development, in the epidermis, and considered that the Merkel cell represented a cell which enters the epidermis from the mesoderm; the possibility of an origin from the neural crest was also raised. However, in 'aneurogenic' *Ambyostoma* larvae, produced by removing the entire nervous system, including the neural crest, Merkel cells still develop.

# 6 Problems in the measurement of kinetic parameters in squamous epithelium

Most of the classical kinetic techniques, described in some detail in Part 1 of this book, have been applied to squamous epithelium, and particularly to the epidermis. However, there are a number of special methods which are certainly worth remarking upon, and several of the more established methods have problems which are peculiar to these tissues.

## 6.1. Proliferative indices

We have already noted the general advisability of relating counts of labelled or mitotic cells to the total nuclear population, and also the potential errors incurred from viewing objects in microtome sections (see Chapter 3, section 1). Several other problems emerge in squamous epithelium, particularly in the case of the labelling index (but some of these considerations also apply to the mitotic index).

### (i) *Correction factors for the labelling index*

The first of these problems is the effect of section thickness upon determination of the flash labelling index. This point has received some notoriety as a result of the current controversy concerning the cell cycle time in the skin disease psoriasis; Weinstein and Frost (1968), using sections cut at 5 $\mu$m, employed a correction factor of 1.4, by which the experimental labelling index should be multiplied, to correct for the short range of $\beta$ particles in tissue sections (Iversen and Evensen 1962); Simnett (1968) calculated a similar factor for use in the analysis of lung sections. However, as Gelfant (1976) points out, Weinstein and his co-workers have been rather inconsistent in applying this correction factor to their data. Goodwin, Hamilton and Fry (1974) rejected the use of the correction factor, and quoted, in support, the results of Gelfant (1962), who reported no difference in $I_S$ values in mouse epidermis, whether measured in 2 or 7 $\mu$m sections. In a later publication, Gelfant (1976), now considering human epidermis, could find no statistical difference in $I_S$ in *adjacent* sections cut at a nominal thickness of 3, 5, 6, or 7 $\mu$m. Taguchi and Tabachnik (1973), in an unemotive report, compared $I_S$ values in flash labelled guinea pig epidermis in 4 and 6 $\mu$m sections. There was a 7–10 per cent decrease in $I_S$ values in the thicker sections, a difference which, in most studies, would be small compared with the sampling error of the count (cf. Weinstein and Frost 1968). In considering these arguments, the reader should remember the difficulty in cutting thin sections of skin in paraffin embedded material; even so, autoradiographically speaking, infinite thickness occurs at 2 $\mu$m, but increased efficiency compensates for decreased grain counts in thinner sections. The only reason the $I_S$ has been so critical is because of its use in calculating the cell cycle time by a stage-duration calculation. Of course, this is not a cell cycle time, but merely a turnover time for the included nuclear population; the labelling index therefore decreases in importance because of the several problems involved in such calculations.

If, however, labelling index measurements are used *comparatively* in more important measurements, as for example, in a double labelling experiment to measure $r_S$, then these difficulties can be disregarded.

### (ii) *The identification of the proliferative compartment*

In some squamous epithelia, this is a simple matter, since, in accord with classical ideas (see Marques-Periera and Leblond 1965), only the basal layer contains mitoses and labelled cells; this is the situation in, for example, the mouse oesophagus (Marques-Periera and Leblond 1965), the hairless mouse dorsal epidermis (Iversen *et al.* 1968), and also, apparently, the hamster cheek pouch epithelium (but see Møller *et al.* (1979) who disagree with this). However, problems occur when proliferative activity is present above the level of the basal layer; Hoffman (1949) was in no doubt at all that mitotic activity was present well above the basal layer in resting human epidermis, and in resting oral epithelia of several species, Macdonald (1971) has remarked on the occurrence of flash labelled cells in suprabasal layers. However, we should be cautious in accepting such observations, since a suprabasal labelled cell in one section may in fact be in the basal layer if viewed in the next serial section. But Penneys *et al.* (1970), who photographed serial autoradiographs of normal human epidermis (thus avoiding this latter error), noted that no less than 32 per cent of labelled cells were in suprabasal layers.

However, it is in hyperplastic states that this problem acquires real significance. During the proliferative response to wound healing, we shall see that numerous observers have reported the presence of suprabasal proliferative activity (Chapter 10), and in hyperplastic epidermal states such as psoriasis, it is usual to find as many as three or four layers of proliferative cells; that this was not artefactual was shown by Penneys *et al.* (1970), who reported that the proportion of labelled suprabasal cells was increased to 49 per cent in psoriatic epidermis. A further problem was raised by Schellander and Marks (1973), who ascribed the 13–14 per cent of labelled cells observed in 'high epidermal levels', in epidermal hyperplasia induced by dermal granulomatous inflammation, to labelled Langerhans cells, but since ATPase with concomitant [$^3$H]-TdR labelling was not carried out on the same section, we cannot be sure of this conclusion (see Chapter 5, section 3).

There remains the question of which population to include in the count when estimating proliferative indices; this will, of course, depend upon what is wanted. If the turnover time for the viable cell population is required, then we should include all nucleated cells in the count, but if, for example, the labelling index of only the potentially proliferative population is needed, to use in a stage-duration calculation of the cell cycle time, then such a counting procedure would incorporate large numbers of non-proliferating cells in the count. On the other hand, Penneys *et al.* (1970) considered that confining the count to the basal layer only may give artificially high readings in hyperplastic states, presumably because either more basally sited cells are proliferative, or because of their shortened $T_C$. There are two further counting alternatives: some workers (Weinstein and Frost 1968; Duffill *et al.* 1976) have laid down morphological criteria for what are euphemistically called 'germinative cells' in human epidermis. For example, Weinstein and Frost (1968) defined the germinative cell compartment as 'the area of labelling and mitosis'(!) but noted that such cells also have an increased nucleocytoplasmic ratio. Whatever the definition used, it is evident that the criteria for recognition will differ between workers, with the consequence that the comparison of absolute values for the labelling index between groups (Weinstein and Frost 1968; Goodwin *et al.* 1974) is a singularly fruitless exercise.

The other alternative, which in our opinion exists only to be deplored, is the practice of counting mitoses or labelled cells per unit length of epidermal surface, or of basement membrane. Admittedly this avoids the problem of the definition of the proliferative compartment, but automatically introduces the further problem of the constancy of the nuclear density. In comparison of proliferative indices, if the particular experimental conditions lead to an increased length of the basal lamina (as, for example, in a change from the Type I epidermis of Bullough and Deol (1975) to their Type 2 epidermis; see Chapter 11, section 2), or to alteration in cell dimensions, then indices will change in the absence of kinetic differences. Stringent nuclear counts then become necessary (see Chapter 3, section 1). Furthermore, without relating the index count to the nuclear population in some way, kinetic calculations of phase duration become largely non-tenable.

In this situation it is difficult to know what to advise: in experimental comparison of indices, then the problem becomes unimportant so long as the same population is counted each time. If turnover times are required, count the population in question (basal cells, viable cells, etc.). But when absolute measurements of $I_S$ are needed for the calculation of $T_C$ from, say

$$T_C = t_S/I_S$$

or
$$I_S = \exp(t_S \ln 2/T_C - 1) \cdot \exp(t_2 \ln 2/T_C)$$

then, unless proliferative activity is confined to the basal layer (see above), assessment is difficult. We should define our population as well as we are able, and calculate $T_C$ for this population: we should not expect, however, that our results will be directly comparable with those of other investigators, using different criteria; and we should not represent our indices, and resultant calculations, as *absolute*.

### (iii) *The use of intradermal injections*

Because of ethical constraints, in human studies, or because of expense in animal work, many investigators have employed the local intradermal injection of [$^3$H]-TdR or of metaphase arrest agents (see for example, Hell and Hodgson 1966; Weinstein and Frost 1968; Goodwin *et al.* 1974; Duffill *et al.* 1976). Naturally a first proviso must be that there is no difference between the results by local as compared with systemic administration. This problem has been analysed most recently by Taguchi *et al.* (1975), who showed, for [$^3$H]-TdR at least, that there was no difference in the $I_S$ of guinea pig epidermis when intradermal, intraperitoneal or subcutaneous routes were used. Furthermore, there was no change in the $I_M$, or in the FLM, for up to seven hours after injection.

A second reservation would be whether the local procedure itself would have an effect on the proliferative status of the epidermal cells. This question was also considered by Taguchi *et al.* (1975), in the guinea pig, and by Grove (1980), in human epidermis; Taguchi *et al.* showed that a single needle puncture of guinea pig epidermis caused a local increase in $I_S$ at 12 hours after wounding, observable for up to 5 mm from the wound, and peaking at 24 hours. Injection of a small amount of fluid caused a similar response, but with a higher peak value. Grove (1980) has investigated the problem in the human skin, using integrating microdensitometry, in both normal, non-lesional, and lesional psoriatic epidermis; although there was no difference in the fraction of hyperdiploid cells at 30 hours after an intradermal injection of normal saline in psoriatic epidermis, and also little difference in normal epidermis, there was evidence of a prominent proliferative response in the non-lesional epidermal cells.

In some ways, these results are similar to those described in the epidermal response to wound healing (see Chapter 10), and indicate that, after a period of about 12 hours, numbers of cells enter DNA synthesis in response to the stimulation; although the results may not affect the measurement of the flash labelling index, or similar short-term procedures, if we were doing an FLM experiment, this effect might lead to an influx of unlabelled mitoses at about the time of the descending limb of the first peak, and an underestimate of $t_S$ in the first instance. The transit time through the epidermis would also be affected by the ensuing proliferative response, and, in this respect Taguchi *et al.* (1975) reported nine days from basal layer to granular layer with systemic exposure to [$^3$H]-TdR, but only seven days after intradermal injection.

Thus, although intradermal injection seems reasonable for the determination of the $I_S$, because of the induced proliferative response some doubt exists as to the accuracy of measurements in more prolonged procedures, as in FLM or transit time experiments. Added to this, there is also the problem of a variable grain count: Taguchi *et al.* (1975) reported a decreasing grain count with distance away from the injection site, making grain count experiments unreliable. Nevertheless, it should be said that intradermal injection is a convenient method of performing short-term kinetic experiments in situations, such as human studies, where there is a need to limit the amount of [³H]-TdR given; however, total recovery of [³H]-TdR, often pleaded in mitigation for this procedure, may be untenable, since Arndt and Friedberg (1971) showed that only 1.1 per cent of injected radioactivity is removed by a 4 mm punch of guinea pig epidermis at one hour after injection.

### (iv) *Does [³H]-TdR label all cells in DNA synthesis in squamous epithelium?*

We have mentioned this problem before (Chapter 3, section 1); we usually assume that all cells are detectable by the incorporation of [³H]-TdR, and are then seen in autoradiographs. We of course can appreciate that the amount of [³H]-TdR given, its specific activity, and the emulsion exposure time can modify the labelling index (Simpson-Herren *et al.* 1976), but in the work of Clausen and Lindmo (1976), Møller and Larsen (1978), Clausen (1979), and Clausen *et al.* (1979) there are definite indicators that there are cells in the *S* phase which do not take up [³H]-TdR; the problem was investigated in some detail by Clausen *et al.* (1980), who compared the fraction of hairless mouse epidermal cells in the *S* phase by both flow cytophotometry and autoradiography at 30 minutes after the administration of various amounts of [³H]-TdR; at low [³H]-TdR dosage levels (i.e. 0.5–1 µCi/g body wt), it was found that the autoradiographic method significantly underestimated the *S* fraction, compared with flow data, by 40–50 per cent! At higher [³H]-TdR dose levels, the degree of underestimation was smaller, but even at doses as large as 10 µCi/g body wt, flow measurements indicated a higher *S* fraction than autoradiography; although increasing the exposure period decreased the deficit slightly, this did not make a major difference. We have, however, seen that a problem in the interpretation of flow data is the method used to measure the *S* fraction from the DNA histogram. Clausen *et al.* used the planimetric method, but of course the objection could still be raised that the method overestimated the *S* phase fraction; this was effectively countered by using the cell sorter facility to sort [³H]-TdR labelled cells, and to isolate cells from the middle of the *S* phase; this showed that 1 µCi/g body wt labelled about 50–80 per cent of mid-*S* phase cells, while 80 per cent were labelled by 4 µCi/g body wt. Interestingly enough, 93 per cent of mid-*S* phase cells were labelled in epidermis which was regenerating after damage.

The reasons for this discrepancy are presently unknown: Møller *et al.* (1982) have also detected *S* phase cells which do not label in hamster cheek pouch epithelium, so the phenomenon is presumably widespread. There are several possible mechanisms; cells could be stopping DNA synthesis completely at some point in the *S* phase, in the manner indicated by the experiments of Howard and Dewey (1961) using metaphase grain count analysis in some plant meristems;

alternatively there could be some metabolic reason for the $[^3H]$-TdR not labelling the $S$ phase nucleus. Møller and Keiding (1982) have claimed that the endogenous TdR pool varies in size throughout the day in hamster cheek pouch squamous cells; if the TdR pool was larger in some cells than others, this might account for them not labelling. Furthermore, Møller *et al.* (1982) also considered that ordinary basal cells were capable of storing $[^3H]$-TdR in the intracytoplasmic pool, which was later incorporated, possibly over the next three or four hours; this could also lead to the presence of unlabelled cells in $S$ after only one hour's exposure. The small fraction of cells showing late labelling up to 17 hours after injection, described by Potten *et al.* (1978; see Chapter 8, section 4) are probably too few in number to really contribute to the non-labelled $S$ fraction. In this context, Imondi *et al.* (1969) and Salser and Balis (1973) have detected a correlation between low TdR kinase levels and increased levels of differentiation, and Clausen *et al.* (1980) have speculated that the unlabelled $S$ phase cells in mouse epidermis are those which are destined to differentiate after the next cell division, and the low TdR kinase levels restrict $[^3H]$-TdR incorporation in these cells, which must therefore continue to synthesize DNA by the *de novo* pathway.

There are thus several candidate mechanisms, but whatever the cause, it is mandatory to recall this important limitation to the use of the flash $[^3H]$-TdR labelling index. We have already stated our position on the advisability of assessing proliferative rates by the incorporation of $[^3H]$-TdR into TCA precipitable material (see Chapter 3, section 2, where our general reservations are set out), but we should also note that specific objections to this practice emanate from the work of Davidson *et al.* (1979), who reported a consistent failure of $[^3H]$-TdR incorporation to correlate with direct measurements of cell number or of DNA content in keratinocytes growing in culture; this was particularly evident after cells had been growing for some time, and was apparently correlated with a decrease in TdR kinase activity. On the other hand, studies with $[^{14}C]$-aspartate, another pyrimidine nucleotide precursor, indicate an increase in *de novo* synthesis at this time. Thus the utmost care must be taken when using this measurement, and we must counsel strongly against using it as the sole determinant of proliferative rate.

### (v) *The influence of circadian variation*

This topic is discussed in some detail in Chapter 9, but here we should note that the flash labelling index is remarkably dependent on the time of day it is measured. There are very large circadian swings in $I_S$, and any attempt to obtain a value for the labelling index without taking such variation into consideration will naturally lead to serious error in any stage-duration calculation; there can be little doubt that, for accurate work, a 24 hour average should be used if such calculations are contemplated (Potten 1980).

## 6.2. The measurement of the cell cycle time

Many authors have tried to measure, by direct FLM analysis, the cell cycle time in resting stratified squamous epithelium, but few authors have succeeded in obtaining a defined second peak. The reasons for this were discussed by Brown and Oliver (1968), who noted that, in systems where $T_C > 100$ hours, a second FLM peak was not defined; we might now suggest that the basis for this argument lies in the potentially wide distribution of phase durations which might be found in such a long cycling population, leading to rapid randomization of cells throughout the cell cycle, loss of synchrony of the labelled cohort, and no second peak. However, we should note that Potten (1976) considers that one reason for the absence of a second peak is that most basal layer mitoses, in the stacked mouse epidermis at least (see Chapter 8, section 2), are the final mitosis before migration and differentiation in the suprabasal layers ensues; thus most labelled basal cells only divide once, and no second peak is discernible.

There are several problems involved with the application of the FLM method in squamous epithelium, not wholly related to the lack of a second peak; the presence of a circadian rhythm in phase fractions is only one of these; one basic assumption of the FLM method is asynchrony of the proliferating cells, a problem considered in Chapter 9, section 3. One problem more related to the length of the cell cycle is the possibility of $[^3H]$-TdR reutilization (see Chapter 3); upper granular layer cells undergo nuclear degeneration, with release of $[^3H]$-TdR at perhaps 3–5 days after injection in mouse epidermis (Cutright and Bauer 1967), thus releasing labelled TdR during the period when, with a $T_C$ approaching 100 hours (see Fig. 7.2, p. 254), a second peak might be expected (Potten 1981).

Other problems are more related to the properties of the epithelium, and particularly of the stem cells; one of these is *delayed labelling*; there are some cells which are thought to be able to store $[^3H]$-TdR intracellularly, and only incorporate it at a later time; consequently the duration of the pulse could be increased for these cells. Although Potten *et al.* (1978) have ascribed this property to only a small epidermal subpopulation, Møller *et al.* (1982) have shown that all cells in the epithelium of the hamster cheek pouch have the potential to incorporate $[^3H]$-TdR over a period of 3–4 hours (see section 6.1). Again, this negates one more assumption of the FLM method: that the labelling period is short (see Chapter 3, section 4). Potten (1981) has also suggested that the non-random segregation of DNA at mitosis in certain putative stem cells may affect the FLM curve (see also Chapter 8, section 4); however, the number of these cells is small, their cell cycle time supposedly long, and the influence of this factor is likely to be of minor importance only.

Having said this, there are a number of FLM curves in squamous epithelium which do achieve a second peak, and their interpretation is of considerable interest (see Chapter 7, section 2, where these and other problems of the FLM curve in squamous epithelium are further discussed).

A practically ingenious and theoretically interesting method for deriving $T_C$ in stratified squamous epithelium was proposed by Brown and Oliver (1968), and is worthy of some attention; this method is based on the premise that (hamster cheek pouch) stratified squamous epithelium is divided into two well-defined compartments, the basal and differentiating cell layers, and that cell loss from the basal layer only occurs by migration into the differentiating layer. Brown and Oliver (1968) flash labelled hamsters with [³H]-TdR, and killed animals serially over 100 hours after injection (see Fig. 6.1A). The rate at which

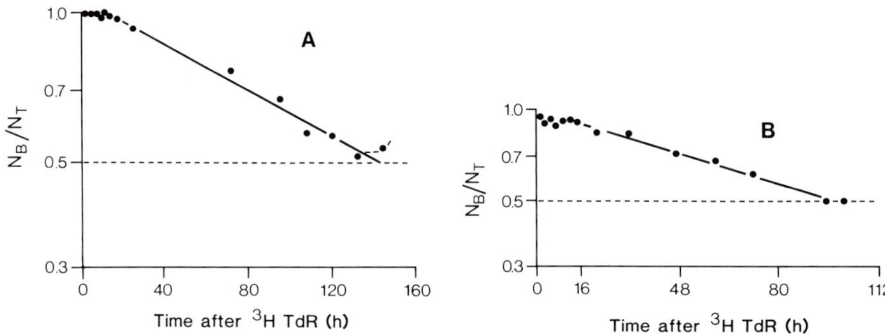


Fig. 6.1. Changes in the value of $N_B/N_T$ with time after injection of [³H]-TdR in (A) the hamster cheek pouch squamous epithelium, and (B) the mouse dorsal epidermis. The slope of the line gives an estimate of $T_C$ if $I_P$ is known. In (A) the line passes through the unit ratio at 12 hours after [³H]-TdR, for reasons explained in the text, and thereafter the line is of best fit (least squares) up to 132 hours. (Redrawn from (A) Brown and Oliver (1968) and (B) Hegazy and Fowler (1973*a*).)

labelled cells migrated from the basal layer was assessed by plotting the ratio of labelled basal cells ($N_B$) to total labelled cells ($N_T$); this is then drawn on a semilogarithmic plot. The ratio $N_B/N_T$ remains constant for about 12 hours, in accord with the observation, of Marques-Periera and Leblond (1965), that labelled cells do not begin migration until the first mitosis is completed (see Chapter 8, section 6). However, even if basal cell migration was a completely random process, there would tend only to be a very small fall in the ratio $N_B/N_T$ since, as nearly all cells divide in the plane of the basal layer (Brown and Oliver 1968), the increase due to cell division would compensate for any labelled cells lost as a result of the random removal process. However, after 12 hours, the ratio starts to fall in an exponential manner, as indicated by the straight line, and supporting a random migration mechanism. The slope of the line is then the birth rate, $k_B$, and with an exponential age distribution (see section 6.7),

$$k_B = \ln(1 + I_P)/T_C. \tag{2.13}$$

There are then two factors affecting the limiting value of $N_B/N_T$; first, at the end of one cell cycle time, the remaining basal cells will divide again, doubling the value of $N_B$. Thus a discontinuity should appear at time = $T_C$. Assuming a random removal process, after the first division of labelled cells (i.e. $\sim$12 hours), $N_B$ after a further time $t$ is

$$N_B = N_T e^{-k_B t} \tag{6.1}$$

then, from eqn 2.13 above

$$N_B/N_T = e^{-k_B t} = \exp[-\ln(1 + I_P)/T_C]\,t \tag{6.2}$$

and when $t = T_C$

$$N_B/N_T = \exp[-\ln(1 + I_P)]. \tag{6.3}$$

If $I_P = 1$, and all cells are proliferating, then the discontinuity at $t = T_C$ should therefore occur at $N_B/N_T = 0.5$ (by substituting $I_P = 1$ into eqn 6.3). Fig. 6.1A shows such a discontinuity, and, furthermore, so does the data of Hegazy and Fowler (1973$a$) for mouse epidermis (see Fig. 6.1B). It would appear then, in these tissues $I_P$ is unity.

The second factor modifying the limiting value of $N_B/N_T$ is the mean transit time $T_T$ through the differentiating or spinous layer. At $t = T_T$ cells will be lost into the stratum corneum. If $N_S$ is the number of cells in the basal layer, the number of cells entering the spinous layer per unit time will be $N_S k_B$; similarly the total number of cells in the spinous layer $N_D$ will be $N_S k_B T_T$. Then

$$k_B T_T = N_D/N_S = T_T/T_C \ln(1 + I_P).$$

From eqn 6.1, at $t > T_T$ the ratio $N_B/N_T$ will be a constant $e^{-k_B T_T}$, where $-k_B T_T = N_D/N_S$. Brown and Oliver found this value to be 0.99 from direct observation; now the limiting value of $N_B/N_T = e^{-0.99} = 0.37$. However, the 0.5 discontinuity has intervened in each of the experiments shown in Fig. 6.1. Consequently, $I_P$ is unity, and gives, for the cheek pouch, $T_C = 130$ hours, and for mouse epidermis, 107 hours; both values are in good agreement with alternative techniques. Note that a value for $T_T$ can be obtained by extrapolation to $N_B/N_T = 0.37$.

This method apparently gives values for $T_C$, $I_P$, and $T_T$, and would appear useful in squamous epithelia. However, there are several limitations which compromise its applicability generally: (i) the presence of proliferative activity above the basal layer, commonly found in both human and animal (Penneys *et al.* 1970; Møller *et al.* 1979) epidermis would complicate the model; (ii) in many epithelia, cells do not divide in the plane of the basal layer (Duffill *et al.* 1977, and see section 6.7), and the presence of vertical mitoses would mean the migration of *labelled cells by division*, rather than by a squeezing out process by horizontal mitosis. These two values would both have the effect of modifying the initial values of $N_B/N_T$, and clearly, migration closely associated with a cell cycle event cannot be said to be random. This leads to our third consideration: (iii) many stratified squamous epithelia have a highly ordered structure (see Chapter 8, section 2), with a distinct columnar arrangement. Repeated reports have shown that migration occurs selectively at the periphery of these columns, and recent theory maintains that the migrating cells are post-mitotic cells which are already in the process of differentiating. This would be inconsistent with random removal (see Chapter 8, section 6). Slowly proliferating mouse epidermis shows such structure, but we should note that Hegazy and Fowler (1973$a$) obtained a straight line of $N_B/N_T$ decrease, support for a random removal process; (iv) the discrete discontinuity of $N_B/N_T$ in both graphs at $t = T_C$ (see Fig. 6.1), indicates a narrow distribution of cell cycle times. The effect of an increasing variance of $T_C$ would be to blur the discontinuity and make the measurement of $T_C$ difficult. However, we should note that the form of the FLM curve obtained by Hegazy and Fowler (1973$a$), for resting mouse epidermis, does indicate such a narrow distribution of cell cycle times (see Fig. 7.2A, p. 254); (v) the ratio of the numbers of differentiating cells to basal cells, $N_D/N_S$, is important. In Brown and Oliver's results $N_D/N_S = k_B T_T = 0.99$, and the plateau value at $e^{-0.99}$ or 0.37 could be distinguished from the 0.5 expected from the $I_P = 1$ argument. However, if $N_D/N_S = 0.69$, we would have $e^{-0.69}$ or 0.5, and be unable to distinguish the $T_T$ and $T_C$ effects, which would lead to some confusion; in fact, Møller (personal communication) has reported $N_D/N_S$ to be 0.7 in her strain of hamsters, and in many other, even normal, squamous epithelia this situation could well obtain; (vi) all that we can really say about the growth fraction is that all cells in the basal layer are potentially proliferative, i.e. will enter the cell cycle at some time in time $t = T_C$. This does not of course exclude a subpopulation in $G_0$ with a turnover time $<T_C$ (see Chapter 7, section 4). A similar argument applies to $I_P$ estimates from continuous labelling procedures (see Chapter 7, section 4).

Finally, we should note that Hegazy and Fowler (1973$a$) have applied this method to plucked epidermis, where a hyperproliferative state is induced, and steady state conditions no longer apply. In addition to the prospect of an induced wave of proliferation, there is also the distinct possibility that the suprabasal layers will contain proliferative cells (i, above); that more mitoses will have vertical axes (ii, above; and see Bullough and Mitrani 1978), and that the migration pattern will become disordered (Iversen *et al.* 1968).

Thus we should beware the indiscriminate application of this potentially powerful technique to systems without rather stringent criteria, particularly in perturbed conditions. We should note that not all attempts to use this method have been deemed successful; Olsson (1976), in mouse epidermis, could not record an upswing when the ratio $N_B/N_T$ had reached 0.5; in fact, careful inspection of the data shows that the ratio 0.5 was not in fact reached.


## 6.3. Turnover times and transit times

Workers in epidermal cell kinetics particularly have been fond of calculating turnover times, and much of the present controversy about the kinetic situation in human psoriatic epidermis stems from confusion about what constitutes a turnover time and what a cell cycle time (Wright and Camplejohn 1983). In a typical squamous epithelium, with proliferation confined to the basal layer, the basal layer turnover time $T$ will typically be $1/k_B$ where $k_B = \ln2/T_C$ or $\ln(1 + I_P)/T_C$ where $I_P < 1$. The choice of age distribution is an important step in squamous epithelium, and is discussed in section 6.7. In many situations in squamous epithelium we do not have values for $T_C$ or $I_P$, and the best that can be done is to calculate $k_B$ from $I_S$ and $t_S$, or $I_M$ and $t_M$, and obtain an estimate of the basal layer turnover time, but we should appreciate it as *a turnover time only*. Of course, because the basal layer contains the stem cells of the system, we cannot have a transit time *per se*, only a rate of turnover and a turnover time.

In the spinous layer, things should be simpler, since it classically represents a simple transit compartment, where $T = T_T$. Many investigators have simply labelled basal cells with $[^3\mathrm{H}]$-TdR, and measured the time taken for labelled cells to reach the granular (or corneal) layer (e.g. Epstein and Maibach 1965); this was then the transit time through the spinous layer, and equal to the turnover time. Halprin (1972) and Potten (1975*a*) have critically examined this concept, and have exposed several problems: (i) at the level of the stratum corneum, the cells are attached firmly, and consequently probably move at the same rate, with the result that the minimum, mean, and maximum transit times are all the same; migration from the basal layer may well be a random process (Marques-Periera and Leblond 1965), and consequently the spread (in time) of labelled cells arriving at the granular layer will be wide. Thus we must distinguish between the *minimum transit time*, or the time taken for the first labelled cell to reach the granular layer, and the *average or mean transit time*, which is much more difficult to derive. Nearly all published measurements are minimum transit times; (ii) labelled cells are known to stay in the basal layer for some time before migration; Leblond *et al.* (1964) and Marques-Periera and Leblond (1965) reported that the minimum delay for labelled cell migration was 12 hours in mouse oesophagus, but Potten (1975*a*) has calculated that labelled cells, passing through a post-mitotic phase, may stay in the basal layer for up to two days in mouse dorsal epidermis, and as long as five days in tail epidermis. This *post-mitotic maturing phase* will compromise any measurement of transit time made using $[^3\mathrm{H}]$-TdR by this particular method, and should be subtracted from the measured $T_T$ to the granular layer; (iii) errors are also incurred if labelled cells are present above the level of the basal layer, which is, in fact, the normal arrangement in human epidermis. This is really one of the problems in defining the proliferative compartment, since, if there are flash labelled suprabasal cells, then the time of appearance of labelled granular layer cells is not, of course, the minimum transit time from the basal layer, even ignoring the problems set by post-mitotic maturing cells.

We have seen above that transit time through the stratum corneum is a simpler concept, since minimum, mean, and maximum transit times are held to have the same value. Many investigators have measured the transit time through the stratum corneum by studying the disappearance of radioactively labelled amino acids, or of fluorescent compounds, from the epidermis; the labelled amino acid method depends on the uptake of the label by viable epidermal cells upon systemic or intradermal injection, and on passing through the stratum corneum, label appears in superficial squames and can be detected in tape strippings of the surface (see Chapter 7, section 6; Porter and Schuster 1961; Rothberg *et al*. 1961). Now these methods are reasonable if label is incorporated regularly into epidermal cells, particularly in the granular layer. However, if label is incorporated into spinous cells and not granular layer cells, then the transit time will be overestimated; in this respect, Fukuyama and Epstein (1966), localizing a wide selection of amino acids in newborn rat epidermis, concluded that all amino acids were incorporated into all layers, but leucine, lysine, and methionine were concentrated in the basal and spinous layers, arginine, histidine, and glycine in the granular layer, while proline and tyrosine were evenly distributed in all viable layers; so, if these results apply more generally, then the transit times obtained will largely depend on the amino acid used. The loss of fluorescent compounds, such as dansyl chloride, which stains corneocytes only, is followed daily, usually in the human epidermis, using a UV source, to give the stratum corneum transit time (Baker and Kligman 1967; Jansen *et al*. 1974).

Potten (1975*a*) has attempted to overcome some of the problems implicit in transit time measurements, and particularly that concerning post-mitotic cells in the basal layer, by combining studies of stratum corneum transit time with $[^{35}S]$-methionine with DNA labelling methods, employing $[^{3}H]$-TdR and also $[^{125}I]$-UdR, to overcome the problem of TdR re-utilization in longer-term labelling experiments in mouse epidermis. Making the assumption that the loss of total radioactivity in epidermal punch biopsies after $[^{3}H]$-TdR or $[^{125}I]$-UdR was due to destruction of DNA in the granular layer (see Chapter 4 and Iversen *et al*. 1971), the transit time from basal layer to granular layer, including the post-mitotic maturation time, was estimated; the total transit time (including the stratum corneum) was found from the appearance of label in tape strippings of the surface, which would eventually yield $^{3}H$- or $^{125}I$-labelled corneocytes; finally the transit time through the stratum corneum was found by following the appearance of $[^{35}S]$-methionine-labelled corneocytes in tape strippings after incorporation by viable epidermal cells. The data obtained by these measurements are summarized in Table 8.1 (p. 294), but here we might note Potten's method for calculating the post-mitotic maturing time in the basal layer. Knowledge of the number of cell layers in the stratum corneum, together with its transit time, allows a calculation of the transit rate per layer; assuming that this transit or flow rate also applies to viable cell layers, Potten calculated the *theoretical* transit time from the basal layer. The difference between this value and the measured total transit time would reflect the time spent by labelled cells, still in the basal layer, to pass through the post-mitotic maturing stage and migrate. In the mouse epidermis, this time varied from two days in the dorsal epidermis to 3.8–4.8 days in tail epidermis.

Difficulties in this approach include the acceptance of the various assumptions that are made: that decreased radioactivity in whole punch biopsies reflects the transit time to the granular layer; that the appearance of radioactivity in strippings reflects the total transit time, and that the migration rate in the stratum corneum accurately reflects the transit rate in the viable cell layers. The argument for equivalence of transit rates in the viable and corneal cell layers would appear to be reasonable, and was also made by Bergstrasser and Taylor (1977), in their calculation of transit times. However, Iversen *et al.* (1971) considered that the appearance of radioactivity in strippings was questionable when used as an indicator of transit time (and see Chapter 4); moreover, it is evident that the estimates obtained will be on the minimum side. Nevertheless, this approach, although approximate, is interesting because of the light cast on the kinetics of post-mitotic cells in the basal layer.

## 6.4. Cytophotometry

The general principles of this technique have already been discussed (see Chapter 3, section 8). There have been applications of the two main methods, i.e. integrating microdensitometry in sections (Grove and Graham-Smith 1976; Grove 1979), where there are of course problems of sectioned and overlapping nuclei to contend with, and flow methods, and the limitations of this method, in terms of its application to epithelia, where variations in kinetic microarchitecture can be extreme, have already been mentioned. However, useful information has been obtained; earlier workers have studied the microdensitometric properties of epidermal cells; Alavaikko (1971) in mouse cell suspensions prepared by enzyme digestion, and Grove *et al.* (1976) in cell squashes prepared from psoriatic epidermis. From the point of view of obtaining phase population density, this method suffers from the drawbacks of small sample size, and from the difficulties in distinguishing 'germinocytes' or potentially proliferative basal cells from differentiating spinous cells. Several authors have now described separation techniques for use with the flow cytophotometer, which effectively removes the objection to small sample size experienced with the microdensitometric method; Haag *et al.* (1975), Clausen *et al* (1976), Bauer and de Grood (1975, 1976), Møller and Larsen (1978), Frentz *et al.* (1980), and Bauer *et al.* (1980, 1981) have all described methods of isolating single cell suspensions; in some of these studies (for example, Bauer and de Grood (1976), in human epidermis), no attempt was made to separate potentially proliferative cells from differentiating cells, and consequently the $G_1$ peak will be artificially inflated by numbers of differentiating keratocytes. This means that the only useful parameter one has in discrimination is the hyperdiploid fraction, i.e. $S + G_2$ cells, which to some extent is a serious drawback (Frentz *et al*. 1980). However, Clausen *et al*. (1976), for example, have adapted the technique described by Laerum and Boyum (1969), and used by Elgjo *et al*. (1971, 1972), for separating mouse epidermal basal cells from differentiating cells, whereby spinous cells are removed by a trypsinization procedure, leaving basal cells which are then mechanically removed from the basal lamina. Clausen *et al*. (1976) calculated the purity of the basal layer fraction at 87 per cent, indicating only a minor contamination with differentiating cells; cells in mitosis are lost by this procedure, but, as the $I_M$ is low this gives only a small error; Sasai *et al*. (1979) have described a method for human epidermis. Admittedly, this method will still incur errors because of the inclusion of non-proliferating cells in the $G_1$ peak, but this could be rectified by using a two-parameter method for distinguishing, not only non-proliferating post-mitotic cells, but also cells in a $G_0$ compartment (Darzynkiewicz *et al*. 1978, and see Chapter 3, section 8). Despite the problems still remaining in data analysis, this method is already making its impact on epithelial cell kinetics (see, for example, Clausen and Lindmo 1976; Aarnes *et al*. 1981).

### 6.5. *In vitro* methods

Numerous workers have employed *in vitro* methods of analysing cell proliferation in squamous epithelium, notably epidermis; those methods which have found favour include organ culture of skin or of isolated epidermis (Bullough and Laurence 1961*a*; Friedmann *et al*. 1966; Kierney and Rothberg 1969; Hall 1969*a,b*; Marks *et al*. 1971*a*; Levine 1972; Hell and Maibach 1972; McGuire and Arnesen 1972; Macdonald and Los 1973; Bishop and Cox 1974; Pullman *et al*. 1974; Gaylarde and Sarkany 1975; Haustein 1979); primary epidermal outgrowths (Flaxman and Chopra 1972); skin plugs (Young *et al*. 1975); *in vitro* clonal growth (Green 1979, Pechl 1980), and isolated epidermal cells growing in culture (Laerum and Boyum 1969; Olvey and MacDonald 1972; Vaughan and Bernstein 1971; Elgjo *et al*. 1976; Delescluse 1977, 1979; Liu 1978, 1979). Some of the kinetic results of these and other studies are discussed in Chapter 7, but there has been a disturbing tendency for investigators, using these methods, to equate kinetic measurements *in vitro* with the *in vivo* situation; for example, Flaxman and Chopra (1972) compared the cell cycle times of normal and psoriatic epidermal cells in explant culture, using the FLM method, and Pullman *et al*. (1974) concluded that circadian variation in $I_S$ and $t_S$ occurred in psoriatic epidermis, and not in normal epidermis, again on the basis of *in vitro* studies alone. With a few notable exceptions (see, for example, Elgjo *et al*. 1976) authors have not bothered to stringently categorize the experimental system in terms of the kinetic changes which occur over the time period used for the procedure, and many have relied on the demonstration of linearity of [³H]-TdR incorporation over a short period (see for example Kierney and Rothberg 1969), or comparison of $I_S$ values *in vivo* and *in vitro* at a single time point (Hell and Maibach 1972), as the sole criterion of validation. However, there is evidence, in some studies, that the DNA synthetic rate is certainly not linear over even a short period after explantation (see for example, Schmid and Hornstein 1974); moreover, there may be marked changes in phase distribution when tissues are explanted (Simnett 1972), with accumulation of cells in $G_2$, and, additionally, marked synchrony is frequently observed when cells are explanted or cultured; for example, Elgjo *et al.* (1976) showed that isolated newborn mouse epidermal cells, grown on glass, showed numerous peaks of $I_S$ and $r_M$ evidently due to synchronization rather than the emergence of sub-populations with different transit times from a $G_0$ compartment. Similarly, MacDonald and Los (1973), in primary explants of guinea pig ear, reported $I_S$ peaks at 8 and 30 hours after explantation, while Olvey and MacDonald (1972) and Young *et al.* (1975) found early bursts of proliferative activity in disaggregated guinea pig epidermal cells and rat skin plugs, respectively. Elgjo *et al.* (1976) considered that the phenomenon was unrelated to medium change or to the resting *in vivo* condition, and proposed that, because of the similarity of the response to kinetic changes which occur after wounding, where synchron-ized peaks of labelling activity are commonly found (see Chapter 10), the *in vitro* proliferative pattern reflects a wound response to the explantation or separation procedure (the mechanism of which is explored in Chapter 10). Whatever the reason for these changes, unless actively looked for, they could introduce enormous complications into say, an *in vitro* assay for a growth inhibitor (see Chapter 14, section 1) using explanted epidermal cells, or attempts to measure $T_C$ by the FLM method (Chopra and Flaxman 1974).

Nevertheless, reports are available which do show that no significant differ-ences are evident in flash labelling indices as measured in the same epidermal tissue *in vitro* and *in vivo* (Hell and Maibach 1972; Lapachelle and Gilman 1969; Schmid *et al.* 1974). However true this may be for $I_S$ values realized in immediate explants, much careful work is required before one is justified in using the *in vitro* method for more prolonged kinetic procedures.

Thus, before accepting *in vitro* results at face value, we must ask: has the system been adequately investigated in terms of the kinetic changes which can accompany explantation? If it has, how far does the system reflect the *in vivo* situation, particularly when *in vitro* systems are used as the basis of physiological or pharmacological experiments, as in the numerous chalone assays (see Chapter 14, section 1)? In this respect, we must agree with Elgjo *et al.* (1976) who urged considerable caution in accepting the results of *in vitro* assays for proliferation control agents. We should also be slow in accepting the equation of *in vitro* deter-minations of kinetic parameters with their *in vivo* counterparts (see Chapter 7).


## 6.6. Cell kinetics and three-dimensional structure

In recent years there has been a movement away from pure kinetic analysis towards the correlation of kinetic organization with the three-dimensional arrangements in squamous epithelia, notably epidermis (see, for example, MacKenzie 1972; Allen and Potten 1974*a*). The techniques which have been used in this rapidly advancing field are fully discussed in Chapter 8, section 2.


### 6.7.   The age distribution in stratified squamous epithelium and its influence on measurements of proliferative rate

We have seen that the age distribution of a population can have profound effects on the estimation of kinetic parameters, particularly when indirect calculations are carried out; the age distribution which is eventually decided upon for any population is usually the result of assumptions, which are normally based on little in the way of experimental data. We have also made the point that the prevailing idea of a steady state population with the classical rectangular age distribution is a gross oversimplification (Aherne *et al.* 1977*a*); moreover, the other limiting case, beloved of mathematical modellers, of a population whose age distribution is exponentially distributed, is equally an oversimplification.

Epithelial tissues which have an ordered kinetic architecture are subject to specific migration processes which will markedly affect the age structure, and the particular mode of migration will determine this effect. The ensuing discussion concerns the dynamics of migrating cells and their effect on the age distribution and on the measurement of proliferative rate (Appleton *et al.* 1977; Duffill *et al.* 1977).

Consider an ordered, layered arrangement of cells as indicated in Fig. 6.2B; for the moment, all cells are proliferating, and have a constant cell cycle time; let us also agree that cells can divide in any plane with respect to any layer. It is evident that the mode of migration will depend on *the direction of the mitotic axis*. In any layer $k$, at each mitosis, at least one daughter cell remains in that layer; the probability that the other cell also remains is $p_k$, but with probability $1 - p_k$ it moves into layer $k + 1$, and in doing so it displaces a cell from that layer into the consecutive layer $k + 2$; the cell displaced does so at random age. If both cells remain in layer $k$, an adjacent non-mitotic cell in layer $k$ is displaced into the layer $k + 1$, with each cell in turn displacing a cell in the layer above. It should be noted that the value $p_k$ can be obtained by observing the proportion of mitotic figures in layer $k$ which have their axes, here defined as a line joining the two daughter nuclei, in the plane of the layer, and consequently it becomes possible to make measurements which reflect directly on the age distribution (Fig. 6.2A).

We will recall that the age distribution of any population is a useful function which allows us to calculate the probability of finding any cell, chosen at random, at a particular age; let the age distribution in our layer $k$ be $y_k(x)$; if we allow $y_0(x) = 0$, and reserve very young cells and those nearing the end of their mitosis until later, the proportion of cells in layer $k$ at age $x + dx$ is

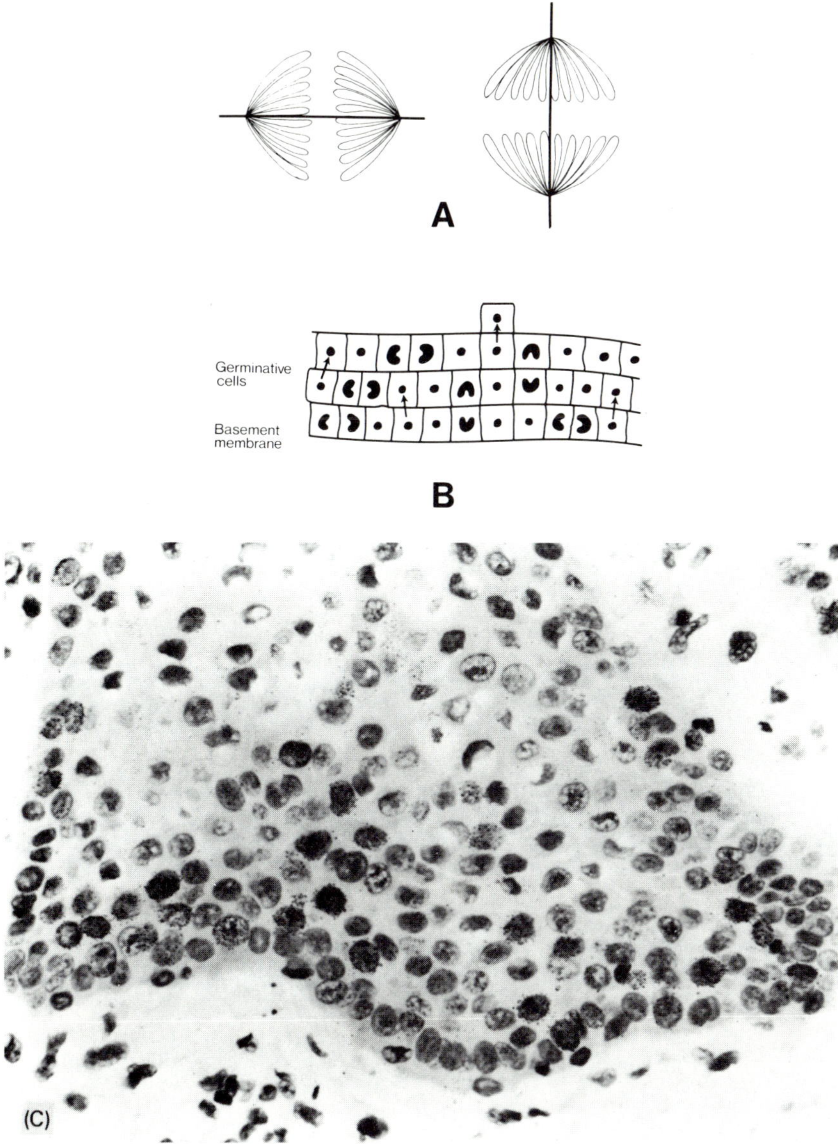



Fig. 6.2.  (A) The orientation of mitotic axes as defined herein: on the left is a metaphase with *horizontal axis*, which is dividing in the plane of the basal layer; on the right is a metaphase with a *vertical* mitotic axis, contributing cells both to the basal and suprabasal layers at division.  (B) A diagrammatic representation of a multilayered array of cells with their mitotic axes orientated at random with respect to the plane of the basal layer.  (C) A rete peg from psoriatic epidermis which has been injected with [³H]-TdR ($10\,\mu$Ci/site) and exposed for 28 days. Note the numerous labelled nuclei, many of which are in suprabasal cell positions, some 4 or 5 cell positions above the basal layer. ($\times 420$.)

$$y_k(x + dx)$$

which equals the proportion of cells at age $x$, i.e. $y_k(x)$; however, we should subtract those cells, at the same age, which are pushed out by the horizontal mitoses occurring in that layer in time $dx$, i.e.

$$y_k(x) - y_k(x)y_k(T_C)p_k\,dx.$$

We should subtract those cells pushed out by any mitoses in any of the $k - 1$ layers

$$-\sum_{i=1}^{k-1} y_k(x)y_i(T_C)dx$$

and of course we must add those cells pushed upwards into $k$ by horizontal mitoses in layer $k - 1$

$$+ y_{k-1}(x)y_{k-1}(T_C)p_{k-1}\,dx$$

and also cells pushed up by any mitosis in the $k - 2$ layers below that

$$+\sum_{i=1}^{k-2} y_{k-1}(x)y_i(T_C)dx.$$

Now, if we put

$$C_k = p_k y_k(T_C) + \sum_{i=1}^{k-1} y_i(T_C)$$

we can rewrite this as the differential equation

$$y'_k(x) + C_k y_k(x) = C_{k-1} y_{k-1}(x)$$

which can be solved

$$y_k(x) = A_k e^{-C_k x} + C_{k-1} e^{-C_k x} \int_0^x y_{k-1}(x) e^{C_k x}\,dx. \qquad (6.5)$$

Here we have chosen $A_k$ as a constant which satisfies the boundary conditions which are given as

$$\int_0^{T_C} y_k(x)dx = 1$$

or, considering the youngest cells

$$y_k(0) = (1 + p_k)y_k(T_C) + (1 - p_{k-1})y_{k-1}(T_C). \qquad (6.6)$$

Let us now consider the basal layer only; of course, in most squamous epithelia, this will be the only layer of proliferating cells and can be considered as such, or alternatively thought of as the initial layer in our layered array (Fig. 6.2B).

Here eqn 6.5 simplifies to

$$y(x) \;=\; A\, e^{-cx}$$

where $c = py(T_C)$; eqn 6.6 then reduces to

$$y(0) \;=\; (1+p)y(T_C)$$

which can be solved and written in the form

$$y(x) \;=\; \frac{1+p}{p}\;\frac{\log_e(1+p)}{T_C}\;\exp\left\{\frac{-\log_e(1+p)}{T_C}\,x\right\} \qquad (6.7)$$

which is similar to eqn 2.6, derived by Steel (1968), apart from the addition of $p$, the probability of a cell remaining in a layer after mitosis, which of course is equivalent to the proportion of cells with mitotic axes in the plane of the layer.

In practice, mitotic axes are found in both planes, in varying proportions; some axes are vertical, with immediate loss of one cell into the Malpighian layer, and some horizontal, with retention of both young cells. Figure 6.3A shows the arrangement diagrammatically, with the resulting age distribution for $p$ set at various values. With $p = 1$, the age distribution is exponential, and the initial ordinate falls with decreasing values of $p$, until, with $p = 0$, the age distribution is again rectangular.

This arrangement, with a distribution of mitotic axes, is seen in many squamous epithelia, for example normal human epidermis (Pinkus and Hunter 1966), and, in fact, evidence will be given in Chapter 8, section 7, that most mouse squamous epithelia have this sort of kinetic organization.

Now, one limiting case is where all axes in the single basal layer are horizontal, i.e. $p_k = 1$; then

$$y_k(x) = 2\,\frac{\log_e 2}{T_C}\,\exp\left\{\frac{-\log_e 2}{T_C}\,x\right\}$$

which is, needless to say, eqn 2.6 (p. 26) (Steel 1968), and we have a simple exponential age distribution; this kinetic situation is shown diagrammatically in Fig. 6.3B. At each mitosis the two newborn cells remain in the basal layer, with a resulting preponderance of young cells in that layer, and an adjacent cell is ejected (assumed to be at random with respect to age in the cell cycle) to differentiate; the population density will then fall exponentially with age in the cell cycle as basal cells undergo the random removal process. This kinetic arrangement, with all mitotic axes horizontally orientated, is said to occur in the hamster cheek pouch (Brown and Oliver 1968), and also in the epidermis of mature mice

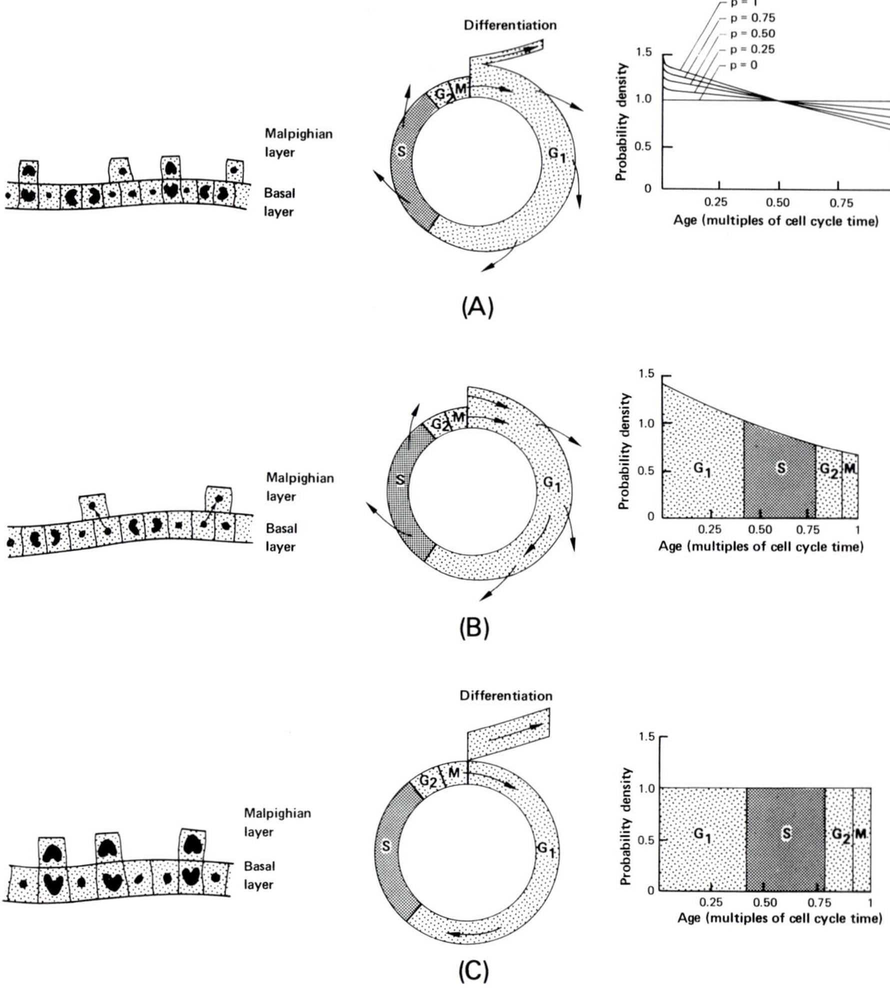


Fig. 6.3. The age distribution in the basal layer of stratified squamous epithelium and its dependency on the direction of the mitotic axis in the basal cells. In (A) is shown the situation where the cells show a distribution of mitotic axes, i.e. cells divide both vertically and horizontally; thus the vertical fraction will contribute one daughter cell immediately to the Malpighian layer, while the daughters of the horizontal mitoses are ejected at random with respect to age in the cycle. The resulting age distribution thus depends upon the value of $p$, the proportion of cells which divide with horizontal axes. In (B), when $p = 1$, all cells divide in the plane of the basal layer and are *all* removed at random through the cell cycle with the resulting exponential age distribution. In (C), all cells divide at right angles to the plane of the basal layer, i.e. with vertical axes, and one cell remains to proliferate, while the other daughter is removed to the Malpighian layer and differentiation; by definition, this gives a rectangular age distribution. (From Duffill *et al.* (1977).)

(Smart (1970*a*). We shall see that, however, the assumption of a random removal process may be an oversimplification in some squamous epithelia, even though advocated by many authors (see, for example, Marques-Periera and Leblond 1965) in the mouse oesophagus, but see Chapter 8, section 7).

At the other extreme there is the case where all mitoses occur perpendicular to the plane of the basal layer, and we have

$$y_k(x) = 1/T_C. \tag{6.8}$$

Here one cell remains in the basal layer to proliferate, and the other migrates; by definition we have a rectangular age distribution (see Fig. 6.3C). This kinetic arrangement is very rare, but has been reported in embryonic mouse epidermis (Smart 1970*a*).

Let us now return to our layered array of proliferative cells; having considered the basal layer, we must now calculate the age distribution for higher layers containing proliferating cells; this does occur in squamous epithelia, for example in hyperplastic states (Bullough and Mitrani 1978), and in the skin disease psoriasis, several layers of proliferating cells are present (Penneys *et al.* 1970), and incorrect assumptions concerning the age distribution have led to several misconceptions in indirect calculations of the cell cycle time.

One limiting case will be where all mitotic axes, in all layers, are in the plane of the basal layer, i.e. $p_k = 1$, for all $k$; here we may show that eqn 2.6 applies for all $k$, and the age distribution for the basal layer, and consequently for the tissue as a whole, will be exponential.

However, consider the other extreme, where all mitotic axes are perpendicular to the basal layer, i.e. $p_k = 0$ for all $k$. In layer 1, the basal layer,

$$y_1(x) = 1/T_C \tag{6.8}$$

and applying this to eqn 6.5 for the second layer,

$$y_2(x) = \frac{\kappa_2}{T_C} \, e^{-x/T_C}$$

where

$$\kappa_2 = e/(e-1) \simeq 1.582, \text{ i.e. } y_2(x) = 1.58/T_C \, e^{-x/T_C}.$$

Now, for the third layer

$$y_3(x) = \frac{\kappa_3}{T_C} \, e^{-\kappa_2 x/T_C} + \frac{1}{T_C} \, e^{1-x/T_C}$$

where

$$\kappa_3 = \kappa_2(2-e)/(1-e^{-\kappa_2}) \simeq 1.430,$$

i.e.

$$y_3(x) = \frac{1}{T_C} \left\{ 2.72 e^{-x/T_C} - 1.43 e^{-1.58x/T_C} \right\}.$$

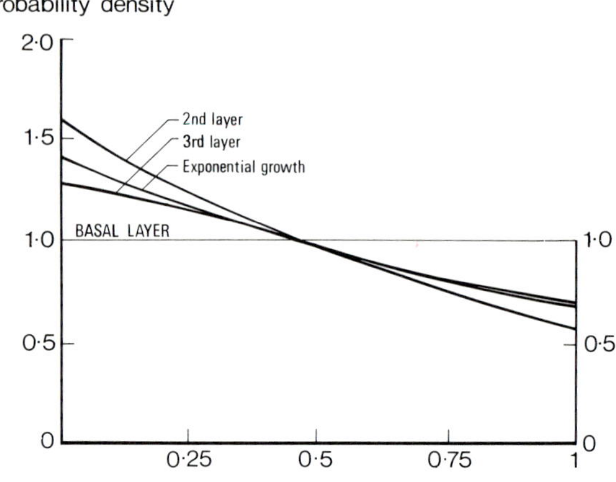


Fig. 6.4.  The age distribution for the first three layers of cells in an array where all mitotic axes are perpendicular to the plane of the basal layer. The first layer is of course rectangular, but the second and third layers are exponential in type, and, importantly, the sum of the three distributions will also be close to an exponential. (From Duffill *et al.* (1977).)

These distributions are shown in Fig. 6.4. An addition of these three distributions would give an age distribution close to exponential, which, for mitotic axes all perpendicular to the plane of the basal layer, is quite a surprising result. Consequently, we are forced to the important conclusion that the age distribution for a many layered structure of proliferating cells, taken as a whole, is close to an exponential age distribution, irrespective of $p_k$ (Appleton *et al.* 1977).

The analysis, thus far, has unrealistically assumed that $T_C$ is constant; let us introduce variability by assuming that the cell cycle time distribution $q(T_C)$ has a gamma distribution with mean $n/\lambda$ and coefficient of variation $n^{-\frac{1}{2}}$, i.e.

$$q(T_C) = \frac{1}{\Gamma(n)} \lambda^n T_C^{n-1} e^{-\lambda T_C}. \tag{6.9}$$

Now, if $y(x/T_C)$ is the age distribution for cell cycle time $T_C$, the overall age distribution is

$$f(x) = \int_x^\infty y(x/T_C)q(T_C)\mathrm{d}T_C.$$

But, for our present purpose, where we mean to consider metaphase arrest experiments, Appleton (Appleton *et al.* 1977) has introduced a most useful concept of *the time remaining to mitosis*, here called $g(x)$

$$g(x) = \int_x^\infty y(T_C - x/T_C)q(T_C)\mathrm{d}T_C = \int_0^\infty y(t/t + x)q(t + x)\,\mathrm{d}t.$$

Returning to the basal layer, i.e. reverting to the usual age distribution diagram, and writing $\theta = \log_e(1 + p)$, combining eqns 6.7 and 6.9

$$g(x) = \frac{1+p}{p}\,\theta\,\frac{\lambda^n}{\Gamma(n)}\int_0^\infty (t+x)^{n-2}\exp\left\{\frac{-\theta t}{t+x}\right\}\exp\{-\lambda(t+x)\}\,dt.$$

Figure 6.5 shows the distribution of times to mitosis for $p = 1$ in the basal layer, for various values of the coefficient of variation of the cell cycle time distribution;

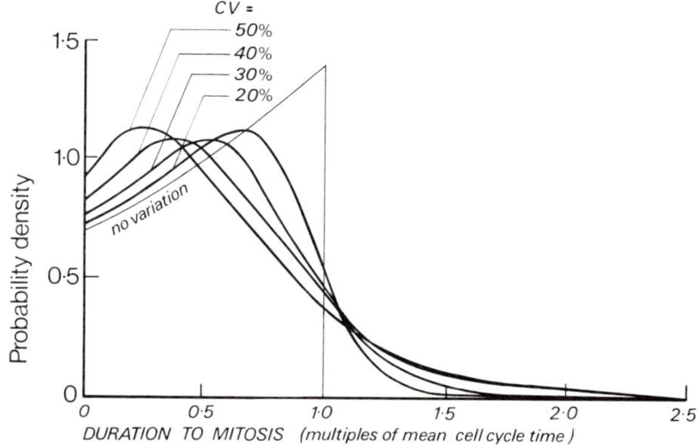


Fig. 6.5. The distribution of times to mitosis in the basal layer when all mitotic axes are in the plane of the layer for values of the coefficient of variation of the cell cycle time of 20, 30, 40, and 50 per cent; a gamma distribution is assumed for transit times through the cell cycle. (From Appleton *et al.* (1977).)

note the differences from that obtained assuming no variation; these differences affect the analysis of metaphase arrest experiments, since the rate of entry into mitosis at the beginning of an experiment is

$$g(0) = \frac{n}{n-1}\,\frac{\log_e(1+p)}{p}$$

where the values of the ratio $n/(n-1)$, which relates to coefficients of variation of 20, 30, 40, and 50 per cent (see Fig. 6.5) are 1.04, 1.10, 1.19, and 1.33 respectively. These can be thought of as correction factors by which the initial value of $T_C$ should be multiplied to obtain a better estimate. On the other hand, we are rarely in a position to quote a coefficient of variation of $T_C$, unless we have a good FLM curve with an analysis by, say, the Gilbert (1972) method. But, in practical terms, the reader may ask why, if we have an FLM curve, we are doing a stathmokinetic experiment as well? This is considered in Chapter 3.

## 6.8.  The effect of the age distribution in squamous epithelia on measurements of the proliferative rate

In Chapter 2, section 3 we agreed that the age distribution could be a critical factor in the calculation of the cell cycle time by indirect methods, cf. the stage-duration calculation. Let us consider what effects the conclusions reached above will have on the measurement of the birth rate and the cell cycle time in a squamous epithelium, using the metaphase arrest technique (after Appleton *et al.* 1977). Let $M_{(k)}(t)$ be the mitotic index in layer $k$ after time $t$ in our experiment. Then

$$M_k(t) = \int_{T_C-t_m-t}^{T_C} y_k(x)\,dx$$

using the usual notation. From eqn 6.5

$$M_k(t) = A_k \int_{T_C-t_m-t}^{T_C} e^{-C_k x}\,dx + C_{k-1} \int_{T_C-t_m-t}^{T_C} e^{-C_k x}\,dx \int_0^x y_{k-1}(\xi) e^{C_k \xi}\,d\xi$$

and rearranging the double integral

$$M_k(t) = \frac{1}{C_k} \left\{ y_k(T_C-t_m-t) - y_k(T_C) + C_{k-1} M_{k-1}(t) \right\} \tag{6.10}$$

or approximately

$$M_k(t) \simeq (t_m + t)\, y_k(T_C) \tag{6.11}$$

and note that the native mitotic index is

$$M_k(0) \simeq t_m y_k(T_C)$$

as we would expect (see Chapter 2).

Now, remembering that $c = py\,(T_C)$, eqn 6.10 becomes

$$M_{(t)} = \frac{1}{c} \left\{ y(T_C - t_m - t) - y(T_C) \right\}$$

and so

$$\log_{1+p} \left\{ 1 + pM_{(t)} \right\} = \frac{1}{T_C}\, (t_m + t). \tag{6.12}$$

At the limit, where $p = 0$, all mitoses are vertical to the plane of the basal layer (cf. eqn 6.8), and for a rectangular age distribution

$$M_{(t)} = \frac{1}{T_C}\, (t_M + t) \tag{6.13}$$

and when $p = 1$, i.e. all mitoses are horizontal, we obtain the collection function of Puck and Steffen (1963)

$$\log_{10}\{1 + M_{(t)}\} = \frac{\log_{10} 2}{T_C}(t_m + t) \tag{3.14}$$

and the native mitotic index, $M_{(0)}$

$$M_{(0)} = \frac{t_m}{T_C} \frac{\log_e(1 + p)}{p}$$

Consequently, if we know $p$, the correct line to fit to the mitotic accumulation data is eqn 6.12, which will compensate for the form of the age distribution. Look again at eqn 6.11; this is an approximation which will be valid if $(t_m + t)$ is small compared with $T_C$, and, at least in squamous epithelium where the experimental period is always (or should be!) short compared with $T_C$, this approximation is good. Then, from eqn 6.7 for the basal layer

$$M_{(t)} \simeq \frac{\log_e(1 + p)}{p} \frac{1}{T_C}(t_m + t) \tag{6.14}$$

and for $p = 0, 0.25, 0.5, 0.75,$ and $1$, the values of $[\log_e(1 + p)]/p$ are $1.00, 0.89,$ $0.81, 0.75,$ and $0.69$ respectively, by which the $T_C$ estimate from fitting eqn 6.14 should be multiplied.

We can also derive suitable values for layers 1, 2, and 3, i.e. $M_1(t), M_2(t),$ and $M_3(t)$, viz.

$$M_1(t) = \frac{1}{T_C}(t_m + t)$$

$$M_2(t) = \frac{1}{e - 1}\left\{\exp\left[\frac{t_m + t}{T_C}\right] - 1\right\}$$

$$M_3(t) = \left\{\exp\left[\frac{(t_m + t)}{T_C}\right] - 1\right\} + \frac{\kappa_3}{\kappa_2} e^{-\kappa_2}\left\{\exp\left[\frac{\kappa_2(t_m + t)}{T_C}\right] - 1\right\}.$$

From eqn 6.11, again ignoring terms higher than $(t_m + t)/T_C$, $M_k(t) = (\gamma_k/T_C)(t_m + t)$, where $\gamma_k$ for $k = 1, 2,$ and $3$ is $1.00, 0.58,$ and $0.71$ respectively; as $k \to \infty$, so $\gamma_k \to 0.69$, i.e. for higher layers we become closer and closer to an exponential age distribution. Note that if $\gamma_k$ is high then layer $k + 1$ receives many young cells, and $\gamma_{k+1}$ is low. Since $\gamma_k$ varies, this may indicate that the native mitotic index may be a poor estimator of the proliferative rate, since, in this situation, layer 2 will have $< 60$ per cent of the mitoses which are present in the basal layer, despite the fact that our initial assumption was that all cells in the array are proliferative.

Consequently, if we were to analyse experimental data for any succeeding layer, employing a rectangular age distribution, we would incur errors, notwithstanding the fact that $p = 0$ for the basal layer. On the broader front, we will now see what sort of error is incurred by applying the incorrect age distribution to single layer data. Figure 6.6 shows a metaphase arrest experiment for the

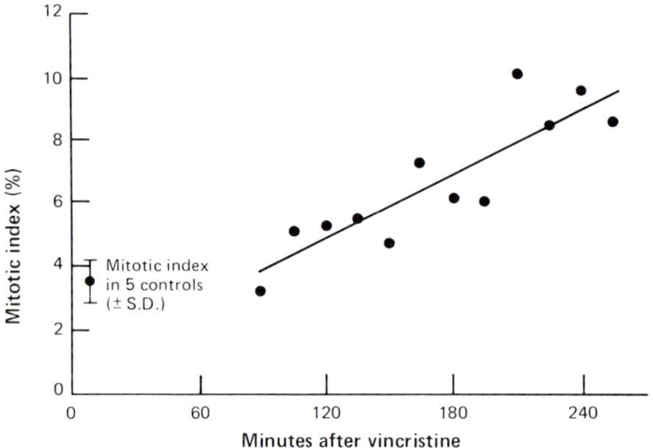


Fig. 6.6. A metaphase arrest curve in the mouse oesophageal squamous epithelium after injection of vincristine (1 mg/kg). Note the 75 minute delay period before the curve begins to rise.

mouse oesophagus using vincristine. The direction of the mitotic axis was determined from a count of 155 mitoses, of which 58 were horizontal, 82 were perpendicular, and 15 were intermediate, and the latter were treated as if they were vertical, since one daughter probably left the basal layer; thus $p = 0.37$, and the line in Fig. 6.6 was fitted by putting this value into eqn 6.12. This gives a value of 43 hours for $T_C$ (remember we assume that the growth fraction for the basal layer is unity); if $p$ is put at 0, or 1 (i.e. the age distribution becomes rectangular or exponential), $T_C$ becomes 49 and 36 hours respectively: even recalling the confidence intervals likely in a metaphase arrest experiment, this error appears worthwhile eliminating. The other regression lines for $p = 0$ and 1 fit the data equally well, and it is plainly impossible to use stathmokinetic data to determine the age distribution as Steel (1977) has suggested, no matter how good the experimental data.

Several groups of workers have determined the direction of the mitotic axis in squamous epithelia (see Chapter 8). For example, a value of $p = 0.34$ can be derived for normal human epidermis from the results of Pinkus and Hunter (1966), and this has the important consequence that the age distribution in human basal cells is not rectangular; there are similar implications for the calculation of $T_C$ from measurements of $t_S$ and $I_S$, and this will mean that the 459 hour cell cycle time calculated by this method for human epidermis by Weinstein and Frost (1968), who assumed a rectangular age distribution (i.e.

that $p = 0$), will be shortened. Similar considerations apply for the cell cycle time in psoriatic epidermis.

This discussion may explain an anomaly reported by Hegazy and Fowler (1973$a$) in adult mouse epidermis; in the equation

$$I_S = \mu I_P t_S / T_C$$

$\mu$ is a constant, which, for an exponential age distribution, should equal 0.73 (Steel 1968; and see Chapter 2); in fact, in Hegazy and Fowler's study, $\mu$ was calculated at 1.28 for mouse dorsal epidermis. This discrepancy could be explained by the presence of a fraction of cells, with vertical axes, lowering the initial ordinate on the exponential age distribution by ejecting a cell directly into the Malpighian layer at mitosis (see section 6.7).

We may conclude by saying that the age distribution can affect the measurement of proliferative rates in squamous epithelia. However, we shall see, when we come to discuss the growth fraction and the kinetic organization of the basal layer (see Chapter 8), that some of the assumptions we have made above are questionable, and in particular, our proposal of random migration from the basal layer. There is now good evidence that cells in $S$ and $G_2$ do not migrate, so the random component will be lost from the exponential age distribution. However, if migration is random in $G_1$, because of the possibly convenient fact that $G_1$ occupies the greater part of the cell cycle, this could mean that migration was effectively random. On the other hand, anticipating the discussion in section 8.6, there is evidence of an ordered migration sequence, with possibly the oldest $G_1$ cell nearest a horizontal mitosis migrating (Iversen *et al.* 1968), and this would, of course, further complicate the age distribution diagram. We have also assumed asynchrony of cells in the cell cycle, and, in Chapter 9, we shall see that this is anything but the case; circadian variation in proliferative indices is a prominent phenomenon in squamous epithelia, and takes the form of a partial synchrony; hence the age distribution diagram is affected (Izquierdo and Gibbs 1972, 1974). Potten (1981) has also suggested that the presence of resting non-proliferating cells in the basal layer will further complicate the age distribution, but we shall argue that there is no firm evidence for such resting cells in epidermis, at least (see Chapter 7, section 4, and Chapters 8 and 10).

Despite these factors, which of course horribly complicate the age distribution diagram, we consider that the age distribution does have an impact upon both the measurement and the understanding of cell kinetics in squamous epithelium, and its importance must be appreciated. In fact, it is essential to the understanding of migration patterns, as we shall see in Chapter 8.

# 7 Kinetic parameters in stratified squamous epithelia

There have been many measurements of the various kinetic parameters in squamous epithelia, and particularly in the epidermis. It is not really the purpose of this chapter to provide a detailed list of all the published measurements, but more to sift the more reliable estimates to obtain a global picture of the cell population kinetics, and *en passant*, to note the reasons for regarding some measurements as unsatisfactory, and perhaps to explain discrepancies between different estimates.

## 7.1. Proliferative indices

We have already referred to the conceptual and technical difficulties encountered in determining proliferative indices, and particularly the labelling index, in epidermis (see Chapter 6, section 1). In both the experimental and human fields, there are considerable problems in quoting a value for *the* labelling or mitotic index, even when we can define our population, because of the phenomenon of circadian variation in proliferative indices, which is found in almost all squamous epithelia (see Chapter 9). Potten (1975*b*), in a detailed study of the literature, has called attention to the fact that values for the labelling index in mouse epidermis, when averaged over a 24 hour period, are almost twice that of single $I_S$ values usually quoted for the hours of daylight (or of the light component of an artificial light regime). In mouse dorsal epidermis, estimates of $I_S$ (for daylight hours) vary from 2.1 to 6.8 per cent for the basal cell compartment, with an average value of 4.0 per cent, whereas the 24 hour average, from published and personal observations, ranged from 4.4 to 10.7 per cent, with a mean value of 7.0 per cent (Grube *et al.* 1970; Tvermyr 1972; Hegazy and Fowler 1973*a*; Potten 1975*b*). This difference was largely due to the high $I_S$ noted during the dark hours. Similarly, in the mouse ear epidermis, daytime $I_S$ values showed a mean of 4.1 per cent, again for the basal layer, while Potten found a 24 hour average of 8.9 per cent, again almost twice the daytime $I_S$ value.

When readings are obtained from other anatomical sites, it is clear that a good deal of variation can be attributed to site and the presence of ordered epidermal structure (see Chapter 8, section 2). The dorsal and ear epidermis of the mouse, referred to above, shows a well-defined stacked columnar arrangement, whereas the mouse plantar epidermis, without ordered arrangement, shows a 24 hour average of 10.7 per cent (Potten 1975*b*). These differences are due to higher cell production rates in non-stacked epidermis.

In addition to variation in the labelling index from region to region, there is also considerable variation within region; for example, in the gingival epithelium are several now well-recognized anatomical sites (Skougaard 1965a,b, 1970), which are classified with respect to the enamel of the adjacent tooth, and although the detailed structure does vary between different species, the general morphology is shown in Fig. 7.1; there are evident differences in thickness

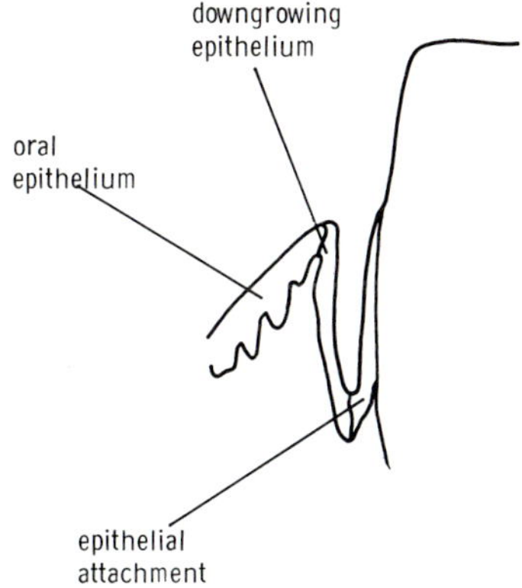


Fig. 7. 1.  A diagram of the gingival epithelium in the oral cavity at the site of the epithelial attachment to the tooth, showing the anatomical relationships between the various zones referred to in the text.

between the several components of the epithelium, and opinion is sharply divided concerning the proliferative rate in each site (see Table 7.1). Earlier authors considered that the epithelial attachment (EA) was a non-renewing, non-keratinizing degenerating zone, with cells being fed into this area from the down-growing oral epithelium (Dimassimo 1963; MacHugh 1959, 1960; MacHugh and Zander 1965; Engler *et al.* 1965), but other workers have reported $I_S$ values in the EA which are as high as in the adjacent areas (Trott and Gorenstein 1963; Randers Hansen 1967; Beagrie 1963; Beagrie and Skougaard 1962; Skougaard 1965a,b).

Generally speaking, labelling indices in oral squamous epithelium are comparable with values reported in the epidermis, although, in some areas, values are much higher; for example, in the interpapillary epithelium of the rat tongue $I_S$ values of 17.5 per cent are evident (Hume and Potten 1976), while in the basal layer of the rat filiform papilla, Cameron (1966) has found an $I_S$ of 50 per cent; it also appears that, at certain times of the day, nearly all cells in basal cell positions are in DNA synthesis, because of the extreme synchrony induced by circadian variation (Potten *et al.* 1978).

It is obviously a mistake to assume that squamous epithelium is homogeneous with respect to labelling index where site is concerned, and we should also

Table 7.1. Kinetic parameters in oral stratified squamous epithelium

| | $I_M$ (%) | $r_M$ (%/day) | $I_S$ (%) | $T$ (d) | $t_S$ (h) | $t_M$ (min) |
|---|---|---|---|---|---|---|
| **Oral epithelium** | | | | | | |
| Beagrie and Skougaard (1962) mouse | | | 7.0 | 4.2 | | |
| Trott and Gorenstein (1963) rat | | 2.27 | | 44.1 | | |
| Randers Hansen (1967) rat | | 4.04 | | 24.8 | | |
| Engler *et al.* (1965) Rhesus | | | 1.9 | 15.4 | | |
| McHugh and Zander (1965) Rhesus | | | 3.1 | 9.4 | | 57–72 |
| Skougaard (1965*b*) marmoset | | | 3.7 | 7.9 | 8.0 | |
| Marwah *et al.* (1960) man | 0.05 | | | 83 | | |
| Hayes *et al.* (1964) | 0.04 | | | 104 | | |
| **Downgrowing oral epithelium** | | | | | | |
| Beagrie and Skougaard (1962) mouse | | | 7.5 | 3.7 | | |
| Trott and Gorenstein (1963) rat | | 6.21 | | 16.1 | | |
| Randers Hansen (1967) rat | | 9.59 | | 16.1 | | |
| Dimassimo (1963) Rhesus | | | 2.8 | 10.1 | | |
| McHugh and Zander (1965) Rhesus | | | 6.0 | 4.9 | | |
| Engler *et al.* (1965) Rhesus | | | 4.9 | 6.0 | | |
| Skougaard (1965*b*) | | | 7.2 | 4.1 | 8.0 | |
| **Epithelial attachment** | | | | | | |
| Beagrie and Skougaard (1962) mouse | | | 5.5 | 5.3 | | |
| Trott and Gorenstein (1963) rat | | 9.5 | | 10.5 | | |
| Randers Hansen (1967) rat | | 9.4 | | 10.9 | | |
| Dimassimo (1963) Rhesus | | | 0.72 | 40.5 | | |
| McHugh and Zander (1965) Rhesus | | | 0.2 | 145.8 | | |
| Engler *et al* (1965) Rhesus | | | 0.3 | 97.2 | | |
| Skougaard (1965*b*) | | | | | 7.1 | |

beware very local variations in both space and time. Similarly with the mitotic index; when values are averaged over a 24 hours period, for example in the dorsal mouse epidermis, values varying from 1.2 to 1.6 per cent are found (Tvermyr 1969, 1972; Hegazy and Fowler 1973a), whereas single readings, taken mainly in the morning, realize only 0.5 to 0.6 per cent (Iversen and Evensen 1962; Evensen and Heldaas 1964). Other mouse epidermis with ordered structure shows similarly low values; reports for the ear epidermis, for example, vary from 0.12 per cent (Sherman *et al.* 1961) to 0.7 per cent (Laurence 1973), for single readings. Observations in non-stacked epidermis are not as frequent, but Laurence (1973) noted an $I_M$ of 1.8 per cent in Swiss mouse foot epidermis, more than twice that in ear epidermis. In oral tissues, such as that of tongue and palate, values seem to resemble those for non-stacked epidermis (Dhawan and Toto 1965), with a tendency to higher values, as seen in the mouse oesophagus (Gruelich 1964) and squamous stomach (Frankfurt 1967).

Thus the main factors which influence epidermal proliferative indices in resting experimental animals are the time of day and the anatomical site; other influences such as sex and strain may have an effect (Potten 1975b), but as yet have not been rigorously documented. One influence which has been explored is the effect of the age of the animal (Thrasher 1971), and this is discussed in section 7.9).

## 7.2. The cell cycle time

We have already discussed the several reasons why few workers have been able to measure the cell cycle time in squamous epithelia (Chapter 6, section 2), and it is in this field where many have fallen into the elementary but important error of equating the cell cycle time with the turnover time, both in animal (see for example, Thrasher 1971) and in human (Weinstein and Frost 1968) tissues; if the basal layer turnover time is calculated by any reasonable method, for example $1/k_B$, or $t_S/I_S$, but *called* the cell cycle time, we incur possible errors, not only due to an $I_P$ of less than unity, but also because of potential errors due to the choice of the wrong age distribution, and the use of a single $I_S$ or $I_M$ reading (or $k_B$ measurement for that matter) when there is diurnal fluctuation in the respective measurements (Duffill *et al.* 1977).

We can at once note that the only really direct measurement of $T_C$ in epidermis was made by Hegazy and Fowler (1973a) with an FLM curve of mouse dorsal epidermis, which is quite unique and is shown in Fig. 7.2A. After an almost trapezoidal first wave of labelled mitoses, the FLM remains at zero until about 96 hours, when a second well-defined peak becomes evident. Such was the unusual nature of this FLM curve that the experiment was repeated three times: the first two experiments were carried out on mice who were injected with [³H]-TdR at a fixed time (0900 h) and killed at different times of the day, whereas in the third experiment animals were injected with [³H]-TdR at various

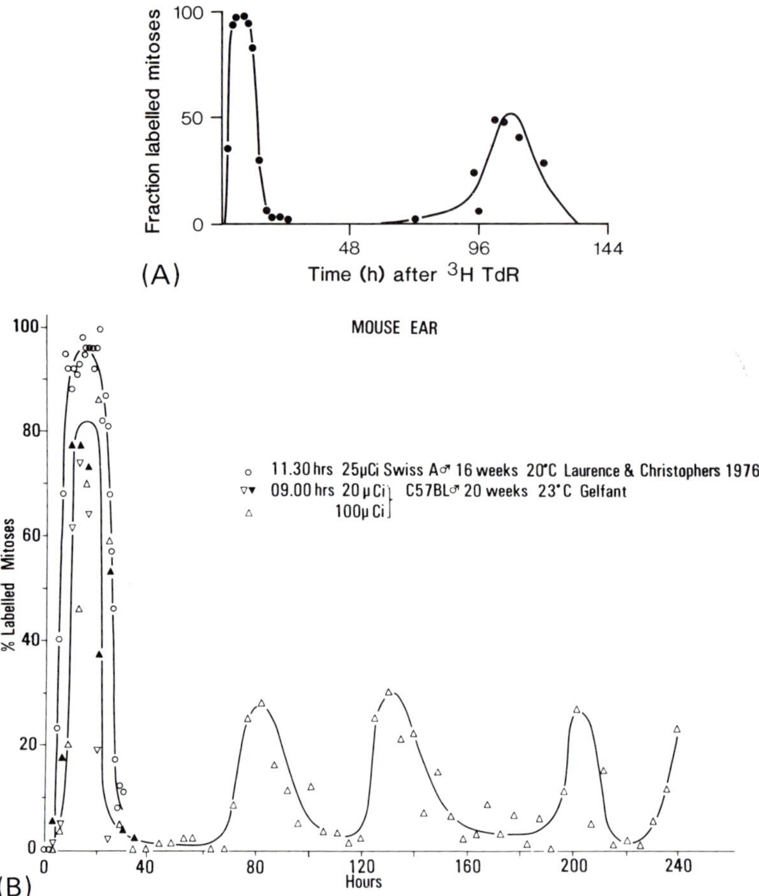



Fig. 7.2. FLM curves in stratified squamous epithelium: (A) mouse dorsal epidermis. [³H]-TdR was injected at 0900h. (Redrawn from Hegazy and Fowler (1973a).) (B) In the mouse ear epidermis, with data from Laurence and Christophers (1976), and some unpublished data of Gelfant. [³H]-TdR was injected at the times shown. (By courtesy of Dr C. S. Potten.) (C) From the mouse dorsal epidermis. The times of [³H]-TdR injection were

times of the day and killed at the time when the native $I_M$ was at its diurnal peak (1100 h). This curve gives a median $T_C$ of 101 hours, and a $T_C$ of 101 ± 8 (SE) hours when fitted by the Barrett (1966) model (Steel and Hanes 1971), which, it will be remembered, assumes a log normal distribution of cell cycle times. The undoubted partial synchrony of the basal cell population induced by circadian rhythm might have some effect on this curve (see Chapter 9), although it is interesting to note that varying the times of the injection of [³H]-TdR had no apparent effect on the format of the curve. We may also note that Dhawan and Toto (1965), who carried out a protracted FLM analysis in mouse oral tissues, reported a $T_C$ of 98 hours, close to that of Hegazy and Fowler (1973a).

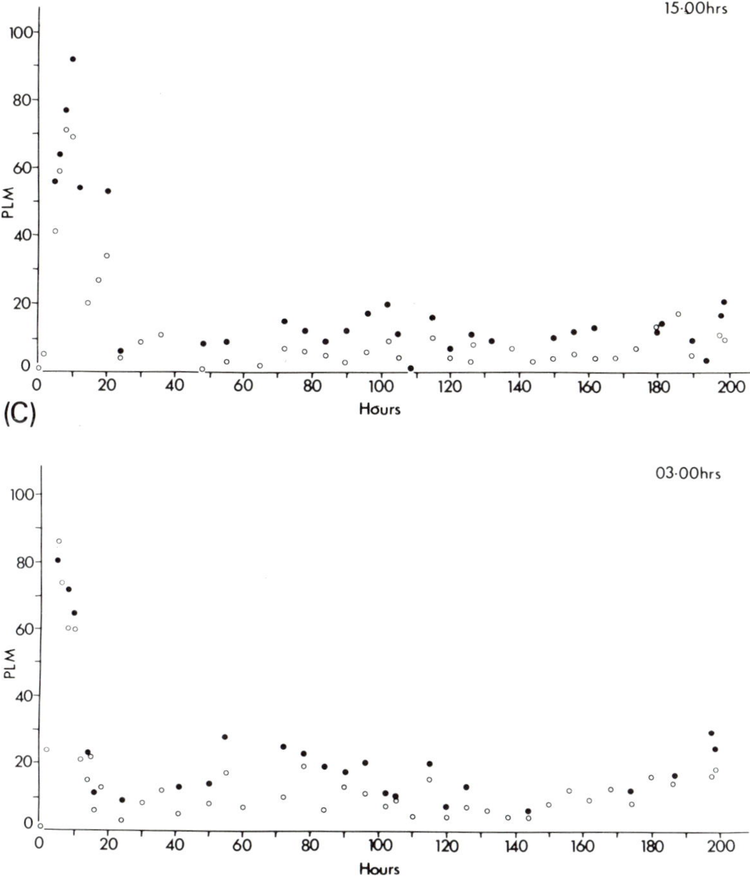



0300 h and 1500 h. See Fig. 8.24 (p. 342) for a computer model fit to this data. The open circles are points obtained using a two week exposure period with a threshold of three grains per nucleus for a labelled cell. Closed circles are points obtained for longer exposure times or a lower threshold (two grains) for labelled cells. (By courtesy of Dr C. S. Potten.)

On the other hand, it should be understood that such well-defined peaks are unusual; there have been other attempts to carry out a prolonged FLM curve in squamous epithelia; Gelfant (quoted by Potten 1981) obtained an FLM curve for the mouse dorsal epidermis with a second peak indicating a median $T_C$ of about 80 hours. Experiments with ear epidermis give FLM curves with multiple peaks, one at about 80 hours, coinciding with the dorsal epidermis peak, and several smaller ones (Gelfant, quoted by Potten 1981), and shown in Fig. 7.2B. Similarly, Potten (1981), in dorsal epidermis, found two smaller peaks following the first well-defined one, one at around 100 hours and the other at about 190 hours (Fig. 7.2C). Hamilton and Blackwood (1974) have also reported a succession of smaller peaks occurring in mouse oral squamous epithelium.

In an extensive study, Potten *et al* (1982*b*) have studied the FLM curve after [³H]-TdR injection at both 0300 h and 1500 h. At 0300 h, the first peak did not reach 100 per cent, and at 196 hours after injection there was a suggestion of a further peak; however, if longer exposure times (or lower grain counts) were used, a further possible peak was discernible at 80 to 85 hours, which was again interpreted as meaning that a subpopulation of cells existed with a cell cycle time of 60 to 80 hours, with a lower [³H]-TdR uptake because of a larger TdR pool, or because of a low DNA synthetic rate; other subpopulations may have a longer cell cycle time. In fact, the curve obtained after injection of [³H]-TdR at 1500 h showed essentially similar features, with putative peaks at 190 hours and a grain count dependent peak at 80 to 90 hours after [³H]-TdR.

Potten *et al.* (1982*b*) did statistical analysis to show that the variation in the FLM over the relevant periods where the peaks were apparent could be distinguished from random noise, but it is difficult to be sure that these peaks are meaningful. On the other hand, these latter peaks seriously complicate the picture presented by the Hegazy and Fowler (1973*a*) FLM curve; it may be that there are subpopulations of cells whose complex cell cycle time structure produce such complications, but there are several problems which mitigate against definitive characterization; (i) *the presence of circadian variation*, notwithstanding the minor differences in the curves obtained for the two injection times; the FLM method does assume that there is total asynchrony, but, as we shall see (Chapter 9), there is considerable circadian synchronization in squamous epithelia. The changes which constitute these peaks are so minor that partial synchronization, albeit of a subpopulation, could produce small, but statistically evident fluctuations in the FLM, and it does remain to be seen if these changes are, in fact, related to the cell cycle time. The influence of such synchronization on the FLM curve is discussed in Chapter 9, section 3; (ii) *the influence of [³H]-TdR on the FLM curve*; although conclusions on its actual effect and mechanisms of action differ, there is evidence that [³H]-TdR itself does modify the mitotic index in squamous epithelium (see Chapter 9, section 3). For example, Møller *et al.* (1974) considered that [³H]-TdR injection caused a shortening of $t_{G_2}$ by causing unlabelled cells to move rapidly into mitosis, which would affect $t_{G_2}$ values as seen in the FLM curve, and this conclusion was supported by a theoretical analysis of Møller's data by Hopper and Brockwell (1978); (iii) *the variance of the cell cycle time*: in some systems, for example the squamous epithelium of the hamster cheek pouch, there is no evidence of a second peak (cf. Brown 1970); instead, values for the FLM fluctuate between 0 and 7 per cent from 49 to 173 hours after injection. Brown (1970) considered that the reason for this was the large variance of $t_{G_1}$ in the population, and in the Hopper and Brockwell (1978) model, the variance of $t_{G_1}$ resulted in the second wave being practically undetectable. This consideration could be the main reason for the lack of a second peak in many studies; (iv) a further problem is *the presence of decycling divisions*: if many cells are decycling after

the first peak, they will not contribute to succeeding peaks. In the EPU model
(Chapter 8, section 2), most divisions are decycling, and this would be another
reason for the lack of a defined second peak; (v) *reutilization of DNA degradation products*: if granular layer cells are providing DNA products, which include
[$^3$H]-TdR, these would be available at about three to five days after injection,
i.e. at about the time of the second peak in mouse dorsal epidermis; (vi) Potten
(1981) considered that the effects of *delayed labelling* would only be minor,
since, on the basis of his own observations (Potten 1971*a*) only a small fraction
of the cell population are late labelling (see Chapter 6, section 1 and Chapter 8,
section 4). However, in some tissues (cf. hamster cheek pouch) most cells
incorporate [$^3$H]-TdR over a 3–4 hour period (Møller *et al.* 1982), and this would
have the effect of lengthening the labelling period, which is usually assumed to
be negligible (see Chapter 3, section 4); (vii) it is of course possible that there are
*proliferative subpopulations*: if such subpopulations exist, with differing cell
cycle times, then this would indeed complicate any FLM curve, tending to smear
the distribution. However, it does appear that, if subpopulations such as these are
present, they have very discrete cell cycle time distributions, as indicated by the
sharp definition of the second peak in some instances (Fig. 7.2A, B).

Hegazy and Fowler (1973*a*) supported their FLM derived $T_C$ estimate by the
Brown and Berry (1968) method (see Chapter 6, section 2), obtaining excellent
agreement between the two techniques (see Table 7.2). An estimate derived
from following the continuous labelling pattern of basal cells (see Fig. 7.4,
p. 270), with a plateau beginning at 96 hours, gave a $T_C$ value of 107.5 hours
(96 + $t_S$ (11.5 hours)), again in good agreement. Additionally, Hegazy and
Fowler calculated $T_C$ from the equation

$$I_S = \mu I_P t_S / T_C$$

where $\mu$ is the correction factor for a non-linear age distribution (Steel 1968;
and see Chapter 6, section 7). Estimating $\mu$ at 0.92 from the FLM data and
assuming (from the continuous labelling data) that $I_P$ was unity (but see section
7.4): $T_C$ was calculated at $110 \pm 14$ (SEM) hours. Despite the fact that a continuous labelling index of 100 per cent does not necessarily reflect an $I_P$ of
unity (in fact, it almost certainly does not in squamous epithelium), and that
only a single value for $I_S$ was used (circadian variation in $I_S$ would compromise
this; see Chapter 9), this estimate is in good agreement with the other more
direct values. It is of interest to note that Hegazy and Fowler (1973*a*) are among
the very few authors to even have considered the age distribution and its effect
on stage-duration calculations, although Brown and Oliver (1968), having
measured $T_C$ in hamster cheek pouch epithelium at $130 \pm 7$ (SEM) hours by their
basal cell migration method (see Chapter 6, section 2), calculated $T_C$ from

$$I_S = \ln 2 t_S / T_C$$

where $t_S$ was measured from an FLM curve; this is only an approximate formula

Table 7.2. *Kinetic parameters in stratified squamous epithelia*

| Tissue and reference | Parameter | $I_S$ (%) | $I_M$ (%) | $T_C$ (h) |
|---|---|---|---|---|
| **Mouse epidermis** | | | | |
| Sherman *et al.* (1961) | ear | 2.8 | 0.12 | 583[7] |
| Blenkinsopp (1968*a*) | ear | | | 532[14] |
| | abd. | | | 156 |
| Hegazy and Fowler (1973*a*) | | | | 101[4] |
| (dorsum) | | | | 107[5] |
| | | | | 96–102[6] |
| | | | | 110[7] |
| Clausen *et al.* (1979) | | | | 53.6 |
| (dorsum) | | | | |
| Iversen *et al.* (1968) | | | | 84[5] |
| (dorsum) | | | | |
| Hamilton and Potten (1972) | | | | 107 |
| (dorsum) | | | | |
| Potten *et al.* (1974) | | | | 120[5] |
| (dorsum) | | | | |
| Cameron (1970) | dorsum | | | |
| | ear | | | |
| | abd. | | | |
| | plantar | | | |
| Olsson (1976)   (dorsum) | | | | 65 |
| Clausen *et al.* (1981) | | | | |
| | | | | |
| **Rat epidermis** | | | | |
| Storey and Leblond (1951) | | | | |
| (plantar) | | | | |
| Bertalanffy *et al.* (1965) | plantar | | | |
| | abd. | | | |
| | ear | | | |
| Fukuyama and Epstein (1966) | | | | |
| | | | | |
| **Mouse oesophagus** | | | | |
| Leblond *et al.* (1964) | | 6.5 | 0.3 | 70[14] |
| | | | | |
| Blenkinsopp (1968*a*) | | | | 58[14] |
| Blenkinsopp and Gilbert (1969) | | | | 64[14] |
| Blenkinsopp (1969) | | | | 41[7] |
| Thrasher (1971)[16] | | 6.6–12.6 | | |
| | | | | |
| Duffill *et al.* (1977) | | | | 39[10] |
| Greulich (1964) | | 3–11 | | |
| | | | | |
| **Rat oesophagus** | | | | |
| Leblond *et al.* (1964) | | | | 54.3[14] |
| | | | | |
| **Mouse squamous forestomach** | | | | |
| Wolfsberg (1964) | | | | 30[4] |
| Frankfurt (1967) | | | | 55[4] |

| $t_S$ (h) | $t_{G_2}$ (h) | $t_2$ (h) | $t_M$ (h) | $k_B$ (cells/ 1000 cells/h) | $T_T$ (h) | $T$ (h) | $I_P$ (%) |
|---|---|---|---|---|---|---|---|
| 30[4] | 5.9[4] | 6.5[4] | 1.3[4] | 9.3[12] | | 489[9] | 54.5[2] |
| 34[13] | | | | | | | 68[3] |
| 8.4 | | | | | | | 68[3] |
| 11.6[4] | | 2.0[4] | 1.4[4] | 6.9[12] | | | 100[5] |
| | | | | | | | 100[6] |
| 8.8 | 3.2 | | 1.0 | 4.0 | | 144[8] | |
| | | | | | | 96[9] | |
| | | | | | | | 58[3] |
| | | | | | | 187[8] | 57[15] |
| 14[5] | | | | | | | 90[5] |
| | | | | | | 7–14[5] d | |
| | | | | | | 14–21  d | |
| | | | | | | 7–14  d | |
| | | | | | | 10    d | |
| 8.5 | | 3.2 | | | | | |
| 8.9–11 | 1.2–2.4 (1.7) | | | | | | |
| | | | | 2.2[10] | | 16.9[8] d | |
| | | | | 1.9[10] | | 24.4[8] d | |
| | | | | 2.6 | | 15.9  d | |
| | | | | 1.4 | | 29.9  d | |
| | | | | | 3 d[1] | | |
| 7[4] | | | | 12[10] | | 105[10,11] | 70[3] |
| | | | | | | 175[8] | |
| 9.1[13] | 2.8 | | | | | | 71[3] |
| 8.5[5] | 2.8 | | | | | | |
| 7.6[4] | 1[4] | 1.6[4] | | | | 64–120[11] | |
| 8.0 | 1.75 | 2.0 | | | | | |
| 8.2[4] | | | | | | | |
| | | | | | | 81.5[10,11] | 58 |
| | | | | | | 131.8[8] | |
| 13.5[4] | | 2–2.5[4] | | 3.9[12] | | 260[11] | 11 |
| 7[4] | | | | 6.5[12] | | | 100[5] |
| | | | | | | | 57[3] |

| Parameter | | $I_S$ (%) | $I_M$ (%) | $T_C$ (h) |
|---|---|---|---|---|
| **Tissue and reference** | | | | |
| **Hamster cheek pouch epithelium** | | | | |
| Brown and Oliver (1968) | | | | $130\pm7$[6] |
| | | | | $135\pm17$[7] |
| Brown and Berry (1968)[17] | | | | |
| Gibbs and Casserrett (1969) | | | | |
| Izquierdo and Gibbs (1974) | | | | |
| **Oral epithelia** | | | | |
| Blenkinsopp (1968a) | Mouse tongue | | | $53$[14] |
| Blenkinsopp and Gilbert (1969) | | | | $84$[14] |
| Cameron (1970) | Filiform papilla | | | |
| Cameron (1966) | | | | 21–35 |
| Potten and Hume (1976) | | 6.9 | | |
| Beagrie (1963) | Mouse gingiva | 5.5–7.9 | | |

1. As measured from the basal layer.

2. This is the fraction of the population occupied by the basal layer, and assumes that all basal layer cells are proliferative.

3. This is calculated from the progenitor:non-progenitor cell argument detailed in Chapter 8, section 5.

4. As measured by the FLM method.

5. As measured by the continuous labelling method.

(see chapter 2, section 7), but Brown and Oliver did obtain a value of $135 \pm 17$ (SEM) hours. The SEM estimates in these measurements were derived from the least squares regression of $N_B/N_T$ (Fig. 6.1, p. 230), and from SD $t_S$ and SD $I_S$ for the two methods used respectively. These values are obviously close, especially when the confidence interval is taken into consideration. In this respect, it is perhaps worth emphasizing that the format of the FLM curve obtained by Hegazy and Fowler (1973a) shows such a square wave course that there is likely to be a very tight distribution of cell cycle times for those cells which are contributing to the second peak.

The above mentioned studies are the only ones published where independent corroboration from more than one method has been forthcoming, but there are several other published values for $T_C$ which are worth considering, and these are summarized in Table 7.2. The continuous labelling method has been employed to measure $t_{G_2} + t_M + t_{G_1}$ from the inflexion point on the curve and then adding the duration of $t_S$, usually found from an FLM analysis. Frankfurt (1967) reported a value of 55 hours for mouse squamous forestomach, while Iversen *et al.* (1968) obtained 84 hours for mouse dorsal epidermis; Blenkinsopp (1968a; 1969) calculated values of 41 and 64 hours for mouse oesophagus, and of 84 hours for the mouse ventral tongue, all obtained using the continuous labelling technique. In mouse dorsal epidermis, Potten *et al.* (1974) reported a value of 120 hours,

| $t_S$ (h) | $t_{G_2}$ (h) | $t_2$ (h) | $t_M$ (h) | $k_B$ (cells/ 1000 cells/h) | $T_T$ (h) | $T$ (h) | $I_P$ (%) |
|---|---|---|---|---|---|---|---|
| 7.6[4] | | | | | | | |
| 8.6±0.8[4] | | 2.4[4] | 0.8–1.2[10] | 4.1[10] | | 113[11] | |
| | | | | 5.1[10] | | 120[11] | |
| 10[4] | | 2[4] | 53–77 min | | | 72–96[1] | |
| | | | | 0.12 | | 8.4 d | |
| | | | | 0.26 | | 3.8 d | |
| 8[13] | | | | | | | 71[3] |
| | | | | | | 4 d[5] | |
| 7 | | | | | | 53 | |
| | | | | | | 65–86 | |
| 7 | | | | | 24h–10d | 24h–10d | |

6. Using the Brown and Oliver method (see Chapter 6, section 2).
7. A stage-duration calculation.
8. Total turnover time (including basal cells).    9. Differentiating layers only.
10. Stathmokinetic method.    11. Basal layer only.
12. From $k_B = I_S/t_S$.    13. From $t_S = I_S/r_S$, assuming $r_S = r_M$.
14. From calculated life of progenitor cell in basal layer.    15. See text (section 7.4).
16. Depending on age.    17. Depending on time of day.

again from the continuous labelling method, while the required $t_S$ value was calculated from the continuous labelling curve. One of the several drawbacks of this method is the difficulty in establishing the point of inflexion (see Chapter 3, section 5), but from the format of the published curves (Fig. 7.4, p. 270) the inflexion point for the germinative cells is usually quite discrete, and consequently no great difficulty is experienced in reading off the value of $t_{G_2} + t_M + t_{G_1}$.

Indirect measurements from a foreshortened FLM curve have also been popular; Sherman *et al.* (1961) did as several authors have done, and calculated cell cycle time values from $I_P/k_B$, assuming that $I_P = 54$ per cent (see section 7.4), calculating $k_B$ from $I_S/t_S$, and assuming a rectangular age distribution, to give a value of 583 hours for the mouse ear. Thrasher (1971), in the mouse oesophagus, obtained values ranging from 64 to 120 hours, depending on the age of the animal; Cameron (1966) reported 21 to 37 hours for the mouse filiform papilla; these values would of course more properly be called *basal layer turnover times* rather than cell cycle times (see Chapter 6, section 3). On the other hand, in the mouse squamous forestomach, Wolfsberg (1964) fared rather better; her curve for cells with $> 6$ grains per nucleus shows a distinct second peak, and indicates a median cell cycle time of about 30 hours.

Some investigators have calculated the cell cycle time from metaphase arrest experiments; Blenkinsopp (1968*a*) used a relationship which in fact reduces to

$$T_C = I_P/k_B$$

where $k_B$ was derived from a colchicine experiment, and also

$$T_C = t_S \cdot I_P/I_S$$

to calculate $T_C$ in mouse oesophagus, tongue, and skin, which methods gave results varying from 58 hours in the oesophagus to 532 hours in the ear epidermis. In a later publication, proposing that cell loss from the basal layer was random, and that consequently the age distribution was exponential, Blenkinsopp (1969) used the relationship

$$T_C = t_S L \cdot \ln 2/I_S$$

(where $L$ is the duration of the labelling period), and derived a shorter value of 41 hours in the mouse oesophagus; here $t_S$ was calculated from

$$t_S = I_S/r_S, \text{assuming } r_S = r_M;$$

$r_M$ was measured from a metaphase arrest experiment. If the age distribution is exponential, then of course $r_S > r_M$, but this may not matter so long as $T_C$ is long and $S$ occurs towards the end of the cell cycle. This value of 41 hours for the mouse oesophagus is similar to that obtained directly from a metaphase arrest experiment reported by Duffill *et al.* (1977) of $39 \pm 4$ hours, again for the mouse oesophagus; Duffill *et al.* assumed a growth fraction of unity, but calculated the age distribution from a consideration of the direction of the mitotic axes in the basal layer (see Chapter 6, section 8).

A more convoluted way of calculating $T_C$ is due to Hamilton and Potten (1972) who proposed that plucking of the hair from mouse dorsal epidermis removed all the post-mitotic maturing cells (see Chapter 8, section 4) from the basal layer by migration (see Chapter 10), and these numbered 4800 cells/mm$^2$. Responding cells entering $S$ during the ensuing proliferative response were considered to be in the $G_0$ phase, and numbered 2500 cells/mm$^2$. If 500 cells/mm$^2$ are melanocytes or Langerhans cells, then 7800 out of 18 200 cells/mm$^2$ are non-proliferative, giving an $I_P$ of 0.57. The basal layer turnover time was calculated at 187 hours, and, using the $I_P$ value, $T_C$ was then 107 hours for proliferating basal cells, with a range of 92 to 138 hours.

It will be seen from Table 7.2 how few point estimates of $T_C$ actually give any form of confidence interval. This is specially relevant in the instances where a quotient has been used to calculate $T_C$. Because of this we should perhaps hesitate to compare values from one anatomical site with another. However, we might note that the non-keratinizing epithelia of the tongue, oesophagus, and squamous stomach tend to have the shortest cell cycle times, with values ranging from about 30 to 70 hours, while most estimates in mouse dorsal epidermis and hamster cheek pouch are around 100 hours. On the other hand, the few values that are available for the mouse ear epidermis indicate a far longer cell cycle time (see Table 7.2). It is possible that these differences may be related to the low temperature of the exposed ear epidermis, compared with the other, less exposed epidermis, and with internal epithelia such as the oesophagus, which will be closer to the axial temperature; Gelfant (1975) has reported increased proliferative rates in epidermis when the temperature is elevated.

Neither can we really draw any worthwhile conclusions as to the comparison of methods, but where investigators have compared different methods in the same model, and at the same time, agreement has been generally good (see Hegazy and Fowler 1973*a*).

In interpreting these values of the cell cycle time, we should remember that many of the mentioned epithelia show an ordered arrangement, and at least one modern hypothesis of cell lineage in epidermis (Potten 1974) proposes a very heterogeneous kinetic structure for the basal layer, with much variation in $T_C$. Indeed, in the context of the *epidermal proliferative unit* (see Chapter 8, section 2), Potten (1975*b*) has doubted the 'significance' of these estimates; we cannot really doubt their 'significance', but we can question their precision; epidermal cells, albeit in an ordered arrangement, will still have a modal cell cycle time, and it will be true to say, for the FLM estimates anyway, that our measurement will be close to $T_C$ for the most mitotically active cell (?in cell position 4, see Fig. 8.14, p. 313). The kinetic heterogeneity will then contribute to our confidence interval. Indeed, by constructing differential continuous labelling curves for different cells in the epidermal proliferative unit, Potten (1974) has identified a subpopulation of cells which label more slowly, and consequently have a prolonged cell cycle time; it is these cells that will enlarge our confidence interval, but our estimates will be none the less valid, as point estimates for the mean cell cycle time, with, it must be hoped, an error bar.

### 7.3. Cell cycle phase durations

#### (i) *The duration of mitosis*

We have already concluded that the available methods of determining the mitotic duration in the *in vivo* situation are fraught with difficulty, and numerous over-optimistically precise measurements have been published for squamous epithelium, often to a single minute, or to two places of decimals in the hour, but rarely with any confidence intervals. Most measurements have been made using the metaphase arrest method, and of course many of these have incurred additional errors due to faulty technique (see Chapter 3, section 3). Thus Iversen and Evensen (1962) and Evensen and Haldaas (1964) reported values for $t_M$ in hairless mouse epidermis of close to 1.0 hour for a single time period, whereas, in the same system, Tvermyr (1969) obtained a value of 2.66 hours over a 24 hour period. In mouse ear epidermis, stathmokinetic estimates have again been longer, varying from 78 minutes to 222 minutes (Sherman *et al.* 1961;

Bullough 1962; Bullough and Laurence 1966*a,b,c*; Laurence and Randers Hansen 1971; Laurence 1973), although longer values have been found in oral squamous epithelia (Dhawan and Toto 1965). But we should emphasize that these readings were mostly obtained with decidedly suspect technique, and we have no way of knowing the statistical worth of the data; hence real comparisons between sites become invidious.

Because of the difficulties involved, few workers have read $t_M$ values directly from the FLM curve, but Sherman *et al.* (1961) did calculate $t_M$ from the usual $I_M/k_B$, where $k_B = I_S/t_S$, and $t_S$ was derived from an FLM curve; they calculated a value of 1.3 hours. Hegazy and Fowler (1973*a*) calculated $t_M$ from the

relationship $\qquad\qquad t_M/t_S \;=\; I_M/I_S$

corrected to $\qquad\qquad t_M/t_S \;=\; 1.07\,I_M/I_S$

to allow for the postulated non-linear age distribution, which gave a value of 1.4 hours for mouse dorsal epidermis.

Values for $t_M$ based on the disappearance of mitoses after irradiation (Knowlton and Widner 1950) have given much shorter values for $t_M$ for mouse ear epidermis (0.5 hours) than the other methods.

The only reasonable conclusion we can draw from these studies is that, in ear epidermis, mitosis takes approximately three hours to be completed, whereas in mouse dorsal epidermis $t_M$ is closer to one hour. In the hamster cheek pouch epithelium, stathmokinetic measurements vary between 0.8 and 1.2 hours, according to the time of day (Brown and Berry 1968), but we should note that Møller *et al.* (1979) maintain that the duration of mitosis varies considerably throughout the day, and that metaphase arrest periods as short as one hour are necessary to detect these rapid changes; this concept is fully discussed in Chapter 9.

### (ii) *The duration of DNA synthesis*

The main methods which have found favour for measuring $t_S$ in squamous epithelium are the FLM method and the double labelling technique. Calculations from $I_S$, $I_M$, and $t_M$ have also been reported, and occasionally measurements have been made with the continuous labelling method (see, for example, Potten *et al.* 1974). In the mouse dorsal epidermis, most measurements, with either technique, give values ranging from 5 to 8.5 hours for a single time point (see Table 7.2), although continuous labelling (measuring $r_S$ from the initial rise in $I_L$, before labelled cells have divided, and $t_S = I_S/r_S$), gives a longer value at 14.0 hours. When values are quoted for a 24 hour period (Potten 1975*b*), longer values are also apparent, also approaching 14 hours, because of the longer $t_S$ estimates in the night. The converse applies in the mouse ear epidermis; single estimates in daylight hours are 30 hours or more (Sherman *et al.* 1961; Blenkinsopp 1969; Laurence *et al.* 1979), but, using a double labelling method,

Potten (1975*a*) reported a mean value of 11.3 hours for a 24 hour period, evidently due to shorter $t_S$ values in the hours of darkness. Similarly, in the mouse plantar epidermis, $t_S$ is 17.2 hours, as measured by the FLM method (Laurence 1973; Laurence *et al.* 1979), whereas Potten's 24 hour average estimate was 7.3 hours.

In other species, epidermal $t_S$ values are similar (see Table 7.2): for example, in the guinea pig epidermis, Song and Tabachnick (1968), using double labelling with $[^3H]$-TdR and $[^{14}C]$-TdR, obtained a value of 9.3 hours. In other, non-keratinizing squamous epithelia, such as oesophagus, tongue, gingiva, and squamous stomach, $t_S$ values are remarkably constant, with most estimates being around 8-9 hours (Skougaard 1965*a,b*; Skougaard and Beagrie 1962; Greulich 1974; Toto and Dhawan 1966; Dhawan and Toto 1965), although values derived *in vitro* appear to be longer (Shand and de Sousa 1974). We might note that Wolfsberg's (1964) estimate for the mouse squamous forestomach, as measured at the 37 per cent peak value on the FLM curve (Quastler and Sherman 1959; and see Chapter 3, section 4) does give a $t_S$ of 11.0 hours. In the mouse oesophagus at any rate, there does not appear to be any marked change in $t_S$ with age: Thrasher (1971) obtained FLM estimates, all of about eight hours, in the four groups of mice he studied, which were of widely divergent ages.

We have already discussed the problems involved in the measurement of $t_S$ using the present techniques (see Chapter 3, section 4), but in squamous epithelia we should remember that the estimates depend to a great extent on the time of day; the FLM method, particularly, depends on a steady and constant flow of cells through the cell cycle, which of course does not occur in this tissue, because of the considerable degree of partial synchrony of cycling cells which are responsible for the remarkable circadian swings in $I_S$ and $I_M$ (see Chapter 9). Similarly, relatively rapid changes in the flux of cells into and out of the $S$ phase will seriously modify the estimates of $t_S$ derived from the double labelling method, since $t_S$ here is usually estimated from $I_S$ and $r_S$, the latter being the difference between two $[^3H]$-TdR labelling indices (Potten 1975*a*) or between $[^3H]$-TdR and $[^{14}C]$-TdR indices (Song and Tabachnick 1968).

Other potential errors with double labelling in squamous epithelium include the possibility of migration of labelled cells during the experiment, if the basal layer alone is being considered; this may be insignificant in some systems (Marques-Periera and Leblond 1965; Brown and Oliver 1968), but could be important in circumstances where there are a significant proportion of vertically orientated mitotic axes (Duffill *et al.* 1977). Other problems include the choice of the wrong age distribution (particularly rife, see Chapter 6, section 7), and the necessity to ensure that cells have not divided before the second label; in epidermis, $t_2$ is probably long enough to allow a reasonable accumulation period, but in oesophagus $t_2$ is sufficiently short for this to be a problem. More general points, such as the recognition of the secondarily labelled cohort, have already been discussed.

We may conclude that, apart from the longer $t_S$ values reported from single time point readings in mouse ear epidermis (which in fact become shorter when a 24 hour mean is taken), $t_S$ measurements in squamous epithelium appear remarkably constant.

### (iii)  $t_{G_2}$ and $t_2$

These have been measured almost exclusively by the FLM method in squamous epithelia, and are usually a minimum $t_{G_2}$ estimate, or a median $t_2$ measurement at the 50 per cent level on the ascending limb of the first peak. The minimum $t_{G_2}$ is particularly subject to error, since the observation of the first labelled mitosis will, to a great extent, depend on the number of readings (cf. Dhawan and Toto 1965). The estimate of $t_{G_2}$ of Sherman $et$ $al.$ (1961) at 6.5 hours for mouse ear epidermis is long compared with other published values for mouse dorsal epidermis, oesophagus, and hamster cheek pouch, where $t_{G_2}$ and $t_2$ are both relatively short (see Table 7.2).

Clausen $et$ $al.$ (1981) have recently reported some very interesting results on the possibility of subpopulations of cells in mouse epidermis with differing transit times through $G_2$. A family of short-term FLM curves, initiated at various times throughout the day, showed that some 10–20 per cent of labelled mitoses were evident at 30 minutes, and, using the usual methods, the estimated mean $t_{G_2}$ varied from 1.4 to 2.5 hours; however, using $[^3H]$-TdR double labelling, both Clausen $et$ $al.$ (1979) and Tvermyr (1972) have calculated much longer times in the afternoon, although values of one to two hours were apparent in both studies in the morning period. Clausen $et$ $al.$ (1979) also calculated the circadian variations in $t_{G_2}$ (see Chapter 9) by measuring $G_2$ efflux from meta-phase arrest studies, and the $G_2$ fraction from FCM studies, and found $t_{G_2}$ values of about two hours from 0400 to 1200 h, rising rapidly to a maximum of seven hours at 1600 to 2000 h. Clausen $et$ $al.$ (1981) calculated that, for mean $t_{G_2}$ values of one and seven hours, 2 and 56 per cent of cells respectively had a $t_{G_2}$ in excess of four hours (see Fig. 7.3); with a mean $t_{G_2}$ of two hours, 14 per cent have a $t_{G_2}$ of more than four hours, while only 3 per cent have a duration of seven hours or longer. It is difficult to see how we can project a continuous distribution of $G_2$ transit times, with a mean of 1.4–2.5 hours as indicated by the FLM studies, and maximum values of seven hours; thus it would appear that some cells are passing through the $G_2$ phase very slowly, or are in a resting state in this phase, and indeed this was the conclusion reached by Clausen $et$ $al.$ (1981). However, it is possible that we are dealing with a $t_{G_2}$ distribution of peculiar shape; it is also interesting to note that Clausen $et$ $al.$ (1981) thought that, if such transiently blocked cells did exist, they should be distinguished from the postulated $G_2$ block of Gelfant (1975, 1976). These latter cells are envisaged as being blocked indefinitely in $G_2$, awaiting a suitable trigger; the delay in $G_2$ postulated by Clausen $et$ $al.$ is of strictly limited duration. It will be interesting to see how these studies develop: Clausen (personal communication)

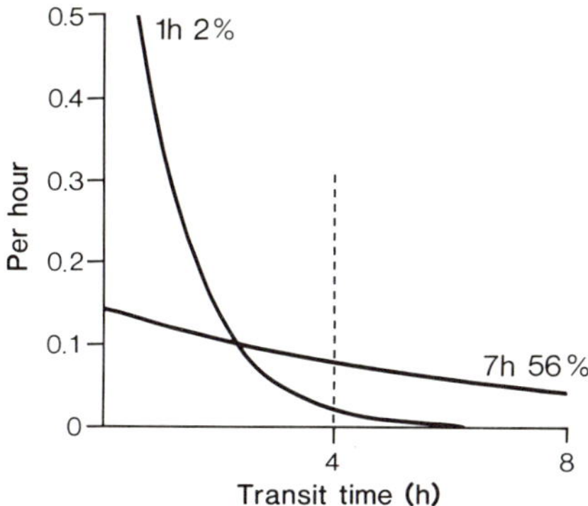


Fig. 7.3. Variation in the duration of $t_{G_2}$ in hairless mouse epidermis. Two distributions are shown where the mean transit times are one and seven hours, in which 2 and 56 per cent of cells have $t_{G_2} >$ four hours. (Redrawn from Clausen *et al.* (1981).)

has supported these proposals by performing continuous labelling experiments on mouse epidermis, and then sorting the cells into $G_1$, $S$, and $G_2 + M$ fractions, and counting the proportion of labelled cells in each phase; early analysis indicated a maximal $t_{G_2}$ as long as 17 hours.

Values for $t_{G_1}$ will of course be reflected in the respective $T_C$ values, where available (see Table 7.2).


## 7.4. The growth fraction

We now come to the vexatious topic of growth fraction, which, it is probably true to say, has produced more confusion among those interested in squamous epithelium than any other kinetic parameter. Proliferative cells are found in the basal layers, and reproductive capacity is lost when cells leave this proliferative compartment to differentiate. When speaking of the 'growth fraction' in squamous epithelium, most workers are concerned with the proportion of the proliferative compartment that consists of cells actively cycling; however, we should note that Sherman *et al.* (1961), observing that basal cells occupied 54.5 per cent of the total epidermal cell population in the mouse ear, regarded this as the maximum growth fraction for the whole epithelium.

We have seen how growth fraction/transition probability discussions often dissolve into semantic arguments (Chapter 2, section 7), but in squamous epithelium the question 'How many basal cells are proliferating?' can be meaningful. This is because of the heterogeneous nature of the basal layer, especially in epidermis. In epidermis, we know that, in the mouse, some 10–15 per cent of basal cells are non-keratocyte, and cannot usually be included in the basal cell proliferative pool, but a more general statement of the problem in squamous epithelium is that some cells may have completed their proliferative sequence, are preparing to migrate, and undergo early differentiation change *while still in the basal layer* (Potten 1981). These *post-mitotic maturing cells* are considered in more detail in Chapter 8, where the kinetic organization of squamous epithelium is discussed, but here we might note that the presence, in the basal layer, of cells predestined to migrate, is not in accord with the simplistic random migration process we have hitherto assumed, and which is central to the Brown and Oliver (1968) method of measuring the cell cycle time (see Chapter 6, section 2). Thus the manner in which investigators regard the proliferative status of cells in the basal layer becomes operationally important, and not merely a semantic argument; if there indeed is a separate subpopulation of cells in the basal layer which have undergone a quantal mitosis, lost reproductive capacity, and are awaiting a signal to migrate, then it certainly appears worthwhile to talk in terms of a *growth fraction*, with the post-mitotic cells occupying the non-growth fraction. However, if we propose a homogeneous basal layer with a random migration process, then we are in semantic difficulties; the word *homogeneous* implies that there is proliferative equality between cells, and, as we shall see, most continuous labelling curves result in effectively all basal cells being labelled, so that they are all *potentially* capable of proliferation, and the interpretation of the homogeneous model is that they are all proliferative *at one and the same time*. However, even at that point in time, there will be a proportion of cells which are destined to progress through another cell cycle, while the remaining proportion are destined to migrate, and be lost from the basal layer during that cell cycle time, or more properly, $t_{G_1}$ (see Chapter 8, section 6). These groups of cells have been termed the *progenitor* and the *non-progenitor* cell populations respectively; the manner in which the proportions have been estimated are basically the same, and is by comparison of the cell cycle time (i.e. the residence time of the *progenitor* cells in the basal layer) with the residence time of the *non-progenitor* cell population; if a kinetic steady state exists, then the ratio of these times will be equivalent to the numbers of cells in the respective compartments. Such calculations have been made by Leblond *et al.* (1964), Frankfurt (1967), Iversen *et al.* (1968), and Blenkinsopp (1969); their methods are reviewed in Chapter 8, section 6, when we consider mechanisms of migration, and the values are given in Table 7.2, and summarized at the end of this section.

More important, for our present purpose, is that the presence of post-mitotic maturing cells in the basal layer means that the remaining cells are proliferative, and these form our growth fraction, even if we do embrace the concept of a transition probability model, and exclude a formal $G_0$ state. The possible presence of real $G_0$ cells in the proliferative compartment is another, more difficult question, but we shall see that there is only scanty evidence for a classical $G_0$ compartment in normal squamous epithelia, a situation which also exists in the small intestinal crypt; as we shall see, however, post-mitotic maturing cells are found in the crypt, albeit in a morphologically more distinct arrangement, at the top of the crypt (Chapter 19).

Most confusion about the growth fraction in squamous epithelium has come from the consideration of continuous labelling curves, and we should, therefore, study these in some depth. We have already made our routine objections to this method of measuring the growth fraction, which apply in any kinetic situation, but there are additional problems which are specific to squamous epithelium. It is the usual practice in most epidermal experiments to give the $[^3H]$-TdR by repeated injections, with the consequent problems of stress induced by repeated animal handling during some 10–40 injections; handling directly stimulates cell proliferation in the epidermis (see Chapter 10), to say nothing of the radio-toxicity of such a large amount of $[^3H]$-TdR; the resultant of these effects might be thought to be an undoubted perturbation of the system from the assumed steady state (Potten and Major 1980), and indeed, although there were only minor perturbations detected over the first four days of labelling in mouse epidermis, by the stathmokinetic and flow cytophotometric methods, thereafter proliferative activity was seen to decrease (Thorud and Clausen 1982). Alternatively, the $[^3H]$-TdR can be given by long-term infusion, or in the drinking water (Cameron 1972a), in which case it is difficult to be sure that all cells are being labelled when they should be, and the actual dosage of $[^3H]$-TdR remains unknown.

A selection of continuous labelling curves are reproduced in Fig. 7.4. The basal layer curve in each case is seen to reach 100 per cent, and the natural reaction, emphasized particularly by Hegazy and Fowler (1973a) and Fowler and Denekamp (1976), is to regard the growth fraction as equal to one, and conclude that all basal cells are proliferating. Denekamp *et al.* (1976) published continuous labelling curves for both ventral and plantar aspects of mouse feet; the data is shown in Fig. 7.4C. The plantar data shows a plateau value of about 97 per cent, which was interpreted as indicating an $I_P$ of unity, with a few unlabelled cells being accounted for by false-negatives due to self-absorption of $\beta$-particles. On the other hand, they could not account for the 75 per cent plateau value in the ventral aspect data, which suggested a non-growth fraction of 25 per cent. Of course these could be classical $G_0$ cells, although it was noticed that, during a proliferative response to irradiation, the putative $G_0$ cells did not contribute. This latter result is not in accord with their statement, published in the same year (Fowler and Denekamp 1976) to the effect that all cells are proliferating. Moreover, Gelfant (Gelfant 1976; Gelfant *et al.* 1983) has reported an $I_P$ value of $\sim$20–40 per cent in normal human epidermis based on biopsies taken after nine days continuous labelling; this introduces special problems in interpretation, which are discussed in section 7.8.

Internal squamous epithelia submitted to continuous labelling conditions also realize high labelling indices; after five days in mouse oesophagus, Greulich (1964) reported 100 per cent basal layer labelling, while Frankfurt (1967), in mouse squamous forestomach, also believed that the 7.7 per cent unlabelled nuclei present after 48 hours labelling represented false-negatives; Cameron

(1972*a*) also reported 100 per cent labelling in the basal layer of mouse tongue with [³H]-TdR given in the drinking water for periods of up to five days. Similar data, showing 100 per cent labelling under continuous labelling conditions, were seen in the squamous epithelium of the hamster cheek pouch (Brown 1970; Izquierdo and Gibbs 1974). Additionally, it will be remembered that Brown and Oliver (1968), in the cheek pouch, and Hegazy and Fowler (1973*a*), both using the $N_B/N_T$ method, have reported a growth fraction of unity from the inflexion point at the 0.5 level.

There are several potential objections of some power to this point of view: there is now considerable evidence, at least for epidermis, for the presence of *post-mitotic maturing cells* in the basal layer, which will not, in normal circumstances, re-enter the cell cycle (see Chapter 8, section 6). This will mean that a proportion of basal cells will be non-proliferative *at any point in time*, but these will not be, and *cannot be*, detected by the continuous labelling or $N_B/N_T$ method. In the case of the continuous labelling method, since post-mitotic maturing cells are held to have a residence time in the basal layer of, for example,

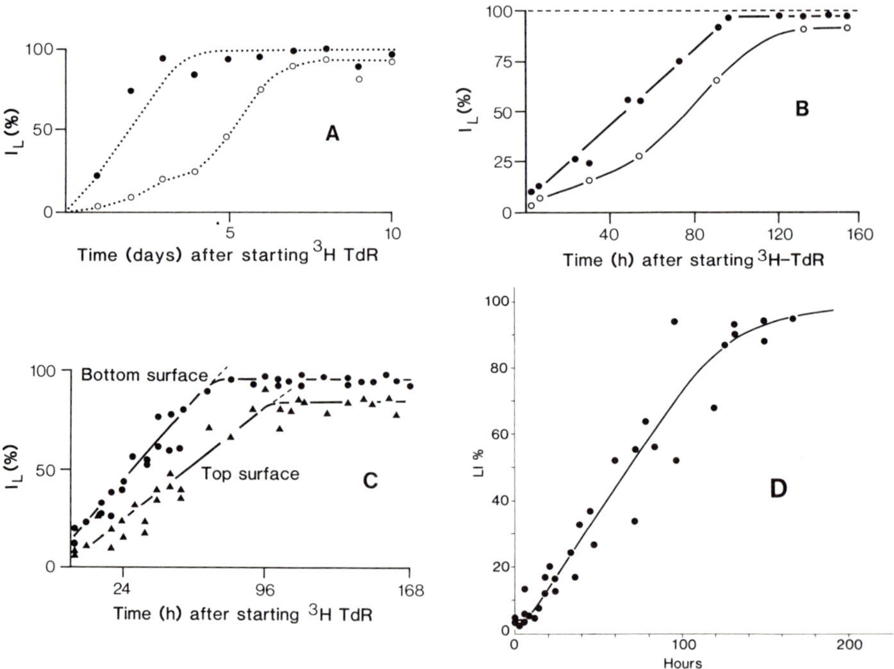

Fig. 7.4.  A selection of continuous [³H]-TdR labelling curves for basal cells (●) and differentiated cells (○) in stratified squamous epithelium:  (A) hairless mouse dorsal epidermis. (Redrawn from Iversen *et al.* (1968).)  (B) Mouse dorsal epidermis. (Redrawn from Hegazy and Fowler (1973*a*).)  (C) The ventral (▲) and plantar (●) epidermis of the mouse foot. (Redrawn from Denekamp *et al.* (1976).)  (D) Mouse dorsal epidermis (counted in sheets). (Courtesy Dr C. S. Potten).)

two days in the mouse dorsal epidermis, they will all have migrated by the time the continuous labelling curve reaches a plateau, and their place taken by labelled post-mitotic maturing cells; at this time we see labelled cells in the suprabasal layers, which have already passed through the post-mitotic maturing phase, and migrated. Hence all basal keratocytes will label, and the growth fraction will appear to be unity even though, at the beginning of the experiment, there will be unlabelled, non-proliferative basal layer cells, and at the end of the experiment there will be labelled non-proliferating cells.

The problem of continuous labelling curves in squamous epithelium has been considered for mouse epidermis by Potten (Potten 1981; Potten and Major 1980). Analysing the several curves which are available (Iversen *et al.* 1968; Hegazy and Fowler 1973*a*; Potten *et al.* 1974; Denekamp *et al.* 1976; Fukuda *et al.* 1978), all obtained using the serial injection procedure, Potten and Major produced the composite curve shown in Fig. 7.5, and from this made several

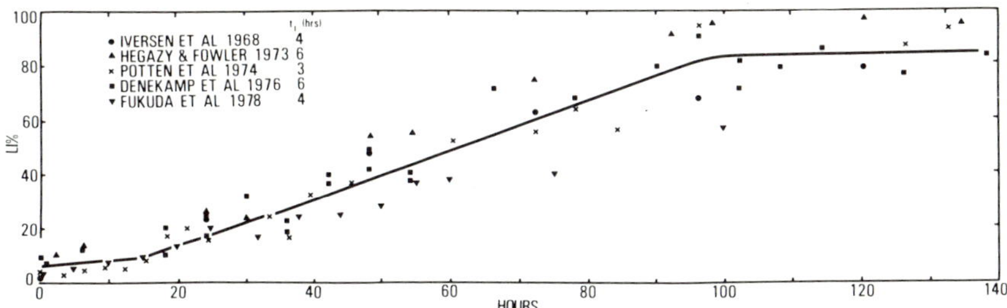


Fig. 7.5. A composite continuous [³H]-TdR labelling curve for the basal cells of the mouse epidermis, constructed from the data indicated. The time $t_i$ is the time between injections of [³H]-TdR. (Courtesy of Dr C. S. Potten.)

interesting points: (i) there is no real difference between the curves, despite the different strains of mice and the different injection schedules used; (ii) there is an early shallow slope, probably related to the length of $G_2 + M$; (iii) that there is considerable variation about the hand-drawn line, and no real plateau was evident at labelling indices less than 100 per cent. Potten and Major (1980) unsuccessfully attempted to discriminate between several different models of basal cell organization on the basis of this data, and reached similar conclusions concerning the continuous labelling method in squamous epithelium to those of Wright (1980*c*), which are rehearsed above and in Chapter 3, section 5, but added the observation that the inherent scatter of the data points obviates the definition of the plateau value; non-proliferating cells may in fact *not* be turning over, and the continued stress of the labelling procedure could stimulate cells from $G_0$ back into the cell cycle, and, as we have previously observed (see Chapter 3, section 5), most curves, if continued for long enough, would reach 100 per cent. Simulation studies indicated that the timing of the injections was quite critical in defining the slope of the curve; a three hour schedule gave a steeper slope than a 12 hour schedule, with the peak reached at 90 as opposed to 130 hours, emphasizing the importance of labelling all cells. The influence of circadian variation was found to be small for short injection schedules, with minor displacements depending on when the first injection was given with respect to the circadian dependent $I_S$ peaks or troughs.

Thus we must have serious reservations regarding the measurement of $I_P$ using the continuous labelling method in squamous epithelium; indeed, we should be quite frank, and say that, because of the arguments put forward, we do not believe that the growth fraction is approachable using this technique.

In Brown and Oliver's (1968) $N_B/N_T$ experiment, based on migration, the unlabelled post-mitotic cells will migrate early in the experiment, and will not contribute to the graph, since here we are concerned only with the ratio of labelled basal cells to labelled total cells. As the experiment proceeds, post-mitotic maturing cells become labelled in their turn and migrate, apparently in a random manner, as evidenced by the exponential decline in $N_B/N_T$ (see Fig. 6.1, p. 230). Nevertheless, the non-proliferative cells will not be picked up.

A further, more important objection to the proposition that all basal cells are proliferating comes from Potten (1975$b$), who claimed that the published values for $T_C$ from FLM curves (cf. Hegazy and Fowler 1973$a$) of about 100 to 130 hours were not really consistent with the published values for cell production rates and observed transit times; for example, a calculation of the transit time in the mouse ear epidermis from published $k_B$ and $T_C$ data, assuming an $I_P$ of unity, realized 56 days, compared with an observed total transit time of 11–24 days; a smaller growth fraction would contribute to a shorter transit time value (assuming that the $T_C$ estimates are reliable, which may not be so; see section 7.2).

We can also criticize the concept of a growth fraction of unity from a consideration of the time basis of such experiments as the continuous labelling and $N_B/N_T$ methods; in continuous labelling, the plateau is reached in time $t$ assumed to equal $t_{G_2} + t_M + t_{G_1}$; a limiting value of 100 per cent would also be consistent with cells, after mitosis, entering a $G_0$ compartment with constant influx and efflux, and a residence time related to the time taken to achieve the plateau, which can be as long as five days. Such an alternative model was considered by Brown and Oliver (1968), but the criticism would appear to be excluded by Hegazy and Fowler (1973$a$), who independently confirmed the $T_C$ value by direct measurement of the FLM curve.

We could argue, with some justification, that we are now entering the sphere of semantics again, equating $G_0$ with a long $G_1$ or even with a putative $A$ state, as discussed in Chapter 2, section 7, but several authors do consider that true $G_0$ cells occur in epidermis, and it is worthwhile exporing the possibility. Wolfsberg (1964) was firmly of the opinion that considerable numbers of $G_0$ cells were present in mouse squamous forestomach epithelium, based on a consideration of the cell kinetics; an FLM curve for basal cells, with a grain count threshold of 10 grains, showed a small second peak (see Fig. 7.4A), indicating a $T_C$ of about 30 hours. The birth rate, calculated from $I_S/t_S$, was 0.0039 cells/basal cell/hour, giving an apparent cell cycle time, or turnover time, of 260 hours (10.8 days). Wolfsberg estimated that the second peak of the FLM curve was about 33 per cent of the area of the first peak, and defined an $x$ which equalled the probability of a newborn cell's entry into $G_0$/probability of its re-entry into $G_1$; this is a decycling/recycling ratio, and its similarity to Bresciani's $d$ ratio of $\overleftarrow{\eta}/\overrightarrow{\eta}$ should be recognized (see Chapter 2, section 7). For Fig. 7.6A $x$ was 2, so that twice as many cells enter $G_0$ as continue in the $G_1$ phase. If the grain count threshold was decreased to six grains per nucleus, the second wave of

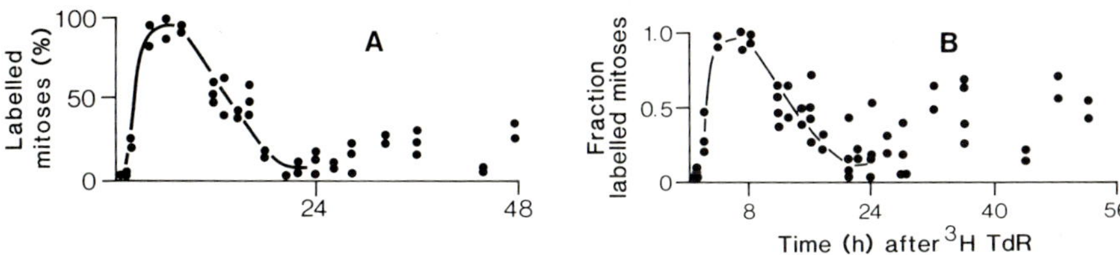


Fig. 7.6. FLM curves in the mouse squamous forestomach. The labelling criterion is (A) 10 and (B) 6 or more grains per nucleus. The symbols refer to different experiments performed on the same tissue and mouse strain. (Redrawn from Wolfsberg (1964).)

labelled mitoses increased to about two-thirds the size of the initial wave; $x$ became $\frac{1}{2}$, and twice as many cells recycled as entered $G_0$. Depending on the value of $x$, Wolfsberg estimated a '$t_{G_0}$', which varied from two to four weeks.

This data does appear to show that many cells are decycling, to re-enter the cell cycle after a variable residence time in $G_0$. The problem in accepting this is highlighted in Fig. 7.7, redrawn from Wolfsberg's paper, which included a 'transitional' phase. These were cells which had completed mitosis and were waiting to migrate: our *post-mitotic maturing cells*. Such cells will contribute to the value of the birth rate, as they are included in the population for the determination of the labelling index, in the expression $k_B = I_S/t_S$; they will not appear in the waves of labelled mitoses (Fig. 7.6), since they are, by definition, post-mitotic, and will therefore contribute to the population of decycling cells. On the basis of the Potten (1975b, 1976) model, where 33 per cent of basal cells were held to be post-mitotic maturing cells, we could explain the results of the six grain FLM curve, at least.

We should also note that, as a result of studies on circadian variation in hamster cheek pouch epithelium, Izquierdo and Gibbs (1974) and Izquierdo (1977) considered that basal cells entered a $G_0$ phase, where synchronization occurred, and then cells were released into $G_1$, accounting for the circadian variation in proliferative indices; however, these observations can readily be explained by a variable $t_{G_1}$ with synchronization in that phase, without recourse to a $G_0$ phase.

In addition to the present comments, the concept of a large subpopulation

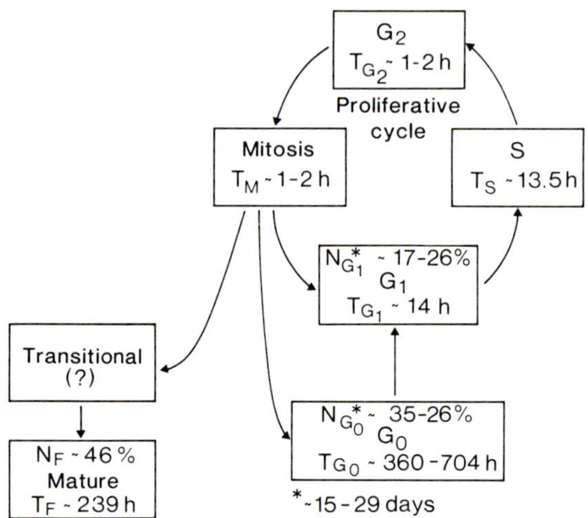

Fig. 7.7.    A model for the organization of cell proliferation in the stratified squamous epithelium of the mouse forestomach. Superscript * means that more than two values were available, proportional to the grain count threshold used. $N$ = number of cells in any phase; $T$ = transit time; $F$ = functional or mature compartment. (Redrawn from Wolfsberg (1964).)

with a definite residence time in $G_0$ was not well received by Brown (1968) or by Frankfurt (1967). Brown (1968), in a little quoted but salient paper, emphasized that such a concept was indistinguishable, both by definition and in practice, from a long $G_1$ phase. Brown (1968) proposed that a long $G_1$ phase could be distinguished from a classical $G_0$ phase in which the length of the stay was not definite, but *random* (that is, random with respect to the length of time already spent in $G_0$). These alternatives are shown in Fig. 7.8A, and the relationship to the general cell cycle models proposed by Burns and Tannock (1970) and by Smith and Martin (1973) will be readily appreciated (see Chapter 2, section 11). Making the rather important assumption of random migration from the basal layer, $I_L$, the continuous labelling index, will be, for model (1):

$$I_L = 2 - 1.950e^{-\lambda t} \text{ (for } t < T_C, \text{ as when } t > T_C, I_L = 1) \tag{7.1}$$

and for model (2):

$$I_L = 1 - 0.95e^{-2\lambda t} \text{ (for all } t) \tag{7.2}$$

where $\lambda$ = rate constant for cells entering the $S$ phase, and equivalent to the birth rate, or $I_S/t_S$, and equal to 0.0051 cells/cell/hour. The values 1.95 and 0.95 arise because $I_S = 0.05$.

The models were compared by (a) fitting the experimental $I_L$ curve, and (b) transforming eqns 7.1 and 7.2 above to a linear form (i.e. eqn 7.1 becomes $\ln(1.95/2 - I_L) = \lambda t$ and doing a least squares regression. Figure 7.8B shows the predicted curves together with the experimental data for the hamster cheek

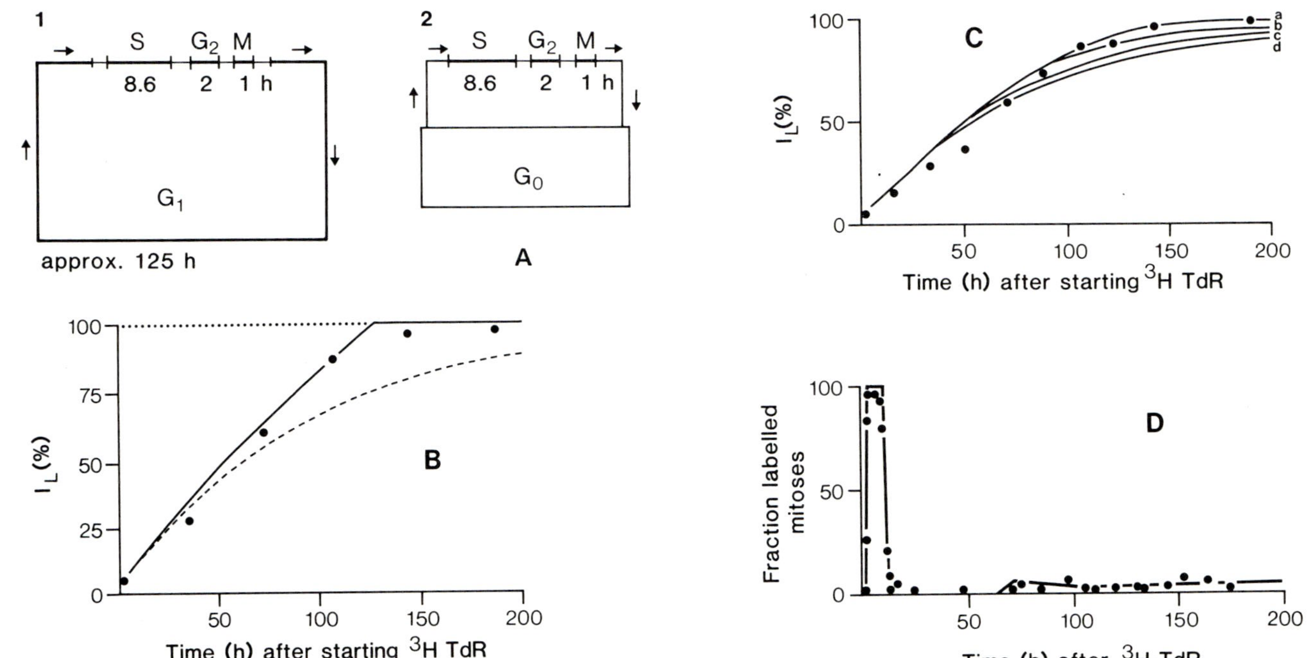



Fig. 7.8. (A) Alternative models for the cell cycle in hamster cheek pouch squamous epithelium: Model (1) the long $t_{G_1}$, Model (2) the $G_0$ state. (B) Predicted continuous labelling curves for the two models. The points show the experimental values, and —— the long $t_{G_1}$ model, ----- the $G_0$ model. (Redrawn from Brown (1968).) (C) The same continuous labelling data fitted by the Burns and Tannock (1970) $G_0$ model (see Fig. 2.25A, p. 91), assuming that $K_p$ (the birth rate constant) = 0.51 per cent/hour, with $T$ values (the $C$-phase transit time) of 100, 80, 60, and 40 hours for curves a, b, c, and d respectively. Also shown (D) is the FLM curve for the hamster cheek pouch (Brown, unpublished) fitted by the Burns and Tannock model, with $K_p$ = 0.51 per cent/hour, $T$ = 60 hours, and $t_S$ = 8.6 hours. (Redrawn from Burns and Tannock (1970).)

pouch, and model (1) evidently gives the best linear relationship from the transformed equations. Hence the long $G_1$ model was considered the most realistic of the two, and few, if any, cells were thought to be in $G_0$.

In this analysis, Brown (1968) considered that the $T_C$ value of 135 hours gave the best fit, and although he maintained that there was no $G_0$ phase, favouring model (1), there was no evidence of a second peak on the FLM curve. Figure 7.8C shows Brown's experimental points and theoretical curves derived by Burns and Tannock (1970), on the basis of their $G_0$ model, using an initial $I_S$ of 2.0 per cent and $T_C \sim 60$ hours; the fit is reasonable, and the $G_0$ argument could explain the absence of a second FLM peak. However, we will recall that Hegazy and Fowler (1973$a$) did obtain second peaks, and we have already seen how their application of the Brown and Oliver (1968) method to mouse epidermis indicated $I_P = 1$ (even if insensitive to post-mitotic maturing cells, the method should pick up resting cells!).

Burns and Tannock (1970), in their turn, and using the same continuous labelling data, claimed that the $G_0$ model could not be disregarded for squamous epithelium. We have already discussed this model (see Chapter 2, section 11). It will be remembered that the model (see Fig. 2.25A, p. 91) proposes that flux into the $C$ state from $G_0$ occurs with a constant probability $K_{G_0}N_{G_0}$. Cells are produced from the $C$ state at a rate $2k_PN$. The continuous labelling function is derived by integrating the age distribution diagrams for the $G_0$ and $C$ phase cells. This was done at $t \leqslant t_2$, $t_2 \leqslant t \leqslant t_1 + t_2$, and at $t \geqslant t_1 + t_2$. Initially $I_L = I_S$. At zero the slope is $k_P e^{k_P t_2}$, which changes to $k_P$ at $t = t_2$. At $t > t_2$ the slope is $2k_P$ and will then decline as the curve approaches 100 per cent asymptotically. Figure 7.8C shows the best fit to the experimental data using different values for the length of time spent in the $C$ phase. The relatively poor fit of the curves up to the 50 per cent level was considered to be due to the circadian variation so prominent in the hamster cheek pouch epithelium (see Chapter 9); after the 50 per cent level, however, all curves with a $C$ phase transit time $> 60$ hours gave a reasonable fit. Thus, if we assume that cells leave $G_0$ randomly with a constant probability per unit time, we can fit continuous labelling curves satisfactorily. On the other hand, it must be realized that the fitting of continuous labelling curves is very sensitive to the mode of cell loss from the population under consideration (Steel 1977; Wright 1980$c$; Iversen $et\ al.$ 1968) and some assumptions about removal mechanisms must be made before information can be derived from the curve. In the present situation, alternatives include the simplistic random removal mechanism favoured by some (Brown and Oliver 1968), cell loss from the oldest $G_1$ cell nearest a mitosis (Iversen $et\ al.$ 1968), or selective loss from post-mitotic cells (Potten 1976); all these would modify the format of the continuous labelling curve, but we should note that a recent attempt to descry different models from such curves proved singularly unsuccessful, mainly on account of the scatter of the data (Potten and Major 1980).

However, this sort of model, with the $G_0$ phase as envisaged by Burns and

Tannock (1970) and by de Maertelaer and Galand (1977) and Izquierdo (1977), again introduced the semantic problem, compounded by Smith and Martin (1973), about what $G_0$ really means. While the $G_0$ concept of Wolfsberg (1964) of a protracted $G_0$ phase with a definite, constant sojourn is probably not tenable, we cannot really exclude, on experimental evidence, a $G_0$ phase as envisaged by Burns and Tannock (1970).

Frankfurt (1967) made several observations very relevant to this discussion; again studying mouse forestomach squamous epithelium, Frankfurt could find no evidence of the protracted $G_0$ phase proposed in Wolfsberg's investigation. A continuous labelling experiment plateaued at 93 per cent at about 48 hours; with $t_S$ at seven hours (from an FLM curve) a maximum value for $T_C$ was about 55 hours. The 7 per cent of cells which did not label were assumed to represent non-labelled cells which had not migrated from the basal layer at 48 hours after commencing $[^3H]$-TdR injections, or possibly false-positive labelled cells (sections were 5 $\mu$m in thickness, and we have seen how estimates of the distance $\beta$-particles travel in tissue sections vary; Chapter 3, section 1). If, as Wolfsberg suggests, 33–60 per cent of cells were in a $G_0$ phase which lasted as long as 2–4 weeks, naturally such a plateau in the continuous labelling curve would not be realized. Frankfurt's FLM findings also differed; using a grain count threshold of three grains per nucleus, the second peak only reached 16 per cent at 28 hours after $[^3H]$-TdR. This would presumably reflect recycling cells at the lower end of the distribution of cycle times with a mean of 55 hours. Frankfurt did note that this second peak was not associated with any change in the grain count.

This is not to say that all basal cells were proliferative at the outset; Frankfurt (1967), from a consideration of the rate of migration of labelled cells, calculated that the *mean maturing time in the basal layer* was 41 hours (actually this is a median value). Proceeding from the premise that the average size of each sub-population is proportional to the transit time through that population, Frankfurt compared the 55 hour cell cycle time with the maturation transit time (i.e. 55:41) to show that 43 per cent of basal cells are in the maturing phase. This evidence goes some way to explaining Wolfsberg's results, especially if we note that 43 per cent could be a minimum value, since Frankfurt compared a near maximum $T_C$ value with a median maturation time. In this context, about 70 per cent of basal cells are found to be *progenitor* or proliferating cells in a variety of squamous epithelia, while the remaining 30 per cent were thought destined to migrate, with a basal layer residence time that varies between 12 hours in the mouse tongue to about 250 hours in the ear (see Chapter 8, section 6). Somewhat anticipating the discussion to be taken up in Chapter 8, the use of the term 'maturing time' by Frankfurt and 'post-mitotic maturing cells' by others (e.g. Potten 1975*b*, 1976) implies an ordered migration sequence from the basal layer, whereas other authors (e.g. Brown and Berry 1968; Blenkinsopp 1968*a*; Iversen *et al.* 1968) regard any sojourn of cells in the basal layer prior to migration as a byproduct of a random removal mechanism. This distinction is discussed in Chapter 8, sections 4 and 6.

This far we have introduced several concepts which modify any simple model of stratified squamous epithelium; the presence of circadian variation, post-mitotic maturing cells and a possible $G_0$ compartment has rather battered the initial simple assumptions upon which, for example, we based our consideration of the age distribution in Chapter 6, section 7; however, the situation may be even more complex: in the stratified squamous epithelium as found in the epidermis in certain anatomical sites, there is a definitive ordered structure, discussed in Chapter 8, section 2. Potten (1975a) has made a critical analysis of the kinetic requirements of maintaining such an ordered structure, including the implications for the epidermal growth fraction, and Table 8.2 (p. 311) shows the various values for $I_P$ which would be consistent with the several possible models permitting the maintenance of such ordered structure; however, here we may note that the favoured concept is one where 60 per cent of basal kerato-cytes are proliferating, while the remaining cells are post-mitotic maturing cells waiting to migrate.

So what conclusions can we make regarding the measurement of proliferating populations in squamous epithelium? Purely cell kinetic measurements, using classical methods, as applied by Brown (Brown and Oliver 1968; Brown 1968), Hegazy and Fowler (1973a) and others do indicate that all cells are proliferating; however, this does imply a simplistic idea of kinetic organization, and when we introduce the concept of basal layer non-proliferative subpopulations, consider-able problems become apparent in interpretation. Wolfsberg's (1964) concept of a classical $G_0$ subpopulation can probably be dismissed, but the data are not inconsistent with the Burns and Tannock (1970) $G_0$ proposal; when we have discussed the kinetic organization of the basal layer (Chapter 8) a more definitive summary can be made.

### 7.5. The birth rate in squamous epithelia

There are numerous values for the birth rate in many squamous epithelia, and these have been summarized by Potten (1975b), with particular reference to ordered epidermal structure. Values are given in Table 7.2 (pp. 258-61), and the relevance of such measurements to ordered structure is discussed in Chapter 8, section 2. Many workers have calculated the birth rate from $I_S/t_S$, but there are also direct measurements available from stathmokinetic experiments; for example Bertalanffy *et al.* (1965) reported birth rate values varying from 3.4 to 6.3 cells/100 cells/day in rat epidermis, depending on the site, and Storey and Leblond (1951) obtained a value of 5.2 cells/100 cells/day for the rat plantar epidermis, while estimates for various sites in the rat oral squamous epithelium vary from 2.27 to 9.59 cells/100 cells/day (Trott and Gorenstein 1963; Randers Hansen 1967). Blenkinsopp (1968a, 1969, Blenkinsopp and Gilbert 1969) has given $k_B$ values for many squamous epithelia as measured by the stathmokinetic

Table 7.3. Birth rate and turnover time values in the basal layer of mouse squamous epithelia

|  | $p*$ | Birth rate † (cells/cell/h) | Turnover time ‡ (h) |
|---|---|---|---|
| Oesophagus | 0.37 | 0.026 | 38 (27, 67) |
| Squamous stomach | 0.45 | 0.020 | 50 (38, 72) |
| Epidermis |  |  |  |
|   Ear | 0.94 | 0.014 | 73 (43, 96) |
|   Ventral | 0.89 | 0.0095 | 105 (81, 122) |
|   Dorsal | 0.95 | 0.0092 | 109 (88, 126) |
|   Plantar | 0.73 | 0.017 | 59 (42, 76) |
|   Tail | 0.65 | 0.0083 | 120 (99, 143) |

*$p$ refers to the proportion of basal layer mitoses with axes in the plane of the basal layer.
†This is calculated from the regression line of the metaphase arrest data after vincristine injection, fitting eqn 6.12 and substituting $p$.
‡The figures in brackets are the 95 per cent confidence intervals from the regression analysis.

method (colchicine), but none of these measurements have been corrected for the age distribution changes that will occur because of the variation in the direction of the mitotic axes, which, as we have seen in Chapter 6, section 7, is a source of error in the procedure. Table 7.3 shows $p$ values (the proportion of horizontal mitoses) in a number of squamous epithelia in the mouse, together with the birth rate and turnover time data for the basal layer only. While most epidermal sites have a majority of horizontal mitoses (but see plantar and tail epidermis) there are considerable numbers of vertical mitoses in internal squamous epithelia such as the oesophagus and the squamous stomach. It is interesting that Bullough and Mitrani (1978) reported an increase in the proportion of vertical mitoses in the epidermis of induced hyperplasia, and note that most of the epithelia with increased numbers of vertical mitoses in Table 7.3 have relatively high birth rates and no ordered structure (see Chapter 8, section 2).

## 7.6. Transit times in squamous epithelia

We have already noted the difficulties which are implicit in the concepts of transit times and turnover times in squamous epithelia (see Chapter 6, section 3). In the main, viable compartment transit times have been measured from the injection of [³H]-TdR to its appearance in the granular layer in keratinizing squamous epithelium. We have seen that this is problematical in that labelled cells remain for a short but variable period in the basal layer before migration, a drawback which has in fact been made use of by Potten (1975$a$) to measure the residence times of post-mitotic cells in the basal layer (Chapter 8). To this caveat we must add the semiquantitative nature of transit time measurements, the possibility of [³H]-TdR reutilization which is always present in such long-term

Table 7.4. Kinetic parameters in normal human epidermis

| $I_P$ (%) | $t_S$ (h) | $T$ (germinative cells) (h) | $I_S$ (%) | $k_B$ (cells/1000 cells/h) | $T_T$ (minimum) (viable cells) (days) |
|---|---|---|---|---|---|
| – | 16 | 457 | 3.5 | – | 12–14 |
| – | 7.3 | 206 | 3.8 | – | – |
| – | – | – | – | – | – |
| – | – | – | – | – | – |
| – | – | 152 | 4.9 | – | 10–14 |
| – | – | – | – | 0.02 | – |
| – | – | – | – | – | – |
| – | – | – | – | – | – |
| 20–40* | – | – | – | – | – |
| – | – | 50† | 6 | – | – |
| *In vitro* | | | | | |
| – | 6.6 | 282 | 2.4 | – | – |
| – | 7.6 | 170 | 4.5 | – | – |
| – | 10.3 | 184 | 5.8 | – | – |
| – | 11.5 | 59‡ | 10.5 | – | – |

*This is a value derived from a continuous labelling curve carried out over 15 days; only 17 per cent of basal cells were labelled, and with a cell cycle time of 200 h, all proliferating cells should be labelled. Also, the ratio of the available cell cycle time values (Bauer and de Grood 1975; Chopra and Flaxman 1974) to the turnover time i.e. 60/200, will give a value for the growth fraction (40 per cent).

experiments, and that the measurements obtained are minimum rather than mean values. Nevertheless, a considerable number of values for the minimum transit time for viable or total epidermal cells are available in the literature (Cameron 1970), some of which are given in Table 7.2. Naturally, in comparing different sites, it is important to remember that the thickness of the epithelium, as well as its cell production rate, will define its transit time; thus the thin epithelial attachment area of the gingival epithelium has a transit time (viable cells) of only 24 hours, whereas cells in the thicker oral epithelium take 10–12 days to traverse the viable cell compartment (Beagrie 1963).

The stratum corneum transit time has been measured by following the disappearance of marker substances such as amino acids or fluorescent markers, which are supposedly incorporated into the stratum corneum only. For example Rothberg *et al.* (1961) injected [$^{14}$C]-glycine into humans, and detected radioactivity in surface scrapes on the 13th day, which reached a peak at day 26–28; they therefore claimed that the minimum stratum corneum transit time was 13–14 days; this, plus the time taken for cells to reach the stratum corneum (12–15 days), gave the total transit time, 26–28 days. Porter and Shuster (1967) reported similar timings for peaks of [$^{14}$C]-glycine and methionine in surface strippings from human epidermis, while Downes *et al.* (1967) reported peaks of [$^{35}$S]-cystine in un-named sites of mouse epidermis at 5–6 days, and considered

| $T_T$ (S. corneum) (days) | $T$ (viable cells) (h) | $T$ (Total) (days) | Reference |
|---|---|---|---|
| – | – | – | Weinstein and McCullough (1973) |
| – | – | – | Allegra and De Panfilis (1974) |
| – | 61 | 71 | Halprin (1972) |
| 6–20 | – | – | Baker and Kligman (1967) |
| – | 17.7 | – | Epstein and Maibach (1965) |
| – | – | – | Fisher (1968*a*) |
| 13–14 | – | – | Rothberg *et al.* (1961) |
| – | 27 | – | Pinkus (1952) |
| – | – | – | Gelfant (1976) |
| – | – | – | Bauer and de Grood (1975) |
| – | – | – | Pullman *et al.* (1974) |
| – | – | – | Heenen *et al.* (1975) |
| – | – | – | Heenen *et al.* (1973) |
| – | – | – | Chopra and Flaxman (1974) |

† This is a cycle time value derived from pulse cytophotometry, also dependent on $I_P$ assumptions.

‡ This is a cell cycle time derived from a two-peak FLM curve.

that the stratum corneum transit time was 4–5 days. However, Fukuyama and Epstein (1966) reported variable concentrations of different amino acids in the different layers of newborn rats after intraperitoneal injection, which must cast doubt on the basis of this method, especially as the levels of amino acids obtained by scraping or stripping the surface may be very low (Downes *et al.* 1967). The fluorescent methods depend on the ability of certain compounds to combine selectively with the keratin in the stratum corneum, and their disappearance is followed by noting the time taken for the ultraviolet-induced fluorescence to disappear. Previously used compounds, such as those used by Baker and Kligman (1967) were held to be potent skin sensitizers by Jansen *et al.* (1974), who suspected that the induced inflammation could lead to increases in the $I_M$ and a decrease in the transit time; they introduced the use of 5-dimethylamino-1-naphthalenesulphonyl chloride (dansyl chloride), which combines with amino groups and is non-irritating. Penetration is complete in human epidermis by 24 hours, and the method gives stratum corneum transit time values of around 14 days (see Table 7.4).

## 7.7. Turnover times in squamous epithelia

Most estimates are derived from measurements of the cell production rate, which in turn is found from the direct measurement of the birth rate by the metaphase arrest method (Blenkinsopp 1968$a$, 1969), or by calculation from $I_S/t_S$ or $I_M/t_M$ (Sherman *et al.* 1961), and relating this value to the population under consideration (e.g. basal layer, viable cell compartment) by including the appropriate cells in the interphase count for the respective index; the reciprocal of the birth rate is the turnover time, assuming a rectangular age distribution, where appropriate (see Chapter 6, section 7). Naturally these values are very dependent on the value chosen for $t_M$ or $t_S$, particularly if this is assumed (Pinkus 1952; Epstein and Maibach 1965), and of course, in the presence of circadian variation in proliferative indices, a single value for $I_M$ or $I_S$ may lead to the turnover time being erroneous. Consequently there are widely varying times reported for turnover times in the various sites shown in Table 7.2.

An interesting approach to the analysis of turnover times was made by Bergstrasser and Taylor (1977) who proposed, for individual surface corneocytes in steady state, that the desquamation rate $(r_1)$ equalled the rate of keratocyte maturation $(r_2)$ which was also equal to the birth rate, i.e.

$$r_1 = r_2 = k_B.$$

Now the turnover time $(T) = N/k_B$, where $N$ is the number of cells in any epidermal compartment. In the human epidermis considered, there were 20 corneocyte layers, and since the area of a corneocyte was $\sim 937\ \mu m^2$, and $N_{\text{keratocyte}}$ was $4.7 \times 10^4/mm^2$, there are 44 keratocytes per surface corneocyte. $T_{\text{corneocyte}}$ was 14 days (Baker and Kligman 1967; Jansen *et al.* 1974), and so $r_1 = 20$ layers/14 days or 1.4 layers per day (or in the context of a surface corneocyte, 1.4 cells per day). Now, $T_{\text{keratocyte}} = N_{\text{keratocyte}}/r_2$, and as $r_1 = r_2$, this is 44/1.4 or 31 days. Hence $T_{\text{total}} = 31 + 14 = 45$ days. For collateral, Weinstein and Frost's (1969) value of 19 days can be taken as the $T_{\text{basal}}$, and with the 14 day value for the suprabasal keratocytes obtained by Weinstein and Van Scott (1965), this gives a $T_{\text{keratocyte}}$ of 33 days + 14 days, and $T_{\text{total}} = 47$ days.

Now this is against the classical values for the turnover time in human epidermis: Pinkus (1952) had previously reported a $T_{\text{total}}$ of 26.7 days from an $I_M$ value of 0.00156 and an assumed $t_M$, while Epstein and Maibach (1965), also using an assumed $t_S$ value, found a mean $T_{\text{viable}}$ of 17.7 days with a measurement of the basal layer birth rate from $I_S/t_S$ ($T_{\text{basal}}$ was 152 hours); this, with a $T_{\text{corneum}}$ of $\sim 14$ days, would give a total turnover time of around 30 days. Moreover, following the migration of labelled cells after injection of [³H]-TdR, Epstein and Maibach (1965) and Weinstein and Van Scott (1965) provided estimates of 10–13 days for cells to reach the stratum corneum, and in these, the suprabasal and cornified layers, the transit time should equal the turnover time, since both are notionally simple transit compartments. This would leave about six days for the turnover time of the basal layer, but our best *in vivo* estimate is about 19 days (Weinstein and Frost 1969).

However, Halprin (1972) has given some support for the concept of longer turnover times in the human epidermis; noting that, since migration was probably random, these estimates of Epstein and Weinstein were likely to be minimal (see Chapter 6, section 3), Halprin regarded the 457 hours measured by Weinstein and Frost (1969), and reported by them as a cell cycle time, as the $T_{\mathrm{basal}}$. However, from Penneys *et al.* (1970), some 32 per cent of labelled cells are found above the basal layer, and since there are usually about six viable cell layers in human epidermis, Halprin calculated a $T_{\mathrm{Malpighian}}$ of about 42 days; this, plus the $T_{\mathrm{basal}}$ and $T_{\mathrm{corneum}}$ gave a $T_{\mathrm{total}}$ of 75 days. However, this estimate is heavily dependent on the $T_{\mathrm{basal}}$ value, and, as we shall see, this estimate is not very robust.

## 7.8. Cell proliferation in normal human epidermis

Despite its importance in baseline studies for the analysis of cell proliferation in disease states, there has, comparatively speaking, been little work in normal human epidermal cell kinetics, and the methodology of existing studies leaves a good deal to be desired (Wright 1980*c*). Pinkus (1952) observed a mitotic index of 0.156 per cent of 'epidermal cells', while *in vivo* labelling index values for basal cells vary between about 2 and 5 per cent; Epstein and Maibach (1965) reported a mean value of 4.9 per cent, Allegra and de Panfilis (1974) 3.8 per cent, and Weinstein and his colleagues, in a number of studies, obtained values which varied between 3.0 and 5.6 per cent (see Gelfant 1976). Lower values of 2.5 per cent have come from the *in vivo* work of Hell and Maibach (1972) and of Lapachelle and Gillman (1969).

Unfortunately, there has been no direct measurement of the cell cycle time in normal basal cells. Weinstein and Frost (1969) obtained a first wave of labelled mitoses, and calculated a 'cell cycle time' of 457 hours from $I_S/t_S$, but the experiment was not prolonged to look for a second peak. Of course, as previously noted, this is not a cell cycle time, but a turnover time; the age distribution is probably exponential in type (see Chapter 6, section 7), and the growth fraction is almost certainly less than unity, and both these factors will decrease the magnitude of this estimate.

Thus we simply do not know the intermitotic time of normal human basal cells. Bauer and de Grood (1975; Bauer *et al.* 1980) derived values varying from 50 to 137 hours from flow data, but these values also depend on growth fraction assumptions, although there is some support from the *in vitro* results of Chopra and Flaxman (1974), who measured the cell cycle time at 59 hours by the FLM method in outgrowths of normal human skin. Should these results be substantiated, it will mean that large numbers of basal cells in human epidermis are non-proliferative; if we take the value of 450 hours for the turnover time and 60 hours for the cell cycle time, only about 13 per cent of the basal cells would be proliferating, and the remaining 87 per cent would be post-mitotic maturing cells, resting cells in $G_0$, or non-proliferating (?slowly proliferating dendritic cells). Naturally, the value for the growth fraction will depend on what estimate we take for the turnover time. Table 7.4 shows that several estimates are closer to 200 hours, in which case about 60 per cent of the basal cells would be non-proliferative. Gelfant (1976) and Gelfant *et al.* (1983) have presented some interesting data in this respect: patients were infused with [$^3$H]-TdR for periods up to 15 days, but only 40 per cent of the normal basal cells were labelled. With a cell cycle time of $\sim$200 hours, all cells, if proliferative, should have incorporated [$^3$H]-TdR at least once (but see the several problems with this method, section 7.4). Thus a relatively large fraction of the population may be non-cycling, in accord with the above calculation.

This is about the sum total of our knowledge of proliferative cell number in the normal human epidermis; however, Penneys *et al.* (1970), have, by careful studies in serial sections, shown that about 30 per cent of flash labelled cells are in the suprabasal layers. It is possible that these suprabasal cells are born in the basal layer; if migration from the basal layer is random, or if, as Iversen *et al.* (1968) suggest, the oldest $G_1$ cell next to a mitosis migrates, these cells may migrate before commencement of the $S$ phase, to complete their cell cycle in the suprabasal layers. Alternatively, the cell may migrate at mitosis and spend a whole cell cycle time in the suprabasal layers. Whatever the mechanism, from a practical viewpoint it is important to remember that proliferative cells are found in suprabasal layers.

As far as the other phases of the cell cycle are concerned, the only FLM derived value for the duration of DNA synthesis has been reported by Weinstein and Frost (1969); this is 16 hours, but this value does not compare well with results from other methods, and in particular the double labelling method. *In vitro*, Allegra and de Panfilis (1974) found $t_S$ to be 7.3 hours, while other investigators have presented values ranging from 5.9 to 10.3 hours, also using the double labelling method *in vitro* (Christophers and Schaumloffel 1967; Pullman *et al.* 1974; Heenan *et al.* 1975). It is possible, of course, that differences in technique could play some part here, and note that the $t_S$ values obtained by this method give turnover times for the basal layer of the order of 200 hours, less than half of Weinstein and Frost's (1969) estimate.

Although measurement of the birth rate in psoriatic epidermis, using the metaphase arrest method with intradermal injection of vincristine, has met with some success (Duffill *et al.* 1977; Ralfs *et al.* 1981*a*) there are problems when the technique is applied to normal human epidermis; for example Camplejohn *et al.* (1981) failed to obtain any increase in the metaphase index for 2.5 hours after vincristine injection, an effect which we have also encountered (Moffat and Wright, unpublished). Fisher (1968*a*) was able to obtain metaphase accumulation in normal epidermis, but only by extending the experimental period to as long as 12 hours, and reported a birth rate of 0.027 cells/1000 cells/hour. However, as we have seen, there is a grave danger here of incurring errors due to metaphase degeneration (Aherne and Camplejohn 1972), and if the four hours reported by Ralfs *et al.* (1981*a*) for the minimum lifetime of an arrested metaphase in psoriatic epidermis is also applicable to normal epidermis, then from the correction factor proposed by Aherne and Camplejohn (1972), this value of Fisher's could be increased to 0.17 cells/1000 cells/hour.

We have already discussed the rather controversial topics of turnover time and transit time with reference to human epidermis (see section 7.7). Many kinetic parameters for human epidermis are given in Table 7.4.


## 7.9. Kinetic parameters *in vitro*

The techniques used to cultivate epidermal cells have been mentioned briefly in Chapter 6, section 5, and while the use of culture methods in epidermis have given some insights into its cell biology (e.g. Vaughan and Bernstein 1971), few workers have concerned themselves with the critical examination of proliferative rates (but see Chopra and Flaxman 1974). It is always difficult to know how much weight to put on *in vitro* measurements of proliferative rates in terms of their relationship to *in vivo* relevance, and, in their interpretation, the caveats listed in Chapter 6, section 5, should be recalled.


## 7.10. Changes in proliferative rates with aging

Probably because of the difficulties involved in the measurement of proliferative rates in slowly renewing populations, hard kinetic information in squamous epithelium during the aging process is limited. There are several reports in the older literature which suggest that mitotic indices and daily mitotic rates increase with age in both rodent and human tissues. Bullough (1949) in the mouse, and Cooper *et al.* (1951) in man reported rising mitotic indices until middle life, followed by a falling (Bullough 1949) or a constant (Cooper *et al.* 1951) mitotic index. These changes are occurring in the presence of a decreasing density of the epidermal cell population (Thuringer and Katzberg 1951; Katzberg 1952; Bertalanffy *et al.* 1965; Sharav and Massler 1967), and a declining DNA content in individual epidermal cells (Goerttler *et al.* 1970). Katzberg (1952), assuming a constant $t_M$ at all ages, calculated a reduction in the total viable cell turnover time in human epidermis from 101 days in the first decade to 46 days in the seventh decade.

Other workers, on the other hand, have reported a decrease in mitotic index with age, in mouse ear epidermis (Whiteley and Horton 1963), in rat gingival epithelium (Sharav and Massler 1967), and also in human gingiva (Marwah *et al.* 1960). Studies on the corneal layer transit time in human epidermis, using the fluorescent method, are also consistent with a decrease in proliferative rate in old age (Baker and Blair 1965; Jansen *et al.* 1974), although, oddly enough, elderly women do not show the same increase in transit time as men. Gelfant and Smith (1972) considered that the decline in proliferative rates with age in mouse epidermis was due to an increase in the proportion of non-cycling cells;

the proportion of $G_2$ blocked cells was decreased in three-month-old mouse epidermis, compared with that of 20-month-old epidermis. The techniques used were the same as those described by Gelfant (1966) which are discussed in Chapter 2, section 7. Some work of Cameron (1972$a$) could also be interpreted as meaning a reduction in the size of proliferating populations with age; Cameron subjected mice to continuous labelling in the drinking water for as long as 90 days, and noted a reduction in the maximum $I_L$ attained in the basal cells of the ear epidermis and the oesophagus with increasing age. From the form of the curves Cameron considered that there was a reduction in the proliferative rate with age. Assuming that the continuous labelling procedure encompassed the $T_C$ of the slowest proliferative cell in each case, and that the administration of [$^3$H]-TdR in the drinking water is a reliable procedure (see section 7.4) – we could suggest that there is a decrease in the growth fraction of the basal cells with age. Interestingly enough, there was no such reduction in the plantar epidermis, or in the epithelium of the ventral surface of the tongue. Using the ventral surface of the tongue as a reference population, Cameron (1970$b$), having found total basal cell labelling within five days at 3, 13, and 19 months of age, and therefore assuming a maximum $T_C$ of five days, calculated that there would be some 146 doublings within two years, the mean life-span of the animal; this, of course, is not in accord with the Hayflick (1965) concept of limited cell generations.

Thrasher (1971) also reported a reduction of $I_S$ from 12.6 per cent in the oesophagus of 10-day-old mice, to 6.6 per cent in 600-day-old mice. There was no effective change in $t_S$ as determined by an FLM curve, but Thrasher calculated that the turnover time (from $t_S/I_S$) was increased from 64 hours in the infant to 120 hours in the senescent animals. This data does not allow a distinction to be made between the role of the cell cycle time and the growth fraction, but Sharav and Massler (1967), noting a reduction in $I_S$ over 27 months in the rat palate, concluded that the 'progenitor' population had decreased, although they did not measure $t_S$ in support of this contention.

In conclusion, it would appear as if the bulk of the evidence favours an increase in proliferative rate with age in human epidermis (if $I_M$ readings only may be relied upon), while in experimental animals rates appear to decline with age in some squamous epithelia, and remain constant in others. There also may be a decrease in the total population with age, which of course will complicate comparisons of turnover time values between different age groups. The cause of these changes is not yet clear, but there is some evidence to suggest that the decreased rates observed in murine tissues with age may be associated with a decrease in the proliferative population size.

# 8 The kinetic organization of squamous epithelium

We have now discussed the various kinetic parameters which apply to stratified squamous epithelium, and the vagaries of their measurement. Because basal cells are disposed in what amounts to a flat sheet, there has been a tendency to regard them as a distinctly homogeneous population, with the same reproductive capacity, and with identical kinetic status; migration was by a random removal process; all basal cells were proliferative, and cells lost the capacity to divide on entering the suprabasal layers. Migration through the epithelium then followed first-order kinetics, and desquamation naturally followed the same process. However, there is now accumulating evidence that the basal cells are, in fact, a cell population with a definite kinetic organization, which is rather complex; indeed, in certain circumstances, such as the thin epidermis of the mouse, the epithelium is highly organized, with corneocytes and suprabasal keratocytes arranged in ordered columns (MacKenzie 1969, Christophers 1971*b*), and recent theory maintains that this organization extends into the basal layer, so that proliferation is orientated towards the maintenance of these ordered columns (Potten 1975*b*, 1976). In other thicker and faster proliferating epidermis, while there is no apparent structural organization, there is some evidence that an ordered migration process operates.

A former simplistic viewpoint was that all basal cells are reproductively homogeneous, and they were all regarded as *functional* stem cells (Cairnie 1976); this is now regarded as being unlikely, and it is probable that functional stem cells exist as a small basal cell subpopulation, and that the great majority of observed basal layer mitoses are in fact amplification divisions in reproductively doomed cells (Potten 1975*b*). Moreover, clonogenic cells are few in number in epidermis compared with total proliferative basal cells (Potten and Hendry 1973), further evidencing that kinetic subpopulations are a reality in squamous epithelium.

In the following pages we will examine the evidence for these intriguing proposals, which are, we believe, beginning to provide real insight into how this deceptively simple tissue is put together. The first of these pointers came from the observation that the spatial distribution of mitoses in squamous epithelium was non-random in nature.

## 8.1. The spatial distribution of proliferative activity in squamous epithelium

For many years there has been speculation that mitotic activity in squamous epithelium showed some degree of clustering or non-randomness. For example, Thuringer (1928) described 'clustering growth waves', measuring 100 $\mu$m in diameter, with as many as 10 to 14 mitoses in one high-power field, in human epidermis; other, basically qualitative observations confirming this impression were made by numerous other authors in both human and animal epidermis.

However, putting the work on a quantitative footing Meyer *et al.* (1960) counted the number of mitoses per section in the lateral palate epithelium of the rat, and reported conformity to a Poisson distribution, and concluded that mitotic activity was randomly distributed. On the other hand, Frei *et al.* (1963), with a more sophisticated approach, used a grid to study the distribution of mitoses in isolated sheets of epidermis from the mouse ear; when checked against a Poisson distribution with the $\chi^2$ test for the goodness of fit, the probability was $>0.5$ that the distribution differed from a Poisson, while Rowe and Dixon (1972), who used sections and Poisson statistics to study the distribution of mitoses in human sacral skin at different times of day, found evidence of mitotic clustering, not only in the same biopsy specimen, but also within different biopsies in the same area; a non-random distribution of both labelled and mitotic cells has also been reported in the epithelium of the mouse tongue by Löe *et al.* (1972).

A new approach was made by Leblond *et al.* (1964), and by Cameron *et al.* (1965), who mapped the number of non-labelled and non-mitotic cells between labelled and mitotic cells respectively, in the mouse oesophagus after tritiated thymidine labelling and metaphase blockade. In one case, the frequency of non-active cells did not fit a geometric distribution, considered by Cameron *et al.* to be the appropriate distribution if labelled or mitotic cells were randomly distributed (see Fig. 8.1). However, if the zero value was excluded, then the $\chi^2$ test for goodness of fit for the geometric distribution became satisfactory, so that the distribution of proliferative activity was almost random, apart from the tendency for dividing cells to occur in adjacent positions; Leblond *et al.* (1964) did not detect this clustering, and concluded that their data supported a random distribution.

These findings were extended by Gibbs and Cassarett (1972), in hamster cheek pouch epithelium, where the distribution of labelled and mitotic cells was tested against a Poisson distribution, and of non-active cells against a geometric distribution, and in each case, significant deviation from randomness was apparent for both labelled and mitotic cells, although animals not injected with [$^3$H]-TdR did show a Poisson distribution for mitoses, an effect which was attributed to the [$^3$H]-TdR injection (although any mechanism must be speculative). The presence of uniform observations of fewer runs than expected from random sequences indicated that, when high or low observations occur, they occur in uninterrupted sequences, with consequent high indices occurring together, and grouping of labelled and mitotic cells.

Consequently the probability of finding mitoses or labelled cells is not

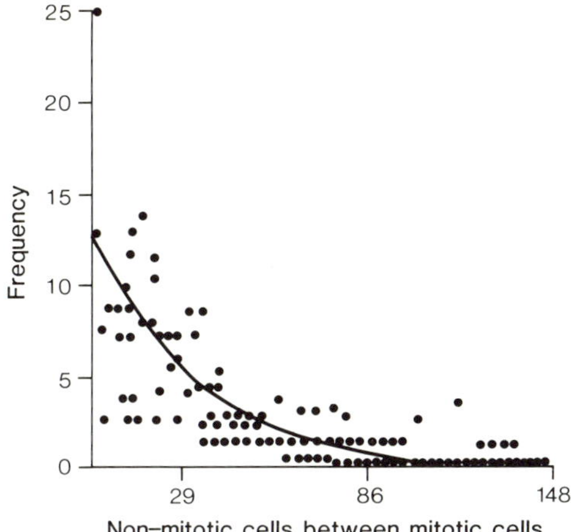


Fig. 8.1. The distribution of the number of non-mitotic cells between mitotic cells in the basal layer of the mouse oesophagus. The line indicates a geometric distribution. (Redrawn from Cameron *et al*. (1965).)

independent of neighbouring cells, and these results then support the concept of localized areas of proliferative activity. The possible mechanisms behind these interesting results vary from suggestions regarding some ill-defined humoral effect (Cameron *et al*. 1965) to non-randomness in time, i.e. a form of spatial cycle synchrony, with sister cells in phase with each other (Cameron *et al*. 1965; Gibbs and Cassarett 1972); however, Rowe and Dixon (1975) excluded all touching sister cell clusters to try to correct for paternity effects of cells taking origin from the same mitosis, with possibly the same $T_C$ value, but still found the degree of clustering significant in human epidermis; they considered that the phenomenon could be due to some sort of spreading mitotic stimulus, as yet undefined.

The possibility of synchrony as an explanation is worth pursuing; it is unlikely that paternity alone can have such an effect, because although there is a negative mother–daughter correlation in cycle times in prokaryotes (Mitchison 1970), there is little evidence of such correlation in mammalian cells. But there is substantial evidence of synchrony in squamous epithelia taken as a whole (cf. Chapter 9), and of course it is possible that such synchrony could lead to a non-random distribution of proliferating cells at a local level, which would naturally depend on the timing of the synchrony (Rowe and Dixon 1975).

In thin epidermis there are now indications that the proliferative activity in the basal layer is concentrated towards the periphery of the *hexagonal units* called by Potten (1975*b*) 'epidermal proliferative units' (EPUs) (MacKenzie 1970; Karatchai *et al*. 1971). While cell proliferation and its relationship to ordered structure is really the province of section 8.2, we might note here that MacKenzie (1970, 1975*a*), who plotted the distribution of colchicine-arrested mitoses in mouse ear epidermis, using interference microscopy, observed a non-random distribution, with mitotic activity concentrated towards, but not actually at, the periphery of the hexagon. As we shall see, Potten (1975*b*, 1976) would have us believe that this arrangement is essential for the maintenance of ordered structure (see section 8.2), but MacKenzie (1975*a*) could not detect any morphological evidence for functionally independent groups of basal cells beneath each cell column, and considered that there was no *a priori* reason for proposing a functional proliferative unit, despite the fact that supportive evidence was available in that peripheral basal cells in the EPU labelled faster than centrally placed cells under continuous labelling conditions (Potten *et al.* 1974); be that as it may, while the EPU argument provides a beguiling explanation for the non-randomness of the distribution of cell proliferation in the basal layer of thin epidermis, there is, as yet, no convincing reason for its occurrence in thick, non-stacked squamous epithelium.

## 8.2. Ordered structure in mammalian epidermis

### (i) *Epidermal stacking*

Prior to 1969, the stratum corneum was regarded merely as numbers of layers of very flattened keratinized cells. Because fixation, dehydration, and embedding produce much distortion, and also because of the limits of resolution of the optical microscope, the ordered structure which has recently been described (Mackenzie 1969; Christophers 1971*b*) was not appreciated until specialized methods became available. If skin sections are expanded in 0.1 M NaOH (Christophers and Kligman 1964), the flattened, keratinized cells will be seen to be *stacked* one above the other in ordered columns of cells. Suitable modifications of the method, such as treating frozen sections with dilute alkali, and staining with aqueous methylene blue, have allowed direct visualization of cell proliferation and migration in respect to these columnar arrangements.

The geometrical regularity of stacking varies with respect to site (Allen and Potten 1976*a*), species (Mackenzie 1975*c*), and age (Mackenzie and Zimmerman 1981); for example, primate (including human) epidermis shows a columnar arrangement, but there is considerable variation in the extent to which individual columns interdigitate with each other (Mackenzie 1975*c*). On the other hand, palmar and plantar epidermis, including rodent foot pad, shows no ordered structure; indeed, an ordered arrangement could not be demonstrated in oral mucosa, lip, nipple or vagina (Mackenzie 1972; Mackenzie and Linder 1973). Despite claims to the contrary (Bergstrasser and Chapman 1980), Menton and Eisen (1971*a*) and Mackenzie *et al.* (1981) were able to show a columnar pattern in human stratum corneum from abdomen, scalp, chest, and arm, but this could not be traced beneath the granular layer. Furthermore the overlap zone between contiguous squames was larger in human epidermis, although stratum granulosum cells can show a highly ordered structure (Hume and Potten 1983).

Ordered stacking is seen typically in thin rodent skin, for example ear epidermis (Fig. 8.2), where, after expansion, the columns of squames are readily seen. Each squame is in contact with its neighbour above or below it, whereas at the margins there is a degree of overlap or interdigitation with adjacent epidermal columns. Observation with the electron microscope (Mackenzie 1972; Menton and Eisen 1971a) indicates that the junctional regions are in fact complex, with complementary projections and depressions which are seen to interlock with the squames of the adjacent column. Inspection of Fig. 8.2, which depicts an alkali-expanded ear epidermis stained with methylene blue, shows that cells in the basal layer have no definite positioning with respect to stacking, whereas all suprabasal cells tend to be stacked beneath the overlying columns of squames.

Viewed from the surface, squames appear as flat, hexagonal plates (Gold-schmidt and Kligman 1967; Marks and Boghel 1973), and thus are readily appreciated by scanning electron microscopy, or more simply by staining epidermal sheets with silver impregnation methods (Mackenzie 1972; Mackenzie and Linder 1973); this latter technique (Fig. 8.3A) shows that mouse ear squames have well-defined junctions, and here the interdigitations appear as parallel bands surrounding each column.

Mackenzie (1972) has asked what particular advantage such a structured organization plays; he had previously proposed that epidermal columns were indeed independent proliferative units, and this concept has been taken up, and considerably expanded, by Potten (1974). Potten dignified this concept with the appellation 'epidermal proliferative unit' or *EPU*. Allen and Potten (1974a) have analyzed ordered epidermal structure in the mouse ear at the light and electron microscopical level. The EPU (Fig. 8.4) possesses a constant number of cells, some 20 in all; there are 10-11 basal cells, arranged in a peripheral ring, with 3-4 central nuclei, one spinous cell, two granular cells, and 5-7 cornified cells. The lowermost cornified cell is called the *basal cornified cell*, which is a good guide to the limits of the column, since, like the granular and spinous cell, it runs the whole length of the column.

Christophers and Laurence (1973), combining a fluorescein isothiocyanate (FITC) method for staining keratin (Fig. 8.2B), with autoradiography in the same site, have quantified the cellular content of epidermal columns in the ordered mouse epidermis, and compared this with the number of epidermal cells associated with one surface corneocyte in the foot epidermis of the mouse, which does not show stacking. These values are shown in Table 8.1 (p. 294), which also gives values for the cellular content of other stacked, partially stacked, and non-stacked epidermis. In the mouse ear epidermis, although Christophers (1972a) values for basal cells and cornified cells are comparable with the figures of Allen and Potten (1974a), there are many more keratocytes. It should be noted that the non-stacked foot epidermis is much thicker, with double the number of cornified cells; furthermore, some 73 per cent of the total cell population in the foot are cornified cells, compared with only 25 per cent in the

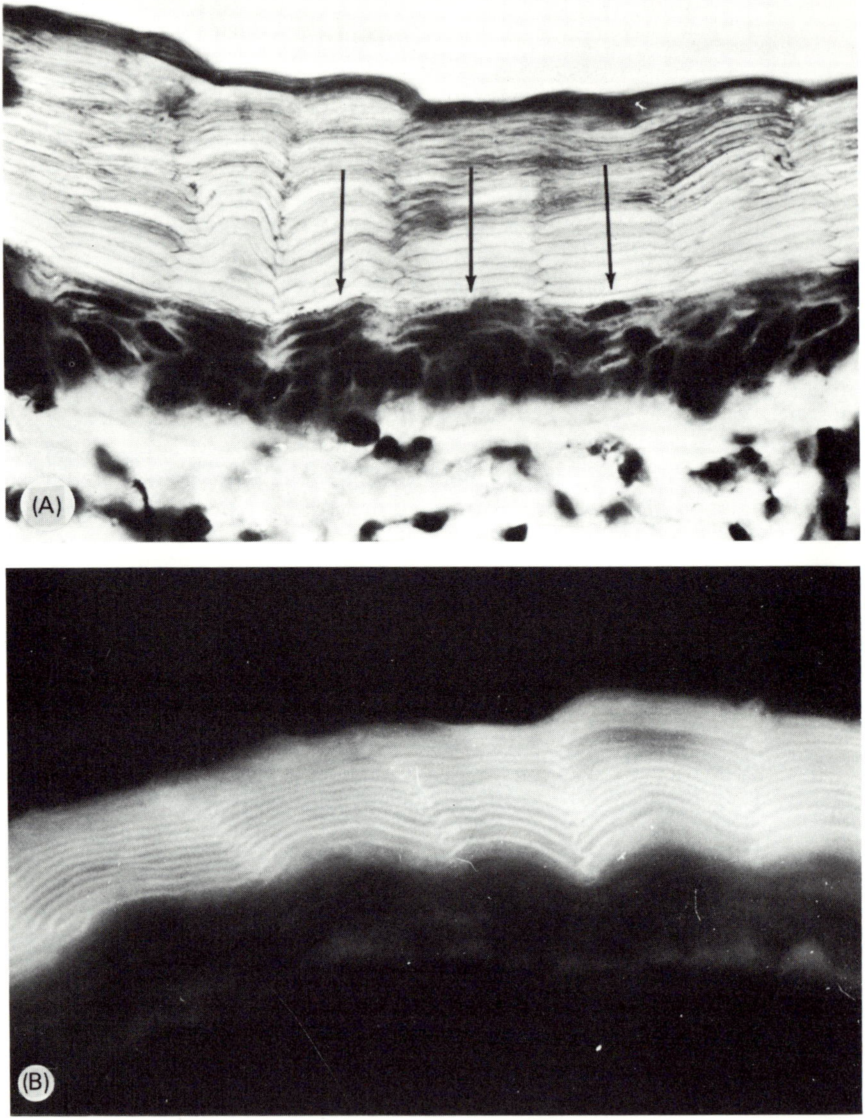

Fig. 8.2. (A) Mouse ear epidermis, expanded by treatment with O.IM NaOH, showing the typical arrangement of corneocytes in columns, overlapping at the margins. The arrows point to three adjacent stacked columns, where the Malpighian layer cells themselves can also be seen layered one above the other. (×252.) (From Menton (1976*b*), by courtesy of Dr D. Menton.)  (B) Expanded mouse dorsal epidermis stained with fluorescein isothiocyanate and photographed under ultraviolet light: again the columns in the stratum corneum are easily visible. (×180.) (By courtesy of Dr C. S. Potten.)

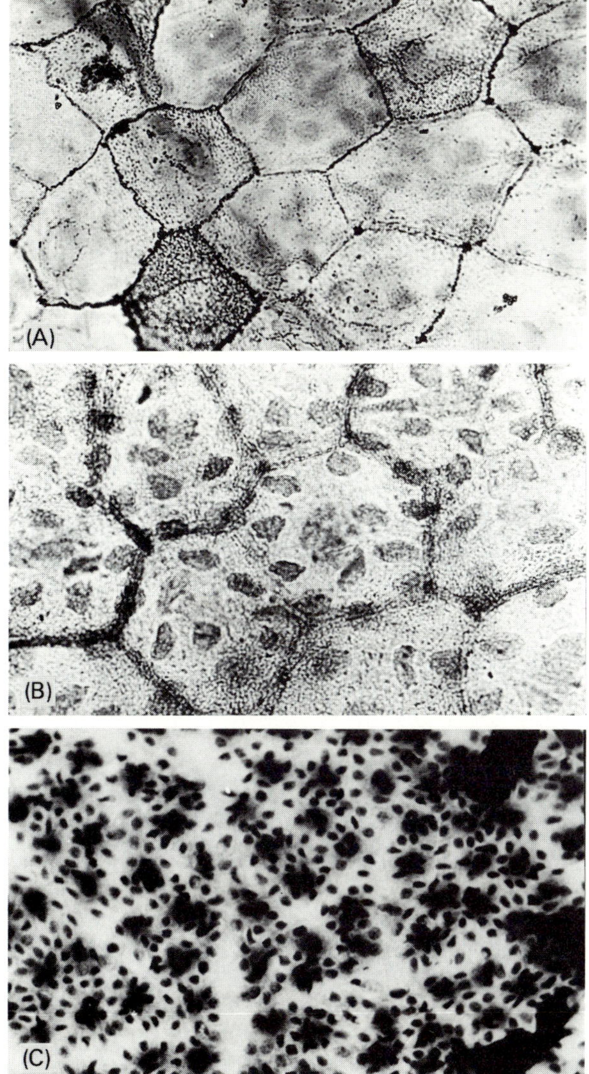

Fig. 8.3. (A) A sheet of mouse ear epidermis stained by a silver impregnation method. The surface squames are roughly hexagonal in outline, and the parallel lines present at the boundaries of some represents the area where each column of squames interdigitates with its neighbour. (B) The same specimen photographed in a different optical plane to show the relationship of the underlying basal cells to the hexagonally shaped units. The large central cell, in the mouse a Langerhans cell, is evident, surrounding which are at least two central epithelial cells; the remaining basal cells associated with the column are disposed around the margin of the hexagon. The interdigitating area between adjacent columns is better seen in this plane. (C) An epidermal sheet from the mouse ear photographed from beneath, showing the grouping of the marginal basal cells around the central component. (By courtesy of Dr C. S. Potten.)

Table 8.1.  Cellular content of individual epidermal columns in stacked epidermis, values per surface corneocyte in non-stacked epidermis, and the relationship between epidermal cell production rates and ordered epidermal structure

| | Mouse dorsum | Hairless mouse dorsum | Mouse ear | Mouse foot | Mouse tail | Guinea pig ear | Human |
|---|---|---|---|---|---|---|---|
| Basal cells/EPU[a] | 10.6[1] | 11[1] | 10–11[5], 9.3[2] | 14.1[2] | 25[1] | – | 20–28[1] |
| Keratocytes/EPU[a,b] | 3.0[4] | 3.5[4] | 3[5], 15.3[2] | 35.7[2], 12[4] | 5–6[4] | – | ~5[1] |
| Corneocytes/EPU[a,b] | 6[4] | 7[4] | 6–8[5], 10[2] | 37[2], 40[4] | 32[4] | – | 10–28[1] |
| Total epidermal cells/EPU[a] | ~20[1,4] | ~21[1,4] | ~20[5], 25.3[2] | 72.7[2], 52[4] | ~80[1,4] | – | 35–60[1] |
| Proliferative cells/EPU[a] | 10.6[1] | 11[1] | 9.7[1] | 20[1] | 25[1] | – | 20–28[1] |
| Cell production rate/EPU/day[a] | 1.3[1] | 2.4[1] | 1.0[1], 0.6[3] | 1.7[3], 6.7[1] | 4.1[1] | 1.0[3] | 2.6[1] |
| Observed transit time (days) | 9[1] | 4–5[1] | 11–24[1], 20–24[3] | 6–7[3], 7[1] | 9[1] | 12–14[3] | 13.5[1] |
| Labelled cells/EPU[a] | ~0.4[1] | ~0.4[1] | 1[4], 0.48[2] | 3.81[2] | – | – | – |
| Organization | Stacked | Stacked in places | Stacked | Non-stacked | Stacked with maximal overlap | Stacked in places | Stacked in places |

1. Potten (1975*b*) from a review of the literature.    2. Christophers (1972*a*).    3. Christophers *et al.* (1974).
4. Allen and Potten (1976*b*).    5. Allen and Potten (1974*a*).

(a)  or per surface corneocyte in non-stacked epidermis;
(b)  where no actual counts/EPU or surface corneocyte are available, inferred from measurements of the number of layers in the respective compartment.

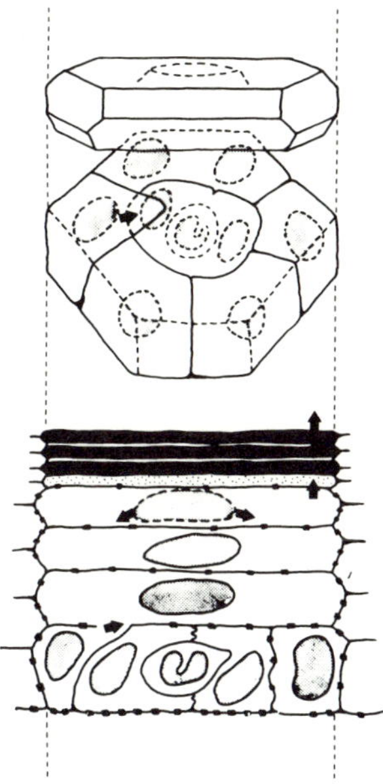

Fig. 8.4.  A diagrammatic representation of the epidermal proliferative unit. This is a mouse EPU, with a central dendritic Langerhans cell. Migration is peripheral, and there are three nucleated suprabasal cells beneath the stratum corneum. (Redrawn from Potten and Allen (1974a).)

ear. Finally, the number of labelled cells per cornified surface cell is higher in the foot, indicating a faster rate of proliferation. Christophers *et al.* (1974) believe that the association of a slowly renewing epidermis with ordered structure is not merely fortuitous, but that any tendency to increase proliferative rates and thicker epidermis leads directly to a loss of epidermal stacking (see iii below).

However, hairless mouse dorsal epidermis, which shows a similar cellular complement to the stacked mouse ear and dorsal epidermis, has a variable structure, with stacking only in places. On the other hand, mouse tail epidermis, with a large cellular complement, shows stacking with maximal overlap (Allen and Potten 1976a), in a manner resembling a tiled roof, also seen in guinea pig epidermis that has undergone mild hyperplasia (Bullough and Stolze 1978), while human epidermis, also with many cells per surface corneocyte, also shows stacking in some areas (Mackenzie 1975c; and see above).

Allen and Potten (1974b) have considered the intercellular desmosomal contacts in the EPU. Normal desmosomal contacts are seen up to and including the lower margin of the basal cornified cell. Above the basal cornified cell, where normal desmosomes are absent, Allen and Potten described a *squamosome*, which runs around the squame column, forming an unbroken ring (see Fig. 8.5). These squamosomes are present in each cornified layer, and Allen and Potten speculated that the decrease in mechanical bonding caused by this arrangement

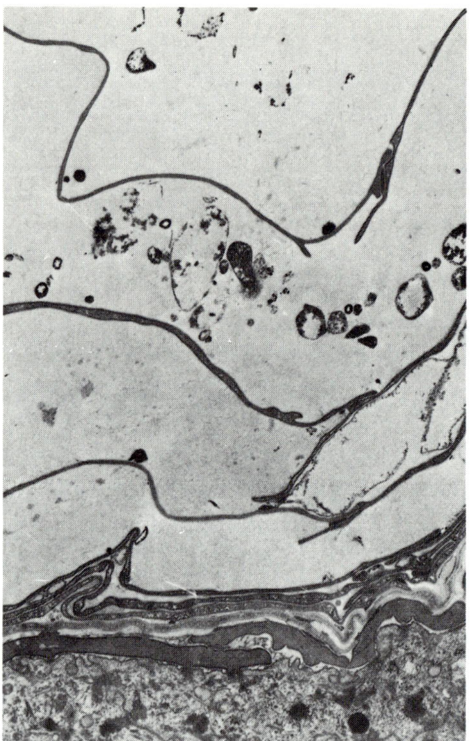

Fig. 8.5.  An electron micrograph of the squame overlap region in the stacked dorsal epidermis of the mouse. Here the squamosomes are cut in transverse section, and appear as electron-dense regions holding the squames together. (From Allen and Potten (1974*b*), by courtesy of Dr C. S. Potten.)

is responsible for desquamation. The question then arises how desmosomes suddenly become squamosomes at the level of the basal cornified cell, and it was suggested that desmosomes migrate to the periphery of the cell as the basal cornified cells are replaced by the next granular cell in sequence.

Stacking will obviously have important consequences for cell-shedding mechanisms, but a further possibility involves the contribution of the arrangement to the stability of the epithelium. An ordered, stacked epidermis would be resistant to the formation of planes of cleavage, and prevent scale formation. Allen and Potten (1974*a*) considered that each stacked squame is in contact with 12 other cells, six above and six below; adjacent columns are not joined, but two vertically adjacent cells are in contact with the intercalated cell. Consequently the cornified layer is extensively interconnected, and forms an excellent barrier. Such considerations do not explain why plantar and palmar epidermis, which is not stacked, can withstand considerable trauma. Thick epidermis, with a higher proliferative rate than thin epidermis (e.g. ear epidermis), has thicker individual squames in its prominent stratum corneum, and may rely upon these factors for resistance to trauma.

(ii) *The cellular origin of ordered structure, and its relationship to proliferation*

Since the surface corneocytes are dead, stacking must be determined by the lower, viable epidermal cells. During establishment of an ordered epidermal structure, cells, formed in the basal layer, migrate into the Malpighian layer, and, flattening and moving laterally, become vertically aligned into columns of hexagonal cells. Presumably such organization could be due to cell pressure produced during keratinization, to cell-contact phenomena, or be the result of columnar organization at the deeper level.

While it is true that hexagonal packing of spherical bubbles can be demonstrated in a single layer of soap bubbles on the surface of the water, a second layer produces a regular overlap of about half a diameter of a bubble (Mackenzie 1972); consequently the hexagonal shape is more likely to be due to control of the lateral expansion of the cells in certain directions, rather than simple mechanical packing of spheroids (but see Menton 1976*a,b*; and below). Since epidermal cells possess desmosomes at the cell interfaces, for lateral expansion to occur desmosomal contact must be broken, and later re-established. Mishima and Pinkus (1968) have described disruption and reformation of desmosomes in tape-stripped epidermis, but, as Mackenzie (1972) points out, the rate of membrane change during renewal of normal epidermis is more difficult to explain. Mackenzie (1970) realized that cells became flattened immediately after leaving the basal layer, increasing their surface area and eventually covering an area of some 10 basal cells. Potten (1975*b*; Potten and Allen 1975*a*) has calculated that about one cell is produced in each EPU per day, and thus epidermal cells are able to change their spatial relationship and synthesize new epidermal cell contacts every 24 hours.

The forces responsible for flattening and increasing surface area are uncertain. It is apparent from Fig. 8.4 that cells expand and become arranged into columns as soon as they leave the basal layer; this could be the result of population pressure, although the observation that cell flattening occurs during keratinization in epidermal outgrowths *in vitro* (Friedmann-Kien *et al.* 1966) might suggest that flattening is an inherent, active property of individual epidermal cells, possibly related to microtubule assembly (Branson 1968).

Christophers (Christophers 1971*c*; Christophers *et al.* 1974; Christophers and Laurence 1976) has proposed that epidermal columns are maintained by the order of addition of cells to each column (see Fig. 8.6). Having moved into a column, a cell interdigitates with adjacent cells, leaving a step-like boundary (at the site of contact). This step determines the position of the next cell in sequence, which flattens into line beneath an adjacent column, leaving a suitable site for the addition of the next cell to the original column. However, in Fig. 8.6 the difficulties of a necessarily three-dimensional configuration are also

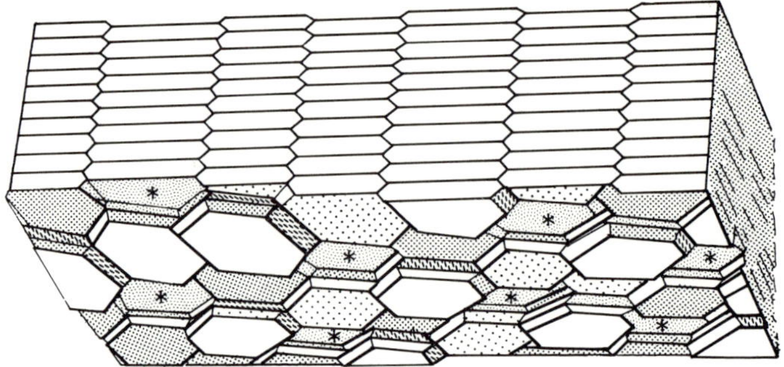

Fig. 8.6. A stereogram of the undersurface of the stacked stratum corneum of mouse ear epidermis, demonstrating the stepped position of the corneocytes; the cells in the 'up' position are denoted by asterisks. (Redrawn from Christophers *et al.* (1974).)

shown; a hexagonal column has six potential horizontal planes for migration into adjacent columns, rendering this arrangement very complex.

At this point some observations of Christophers (Christophers *et al.* 1974; Christophers and Laurence 1976) are germane: from Fig. 8.6 it is clear that the basal cornified cell is either above or below its immediate neighbours, and of course there will be more cells in that column which has its basal cornified cell in the superior position. In this context, Potten and Allen (1975*a*) found that 38 per cent of columns in mouse ear epidermis have less than the modal value of four viable keratocytes; these columns are thus in a position to receive a migrating basal cell. Figure 8.6 shows that the arrangement of the basal cornified cell in the 'up' or 'down' position is not regular. Evidence from [$^3$H]-proline uptake indicated that cell maturation occurred in an alternating fashion from one column to the next. In stacked epidermis, subcorneal granular cells showed alternating uptake, and this difference in the degrees of differentiation of cells in adjacent columns was confirmed by ultrastructural studies, using cell organelles as markers. Thus cells at the same level in the epidermis have different ages in adjacent columns, and moreover, this arrangement was not found in non-stacked epidermis. Christophers envisaged that subcorneal keratocytes move into columns in an alternate manner, and only cells with enough space will leave the basal layer, to be positioned beneath the column with the least corneocytes. Since cells are locked together, entry is barred until cornification of another corneocyte occurs; lateral movement is inhibited by the stacking arrangement. Moving into the stratum corneum, a granular cell becomes a hard hexagonal plate of keratin, and Potten and Allen (1975*a*) suggest that the framework of six similar cells is an essential component of the final stages of keratinization, holding the cell firmly in position.

Such a model of alternating cornification and migration in adjacent columns would lead to a sequential desquamation process; and Potten and Allen (1975*a*) have argued strongly for a control process operating *between* epidermal prolifer-

ative units. The existence of a tight-ordered structure, taken together with the
fact that stacking is re-established after injury (see Chapter 10), points to the
conclusion that EPUs are autonomous, but the rigid control of epidermal
thickness, with no localized fluctuations in the number of cell layers, indicates
that a control system exists between EPUs. Even more persuasive is the consequence of the concept of Christophers *et al.* (1974) of alternate overlapping in
adjacent columns: with this pattern, desquamation can only occur when all
margins of the corneocyte are free. Since desquamation necessarily reflects cell
production, such a system can only work by some form of synchronous cell
production between neighbouring EPUs, as a stack cannot add a second cell to
its column before all six adjacent columns have done so. This arrangement is
only consistent with two models; one with 25 per cent of EPUs synchronized
in pairs, and the other with 33 per cent of EPUs synchronized in groups of three;
the latter is simpler and more symmetrical, and more adaptable to the constraints
of diurnal variation. That such inter-EPU control mechanisms must exist is seen
by the consideration that, if EPUs acted independently of each other, there
would only be a 3 per cent chance of desquamation, whereas CPR measurements
indicate a desquamation rate which is ten times more than this. The concept of
synchrony in trios of EPUs, in which EPUs have a limited amount of time in
which to add a cell to the base of the column, is shown diagrammatically in Fig.
8.7; groups of three columns about any given column will be synchronous.

Mackenzie (1972) has an alternative proposal. Columns are not maintained
by channelling of migration by different numbers of cells in adjacent columns,

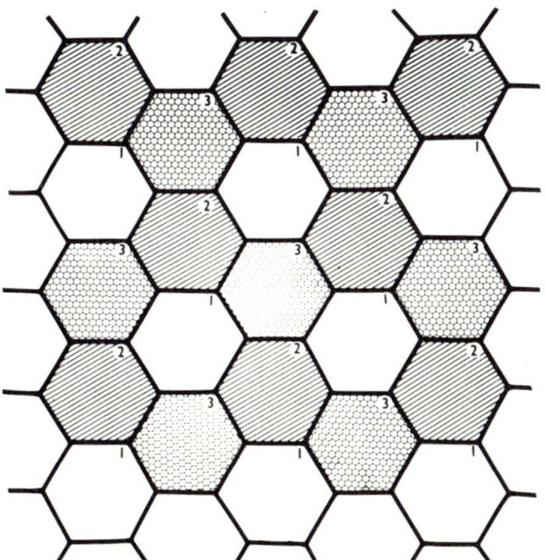

Fig. 8.7.    A diagram showing the possible synchronization of neighbouring epidermal
columns as a necessary consequence of desquamation occurring in stacked epidermis. A
third of EPUs are synchronized; columns numbered 1 add or lose a cell, followed in turn
by columns numbered 2 and 3. (From Potten (1976), by courtesy of Dr C. S. Potten.)

but by the same junctions between previously aligned cells limiting the lateral expansion by some form of contact inhibition. In this case, it would only be necessary to postulate that inhibition occurs at contacts with more mature cells, while cells in the process of flattening move freely against non-flattened cells.

These arguments have attracted a powerful attack by Menton (1976*a,b*). Menton saw three possible alternatives for a stacking mechanism: (i) *non-random mitotic activity* (cf. Karatschai *et al.* 1971; Mackenzie 1975*a*). Menton was not clear how mitotic control itself could provide a mechanism for stacking; mitosis occurs in the basal layer, which is often (e.g. in guinea pig or primate epidermis) separated from the overlying columns by randomly arranged cells. Menton considered that migration from the basal layer, and subsequent movement, was random so that cell age varies when cells reach the columns. However, we might note that the basal layer migration process envisaged by Potten (1975*b*, 1976) is anything but random, and is based on a sequential aging idea (see section 8.6); (ii) *active vectorial movement into alignment*. There is no direct evidence suggesting this; alignment occurs only during flattening, a change in shape brought about by the assembly of cortical microtubules which appear along the axis of extension (Branson 1968); (iii) Menton favoured the view that stacking was the result of inherent geometrical organization. In a simple epithelium, cells become hexagonal with *trihedral* junctions, which arrangement gives the least perimeter in relationship to the base. In fact trihedral junctions are most stable between semi-liquid structures. Moreover, this shape is soon restored after cell division (Lewis 1926). Viewed in three dimensions, these are then 14-sided *orthic (or regular) tetrakaidecahedra*, the only geometric shape which stacks, without interstices, in columns of cells which regularly interdigitate with six surrounding columns, and provides the smallest surface:volume ratios. Such a shape is mandatory in stacked epidermis, where most columns are hexagonal in shape.

These structures are shown in Fig. 8.8. By flattening these tetrakaidecahedra to a truncated form, Menton (1976*a*) succeeded in building a physical model which distinctly resembled stacked epidermal cells in an alkali-swollen preparation (see Fig. 8.8). All junctions in such an arrangement are of the stable trihedral type, and Menton proposed that epidermal cells have an intrinsic tendency to form this shape, as was proposed for cambium by Lewis (1928). A single layer of bubbles blown through a layer of liquid forms such orthic tetrakaidecahedra, and when multilayers are formed the bubbles form columns, *but only when viewed at an angle of 60° to the liquid surface*, which would account for Mackenzie's (1972) failure to show vertical stacking in multilayered bubbles. Extrapolating to the epidermis, Menton considered that the 60° tilt could be due to the nature of the basal cells not being an aqueous substratum, and proposed a mechanism whereby non-aligned cells were added to the deep surface, or 'forming-face', of the stacked epidermis—such a face would have three levels (see Fig. 8.9); a flattening cell would occupy the most superficial position available on the forming face (position 1, Fig. 8.9), and continue the flattening

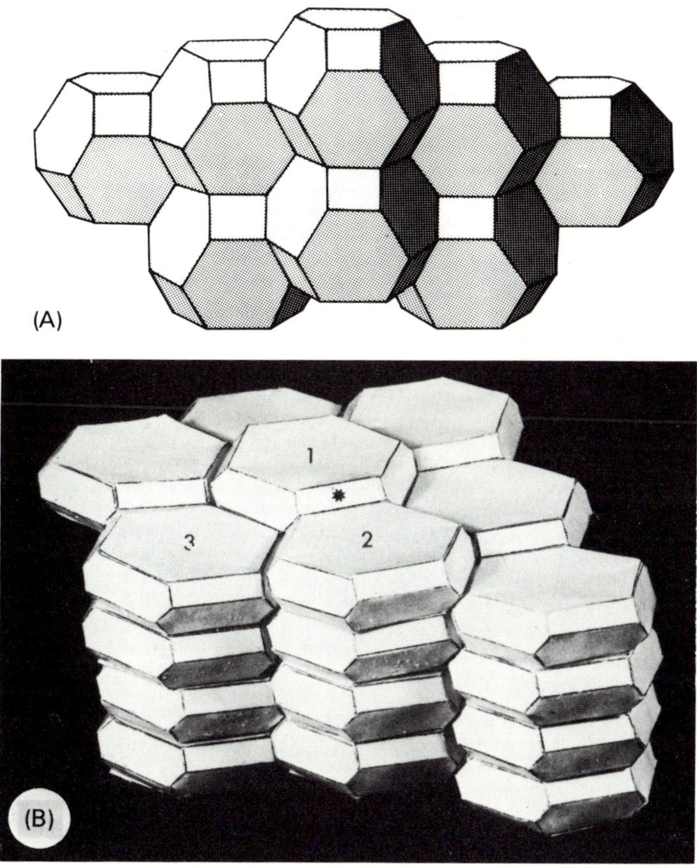

Fig. 8.8.  (A) Orthic tetrakaidecahedra stacked without interstices.  (B) A stacked array of flattened or truncated tetrakaidecahedra, again stacked without interstices, and distinctly resembling stacked corneocytes in the epidermis. (From Menton (1976*a*), by courtesy of Dr D. Menton.)

process until contact is achieved with cells in adjacent columns (?contact inhibition, see Mackenzie 1972, and above). It should be noted that each column numbered 1 has three columns numbered 2 and three numbered 3 arranged in an alternate manner around it. The same arrangement is depicted in Fig. 8.7, which shows a possible synchronization in trios of adjacent EPUs, although here the view is from the surface. If the mechanism proposed by Menton (1976*a*) is tenable, then it may not be necessary to postulate any special inter-EPU control, as the trilevel arrangement possesses an inherent synchronization sequence in trios of EPUs.

These proposals have not found favour with Allen and Potten (1976*b*). Allen and Potten agree with the proposal that ordered structure can be simulated by an assembly of truncated tetrakaidecahedra, and accepted the concept of a

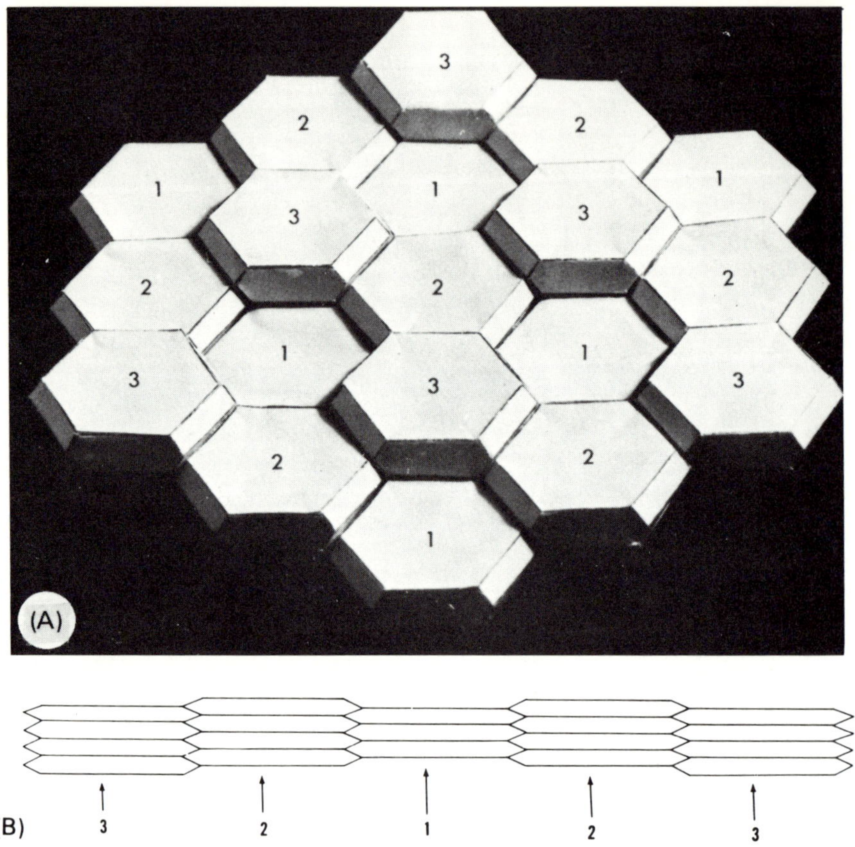



Fig. 8.9. (A) The forming face, or deep aspect, of the stratum corneum in stacked epidermis, represented by stacked tetrakaidecahedra, showing the trilevel arrangement visualized by Menton (1976*a*). Cells are added first to position 1, then 2, then 3, thus ensuring regular interdigitation and perpetuating the trilevel arrangement. Note the similarity to Fig. 8.7. (B) Stacking as seen in two dimensions: cells are added consecutively to the forming face in the order shown. (From Menton (1976*a*), by courtesy of Dr D. Menton.)

trilevel forming face providing a partial self-assembly system which predetermines the column entered into by the next flattening cell. However, Allen and Potten maintained that the trilevel columnar arrangement (Fig. 8.9) imposed a sequence of migration and cell production on the *basal layer*, since, in mouse epidermis anyway, ordered structure is visible down to the spinous layer. Migration into each respective superficial position (position 1, Fig. 8.9) was from the basal cells beneath that column (i.e. from the same EPU); migration from outside the EPU was considered to involve unnecessarily extensive lateral cell migration. Furthermore, the epidermis was seen as a tightly arranged cell structure, bound by the desmosomes and hemidesmosomes; migration involves extensive breaking and remaking of desmosomal contacts, a complex process reviewed by Allen and Potten (1974*b*). In a bubble chamber, with free lateral movement and jostling, and no interbubble connections, bubbles can jostle into columns regardless of origin. Epidermal columns were considered to be produced only by control of cell proliferation, which occurs not only within individual EPUs, but also between EPUs; random basal cell migration and mitosis are consequently precluded.

It does then appear that a highly organized pattern, with peripheral migration from the basal layer (Christophers 1971*c*) and the complex basal layer kinetic organization about to be described, is unlikely to be wholly the result of an intrinsic minimum surface area argument.

### (iii) *Ordered structure and cell proliferation*

As mentioned above, an ordered epidermal structure is normally found only when proliferative rates are low, as in mouse ear epidermis. In situations where the proliferative rate, and consequently the flux into the stratum corneum is increased, then column formation is usually lost. Thus stacking is not a feature of palmar and plantar epidermis, where proliferative rates are relatively high (Goldschmidt and Kligman 1967), or following the induction of increased proliferative activity in the ear epidermis of the guinea pig or mouse after localized injury (Christophers and Braun Falco 1970*a*; Potten and Allen 1975*b*; see Chapter 10). However, when the normally slow renewal pattern is established, then stacking returns. Mackenzie (1974*b*) has also reported a loss of orderly columnar arrangement in epidermis showing a thickened corneal layer and an increased proliferative rate as a result of a chronic frictional stimulus.

Christophers *et al.* (1974) believe that a direct relationship exists between epidermal cell production and the presence of an ordered structure. This is indicated in Table 8.1. In mouse ear, where measurements indicated that only one new epidermal cell is produced per surface corneocyte in two days, stacking is present: in mouse plantar epidermis, which produced two new cells/corneocyte/day, an ordered structure is not evident. Consequently Christophers *et al.* (1974) postulated that ordered epidermis is only found below a certain proliferative rate. Loss of ordered structure in more rapidly renewing epidermis may be the result of lack of time for flattened cells to mature before other cells begin flattening beneath them (Christophers and Braun Falco 1970*a*); similarly, if insufficient time is available for maturation, lateral steps would not form, thus inhibiting the lateral expansion of younger cells in the process of flattening.

In further support of this contention, Christophers (1971*a*) has shown that the hyperplastic epidermis which results from scratch-wounding mouse ear epidermis is no longer stacked. Stacking reappears at the wound margins, and gradually moves in to establish ordered structure only after 15–21 days, when the epidermis reverts to its slowly proliferating, thinned state. The situation is evidently more complex than a simple relationship between a low proliferative rate and ordered structure, because compression of the ear resulted in an increase in cell production, but loss of stacking occurred only in the lower epidermis.

Enhancement of proliferative rate may not be the sole factor which leads to loss of stacking in ordered epidermis; Potten and Allen (1975*b*) have made the salient observation that, in the initial period following wounding, when migration into differentiating layers is high, stacking is retained. Later in the response, stacking is lost at a much lower migration rate. This indicates that a stacked epidermis needs a certain post-mitotic basal cell maturation period, and if this is permitted, $k_B$ could be high but stacking remains. Consequently stacking may be better correlated with basal cell maturation time rather than $k_B$ or $k_L$. In this context, the reversion of non-stacked, rapidly proliferating embryonic epidermis and regenerating post-wounding epidermis to an ordered structure could be a modification of the unstacked situation, where most constraint exists (Potten and Allen 1975*b*). In this context it is interesting to note that in mouse tail, and in some human epidermis, where proliferative rates are relatively high, stacking, albeit with maximal overlap, is still present.

However, in more physiological circumstances, stacking is associated only with slowly renewing epidermis. Thus Christophers (1972*a*), studying guinea pig epidermis, which varies in thickness from animal to animal, reported that ordered structure decreased and was lost as the epidermis increased in thickness, with a critical thickness of 38–46 $\mu$m, above which normal stacking was not evident. Furthermore, there was a linear relationship between $I_S$ and epidermal thickness (see Fig. 8.10), emphasizing the correlation between stacking and proliferation. Recalling the possibility of sequential surface desquamation and alternate entry of migrating cells into columns containing least corneocytes, an increase in the birth rate in stacked epidermis would quickly lead to loss of structure as many cells of the same age compete for the same position. A similar situation obtains in the human epidermis in disease; Menton and Eisen (1971*b*) found that in psoriasis and lamellar icthyosis, both diseases associated with an elevated birth rate, regular stacking of corneocytes was absent, but in icthyosis vulgaris, with a normal birth rate, the stratum corneum showed the usual ordered arrangement.

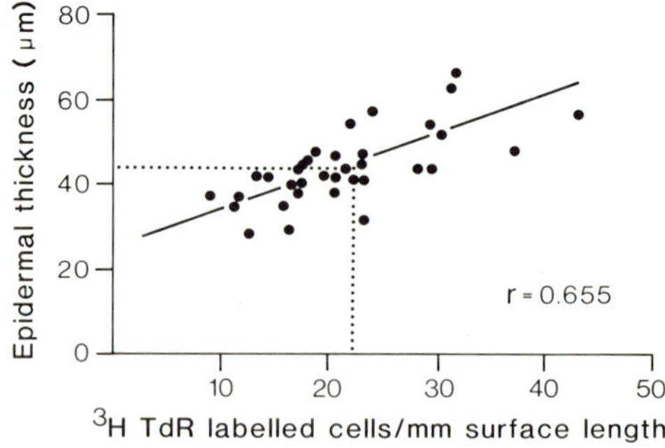


Fig. 8.10. Correlation between epidermal thickness and the number of flash-labelled nuclei in guinea pig ear epidermis. (Redrawn from Christophers (1972*a*).)

Mackenzie (1970, 1972, 1975*c*) has reported an interesting association between the spatial arrangement of mitotic activity and epidermal columns. Using methylene blue stained sections and acid expansion to visualize stacking of suprabasal cells after a six-hour collection of metaphases with colcemid, the distance of metaphases from the columnar boundary was measured. More metaphases were seen beneath the margins of the column. This arrangement was more pronounced in wide columns than in narrow columns; in narrow columns the distribution of metaphases was random (see Fig. 8.11), and these

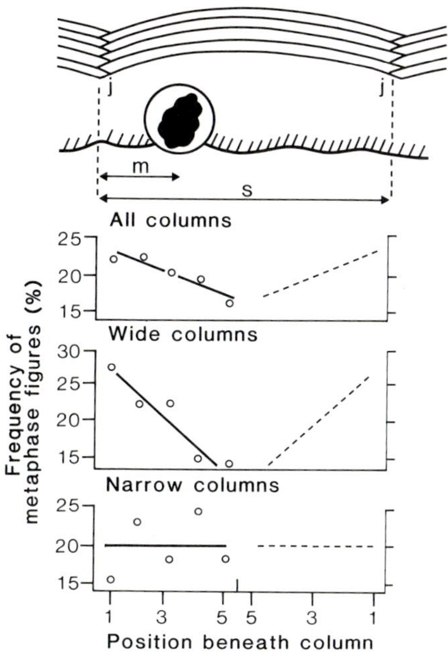


Fig. 8.11.   The distribution of arrested metaphases in relationship to the EPU; *s* is the width of the column, and *m* the distance of a metaphase from the boundary. The distribution is plotted in each of five equal divisions from the boundary to the centre. Note that the tendency for metaphases to occur peripherally, with fewer centrally, is more prominent in central sections, and absent in peripheral sections. (Redrawn from Mackenzie (1972).)

were probably columns sectioned through their periphery. The same author (Mackenzie 1975*c*) further examined the problem of the spatial distribution of mitoses in relationship to ordered structure, using interference microscopy on epidermal sheets, which avoids difficulties experienced with the orientation of sections with respect to the cell columns. Dividing each hexagonal column into 10 smaller, concentric hexagons, and measuring the distance of colchicine-metaphases from the centre of the hexagon, Mackenzie found that the distribution of mitoses was non-random and the largest number of mitoses was located just inside the periphery of the epidermal column (see Fig. 8.12). Basal cells just

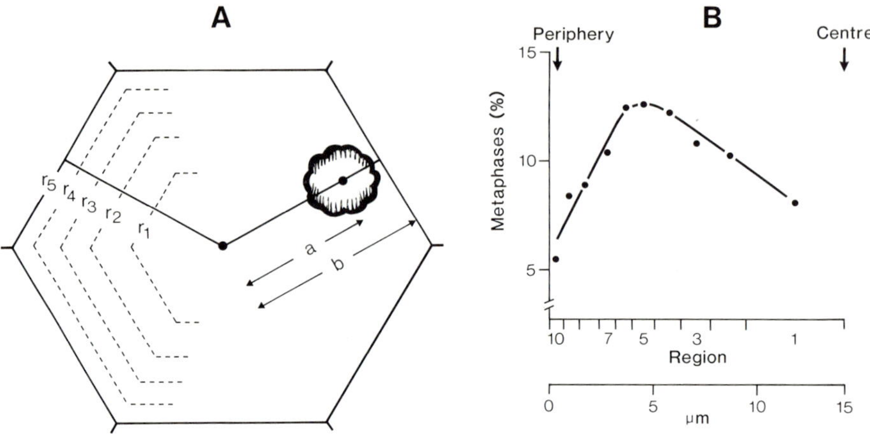


Fig. 8.12. (A) Showing the method used by Mackenzie (1975*a, c*) to divide the hexagonal EPU into concentric regions of equal area by division of the distance (*r*) into appropriate lengths $(r_1, r_2, r_3)$ so that $r_1 = r_2\sqrt{2} = r_3\sqrt{3}$. The mitotic distribution in 10 such regions is then calculated from the ratio $a/b$. (B) The distribution of metaphase (●) nuclei in the EPU. Whereas the interphase nuclei are randomly distributed, the distribution of mitotic nuclei differs significantly ($p < 0.001$) from a random distribution. (Redrawn from Mackenzie (1975*c*).)

within the periphery have the greatest growth potential; such a situation may explain why other authors (Frei *et al.* 1963) suspected a random distribution of mitoses but could not explain the greater frequency of division in adjacent cells (see section 8.1). The peripheral localization of mitotic activity has also been confirmed by Goerttler *et al.* (1973) and by Karatschai *et al.* (1971), who showed that 85–90 per cent of colcemid-blocked metaphases were found at the margin of the EPU.

The kinetic organization of the EPU is central to any discussion we may have concerning the mechanism of basal cell migration. As is more fully discussed in section 8.6, a very reasonable suggestion is that, if the basal cells divide in the plane of the basal layer (that is, if the mitotic spindle axes are *horizontal*), then it is the oldest $G_1$ cell nearest a basal layer mitosis which migrates (Iversen *et al.* 1968). Mackenzie (1970, 1972) proposed that the marginal rings of proliferative cells produced progeny which moved centrally into line beneath the overlying epidermal cell column; the (now late $G_1$) cell then migrated from the basal layer in a *central* position, to flatten itself over the adjacent marginal basal cells. However, this mechanism of migration needed revising to account for Christophers' (1971*c*) observation, using the FITC technique for distinguishing between cells of differing maturity, that the basal cells underlying *lateral junctional regions* of the overlying columns acquire a pedunculated shape and migrate from here into the above column. This latter mode of migration has been incorporated by Potten and his colleagues (Potten 1974; Potten and Allen 1975*a*; Potten 1976) into their model of kinetic organization in ordered epidermis (see v below).

This model grew from observations made by Potten (1974) on some kinetic responses, particularly to hair-plucking, of mouse epidermis. The responses were quantified in terms of the epidermal proliferative unit, which it will be remembered, contains about 10 basal cells. Of these 10 cells, seven are peripheral, two are intermediate, and one is central. Since there are some 15 000 basal cells/mm$^2$, there are 1400 EPU/mm$^2$ and consequently a similar number of single central cells. Under continuous labelling conditions with [$^3$H]-TdR, the labelling of the central cells rises more slowly than the other peripheral cells in the EPU. After 150 h of continuous labelling, 50 per cent of the unlabelled cells were central cells, while the remaining unlabelled basal cells were regarded as late $G_1$ cells or melanocytes, etc. Hence the central cells are either cycling at a relatively slow rate, or are out of cycle.

In a resting EPU, there is about one labelled cell, and 10 per cent of all labelled cells are in the central position (Potten 1974). Stimulating the epidermis by plucking the skin leads not only to an increase in labelled EPU, but also to an increase in the labelling of central cells, which is more prominent than in peripheral cells; this was shown by the increase in the ratio of labelled central cells to all labelled cells, which led Potten to suggest that central cells were more responsive than peripheral cells in this kinetic situation. Potten is firmly of the belief that the central cell in the EPU is of major importance in the control of epidermal cell proliferation and homeostasis in mouse ear epidermis. In mouse ear epidermis the Langerhans cell is found in the central position, and Potten and Allen (1976) have built a hypothesis deeply implicating the Langerhans cell in keratinocyte proliferation control (see iv below). Furthermore, measurement of clonogenic cell number in the mouse epidermis gives estimates which were considered close to the total number of central cells (i.e. about 1000/mm$^2$; but see section 8.4 and Potten and Hendry 1973) and the same number show late metabolism of [$^3$H]-TdR, which is possibly a feature of epithelial stem cells (see Chapter 2, section 9). These findings have led Potten and his colleagues to a proposal of equivalence between the central (or Langerhans cell) and the epidermal clonogenic cell population. Because of the relationship between Langerhans cell number and granularity on the one hand, and epidermal cell production rates on the other, it was proposed that the Langerhans cell controls the cell production rate by a local negative-feedback mechanism (see Chapter 2, section 10).

However, the link between central cells and clonogenic cells remains circumstantial, and is based largely on their numerical equivalence; the potential weaknesses of this arrangement have already been explored (see Chapter 2, section 9). The kinetic evidence for the importance of central cells is more impressive: the facts that central cells are stimulus responsive, and slowly cycling, possess characteristic thymidine pools, and have a strong circadian rhythm (Potten *et al.* 1977), by the analogy with known *functional* stem cells in the small intestinal crypt, may indicate that central cells are involved in cell production in normal circumstances at least. This point is considered further in section 8.4.

(iv) *The possible role of the Langerhans cell in the maintenance of ordered epidermal structure*

Although the definite function of the Langerhans cell remains to be decided, recent evidence inclines to the view that it is of bone marrow origin, and may be related to the dermal histiocyte, forming an intraepidermal macrophage system (Hashimoto and Tarnowski 1968; Mackenzie 1975b; Rowden 1980; see Chapter 5). Wolff and Winkelman (1967a,b) have reported that, in guinea pig epidermis, Langerhans cells are spatially regularly arranged, and proposed that a special, possibly functional relationship might exist between a Langerhans cells and the surrounding squamous cells. The morphological aspects of the Langerhans cell have already been briefly described (see Chapter 4, section 2). In the mouse epidermis, Langerhans cells are always basal, and despite lack of desmosomal contacts with adjacent keratocytes, Allen and Potten (1974a) have described how they are firmly anchored, above and below, by the cytoplasm of adjacent keratocytes, which effectively isolate the Langerhans cell from the basement membrane below and the spinous layer above, holding it in the basal layer. In the mouse Wolff and Winkelman (1967a,b) counted some 920 Langerhans cells/mm$^2$ and Mackenzie (1972) has calculated that each associated group of keratocytes would approximate to the size of an epidermal column. Similarly Allen and Potten (1974a) found that Langerhans cells formed 10.6 per cent of the basal cell population in the mouse ear epidermis; as there are 14735 basal cells/mm$^2$, some 1400 Langerhans cells are present/mm$^2$, in reasonable agreement with Wolff and Winkelmann (1967a,b). It will be remembered that there are 1400 EPUs/mm$^2$ (Potten 1974), and thus Langerhans cells are 1:1 with epidermal columns; in non-stacked epidermis, the ratio is also one Langerhans cell/surface corneocyte (Potten and Allen 1976).

More direct studies by Mackenzie (1972, 1975b), demonstrating mouse Langerhans cells by their large content of ATPase, and showing up the hexagonal columns with Sudan Black B, indicated that the number of ATPase-positive dendritic cells was similar to the number of epidermal columns, and furthermore that dendritic cells tend to lie beneath the centre of the overlying columns (see Fig. 8.13). Quantifying this observation by a superimposed template of concentric circles to find the centre of the hexagonal column, Mackenzie showed that the spatial relationship of the Langerhans cell to the epidermal column was non-random. In 60 per cent of cells examined, the Langerhans cell was in the central zone of the column. In the face of some criticism of the ATPase method for Langerhans cells (see Chapter 5, section 3), Mackenzie quoted additional, personal ultrastructural evidence showing Langerhans cells beneath epidermal columns of mouse ear epidermis. This has been confirmed and extended by Potten and Allen (1976), who added the point that the dendritic processes do not extend beyond the margins of the epidermal column. Mackenzie (1975b) concluded that Langerhans cells may have a function related to the establishment of a columnar

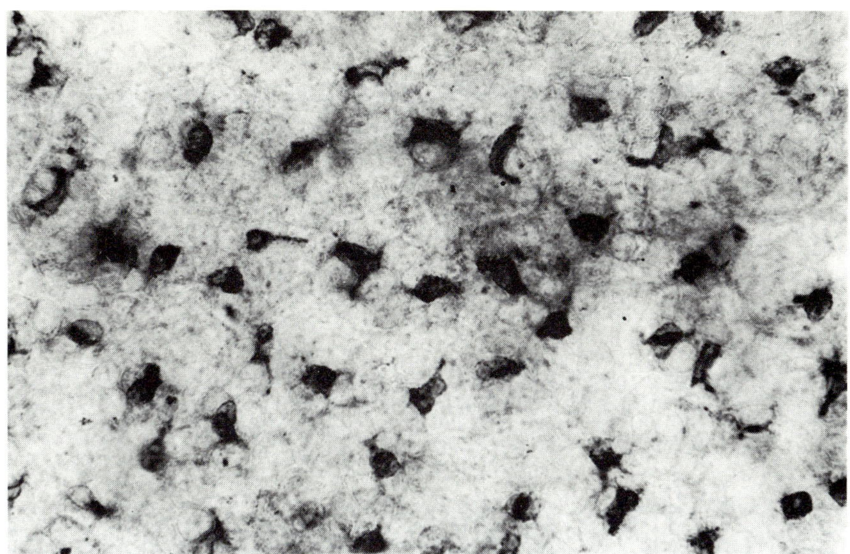

Fig. 8.13. Sheet preparation from mouse ear epidermis showing the regular arrangement of the dendritic ATPase-positive Langerhans cells; these are in fact at the centre of the hexagonal EPU. (By courtesy of Dr C. S. Potten.)

structure. Of the several postulated functions of the Langerhans cell, its possible role in assisting keratocyte movement (Breathnach and Wyllie 1966), and in the control of cell production (Prunieras 1969; Hashimoto and Karnowski 1968; Wong and Buck 1971), both might be expected to lead to a non-random, possible central arrangement with respect to cell columns. By analogy with the epidermal melanin unit, there may exist an *epidermal Langerhans unit*, composed of a Langerhans cell and its surrounding clone of keratocytes (see Chapter 5, section 3).

We have seen above that Potten (1974) considered that the central cell in the EPU was slowly proliferating, and could be the Langerhans cell, on the basis of Mackenzie's (1972, 1975*b*) findings on the central position of the Langerhans cell, and on his own observation of equivalence between the number of Langerhans cells and of EPUs per unit area. However, as we have already seen (Chapter 5, section 3), opinion is divided, not only on the function of Langerhans cells in stratified squamous epithelium, but also on the question of their origin. Most of the various theories do involve the proliferation of Langerhans cells, and of course Mackenzie (1975*b*) claims direct evidence for Langerhans cell proliferation; such proliferation is essential for Potten's hypothesis to be tenable. However, Potten and Allen (1976) went even further—that the Langerhans cell is at least a functional, if not a potential stem cell for the epidermis: being in a central position it is suitably placed for such a role. On division of the Langerhans stem cell, one daughter remains central and enclosed by basal cells, while the other, largely devoid of Langerhans granules as a result of asymmetrical division, starts a division–differentiation sequence towards a mature keratocyte. Certainly the evidence does point towards Langerhans cells being proliferative (reviewed in Chapter 5, section 3), and Langerhans granules have been seen in keratocytes (Potten and Allen 1976), supporting the concept of an asymmetric division of some sort. Allen and Potten (1974*a*) alternatively proposed a 'managerial' role for the Langerhans cell (Bruce 1976), in which they see the cell having a controlling role, possibly transferring information to keratocytes. They also speculated that, since the central keratocytes are relatively differentiated, and as further differentiation and migration is a peripheral event, the Langerhans cell might maintain the central keratocyte in an undifferentiated state until movement to the margin of the unit occurs. Studies of the response of Langerhans cells to wounding may also indicate a controlling role in cell production; measurements of the frequency of clear cells in tape-stripped epidermis (corrected for melanocytes and indeterminate forms) show an inverse relationship between the density of putative Langerhans cells and the cell production rate. Moreover, counts of granule density in electron micrographs reveal a fall in count with increasing cell production rate. Later on in the recovery sequence, the granule density increases and cell production rate falls. This admittedly circumstantial evidence seems to point to a negative-feedback control of keratocyte division by Langerhans cell granules (are the granules a chalone or chalone precursor?). However, this proposal was opposed by Schweizer and Marks (1978), who could find no difference in proliferative rates in areas of mouse tail epidermis where Langerhans cells were present or absent; furthermore, epidermal chalone acted on both areas. There is also other confounding data: Mackenzie (1975*c*) has shown that Langerhans cells are by no means centrally placed in columns in the epidermis of hamster, guinea pig, and monkey, and considers that the relationship to cell columns in mouse epidermis is atypical and 'not important to the establishment of cell columns'. The labelling indices of Mackenzie (1975*b*) and Potten (1974), of Langerhans cells and EPU cells respectively, would indicate that all that is central is not a Langerhans cell, since Mackenzie (1975*b*) reported an $I_S$ value of $<0.01$ per cent for mouse Langerhans cells, whereas the labelling index data of Potten (1974) would indicate that the $I_S$ of centrally placed nuclei in the EPU is higher than this. It is also clear that the hypothesis is at variance with the role of Langerhans cells as the afferent arm of the immune system in the epidermis (see Chapter 5, section 3). However, this does not mean that Langerhans cells cannot interact with the keratocyte population so as to affect keratinization (Schweizer and Marks 1978) or by sensing antigenic changes on the cell membrane of adjacent keratocytes (Hume and Potten 1983). Any global hypothesis will have to account for the central position of the Langerhans cells in the epidermal cell columns of the mouse epidermis.

(v) *Kinetic organization and the epidermal proliferative unit*

It is probably true to say that some of the most intriguing and exciting ideas on organization in epithelia have come from the work of Potten and his colleagues in Manchester. In the epidermis their particular contribution has been the correlation of ordered epidermal structure with the presently available information on kinetic parameters in the epidermis (Potten 1975b; Potten and Allen 1975a; Potten 1976; Potten *et al.* 1977; Potten 1978a; Hume and Potten 1983).

Central to this correlation is the concept of the heterogeneous kinetic nature of the basal layer. Disregarding for the moment the 15 per cent of the basal cell population made up of melanocytes, Langerhans cells, and Merkel cells (in the mouse anyway), basal cells are either proliferative or non-proliferative. Of the non-proliferative cells, we have $G_0$ cells (for which the supporting information is dubious, see Chapter 7, section 4), and cells which are variously described as post-mitotic maturing or basal differentiating cells. These can be seen as FITC-positive, keratin-containing cells in the basal layer; their presence can be inferred from the time of appearance of the first suprabasal [$^3$H]-TdR-labelled cells after flash labelling, or from patterns of cell migration after wounding (see Chapter 10). Potten (1975a,b, 1976) compared the measured time taken for [$^3$H]-TdR-labelled cells to migrate to the surface with estimates for the rate of transit through the given number of cell layers, and noted a discrepancy which could be explained on the basis of some 20–30 per cent of non-mitotic basal cells spending some 2–3 days in the basal layer prior to migration. At the same time, some cells are available for rapid transit into the Malpighian layer after wounding; post-mitotic cells would be of sufficient maturity to migrate in such circumstances.

Considerable confusion exists as to the concept and size of the growth fraction in the basal epidermal cells (see, for example, Fowler and Denekamp 1976; and Chapter 7, section 4) and Table 8.2 (after Potten 1975b) itemizes some $I_P$ values and their connotations for the kinetic organization of the EPU. The bulk of the evidence presented favours the value of about 0.6, or, in the context of the EPU, about six out of the 10 basal cells are cycling. Potten (1975b) has exhaustively

Table 8.2. Some values for growth fraction in the epidermis and their implications for the concept of the epidermal proliferative unit. (After Potten (1975b))

| $I_P$ | Implications |
| --- | --- |
| 1.00 | All basal cells proliferative. Continuous labelling gives labelling of all basal cells with time, thus all basal cells *eventually* pass through S; this does not mean that all cells are in cycle *at any point in time*. Does not allow for post-mitotic maturing cells |
| 0.6 | Calculated from literature (see Chapter 7, section 4). Allows for four post-mitotic maturing cells per EPU, and six proliferating cells. Does not incorporate any $G_0$ cells blocked in $G_1$ or $G_2$. Necessitates postulation of precisely phased cell cycles in proliferative cells to account for CPR data of 1 cell/EPU/day. Meets FLM data requirement of $T_C \sim 100\,\mathrm{h}$ |
| 0.1 | Only one basal cell in cycle, with $T_C$ of 24 h. Allows for post-mitotic maturing cells and $G_0$ cells and meets requirement of 1 cell/EPU/day. Does not meet the FLM data requirement of $T_C \sim 100\,\mathrm{h}$ |

Table 8.3. Estimates of epidermal cell production rates per EPU per day in mouse epidermis. (Summarized from Potten (1975$b$, 1976))

| | CPR (cells/EPU/day) | |
| --- | --- | --- |
| | Dorsal epidermis | Ear epidermis |
| Average from measurements of transit time | 1.0 | 1.0 |
| Average from $I_S$ and $I_M$ measurements throughout a 24 h period | 1.5–1.8 | 1.8 |
| Average of single reading $I_S$ and $I_M$ data | 1.4 | 0.5 |
| Average from $T_C$ measurements, with $I_P = 0.6$ | 1.2 | 0.9 |
| Overall average | 1.3 | 0.9 |

analysed the published data for epidermal cell production rates, and Table 8.3 is summarized from this analysis (see also Potten 1976). In the ordered structure of the mouse dorsal and ear epidermis, each EPU receives about one cell per day from the basal layer. With a cell production rate of one cell/EPU/day, we can have a kinetic situation where there is only one basal cell in cycle in each EPU, with a $T_C$ of 24 h and an $I_P$ of 0.1, or in accord with the kinetic data, which demands an $I_P$ value of 0.6 and a $T_C$ in excess of 100 h (Hegazy and Fowler 1973$a$), six out of the 10 basal cells/EPU must be in cycle (see Table 8.2). In this case the cell cycle time would be about five days, and the proliferative cells must be precisely phased so as to give one cell/EPU/day. We can refine the proliferative population even further by remembering that continuous labelling data (Hamilton and Potten 1972; Potten *et al.* 1974) shows that the central cells in the EPU are late labelling compared with peripheral cells, and hence are likely to have a longer cell cycle time. These considerations gave rise to the concept of *age differences in EPU cells*: a slowly cycling central cell which is stimulus-responsive and also a strong candidate for the one clonogenic cell/EPU (see Potten and Hendry 1973; Potten 1976), a more peripheral group of faster cycling cells, five or six in number, and about four peripherally situated post-mitotic maturing cells which are destined for early migration.

With these ideas in mind, Potten (1976) delineated the numerous experimental observations which any global model of epidermal kinetic organization must incorporate: (i) the complex age structure of the basal cells in the EPU, described above, with a slowly cycling, stimulus-responsive central cell, more peripheral faster cycling cells, and marginal post-mitotic maturing cells; (ii) that most of the cell division and migration occurs at the marginal part of the EPU basal cell complement; (iii) the measured kinetic parameters: (a) cell production rate (Table 8.3); (b) constancy in basal cell number per EPU (8–12 per EPU, see Table 8.1); (c) continuous labelling data (see Table 8.2; Potten 1974; Iversen *et al.* 1968); (d) FLM measurements of the cell cycle time (see Table 7.2 and Hegazy and Fowler 1973$a$).

Potten (1975, 1976) has proposed such a model, and this is shown as a flow

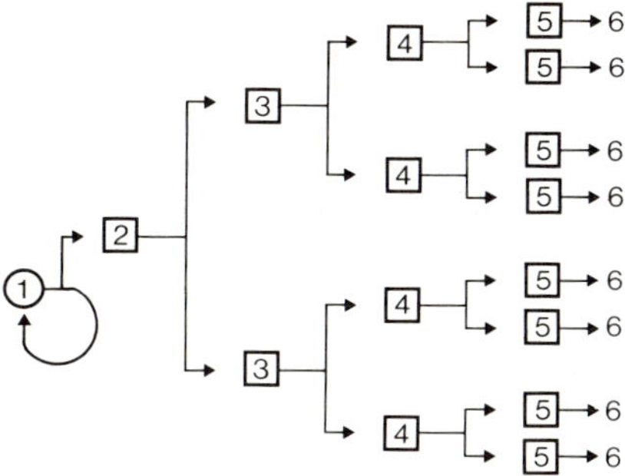

Fig. 8.14. The sequential division process in the EPU as visualized by Potten (1976). The stem cell (1) begins the sequence and there are a series of amplification divisions with irreversible differentiation until post-mitotic maturing cells are reached. Cells in position 6 migrate at the periphery. (Redrawn from Potten (1976).)

diagram in Fig. 8.14. A stem cell (numbered 1) gives rise to progeny which move by successive divisions irreversibly towards the post-mitotic maturing cell (5) and migration to suprabasal layers (6). As the proliferative potential of the cells becomes decreased, cells are more committed towards differentiation. The stem cell, which is centrally placed in the EPU, has a cell cycle time of about 8 days. The progeny are capable of three transit divisions, most of which occur towards the periphery of the EPU, with a cell cycle time of around 4–5 days. Such an arrangement is shown diagrammatically in Fig. 8.15; cells in position 6 migrate,

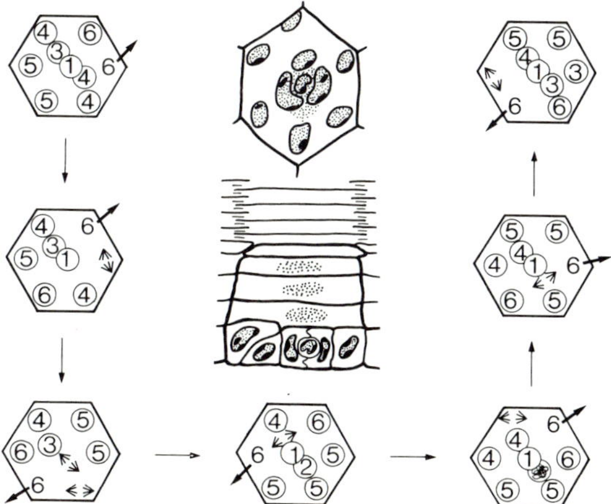

Fig. 8.15. The progressively aging sequence of cell divisions depicted in Fig. 8.14 brought into the context of the EPU. Cells 1–4 can divide, cell 5 is a post-mitotic maturing cell which migrates in position 6. Migrating cells leave a space which is filled by mitosis or lateral movement of an adjacent cell. Cellular depletion in any area, peripheral or central, is seen as the stimulus for an amplification or stem cell division respectively. (Redrawn from Potten (1976).)

leaving a space which is filled by division of a neighbouring cell (numbered 4). When we consider wound healing, for example after repeated stripping of the epidermis (see Chapter 10 and Potten and Allen 1975*b*; Etoh *et al.* 1973, 1977) we conclude that the stimulus for basal cell proliferation is depletion in basal cell numbers, Similarly, here in the normal situation, basal cell depletion could be the signal for local cell division.

It can be seen that this model does fulfil the requirements laid down above. In addition, we may note that (i) the system will be in a steady state with production of one cell/EPU/day, meeting the CPR data (Table 8.3); (ii) a reserve division potential in times of cell need is accommodated by the amplification compartment (cells 3, 4), and a migration potential by the peripheral post-mitotic cells (cells 5, 6); (iii) cells migrate from the basal layer in order of age, with precise sequential senescence directed towards the peripheral post-mitotic cells; (iv) the central stem cell unit is seen as unstable, after for example wounding, and capable of repopulation; (v) as far as the available kinetic data goes, the model would allow complete labelling under continuous labelling conditions with later labelling of the central cell. Because the divisions of cells 3 and 4 are more numerous than the slower stem cell division, these would predominate in any FLM curve and thus give a $T_C$ of around 100 h (Hegazy and Fowler 1973*a*). Divisions in cell 4 are most numerous (see Fig. 8.14) and are by definition terminal cell divisions; they would thus account for the well-known difficulties encountered in obtaining well-defined second peaks in epidermal FLM curves. It should be noted that, because only one cell can occupy position 1, stem cell divisions are perforce asymmetric; thus the model does allow the deletion of potentially damaged DNA, as Cairns (1975) proposed for stem cell systems. However, Mackenzie (1975*a,c*) was extremely doubtful about the advisability of relating cell proliferation directly to ordered structure, and questioned the whole concept of the EPU. He raised two points: (i) peripheral displacement of mitotic activity could be due to the position, centrally, of a dendritic (Langerhans) cell of very low proliferative activity, with subsequent localization of division in the periphery, without any consequence for basal cell proliferative control, and (ii) that basal cells functionally related to an EPU could be aligned within the periphery of the aforesaid column, thus influencing the observed position of mitosis. Illustrating these points Mackenzie (1975*c*) did two further experiments: (a) he mapped the position of vinblastine metaphases with respect to the EPU as previously described (see Fig. 8.11) in *hamster* epidermis, where Langerhans cells are not centrally dispersed with respect to the EPU; a mitotic distribution curve was obtained similar to that in the mouse. This finding indicates a relationship to the control of cell proliferation rather than simple exclusion from the central area; (b) Mackenzie repeated his previous (1975*a*) observation that peak proliferative activity is found just within the boundary of the mouse EPU, with a low incidence centrally and beneath the actual EPU junctions; however, the pattern of distribution for all basal cells did not differ from a random distribution.

Consequently there is no pre-existing alignment of peripheral interphase cells to account for the observed mitotic distribution, and Mackenzie considered that the reason for the observed position of mitosis was obscure, but there was no evidence for grouping of basal cells into EPUs.

But this evidence in no way disturbs the EPU hypothesis; the finding of peripheral mitoses in hamster epidermis merely indicates that central cells are slowly cycling but are not Langerhans cells, in accord with the EPU arguments (see Fig. 8.14), and studies of continuous labelling and response to wounding in hamster epidermis would be of interest. Similarly the finding of peak mitotic activity towards the margin of the EPU without concentration of basal cell number in that location, taken together with the observation of Christophers (1971c) that migration is also found exclusively at the margins of the EPU, would argue for organization of proliferative activity on the basis of epidermal proliferative units. Of course, such a peripheral concentration of mitotic and migratory activity is a basic tenet of the kinetic organization of the EPU model of Potten, described above.

Perhaps we should emphasize that this scheme is proposed only as being applicable to epidermis with ordered structure, and the kinetic organization of the basal layer in non-stacked epidermis and squamous epithelia such as oesophageal or vaginal mucosa may be quite different, as is discussed in section 8.3. Furthermore, it is difficult to incorporate the ideas of Gelfant (1966, 1977; see Chapter 7, section 4) on a subpopulation of 5–10 per cent stimulus-responsive $G_2$ blocked cells. The model also begs the question on the identity of the stem cell; also, what role does the Langerhans cell play? It will be interesting to know.

## 8.3. Kinetic organization in other squamous epithelia

We have now considered the situation as it affects thin stratified squamous epithelium without significant rete ridge formation, and with ordered structure, what Bullough (1972; Bullough and Mitrani 1976) has called *Phase 1 epidermis*. Not all squamous epithelia are like this (see Chapter 4), and in fact, most human epidermis, and internal epithelia such as line the oral cavity in many species, is thickened, non-stacked, and shows rete ridges (Fig. 4.1C, p. 206). Additionally, there are specializations in surface epithelia, such as the filiform papilla in the mouse tongue, which has given considerable insights into the behaviour of squamous epithelial cells, and warrant our attention.

### (i) *The mouse filiform papilla*

This is shown in Fig. 8.16A and diagrammatically in Fig. 8.16B. These papillae are highly ordered structures (Cameron 1966), formed of two large and two small tilted columns of corneocytes; the basal layer of cells is folded over a dermal papilla, and in thin sections, the column boundaries can be traced from the surface to the basement membrane, and the cell positions of the basal cells can

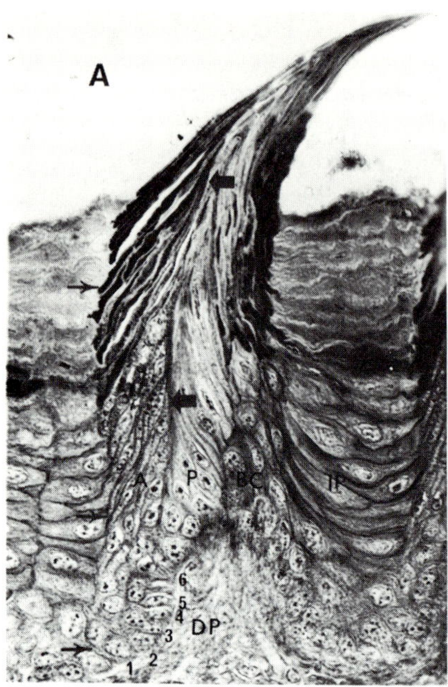

Fig. 8.16. (A) A longitudinal section of a filiform papilla from the mouse tongue, showing the typical columnar structure. (Palmgrens stain. × 163.)

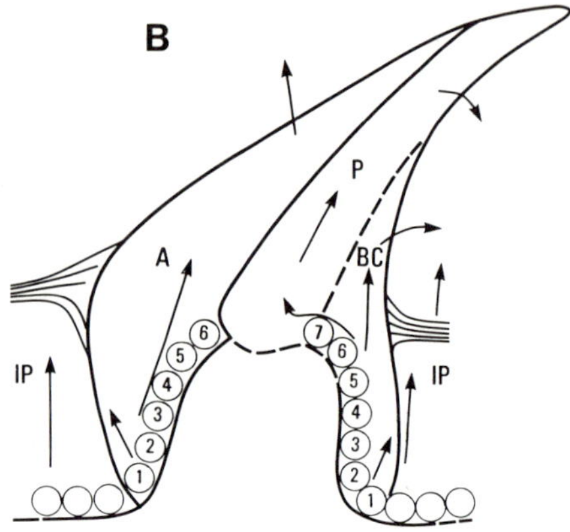

(B) A diagrammatic representation of the several columns and of the proposed directions of cell flux in the papilla. The boundaries of the anterior (A) and posterior (P) can be traced to the surface from the basement membrane (the anterior–posterior column boundary is arrowed in the photograph, as is the junction between the anterior column and the interpapillary epithelium (IP). The basal cells of the anterior column can be numbered according to their position on the connective tissue papilla. From the stem cell position (cell number 1) the basal cells move up through successive positions towards the apex of the connective tissue core before migrating suprabasally; there is also some migration of cells suprabasally from lower cell positions (indicated by the arrow in the diagram). The buttress column (BC) is less well-defined, but the cells here can also be numbered in the same way, and their flux directions are indicated in the diagram. (By courtesy of Dr W. J. Hume.)

be accurately delineated (Hume and Potten 1976, 1979). These cells show differing patterns of circadian variation in $I_S$ with position from position 1 (see Fig. 8.20, p. 325), but Hume and Potten (1979) have shown, by chasing a flash [³H]-TdR label autoradiographically, that there is a significant flow of basal cells along the basement membrane, from low to high positions in both columns, and cells at positions 4 and 5 have little proliferative potential, whereas cells in position 1 show high $I_S$ values (Cameron 1966), and display 100 per cent labelling with only 24 hours serial [³H]-TdR injections; it is likely that cells in lower cell positions have a high $I_P$, and that the cause of the falling $I_S$ with cell position is a decreasing $I_P$ (since FLM data evidently showed no change in $T_C$), with the result that cells overlying the dermal papilla (i.e. cells above positions 5 or 6), which do not label or divide, are post-mitotic maturing cells. It is also likely that some labelled cells leave positions 1 and 2, and pass suprabasally at early times after labelling.

In this model there is no doubt that these basal cells have a distinct age structure, and that migration of basal cells occurs along the basement membrane. As we shall see, cells in position 1 show evidence of selective segregation of DNA (Potten *et al.* 1978), and since the origin of the flow must be in cell position 1, these cells are important candidates for stem cells in this tissue. Moreover, it is possible that there is functional interdependence between groups of these columns, so that cell production and desquamation are co-ordinated, as is envisaged in the EPU model (Hume and Potten 1979). Thus we have a good example of a thick epidermal derivative with a highly ordered structure.

(ii) *The organization of thick epidermis*

Figure 4.1C (p. 206) shows a representative portion of thick epidermis, with its well-formed rete ridges, and illustrates the thin epidermis found over the dermal papillae; this is what Bullough (1972) calls *phase 2 epidermis*, which is characteristic of the more rapidly proliferating epithelia, and also, in even more pronounced form, in the various hyperplastic states. We shall consider the changes from phase 1 to phase 2 later (see Chapter 11) but here we merely consider how this epithelium is organized. Figure 8.17 (after Hume and Potten 1979), illustrates several models for the kinetic organization of thick epidermis. We might begin with the premise that the rate of cell loss by desquamation from the surface overlying both papillae and rete ridges is equal, which seems a reasonable assumption; additionally, for steady state conditions, this cell loss must be balanced by the birth rate (or cell production rate in the 'germinative' layers). Hume and Potten (1979) considered three possible models for this epithelium: (a) a *homogeneous* basal layer, with all cells as stem cells, all actively cycling, but without movement along the basement membrane; if this were the case, the cell production rate per cell in the thin epidermis above the papillae would have to be higher than in the rete ridges, and hence proliferative indices would be higher above the papillae. In fact, the converse is found in this type of epithelium; e.g. in psoriatic epidermis, more labelled cells are seen in the rete pegs than the papillae, and Löe *et al.* (1972) reported that mitoses tend to be localized in the lower part of the rete pegs in thick gingival epithelium. So, in effect, most cell production occurs in the rete pegs; (b) again with a homogeneous basal layer, with a high $I_P$, but here cells of equal proliferative potential *flow along the basement membrane*, so that the cell production rates over the papillae and pegs are equal; (c) a *stem cell subpopulation* at the base of the peg, with cell flow along the basement membrane, with amplification by dividing transit divisions, and although the cell production rates will be equal in peg and papilla, clonogenic cells are confined to the pegs; the papillary epidermis is occupied by the oldest basal cells of lower proliferative potential and/or by post-mitotic maturing cells.

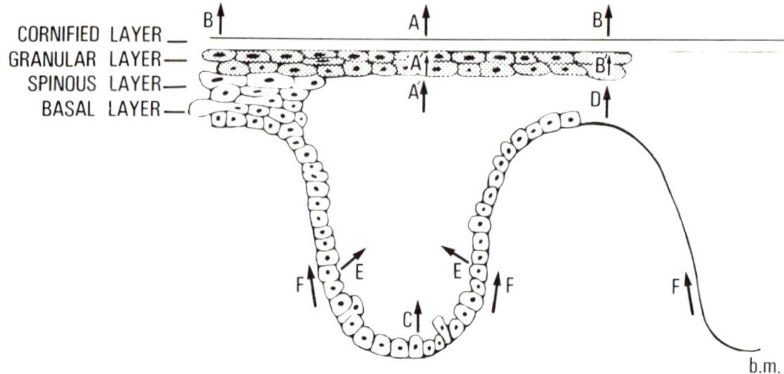


Fig. 8.17. Alternative models for the organization of cell proliferation in stratified squamous epithelium with rete ridges. In the steady state, the rate of surface desquamation at A and B are equal, and thus the cell production rates at A' and B', and therefore at A' and D should also be equivalent. Model (a) proposes a homogeneous basal layer, all of which are stem cells, with a high $I_p$ and no net movement of cells along the basement membrane (b.m.). The area A' will receive cells from both C and E, and to balance the fluxes, the birth rate per proliferative cell must be higher at D than at C; this would suggest that proliferative indices would be higher at D than C. Model (b) also has a homogeneous basal layer, high $I_p$ and a significant flux of cells along the basement membrane. Cells from E are diverted to D so that the cell production rate at D will equal that at A'. Model (c) has a permanent stem cell zone towards the bottom of the rete peg, cell flux occurs along the basement membrane with amplification divisions occurring in transit. Thus the cell production rate at D equals that at A', but, since cells are at the end of the dividing transit sequence, proliferative potential is lower at D, and hence proliferative indices will be lower at D than at C, due to a falling $I_p$ as cells migrate, with or without a lengthening of $T_C$. (By courtesy of Dr W. J. Hume and Dr C. S. Potten.)

Although Hume and Potten (1979) favoured (c), it should be noted that both (b) and (c) will fit what few observations there are for this type of epithelium, and (c) was only really favoured by analogy with the EPU and the filiform papilla, where clonogenic subpopulations and basal cell flow are more likely; for (c) to operate, Hume and Potten (1979) conceived a 'functional' proliferative unit, based on the rete peg; however, evidence for this is lacking, and it must be confessed that we are really ignorant about this type of epithelium, which is arguably the most common type, particularly in humans, notwithstanding our advances in the understanding of the organization of thin epithelium.

## 8.4. The kinetic organization of the basal layer

### (i) Epidermal stem cells and clonogenic cell number

We have already discussed the nosological and semantic problems which beset discussion about stem cells in renewing populations, and we have previously noted that there has, hitherto, been a tacit assumption that the epidermis is a simple one-tier structure, with a homogeneous basal cell population, all of which are regarded as stem cells. There is, however, little supporting evidence for this simplistic concept. There are accumulating observations which support the proposal that epidermal stem cells are a small subpopulation within the basal cell population, and in a provocative survey of the available evidence, Potten (1981) has pointed to what he considers to be several large inconsistencies in the older, orthodox simplistic concept: these include (a) observations on epidermis exposed to irradiation (see Chapter 12); (b) the effect of putative cell line specific proliferation control factors (see Chapter 14, section 1); (c) the achievement of higher plateau values on the continuous labelling curve than indicated by speculations on the possible size of the proliferative pool (but see Chapter 7, section 4); (d) the presence of small peaks following the initial major peak on the FLM curve for squamous epithelium (but see Chapter 7, section 2); and (e) the effect of [$^3$H]-TdR on the timing of FLM peaks (but see Chapter 9, section 3). It will be seen, however, that in some of these instances we do find ourselves in disagreement with the interpretation of much of this data.

Potten (1981) also speculates on the more teleological reasons for divorcing a keratocyte stem cell population from a dividing transit population, largely on the basis of analogy with the intestinal crypt and other tissues (but here we will see that there is anything but agreement that intestinal stem cells are a defined subpopulation), and the likelihood that 'stem cell DNA would be wasted'. However, there are compelling reasons for critically examining these data.

We have seen that, as yet, in epithelial cell systems the main methods of stem cell identification and enumeration are radiobiological, so-called *clonal regeneration studies*. We shall examine epidermal regeneration after irradiation in Chapter 12, but germane to our present argument is the dose-survival curve for epidermis, obtained by the macroscopic clonal regeneration assay (Withers 1967a,b; Emery *et al.* 1970; Denekamp *et al.* 1971; Leith *et al.* 1971), or by the microcolony epidermal clonogenic assay (Al-Barwari and Potten 1976). These curves are constructed by plotting the clonogenic survival against the increasing dose, on a semi-logarithmic plot (see Chapter 2, section 9). These curves are reproduced in Fig. 8.18 (from Potten 1981), and all appear to have a common slope with a $D_0$ (see Chapter 2, section 9) of 1.35 Gy; the extrapolation number is $1 \times 10^6$ cells/cm$^2$ at zero dose (Table 12.1, p. 413). Now, in theory, this extrapolation number should give information about the total number of stem cells; there are, roughly speaking, about $1.5 \times 10^6$ basal cells/cm$^2$ in mouse interfollicular epidermis; to this we should also add hair follicular cells, which almost certainly include at least some cells capable of regenerating (Bishop 1945; Oduye *et al.* 1975; Argyris 1976; Al-Barwari and Potten 1976), which are readily demonstrated in and around the hair follicle after irradiation. Potten (1981) has interpreted these findings as indicating that such clonogenic cells differ from those in the interfollicular epidermis; although they would appear to have a common embryological lineage, their radiosensitivity may differ (Hendry *et al.* 1980), and Potten (1981) has speculated that 'follicular' stem cells are more 'efficient' clonogenic units than the interfollicular stem cells, and that they may form a 'stem cell reserve', available if excess death of interfollicular stem cells occurs.

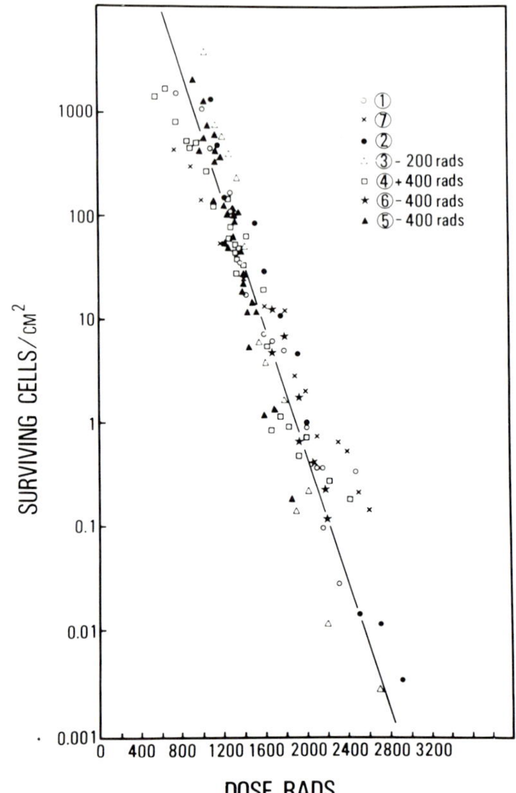



Fig. 8.18. A composite survival curve for clonogenic cells in mouse epidermis. The quality of the radiation in the studies used to compile the curve varied, so the results have been adjusted to fall about a common line, and the modifications necessary are shown on the figure. The common $D_0$ is 1.35 Gy, and the extrapolation number is $1 \times 10^6$ cells/cm². (1) Withers (1967*a*); (2) Withers (1967*b*); (3) Withers (1967*b*); (4) Emery *et al.* (1970); (5) Denekamp *et al.* (1971); (6) Leith *et al.* (1971); (7) Al-Barwari and Potten (1976). (By courtesy of Dr C. Potten.)

The effect of the 'follicular' stem cells would be to increase the total basal cells to some $3 \times 10^6$ cells/cm²; consequently, if there is a simple exponential relationship between clonogenic survival and dose, only about 0.33 of basal cells should be clonogenic. However, there are several drawbacks to this approach, which are set out in Chapter 2, section 9; here we may recall that the curve can, because of the repair of sublethal damage and multiplicity (see Chapter 2, section 9) only be followed over a strictly limited dose range, and the behaviour at lower doses can only be surmised; moreover, the fitting of a single curve implies that all cells in the population have equal radiosensitivity, which appears intrinsically unlikely (see Chapter 2, section 9, and Chapter 12). There is also the possible selection of the most radioresistant cells at high doses, and the likelihood of the formation of transitory colonies from the non-stem dividing transit cells.

Consequently we cannot reach any firm conclusions, except to say that if a shoulder region on the curve exists, similar to that evident for other cells, for example in culture (e.g. Nias and Fox 1968), then the true origin of the curve will be $< 1 \times 10^6$ cells/cm$^2$, and it is possible that clonogenic basal cells capable of regeneration may only constitute a fraction of the basal cell population; what fraction of the total clonogenic population is found in the follicles will, of course, depend on our estimate of the total clonogenic cell population, which remains speculative. This follicular contribution will not be negligible, and it will become important to investigate their properties *vis à vis* interfollicular clonogenic cells.

If doses are split to allow maximal radiation recovery, then making the usual assumptions, we could calculate the clonogenic cell number; Fig. 2.20 (p. 64) shows a diagrammatic representation of a split-dose experiment in mouse epidermis (Potten and Hendry 1973). Here $A$ is the initial clonogenic cell number, while $N$ is the extrapolation number; if $N$ is known then we can calculate the number of clonogenic cells. Now $N$ is the same, or $\geqslant R$, the recovery factor, where

$$R = \log_{10}N = 0.4343(D_2 - D_1)/D_0 \tag{8.1}$$

where $D_2 - D_1$ is the difference between the doses and the extrapolation number is calculated from

$$\log_{10}AN = \log S' + 0.4343D'/D \tag{8.2}$$

where $S'$ is the survival value at dose $D'$. It can be calculated that there are between 309 and 1107 clonogenic cells/cm$^2$, or that between 3 and 7 per cent of the basal cell population are clonogenic (Potten and Hendry 1973; Potten 1981). We can see from Table 8.1 that, on the principle of 'numerical equivalence' (see Chapter 2, section 9) there might be a single clonogenic cell per EPU. In support of this concept, Potten (1981) cites the rigidly ordered structure of stacked epidermis as evidence that each EPU must contain at least one long-lived, self-perpetuating cell, notwithstanding the relative ease with which the ordered structure can be lost and regained after injury; moreover, it is conceivable that the second cell in the envisaged proliferation sequence (close to the central area, see Fig. 8.3C) may be of greater maturity, but retains some regenerative capacity.

We might conclude that, although the data are not yet as hard as we would wish, there are distinct indicators that the basal cell population is composed of a separate stem cell subpopulation and a dividing transit population; clonogenic cells are also found in follicular epithelium, and it is possible that these two clonogenic compartments may differ in their clonogenic behaviour.

(ii) *Properties of epidermal stem cells; topography and cell kinetics*

Potten (1976) has proposed that epithelial stem cells may have discrete properties which set them apart as a proliferative subpopulation, particularly in the context of epidermis with ordered structure, where the EPU argument maintains that the stem cell occupies a central position in the basal layer, and is slowly cycling (or even in a resting phase) compared with its surrounding ring of amplifying proliferative cells (see section 8.2). There are no apparently distinct morphological features which are associated with the stem cell, although at one time there was speculation that a centrally placed dendritic cell was the EPU stem cell (Potten and Allen 1976), although, with the advent of the marrow origin of Langerhans cells and their role in antigen presentation, this hypothesis has been largely negated.

However, there is little or no information available concerning the nature of stem cells in other non-stacked squamous epithelia, apart from the filiform papilla in the mouse tongue, where some evidence for a 'presumptive stem cell compartment' is available (see section 8.3).

(iii) *Endogenous thymidine pools*

There have been a long series of rather peculiar observations, largely accruing from work involving the rate of [³H]-TdR incorporation into DNA, which now indicate that some cells in the presumptive stem cell compartment, in common with some other cell systems such as frog and rat lens epithelium (Rafferty and Gfeller 1970; Miller 1972), have singularly unusual endogenous TdR pools. Moffat and Pelc (1966) and Moffat (1968) reported that some resting hair follicle cells showed [³H]-TdR labelling *in frozen sections*, but not in paraffin sections after acid/aqueous fixation; hair plucking after [³H]-TdR administration led to persistence of the label, which was thus transferred from a fixative-soluble to a fixative-insoluble state. It later became evident that plucking induced a large increase in the acid-soluble pool size (Potten and Croxson 1971), which decreases slowly thereafter, although it was not clear whether this increase in pool size was due to decreased incorporation of [³H]-TdR or an actual increase in pool size; this naturally causes an immediate fall in $I_S$ and grain count, although the mitotic index also fell (Potten 1972), which does support a real decrease in proliferative rate. Potten (1971$a$) showed that a similar situation was evident in mouse interfollicular epidermis, where 'late labelling', or delayed uptake of [³H]-TdR was demonstrable if interfollicular basal cells were stimulated into division as long as 24 or 48 hours after flash [³H]-TdR labelling, probably before [³H]-TdR reutilization could confuse the issue, since the period concerned was shorter than the epidermal transit time (but not intestine?). This late labelling was hydroxyurea-sensitive, so it was probably associated with DNA (Potten 1973), and thus some basal cells are apparently able to maintain an endogenous cytoplasmic TdR pool, which, at a later time, and possibly only after a stimulus, is incorporated into DNA. Since Potten originally interpreted the response after plucking in terms of a stimulation of $G_0$ cells (see Chapter 10), the phenomenon was regarded as being $G_0$ specific, so that when $G_0$ cells were ejected into cycle, the [$^3$H]-TdR was incorporated.

The significance of this finding has yet to be shown, although Potten (1981) believes that this capacity may remain a property of stem cells, since centrally placed EPU cells, the 'stimulus-responsive cells' (Potten 1971$b$, 1973, 1974) showed some localization of delayed TdR incorporation, and, moreover, the number of cells showing delayed labelling may be numerically equivalent to the stimulus-responsive cells (Potten 1973). Additionally, there is now good evidence that putative stem cells in the presumptive stem cell zone of the mouse filiform papilla undergo delayed [$^3$H]-TdR labelling (Hume and Potten 1982); after flash [$^3$H]-TdR labelling, the initial doubling of the labelling index due to cell division in the stem cell zone was followed by a *second* rise of 14–16 per cent at about 16 hours after injection; that this rise was not dependent on cell division was indicated by its insensitivity to vincristine, but, unlike the situation in interfollicular epidermis, although the initial doubling of the $I_S$ and the *initial* incorporation of [$^3$H]-TdR was hydroxyurea-sensitive, HU had no effect on the secondary rise in the $I_S$. Changes due to migration into the stem cell compartment were considered unlikely, and the favoured explanation is that the rise in $I_S$ at 16 hours occurs because an average of one cell/column incorporated [$^3$H]-TdR from its long-lived pool; we shall also see that one cell/column can selectively segregate DNA at mitosis (Potten *et al.* 1978) (see below), and of course they may be common properties of the same cell. Be that as it may, the specificity of delayed labelling for stem cells must remain somewhat speculative at this time, as must its metabolic and cell kinetic role; Hume and Potten (1982) have speculated that such a pool may protect the stem cell from the known mutagenic effect of nucleotide imbalance (Bradley and Sharkey 1978) by, as yet, unknown mechanisms. In practical terms, the emergence of labelled cells at some time after flash labelling will certainly complicate the FLM curve, particularly in perturbed situations (cf. Hegazy and Fowler 1973$a$; and see Chapter 2, section 6).

An even more complex situation *vis à vis* routine thymidine labelling is indicated by the work of Møller and Keiding (1982), who demonstrated a diurnally varying endogenous acid-soluble pool in the hamster cheek pouch (see Fig. 8.19A,B); this resulted in a circadian-dependent delay in the incorporation of [$^3$H]-TdR into DNA, with a consequent pronounced circadian variation in the influx into and efflux from the $S$ phase, as indicated by double-labelling techniques (see Chapter 9); Møller and Keiding ascribed this effect to possible cell-cycle-dependent fluctuations in TdR kinase activity (Bray and Brent 1972). These changes did not appear to be confined to any subpopulation, and consequently will markedly affect the interpretation of routine $I_S$ measurements in squamous epithelium, particularly in studies of circadian variation; here it is usually assumed that where $I_S$ values are compared in the same tissue but at different times, they are independent of the pool size problems that can beset

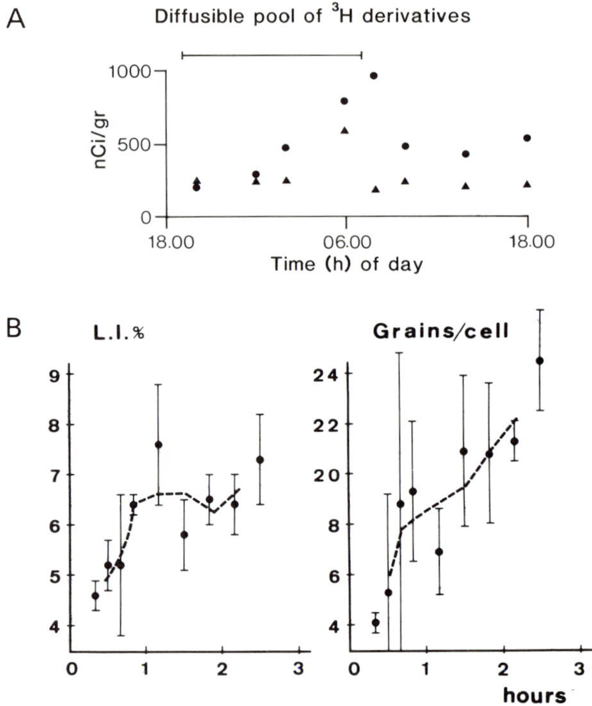



Fig. 8.19. (A) Circadian variation in the endogenous TdR pool size, as reflected in the radioactivity of the acid-soluble washings, in the stratified squamous epithelium of the hamster cheek pouch. (B) The variation in the labelling index and in grains per labelled cell after [$^3$H]-TdR in the hamster cheek pouch epithelium. There is a progressive rise in the grain count for at least 2.5 hours after labelling, reflecting [$^3$H]-TdR incorporation probably from a cytoplasmic pool. (By courtesy of Dr U. Møller.)

kinetic comparison of [$^3$H]-TdR incorporation rates. This can be important in, for example, chalone research, where pool problems between control and experiment are now more widely appreciated (Laurence *et al.* 1979).

### (iv) *Circadian variations*

Potten *et al.* (1977) have called attention to the large changes in proliferative indices which occur in several presumptive stem cell zones, cf. the central cells in the EPU, the bottom cell positions in the anterior and posterior aspects of the mouse filiform papilla, and the basally sited crypt cells in the small intestine; the changes in amplitude, and the ratio between peak and trough are particularly high for the cells in the lowest positions (i.e. the 'stem cell') in the filiform papilla (see Fig. 8.20), and get progressively smaller as we move up the column (see section 8.3 for the organization and numbering of cells in the column). It is also suggested that the central areas of the EPU have higher amplitudes

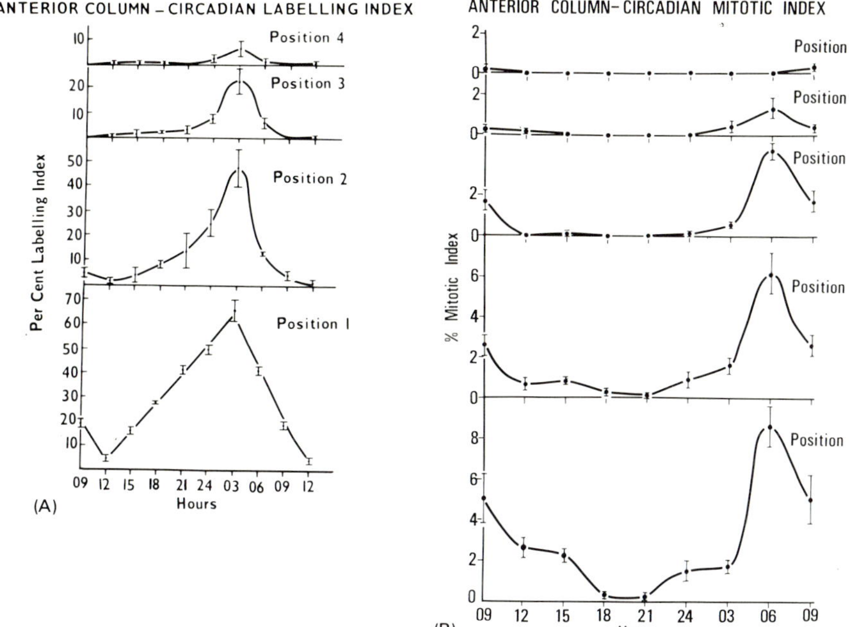

Fig. 8.20. The variation in (A) labelling and (B) mitotic indices with time of day in the basal cells of the anterior column of the mouse filiform papilla. Note that the peaks of $I_S$ and $I_M$ are about 3–4 hours out of phase, and also the very high flash-labelling indices which can be obtained in cell position 1 at 0300h. (By courtesy of Dr W. J. Hume.)

in proliferative index variation than the peripheral portions.

The occurrence of strong circadian rhythms in stem cell compartments could result in entrainment of the remaining proliferative cells, and account for the presence of circadian variation in these tissues (see Chapter 9); again the reasons for such prominent variation in putative stem cells is obscure.

### (v) *Conservation of stem cell DNA*

We have discussed the general argument concerning the advisability of maintaining 'immortal strands' of DNA through successive stem cell divisions (Cairns 1975; and see Chapter 2, section 9). The major problem in testing this hypothesis is, of course, the difficulty in recognizing stem cells morphologically; however, in the mouse filiform papilla there is, at least, presumptive evidence that stem cells are found in position 1 of the anterior and posterior columns (Hume and Potten 1976, 1979). Because of the extreme circadian synchrony which occurs at cell position 1 (see Fig. 8.20 and Chapter 9), it is possible, after a series of judiciously timed injections of [³H]-TdR, to label all cells in cell position 1 (see Fig. 8.21; Potten *et al.* 1978). After 24 and 36 hours, all cells at cell position 1 remain labelled, but at 48 hours there is an abrupt fall in the $I_S$ to 0.85; in

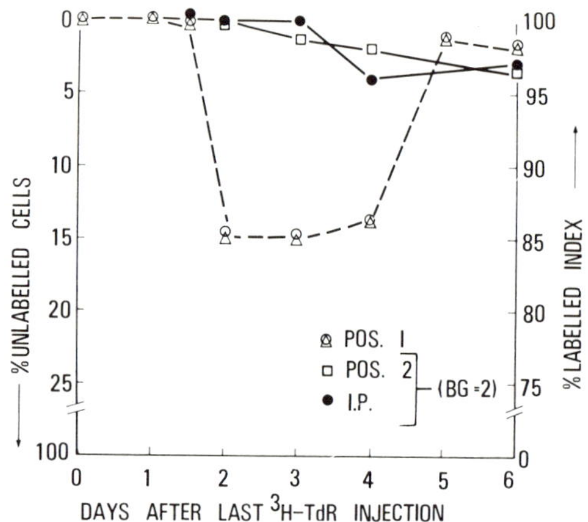



Fig. 8.21. The changes in the fraction of *unlabelled* cells with time after total labelling in the basal cells of the anterior column of the mouse filiform papilla. Three injections of [³H]-TdR given five hours apart (at 2200, 0300, and 0800h), or six four-hourly injections are sufficient to label virtually all cells at cell position 1. (By courtesy of Dr W. J. Hume.)

contrast, cells in neighbouring positions showed $I_S$ values of between 0.96 and 1.0 for at least five days after the labelling procedure. However, after five days, the proportion of labelled cells at cell position 1 is again increased to 0.98.

Potten *et al.* (1978) interpreted these results as indicating that about one out of every six cells at the column base is selectively segregating its DNA during *S*, with a $T_C$ approaching one day. Naturally, there are problems in proposing such a view; the grain count threshold for determining a cell as labelled is critical, since a lightly labelled cell which halved its grain count by division, between 0 and 48 hours after labelling, could fall below the level used for recognition, and be counted as an unlabelled cell. However, Potten *et al.* (1978) used an initial orthodox threshold of five grains, apparently based on background counts, but at 48 hours, the criterion for regarding a cell as unlabelled was only two grains, supported, in one experiment, by an exposure period of 365 days. Moreover, the results have an internal control, in that such observations were confined to cell position 1, and higher cell positions did not show any such change. But one would, on the basis of the proposed explanation, predict that the number of *unlabelled* cells at cell position 1 would increase as the unlabelled stem cell and its progeny continue to divide; this, in fact, does *not* occur, possibly due to reutilization of [³H]-TdR released by nuclear degeneration in the granular layer, a hazard of all longer-term experiments in squamous epithelia (Cutright and Bauer 1967; Iversen *et al.* 1971); that this was a reasonable explanation was the observation that the administration of cold TdR (to increase the pool size) over the post-labelling period led to a further decrease in the labelling index to 0.70 at day 3.

Despite, therefore, the caveats and inherent difficulties in interpreting these complex experiments, the Cairns hypothesis does appear to offer an explanation for these observations, and should be pursued with vigour.

We may conclude this survey of the nature and properties of stem cells in squamous epithelium by inferring that current work supports the concept of a small group of cells, probably slowly cycling, and, in stacked epidermis at least, of discrete topography or 'focal point' (Potten *et al*. 1979); there is evidence of singular endogenous nucleotide pools, the significance of which is obscure, of pronounced circadian rhythms, and some evidence of selective DNA segregation. As yet there is little to denote differential 'shells of stemness' (Potten *et al*. 1979), and no definitive evidence that such cells are in a discrete $G_0$ compartment (see section 8.5, Chapter 7, section 4, and Chapter 10), so that only some of the criteria of the Potten *et al*. (1979) model are fulfilled for squamous epithelium. A major problem would appear to be the difficulty in studying stratified squamous epithelium which is unstacked (which, after all, is in the great majority) and the above conclusions would appear to be valid only for epithelia with a highly ordered structure (cf. the EPU and the filiform papilla). We do, evidently, need new methods for the analysis of stem cell characteristics and numbers in other, less well-organized and more ubiquitous epithelia.

## 8.5. Proliferating and non-proliferating cells in the basal layer

We have already discussed the experimental problems which accompany any consideration of the growth fraction in squamous epithelium, and conceptual difficulties about what the proliferative compartment actually means.

We could use the term growth fraction in a loose sense to denote those cells which are actively cycling; however, we might ask where we would place any cells which are in the $G_0$ compartment, *vis à vis* those cells which have irretrievably differentiated, but are still in the basal layer; sure enough, these are both groups of non-proliferating cells, but not quite of the same type. We will see how some authors have calculated the proportion of progenitor cells (i.e. those cells which will migrate) (Marques-Periera and Leblond 1965; Frankfurt 1967; Iversen *et al*. 1968; Blenkinsopp 1969; and see Chapter 7, section 4), in which case the ratio could be regarded as the growth fraction value. However, this is a special case, since these authors start with the premise (based on continuous labelling) that all keratocyte basal cells are at least potentially proliferative, and remain so as long as they abide in the basal layer; however, on this view (but see Potten 1975*b*; and section 8.6), the non-progenitor cells are not in any way 'destined' to migrate, and anyway, their migration is ordered by random mechanisms. Consequently, this definition of growth fraction is largely a statistical argument.

However, from Chapter 7, section 4, it does appear that, notwithstanding the assertions of Potten (1981), there is no substantiative experimental evidence for a classical $G_0$ state in epidermal cells, and the only pointers we have towards this conclusion are certain mathematical models, which incorporate a $G_0$ state, and simulate experimental data in a more or less satisfactory manner (Burns and Tannock 1970; DeMaertelaer and Galand 1977*a*); even so, applying other models (cf. Smith and Martin 1973; Shields and Smith 1977), we can, in fact, do without any $G_0$ compartment.

However, despite the fact that we have argued quite strongly against the notion of a classical $G_0$ compartment in epidermis, there are other assertions regarding proliferating cells in epidermis which warrant attention—these are (a) the concept of cells temporarily arrested in $G_1$ or $G_2$, introduced largely by Gelfant (1963*a,b*); (b) the location of proliferating cells; and (c) the possible presence of a discrete compartment of post-mitotic maturing cells in the basal layer.

## (a) *Arrested cells in $G_1$ and $G_2$*

We have already discussed the general concepts and techniques introduced by Gelfant (see Chapter 2, section 7), in which a general system of proliferative control is advanced, which accounts for such phenomena as aging (Gelfant and Smith 1972) and neoplasia (Gelfant 1977) in terms of increased numbers of resting cells (see Chapter 7, section 9); some of these concepts have also been considered in mechanisms of wound healing (see Chapter 10); Gelfant (1977) has interpreted epidermal cell behaviour in wound healing, with its variable lag or delay period before the onset of DNA synthesis, in terms of a classical $G_0$ phenomenon with a characteristic prereplicative period (see Chapter 24). In Chapter 10, section 4, we give reasons why we consider such a conclusion to be at least premature, if not erroneous. However, Gelfant (1977) has pointed to certain rather isolated reports (Gelfant 1962, 1963*a,b*; Block 1963) which indicate that a small proportion of basal cells in mouse ear or rat epidermis or tongue enter DNA synthesis *immediately* after wounding by cutting, or at least within the first hour thereof; this would indicate that some cells are blocked in late $G_1$, and have the ability to prepare for DNA synthesis remarkably quickly. As yet, the biochemical series of events which presage the appearance of these early labelled cells has not been itemized in the way in which preparation for DNA synthesis in $G_0$ cells has (see Chapter 24); if these claims are substantiated, it does appear as if synthesis of the necessary enzymes and other proteins essential for DNA synthesis initiation can occur very quickly, and that in these cells the control point in late $G_1$ may be much closer to the $G_1$-$S$ boundary than is usually thought (Prescott 1976). After these cells have entered DNA synthesis, the response to injury then follows a biphasic pattern, with the main wave of proliferation coming on a variable time later. The number of cells with this capability is apparently small, and furthermore, most other authors have emphatically stated that cell migration ensues *before* the onset of DNA synthesis in wounded epidermis, and in most reports the delay period is much more tangible (see Table 10.1, p. 374).

Gelfant has also proposed that a small number of basal cells are arrested in $G_2$, the evidence for which comes from two main observations: (1) the appearance of mitoses after a stimulus, cf. wounding (Storey and Leblond 1957; Gelfant 1966, Gelfant and Candelas 1972), high temperature (Gelfant 1975) or adrenaline administration (Evensen 1964) at a time before cells, on theoretical grounds, could have traversed $S$ and $G_2$; and (2) the appearance of unlabelled mitoses in stimulated epidermis after a flash or continuous labelling procedure which was designed to label all cells, and where stimulation is carried out after the labelling procedure. Such mitoses were considered only to have originated from cells, arrested in $G_2$ during the labelling procedure (and therefore unlabelled) being ejected from $G_2$ into mitosis, sometimes as long as 48 hours after wounding (Gelfant 1966).

It is, of course, difficult to argue with the first of these methods; moreover, this early mitotic burst was evidently not prevented by the DNA synthesis inhibitor, hydroxyurea, or by inhibitors of transcription such as actinomycin D, but was decreased by translation inhibitors such as cycloheximide (Gelfant and Candelas 1972); however, with regard to the second proposal, there is accumulating evidence that some basal cells with $S$ phase DNA content do not label with [${}^3$H]-TdR (Clausen $et$ $al$ 1979; Møller and Keiding 1982), and previous studies, in numerous tissues and cell types, have shown that cells at the very beginning and end of the $S$ phase have very low grain counts (see Chapter 3, section 1); moreover, there is always the problem of self-absorption of $\beta$-particles, which can arise both in sections, which in Gelfant's work were as thick as 7 $\mu$m (Rogers 1973), and in smears (Quastler 1963). In addition, there are several experiments, often cited in support of a $G_2$ block (cf. Gelfant 1962, 1963$a,b$) where the epidermis has been labelled serially at 12, 32, and 48 hours after wounding, followed by a five hour collection of metaphases with colchicine; the presence of 'many unlabelled mitoses' is taken as adductive evidence of cells, arrested in $G_2$, which enter mitosis for the first time at 48 hours after wounding. However, $T_C$ in wounded mouse epidermis could well be $< 20$ hours (Hell and Cruikshank 1963), and $t_S$ is certainly shorter than the labelling period used. Thus it is hardly surprising that unlabelled cells resulted. Later, however, experiments using a more appropriate continuous labelling procedure were carried out (Gelfant 1966; Pedersen and Gelfant 1970), but again we might note that Potten (1981) considers that some basal cells may have extremely long cell cycle times, as evidenced by continuous labelling curves which do not reach 100 per cent labelling even after a long period; on the other hand, we should also note that Potten (1974), observing the intense staining of central EPU cells with Feulgen, speculated that these cells, which may constitute the 'stimulus-responsive' cohort (see section 8.2 and Chapter 10), and possibly be stem cells, might be in $G_2$. However, a microdensitometric search for such a compartment was unsuccessful (Sauerborn $et$ $al.$ 1978).

Consequently, there are several reasons for regarding the basis of the second method (cf. (2) above) with some suspicion; how much weight to put on the first series of observations is problematical, since it certainly weighs contrary to most evidence on the kinetic sequence after wound healing (see Table 10.1, p. 374). In conclusion, we might state that, while there might be a number of cells that remain blocked in $G_2$, the number of these is small, and they do need further characterization.

## (b) *The location of proliferating cells in squamous epithelia*

It will be recalled that, in the simple model of squamous epithelium, cell proliferation was confined to the basal layer (see Chapter 4). Many workers have asserted that this is, in fact, the situation as it exists in reality (Marques-Periera and Leblond 1965; Brown and Oliver 1968; Hegazy and Fowler 1973*a*), but we have already remarked that Penneys *et al.* (1970), in serial sections, have shown that some 32 per cent of labelled cells are in suprabasal positions in human epidermis (see Chapter 6, section 1). Indeed, there have been conflicting reports on the specific location of proliferating cells in squamous epithelia; Thuringer (1924; Thuringer and Katzberg 1959) and Hoffman (1947) were in no doubt that the human epidermis showed a mitotic gradient, declining from basal layer to upper Malpighian layer; indeed, Thuringer (1928) considered that maximum mitotic activity was present in the lower third of the Malpighian layer in human prepucial epidermis. Where the oral and oesophageal mucosae are concerned, there are numerous reports which indicate that labelled and mitotic cells are found well above the basal layer (Henry *et al.* 1952; Skougaard 1965*a,b*, 1970; Greulich 1964). Even in the apparently well-investigated epithelium of the hamster cheek pouch, Møller (Møller *et al.* 1979; Møller and Keiding 1982) has reported that 1 per cent of labelled cells are in suprabasal positions, while some 3 per cent of colcemid-arrested metaphases are above the basal layer.

The phenomenon of proliferating suprabasal cells is one which has not attracted much attention in the literature, apart from its occurrence in induced hyperplastic states (Rohrbach *et al.* 1968) and in hyperplastic skin diseases such as psoriasis (Penneys *et al.* 1970; Duffill *et al.* 1977), and in these cases this is probably due to an absolute expansion in the size of the proliferative compartment, with consequent increase in net cell production. Perhaps we should emphasize that these reports of suprabasally sited proliferating cells are confined to non-stacked and therefore, by definition, relatively rapidly proliferating epithelium, as the presence of suprabasal mitoses is largely inconsistent with the EPU model (see section 8.2).

As previously noted, we do not yet know the nature of suprabasal proliferation, whether a cell is ejected from the basal layer towards the end of its cycle in the basal layer, or whether its cycle is both initiated and completed in suprabasal layers. However, anticipating the discussion of section 8.6, we might note that Leblond *et al.* (1964) and Iversen *et al.* (1968) considered that cells in $S$ and $G_2$ were bound to the basement membrane and could not migrate; on this viewpoint, cells in suprabasal layers presumably migrate from the basal layer some time in $G_1$, but DNA synthesis is not repressed as occurs in most cells

(Vaughan and Bernstein 1971; Vaughan *et al.* 1976), and the cycle is completed in the suprabasal layers.

### (c) *Post-mitotic maturing cells*

We have already mentioned the possibility that some basal layer cells have completed their reproductive cycle, and have left the proliferative pool; these can then be regarded as cells which have initiated differentiation while still in the basal layer, have progressed beyond the stage where cell division is still possible, and are awaiting their appropriate time for migration (Potten 1981); this argument is an essential part of the EPU model (section 8.2). In the next section we shall examine the evidence for this concept, as we consider the mechanisms of basal cell migration, and we shall see that it is, in fact, experimentally very difficult to distinguish between a post-mitotic maturing cell compartment with a basal layer residence time of 48 hours (Potten 1975*a*), and a kinetic situation where migration occurs at a fixed cell age, say 48 hours, into $G_1$ (Iversen *et al.* 1968).

## 8.6. The mechanism of basal layer migration

The simplest assumption concerning the mode of migration of basal cells is that it is a random removal process, with each cell on the basal layer having an equal chance of migrating; this has, in fact, been the favoured hypothesis in many analyses of basal cell proliferation.

Probably the main impetus for regarding migration as a random process, and indeed for visualizing the basal layer as a homogeneous population with equal proliferative and regenerative capabilities, came from the work of Leblond and his colleagues (Leblond *et al.* 1964; Greulich 1964; Marques-Periera and Leblond 1965; Leblond *et al.* 1967), worked out largely for the mouse and rat oesophagus.

The progeny of basal cell divisions are especially relevant to this argument, and Fig. 2.17 (p. 60) shows the several possibilities which have already been outlined in our general discussion (see section 2.9). Figure 2.17(i) shows the *asymmetrical or differential* division, originally introduced by Rolshoven (1957) for spermatogonia, which of course is also the statistical argument (Greulich 1964), essential, on the average, for maintaining the steady state size proposal, since one of the progeny of the division will migrate and desquamate, and the other will remain to proliferate (again, on the average); Fig. 2.17(ii) shows symmetrical divisions, called *stochastic* by Hume and Potten (1979), whereby two $P$ or two $Q$ cells are formed at each mitosis, and, with equal numbers of each (Cairnie *et al.* 1965*b*), a steady state situation is still admissible; Fig. 2.17(iiic) indicates that all divisions produce $P$ cells only, but that one or both of these become committed to differentiation, without further mitosis.

Leblond *et al.* (1964) noted that, at the conclusion of the first peak of the

FLM curve for the rat oesophagus, at 12 hours after [³H]-TdR injection, only 3 per cent of suprabasal cells were labelled, and Greulich (1964) also reported that, whereas all flash-labelled cells had divided by 15 hours after injection, continuous labelling studies failed to show labelled suprabasals until 20 hours after the beginning of the labelling procedure, time values which were largely confirmed by Thrasher (1966). Combined with the observation that all basal cells could be labelled by a protracted continuous labelling procedure, and were therefore, by definition, all proliferative (but see Chapter 7, section 4, etc.), this indicated that the progeny of all mitoses contributed daughter cells to the basal layer, and remained, again by definition, proliferative; consequently it became possible to disregard proposals (i) and (ii) in Fig. 2.17 (p. 60), in favour of proposal (iiic), a view shared by Thrasher (1966). It then became possible to postulate that (i) migration of a cell occurs in a random fashion, and is independent of the mitosis from which it arose; cells are therefore ejected from the basal layer by population pressure produced by mitoses in adjacent cells; and (ii), the local microenvironment is such that all cells in the basal layer retain a proliferative habitus, and on leaving the basal layer, conditions change so that proliferative capacity is lost, and differentiation ensues; thus the process of migration itself was seen as inducing differentiation (Greulich 1964).

As to what these implicit changes in the microenvironment are, no-one is very clear; it is unlikely to be merely an oxygen tension or nutrient diffusion effect; as we shall see (Chapter 14, section 1), Bullough and Mitrani (1976) have suggested that a concentration gradient of a mitotic-inhibiting chalone exists across the epidermis, allowing proliferation in the basal layer, but ensuring that suprabasal cells remain post-mitotic. There is, however, no experimental evidence for this viewpoint, as yet. Vaughan and Bernstein (1971) have isolated basal cells and differentiating suprabasals in culture, and reported a hundredfold difference in [³H]-TdR incorporation in the basal cells, compared with the more differentiated cells, and proposed that this was direct evidence that loss of reproductive power was related to changes in the cell itself, rather than in its microenvironment; furthermore, while whole spinous cells do not incorporate [³H]-TdR, isolated nuclei from these cells apparently do (Vaughan *et al.* 1976), indicating that replication is possible but inhibited, in whole cells. If we can be sure that there is no contamination with basal cells, and Vaughan *et al.* went to some pains to insure against this, it would appear that DNA synthesis is repressed by some cytoplasmic constituent, as an active process indigenous to the cell itself; consequently, there may be no need to postulate changes in the microenvironment or in putative inhibitor concentration.

Naturally, these proposals concerning basal cell migration warrant some further clarification; the statement that all (or most) progeny are retained in the basal layer for at least some time after mitosis implies that practically all mitotic axes are orientated in the plane of the basal layer, and indeed Marques-Periera and Leblond (1965) reported that 96 per cent of mitoses had horizontal axes, and 4 per cent were vertically orientated (i.e. $p = 0.96$, and see Chapter 6, section 7 for definition of mitotic axis directions) for the oesophageal epithelium; this, of course, would be consistent with some 3 per cent of labelled cells being in the suprabasal layers at 15–20 hours after flash labelling. Thus, there will, on the basis of the microenvironment model, presumably be a small proportion of either asymmetrical (i.e. $P \to P + Q$) or possibly $P \to Q + Q$ (see below).

How general such an argument will be is debatable; these data are for the rat oesophagus, but in the oesophagus of the mouse, only 39 per cent of mitoses are horizontal (see Chapter 6, section 7; and Duffill *et al.* 1977). Moreover, Smart (1970*b*) has found that, although some 90 per cent of mitotic axes were horizontal in early embryonic mouse oesophagus, with the achievement of adult habitus only 66 per cent remained horizontal, and we have seen how the proportion of axes can vary in different types of stratified squamous epithelium (Table 7.3, p. 279), and also that the proportion may vary during the development of epidermal stratification (Smart 1970*a*). In adult human epidermis, some 54 per cent of mitoses have vertical axes, and so contribute at least one partner directly into the suprabasal layer (Pinkus and Hunter 1966). Consequently, in some epithelia, as many as 60 per cent of divisions may follow the pattern indicated in Fig. 2.17(i) (p. 60), and the hypothesis put forward by Leblond *et al.* (1964) will thus need adjustment. Similarly, on the basis of the EPU model, *all* mitoses must, of perforce, be in the plane of the basal layer, since the model envisages a consecutive series of mitoses with lateral migration of post-mitotic cells; however, Table 7.3 does show that, even in stacked epidermis, some mitoses have vertical axes.

In support of the random migration argument, Marques-Periera and Leblond (1965), by means of serial sections, mapped the distribution of the daughter progeny of labelled cells, at 24 and 48 hours after labelling (Fig. 2.18, p. 61); epidermal sheets can be used to observe the same phenomena, as Karatschai *et al.* (1971) have shown. Marques-Periera and Leblond, with the aid of grain counts, were able to match daughter pairs of 256/305 cells counted. At 24 hours after [$^3$H]-TdR labelling, 69 per cent of pairs were both basal, 5 per cent were both in suprabasal positions, and 26 per cent were mixed (i.e. one basal, one suprabasal); at 48 hours, only 29 per cent of pairs were basal, 29 per cent were suprabasal, and 42 per cent were mixed. These observations are again contrary to the predictions of scheme (i) in Fig. 2.17 (p. 60), since we would expect that $>50$ per cent would be in the suprabasal layers at 24 hours, as all cells would be mixed pairs; in fact, only 26 per cent were mixed pairs, and with only 3 per cent vertical mitoses, on the basis of the preferred explanation (Fig. 2.17(iiic)), this was due to the migration of one member of a labelled pair which was initially in the basal layer, and was ejected at random by mitotically induced population pressure; in a similar manner, in 5 per cent of pairs both daughters had been ejected. On this argument, the two *basal layer* daughters arising at mitosis are equally likely to migrate, and to test this hypothesis Marques-Periera and Leblond compared the distribution of pairs observed experimentally with those that would be expected on a random basis,* and found no difference at either 24 or 48 hours after labelling; this supports the concept that all labelled cells have an equal probability of migrating, with the net result that one or both daughters in a pair can migrate; all basal cells are the same, i.e. are stem cells, removal is random, and caused by population pressure due to adjacent mitoses. One or two *provisos*; cells in mitosis do not migrate (apart from the 3-4 per cent with vertical axes in the oesophagus, but see Table 7.3 (p. 279) for other tissues), and because labelled cells are not seen in suprabasal layers for at least 12 hours after labelling (when even cells at the beginning of $S$ at the time of labelling will have completed DNA synthesis), Leblond *et al.* (1964) concluded that cells in $S$ also did not migrate; consequently, cells migrate randomly in $G_1$.

Since, on this hypothesis, all cells are retained in the basal layer at each mitosis, on average 0.5 will migrate with equal probability, and the mean time spent in the basal layer will be $0.5T_C$; therefore each mitosis gives rise to cells which remain in the basal layer for an average time of $1.5T_C$. Now this will be equal to the turnover time for the basal layer, i.e. $T_{basal} = 1.5T_C$ (Leblond *et al.* 1964), since $T_{basal} = 1/k_B = 1/1/T_C = 1.5/1/T_C = 1.5T_C$; consequently, in the mouse, $T_{basal}$ was 105 hours, and 81 hours in the rat; $T_C$ for the mouse oesophagus is 70 hours, and 54.3 hours in the rat (see Table 7.2, pp. 258-261). Consequently,

---

*This analysis is due to Nadler (see Marques-Periera and Leblond 1965). Consider $N$ pairs of labelled cells ($2N$ cells in all). After, say, 24 hours, a proportion $f$ of the $2N$ cells migrate. If this is a random process, the probability of migration is $f$, and of not migrating is $(1-f)$.

Then, (i) the probability that both daughter cells will migrate $= f \times f = f^2$, or $f^2N$ pairs;

(ii) the probability that the first cell only will migrate $= f(1-f)$, or $f(1-f)N$ pairs;

(iii) the probability that the second cell only migrates $= f(1-f)$, or $f(1-f)N$ pairs.

Therefore the probability of finding mixed pairs should be $2[(1-f)N]$, if migration is random.

(iv) the probability that neither migrates $= (1-f)(1-f)$, or $(1-f)^2N$ pairs; $f$ can be found by observation, and will be equal to

$$f = \frac{\text{(mixed pairs)} + 2\,\text{(both suprabasal pairs)}}{\text{cells counted}}$$

The expected number of pairs in each category can then be compared with those actually found by observation, using a $\chi^2$ test. It should not, however, be construed from this argument that the daughters of a mitosis continue to behave in pairs. Although they will initially have the same age, the transition probability hypothesis predicts that they would soon lose synchrony. We have seen that this desynchronizing effect may not be large, viz. the achievement of a second peak on the FLM curve (Hegazy and Fowler 1973*a*; Potten 1981; and see Fig. 7.2), but Potten (1981) has wondered how the mechanisms controlling proliferation prevent daughter pairs from responding synchronously to any trigger or stimulus; daughter cells do *not* behave as a pair (apart from post-wounding synchrony); the random elements in the transition probability hypothesis would in fact predict this (Bullough and Mitrani 1976). But this should not be seen as proof of the transition probability hypothesis as it applies to squamous epithelium, as implied by Bullough and Mitrani (1976) and Mitrani (1978). If there were no random transition point in the cell cycle, since there is no convincing evidence of mother–daughter inheritance of cycle times in mammalian cells, so long as the cell cycle time was not exactly deterministic, but conformed to say, a gamma distribution, or a log-normal distribution, there would be no reason to predict synchrony.

the average residence time from mitosis to migration is 35 hours in the mouse and 27.1 hours in the rat. A similarly directed study of the mouse squamous forestomach by Frankfurt (1967), using essentially the same methodology, showed that migration of labelled cells began at 32 hours, and 50 per cent of labelled cells had migrated at 48 hours after injection of [³H]-TdR (all mitoses were presumably horizontal); Frankfurt calculated what would be a *median residence time* for migrating cells of about 41 hours, and that 'progenitor: non-progenitor' basal cells (Blenkinsopp 1969 and see Chapter 7, section 4, and Chapter 8, section 5) were present in the ratio 57:43.

Thus this penetrating analysis, due to Leblond, gives us insight, not only into the mode of migration, but also an estimate of the average time taken for a cell to migrate after completing mitosis in the basal layer. However, we are still faced with the problem of deciding whether these cells are ejected at random at some time in the $G_1$ period, or whether they form a discrete post-mitotic maturing compartment, as envisaged by Potten (1981).

Iversen *et al.* (1968), however, came to essentially the same conclusions in a study of hairless mouse dorsal epidermis; after about 12 hours, some 7 per cent of suprabasal cells were labelled, and Iversen *et al.* considered that about 86-88 per cent of mitoses had horizontal axes (see Table 7.3). Continuous labelling data showed a delay of between 24 and 48 hours before significant numbers of differentiating cells were labelled, again indicating that most divisions give rise to two new *basal* cells. Using a computer model, Iversen *et al.* then proceeded to simulate the continuous labelling (multiple [³H]-TdR injections) curves to be expected for both basal and differentiating suprabasal cells ($CL_{basal}$ and $CL_{diff}$ respectively) on the basis of the several possible models (see Fig. 8.22), and compared them with experiment (see Fig. 8.22A). When the simple $P \rightarrow P + Q$ model was used (presumably with all cells having vertical mitotic axes), the $CL_{diff}$ curve climbed higher than the $CL_{basal}$, not conforming to the experimental curve, and evidently inconsistent with the data; the Leblond model of random removal in $G_1$ was next tried with good agreement. However, the best fit to the data was obtained when it was assumed that the *oldest $G_1$ cell next to a mitosis* was the cell which was removed; we should perhaps stress that this argument rests solely, as yet, on computer simulation. Potten and Major (1980) have reapplied the methods of Iversen *et al.* (1968), analysing continuous labelling curves in mouse epidermis on the computer by proposing several different models, and analysing the results by isomorphy or goodness of fit. Assuming random loss, and a $T_C$ of 100 hours, there was a very poor fit to the data, but extending the $T_C$ to 140 hours and confining migration to late $G_1$ (i.e. 50-70 hours into $G_1$), produced a somewhat better fit, confirming the impression of Iversen *et al.* (1968), and shown in Fig. 8.23. Potten and Major then tested their own model on a small subpopulation (10 per cent) of cells giving rise to the basal cell populations discussed in the EPU argument (see section 8.2), with the stem cells being long cycling (192 hours), with the

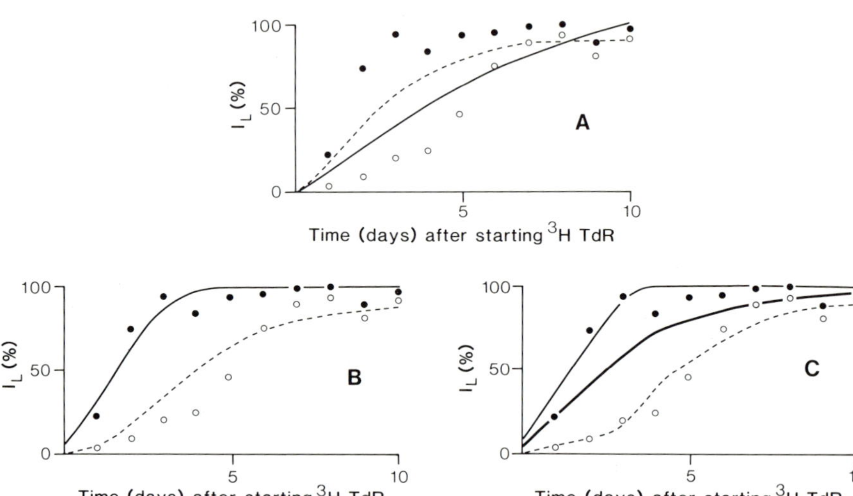

Fig. 8.22.   Simulation of continuous labelling data in the hairless mouse dorsal epidermis. Labelled basal (●) and differentiating (○) cells with time after [³H]-TdR. (A) A simulation using the classical model, i.e. at mitosis one daughter migrates and one remains in the basal layer.   (B) Using a random removal argument.   (C) Assuming that the oldest basal cell nearest a mitosis migrates – this was considered to be the best fit. Total labelled cells are shown by the heavy line, labelled basal cells by the solid line, and labelled differentiating cells by the broken line. (Redrawn from Iversen *et al.* (1968).)

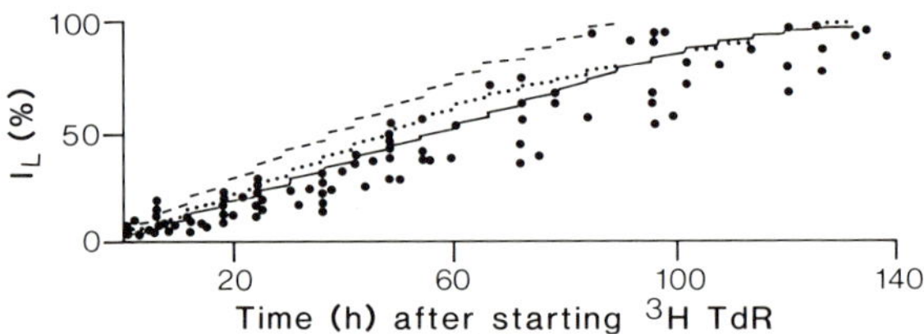

Fig. 8.23.   Theoretical continuous labelling curves in mouse dorsal epidermis. Experimental points are culled from the literature (cf. Fig. 7.5, p. 271);(----) random removal, $T_C$ = 100 hours; (····) loss 50–70 hours into $G_1$, $T_C$ = 140 hours; (——) sequential aging as shown in model 5, Fig. 8.24A, with 10 per cent stem cells ($T_C$ = 192 hours), 18 per cent transit cells ($T_C$ = 144 hours), 18 per cent amplifying cells ($T_C$ = 144 hours), overall $T_C$ = 140 hours, and 41 per cent post-mitotic cells, which gives the best fit. (Redrawn from Potten and Major (1980).)

proliferative differentiating cells more rapidly cycling and producing the post-mitotic maturing cells; the overall $T_C$ was 100–140 hours, and the post-mitotic maturing fraction was 0.4 (see Table 8.2, p. 311). This model fitted the points reasonably well (see Fig. 8.23). However, these results only serve to show that both these models are feasible, but of course do not in any way show which model is applicable; the improvement in fit is very minor.

In this respect, Burns and Tannock (1970) were able to make some calculations with a bearing on the age-dependency of cell loss in the basal layer of the hamster cheek pouch epithelium. It will be recalled, from Chapter 2, section 11, and Chapter 7, section 4, that a $G_0$ phase had been introduced into a model which was proposed for this tissue; with such a model, the possibility exists that cell loss (i.e. migration) occurs only from the $G_0$ compartment (Lajtha *et al.* 1962). If this is so, the cell loss per $G_0$ cell would be greater than the overall cell loss rate, since fewer cells are found in $G_0$. It is possible to draw an age distribution diagram where $G_0$ cells are selectively lost (Burns and Tannock 1970; Steel 1977), and the cell loss per $G_0$ cell is $k_B/1 - k_B R$, where $R$ is the time spent in the $C(G_0)$ phase; where $k_B = 0.51$ per cent/hour and $R = 60$ hours, the cell loss per $G_0$ cell is 0.74 per cent/hour, but for random cell loss, the cell loss per $G_0$ cell is equivalent to the overall cell loss rate, i.e. 0.51 per cent/hour. Brown and Oliver (1968) produced a value of 0.41 per cent/hour for early $G_1$ cells, which Burns and Tannock regarded as more consistent with a random removal process. However, one could argue, semantically, that such evidence supports a random process rather than a migration at a specific age of $G_1$ as proposed by Iversen *et al.* (1968).

Moreover, Hegazy and Fowler (1973*a*), analysing the mouse epidermis by the Brown and Oliver (1968) method (see Chapter 6, section 2), described an initial plateau where $N_B/N_T$ remained at 1.0 for up to 15 hours in both plucked and unplucked epidermis. Now Hegazy and Fowler interpreted this as meaning that mitotic and early $G_1$ cells did not migrate; the presence of a linear fall in $N_B/N_T$ thereafter indicated random removal, however. If we disregard, for a moment, the existence of vertical mitoses (see Chapter 6, section 7, and Chapter 7, section 5), after 15 hours, theoretically $15 \times k_B$ mitoses would have been completed; if $k_B = 0.6$ per cent/hour, then $15 \times k_B$ or about 9 per cent of labelled cells would have divided; if removal was random, then about 4 per cent of labelled cells would have migrated. Whether or not this is beyond the resolving power of the Brown and Oliver method is debatable, but we could well argue that some migrating cells in the suprabasal layers should be seen. In this respect, Hegazy and Fowler (1973*a*) thought that few, if any, mitoses had vertical axes; we could make a similar point about the data of Leblond *et al.* (1964), since they reported that only 3 per cent of labelled cells were in the suprabasal layers at 12 hours after labelling; these are surely accounted for by the 4 per cent of mitoses with vertical axes, but what about the 12 hours worth of cell production which should have been pushing labelled cells, at random, into the suprabasal layers? In this particular investigation, $k_B$ was 1.2 per cent/hour, so there should be about 7 per cent of extra labelled cells in the suprabasal compartment, if removal was random in $G_1$. Thus both sets of data indicate that the daughter cells from horizontal divisions do not migrate in the very early part of $G_1$; so, another caveat to add to the fact that $S + G_2$ cells do not normally migrate, again modifying the random argument. We have already seen that Hegazy and Fowler (1973*a*) could not confirm the expected age distribution experimentally (on the basis of a random removal process), since $\mu$ was calculated at 0.92, whereas values of 0.73 would be required for a true exponential age distribution (Chapter 6, section 7); in Chapter 6, section 7, we ascribed this phenomenon to the presence of at least some vertical mitoses; it is of interest to note that, if cell loss is confined to late $G_1$ (i.e. the oldest $G_1$ cell nearest a mitosis migrates, Iversen *et al.* 1968), then $\mu$ becomes $> 1$; Hegazy and Fowler considered that their data does not conform to such a removal mechanism, and believed that cells migrated randomly after the beginning of $G_1$, a view also embraced by Bullough and Mitrani (1976, 1978).

We should perhaps consider why there should be this particular distribution of mitotic axes; Mitrani (1978; Bullough and Mitrani 1976, 1978) has speculated most interestingly on this topic; confirming that about 10 per cent of mitoses were vertical in resting mouse ear epidermis, the induction of hyperplasia with retinoic acid increased this fraction to some 22 per cent; it was also noted that about 29 per cent of mitoses had vertical axes in the more rapidly proliferating plantar epidermis (see section 8.2), and, in the hair bulb, where cell proliferation is rapid ($T_C \sim 13$ hours, Cattaneo *et al.* 1960), all mitoses were reported to be vertical. Moreover, the relatively rapidly proliferating human epidermis contains considerable numbers of mitoses with vertical axes both in normals (Pinkus and Hunter 1966), and in the frankly hyperproliferative skin disease, psoriasis, the basal layer contains many vertical mitoses (Duffill *et al.* 1977).

Bullough and Mitrani (1978) advanced the general hypothesis that the proportion of vertical mitoses was proportional to the proliferative rate; again the reasons for this were not entirely clear, but they proposed that, with decreasing cycle length, the time available for cells to be in a state where migration was possible would also shorten; cells were held to migrate in an isolated portion of $G_1$ (see above). Consequently, there would be smaller numbers of cells available to migrate, and population pressure would force mitotic axes into a vertical position; when no cells are available to migrate, all mitoses are vertical (cf. the hair bulb); this of course cannot be the whole story, since we will see that, in situations where cell production is increased in hyperplasia, cells can apparently migrate in $S$ and $G_2$ (discussed in Chapter 11), so in hyperplasia the factors, which constrain the migration of cells to some part of $G_1$ in normal circumstances, are alleviated, and cells can migrate in any part of the cell cycle (Iversen *et al.* 1968). Nevertheless, this proposal of Mitrani's warrants further attention.

The most favoured mechanism by which basal cells migrate is that of population pressure caused by basal cell mitoses; Iversen *et al.* (1968), and Bullough and Mitrani (1976, 1978) also subscribed to the population pressure hypothesis, and described racket-shaped cells, presumably in the process of migrating, in sections. These were further described by Christophers (1971*c*), using the FITC technique, as deeply stained 'pedunculated' cells, attached to the basement membrane by a thin stalk, and ballooning upwards; these are presumably cells in the process of leaving the basal layer, and indeed, were located at the stack junctions in epidermis with ordered structure. It is interesting to note that Iversen *et al.* (1968) touched on the post-mitotic maturing cell residence time/ $G_1$ debate by postulating that cells about to migrate (?oldest $G_1$ cell) were physically harder, due to differentiating cell change, and thus easier to remove from the basal layer by pressure. Conversely, however, Bullough and Mitrani (1978) speculated that cells ready to migrate were more 'flaccid' than cells in other phases, allowing the mitotic cell to expand in the direction of the flaccid cell, squeezing it into a columnar shape, and in some way inducing loss of its anchorage to the basal lamina.

In this discussion there has been a tacit assumption that mitotic or division pressure is the motive force behind 'population pressure', and migration. However, Etoh *et al.* (1973, 1977), and Yamaguchi and Tabachnik (1972) have thrown a rather large spanner; [$^3$H]-TdR pre-labelled mouse epidermis was given 3000 rep of $\beta$-rays to inhibit mitosis, and although there was no change in the mean grain count, or indeed, in the labelling index, for three days, in this time 50 per cent of labelled basal cells had migrated. Consequently, migration can occur in the absence of cell division, and is, in fact, an active, radioresistant process, independent of so-called division or population pressure. Moreover, Krawczyk (1972) studied intact and deroofed epidermal suction blisters, and, observing that vinblastine and colchicine had no effect on migration at 12, 18, and 24 hours after wounding (cf. tape stripping, Potten and Allen 1975*b*), concluded that migration was a primary phenomenon, independent of mitosis; we will see that, after epidermal wounding, most reports indicate that migration begins first, and is not caused by mitotic pressure (see Chapter 10).

However, it is rather surprising that little attention has been paid to the biophysical mechanism of basal cell migration. There has, on the other hand, been considerable, and rather loose and speculative, discussion of the 'epidermal grip' on the dermis, based largely on the forces needed to raise suction blisters (Bullough and Mitrani 1976), which indicate that the 'grip' is produced mainly by the viscosity of a fluid interface, although Lenn *et al.* (1974) have postulated the existence of a further weaker adherence, possibly produced by mechanical dermal–epidermal connections, which have been described morphologically (Palade and Farquhar 1965; Campbell and Campbell 1971). Moreover, basal cells are in fact deeply interdigitated into the dermis (Winter 1972; Schmidt *et al.* 1974). However, such considerations are inadequate to explain how a single basal cell is separated from the basal lamina, and migrates, even if in response to a neighbouring mitosis. All we can usefully say is that, depending on the age-dependent migration process that we are prepared to accept, cf. random in $G_1$, late $G_1$, etc., some local, and as yet undefined, change in the region of the basal cell membrane allows individual cell migration (Bullough and Mitrani 1976; Bullough and Deol 1975). In epidermis regenerating after stripping, Potten and Allen (1975*a*) have described a marked reduction in hemidesmosomal contacts, which of course must also occur when basal cells migrate.

It is possible to obtain some idea of how long basal cells spend in the basal layer before migrating; from their analysis in mouse epidermis, Iversen *et al.* (1968) concluded that the mean $T_C$ was 83 hours, and that the mean time between mitosis and migration was 60 hours, giving a $T_{basal}$ of about 140 hours, and we have seen that Leblond, Iversen, and indeed Blenkinsopp (1968*a*) have used such figures to suggest that the 'progenitor' compartment size, i.e. those cells which will divide, is between 58 and 70 per cent, with the remainder being migratory non-progenitor cells (see Chapter 7, section 4). Blenkinsopp (1968*a*) considered 100 basal cells, with $P$ progenitor cells with lifetime $T_C$, and $100-P$ non-progenitor cells with random cell loss in $G_1$, and hence a mean residence time in the basal layer of $t_{G_1}/2$. Since, in steady state conditions, $P$ is constant, on average, after each mitosis, one daughter has a lifetime of $T_C$, and one of $t_{G_1}/2$; in our 100 cells, the proportions are

$$P : (100-P) \; = \; T_C : t_{G_1}/2$$

and
$$2(T_C)/t_{G_1} \; = \; P/100-P)$$

Now
$$P/I_S \; = \; T_C/t_S \qquad (\text{i.e.}\; I_S \; = \; t_S \,.\, I_P/T_C \;\text{approximately})$$

and
$$P/I_{\mathrm{meta}(6)} \; = \; T_C/6 \quad (\text{i.e.}\; 1/k_B \; = \; T_C)$$

where $I_{\mathrm{meta}(6)}$ is the number of metaphases collected after six hours metaphase blockade. These equations are readily solved and give values of about 68 per cent for progenitor cells in epidermis, and about 71–74 per cent in internal squamous epithelia such as oesophagus and tongue.

Thus we have values for the residence times that cells spend in the basal layer after mitosis and before migrating, which are fairly comparable even between different squamous epithelia, albeit in rodent tissues. We have already discussed

Table 8.4. Estimates of the average residence time (hours) spent in the basal layer by basal cells before migration

| | |
|---|---|
| Epidermis (mouse) | |
| Dorsal | 46[1]; 60[2] |
| Ear | 72[1] |
| Tail | 101–115[1] |
| Foot | 48–60[1] |
| Oesophagus | |
| Mouse | 35[3] |
| Rat | 27.1[3] |
| Squamous stomach | |
| Mouse | 41[4] |

1. Potten (1975*a*);  2. Iversen *et al.* (1968);  3. Marques-Periera and Leblond (1965); 4. Frankfurt (1967).

the interesting approach used by Potten (1975) to measure the residence times of post-mitotic maturing cells at several sites in the mouse epidermis. These several values are summarized in Table 8.4.

## 8.7.  Implications for the age distribution inherent in migration models

In the consideration of the age distribution in the basal layer (Chapter 6, section 7), a tacit assumption was that migration was random throughout the cell cycle; for horizontal divisions only, this produced an exponential age distribution.

In the light of our discussions, this will obviously need revamping; early $G_1$, $S$, and $G_2$ cells evidently do not migrate, even though in some epithelia we do not have to compensate for a great number of vertical mitoses (cf. mouse dorsal epidermis). Even so, this would complicate our age distribution, though we could argue that, practically speaking, for epidermis at least, random in $G_1$ is tantamount to random in the whole cell cycle (cf. Hegazy and Fowler 1973$a$; $t_{G_1} \sim 100$ hours, $T_C \sim 110$ hours), but we will shortly add the complication of synchrony through circadian rhythms (Chapter 9). If migration is proved to be non-random, if a $G_0$ compartment exists, or if post-mitotic maturing cells are present, the situation will be even worse with regard to simplicity; if the EPU model obtains, with a proliferative compartment composed of cells of different cycle characteristics, we will have a very complex situation indeed (Potten $et$ $al$. 1982$b$);

## 8.8.  Kinetic organization:  conclusions

It is now evident that any simple model of keratocyte proliferation can be discarded; the EPU argument would appear to indicate that, at least in thin epidermis, the basal layer possesses an identifiable age structure, which is implicated in the maintenance of the ordered arrangement. There is circumstantial evidence that each unit contains a clonogenic cell, the properties of which are debatable, but are becoming defined. However, we do have problems when we come to equilibrate the EPU concept with the organization of non-stacked, more rapidly proliferating epidermis, and even with reconciling some of the observations on stacked epidermis with the EPU concept.

First, is migration random? In non-stacked squamous epithelium such as that of the oesophagus, apart from the now well-recognized caveats that mitotic (for horizontal mitoses, that is), early $G_1$, $S$, and $G_2$ cells do not migrate, there is substantial evidence for a random component to migration (?also in stacked epidermis, Iversen $et$ $al$. 1968; Hegazy and Fowler 1973$a$). However, the EPU argument specifically precludes random migration (cf. Fig. 8.15), and proposes a discrete age structure, in which cells migrate as itemized post-mitotic maturing cells. Where the proposals of Potten (1975$a,b$, 1976) and of Leblond $et$ $al$. (1964) and Blenkinsopp (1969) collide is in the interpretation of the migration pattern; the EPU argument predicts an ordered migration sequence of cells which have finished with proliferation, whereas on the Leblond–Iversen model these cells would have continued proliferating had they not been ejected, more or less at random with respect to age in $G_1$, into the suprabasal layers. It is not really possible to reconcile these propositions, and the concept of Iversen *et al.* (1968) of the oldest $G_1$ cell nearest the mitosis migrating, merely complicates the issue, since, by definition, post-mitotic maturing cells are not in $G_1$.

We consider that, as yet, it is impossible to distinguish between these two possibilities; in fact, the *experimental* evidence favours the random model. However, we have noted that some basal cells stain distinctively with the FITC technique, which may indicate the early stages of keratinization (Christophers 1971*a*,*b*), often with a shape which suggests that they may be leaving the basal

---

Fig. 8.24. (A) Five possible models which explain the organization of cell proliferation in steady state mouse epidermis. Stem cell compartments are denoted by A; dividing transit cells by T, where the suffices 1, 2, and 3 represent successive cell cycles, and P indicates the post-mitotic maturing compartment, from which cells are lost in an exponential manner. (Redrawn from Potten *et al.* (1982*a*).) (B) A comparison of model 5 with experimental curves for changes in (a) the labelling index of the basal layer after a single injection of [³H]-TdR; (*b1*) and (*b2*) with FLM data obtained at 1500 h and 0300 h respectively, and with continuous labelling data (c). The kinetic parameters used in the simulation are shown below – the numbers in parentheses are minimum and maximum values corresponding to the 1500 h and 0300 h data.

|  | $T_C$ | $t_S$ | $t_{G_2}$ | $t_M$ |
|---|---|---|---|---|
| $A$ | 180 | 2.5 (2, 3) | 6 | 3 |
| $T_1$ | 180 | 8  (6, 10) | 3  (4, 2) | 1.5 (2, 1) |
| $T_2$ | 90 | 15  (9, 21) | 2  (3, 1) | 1  (1.5, 0.5) |
| $T_3$ | 90 | 15  (9, 21) | 2  (3, 1) | 1  (1.5, 0.5) |
| Average | 126 | 14.1 (8.5, 19.8) | 2.6 (3.5, 1.8) | 1.3 (1.8, 0.9) |

The half-life in P is 30 hours.

In (a) the minimum (dotted line) and maximum (dashed line) are shown, together with an intermediate situation, because the labelling index is so circadian dependent. Note that, in order to fit the first experimental peaks, the initial labelling indices have to be set at some 1.25 times the experimental values; the initial peak at 20 hours is about twice the initial values while the successively more damped second and third peaks occur at 110 and 200 hours after [³H]-TdR respectively. The sharp reduction in the labelling index between 100 and 200 hours is possible because of the large difference in $t_S$ values chosen for the stem cell and later proliferative compartments.

In (b1) the intermediate curve is shown, while in (b2) the maximum and minimum situations are indicated as for the labelling index curves. The FLM curves have been simulated ignoring the background 'noise', assumed to be between 5–10 per cent. The width of the FLM peaks are mainly determined by the $t_S$ values, while the height of the second and third peaks depends upon the two selected different $T_C$ values and the different $t_S$ values, and the inverse relationship between $t_S$ and $t_M$; $t_{G_2}$ has only a minor influence on the position of the first peak.

In (c) the theoretical curve reaches 70 per cent at 90 hours, with a maximum of 85 per cent at 180 hours (cf. Fig. 7.5). The continuous labelling curve is not markedly affected by the choice of any of the parameters selected (see Fig. 8.23).

The theoretical curves have the same general shape and timing as the experimental data and the fits obtained were considered by Potten *et al.* (1982*a*) to be the most appropriate. (By courtesy of Dr C. S. Potten.)

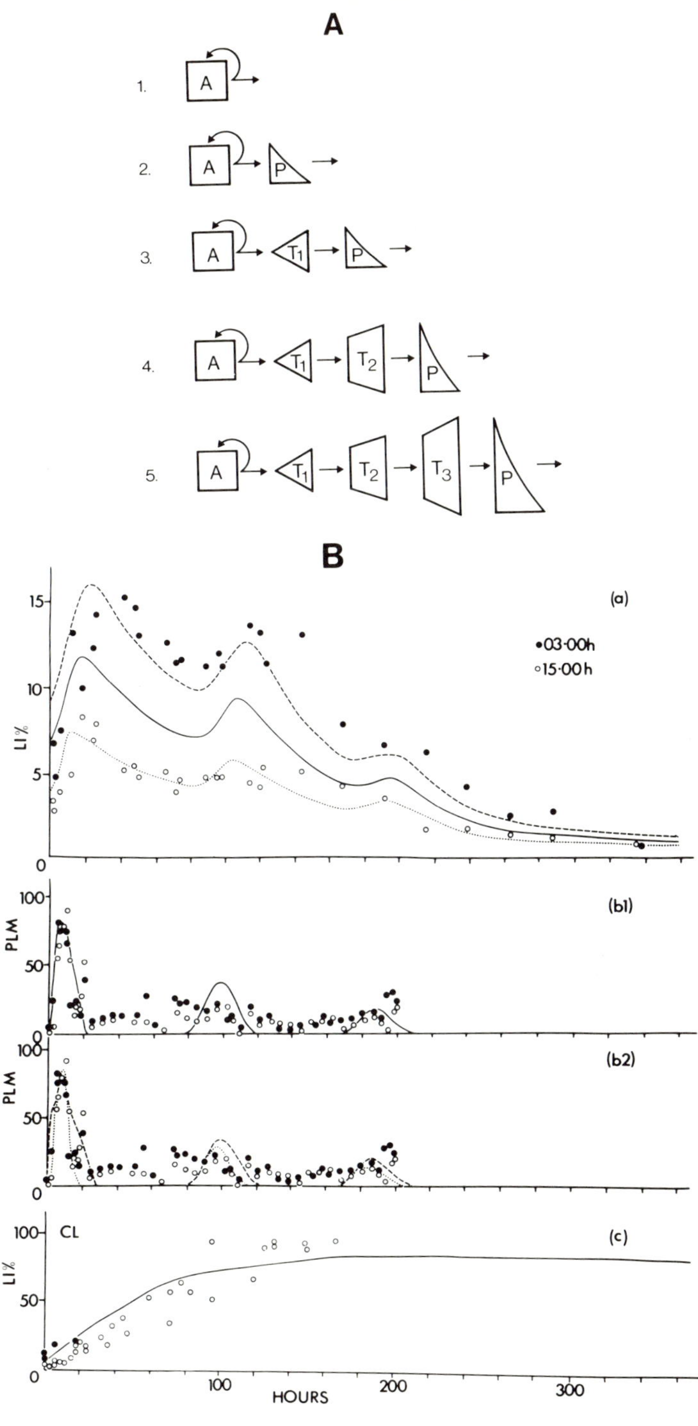
A
1.
A
2.
A
P
3.
A
T1
P
4.
A
T1
T2
P
5.
A
T1
T2
T3
P
B
(a)
15
10
LI%
5
0
03·00h
15·00h
(b1)
100
PLM
50
(b2)
100
PLM
50
0
(c)
100
CL
LI%
50
0
100
200
300
HOURS

layer; they are also seen at the ultrastructural level, and appear to be non-randomly distributed, indeed concentrated at the margins of the columns, in accord with the EPU concept, observations which favour a discrete class of post-mitotic maturing cells (Potten 1981). Moreover, we have seen that mitosis is probably a non-random event when analysed spatially, and that migration may not be dependent upon mitotic activity; moreover, attempts to model continuous labelling data (Iversen *et al.* 1968; Hopper and Brockwell 1978) do support, by isomorphic interpolation, deviation from a random migration sequence; all these factors are circumstantial, adductive evidence for an ordered migration sequence. Recently, Potten *et al.* (1982*b*) analysed protracted FLM curves, continuous labelling curves, and flash [$^3$H]-TdR labelling index chase curves with a series of models by computer simulation (Fig. 8.24 A and B). This analysis was held to support the concept of at least two and probably three (i.e. equivalent to the EPU model) distinct subpopulations of proliferating cells with an exponentially decaying subpopulation of post-mitotic cells with a half-life of about 30 hours; these proliferative subpopulations would have much longer cell cycle times in the stem cell compartment (supported by the differential continuous labelling data of Potten *et al.* 1974), but incongruously, an inverse relationship would be required for $t_S$, with the shortest time in the stem cells. This would be a most unusual kinetic situation, without precedent, and moreover, not supported by experiment. Furthermore, the analysis of these several experiments has not yet included any consideration of circadian variation, which would surely make some difference to the analysis of the various curves. However, on the experimental evidence, rather than on the symmetry of an attractive model, perhaps definitive conclusions are premature.

A further area of conflict is the question of the symmetry of stem cell divisions. Implicit in the EPU model is an asymmetrical stem cell division, whereby one stem cell (position 1) is produced, and a dividing transit cell (position 2, Fig. 8.15) is also produced; such a concept is also supported by the possible selective segregation of DNA (see section 8.4). The experimental evidence on basal cell migration all favours a symmetrical stem cell division with later $P \rightarrow Q$ transition determined by other factors (cf. the microenvironment or genetic programming).

In the 'fixed point' hypothesis advanced by Potten *et al.* (1979; and see Chapter 2, section 9), the local microenvironment effectively decides the asymmetry of the stem cell division, but it is difficult to see how a fixed point, or 'anchorage' is possible in squamous epithelium, even in the context of the EPU. This is presumably essential for the maintenance of polarity in any asymmetrical division; Potten *et al.* (1979) argue that, anyway, an asymmetrical division may not be needed to maintain genetic integrity, which can be achieved by genetic 'housekeeping' during a long $G_0$ state; we have, however, argued strongly that there is no evidence for such a long $G_0$ state in squamous epithelium (see also Chapter 7, section 4 and Chapter 10). However, there may not be more than one stem cell (here stem cell = clonogenic cell) per EPU, by which Potten (1981) denies a stochastic removal to differentiation, in which case the stimulus to stem cell division is removal to differentiation itself, which implies two control mechanisms: one for removal, and one for stem cell division; however, on the fixed point hypothesis, there can only be one control, acting on cell division itself (Potten 1981), and hence the simplest alternative.

We are again in the realm of speculation beyond the data; *ipso facto*, we need more data before firm conclusions on the kinetic organization of the basal layer can be made.

# 9 Circadian variation in stratified squamous epithelia

Hitherto we have largely assumed that basal cell proliferation is synchronous in time, which assumption has coloured our estimates of turnover times based on a single $I_S$ reading (Chapter 6, section 3), our interpretation of FLM curves (Chapter 7, section 2), and our deliberations concerning the age distribution (Chapter 6, section 7). It is an uncomfortable fact (for cell kineticists) that practically all squamous epithelia show pronounced and wide circadian variations in proliferative indices. The phenomenology of this has long been recognized (Cooper and Schiff 1938), and now its kinetic mechanism is in the process of being elucidated (Møller *et al.* 1979), but we are probably as far from explaining its physiological mechanism as ever we were.

## 9.1. Circadian variation in proliferative indices

One problem in analysing the many reports which have dealt with this topic has been the variable conditions in which investigators have kept experimental animals; important variables include the length of the dark period and acclimatization to the dark–light cycle, achievement of a constant temperature, the presence of extraneous noise levels (Møller 1978), and even whether the animals are accustomed to handling during the performance of the appropriate labelling procedure (Møller and Keiding 1982). These variables are rarely constant between investigations, and make comparison difficult; strain, species, and anatomical site variations are superimposed upon these variables.

Nevertheless, Fig. 9.1 shows representative curves for $I_S$ and $I_M$ in the mouse epidermis (Tvermyr 1972; Hegazy and Fowler 1973a; Clausen *et al.* 1979). The major $I_S$ peak is at 1800 to 2200 hours, whereas the $I_M$ peak is at approximately 1000 hours; the nadirs (or troughs) are evident at about 1200 to 1400 hours and 2000 hours, respectively (Pilgrim *et al.* 1963); other, minor troughs are also evident. Other reports show more or less similar timings for mouse epidermis, with peaks of mitotic activity at around midday, and troughs at around 2200 hours (Cooper and Franklin 1940, Bullough and Laurence 1966a) and peaks of cells in the $S$ phase (as detected by both autoradiography and/or flow cytometry) at between 2000 and midnight (Grube *et al.* 1970, Potten *et al.* 1971b; Iversen and Iversen 1976; Clausen *et al.* 1979), and most authors are agreed that labelling and mitotic peaks do not coincide, but differ in their timing, although the extent of this difference varies between reports (see Fig. 9.1).

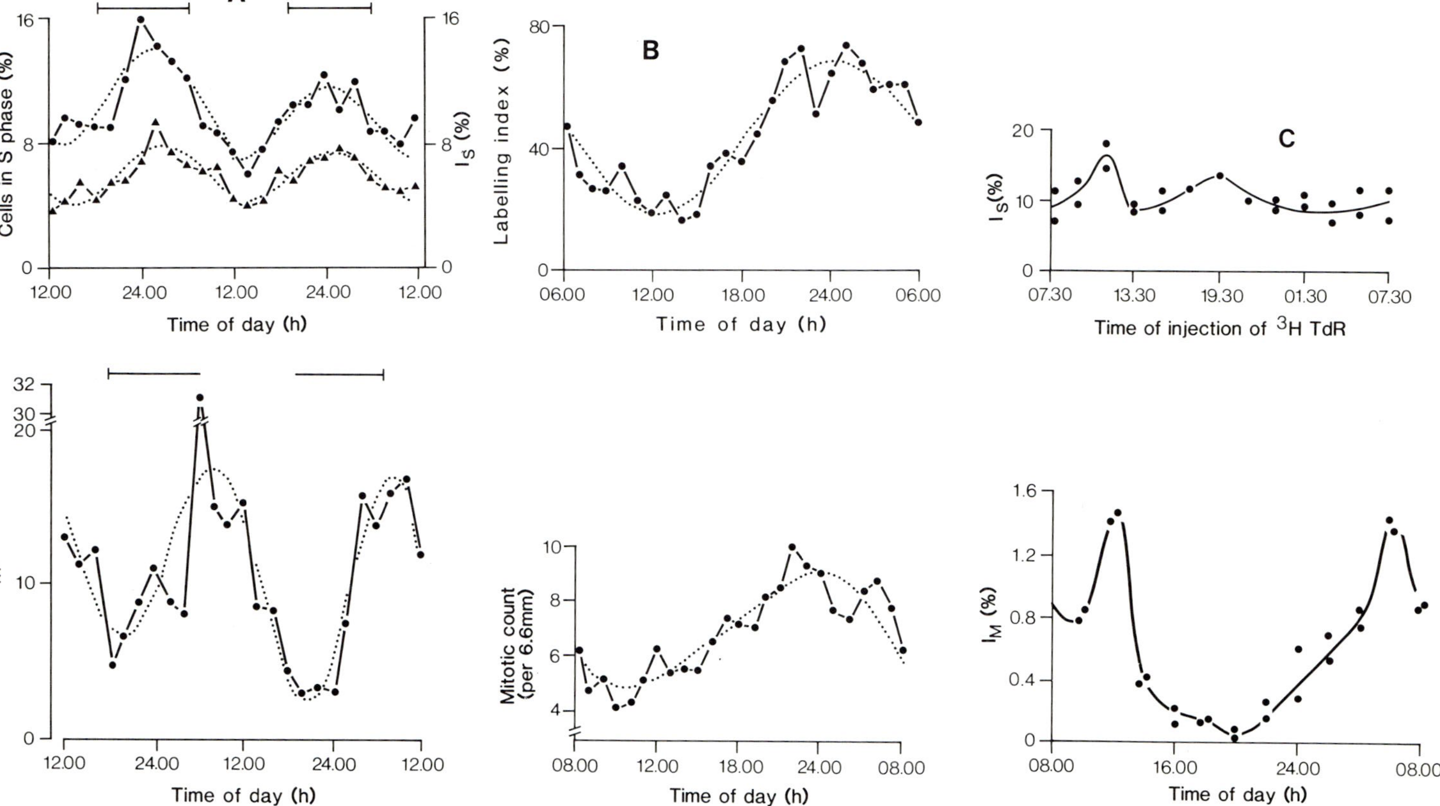



Fig. 9.1.  Changes in labelling and mitotic indices in mouse epidermis with the time of the day: (A) hairless mouse dorsum (Clausen *et al.* 1979). (●——●), cells in *S* phase measured by flow cytometry; (▲——▲), labelling index. (B) Mouse dorsum (Tvermyr 1972). (C) Mouse dorsum (Hegazy and Fowler 1973*a*).

Prominent circadian variation is also found in oral and internal squamous epithelia (England and Burke 1968, 1969, 1971); in mouse tongue and oesophagus Pilgrim (Pilgrim *et al.* 1963; Pilgrim 1967) reported an $I_S$ peak at 1730 hours, with a difference of exactly 12 hours between $I_S$ and $I_M$ peaks; Wolfsberg (1964) and Frankfurt (1967) also noticed $I_S$ peaks occurring in squamous fore-stomach at around 1700 hours. Non-coincident peaks of mitotic activity and [$^3$H]-TdR uptake were also reported for rat tongue by Gasser *et al.* (1972) and for *Ankylostoma* larval epidermis, by Chiakulas and Scheving (1966).

One of the best studied squamous epithelia, from the point of view of circadian variations, is that of the hamster cheek pouch, and the several authors who have studied this tissue have all reported significant circadian variation in proliferative indices, with the $I_S$ and $I_M$ curves being out of phase by about six to eight hours (Gibbs and Casserett 1969; Brown and Berry 1968; Izquierdo and Gibbs 1972, 1974; Møller *et al.* 1979; see Fig. 9.2), although the timing of the individual peaks does vary with report, again probably due to variations in the lighting regime (Møller and Larsen 1978). These fluctuations are also observed when the $S$ and $G_2 + M$ fractions are followed throughout the day by microfluorimetry (Møller and Larsen 1978; and see Fig. 9.2).

However, in an isolated report in corneal squamous epithelium Scheving and Pauly (1967, 1973) have described synchronous peaks of $I_S$ and $I_M$ both occurring at between 2000 and 2300 hours, although the $I_S$ trough tended to occur later, at 1000 to 1200 hours compared with 0600 to 1000 hours for the $I_M$ trough; these times have been confirmed, but only for the mitotic index, by Mantell (1972); corneal epithelium may well be the exception to the general rule of non-coincident peaks of $I_S$ and $I_M$ values throughout the day in squamous epithelium.

Hitherto, it was only possible to measure fluctuations in the number of $S$ and $M$ cells, but flow cytophotometry, when combined with measurements of the mitotic index, allows the fraction of cells in $G_2$ to be followed; in mouse epidermis, Clausen *et al.* (1979) found $G_2$ peaks at 0400 to 0800 hours, with troughs from 1600 to 2000 hours (see Fig. 9.8, p. 359).

In human epidermis, Cooper and Schiff (1938), studying prepucial specimens, showed a marked circadian mitotic rhythm, with a peak at 2245 hours, and a trough at 1030 hours, and similar timings were reported by Scheving (1959); Fisher (1968*b*), in shoulder skin, found a later mitotic peak at 0100 hours. As far as labelling studies go, an *in vivo* examination by Kahn *et al.* (1968) showed only a small peak at around midnight, whereas Pullman *et al.* (1974) could find no evidence of any circadian variation in human epidermis, using a double-labelling method *in vitro*. Singularly, however, Pullman *et al.* did report a circadian rhythm of $I_S$ in lesional psoriatic epidermis, with a peak at 0600 hours; this is against the general feeling that circadian variations tend to be ironed out in hyperplastic states, and indeed an *in vivo* study by Milstein and Cornell (1973), looking at mitoses in psoriatic lesional epidermis, could not confirm a circadian rhythm. A report by Schell *et al.* (1977) points to the difficulties involved in

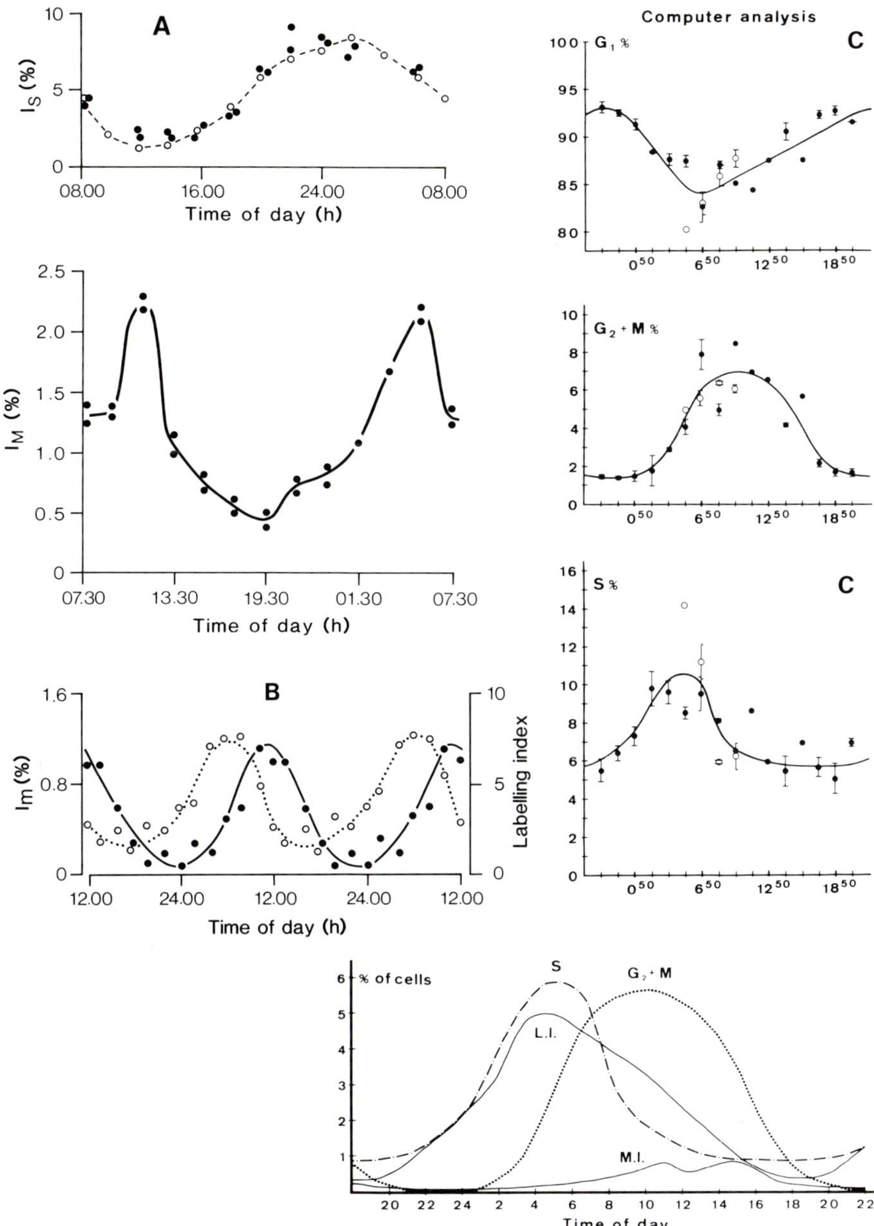



Fig. 9.2. Changes in the labelling and mitotic indices in the stratified squamous epithelium of the hamster cheek pouch: (A) Brown and Berry (1968); (B) Izquierdo and Gibbs (1974) (○ labelling index, ● mitotic index); (C) fractions of $G_1$, $S$, and $G_2$ cells throughout the day measured by computer analysis of DNA histograms derived from flow cytometry, and compared with labelling and mitotic indices in the same tissue. (Courtesy of Dr U. Møller.)

such studies in human epidermis: although variation in $I_S$ was found with time throughout the day, the periodicity of the variation varied with each individual subject, presumably due to the difficulties in the standardization of behaviour patterns; such problems will affect the results when average values between several subjects are sought. Further compounding the error is the observation that irregular seasonal variations are discernible among human subjects (Camplejohn *et al*. 1982).

So, although the results quoted are for various tissues and species, a pattern emerges of significant circadian variation in the fraction of cells in $S$ and $M$, and moreover, the curves describing these variations in $S$ and $M$ fractions appear, in most cases, to be out of phase with each other by a period which also varies from tissue to tissue, but so organized that the $S$ fraction peaks before the $M$ fraction.

## 9.2. The kinetic mechanism of circadian variation in phase fractions

A change in proliferative indices (e.g. $I_S$) can obviously be caused by a change in the mean transit time through $S$ ($t_S$), or a change in the influx into $S$ ($S_{influx}$), or even by a change in the rate in which cells leave the $S$ phase ($S_{efflux}$). There have thus been attempts to explain the phenomena of circadian variation in proliferative indices by primary changes in $t_S$ and $t_M$, in $S_{influx}$ and $S_{efflux}$, and by combinations of these changes.

Perhaps the earliest attempt to provide an analytical explanation for these rather remarkable kinetic phenomena was made by Evensen (1965) and by Kreyberg *et al*. (1965) for mouse epidermis, which involved changes in the duration of mitosis; in this study, $I_M$ peaked at 2400 hours, when $t_M$ was measured at $> 2$ hours, by the metaphase arrest method. At 1400 hours, when $I_M$ was low, $t_M$ was $< 0.5$ hours; $r_M$ remained constant throughout the day, and the conclusion was that the circadian variation in $I_M$ was due solely to changes in $t_M$. Conversely, Bullough and Laurence (1966a) reported an elevated $r_M$ and a short $t_M$ during daylight hours when mice were asleep, but in the dark period $r_M$ fell and $t_M$ lengthened; an inverse relationship between $t_M$ and $r_M$ was also indicated by the work of Tvermyr (1969), who showed that, while the variations in $r_M$ closely followed circadian variations in $I_M$ in mouse epidermis, with a peak at 2400 hours and a trough at 0900 to 1000 hours, $t_M$ itself was longest at 0700 to 0900 hours, and shortest during the dark cycle. However, the significance of these latter findings is uncertain because of the inappropriate application of the metaphase arrest method (Wright and Appleton 1980), and because of our inability to compare $t_M$ values in the absence of quoted interval estimates (see Chapter 3, section 3); moreover, Brown and Berry (1968), and Izquierdo and Gibbs (1972), with a more appropriate application of the method, considered that $r_M$ values mirrored $I_M$ values, and, while there was some circadian variation in $t_M$, this was insufficient to cause changes in the mitotic index; the $r_M$ changes were considered to result from primary variation in the influx into $S$ (see below).

However, the detailed stathmokinetic studies of Thorud *et al.* (1980) clearly demonstrated that circadian variation in $r_M$ does occur in hairless mouse epidermis; support for this viewpoint came from the careful studies of Møller *et al.* (1979), in the hamster cheek pouch, analysed by a useful mathematical model (Hartmann and Møller 1978). The variation in $I_M$ was measured throughout the day, while $M_{influx}$ was measured by *hourly* metaphase arrest experiments with colcemid, allowing $M_{influx}$ values to be quoted at each hour; it was further shown that $M_{efflux}(t)$ was given by

$$M_{efflux}(t) = r_M - \frac{I_M(t+0.5\,\text{h}) - I_M(t-0.5\,\text{h})}{1\,\text{h}}.$$

$t_M$ at any time could then be calculated by incorporating these values into a model which used (a) a 'stationary probability density function' for the transit time, whereby the mean transit time through $M$ remains constant, and (b) a model which allowed the mean transit time to vary throughout the day, what Hartmann and Møller (1978) call the 'non-stationary probability density function model'. The fit produced by the invariate $t_M$ model was poor, but the variable $t_M$ model produced excellent agreement with both the experimental $I_M$ and $M_{influx}$ data (see Fig. 9.3); here $M_{influx}$ changes from 0 to 1.3 per cent/hour over the eleven hours studied; the mean transit time through $M$ varies from 0.3 to 3.3 hours; there was again a tendency for $t_M$ to vary inversely with $r_M$.

Although these values for $t_M$ are not measured but simulated, because of the poor fit of the stationary transit time model, variations in $t_M$ are reasonably to be expected; Møller *et al.* (1979) concluded that, while $M_{influx}$ changes seemed responsible for the $I_M$ variation, it was not possible to say definitively whether $M_{influx}$ or $t_M$ changes were the main causative factor, although this investigation would support the concept of a variable transit time which may contribute to circadian $I_M$ variations.

An important practical point is that $I_M$ varies by the hour; the argument about $r_M$ measurements by the metaphase arrest method when $I_M$ and $t_M$ themselves vary with time is taken up in Chapter 3, section 3, but here we might note that Thorud *et al.* (1980) considered that a four hour metaphase arrest collection was generally sufficient for picking up circadian changes in $r_M$ when $r_M$ was changing slowly, although when very rapid changes occurred, problems were encountered (see Chapter 3, section 3). Using this method, Clausen *et al.* (1979) reported circadian variations in the $r_M$ which showed largely the same periodicity as the $I_M$, with a high rate during the early morning, and a low rate in the evening (see Fig. 9.4); Clausen *et al.* (1979) calculated the influxes and effluxes from each phase from sinusoidal curves fitted to the $I_S$ and $I_M$ data; from these the phase durations are readily calculated, and the changes in $t_M$ are also shown in Fig. 9.4. The results indicate that the duration of mitosis changes markedly throughout the day in the hairless mouse dorsal

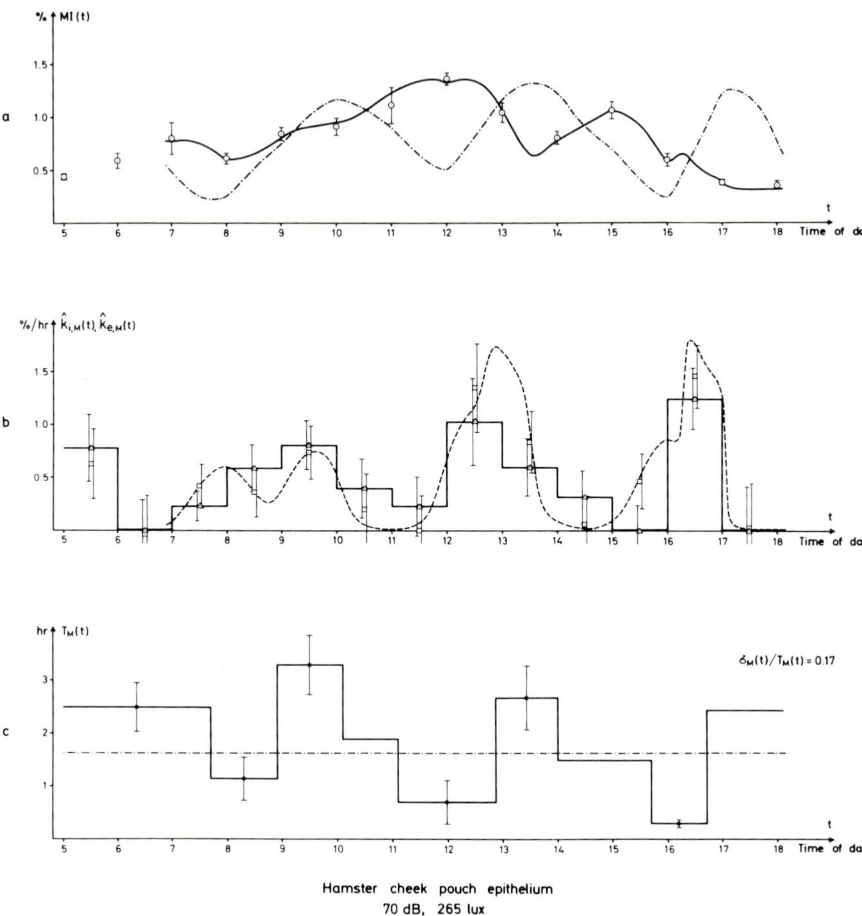



Fig. 9.3. Changes in the (a) mitotic index; (b) in the influx into and efflux from mitosis; and (c) in the mitotic duration in the hamster cheek pouch with time of day. Two theoretical curves are fitted; (———) estimated non-stationary mean transit time; (–·–·–·–) stationary transit time; ⊙ experimental $I_M$; △ experimental influx; ▫ experimental efflux, (– – – –) theoretical efflux. Error bars in a and b are standard errors, and in c standard deviations of the estimated mean transit time. (By courtesy of Dr N. R. Hartmann.)

epidermis, varying between about 0.75 and 2.5 hours; again it is interesting to note that an apparent inverse relationship exists between $r_M$ and $t_M$ as previously reported by Bullough and Laurence (1966a).

A perhaps more popular suggestion has been that the cause of the non-coincident peaks in $I_S$ and $I_M$ was partial synchrony of cells in the $S$ phase (Pilgrim 1967; Scheving and Pauly 1967). Pilgrim (1967), observing that the mean interphase grain count did not vary throughout the day, concluded that there was no variation in mean DNA synthetic rate, and that changes in $t_S$ were

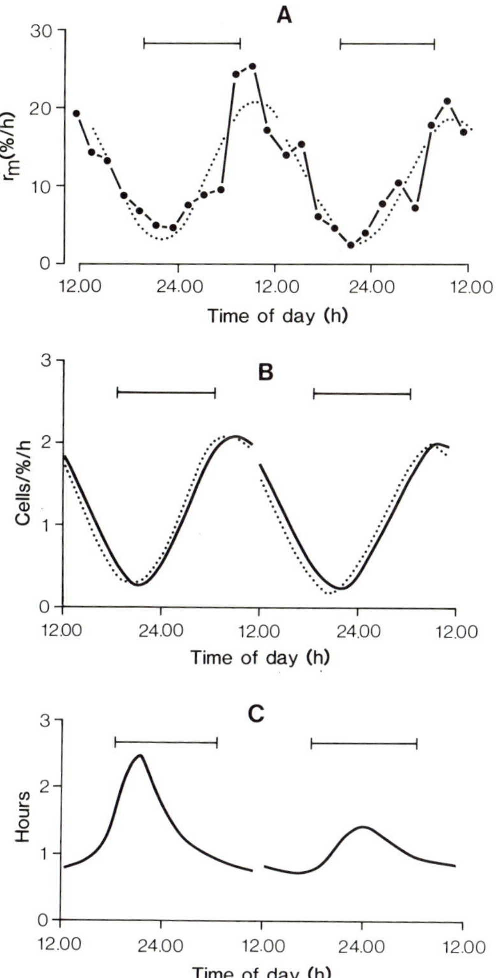

Fig. 9.4.  Changes in the (A) mitotic rate, (B) calculated influx into (····) and efflux from (——) mitosis and (C) the mitotic duration over 48 hours in the hairless mouse dorsal epidermis. The barred lines indicate the dark phase. (Redrawn from Clausen *et al.* (1979).)

therefore unlikely; Pilgrim (1967) proposed that changes in the *influx* into $S$ were responsible for the accumulation of cells in $S$. If $t_S$ remains constant,

$$I_S = \int_{t-t_S}^{t} S_{\text{influx}} \, dt;$$

calculated values of $S_{\text{influx}}$ showed that such a hypothesis was consistent with $I_S$ fluctuations in mouse tongue and oesophagus. On this argument, the accumulation of cells in $S$, induced by $S_{\text{influx}}$ fluctuations, would lead to a rise in $I_M$ in a time related to $t_S + t_2$; for example, Izquierdo and Gibbs (1972) reported that the $I_S$ and $I_M$ peaks in the hamster cheek pouch were six hours apart, and indeed $\frac{1}{2}t_S + t_2 = 6$ hours. Support for this hypothesis came from Brown and Berry (1968): in Fig. 9.2A, if $t_S$, $t_M$ and $t_{G_2}$ remain constant, the fraction of cells in $S$ can be calculated from the $I_M$ curve by subtracting $\{(t_S + t_2) - t_M\} . t_S/t_M$; $t_S$ was known from an FLM curve, and $t_M$ by stathmo-kinetic measurement; assuming that $G_1 - S$ flux is unimpeded, the generated theoretical curve fitted the $I_S$ points very well; Hegazy and Fowler (1973a) also reached much the same conclusion in the mouse epidermis. Changes in the $S_{\text{influx}}$ thus appeared to be an important candidate for the cause of both the $I_S$ changes, and the later $I_M$ changes as the synchronous wave passes into the $M$ phase.

However, we should note that Scheving and Pauly (1967) found a circadian variation in the interphase grain count in rat cornea, which coincided with $I_S$ peaks, and was believed to indicate that, in addition to the accumulation of cells in $S$, there was an associated increase in the rate of DNA synthesis. On the other hand, both Brown and Berry (1968), and Hegazy and Fowler (1973a), were unable to confirm the presence of grain count changes in the hamster cheek pouch epithelium and the mouse dorsal epidermis respectively, so the rate of [³H]-TdR incorporation may not vary in these tissues. More recent studies have favoured grain count changes, with Clausen et al. (1979) reporting circadian variation in interphase grain count in the hairless mouse epidermis, and Møller and Keiding (1982) confirming this in the hamster cheek pouch, in the latter case with increased grain counts from 2200 hours to 1100 hours; although Møller and Keiding attributed this to changes in the DNA synthetic rate, there was also circadian variation in the size of the endogenous TdR pool as indicated by measurements of acid-soluble [³H]-TdR derivatives (see Chapter 8, section 4), in which case interphase grain counts may be more a function of endogenous pool size, rather than DNA synthetic rate (see Chapter 3, section 1). On the other hand, Burns et al. (1976), using metaphase grain count analysis (see Chapter 3, section 1), did show that [³H]-TdR incorporation was maximal at the end of the $S$ phase in the oesophagus, which is dissimilar from most other tissues (cf. Chapter 3, section 1). If, however, pool sizes are so variable, as is indicated by the studies of Møller and Keiding (1982), it could be most unwise (or more unwise than usual) to rely on [³H]-TdR incorporation as the sole criterion of proliferative or $S$ phase activity in squamous epithelium, particularly in studies of circadian variation, whether expressed as activity per 3 mm punch (Potten et al. 1971a), or per μg of DNA (Burns 1981; and see Chapter 3, section 2).

Studies using the double-labelling technique have further tended to confuse the issue (Grube et al. 1970; Tvermyr 1972); Grube et al. (1970), using classical double-labelling techniques with two isotopes, provided important supportive

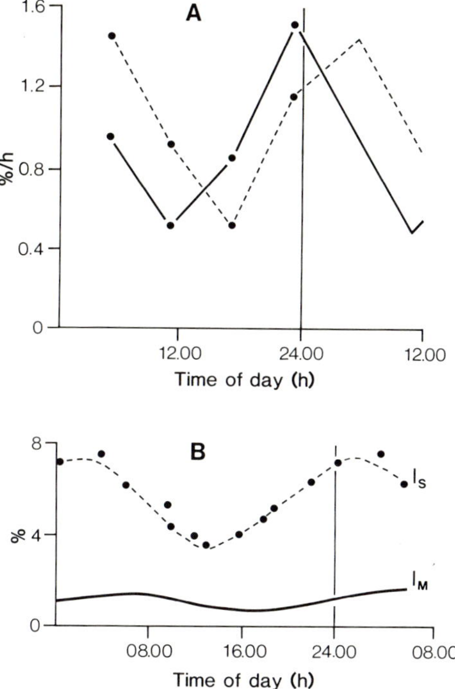

Fig. 9.5. (A) Influx (——) into and efflux from (- - - -) the $S$ phase over a 24 hours period in mouse epidermis, measured by double labelling using [³H]-TdR and [¹⁴C]-TdR. The values given are mean percentage cells/hour. (B) Theoretical $I_S$ and $I_M$ curves (see text), (- - - - -) predicted curves, (●) experimental values. (Redrawn from Grube *et al.* (1970).)

evidence for the $r_S$ fluctuation hypothesis (Fig. 9.5). Tvermyr's (1972) method was to use a double-labelling method with two doses of [³H]-TdR, given three hours apart, and preventing division of cells labelled by the first injection of [³H]-TdR by metaphase arrest with colcemid (see Chapter 3, section 7); this procedure was performed throughout the day in dorsal mouse epidermis (see Fig. 9.6). $S_{\mathrm{influx}}$ per hour was measured in biopsies carried out in the same animal, where $S_{\mathrm{influx}} = (I_{S(2)} - I_{S(1)})/3$; the $S_{\mathrm{efflux}}$ per hour was measured by comparing labelling indices from different animals, i.e. from one animal which received its first injection at time $t + 3$ hours $(I_{S(1)}t + 3)$, and an animal which received its second injection at the same time, $I_{S(2)}t$, having also been labelled three hours previously; this last labelling index will include labelled $G_2$ cells that have completed DNA synthesis during the three hour experimental period, and then $S_{\mathrm{efflux}} = (I_{S(2)}t - I_{S(1)}t + 3)/3$.

The data of Grube *et al.* (1970) show that the $S_{\mathrm{influx}}$ shows exactly similar changes to the labelling index, but anticipating them by about two hours (Fig. 9.5); the changes in the calculated $S_{\mathrm{efflux}}$ also mirrored the $I_S$ changes, but

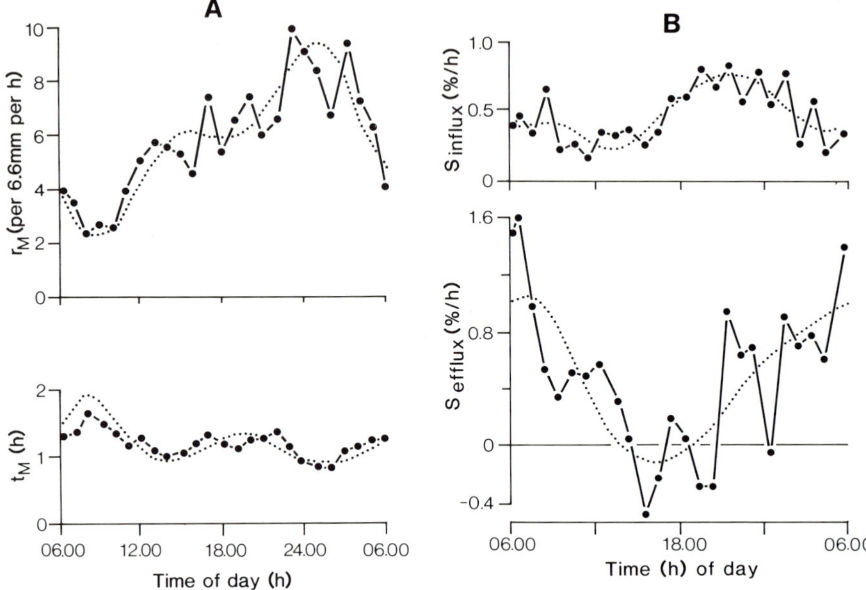

Fig. 9.6.　(A) Changes in mitotic rate and mitotic duration in mouse dorsal epidermis: (——●——) are experimental means and the broken lines are sine-adaptations fitted by computer. In the $t_M$ and $r_M$ curves three hour moving averages are used.　(B) Variation in $S_{influx}$ and $S_{efflux}$ in mouse dorsal epidermis, measured by a double [³H]-TdR labelling method: (——●——) show experimental mean values, and the broken line shows the computer fit. (Redrawn from Tvermyr (1972, 1974).)

about three hours later; consequently $S_{efflux} = S_{influx}$ as it was some 5–7 hours previously, *and $S_{influx}$ is the changing parameter*. Changes in $t_S$ were not considered likely, since a lengthening would maintain the same influx, but decrease the efflux; a theoretical $I_S$ curve calculated from the net changes in $S_{influx}$ gives good agreement with experimental values (Fig. 9.5).

Tvermyr's (1972) results were nowhere so clear cut; Figs 9.1 B and 9.6 show the circadian variation in $I_S$, $S_{influx}$, $S_{efflux}$, $I_M$, and $r_M$. The labelling index showed a peak at 2400 hours and $S_{influx}$ showed similar timings; however, $S_{efflux}$ peaked at 0600 hours, and, during the trough at 1400 hours, negative values were apparent, which were interpreted as zero values, and the negativity ascribed to biological variation. We have however noted (Chapter 8, section 4) that Møller and Keiding (1982) considered that all squamous epithelium may have a relatively long-lived [³H]-TdR pool, and nuclei may not label immediately with [³H]-TdR, even in the $S$ phase (see Chapter 3, section 1); if this is verified, then it could account for the negative $S_{efflux}$ measurements obtained by Tvermyr's (1972) method. Tvermyr (1972) calculated $t_S$ by summing the rates of exit until the $I_S$ at $t$ hours is reached, i.e. by emptying the original $S$ cohort to reach $I_S$; using this method $t_S$ peaked ($\sim$ 10–12 hours in length) at 1000 to 1200 hours, and declined slowly to a trough at 0600 to 0800 hours. Comparison of $S_{efflux}$ and $r_M$ values indicated that $t_{G_2}$ was 1-2 hours in the early morning, and lengthened in the afternoon. Because there was a greater change in $I_S$ than $I_M$ (as a percentage of the mean), Chiakulas and Scheving (1974) also considered that $t_{G_2}$ might vary in *Ankylostoma* epidermis; Clausen *et al.* (1979), in a definitive paper, calculated, from flux considerations, that $t_{G_2}$ varied considerably throughout the day in hairless mouse epidermis, with minimum values approaching two hours between 0800 and 1200 hours, to as long as seven hours at 2000 hours (see Chapter 7, section 3). In Fig. 9.6, note that the lowest $I_S$ values (at 1400 hours), coincide with the $S_{influx}$ low and the $S_{efflux}$ falling, while the highest $I_S$ value coincides with a high $S_{influx}$ and a rising $S_{efflux}$, and Tvermyr thus disagrees with Grube *et al.* (1970) that $S_{efflux}$ is equal to $S_{influx}$ at some time previously.

We consider that a reasonable interpretation to put upon these results is that $S_{influx}$ increases, leading to a rise in $I_S$, this occurring while $S_{efflux}$ and $r_M$ are still low; the later rise in $S_{efflux}$ (as a consequence of the increase in $S_{influx}$) leads directly to changes in $r_M$. Thus the concept of a variable $S_{influx}$ as the prime mover still obtains, although Clausen *et al.* (1979) have reported that, in hairless mouse epidermis, the fluctuations in the influx of cells into the $S$ phase were smaller than the $S_{efflux}$ (see Fig. 9.7), and this could well account for the accumulation of cells in the $S$ phase, with a period of low exit coinciding with unchanged flux into the $S$ phase and consequent synchronous cell cycle traverse through the $G_2$ phase and mitosis. The primary event was seen, by Clausen *et al.* (1979), as a prolongation of $t_S$, resulting in a reduced exit from the $S$ phase. The results of Clausen *et al.* are not therefore really consistent with a primary change in the $S_{influx}$. In this study it was noted that the $S_{efflux}$ reached a peak prior to the peak in $M_{influx}$, and this was considered as adductive evidence that the synchrony was induced near the $S/G_2$ transition.

Further evidence against a primary variation of $S_{influx}$ was claimed by Aarnes *et al.* (1981); formulating a stochastic compartmental model, they introduced periodic variations into the various phase compartments alone or in combination; it proved impossible to fit the experimental data with models assuming variation in $G_1$ alone or in $G_1$ and $S$ (see Fig. 9.8); however, significant improvements were obtained when periodic variations were introduced into the transit time through $S$ and $G_2$, possibly indicating that, in the mouse epidermis at any rate, the origin of the circadian rhythm lies at the $S/G_2$ transition, or within $S$ itself. In this context, it is of some interest to note that Tvermyr (1972) maintained that changes in the mean transit time through $S$ were a contributory factor to the overall rhythm (see also Scheving and Pauly 1967, discussed above). However, there are several problems in measuring $t_S$ from double labelling procedures in squamous epithelium (Gibbs and Cassarett 1969, Møller *et al.* 1974; Olsonn 1976; Møller and Keiding 1982), and we might note that Izquierdo and Gibbs (1972), performing FLM curves at two different times of day (times which

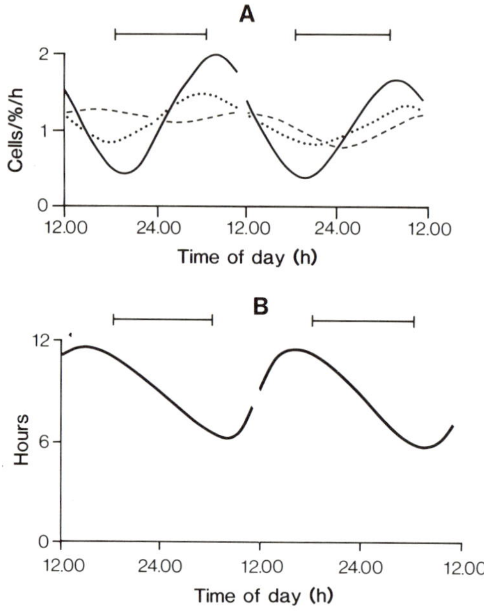

Fig. 9.7. (A) Variation in $S_{\text{influx}}$ calculated from $I_S$ values (····) and $S$ phase DNA content (----), and in $S_{\text{efflux}}$ (——) calculated from sinusoidal curves fitted to experimental data. (B) Calculated values for the variation in $t_S$, from the $S_{\text{influx}}$ and $S_{\text{efflux}}$ data. The barred lines indicate the dark periods. (Redrawn from Clausen *et al.* (1979).)

coincided with the $I_S$ peak and trough), reported no change in $t_S$ or in $t_{G_2}$, and concluded that the circadian rhythm in the hamster cheek pouch was due to synchronous waves of cells entering a relatively constant $t_S$, $t_{G_2}$, and $t_M$, i.e. a deterministic $B$ state (Smith and Martin 1973; and see Chapter 2, section 11) from a variable $G_1$ or $G_0$ state (and possibly mediated via a variable transition probability?). In fact, Izquierdo (1977) considered that some mechanism stimulated fluctuating numbers of $G_1$ cells to migrate suprabasally, and equivalent numbers of $G_0$ cells were stimulated into $S$ and subsequently $M$, giving rise to the respective curves.

Klein and Valleron (1977) also adopted a compartmental approach (see Chapter 2, section 5) to model the data of Izquierdo and Gibbs (1972, 1974), and employed individual transition probabilities for each phase (i.e. $\lambda_S$, $\lambda_{G_2}$, etc.); the model gave a good fit to the $I_S$ and $I_M$ data, and indeed it was found that the results could be simulated with only $\lambda_{G_1}$ varying. Klein and Guiget (1978), using the same model, examined the possibilities of fitting the $I_S$ and $I_M$ data of Brown and Berry (1968; and see Fig. 9.2A), and of Izquierdo and Gibbs (1972, 1974; and see Fig. 9.2B) by variations in the transit times through the various phases; the existence of a long $G_1$ phase in the hamster cheek pouch introduces special difficulties; if there is variation in $t_{G_2}$ or $t_M$ alone, variations

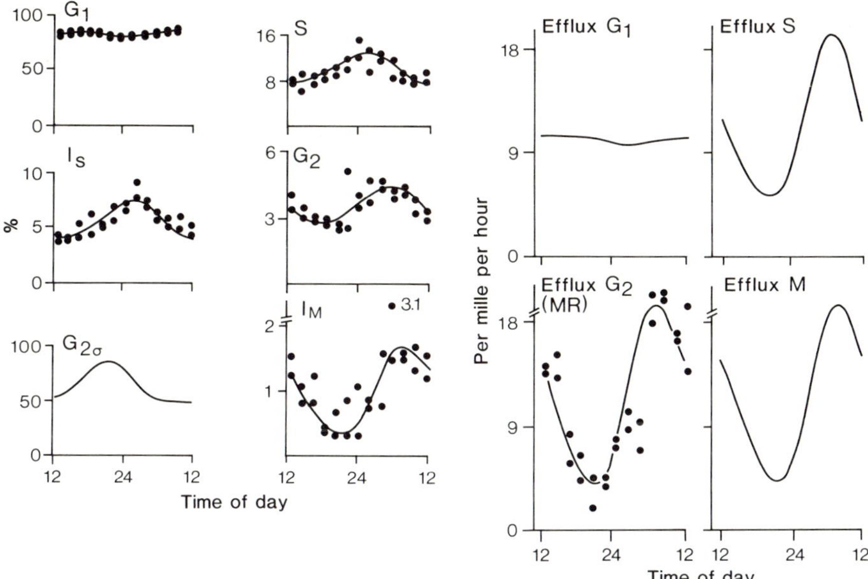

Fig. 9.8. The proportions of cells in $G_1$, $S$, and $G_2$, the labelling and mitotic indices, and the efflux from $G_2$ (mitotic rate) in hairless mouse dorsal epidermis. The data are from Clausen *et al.* (1979) and are fitted by a model making the following assumptions: periodic variations only in the duration of $t_S$; a resting state in $G_2$ ($G_{2\sigma}$). The mean phase durations (with weighted deviations) were: $t_{G_1}$, 38 (0.99) hours; $t_S$, 8.6 hours (0.9); $t_{G_2}$, 1.1 (1.20) hours for cells not in $G_{2\sigma}$; $t_M$, 0.9 hours. The effluxes from $G_1$, $S$, and $M$ are computed from the model. Lines are fitted by least squares. (Redrawn from Aarnes *et al.* (1981).)

in the influx of cells *into* $G_1$ will be damped out through the long $t_{G_1}$, which will result in a nearly constant $S_{\text{influx}}$. Hence, although the $I_M$ curve will be varied, there will be little change in the $I_S$ curve, which, of course, is against all observations. Periodic variation in the $S$ phase transit time in the model caused the labelling index to vary, and, in turn, the mitotic index, but, even though this caused variation in the $G_{1\text{influx}}$, again this periodic influx was found to be damped out in the long $t_{G_1}$, resulting in a constant $S_{\text{influx}}$; this again is not *observed*. However, $G_1$ transit time variations were found to fit the data, and, on these considerations, the origin of the rhythm should be in $G_1$. Moreover, Brockwell (1975), with a random walk model, also fitted the data of Brown and Berry (1968) with purely $G_1$ flux changes. In our opinion, it is difficult to underestimate the importance of the salient observations of Klein and his colleagues in pointing out the critical role of the long $t_{G_1}$ as a modifier of any entrainment introduced after the onset of DNA synthesis; their argument clearly points to the concept of a variable flux from a $G_1$ period of variable length, into a deterministic, or near deterministic, $B$ phase.

On the other hand, a theoretical analysis of the data of Møller *et al.* (1979), of Brown and Berry (1968), and of Izquierdo and Gibbs (1972, 1974) by

Hopper and Brockwell (1978) found it inappropriate to speak of a mean duration of $S$ or of $G_1$ without reference to the time of day under consideration; using a random walk model, Hopper and Brockwell successfully fitted the data of Møller *et al.*, using the assumption of random basal cell migration; cells entering $S$ at 0300 hours average 7.8 hours in passing through $S$, while those entering at 0800 hours average 11.7 hours; $T_C$ was also found to vary with the time of commencement of the cycle in the day: from 130 hours for cells starting the cycle at 1930 hours, to 139 hours for cells beginning at 0430 hours.

Moreover, Hartmann and Møller (1978) applied their successful non-stationary probability density function model to measurements of the $I_S$, $S_{\text{influx}}$ and $S_{\text{efflux}}$, measured using the double [$^3$H]-TdR labelling method, carried out at every 1.5 hours in the epithelium of the hamster cheek pouch. $S_{\text{efflux}}$ was calculated in the same way as the $M_{\text{efflux}}$ (see above), i.e.

$$S_{\text{efflux}} \;=\; S_{\text{influx}}(t) - \frac{I_S(t+1.5\,\text{h}) - I_S(t-1.5\,\text{h})}{3\,\text{h}} .$$

$I_S$ showed marked circadian variation, as does $S_{\text{efflux}}$ (see Fig. 9.9). It is evident that the computer fitted lines to the data are too low for the experimental points; since the computed $I_S$ and $S_{\text{efflux}}$ values are calculated from the experimental $S_{\text{influx}}$, Hartmann and Møller considered that the $S_{\text{influx}}$ measurements were too low. However, the calculations indicated that $t_S$ varied considerably throughout the day, from 4 hours at 2400 hours to about 20 hours from 0600 to 1200 hours. The reasons for the low $S_{\text{influx}}$ values are obscure, but Møller and Keiding (1982) have pointed out some circadian changes in endogenous TdR pools which may modify the $S_{\text{influx}}$ measurements. Although these results do agree with those of Clausen *et al.* (1979) in the sense that the influx amplitude is much lower than the efflux amplitude, Fig. 9.7 shows that, in the data of Clausen *et al.*, the high $S_{\text{efflux}}$ values coincide with a short $t_S$, and low $S_{\text{influx}}$ values with the long $t_S$ values; however, Fig. 9.9 shows no such relationship. Hartmann and Møller (1978) considered that, in order to maintain circadian variation in systems where there are a series of compartments, if at least one compartment has a non-stationary probability density function, this compartment can act as a 'pump' which generates the circadian variation. Thus the $S$ phase may constitute such a pump, although, as we have seen, the mitotic phase also has a non-stationary probability density function (Møller *et al.* 1979), and the question of which one acts as the primary pump then arises. On the other hand, we must remember that any entrainment occurring as early as mitosis will probably not induce much change in the $I_S$ of the following cycle, because of the very long $t_{G_1}$ in this system (Klein and Guiget 1978), and the physiological importance of this putative pumping mechanism is yet to be shown. However important in the mechanism of circadian variation, many reports indicate circadian variation in phase durations: we should also note that Clausen *et al* (1979) reported circadian variations in the

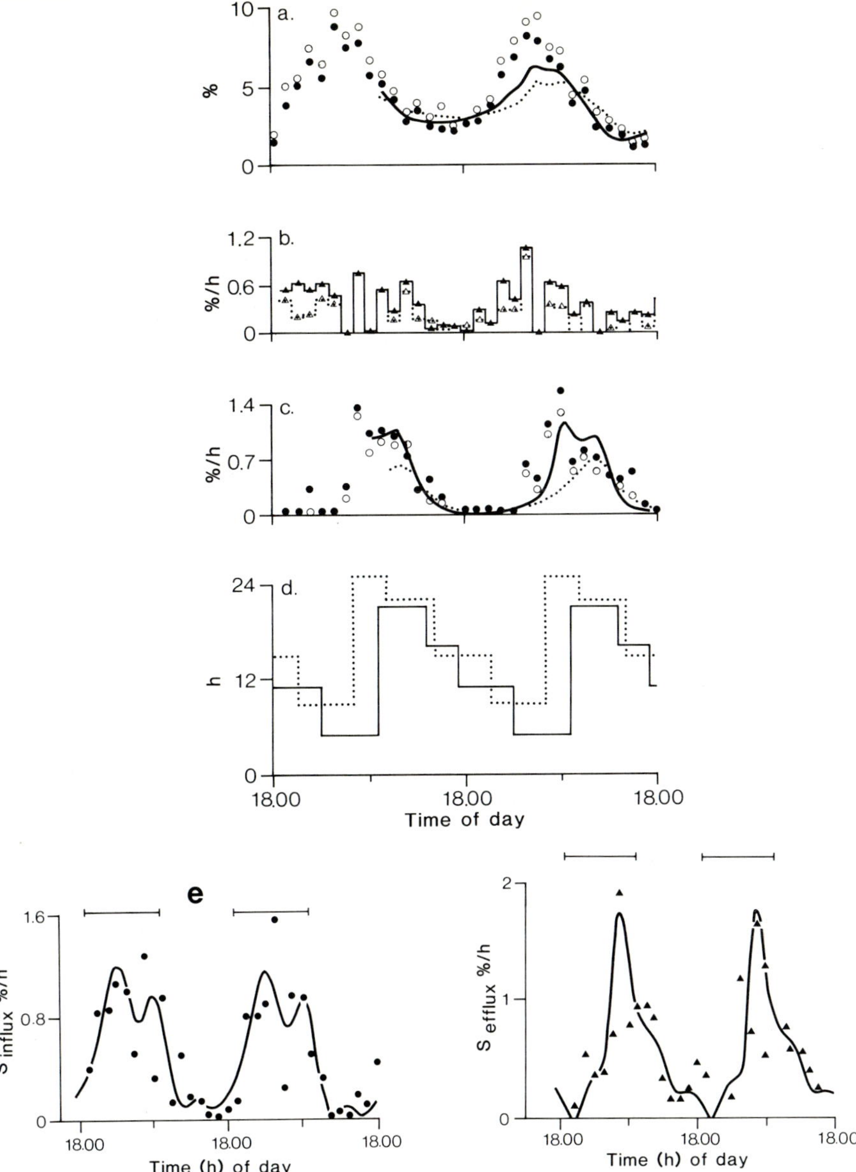

Fig. 9.9. Circadian variation in the $S$ phase of the hamster cheek pouch epithelium: (a) shows the changes in the labelling index. The points indicate the experimental values and the curves are computer generated by the model. (b) Shows the changes in the influx into the $S$ phase, here %/h; the triangles are the experimental values and the curves are model-generated. (c) The efflux from the $S$ phase, also given in %/h. The circles are the experimental data and the curves are from the model. (d) Changes in the duration of the $S$ phase according to the computer model. (Redrawn from Hartmann and Møller (1978).) (e) A later attempt to model the flux data, with values calculated from the double-labelling experiment, where the labelling indices have been corrected for so-called $S_0$ cells in $S$ which do not label. (Redrawn from Møller and Keiding (1982).)

duration of $t_S$, $t_{G_2}$, and $t_M$, calculating the phase durations from phase fluxes, themselves calculated from sinusoidal fits to $I_S$ and $I_M$ measurements.

So where are we? That circadian variation in the form of changes in the fraction of cells in the several phases occurs cannot be doubted; there are also certainly changes in the influx and efflux of both $S$ and $M$, and, by inference, of $G_1$ and $G_2$, and the available evidence would indicate some form of partial synchrony induced upon cells in the $S$ phase; most authors agree that the $S_{\text{influx}}$ is the causative variable (although there are problems in the interpretation of the double-labelling methodology; Møller and Keiding 1982) and that the circadian changes in $S_{\text{influx}}$ are directly reflected in efflux values (cf. Brown and Berry 1968; Grube $et\ al.$ 1970; Izquierdo and Gibbs 1974), although we have seen that Clausen $et\ al.$ (1979) have argued strongly for a primary change in the duration of $S$ itself, causing a reduced $S_{\text{efflux}}$ as the primary event in inducing synchrony; however, there may well be circadian variation in $[^3\text{H}]$-TdR uptake in $S$, as reflected in the interphase grain count, although the available evidence is conflicting on this point. The changes in $r_M$ are usually considered to be secondary to the variable $S_{\text{influx}}$ (Klein and Guiget 1978; but see Møller $et\ al.$ 1979 and Hartmann and Møller 1978), though we should recall the isolated report of Clausen $et\ al$ (1979), who consider that changes in $S_{\text{efflux}}$ are responsible for the synchronization during $S$.

These proposals fit fairly well into the transition probability hypothesis, with periodic variations in the probability of cells passing into a deterministic $B$ phase of $S$, $G_2$, and $M$. Klein and Valleron (1977), and many authors claim to have shown experimentally that $t_S$ and $t_M$ do not show circadian variation, although it must be said that most of these measurements are based on FLM curves, which of course must be suspect in a synchronized system (see below, section 9.3). The reports which maintain that phase durations vary circadianally are based mainly upon double-labelling experiments where the $S$ phase is concerned (Tvermyr 1972), but, even then, not all agree that $t_S$ varies; perhaps the data is not yet good enough. The alternative approach has been to model the system, and derive the values of the various phase durations from other, more easily measured parameters such as the $S_{\text{influx}}$ and the labelling index (Hartmann and Møller 1978; Clausen $et\ al.$ 1979), and we have seen that most of these reports predict considerable circadian changes in $t_S$, $t_{G_2}$, and $t_M$. It is also difficult to decide whether or not $t_M$ varies; again most authors have used the metaphase arrest method to measure $t_M$, and, as previously noted, where $t_M$ is concerned the technique has little intrinsic precision, making interpretation of reported variations in $t_M$ difficult (cf. Bullough and Laurence 1966$a$; and see Chapter 3, section 3). However, the report from Møller and her colleagues (Hartmann and Møller 1978; Møller $et\ al.$ 1979) is singularly persuasive that circadian changes in the mean transit time through DNA synthesis and mitosis do occur, and of course Clausen $et\ al.$ (1979) have made calculations based on the assumption that the changes in $I_S$ and $I_M$ can be described by sinusoidal wave forms, and these results indicate that $t_M$ varies considerably throughout the day. Møller *et al.* (1982) have further developed their model with respect to the double-labelling data; we have already recorded that there may be anomalies in the double-labelling method because of cells in $S$ not labelling, and because of circadianally varying TdR pools. However, Møller *et al.* (1982) simulated both $S_{\text{influx}}$ and $S_{\text{efflux}}$ from the mathematical model, and showed a good fit to the double-labelling data with a model which also introduced a non-stationary probability density function for cells in the $S$ phase (Fig. 9.9e); Clausen *et al.* (1979) also considered $S$ phase variations to be likely.

Møller *et al.* (1974) and Clausen *et al.* (1981) found that $t_{G_2}$ showed only minor circadian variations, as indicated by the FLM method. However, other studies, including double labelling and computer simulation, showed a considerable discrepancy between these and the FLM derived values; FLM values varied from 1.4 to 2.5 hours, whereas computed values, obtained from studies of $G_{2\text{influx}}$ and $G_{2\text{efflux}}$, indicated a variation of 2.5 to 7 hours. Thus Clausen *et al.* (1981) and Aarnes *et al.* (1981) postulated a special state, $G_2\sigma$, where cells rested or passed through $G_2$ very slowly (see Chapter 7, section 3); about 50 per cent of cells are in this phase, and this proportion varies circadianally.

One tacit assumption which has coloured this account throughout has been the implicit suggestion that all cells in the population show equal circadian variation; however, from Potten *et al.* (1977) comes a report that the cells in the 'presumptive' stem cell areas (cf. the cells in the centre of the EPU; cell position 1 in the mouse filiform papilla) show greater amplitudes in their circadian variation than the succeeding proliferative cells; thus it is possible that circadian variation overall could result from a process of 'entrainment' induced by rhythmical cell production rates in the stem cell compartment; why stem cells should have such a strong circadian rhythm is obscure, and further work is needed on both the phenomenology and mechanisms of the effect of entrainment, by a small subpopulation, of circadian rhythms on the remaining proliferative cells.

In conclusion, we might say that a reasonable compromise is to maintain that the most likely kinetic mechanism which accounts for circadian variation in phase populations is the presence of a fluctuating transition probability producing synchronous waves of cells entering $S$, and subsequently, mitosis; this control point is certainly consistent with current theories of cell cycle control in mammalian cells (cf. Chapter 2, section 11). The role of the phase transit times is obscure; there may be circadian variation in $t_S$, and most probably there is in $t_M$, and these changes may contribute to shifts in the phase fraction. At present we know practically nothing about the factors which control phase durations; but that is not, perhaps, a good reason for dismissing such changes as a contributory cause of circadian rhythms.

### 9.3. The influence of circadian rhythms on cell kinetic measurements in squamous epithelia

It is obvious that the induction of partial synchrony transgresses several of the basic assumptions which we employ in cell kinetic methods. These are considered from the general viewpoint in Chapter 3, but here we will make a few points specific to squamous epithelia.

#### (i) *The FLM curve*

In applying the FLM curve we usually assume that all cells are asynchronous with respect to phase in the cell cycle; it is evident that this assumption is inconsistent with the data in squamous epithelia (Møller *et al.* 1974), and Burns, in a series of papers (Burns and Scheving 1973; Burns *et al.* 1975; Burns and Scheving 1975; Burns *et al.* 1976) has explored the results of this conflict. First establishing a strong circadian rhythm in $I_M$ in cornea, oesophagus, and tongue epithelium, with peaks at 1000 hours and troughs at 2100 and 0100 hours, Burns carried out FLM experiments at 0900 hours and 2100 hours in all tissues; in the cornea and the oesophagus, $t_2$ estimates were *doubled* in the 2100 hour curve compared with 0900 hours, while $t_S$ values were *halved*. There are, however, difficulties in accepting this data: for example, only 30 mitoses per point were counted for the corneal epithelium (see Chapter 3, section 4) and the curves were hand-fitted with some highly dubious second peaks in all three tissues. Burns considered that these curves represented possible subpopulations of cells with $T_C$ as short as eight hours in the oesophagus, and on the basis of this short cycle in a fraction of the cells, attempted to explain the singular finding of coincident peaks of $I_S$ and $I_M$ in corneal epithelium. Inspection of this data indicates how easy it is to over-interpret small fluctuations in the FLM curve, and there are evident problems in the estimation of phase durations using the FLM curve in this kinetic situation.

Against this we might set the results of Izquierdo and Gibbs (1972, 1974) in the hamster cheek pouch; initiating FLM curves to coincide with the peak and trough of the $I_S$ values, they were unable to demonstrate any changes in $t_S$ or $t_2$, and it is salutary to note that, whereas the model of Klein and Valleron (1977) fit the data well (cf. Fig. 9.10), the $t_S$ values are the same, although the overall appearance of the curves differ; moreover, the model successfully simulated the $I_S$ and $I_M$ data (Fig. 9.10). In this respect, Møller *et al.* (1974), who initiated five FLM curves at different times of day in the hamster cheek pouch, found no real change in $t_2$, and $t_S$ remained largely constant apart from some shortening at 0500 hours (compare Fig. 9.2). However, we should note that Hopper and Brockwell (1978) compared the median values for $t_S$ from these same curves with the computed $t_S$ values from their random walk model which fitted the circadian $I_S$ and $I_M$ data so successfully; there was little correlation, and Hopper and Brockwell considered that the width of the first wave of

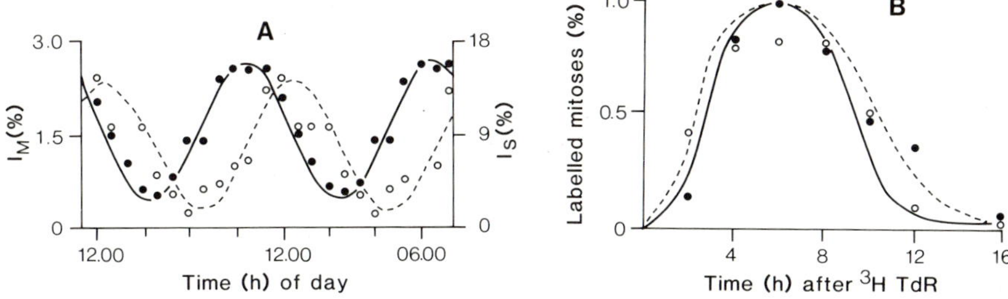

Fig. 9.10. The fit to the $I_S$ and $I_M$ curves and the FLM curves of Izquierdo and Gibbs (1972, 1974) according to the model of Klein and Valleron (1977), which incorporates a time-dependent transition probability in the $G_1$ phase only. In (A), (●) is the $I_S$, (○) the $I_M$, (——) the theoretical fit to the $I_S$ data and (----) to the $I_M$ data. In (B) (●——●) are experimental points and fit at 0800 h, (○---○) experimental points and fit at 2000 h. (Redrawn from Klein and Valleron (1977).)

FLM was a poor indicator of $t_S$ when significant rhythms in $I_S$ and $I_M$ are evident, due to the induced synchronization effect. On the same tack, Clausen *et al.* (1981) compared data culled from FLM curves, initiated at four different times of day, with double labelling, metaphase arrest and flow cytophotometric data, and concluded that the variations in phase duration obtained by the FLM method did not correlate with those derived from the alternative techniques; Clausen *et al*, proposed that the FLM method seriously underestimated the actual $t_{G_2}$, and also that $t_{G_2}$ varied considerably throughout the day; they also considered that circadian variations in $t_S$ cannot be accurately registered by the FLM method. This confirms us in our suspicion that the FLM method is basically non-robust in synchronized systems (see Chapter 3, section 4). In this context, from a practical viewpoint, the questions we would like answered are: does the normal fitting procedure (i.e. an asynchronous steady state model) fit experimental data with circadian variation? And if so, are the generated phase durations close to the real ones? Guiget *et al.* (1978) approached this problem by applying the MacDonald (1970, 1978; and see Chapter 3, section 4) program to four models, where the circadian variation was generated by changes initiated in $G_1$, $S$, $G_2$ or $M$ respectively; the cell cycle time was known, and the FLM curves were simulated at different times of day, namely the maxima and minima of the $I_S$ variation. Although there was no great variation in the $T_C$ values generated, if the circadian variation was initiated in the $S$ phase, errors as large as 25 per cent were apparent if the FLM curve was started at the points of maximum and minimum $I_S$ values; however, interestingly enough, rhythms originating in other phases, cf. $G_1$, $G_2$, and $M$ only gave relatively minor errors. Even so, Guiget *et al.* did conclude that FLM experiments in tissues subject to circadian variation cannot be analysed by standard methods, and specialized techniques should be devised.

## (ii) *The age distribution*

In Chapter 6, section 7, we derived several results for the age distribution in stratified squamous epithelium, based on the assumption of asynchrony; it is evident that these present data will play havoc with our simplistic assumptions, as indicated by Izquierdo and Gibbs (1972, 1974); based on their conclusion that $t_S$, $t_{G_2}$, and $t_M$ did not vary circadianally, and that variable flux from $G_1$ caused the $I_S$ and $I_M$ changes, they constructed age distribution diagrams from the $I_M$ curve. For example, the number of cells entering $S$ (the ordinate on the age distribution curve) should equal the number of cells in $M$ at $(t_S + t_{G_2} + \frac{1}{2}t_M)$, or about 10 hours later, and the ordinate two hours into $S$ would equal the $I_M$ eight hours later.

The resulting age distributions are shown in Fig. 9.11; it can be seen that they are complex; that they are consistent with the $I_S$ data is seen from the fact that integration over $S$ gives $I_S$ values in agreement with the observed circadian curve.

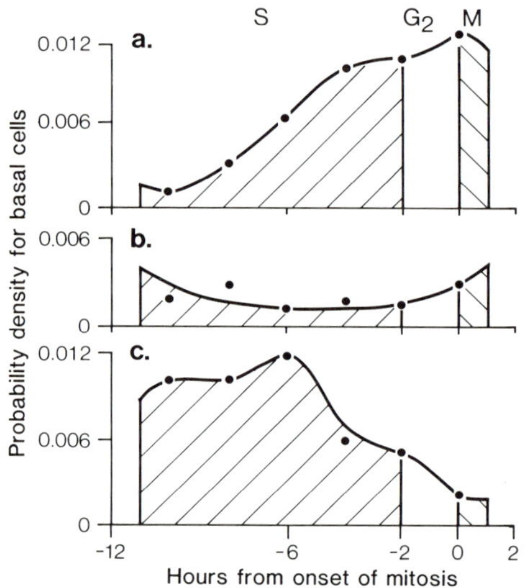


Fig. 9.11. The variation in the age distribution of cells in $S$, $G_2$, and $M$ at (a) 1000 hours; (b) 1800 hours; and (c) 0400 hours, constructed from the $I_M$ curves (see text for explanation). (Redrawn from Izquierdo and Gibbs (1974).)

We may conclude that the age distribution in the basal layer is exceedingly complex, modified as it is by possible migration patterns, mitotic axis directions and the synchronized mode of circadian variation; probably any age distribution diagram that we might draw is a gross oversimplification. In this context we should note that any circadian rhythm in $I_M$ will imply a similar variation in the efflux of cells from the basal layer, which, of course, will be independent of mitotic axis considerations. Theoretically, this should show up on the $N_B/N_T$ experiment of Brown and Oliver (1968) as a step-like decrease in the ratio;

however, perhaps the readings have, as yet, been too far apart to detect this (Hegazy and Fowler 1973*a*).

### (iii) *The influence of [³H]-TdR on the circadian rhythms*

There have been several rather disturbing reports on the putative effects of [³H]-TdR on circadian rhythms, despite the early claim of Blenkinsopp (1967*a*, 1968*b*) that [³H]-TdR had no effect on the $r_M$; in 1969 Gibbs and Cassarett reported that 1.0 $\mu$Ci/g body wt of 6 Ci/mM [³H]-TdR caused an immediate fall in the mitotic index in the hamster cheek pouch epithelium, and although the circadian rhythm recovered, it resumed *in advance* of the control curves; cold TdR had no effect. Similar results were observed by Olsson (1976) in mouse epidermis; 1 $\mu$Ci/g of [³H]-TdR given at 1200 hours depressed $I_M$ at two or three hours later. The $I_M$ rejoined the control circadian curve at 1800 hours, but then overshot with a peak at 1200 hours the next day; cold TdR had no such effect; $r_M$ studies with colcemid showed similar changes, and microdensitometry indicated that $S$ and $G_2$ cells were increased in number at the time of the initial $I_M$ depression; to us the kinetic data indicates a decrease in $G_2$–$M$ flux, although Olsson considered that an inhibition of DNA synthesis was the causative factor.

This mitotic delay could not be confirmed by the detailed studies of Møller (Møller *et al.* 1974; Møller 1978), who reported that the same dose, given at different times, also in the hamster cheek pouch, produced a significant *increase* in $I_M$ within two hours of injection. Counts of labelled mitoses showed that this rise was mostly due to *unlabelled* mitoses, and that the latent period varied with the time of day. The results indicated that [³H]-TdR caused a more or less rapid flux of cells from $G_2$-$M$, possibly caused by a shortening of $t_{G_2}$; Hopper and Brockwell (1978) proposed that, if it is assumed that $G_2$ is shortened by an average of 1.4 hours, it is possible, on the basis of their random walk model, to fit both the $I_M$ and FLM curves described by Møller. The magnitude of the induced peak was found to be independent of the number of $G_2$ cells available (cf. Izquierdo and Gibbs 1972, Clausen *et al.* 1979; and see Figs. 9.4 and 9.8). Cold TdR and saline were inactive, and it does appear singular that unlabelled cells were affected by irradiation from labelled cells.

These reports are of course inimicable, and although they have no deprecatory implications for the flash labelling index, they do caution us (again) not to over-interpret the more time based procedures such as the FLM method; more work is obviously needed on this topic.


## 9.4. The physiological cause of circadian variation in phase fractions

The choice of a physiological mechanism depends, to some extent, on the kinetic mechanism we are prepared to accept (see section 9.2); the presence of rhythmic patterns which vary only slightly in the several studied squamous epithelia argues strongly for some systemic rather than a local control mechanism.

Early investigators, embracing the chalone-adrenaline control of $G_2$-$M$ flux (cf. Chapter 14, section 1), advocated a relationship between blood adrenaline levels and the $I_M$ rhythm (Kreyberg *et al.* 1965; Bullough and Laurence 1966*a*); we shall see that adrenaline is considered a powerful mitotic inhibitor. A parallel was drawn between physical activity (or stress) and adrenaline levels; circadian variation in the blood levels of the stress hormones adrenaline and corticosteroids, both products of the adrenal gland, are well established in both animals and man; adrenaline levels are elevated in the dark phase in rodents (Scheving *et al.* 1979). Kreyberg *et al.* (1965), proposing that all $I_M$ changes were due to variation in $t_M$, showed that intermittent loud noise largely abolished the $I_M$ rhythm, and that $t_M$ remained constant at 1.5 hours. Kreyberg *et al.* (1965) speculated that adrenaline secretion, evoked by physical activity and stress, shortened $t_M$ and thus caused an $I_M$ change, although Møller (1978) reported that intermittent 'white' noise increased the degree of synchrony! However, Bullough and Laurence (1966*a*) found that arousal, and forced exercise, both calculated to increase adrenaline levels, decreased $r_M$ and *lengthened* $t_M$; adrenaline by injection had the same effect, and adrenalectomy, which largely removes the source of endogenous adrenaline, abolished the circadian rhythm; we have discussed elsewhere (Chapter 3, section 3) our reservations about such $t_M$ measurements, but we should note that Fisher (1968*b*) working with human epidermis, reported that physical exercise, in the form of badminton, abolished the circadian rhythm. In view of the imputed relationship between $\beta$-agonists such as adrenaline and the cAMP/cGMP system, it is interesting to note that Marks and Grimm (1972) claimed that the substantial circadian variation in cAMP levels in mouse epidermis was inversely correlated with mitotic activity; this would agree with what was currently held to be the inhibitory role of cAMP in the control of epidermal cell proliferation (but see Chapter 14, section 2). However, Garte and Belman (1980) could not confirm this direct relationship between cAMP and mitotic index, but the levels of cGMP closely followed the mitotic index, and, interestingly enough, the cGMP/cAMP ratio also appeared to correlate with the mitotic index. It is, of course, possible that the cAMP/cGMP system may constitute the final common pathway for the induction of circadian rhythm, and the first messenger could well be, in this case, hormonal.

Scheving and Pauly (1960), observing a mitotic peak at 0400 hours in rat ear epidermis, and a nadir from 0900 to 2300 hours, did not consider an inverse relationship with physical activity appropriate, even though corticosteroids can block cells at the $G_1$-$S$ transition (Frankfurt 1968; Hennings and Elgjo 1971), and there is considerable evidence for entrainment of the partial synchrony by changes in the $G_1$-$S$ flux. These adrenocortical hormones would appear to be good candidates for the cause of the synchrony; moreover, there is a marked circadian rhythm in blood corticosteroid levels in mouse and rat, where peaks are seen in the evening, and troughs in the night and early morning (Halberg *et al.* 1959; Solem 1966). Consequently, it is conceivable that fluctuations in steroid levels would modulate $G_1$-$S$ flux, and cause the partial synchrony, although we could complain that the measurements should be done on the same animal and at the same time. In this context, some interesting data comes from Schell *et al.* (1980), who measured plasma cortisol levels and $I_S$ values in epidermis, throughout the day in normal humans; there is an inverse and time-shifted correlation between $I_S$ and plasma cortisol, which, taken together with the known $G_1/S$ block induced by corticosteroids, is important evidence supporting their role in the induction of circadian variation; a critical study would be to compare this relationship in adrenalectomized patients. However, Scheving and Pauly (1967) and Chiakulas and Scheving (1966) could not confirm that adrenalectomy, or indeed hypophysectomy, abolishes the circadian rhythm; Scheving (Burns and Scheving 1973) reported that isoproterenol (a $\beta$-agonist) advanced the circadian rhythm of mitosis, while Scheving *et al.* (1979) found that epidermal growth factor (see Chapter 14, section 3) stimulated [³H]-TdR incorporation in mouse tongue and oesophagus, an effect which appeared diurnally dependent.

What if we accept the minority view of Clausen *et al.* (1979), that $S$-$G_2$ flux is the primary cause of the induced synchrony? Clausen *et al.* proposed that a rise in $G_2$ chalone (see Chapter 14, section 1) would lead to a decrease in $S_{\text{efflux}}$; however, as we shall see, the demonstrated effect of the $G_2$ chalone, a mitotic inhibitor isolated from basal cells, is to decrease $G_2$-$M$ flux; if this were the operative mechanism which induces the synchrony, there would be a build up of cells in $G_2$, rather than a decrease in $S_{\text{efflux}}$. Although Clausen *et al.* described changes in the number of cells in $G_2$, neither the magnitude of the changes, nor the timing, indicate that this is the primary event; the data do support a modulation in $S_{\text{efflux}}$. This being so, it is unlikely that the $G_2$ chalone is the cause of the synchrony; indeed, it is difficult to suggest a physiological mechanism which could cause such a kinetic event.

Although most workers seem convinced of the critical importance of the lighting schedule (Møller 1978), it is of interest to note that photoreversal apparently has no effect on the circadian rhythm in the frog *Rana pipiens* (Garcia-Arce and Mizell 1972; Morgan and Mizell 1971); pinealectomy appears to abolish the circadian rhythm (a pineal eye?).

The report from Potten *et al.* (1977) that the cells in the 'presumptive' stem cell areas show greater amplitudes in their circadian rhythms than the succeeding proliferative cells could indicate that circadian variation results from a process of 'entrainment' induced by circadianally varying cell production rates in the stem cell compartment.

It is difficult to make anything out of such disjointed and conflicting observations, and we must conclude that the physiological basis of the circadian rhythm, in the cell kinetic field at least, is as yet unknown. In view of our conclusion that the most likely kinetic mechanism is a variable $G_1$-$S$ flux, it is surprising that investigators have not concentrated more on known modulators of this phase, cf. corticosteroids (see Chapter 11), which have a known circadian rhythm, and also affect $S_{\text{influx}}$. Whatever the mechanism, the occurrence of neoplasia in the squamous epithelium of the hamster cheek pouch does not appear to significantly influence the periodicity or amplitude of the circadian variations in $I_S$ and $I_M$, indicating that there is some degree of host physiological modulation persisting through carcinogenesis, and also that such variations are probably induced on a systemic basis (Fukuda *et al.* 1978*b*).

## 9.5. Conclusions

With the advancing study of circadian variation in squamous epithelium, the field appears to get, not simpler, but much more complex; even after quite an exhaustive analysis of the available data, we are only really sure of the phenomenology of the proliferative index shifts; phase duration change data is contradictory, and agreements on kinetic mechanisms or physiological mechanisms does not yet exist, and is a matter for the future. At the lowest level, let us take the existence of such wide variations in proliferative indices as a cautionary tale in our application of such parameters to kinetic calculations, at the very least.

# 10 Cell population kinetics in wound healing

## 10.1. Introduction — experimental models of wound healing

The dynamic events which follow the wounding of skin have intrigued innumerable investigators over many years, and this has naturally led to a large literature; in spite of this, the number of critical kinetic studies in the field is rather small. Three main experimental models have found favour with investigators:

(i) *Removal of the stratum corneum*, produced by repeated applications of adhesive tape (e.g. Scotch Tape, first advocated by Pinkus 1952*a,b*), rubbing with sandpaper (Bertsch *et al*. 1976) or by application of silver nitrate (Block *et al*. 1963). This gives what Christophers (1972*b*) has referred to as a *partial epidermal defect*, and is a powerful stimulus for the induction of basal cell proliferation. The tape is applied repeatedly, usually until a glistening surface is attained, whence the whole stratum corneum has been removed, although the amount of tissue removed can be adjusted by varying the number of applications of the tape (Christophers 1972*b*).

(ii) *Surgical incisions* made by cutting with a scalpel (see, for example, Bullough and Laurence 1957) or scratching with an instrument such as surgical forceps (Christophers 1972*b*). This results in a *deep defect* which may involve the dermis, and the depth of the incision is usually gauged by fitting some sort of guard on the knife (Martinez 1972). Other workers have followed epidermal healing in more open, *shallow wounds*, made by removing a portion of epidermis some 1 cm$^2$ in size (see Winter 1972).

(iii) *Plucking the hair* from hair-bearing mice results in a local depletion of the basal layer, and a prominent ensuing proliferative response (Potten 1971*b*; Hamilton and Potten 1972; Hegazy and Fowler 1973*a*; Potten and Allen 1975*b*).

Other models have more rarely been used; for example Winter (1972) has reported preliminary kinetic data after cutaneous burns in pigs, and Krawczyk (1971, 1972) has studied healing in closed and open suction blisters in mice. Whilst it is probably true to say that the general pattern of epidermal responses in all these models is basically similar, there are some important differences to be observed; Christophers (1972*b*) has convincingly maintained that the behaviour of the epidermis during the healing of both an open wound and after tape stripping basically differs only in that *transverse migration* of epidermal cells occurs in open wounds.

In the kinetic definition of the epidermal wound response, the important aspects which warrant our attention are:

(a) the pattern of cell proliferation;

(b) the mechanisms of cell migration, particularly in relation to cell proliferation;

(c) post-wounding differentiation phenomena, or remodelling (Viziam *et al.* 1961), especially in respect to ordered epidermal structure.


## 10.2. The proliferative response

The pattern of the proliferative response is readily appreciated by study of the proliferative indices, $I_S$ and/or $I_M$ after wounding. Examples of the changes which follow tape stripping, surgical incision, and hair plucking are shown in Fig. 10.1; Fig. 10.1(i) shows the changes in $I_S$ following complete removal of the stratum corneum of hairless mouse epidermis by tape (Clausen and Lindmo, 1976); Fig. 10.1(ii) shows changes in the marginal epidermis following surgical incisions in human epidermis (Rovee *et al.* 1972*a*), and Fig. 10.1(iii) shows the proliferative response occurring in mouse skin after plucking the hair (Hamilton and Potten 1972).

The *morphological changes* after stripping have been described by Mishima and Pinkus (1968) in man, and by Hennings and Elgjo (1970) and Bertsch *et al.* (1976) in the mouse; initially they include cell damage, increase in cell size, hyperaemia, and acute inflammatory changes. In human epidermis a parakeratotic horny layer appears at 24 hours after injury, while a granular layer appears at 48 hours, and an anuclear horny layer is again evident at 72 hours after stripping.

In tape stripping guinea pig epidermis, after a latent period of about 12 hours there are two distinct waves of labelling, occurring at about 35 and 62 hours after injury (Christophers 1972*b*). The waves of DNA synthesis are followed by similar peaks in $I_M$ reflecting the passage of the increased number of $S$ cells into $M$. From our general discussions on the form of $I_S$ curves in conditional renewal systems (see Chapter 24), such a pattern could be due to two populations with differing prereplicative periods (or at least with a differing transit time from a putative $G_0$ or $G_1$ compartment), or to the same population moving synchronously through two consecutive $S$ phases. Although no confirmatory grain count data were proffered, Christophers (1972*b*) favoured the second alternative, inferring from the position of labelled cells in the regenerating epidermis; consequently the cell cycle time would be some 30 hours. But we might note that Clausen and Lindmo (1976) reported several diminishing peaks of proliferative activity at 16, 28, and 38 hours after stripping, which by cytophotometric and grain count analysis were considered to represent waves of decreasingly synchronized cells with a mean $T_C$ of about 12 hours. Similarly, a long series of peaks of proliferative activity were recorded by Schellander (1969), after stripping guinea pig ear, and by Iversen *et al.* (1974*a*) after tape stripping bat web epidermis. Christophers

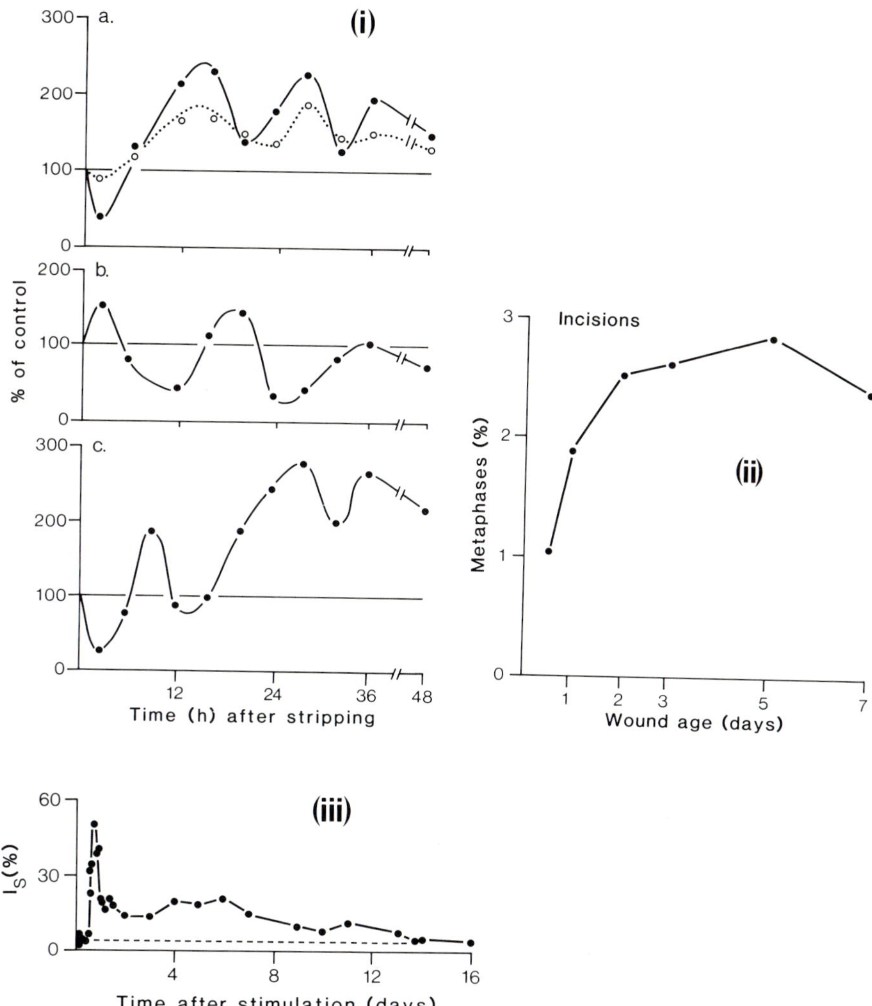

Fig. 10.1. (i) Variations in (a) the labelling index (●), in the number of cells with an $S$ phase DNA content (○), as measured by flow cytometry; (b) in the number of cells in the $G_2$ phase (flow cytometry); and (c) in the colcemid-derived mitotic rate after tape-stripping hairless mouse dorsal epidermis. (Redrawn from Clausen and Lindmo (1976).) (ii) The effect of local incision on the six hour (colcemid cream) metaphase index in human epidermis. (Redrawn from Rovee *et al.* (1972*a*).) (iii) The effect of hair plucking on the labelling index in mouse epidermis. (Redrawn from Hamilton and Potten (1972).)

(1972*b*) noted that the appearance of high $I_S$ values at about 30 hours after stripping coincided with the appearance of keratohyaline granules in the upper epidermis, with possible resumption of tissue specific functions.

Table 10.1. The time relationships of the proliferative response to injury in epidermis. (Modified after Fowler and Denekamp (1976))

| Stimulus | Start $I_S$ | $I_M$ | Peak $I_S$ | $I_M$ | Completion |
|---|---|---|---|---|---|
| **Tape stripping** | | | | | |
| Schellander (1969) (guinea pig) | 24 h | – | 48 h | – | 9 d |
| Hennings and Elgjo (1970) (hairless mouse) | 5–6 h | 8–10 h | 23 h | – | 6 d |
| Potten and Allen (1975*b*) (hairless mouse) | 10–12 h | 18 h | 18 h | 24 h | – |
| Clausen and Lindmo (1976) (hairless mouse) | 8 h | 16 h | 28 h | 36 h | – |
| Pinkus (1952*a,b*) (human) | – | 36 h | – | 72 h | – |
| Williams and Hunter (1957) (human) | – | 24–48 h | – | 48–72 h | – |
| Brophy and Lobitz (1959) (human) | – | 32 h | – | 40–48 h | 72 h |
| Rovee *et al.* (1972*a*) (human) | – | 12–24 h | – | 3–4 d | 10–14 d |
| Bishop and Cox (1974) (human, *in vitro*) | 48 h | – | 96 h | – | 5 d |
| **Incised wounds** | | | | | |
| Block *et al.* (1963) (rat, skin and palate) | – | – | – | 48 h | – |
| Hell and Cruikshank (1963) (guinea pig, *in vitro*) | 4 h | 8 h | 24 h | – | – |
| Giacometti (1967); Giacometti and Montagna (1969) (primate) | – | 36 h | – | 76 h | 92 h |
| Hsu and Hsu (1972) (rat) | 48 h | – | – | – | 6 d |
| Winstanley (1975) (dog) | – | 24 h | – | 6 d | 12 d |
| Epstein and Sullivan (1963) (human) | – | 55 h | – | – | – |
| Rovee *et al.* (1972*a*) (human) | – | 12–72 h | – | 2–5 d | 7 d |
| **Friction** | | | | | |
| Mackenzie (1974*a, b*) (mouse, brush, single application) | – | 42 h | – | 48 h | 60 h |
| Kurman and Argyris (1976) (mouse, abrasion, full thickness) | – | 24–48 h | – | 4–5 d | 14 d |
| Bertsch *et al.* (1976) (mouse, sandpaper) | 8 h | 16 h | 16 h | 24 h | 72 h |
| Argyris (1976) (mouse, abrasion) | – | 3 d | – | 6 d | – |
| **Plucking** | | | | | |
| Hegazy and Fowler (1973*a*) (mouse) | 18 h | – | – | – | – |
| Potten (1971*b*) (mouse) | 10 h | 18 h | – | 20 h | 14 d |

Other methods of removing the stratum corneum, such as rubbing it with sandpaper, or simple stimulation of mouse epidermis by rubbing manually, produce an earlier onset of proliferative activity (Bertsch *et al.* 1976; and see Table 10.1).

Clausen and Lindmo (1976) and Clausen (1979) have produced a most rigorous analysis of the initial events which follow the tape stripping or severe irritation of hairless mouse epidermis. Detailed study of the very early phases showed that $I_S$ had fallen to 40 per cent of control values at two hours after stripping, which was accompanied by a fall in the mean grain count; these changes reflect either a decreased DNA synthetic rate or a partial block of DNA synthesis. In this respect Potten (1971*b*) has also reported decreased incorporation of [$^3$H]-TdR into DNA in the immediate period after wounding. Clausen and Lindmo found that there was an associated increase in the number of $G_2$ cells, as revealed by cytophotometry, and a decrease in $r_M$ as assessed by a four hour colcemid-induced metaphase arrest experiment; similar changes were noted by Hennings and Elgjo (1970), and by Rohrbach *et al.* (1976*c*), also in stripped mouse epidermis. While it is difficult to assess, from their data, how significant the changes are, Clausen and Lindmo (1976) made the reasonable interpretation that, at two hours after stripping, there was a reduction in flux into $M$, while the efflux from $S$ remained unchanged; this resulted in an accumulation of cells in $G_2$, and a fall in $r_M$; moreover, at nine hours after stripping, $r_M$ showed a peak, presumably due to the release of the $G_2$-$M$ flow block (Fig. 10.1(i)).

The proliferative pattern after *incised epidermal wounds* is basically similar, with peaks of proliferative activity occurring after a variable lag phase (see Fig. 10.1(ii); and Epstein and Sullivan 1963; Giacometti and Montagna 1969). In incised wounds of guinea pig ear, the response begins by proliferative activity commencing in cells situated several cell diameters *away* from the wound margin, and cells nearer the margin enter the response as regeneration proceeds (Christophers 1972*b*); in incised wounds of rat palate and abdominal epidermis, Block *et al.* (1963) noted that *labelling* was most prominent at 200-300 cell diameters from the wound margin, up to 48 hours after wounding, but from 72 to 96 hours maximal proliferative activity was evident in the immediately marginal epidermis.

The reasons for this singular distribution of proliferative activity (see Fig. 10.2) have intrigued several authors; with incised or shallow epidermal wounds the initial repopulation or covering is carried out by the migration of initially *non-dividing* cells from the margins (Bullough 1969). In these shallow wounds mitotic and DNA synthetic activity is low or absent in the advancing epidermal tongue, but proliferative rates are high when measured some distance away from the encroaching margin, with a gradient of activity falling over about 1-2 mm from the wound margin (Kobayashi 1976), although in the dog this distance can be larger (Winstanley 1975). Reports on the distribution of *mitotic activity* adjacent to wound margins have differed; Bullough (Bullough and Laurence

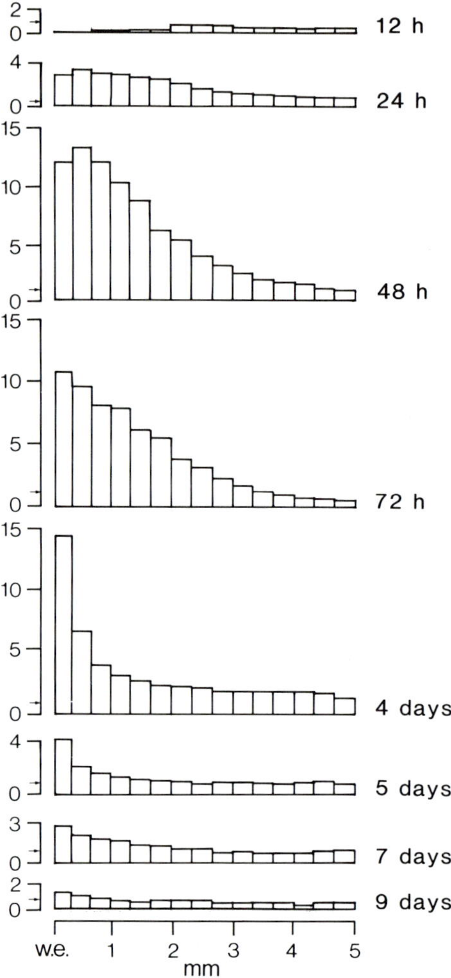



Fig. 10.2. The distribution of mitotic counts in the marginal epidermis next to shallow wounds in pig skin. The arrow indicates normal values. Ordinate: mean number of mitoses per unit length of epidermis (0.33 mm). (Redrawn from Winter (1972).) (w.e. = wound edge.)

1957, 1969) insists that, in the mouse, mitotic activity is maximal at the immediately adjacent wound margin *ab initio*, but this has not met with agreement from other authors; thus Winter (1972), in the pig, maintained that proliferative activity was low at the margins in the early phases of the response, but increased more distally. Proliferative indices in the inner areas increased as the response continued. This view was also propounded by Hell and Cruikshank (1963) in guinea pig ear (explanted after wounding), and by Reeve (1977), in the stratified squamous epithelium of wounded guinea pig tympanic membrane. Reeve observed that, overall, most proliferative activity is found between 35 and 50 cell positions from the wound margin, largely confirming the results of MacMinn and Taylor (1966), also in tympanic membrane.

Thus current opinion favours the view that initially, most cell production occurs at some distance from the wound margin; only later does the process move in to include the innermost wound margin. Figure 10.2 summarizes the results of Winter (1972) in the pig. The reason for this *temporal* distribution of proliferative activity is less clear; Winter (1972) speculated that the commitment of immediately marginal cells to migration could be the cause of the low regional rate of cell division, and in this respect, several authors have commented on the absence of mitoses in migrating cells, and on the apparent mutual exclusivity of migration and proliferation (see MacMinn and Taylor 1966). Thus, Gaylarde and Sarkany (1975), observing no DNA synthesis in migrating epidermal cells during the process of epiboly *in vitro*, noted that hydrocortisone depresses proliferation yet increases migration, while phytohaemagglutinin stimulates mitotic activity but suppresses migration; thus migration and proliferation are probably separate events with different mechanisms. The observations of Etoh *et al.* (1973), who observed migration of [$^3$H]-TdR prelabelled cells, in epidermis where mitotic activity was inhibited by $\beta$-rays, is also relevant in this distinction, while Krawczyk (1972) has reported that neither colchicine nor vinblastine inhibit migration in mouse epidermis healing after the raising of a suction blister. Furthermore, MacMinn and Taylor (1966), in surgically wounded guinea pig tympanic membrane, did not observe migration until 42 hours after wounding, at a time when proliferative activity was already prominent; thus conversely, increased proliferative activity can occur in the absence of migration. While most authors do agree that, in healing epidermis, mitoses are not seen in migrating cells, we should record that Viziam *et al.* (1964), Matoltsky and Viziam (1970) and Winstanley (1975) opine that epithelial cells at the migrating tip divide while moving over the wound surface, and indeed show flash labelling; they consider that only basal cells in the migrating epithelium have proliferative capability, and as migration proceeds, these cells are displaced upwards to begin differentiation. Prolonged treatment with colchicine and busulphan considerably hindered migration, suggesting its dependence upon mitosis. Thus migration, proliferation, and differentiation were seen as contemporaneous processes and mitosis was the cause of migration. But Potten and Allen (1975*b*), studying mouse epidermis after stripping or plucking, report that increased migration from the basal layer occurred *before* enhanced proliferative activity ensued. We may conclude that current opinion favours the independence of mitosis and migration in epidermis; we return to the mechanism of migration in section 10.5.

Once the initial repopulation of incised or shallow wounds is achieved by itinerant keratocytes, proliferation ensues and stratification continues. Christophers (1972*b*) made differential studies of labelling indices in marginal wound epidermis and in the initially repopulating cells (see Fig. 10.3); it is evident that at two days after wounding guinea pig epidermis, the cell production rate in the marginal

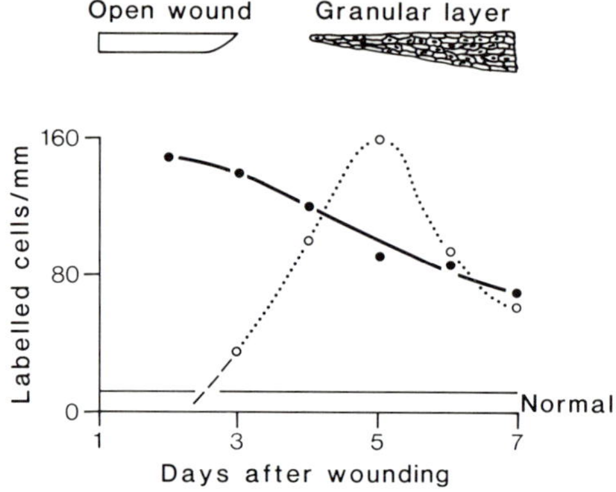


Fig. 10.3. Labelling indices in two portions of a deep wound in guinea pig skin; marginal epidermis (●——●) and regenerating wound floor epidermis (○---○). DNA synthesis in wound floor epidermis is delayed. Wound closure occurred on day 3, but the granular layer only appeared on day 4, both events coinciding with increased DNA synthesis. (Redrawn from Christophers (1972*b*).)

epidermis is high, whereas regeneration in the wound floor keratocytes has only just begun, and peaks at about five days. Note that, as observed in tape-stripped epidermis, peak proliferative activity coincides with the reappearance of the granular layer.

The cells which participate in the proliferative response are both basal and suprabasal cells; thus Argyris (1976) noted a peak of overall mitotic activity at six days after removing the whole of the mouse epidermis by abrasion, whereas the suprabasal mitotic index peaked at three days. Argyris wondered whether such mitoses were initiated in suprabasal layers or merely pushed upwards from the basal layer during mitosis. There is no reason why suprabasal cells should not remain proliferative in a hyperplastic state (e.g. psoriasis) and we should note that Rohrbach *et al*. (1968) claimed that $S$ and $G_2$ cells could migrate in hyperplastic states, so some mitoses in suprabasal cells could have been initiated in the basal layer. However, Pinkus (1952*a,b*), Bullough (1969), Rovee *et al*. (1972*a,b*), and Argyris (1976) are all firmly of the opinion that both basal and suprabasal compartments are proliferatively active, but we should again note the lone dissenting voices of Matoltsky and Viziam (1970), who maintain that proliferation is the province of basal cells only.

Most workers have concentrated on proliferative activity in gauging growth responses to wounding, but Argyris (1976) was able to measure the epidermal wet weight with time after wounding, and reported a recovery to the normal level by four days after removal of the whole of the mouse epidermis by abrasion; this was followed by an overshoot to four times control values by day 5. In this experimental situation, increases in cellularity in the regenerating epidermis appear due to increased numbers of suprabasal cells, while basal cell numbers actually decreased over the regenerative period, only returning to normal levels after 21 days (Argyris 1977).

It is worth noting that the diurnal variation, readily observed in resting epidermis (see Chapter 9), is lost during the epidermal response to injury (Bertalanffy *et al.* 1965; Bullough 1969; Clausen and Lindmo 1976).

In most published proliferative index curves after wounding, there is a distinct lag phase before the onset of increased DNA synthetic activity, which in most publications seems to last about 24–30 hours but is extremely variable in different experimental situations; in rat palatal mucosa and epidermis, for example, the lag phase is as short as one hour, whereas in human epidermis, values as long as 55 hours are reported (Epstein and Sullivan 1963). Thus the response is an evident candidate for induced DNA synthesis in a classical conditional renewal system (but see section 10.4): the situation is, however, complicated by the facts that the evidence for $G_0$ cells in epidermis is scant (see Chapter 7, section 4) and that basal cells are very slowly cycling, but are able to drastically reduce their cell cycle time (see section 10.3; and Hegazy and Fowler 1973*a,b*). Thus we are faced with a familiar semantic problem; is the increased proliferative activity which accompanies epidermal wound healing due to recruitment of resting cells from a $G_0$ compartment, or merely due to an increase in the transition probability of epidermal basal cells? The evidence for these alternatives is assessed below, but here we may note that the lag phase is not immutable (Christophers 1972*b*); if the amount of epidermis removed by the stripping procedure is increased from a few layers to the total removal of the whole stratum corneum, there is a decrease in the $I_S$ measured at 24 hours after stripping. Although this is only a single time point in an ephemeral response, it does point to a longer lag phase produced by increased damage (compare the longer prereplicative period observed after a sub-total hepatectomy with that following partial hepatectomy, see Chapter 25). Moreover, Reaven and Cox (1968) revealed that the lag phase of stripped human epidermis, maintained in organ culture, could be increased from 48 hours in controls to about eight days by lowering the temperature to $31\,°C$, whereas the magnitude of the proliferative response did not change. On the other hand, it is difficult to interfere with the primary proliferative response, once begun; Brophy and Lobitz (1959) stripped human epidermis, followed by a restripping at different times after the initial stimulus, and concluded that the secondary response was essentially the same, and that once initiated, the response to the first stimulus would take precedence and be completed, despite attempts to divert it.

Comparatively little attention has been paid to the factors which control the magnitude of the proliferative response, unlike the situation in, for example, induced DNA synthesis in the prostate, where several of these factors are now recognized (see Chapter 31). However, one modifying influence could be the site of the wound; Kobayashi (1976) has claimed that larger responses are evident in the head region of the mouse, with a declining gradient towards the tail. We should note here that only one time point (24 hours after wounding) was studied, and naturally, this is of no use whatsoever when monitoring an ephemeral kinetic response; consequently this needs repeating.

The time relationships of the response in human epidermis appear to be somewhat longer, particularly the lag phase. Brophy and Lobitz (1959) showed a lag phase of 32 hours (for mitoses) after stripping human epidermis, with a peak at about 40–48 hours, whereas Bishop and Cox (1974), organ culturing human skin after stripping, observed a lag phase of 48 hours, when an increase in $I_S$ occurred with a peak following at 96 hours. These latter observations may of course reflect the artefacts due to explantation, but interestingly enough, Bullough and Laurence (1957) showed no difference in the time relationships of mitotic activity after incising mouse ear, whether the response was followed *in vivo* or *in vitro* after explantation, although higher indices were seen *in vitro*. On the other hand, Pinkus (1952*a,b*) observed a lag phase of 72 hours, and Epstein and Sullivan (1963) and Sullivan and Epstein (1963), pooling data obtained from razor wounds in human leg skin, reported a mean lag phase of $55 \pm 13$ hours (SE). In this respect, Rovee *et al.* (1972*a*) produced evidence that the length of the lag phase and timing of the response in humans varies with the method of stimulation; with linear incisions the lag phase varied in duration between 12 and 72 hours in different individuals, with a variable peak of proliferative activity, whereas with stripping, the lag phase was only 12–24 hours: it is difficult to generalize. Argyris (1976, 1977), has made the important point that there is considerable overshoot in regenerating epidermal mass during the wounding response; in these experiments cellularity could be equated with cell mass. After the induced hyperplasia, remodelling occurred over 21 days in the model Argyris used, which was abrasive removal of the whole epidermis. Argyris speculated that this was due to a feedback overshoot, and noting that overshooting was absent in liver, questioned whether it was a characteristic of renewal systems only (cf. crypt and villus response to cytosine arabinoside, see Chapter 23).

There have been a few reports on the effects of frictional injury on the rate of epidermal cell proliferation (Mackenzie 1974*a,b*); a single application of a motor-driven brush to mouse ear gave no apparent histological change, but a wave of mitotic activity followed which peaked at 48 hours after stimulation. Repeated application of the brush for periods up to 35 days gave thickening in both the stratum corneum and the spinous layers, and a more sustained proliferative response, which Mackenzie in fact interpreted as the emergence of a new steady state, with a 62 per cent decrease in epidermal transit time, despite the increase in epidermal thickness.

The time relationships of the main proliferative responses are summarized in Table 10.1. A general pattern emerges of a variably timed burst of proliferation following a distinct lag phase, and eventual curtailment of the proliferation when normal proliferative rates are again realized.

Having given the descriptive background, it becomes necessary to approach the phenomenon analytically.

## 10.3. Kinetic parameters in epidermal wound healing

Measurement of kinetic parameters during the ephemeral response to wounding is naturally a difficult procedure, since most kinetic procedures are time-based, and hence a rapidly changing situation could lead to confusing results (see Chapter 3). However, in situations where there are two consecutive peaks of labelling or mitotic activity, where synchronous waves of $S$ or $M$ cells are occurring in the same population (Clausen and Lindmo 1976), then a value for $T_C$ can be read from the time between the peaks. Christophers (1972$b$) reported a decrease in $t_{G_1}$ from about five days in controls (a calculated value) to 15 hours after tape stripping in guinea pig ear (see Table 10.2) and $t_S$ at the time of the second labelling peak had fallen from 11.3 hours in controls to nine hours in stripped epidermis. While it is difficult to be sure that $t_S$ has changed, if the $I_S$ curve does reflect movement of synchronous cells, the $t_{G_1}$ and hence $T_C$ are probably much reduced, despite the fact that we do not have an accurately measured $T_C$ in this situation. Also in guinea pig (ear) epidermis, Hell and Cruikshank (1963) reported a $T_C$ of 8–12 hours, based on observation of synchronous $I_S$ peaks.

Table 10.2. Duration (in hours) of the phases of the cell cycle in injured epidermis (for more control values, see Table 7.2)

| Stimulus | $T_C$ | $t_{G_1}$ | $t_S$ | $t_2$ |
|---|---|---|---|---|
| **Tape stripping** | | | | |
| Christophers (1972$b$) (guinea pig) | | | | |
|     stripped | – | 15[1] | 9[1] | |
|     control | – | 120[1] | 11.3[1] | |
| Clausen and Lindmo (1976) | | | | |
| (mouse) | 12[2] | – | 10[3] | |
| | | | (20% decrease) | |
| **Incised wounds** | | | | |
| Epstein and Sullivan (1963) | | | | |
| (human) | 12–18[2] | – | – | – |
| Hell and Cruikshank (1963) | | | | |
| (guinea pig, *in vitro*) | 8–12[2] | – | – | – |
| Devik (1972) (mouse) | 12[1] | – | 6[1] | 1.5–2[1] |
| **Plucked skin** | | | | |
| Hegazy and Fowler (1973$a$) | | | | |
| (mouse) | | | | |
|     control | 101[1], 107[4], 96[5] | – | 11.6[1] | 2.0[1] |
|     plucked | 47[1], 47[4], 48[5] | – | 11.5[1] | 2.4[1] |
| Potten (1971$b$) (mouse) | 45–54[4] | – | 14[4] | – |
| (1972) | 22 | – | – | – |

1. Fraction of labelled mitoses method.
2. Observation of synchronous peaks of $I_S$.
3. Flow cytophotometry.
4. Continuous labelling method.
5. Brown and Oliver (1968) method.

In the hairless mouse, also after stripping, Clausen and Lindmo (1976) measured $T_C$ at about 12 hours (from synchronous peaks of $I_S$), a value to some extent confirmed by the results of Devik (1962) who obtained a value of 12 hours in regenerating hairless mouse epidermis after surgical wounding, using a full FLM curve; $t_S$ was six hours. Application of cantharidin, a powerful skin irritant, also results in synchronous waves of cells traversing the $S$ phase with a cell cycle time of about 12 hours (Clausen 1979). Since $T_C$ in resting hairless mouse epidermis may be about 84 hours (Iversen *et al.* 1969) this reflects a considerable reduction. Using a rectangular age distribution, Clausen and Lindmo (1976) calculated a 20 per cent decrease in $t_S$, and a 60 per cent decrease in $t_{G_2}$, although we might complain that the age distribution is probably exponential (see Chapter 6, section 7). However, Clausen and Lindmo (1976), on the basis of a decreased mean interphase grain count, in association with an elevated $I_S$, suspected that $t_S$ lengthened considerably towards the end of the response. In human epidermis, the time between peaks is about 12–18 hours (Epstein and Sullivan 1963), which also probably reflects a marked decrease in cell cycle time (see Table 10.2).

Potten (1971*a*), inspecting apparently synchronous peaks of labelling and mitotic activity after plucking the hair from mouse skin, proposed a $T_C$ of about 47–54 hours. But perhaps the most extensive analysis of cycle parameter changes has been made by Hegazy and Fowler (1973*a*), who measured $T_C$ by several different techniques, starting at 20–24 hours after plucking mouse skin. The results are summarized in Table 10.2. There was good agreement between $T_C$ as measured by the FLM and continuous labelling methods and by the Brown and Oliver (1968) method. These all showed a decrease in $T_C$ from around 100 hours in control mice to about 47 hours in the plucked animal, agreeing with Potten's (1971*a*) result. However, $t_S$ did not show any change. The achievement of complete labelling of the basal cell population was again interpreted as indicating that $I_P = 1$, as was the appearance of a discontinuity at $N_B/N_T = 0.5$ on the Brown and Oliver curve; we have already discussed the potential sources of error with these methods (see Chapter 6, section 2, and Chapter 7, sections 2 and 4), but here we could note that the FLM curve will be compromised by the synchronous wave of proliferation, and also that the Brown and Oliver method might be untenable because of the appearance of proliferating cells in the supra-basal layers, which would flash label and affect the initial (and later) $N_B/N_T$ values; there is also the distinct possibility that, in the hyperplastic state following plucking, more vertical mitoses would appear, which would affect the mode of migration (see Bullough and Mitrani 1978; and Chapter 6, section 7).

We may conclude that the regeneration which follows stripping, incising or plucking epidermis is associated with a marked reduction in the cell cycle time of the proliferating cells. We might proceed to ask, then, is this the kinetic mechanism of the epidermal response to wounding?

## 10.4.  The nature and control of the proliferative response to wounding

We have considered the phenomenology of the proliferative response to wounding stratified squamous epithelium, notably epidermis, and it is now necessary to comment (and speculate) on the nature of the response, its mechanism and upon factors controlling it. Perhaps the first question we should ask is '*Where do the proliferative cells come from?*'. Unfortunately, practically the only piece of hard kinetic evidence we have to go on is that the cell cycle time of the stimulus-responsive cells is much reduced compared with controls (see Table 10.2), but, even so, there have been three main proposals to account for the appearance of numerous cells in, for example, DNA synthesis after epidermal wounding. The first of these is that epidermal cells are in a classical $G_0$ resting phase (with a $G_1$ DNA complement) from which they are ejected, with some form of distribution of times for leaving $G_0$, giving a synchronous wave of DNA synthesis and later mitosis (Hell 1963; Hell and Cruikshank 1963).

Hennings and Elgjo (1970) found such a mechanism attractive, but Potten (1971$a,b$) presented the basis for such a hypothesis; Potten noted the similarity of the $I_S$ curve to that observed in the liver after partial hepatectomy, with a usually well defined pre-replicative period (see Table 10.1), a phase of cell proliferation and a curtailment period (see Chapter 24). Potten considered that the duration of the pre-replicative period was too long for cells to be merely thrust from $G_1$ into $S$ at a greater rate. Unfortunately, for this present hypothesis to be tenable, we must be sure that $G_0$ cells exist in unstimulated epidermis. We have surveyed the kinetic evidence regarding the growth fraction (see Chapter 7, section 4) and discussed the importance of $I_P$ considerations to the EPU concept (see Chapter 8, section 2), and concluded that the evidence indicated that about 0.6 of mouse basal cells were cycling, while the remainder (0.4) were post-mitotic and destined to migrate, notwithstanding the attainment of 100 per cent labelling in the continuous labelling experiment (Fowler and Denekamp 1976), which in fact is anticipated in the EPU model (see Chapter 8, section 2). Thus there is little room for classical $G_0$ cells. But we have seen (Chapter 8, section 4) that Potten (1971$b$), as a result of a long series of experiments, considered that a subpopulation of epidermal cells were able to store [$^3$H]-TdR in the acid-soluble pool for up to 48 hours; on wounding, stimulus-responsive cells incorporated this [$^3$H]-TdR into DNA. The relationship of these cells to the late labelling cells described by Møller and Keiding (1982) in the hamster cheek pouch is unclear; it appears that *all* basal cells in this epithelium are at least potentially late labelling, rather than just a small subpopulation. Nevertheless, Potten considered that the stimulus-responsive cells were originally in $G_0$, and by this admittedly circuitous argument, calculated, from the extra cells stimulated into $S$ after plucking, that about 8 per cent were in $G_0$ (Potten 1972; Hamilton and Potten 1972). Incidentally the same number showed both late metabolism of [$^3$H]-TdR *and* formed the clonogenic cell population (Potten and Hendry 1973; see

Chapter 8, section 4), and Potten (1973) proposed that $G_0$ cells were restricted to the stem cell compartment. It will be seen that this argument in fact assumes that stimulus-responsive cells are $G_0$ cells, although in fairness we should recall that acid-insoluble [$^3$H]-TdR values in controls were constant, and consequently normally cycling cells cannot draw on the long-lived pool (see Chapter 7, section 1). Furthermore, such an argument really does neglect the continuous labelling index data, which does reach 100 per cent (Iversen *et al.* 1968; Hegazy and Fowler 1973*a*), and does show that the centrally placed cell in the EPU (the putative stem cell, Potten 1974) shows slower labelling under continuous labelling conditions (Potten 1974). In later publications (see for example, Potten 1976) Potten maintains that the central stem cell is slowly cycling; consequently a more reasonable proposal is that this central cell has different TdR pools, and is stimulus-responsive, in which case we must exclude a formal rigorous $G_0$ argument. This is notwithstanding the assertion of Burns and Tannock (1970) that nearly all cells in the basal layer of the mouse epidermis are in $G_0$, a proposal made mainly on the basis of a low cell production rate; we have seen that there are other reasons for a low proliferative rate besides the presence of a large number of $G_0$ cells. Burns and Tannock (1970) showed that, after stripping rat epidermis which had been previously flash labelled 18 hours before with [$^{14}$C]-TdR (which cells had divided, and were in the putative $G_0$ compartment prior to stripping), these [$^{14}$C]-TdR labelled cells appeared to enter $S$ at the same rate as the non-[$^{14}$C]-TdR labelled cells, as indicated by double labelling with [$^3$H]-TdR (see Fig. 10.4). These data were quoted in support of the hypothesis that cells leave the $G_0$ compartment randomly, i.e. with a constant probability per unit time, and that the rate of release of cells from $G_0$ is independent of the

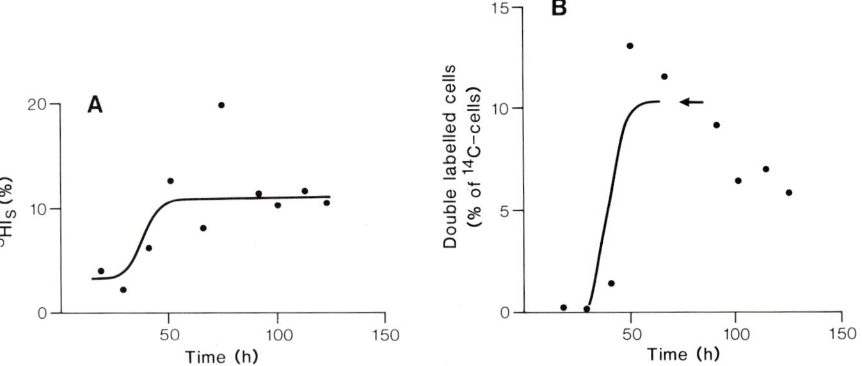

Fig. 10.4.   (A) The effect of stripping on the [$^3$H]-TdR flash labelling index in the rat dorsal epidermis.   (B) The fraction of [$^{14}$C]-TdR labelled cells in DNA synthesis with time after stripping; [$^{14}$C]-TdR was given at zero time, and the presence of [$^{14}$C]-TdR labelled cells in $S$ was detected by double labelling after incubation *in vitro* with [$^3$H]-TdR. The arrow shows the expected level of the curve if $^{14}$C-labelled cells were as likely to leave $G_0$ as the unlabelled cells. (Redrawn from Burns and Tannock (1970).)

time of entry (see Chapter 2, section 11).

However, an alternative explanation would be that, after the rapid migration that occurs after stripping, the remaining cells in the basal layer, and this will include both $[^{14}C]$-TdR and $[^3H]$-TdR labelled cells, have an equal opportunity of being stimulated into $S$. On the Potten (1975$b$) EPU hypothesis (see Chapter 8, section 2), many of the $[^{14}C]$-TdR labelled cells would become post-mitotic maturing cells, and be amongst the first to migrate after stripping, whilst the remaining labelled cells respond with a shortening of their cell cycle. Indeed, the migration of $[^{14}C]$-TdR labelled cells after the stimulus could be the cause of the decreasing trend indicated in Fig. 10.4, in that many $^{14}C$-labelled cells have already migrated (the mean residence time of post-mitotic maturing cells is about 30 hours, see Chapter 8, section 4).

Fowler and Denekamp (1976) aver that the presence of high growth fractions in unstimulated epidermis preclude more rapid proliferation achieved simply by stimulation of more cells into the cell cycle; since Fowler and Denekamp relied mainly on continuous labelling evidence for this statement (which we dispute, see Chapter 7, section 4), we find ourselves in agreement with them, but for the wrong reasons.

That is not to say that changes in the proliferating population are unimportant; Bishop and Cox (1974) considered that the synchronous wave of proliferation after wounding was due to *reversion* of post-mitotic basal cells to the cell cycle, but we will see that Potten and Allen (1975$b$) describe rapid migration of basal cells in stripped epidermis, and it does appear more likely that this is the function of the post-mitotic cells in this situation. Many authors have now described proliferative activity in suprabasal cells in hyperplastic states, and it seems reasonable to propose that there is an absolute increase in the size of the proliferative compartment, with proliferative cells in the suprabasal layers, thus augmenting the cell production rate.

We are left with the possibility that the increase in proliferative indices is due to cells entering $S$ or $M$ at a faster rate from $G_1$ or $G_2$ respectively. Current opinion favours a more rapid $G_1$-$S$ transition (see, for example, Hell and Cruikshank 1963; Kurman and Argyris 1975; Clausen and Lindmo 1976; Bertsch *et al.* 1976) and we would interpret this as an increase in the transition probability of proliferating basal cells, according to the modifications by De Maertelaer (1978) of the original Burns and Tannock (1970) and Smith and Martin (1973) model. If the transition probability remained high for some time, this would account for the synchronous waves of proliferative activity and the FLM finding of a shortened $T_C$ (Table 10.2). The curtailment of the response (Fig. 10.1) would then be due to a gradual decrease in the transition probability. Increased $G_2$-$M$ flux, or release of cells blocked in $G_2$, into $M$, has been put forward by Gelfant (see Gelfant 1963$b$, 1966, 1975; Gelfant and Candelas 1972). We have considered the techniques and arguments put forward by Gelfant for the role of $G_2$ cells in epidermal homeostasis, particularly after wounding (see Chapter 8, section 5) and have set out our reservations: neither has Gelfant's hypothesis found favour with other workers in wound healing: Hell (1963) and Hell and Cruikshank (1963) noted that $I_S$ changes were followed by $I_M$ changes (not the other way around) and the appearance of mitoses at eight hours after wounding was due to labelled $S$ phase cells later entering $M$. Similar complaints were raised by Potten (1971$a,b$) and Clausen and Lindmo (1976), who noted that 94 per cent of mitoses arrested by colcemid at nine hours after stripping had been labelled by a pre-labelling procedure; Bertsch *et al.* (1976) also observed that 90 per cent of mitoses were labelled by a combined continuous labelling and stathmokinetic procedure at 16 hours after stimulation (stripping). These observations are not in accord with $G_2$ cells, or $G_2$ blocked cells, having a major role in epidermal reactions.

How then to account for Gelfant's observations, and for the small fraction of unlabelled mitoses which occur with continuous labelling after wounding? Clausen and Lindmo (1976) noted that after wounding, the proportion of cells with $S$ phase DNA content (as revealed by flow cytophotometry) was larger than the $S$ phase fraction shown by [$^3$H]-TdR labelling; this was explained by some $S$ phase nuclei not labelling or having a grain count indistinguishable from background. It is now well known that cells at the end or beginning of the $S$ phase incorporate low amounts of [$^3$H]-TdR into DNA (see, for example, Wright 1971$b$) and Shackney (1975) has proposed that cells in DNA synthesis may actually have very low grain counts indeed. Consequently, it is conceivable that these cells, in DNA synthesis but not counted as labelled, could be responsible for the presence of unlabelled mitoses in the Gelfant assay system. Other possible reasons for these results are discussed in Chapter 8, section 4, where the phenomenon of late labelling is discussed.

Clausen and Thorud (1980) have also addressed themselves to the reports (see Gelfant 1977) which indicate that mitoses are seen *before* labelled cells appear at the end of the lag phase, which hitherto was interpreted as a movement of cells, blocked in $G_2$, into mitosis. Clausen and Thorud, using flow cytophotometric methods in the main, have shown that tape stripping synchronizes $S$ phase cells in the lag phase, so that they enter mitosis together; this occurs before the major influx into $S$ occurs. The reasons for this sequence of events is unclear, but it would account for the phenomenology.

If we accept the concept of increased $G_1$-$S$ flux, we are left with a further important question: *'Why is there a lag phase before proliferation ensues, which is variable in duration but is nearly always present* (see Table 10.1), *and how do we explain the variability in different experimental situations?'*. Fowler and Denekamp (1976) have introduced a most interesting unifying hypothesis.

In most forms of mechanical injury, viz. stripping, incising, and plucking, the proliferative response is relatively rapid (see Table 10.1), but after irradiation injury or burns, the onset of proliferation is delayed (see Chapter 12, and compare Figs. 10.5 and 12.3). Can we find any reasons which account for this difference? In stripped epidermis Pinkus (1952*a, b*) reported a reduction in basal cell numbers which lasted for up to 72 hours after stripping; after plucking hair, Hamilton and Potten (1972) reported an immediate fall in basal cell density from 17 400/mm$^2$ to 13 000/mm$^2$, confirmed by Potten and Allen (1975*b*), who also recorded a fall in basal cell number after stripping, *which was due to an increased migration rate from the basal layer, independent of mitosis*. With full thickness epidermal wounding, Kurman and Argyris (1975) reported a decrease in the basal cell density in the marginal epidermis 1 mm on either side of the wound, due to increased migration as part of the epidermal response (see section 10.5).

All these models have in common the relatively rapid onset of basal cell depletion; since it takes a finite time for marginal basal cells in a wound to migrate (starting at about 18–24 hours), or for basal cells to migrate after stripping, the regenerative response will be delayed, presumably until basal layer density reaches a critical point. After irradiation, however, loss of basal cells is delayed because of induced mitotic delay (see Chapter 12), and consequently the stimulus for increased proliferation is also delayed (see Fig. 12.3). Just why basal cell depletion should lead to the induction of cell proliferation is unclear, but there are several candidate hypotheses: a decrease in contact or density dependent inhibition, a decrease in the local chalone concentration, or some as yet unknown intercell communicative signal. Whatever the mechanism, it is interesting to note that local basal cell depletion is the envisaged signal for the initiation of cell proliferation which makes good the deficit in cell number, evoked by migration, in Potten's (1976) EPU hypothesis.

Thus the reaction to epidermal stripping cannot be seen simply as negative feedback action from superficial cell loss, as envisaged by Hennings and Elgjo (1970). On the other hand, the Fowler and Denekamp model is not entirely global, since, although there is an immediate fall in basal cell density after plucking, increased [$^3$H]-TdR incorporation did not ensue until 12 hours later (Potten 1971*a*). Of course we could ascribe some of this time to the portion of the $G_1$ phase which is incorporated into the *B*-phase (Smith and Martin 1973), but this may be special pleading.

This brings us to our last question: '*What initiates and controls the proliferative response, leading to its eventual curtailment?*'. Naturally the facts that we are dealing with loss of tissue, and that above we have belaboured the point that basal cell depletion is a good candidate for a proliferative signal, brings negative feedback and the chalone hypothesis into its own. Because many of the basic tenets of the epidermal chalone model come from the study of wound healing, we shall leave detailed consideration until Chapter 14, section 1. Here we might note that stripping hairless mouse epidermis results in a loss of $G_1$ chalone activity in the stripped epidermis lasting for 9–10 days, whereas $G_2$ chalone activity is restored in 5–6 days (Rohrbach *et al.* 1976*c*); this differential sequence is possibly explained by the suprabasal keratocytes (purportedly the source of the $G_1$ chalone) being replaced later than the basal cells, which produce the $G_2$ chalone (Elgjo *et al.* 1971, 1972). Thus, wounding leads to a decreased concentration of chalone (due to loss of and/or damage to cells), with a resulting increase in proliferative rate, in accord with the classical hypothesis of Bullough and Laurence; or does it? Bertsch *et al.* (1976) have made the fascinating observation that epidermal chalones, operative in normal mouse epidermis, *will not act on mouse epidermis regenerating after stripping, for up to 48 hours after injury*, after which the response was restored. Bertsch *et al.* postulated that, with injury, uncoupling of chalone control occurs, with resulting hyperplasia, and drew a parallel with the chalone-unresponsiveness of epidermis treated with tumour-promoting phorbol esters (Bertsch and Marks 1974), and of newborn epidermis (Marks *et al.* 1979). This uncoupling was conceivably brought about by depletion of the chalone-responsive proliferating cell pool by migration, and proliferation is then 'taken over' by chalone-unresponsive committed stem cells, which only acquire chalone-responsiveness by differentiation. This model raises several questions; are Potten's centrally placed stimulus-responsive cells (see Chapter 8, section 2) the chalone-unresponsive committed stem cells? How do chalone-responsive stem cells differ from chalone-unresponsive stem cells? What controls chalone-unresponsive stem cells? These are difficult, and as yet unanswered questions.

We are left with two conflicting hypotheses within the chalone model; is proliferation after wounding due to decreased chalone concentration, chalone-unresponsiveness, or a combination of both? We do not know the answer (see Chapter 14, section 1).

As a final observation, we might note that further evidence for an inter-EPU control mechanism lies in the fact that EPUs marginal to a wound, with no loss of surface corneocytes, also respond to the wounding stimulus (Potten and Allen 1975*b*; and see Chapter 8, section 2).

Our conclusions on the nature of the proliferative response to wounding can be aptly summarized thus: the lag phase is the time necessary for the basal layer to reach the required level of depletion (plus possibly the $G_1$ portion of the *B*-phase?), which, by an as yet unknown mechanism, increases the transition probability of the remaining basal cells, leading to a synchronous movement of cells into DNA synthesis and later mitosis. As the cellular depletion is made good by the elevated cell production rate, proliferation declines until normal rates are re-established. The chalone hypothesis provides an explanation of the initiation of proliferation, either by a reduction in local chalone concentration, or by the emergence of chalone-unresponsive committed stem cells; here curtailment of proliferation would be accounted for by a restoration of chalone concentration to the normal level, or by later differentiation of the unresponsive stem cells. That some form of negative feedback mechanism is operative is supported by Argyris's (1976, 1977) finding that considerable overshooting of the amount of epidermis regenerated occurs at about five days after wounding. Some indirect evidence for this may come from the experiment of Fisher and Maibach (1972) on the effects of surface occlusion on the proliferative response to wounding; after stripping human epidermis, Fisher and Maibach reported that mitotic activity was depressed from days 2 to 6 after stripping, which was interpreted in terms of an increased negative feedback influence from a thickened keratinized layer. Apparently the situation in incised wounds differs: occlusion merely delayed the response, but during the proliferative wave there is an increased migration rate (Rovee *et al.* 1972*a*). Species differences may explain the observation that wound occlusion accelerates the onset of proliferation in pig epidermis (Winter 1972).

We may conclude by noting that the present theory goes some way to explaining the format of the proliferative response (see Fig. 10.1 and Table 10.1), and much of the FLM data (Table 10.2). Fowler and Denekamp (1976) consider that the finding of Block *et al.* (1963) of superficial epidermal damage with silver nitrate inducing basal cell proliferation at 18 hours is not in accord with the basal cell depletion concept; however, the same argument, invoked for basal cell depletion due to an increased migration rate, which replaces lost differentiated cells, *can* apply here, and is not seen as a cogent objection to the theory.

## 10.5. The role of cell migration in wound healing

The onset of cell migration is one of the earliest signs of epithelial activity in open epidermal wounds; for example Gibbons (1969) observed the onset of migration in rat palatal mucosa as early as six hours after wounding, an observation confirmed by Martinez (1972) in rat gingiva. However, in other models, migration occurs much later; not until approximately 18-24 hours after epidermal wounding in mouse (Croft and Tarin 1970), rabbit (Viziam *et al.* 1964), pig (Winter 1972) and man (Odland and Ross 1968; Rovee *et al.* 1972*b*). Consequently, epithelialization is completed much faster (in about 12 hours) in gingival compared with epidermal wounds (Martinez 1972), although of course the exact time will depend on the wound size. The stimulus for the initiation of cell movement is not entirely clear, but reasonably might be expected to be related to release from contact inhibition (Weiss 1961; Winter 1972); this naturally also provides a mechanism for the cessation of migration.

There has been considerable debate on the mechanism of migration; we have seen above (section 10.2) that migration and proliferation are likely to be two independent processes, so direct mitotic pressure, or even indirect lateral pressure due to mitosis, are unlikely to be important mechanisms, despite the contention of Matoltsky and Viziam (1970) that passive movement of epidermal cells occur as the result of mitoses in migrating epidermis. However, several authors have described how advancing keratocytes in the encroaching margin roll or slide over one another during re-epithelialization of the wound surface, as in the movement of a caterpillar track (Winter 1963; Krawczyk 1971, 1972; Rovee *et al.* 1972*a*), both in the formation of the surface monolayer and during stratification. Vaughan and Trinkaus (1966) from their *in vitro* observations, envisaged the epidermal cells migrating as a sheet, and Martinez (1972), while agreeing with previous authors (for example Croft and Tarin 1970; Odland and Ross 1968) that epidermal cells move by amoeboid movement mediated by pseudopodia, could find no evidence for a 'caterpillar track' type of locomotion.

As well as covering an exposed surface, migrating keratocytes also invade the granulation tissue and debris which cover the floor of a healing wound (Croft and Tarin 1970), and Christophers (1972*b*) has shown that prelabelled keratocytes also migrate from the wound margin, over the more deeply placed cells covering the wound base, to enter the granular layer and become keratinized in the wound zone. There has been some controversy about which cells actually migrate, with some authors (for example Viziam *et al.* 1964) holding that only basal cells migrate, while others (Martinez 1972; Rovee *et al.* 1972*a*) maintain that both basal and spinous cells are capable of migration.

As to the speed of migration, Winter (1972) recorded 7 $\mu$m/hour in healing pig epidermis, but as far as the distance migrated is concerned, Winter (1972) considered that cells at the wound margin are displaced only a few cells diameter from their original position; these cells, having migrated, divide within 24 hours of settling on the wound surface; thus Winter (1972) considers that it is the *progeny* of the original migrating cells which end up in the middle of the wound.

It is often said that the stimulus for the initiation of migration in open wounds is loss of contact inhibition of the migrating keratocytes, but the situation in stripped epidermis· is not clear. In this context, Mishima and Pinkus (1968), studying ultrastructural changes in stripped human epidermis, noticed breaks in desmosomal contacts between keratinocytes as long as 72 hours after stripping, and Potten and Allen (1975*b*) reported a reduction in hemidesmosomal contacts before proliferation ensued. The reasons why epithelial cells cease migration are unknown but are likely to be multifactorial (see Croft and Tarin 1970), and may involve contact inhibition and desmosome formation (Martinez 1972).

Thus although cell migration does not strictly come into the sphere of cell proliferation, we cannot appreciate the dynamic nature of the healing process without relating cell migration to the associated proliferative pattern. We conclude that the mechanisms by which migration and proliferation are initiated, maintained, and controlled are largely independent, but since migration may be responsible for triggering proliferation by basal layer depletion, they are undoubtedly complementary processes.


## 10.6. Differentiation after wound healing

For epidermal wound healing to be complete, the keratocytes must form a new granular and corneal layer, and in stacked epidermis, remodel an ordered structure. Christophers (1972*b*) noted that keratohyalin granules and cornification

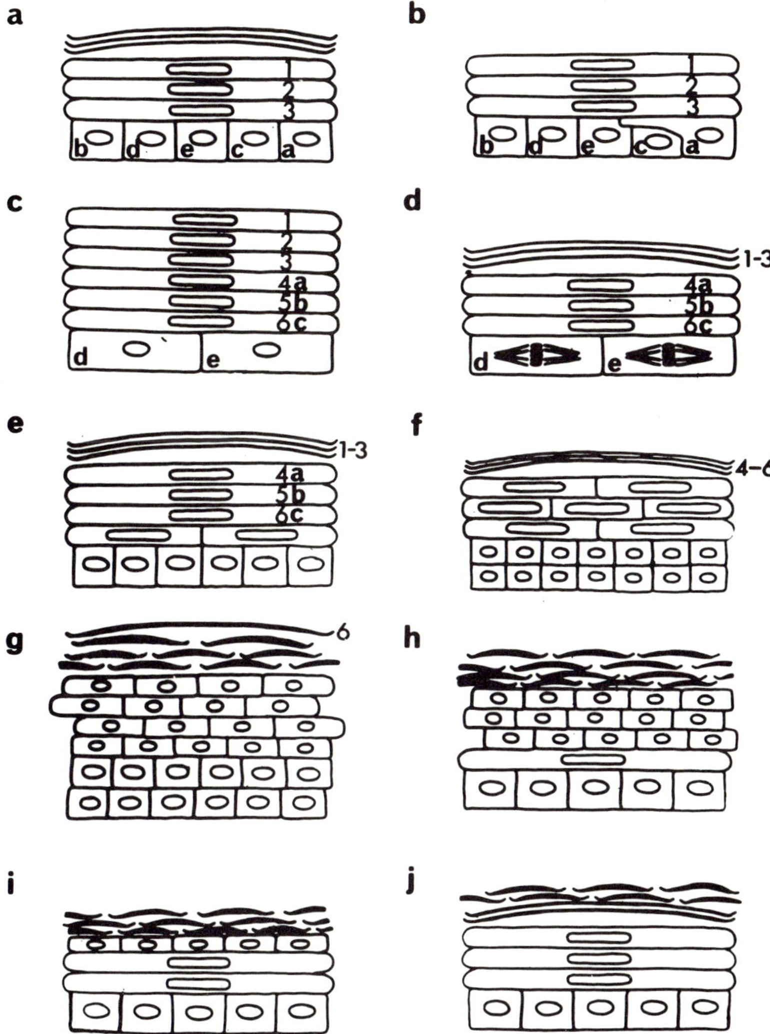

Fig. 10.5.   A summary of the effects of wounding on the ordered structure of the mouse epidermis, after tape stripping: (a) shows a normal profile of the EPU. 1, 2 and 3 are the nucleated layers, and cells a–e indicate the basal cells associated with the unit. (b), (c), and (d) show the *migration* that occurs immediately after stripping; cell a is seen migrating from a lateral position to cover the basal cells of the unit. In (c), cell a is moved to 4a, b to 5b, and c to 6c, replacing the three corneal layers removed by stripping, without the occurrence of mitosis. (d) and (e) are at 18 hours after stripping, and show mitotic activity, with over-production of cells and extra cells shown in the basal and suprabasal layers. (f) shows continued proliferation and loss of stacking, which remains only in the cornified layer. (g) shows the nucleated layer now 6–7 cells thick at seven days after stripping. In (h) the rate of cell production has decreased, the density of the basal cell layer is reduced, the Malpighian layer is thinned. (i) and (j) show the reestablishment of the EPU at 10–15 days after stripping. (By courtesy of Dr C. S. Potten.)

appeared at the incised wound margin at 2–3 days after injury in the guinea pig epidermis, but only later in the central portion of the wound, while a recurrence of the stratum corneum occurs at 72 hours after stripping human epidermis (Mishima and Pinkus 1968). At this time (i.e. 72 hours) after wounding, the wound is covered by several layers of epithelium with no evidence of an ordered arrangement (Christophers 1971*a*), but, starting after about 15 days in mouse ear, stacking reappears at the margins and moves in, until at about 21 days the normal ordered arrangement is seen; thus stacking does not reappear until proliferative rates have normalized again.

The effect of wounding on epidermal ordered structure has been studied in some detail by Potten and Allen (1975*b*). After stripping or plucking mouse epidermis, stacking was maintained during the phase of increased migration of post-mitotic basal cells. However, with the ensuing elevation of proliferative rates, there was synchronous migration of young post-mitotic basal cells into the spinous layer, with associated loss of stacking. Potten and Allen considered that the young cells were unable to flatten so readily, and disordered structure was maintained so long as proliferative rates remained high. With the return of proliferative indices to control levels at 14 days, ordered structure became evident and was complete at 19 days after injury. These changes are summarized in Fig. 10.5.

Thus the elevation of proliferative rates which follows wounding is accompanied by loss of stacking, in accord with the general hypothesis, outlined in Chapter 8, that ordered structure is not consistent with high proliferative rates, and is only found in epidermis with low cell production rates; stacking therefore returns when cell production rates fall to normal levels.


## 10.7. Conclusions

The phenomenological aspects of wound healing in squamous epithelia are now largely agreed, and although kinetic measurements in such a perturbed situation remain problematical, there now exist several conflicting hypotheses which account for the kinetic observations. However, even the favoured hypothesis of basal layer depletion as the proliferative signal is based on largely circumstantial evidence, and awaits conclusive disproof.

# 11 Growth reactions in squamous epithelia

We have already discussed the phenomenology of the reaction to injury, particularly in the epidermis, and examined the several theories which have been introduced to account for these changes. However, after this proliferative response, there is no net change in the number of cells in the epithelium after the cessation of the induced cell proliferation, and the system more or less rapidly returns to the pre-injury state. There are a number of other responses in this epithelial system which do result in a net change in the total number of cells, and these we will examine in this chapter.

## 11.1. Hyperplasia in squamous epithelium

Hyperplasia can readily be induced in squamous epithelium by repeated tape stripping, or even transiently during the recovery from a single episode of tape stripping (Hennings and Elgjo 1970), vitamin A application (see section 11.2), carcinogens (see Chapter 13, section 1), repeated friction (Mackenzie 1974*a,b*), UV irradiation (Argyris 1968), and even by the induction of dermal granulomatous inflammation (Schellander and Marks 1973; Francis and Marks 1977*a*). Increase in cell number is thus a widespread and common general reaction to perturbation, but its mechanism remains poorly understood.

The morphological changes are prominent (see Fig. 11.1): there is marked thickening of the Malpighian layer, and, in some cases, this is associated with elongation of the rete pegs, as occurs in the skin disease psoriasis. Not generally appreciated, but undoubtedly present, is a considerable *hypertrophy* of the individual keratocytes in the Malpighian layer, which of course contributes to the increased epithelial thickness. In most hyperplastic states in epithelia which normally show orthokeratinization, there is a reduction in the thickness of the granular layer and the appearance of parakeratosis, or of nuclear fragments in the stratum corneum. Parakeratosis is probably not caused by a mere increase in cell production rate coupled with a decreased transit time and consequently a decreased amount of time available for differentiation; Christophers and Braun Falco (1970*a*) have shown that although prominent parakeratosis is evident during recovery from tape stripping, after vitamin A or hexadecane application, which causes a similar increase in cell production rate, there is no associated parakeratosis. Thus a simple increase in the cell production rate appears unable to induce parakeratosis by itself, without some defect in cell differentiation.

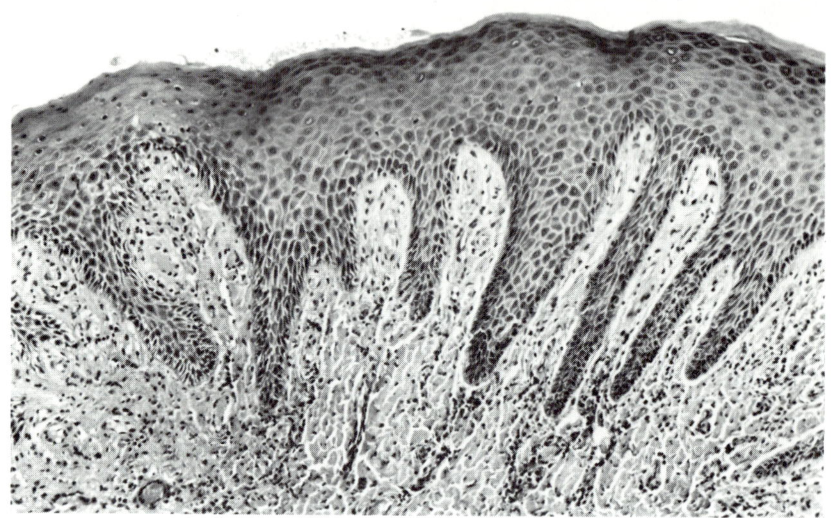

Fig. 11.1. A photomicrograph of hyperplastic epidermis; there is prominent thickening of the Malpighian layer (acanthosis) and marked lengthening of the rete pegs, with reciprocal elongation of the dermal papillae. The granular layer is also thickened. This is taken from human skin near an infected sinus (pilonidal sinus), which accounts for the inflammatory infiltrate in the dermis. (× 100.)

The mechanism by which stripping leads to this defect is not known, but we have seen (Chapter 10) that there is considerable damage induced in the upper epidermal cells, often only revealed at the ultrastructural level (Mishima and Pinkus 1968).

The kinetic basis of these changes are largely speculative, but the kinetic cause of hyperplasia is evidently a *cell production rate which exceeds the desquamation rate*. We are familiar with the concept that basal cells, in normal conditions, spend some time as post-mitotic cells in the basal layer, where synthesis of keratins and other differentiation proteins begins. An increase in the proliferative rate would naturally restrict the time available for such synthesis, both in the basal layer (where even cells in $S$ can migrate in hyperplastic states; Rohrbach *et al.* 1968), and also during the decreased transit time through the Malpighian layer. Consequently, there may be defective keratinization, and, if desquamation is in fact the intensely programmed process envisaged by the EPU model, desquamation becomes abnormal. Banoczy's (1971) conclusion that parakeratotic squames are more readily removed is certainly borne out by the well-known scaling which occurs in the hyperproliferative and parakeratotic epidermis of psoriasis.

The hyperplastic process obviously causes thickening of the viable epidermis, usually called *acanthosis* (Fig. 11.1); a proportion of this thickening is no doubt caused by the need to accommodate increased numbers of proliferative or 'germinative' cells in an expanded proliferative compartment. In psoriasis, for example, there is an evident increase in the number of proliferating cells in the epidermis, as indicated by the presence of mitoses well above the basal layer, and there is also probably a relative increase in the fraction of proliferating cells in the basal layer (i.e. an increase in the growth fraction). It is likely, in most instances of hyperplasia, that increases in the number of proliferating cells contributes to the acanthosis; obviously this is not the whole story, however.

The Malpighian layer is also increased in thickness, evidently due to an increase in the number of post-mitotic cells in the suprabasal layers. In the steady state, we would assume that transit time = size/flux; theoretically, changing only the influx (or basal cell production rate) would lead to a decrease in transit time only, but observations tell us that hyperplastic states are associated, not only with increased flux and decreased transit times, but also with *increased compartment size*. This indicates that there are two controlling parameters of epidermal population size or thickness, the *cell production rate*, and the *transit* or *differentiation rate*. Whether both are operative as causative factors in all hyperplasias is unknown, but we could speculate that an expansion in the Malpighian compartment was required to house all the increased influx from the basal layers in an enlarged maturation compartment. In Chapter 20, section 4, we shall see that, in the enlarged crypt systems encountered in coeliac disease, where a hyperplastic mucosal state also occurs, there is a similar increase in the absolute size of the proliferative compartment, mediated by a three-dimensional expansion in the size of the crypts; *parri passu* there is a large increase in the size of the maturation compartment. Thus it might appear as if absolute increases in proliferative compartment size must be accompanied by concomitant increases in maturation compartment size, for whatever reason, which could conceivably be related to increased efflux from the proliferative compartment. Thus although MacKenzie (1974*a,b*), studying the hyperplastic effect of chronic friction on the mouse ear epidermis, found no change in the *density* of the basal layer, there was an increase in the *number* of suprabasal keratocytes, notwithstanding the decrease in transit time through the viable part of the epidermis; in addition, an increased keratocyte size also added to the hyperplasia.

Bullough (1972; Bullough and Deol 1975; Bullough and Mitrani 1976, 1978) has attempted to explain the evolution of hyperplasia in terms of differences in dermo-epidermal adherence and basal cell pressure. Bullough has usefully conceived two types of epidermis—normal, or *phase 1* (exemplified by mouse ear epidermis, or senile human epidermis), and *phase 2*, or hyperplastic epidermis. Phase 1 epidermis is characterized morphologically by a flat basal layer and few suprabasal layers, with ordered structure, and by a low cell production rate (see Fig. 11.2). The ratio between the cell production rate and the rate of differentiation of maturing or post-mitotic cells is proposed as a constant; similarly the ratio of the number of proliferating to post-mitotic cells is also constant. Thus

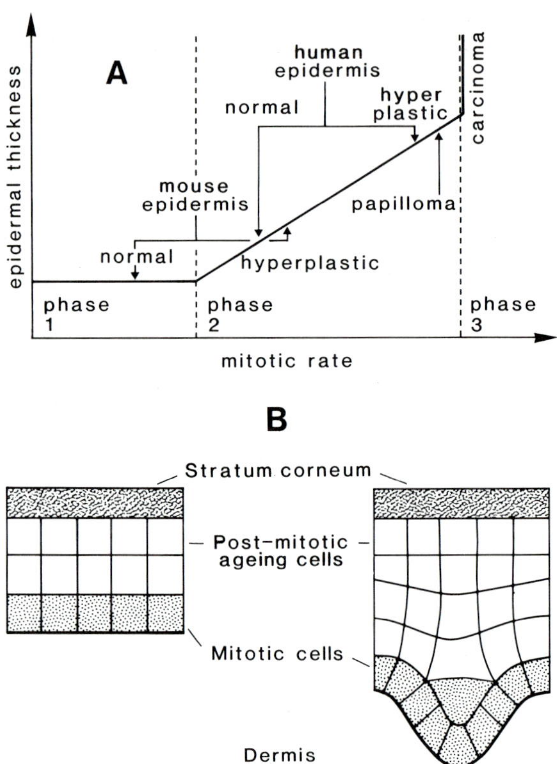


Fig. 11.2. A conceptual view of epidermal changes during hyperplasia. (A) The relationship between the mitotic rate and the thickness of the epidermis. (B) Left is Phase 1 epidermis with changing mitotic rate which does not affect thickness, and right is Phase 2 epidermis with increasing mitotic rate and increasing local pressure, causing increased basal layer folding. (Redrawn from Bullough (1972).)

in epidermis which maintains its phase 1 morphology, an increase in the cell production rate results in an *increase* in the rate of differentiation of post-mitotic cells (i.e. in the viable compartment transit rate) and the epidermis maintains its thickness at a constant level (Bullough and Ebling 1952; Ebling 1957).

Phase 2 epidermis has a folded or undulating basal layer, a relatively high cell production rate, and occurs in mouse foot pad, human epidermis, and in any type of induced hyperplasia (see Fig. 11.1). However, in this instance, the increased cell production rate is accompanied by an increase in the epidermal thickness. With the induction of hyperplasia, all the cells in the basal layer become proliferative; there is exclusion of all post-mitotic cells, which are pushed upwards into the Malpighian layer. On the Bullough hypothesis, this is readily accomplished, since the positive dermoepidermal grip is held to be weak in phase 1 epidermis (Bullough and Deol 1975; and see Chapter 8, section 6).

The outcome of one mitosis is then a net gain of one cell, as no basal layer post-mitotics remain to be pushed out. The basal layer pressure rises, the incidence of vertical basal layer mitoses rises as a result of this pressure (Bullough and Mitrani 1978; and see Chapter 8, section 6), and the layer begins to fold and produce rete ridges, which continue to enlarge in hyperplasia. There is thus an increase in the absolute number of proliferating cells, and in the number of basal cells per unit area; it is a matter of observation that the epidermal thickness increases *parri passu*. The labelling index in normal guinea pig epidermis (a phase 2 epidermis) is positively correlated with epidermal thickness (Christophers and Petzolt 1969), and this is maintained after the induction of hyperplasia with petroleum jelly, vitamin A, and hexadecane (Christophers and Braun-Falco 1970*a*). Thus the increased cell production is associated with an increased number of differentiating cells in the Malpighian layer, notwithstanding the decrease in the viable compartment transit time which accompanies increases in the cell production rate in experimental animals, in humans in psoriasis and in the various forms of icthyosis (Frost *et al*. 1966).

Bullough would have us believe that both factors which control epidermal thickness – viz. the cell production rate and the rate of differentiation and cell loss, are under the control of the epidermal chalone system; the epidermal concentration determines the habitus of the epidermis, i.e. phase 1 or phase 2, with increased concentrations decreasing the cell production rate and slowing the differentiation rate, while decreased concentrations augment the cell production rate and increase the differentiation rate. The *experimental* evidence that chalones control the cell production rate is discussed in Chapter 14, section 1, but the idea that epidermal chalone controls the rate of differentiation is highly conjectural, for which experimental evidence is lacking, and the support for this hypothesis is largely circumstantial. In this context, Marks (1976*a*) found no real need to postulate a separate chalone effect on differentiation; if it is assumed (along with Weiss and Kavanau 1957), that change in the cell production rate is a natural consequence of changes in the maturation rate, the cell production rate becomes coupled to the differentiation rate. Neither does the Bullough hypothesis take cognisance of the fact that two epidermal chalones are proposed, or of the demonstration (Kreig *et al*. 1974; Bertsch and Marks 1978) that the chalone responsiveness of the target cells, rather than the chalone concentration itself, may be the more important factor; proliferating epidermal cells in chemically or physically induced hyperplasia are insensitive to the epidermal $G_1$ chalone.

The relationship between the cell number and the differentiation rate must be nice. During the change from phase 1 to phase 2 epidermis, the ratio of cell production rate to differentiation rate is constant, with resulting uniform basal layer thickness, but this changes in phase 2 epidermis. This would have to be explained by critical levels of chalone concentration, below which a triggering of phase 2 epidermis would occur. Although the rate of differentiation must increase in phase 2 epidermis, cell number also increases; we should have to propose that, at least during the induction phase, desquamation rate < cell production rate until a new steady state is reached. It is not clear, on this hypothesis, how epidermal thickness increases when the *lack* of chalone is said to *increase* the differentiation rate. We should have to say that, for a given level of hyperplastic stimulus leading to unit chalone depletion, the increase in the transit rate or differentiation rate is smaller than the accompanying increase in the cell production rate. Thus, to test the hypothesis, we should measure population size, transit rate, cell production rate, and chalone concentration during the induction of hyperplasia: easier said than done.

In some instances of hyperplasia, particularly in those induced experimentally, the surface and basement membrane lengths remain constant. However, as we have seen, in many pathological states there is an increase in the length of the rete ridges (cf. psoriasis), and this must imply some interaction between the epithelium and the dermal connective tissue. The nature of this interaction is obscure, but its consideration has provoked an entertaining controversy: Pinkus (1970) proposed that the direction of epidermal growth is outwards, in which case augmented proliferation and growth would lead to elongated rete ridges, which would lift the papillary dermal connective tissue upwards, deepening the actual papillae, and accounting for the phenomena. On the other hand, Ryan (1966, 1970) has argued, mainly on a rationalistic basis, that the direction of epidermal growth is inwards, but, because the basal layer is constrained by the basal lamina, the resultant of this growth is an outward migration of cells. Now these hypotheses are not immediately amenable to experiment; as far as the first proposal goes, Hume and Potten (1979) have pointed to the inward migration of the developing pilosebaceous follicle, in embryonic life, which is unassociated with any apparent increase in connective tissue mass. Hume and Potten saw the emergence of a hyperplastic epithelium from epidermis with an undulating basement membrane as the consequence of an increased cell production rate in an epithelium where the distribution of proliferative activity was non-uniform. In squamous epithelia with an undulating basement membrane, cell proliferation is maximal in the rete ridges; an increased cell production rate will result in an augmented Malpighian layer which will be predominantly below the level of the apex of the dermal papillae. The rete ridges then widen and elongate.

Is it then necessary to enquire about the direction of epidermal growth? The induction of acanthosis has two components: an increased cell production rate coupled to some mechanism which delays differentiation so that the Malpighian layer becomes thickened (see above). A combination of an outward growth vector, increased cell production rate and differentiation change could as well explain the hyperplastic phenomena as an inward growth vector and increased cell production. But how do we measure the direction of growth? What is the direction of growth in an intact epithelium? Is there such a kinetic parameter as the direction of growth? Are we giving squamous epithelium a parameter it does not intrinsically possess? We consider that these questions are more semantic–philosophical than biological; in any resting stratified epithelium, with defined basal lamina space and basal cell density, zonally defined cell proliferation confined to the basal layer, and limited beneath by the basal lamina, the resultant vector will always be outwards. During the induction of hyperplasia, so long as the basal lamina remains intact, the vector will remain outwards. In actinic keratoses and squamous carcinoma, where cells do grow downwards as basal buds or invasive tongues, the cells have now assumed new and dangerous properties, and conclusions about the normal epithelium drawn from analogy with these states are probably not substantiable. We remain unconvinced that, however, diverting, such an argument cannot be fruitful in the scientific sense.

## 11.2. Hyperplastic responses in squamous epithelia

Elgjo (1968$a,b,c$) has shown that several substances, for example benzene, will produce a wave of proliferation in mouse epidermis if applied only once, while if repeated applications are made, a protracted hyperplasia will result. Similar effects are discernible with hexadecane (Christophers and Braun-Falco 1970$a$); continuous ultraviolet irradiation (Chopra and Forbes 1974) or $\beta$-irradiation (Yamaguchi and Tabachnik 1972), chronic friction (MacKenzie 1974$a,b$), and many other agents also produce the same effect. In these circumstances the kinetic changes include an elevated cell production rate (Elgjo 1968$a$; MacKenzie 1974$a$), often with a decrease in the mitotic duration (Elgjo 1968$a$) and a decrease in the cell cycle time (Yamaguchi and Tabachnik 1972), and the morphological and organizational changes which are described in section 11.1. Application of carcinogens also results in a more or less protracted hyperplasia, but the kinetic changes may be somewhat different and are discussed in Chapter 13, section 1).

There are one or two experimental models of hyperplasia which have attracted considerable attention, and would perhaps repay closer consideration.

### (i) *The action of vitamin A*

Adequate dietary intake of vitamin A is required for epithelial maintenance, and it is well known that dietary deficiency produces xerophthalmia and follicular hyperkeratosis in the skin, and squamous metaplasia in respiratory and urothelial epithelia. Hypervitaminosis A, on the other hand, induces excessive scaling with an increased cell production rate, hair loss and epidermal thickening. Application of vitamin A to epidermis produces pyknotic nuclei, foamy vacuolization, parakeratosis, and acanthosis with increased DNA synthesis, in fact, all the histological hallmarks of hyperplasia (see Fig. 11.1) and it is thus surprising that the agent has found clinical use in icthyosis and psoriasis (Frost and Weinstein 1969) and in acne vulgaris (Plewig and Braun-Falco 1975). There is some evidence that vitamin A acid has some differences in action from other forms of the vitamin; thus vitamin A acid (retinoic acid or VAA) increased the mitotic index of hairless mouse epidermis within two hours (Sherman 1961; Stefan Zil 1972), compared with 24 hours for the vitamin A alcohol, and Stefan Zil considered that a stimulation of $G_2$-$M$ flux could be partly responsible; even a single application resulted in a considerable increase in cell number. A large increase in $I_S$ is also seen in human epidermis at 48 hours after a single application of VAA (Plewig and Braun-Falco 1975), and Pullman *et al.* (1975) found a decrease in $t_S$, $t_M$, and the turnover time in human epidermis treated with VAA.

In culture, VAA increases early cellular attachment, and induces a wave of DNA synthesis, which starts at about 24 hours, and leads to a marked increase in cell number (Christophers 1974; Chopra and Flaxman 1975; Wilkinson 1978), which is curtailed at 46–50 hours after application, and these latter results would indicate that, whatever the mechanism of initiation, the kinetic mechanism may be similar to that of wound healing, i.e. a decrease in $T_C$ and an absolute or relative increase in proliferative cell number.

VAA not only causes profound changes in the cell production rate, but also acts on differentiation and keratinization, decreasing keratin precursors (tono-filaments), increasing keratinosome number, mitochondria, and endoplasmic reticulum (Christophers and Wolff 1975), and in this way its action does differ from those of the hyperplasiogenic agents mentioned above, which increase the cell production rate, but do not appear to interfere with keratinization. Christophers and Wolff (1975) consider that this latter effect may be the route by which VAA exerts its therapeutic action, in which case it is difficult to understand why such a prominent induction of DNA synthesis and later hyper-plasia accompanies this action. There is no clear indication which biochemical mechanism is involved; whether the hyperplasia is regenerative, and purely in response to the pyknotic cells produced by the compound (Stefan Zil 1972), an action via the cyclic nucleotide system (Christophers and Braun-Falco 1975) or a direct action on chromatin, is unknown. In this respect Prutkin (Prutkin and Bogart 1970; Prutkin 1971) showed that labelled VAA is incorporated rapidly into rabbit ear epidermal cells, and from studies on the time-dependent distri-bution of the label, suggested that direct nuclear activation was involved.

## (ii) *Essential fatty acid deficiency*

Rats, mice, dogs, and pigs fed a diet deficient in essential fatty acids (EFA) develop severe scaly lesions in the skin, which, in the rat, are associated with acanthosis, thickening of the corneal layer, and evidence of increased prolifer-ative activity; MacCullough *et al.* (1978) reported elevated labelling and mitotic indices, a decrease in the turnover time, and a reduced viable compartment transit time. Again the mechanism by which EFA deficiency produces such a change is unclear, but there is a reported reduction in adenyl cyclase activity in EFA-deficient epidermis, which could be a factor (Lowe *et al.* 1980; but see Chapter 14, section 2).

## (iii) *Changes in thyroid status*

Patients with hyperthyroidism are known to have increased epidermal thickness. Holt and Marks (1977) showed that the therapeutic correction of patients with thyrotoxicosis was accompanied by a reduction in the epidermal thickness and the $I_S$, while treatment of patients suffering from hypothyroidism with thyroxine resulted in increases in these parameters; moreover, there appeared to be a linear relationship between the $I_S$ and the epidermal thickness, on the one hand, and the serum triiodothyronine on the other. It appears that thyroid hormones may well be important in the maintenance of normal epidermal homeostasis.

## (iv) *Morphological and kinetic correlates in proliferative skin disease*

There are a group of diseases, of which *psoriasis* and *pityriasis rubra pilaris* are the best recognized, where the epidermis shows the typical appearances of hyperplasia (Fig. 11.3), and where the epidermal state is unequivocally one of excess cell production. There has been considerable debate about the kinetic cause of the hyperproliferation—there are those who consider that the cell cycle time is much shortened in psoriasis (Weinstein and Frost 1968; Weinstein *et al*. 1983), although controversy has raged about the extent of this reduction (Goodwin *et al*. 1974, 1975; Weinstein 1975); the opposing school of thought holds that the growth fraction is elevated (Gelfant 1976; Gelfant *et al*. 1983),

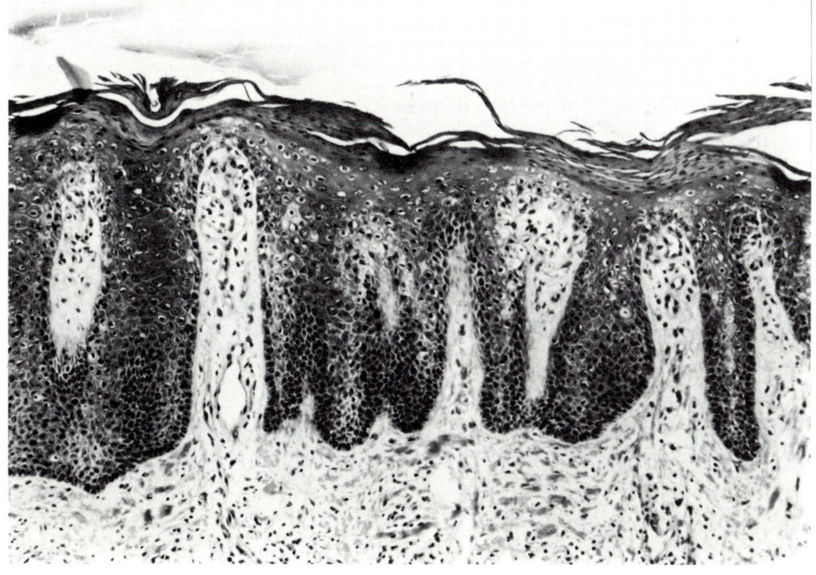

Fig. 11.3. The human epidermis in psoriasis. Note the similarity to the hyperplastic epidermis shown in Fig. 11.1: there is acanthosis with prolongation of the rete pegs, and, in this instance, prominent parakeratosis. Additional features include the vascular and oedematous dermal papillae, and the diffuse mononuclear infiltrate in the dermis. ($\times$100.)

though neither of these concepts withstand critical scrutiny as the only proliferative defect in psoriasis (Wright 1976, 1977, 1980*a,c*). The field has recently been summarized in a volume devoted exclusively to the topic, to which the interested reader is referred (Wright and Camplejohn 1983).

Generally speaking, the presence of parakeratosis (see Fig. 11.3) is correlated with increased proliferative activity in skin disease, and is usually regarded as a result of incomplete differentiation, due to the decreased time available during transit. However, there is some debate on the exact relationship between parakeratosis and cell proliferation: although Cox and Watson (1971) obtained a good correlation between the degree of parakeratosis and mitotic counts, Fry and McMinn (1968) reported that, with certain therapy, such as topical dithranol, the granular layer formed first (i.e. parakeratosis was lost) before the mitotic index fell, indicating a direct effect on the differentiation rate, rather than a direct action via mitosis. Moreover, Christophers and Braun-Falco (1967) induced increased proliferative rate in guinea pig epidermis by both hexadecane (see 11.1) and vitamin A treatment (see (i) above), but noting that neither was associated with parakeratosis, concluded that increased cell proliferation per se does not produce parakeratosis, but that there also has to be an associated defect in differentiation; this might indicate that, in psoriasis, there is an additional defect in differentiation, in addition to the proliferative lesion. However, we ought to note that Johannessen and Hammer (1980), studying lesional epidermis, have shown that the rate of formation of corneocyte layers, as measured with dansyl chloride, is highly correlated with the degree of parakeratosis; this would indicate a direct relationship between cell production rate and parakeratosis, although it does not, of course, exclude a role for a differentiation defect.

That this problem is not simple is indicated by the observation that during treatment with corticosteroids, the granular layer reforms at the same rate that the mitotic index falls, but still, one would expect that if increased mitotic activity was the root cause of parakeratosis, there should be a significant delay between the decrease in $I_M$ and the emergence of a granular layer, and it would appear, on present evidence, that a defect in differentiation, rather than a simple relationship with enhanced cell proliferation, is the cause of parakeratosis in psoriasis.

There is evidence that increased epidermal thickness is associated with an elevated labelling index (Christophers 1972*a*) in guinea pig tissues, and, expressing epidermal thickness as the ratio between the length of the basal lamina to the surface length in sections, Marks and Wells (1973) have found a linear relationship between this ratio and the flash-labelling index in lichen simplex, indicating that a simple relationship exists in human epidermis.

It also seems as if our postulated relationship between proliferative rate and ordered structure may also hold in human epidermis; thus Menton and Eisen (1971*b*) reported good stacking in *icthyosis vulgaris*, a condition associated with a low birth rate, but in conditions where $k_B$ is increased, such as psoriasis and *lamellar icthyosis*, there was much overlapping of surface corneocytes, with no stacking; it seems that elevated birth rates lead to loss of stacking in human epidermis also.

## 11.3.  Atrophy in squamous epithelium

Atrophy can reasonably be defined as a decrease in tissue mass, due either to a fall in the cell production rate while individual cell mass remains constant (hypoplasia), or a decrease in the mass of individual cells, while the cell production rate remains constant (hypotrophy). The resultant epidermis is thin, with a prominent reduction in the thickness of the Malpighian layer (see Fig. 11.4).

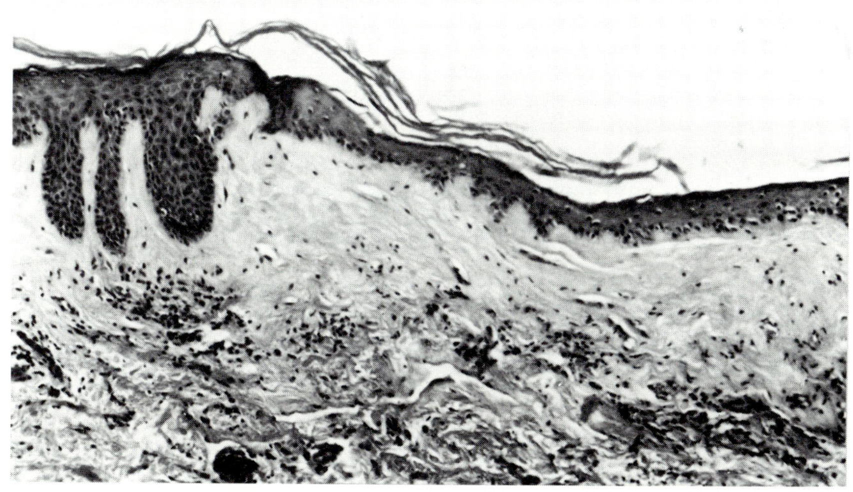

Fig. 11.4.  The junction between a portion of normal epidermis displaying rete pegs (on the left) and an area of atrophic epidermis showing thinning of the Malpighian layer and loss of rete pegs. The tissue is from the human vulva, and shows the features of *lichen sclerosus et atrophicus*; besides the epidermal atrophy, there is homogenization of the upper dermal collagen, and a mononuclear infiltrate in the mid-dermis. (× 100.)

Although not so well studied as hyperplastic responses, there are several good experimental and clinical models in epidermis, and the apparent conflict between putative changes in the cell production rate and cell size is well exemplified by the phenomenon of steroid-induced atrophy in the epidermis.

(i) *Steroid-induced atrophy*

There is good evidence, from work in cell culture (Kollmorgen and Griffin 1971), and also in *in vivo* systems (Wright *et al.* 1973) that glucocorticoids are powerful inhibitors of cell proliferation in epithelial tissues, and most reports are uniform in that they agree that steroids induce a $G_1$-$S$ block. Consequently, it is hardly surprising that reports have appeared which confirm this finding in squamous epithelium *in vivo*, and most workers have ascribed the atrophic effect to this strong inhibitory action on cell proliferation itself: early reports (Chaundry *et al.* 1956) indicated that topical steroids depressed $I_M$ in mouse ear epidermis after a short latent period of four hours, but the more detailed studies of Frankfurt (1968) showed that even low systemic doses (2.5 $\mu$g/g) of hydrocortisone inhibited the flow of cells into $S$ in mouse forestomach, but only if given nine hours before the circadian peak of $I_S$ values, which occurred at 1800 hours; given closer to this peak, i.e. three or six hours beforehand, such low doses were inactive, although high doses (50 $\mu$g/g) given six hours beforehand prevented the peak. These results indicated the existence of a steroid-sensitive point, some nine hours before the onset of the $S$ phase, which can prevent $G_1$-$S$ transition, and of course, such a concept is fully consistent with an action on the transition probability. It is interesting to note that the inhibitory action of hydrocortisone was reduced on chemically-induced hyperplastic epithelium, and largely eclipsed on squamous papillomas; similarly, in the studies of Davis (1964), while subcutaneous cortisol depressed the number of colchicine-arrested metaphases in mouse epidermis, there was no apparent effect on epidermis regenerating after wounding. However, topical steroids are apparently effective in reducing the mitotic index in the hyperplastic epidermis in psoriasis, and cortisone does repress DMBA-induced carcinogenesis when given during the promoting phase (Ghadially and Green 1954; Slaga *et al.* 1975*b*),which would argue for an effect on transformed cells (see Chapter 13).

It thus appears that steroids have a direct action on the $G_1$-$S$ transition: Hennings and Elgjo (1971) interpreted the temporal changes in [³H]-TdR incorporation and in colchicine-arrested metaphases produced by topical hydrocortisone as fully consistent with an inhibition of $G_1$-$S$ flux, and, notwithstanding the negative results of Davis (1964) and Frankfurt (1968) where regenerating epidermis was concerned, Marks *et al.* (1973) introduced the mouse epidermis, regenerating after tape stripping (see Chapter 10), as a model for exploring the antiproliferative effect of steroids; both topical hydrocortisone and betamethasone 17-valerate showed a dose-responsive inhibition of $I_S$ and colcemid-arrested metaphases, while betamethasone reduced $I_S$ in *normal* epidermis, an effect also apparent with the powerful glucocorticoid, dexamethasone (Slaga *et al.* 1975*b*). *In vitro* results with glucocorticoids on mouse epidermis also indicate a dose-responsive inhibitory action on cell proliferation (Bullough *et al.* 1968).

In humans, Marks *et al.* (1971*b*) showed a reduction in [³H]-TdR incorporation in steroid-treated epidermis, and Fisher and Maibach (1971), using their colcemid cream metaphase-arrest method, claimed that both oral steroids and topical fluorinated steroids were powerful mitotic inhibitors, and proposed a relationship between the degree of mitotic inhibition and the therapeutic efficacy in the hyperproliferative skin disease psoriasis; in this context, Baxter and Stoughton (1970) did find a steroid-induced reduction in $I_M$ in psoriatic epidermis.

This evidence would appear to be rather conclusive in indicating that steroids are active inhibitors of cell proliferation; however, we must record a discordant voice; Laurence (Bullough and Laurence 1968$a$; Laurence and Christophers 1976) is not persuaded of the proliferative action of steroids; thus, in mouse epidermis *in vitro*, hydrocortisone had no effect on mitotic activity, although it may have had an additive inhibitory effect when combined with chalone and adrenaline, and was supposed to stabilize the chalone:adrenaline complex (but see Chapter 14, section 1). Surprisingly, Laurence and Christophers (1976) reported that as long as three weeks exposure to hydrocortisone in the drinking water had no effect on $I_S$, $I_M$, $t_S$, or $t_M$ in mouse ear or foot epidermis, although there was a prominent decrease in both epidermal thickness and cell size. Laurence and Christophers therefore claimed that steroids induce atrophy by an acceleration of the rate of differentiation, with a selective action on post-mitotic cells only.

It is difficult to reconcile these two viewpoints; while there is little doubt that steroids do decrease cell size, an effect which correlated with the reduction in epidermal thickness, we consider it rash to discard the quite hard kinetic evidence which argues for a depressive effect on cell proliferation. Few studies have been carried on as long as those of Laurence and Christophers, and, moreover, in their study there was no analysis of the early effects of the treatment; of course, it may be that the initial action of steroids is to inhibit cell proliferation and induce atrophy because of this, but, as a later effect, there is an action on cell size, and, possibly, the maturation rate; more detailed studies would provide the answer to this question, which, because of the widespread use of steroids in dermatological practice, is not without clinical import.

### (ii) *Cytotoxic drugs*

Compared to the intestine, where both the action of, and the proliferative responses to, cytotoxic drugs have been studied in some detail (see Chapter 21), there is a curious lack of information in the epidermis; this could be related to the rarity of cutaneous clinical complications produced by cancer chemotherapy when compared to the gut, but given the important action of these drugs in psoriasis and in squamous carcinomas (Chapter 13, section 4), there is an urgent need for such knowledge. However, Bleomycin, a drug used extensively in the treatment of squamous carcinoma, has been studied in some detail by the Oslo group (Clausen 1975; Iversen *et al.* 1976; Iversen *et al.* 1977; Thorud and Clausen 1978). Bleomycin induces hypoplasia in mouse epidermis, and appeared to depress the fraction of cells in $S$, $G_2$, and $M$, and to inhibit flux through these phases, acting mainly on late $G_1$ and mitosis. An action by delaying maturation was also suspected, and, interestingly enough, this was correlated with an increase in the $G_1$ chalone content of bleomycin-treated epidermis (see Chapter 14, section 1), possibly indicating that mature cells with high chalone content were accumulating in the epidermis.

In epidermis regenerating after tape stripping, bleomycin produces a transient block in the $G_1$ phase, which is soon released, and these cells are then blocked at the $G_2$-$M$ boundary, which action appears to be the main effect of the drug; however, in normal epidermis, this $G_2$ block effect was not a prominent feature (Thorud and Clausen 1978). While there is no doubt that bleomycin kills tumour cells (Terasima *et al*. 1976), there is little indication of this in epidermis; thus even a very high (80 $\mu$g/kg) dose induces only a 10 per cent fall in basal cell density (Thorud and Clausen 1978).

Methotrexate, which, when given systemically, inhibits DNA synthesis in epidermis, probably acts via the *de novo* pathway (White 1971). This agent has found favour in the treatment of psoriasis (Weinstein 1971; Weinstein and Velasco 1972). Although Karesek (1971) was unable to detect a cytostatic effect on human keratocytes growing *in vitro*, Flaxman *et al*. (1977) observed an irreversible mitotic inhibition when cultured human keratocytes were exposed to methotrexate for long periods (24 hours); shorter exposures produced a reversible inhibition of DNA synthesis. Although much has been made of the supposedly selective action of methotrexate on the rapidly proliferating psoriatic cells, Flaxman *et al*. (1977) reported that methotrexate responses in cultured psoriatic keratocytes were similar to those seen in normal cells; however, since, in this culture system, psoriatic and normal keratocytes share much the same kinetic parameters (Chopra and Flaxman 1974), it may be that, since the proliferative advantage (or disadvantage) possessed by psoriatic cells *in vivo* is evidently lost in culture, any selective action due to kinetic differences is lost. It is interesting that the repression of DNA synthesis is reversible: this may indicate that the therapeutic effect of methotrexate in psoriasis is not wholly related to its cell killing ability, and the remaining action may be due to an action on prolonging the cell cycle time, or removing cells from the cell cycle.

On the other hand, *hydroxyurea* and *cytosine arabinoside*, predominantly $S$ phase specific agents, not only depress DNA synthesis in mouse epidermis, but also result in the formation of numerous necrotic cells (Stoughton 1971); hydroxyurea is effective in some cases of psoriasis, and could act mainly by this cell-killing effect. But Hennings and Devik (1971) concluded that the relationship between DNA synthesis inhibition and cell killing was complex; in normal mouse epidermis, cell killing appeared unrelated to the degree of inhibition of DNA synthesis, but possibly more relevant was the duration of the block. In croton oil-induced hyperplasia, even a very large (75 per cent) but transient depression of DNA synthesis was not associated with increased cell kill.

Some other phase-specific agents also block DNA synthesis after *topical* application: 5-fluorouracil, which is used in the treatment of actinic keratoses, inhibits DNA synthesis in both normal and actinic epidermis (Eaglestein *et al.* 1970), and induces prominent ultrastructural degenerative changes (Zelickson *et al.* 1975).

In conclusion, there is evidence that cytotoxic agents produce cell kinetic effects in squamous epithelium which are similar to those observed in other renewal systems, but it is presently unclear whether the epidermal atrophy induced by them is due to a prolonged depression of cell proliferation or to overt necrogenic effects; in this respect, the relationship between the inhibition of cell proliferation and cell death is particularly obscure.

### (iii) *Malnutrition-induced atrophy*

In neonatal animals suffering from severe undernourishment, there is an early appearance of atrophic changes. Mathur and Deo (1976), using a curious modification of the double-labelling method, suspected that cells were held up in the *S* phase, although this would appear to be an unlikely mechanism for the atrophic change.

## 11.4. Conclusions

Like most renewal systems, squamous epithelium has the capacity to respond relatively swiftly to any induced change, and growth reactions in the form of hyperplasia and atrophy are common responses; with the available methods, we can readily follow the progressive kinetic changes which accompany these growth reactions, but we do not yet understand the morphogenetic mechanisms which lie behind the development of the singular appearances that occur in the epidermis in, for example, hyperplasia. It is likely that a combination of changes in the rate of differentiation and in proliferative rate are responsible, and, as yet, we do not have any generally satisfactory method for measuring the maturation rate in quantitative terms on a cellular basis. We would look to the understanding of the process of hyperplasia as one of the pressing needs in the biology of squamous epithelium.

# 12 The response of squamous epithelia to irradiation

Presumably because of its clinical importance, and the relative ease with which studies can be performed, the skin has been the tissue of choice in irradiation studies on squamous epithelium, and much of the ensuing discussion is centred on the epidermis. However, although we are of course concerned exclusively with the epithelial tissue component in this study, it would be a gross oversimplification to presume that tissue reactions in say, the epidermis, are mounted solely by the squamous epithelium; the survival and behaviour of the epithelial cells is strongly influenced by their microenvironment (Devik 1955), and this is indicated, in the skin, by the practice of scoring the response to irradiation by the degree of *erythema*, or transient reddening of the skin. This is, of course, related to the blood vessel changes.

There is a further point to be made: in considering epithelial behaviour after irradiation, we are naturally interested in the proliferative aspects, but this would be less than half of the story. Since the damaging properties of irradiation are, at least partially, related to its ability to kill cells, indivisible facets of the subject include *cell survival* after irradiation (Potten 1981), and subsequent *regeneration*, which makes good the cellular deficit. Thus the post-irradiation phenomena have a good deal in common with wound healing.

## 12.1. Skin reactions to irradiation

If excess irradiation is given to the skin, a burn results. There are a definable series of events which precede this; the *erythematous reaction* can be divided into three stages: an early and variable phase which comes on within a few hours or days; a much more constant and reproducible phase which occurs after 10 to 20 days, and which, because of its reproducibility and dose-dependence, was, and is, used as an effective way of expressing a biologically effective dose (see, for example, Denekamp *et al.* 1966; Douglas *et al.* 1979). However, this 'skin erythema dose' is clearly dependent on many variables, which concern both the irradiation and the skin, and is difficult to standardize as between centres (Potten 1978*b*). Lastly, there is a final, variable phase which develops at a later date. In the early and late phases the cause is probably related to transient dilatation of the dermal blood vessels, and to telangiectasia (a more permanent dilatation and vessel proliferation) respectively. In the second phase, the damage induced to both epidermal and dermal cell populations also probably contributes.

Increasing doses of irradiation induce desquamation and ulceration; in the mouse epidermis, where most work has been done, about 10 Gy of gamma or X-rays are needed before gross skin reactions are evident. A visible reaction occurs at 6-8 days, and peak damage, which might include ulceration if the dose is large enough, ensues on the 12th to 25th day after irradiation. Unless large doses are given, the lesion usually heals within four to eight weeks (Denekamp *et al.* 1969; Hegazy and Fowler 1973*b*; Field *et al.* 1975; Leith *et al.* 1975). If increased cell proliferation is induced by plucking, an earlier reaction occurs (Hegazy and Fowler 1973*b*), and ultraviolet irradiation also causes an early response, which, however, may be more attributable to more prominent vascular changes.

## 12.2. Cell survival in squamous epithelia after irradiation

We have already alluded to experiments on cell survival in our discussion of epidermal clonogenic cells (see Chapter 8, section 4); the investigation of survival phenomena of course demands a suitable end-point, which was provided by Withers (1966), who introduced a *macrocolony method* applicable to epidermis, by defining survivors as those cells capable of sufficient cell divisions to produce a macroscopically recognizable nodule in an otherwise sterilized area of skin. Practically speaking, the test doses are given inside a 'moat' of skin which has been sterilized by a dose of 30 Gy, which minimizes the confluence of clones, and the possibility of migration; the skin is usually plucked between 0 and 24 hours before irradiation, and thus both epidermal cells and follicular cells are probably rapidly proliferating at the time of irradiation. The nodules formed are usually about 2 mm in diameter, and probably contain about $3\text{-}4 \times 10^4$ cells, and take some 12-24 days to develop. This would indicate that a minimum of 15-16 doublings has taken place within this period, and although the doubling time of the cells in these nodules has been estimated at about 20-24 hours (Withers 1967*c*), because some cells are lost to keratinization, Withers (1967*a*) suspects that the cell cycle time in regenerating macrocolonies may be as short as 12 hours, and it may be recalled that this approximates to the minimum $T_C$ values reported in mouse squamous epithelium regenerating after wounding.

Despite the fact that it is only possible to use high doses in the test area, $D_0$ values can be measured, and the macrocolony method usually gives values between 1 and 1.4 Gy with an average value of 1.35 Gy (see Fig. 8.18, p. 320), and the extrapolation number (i.e. the cell number at zero dose) is between $1.2 \times 10^4$ and $3.1 \times 10^7$ cells per $cm^2$ (see Table 12.1). However, different results are obtained if the *microcolony method* of Al-Barwari and Potten (1976) is used; this measurement was introduced to decrease the time period necessary between irradiation and measurement, and it also decreases the chances of confluence and allows smaller dose levels to be studied. Practically, micro-colonies are detected by viewing [$^3$H]-TdR labelled cells in epidermal sheet preparations separated from the basement membrane (Hamilton and Potten 1972). This method gives a $D_0$ of 2.23 Gy, and, peculiarly enough, it is not possible to fit the data with the macrocolony-derived $D_0$ of 1.35 Gy, and the reason for this difference is not yet evident, but may be caused by the macro-colony method selectively favouring the detection of the most rapidly growing colonies in the presence of a considerable range of clonal growth rates (Al-Barwari and Potten 1976; Potten 1978*b*). We have seen elsewhere that one of the first reactions to injury is cell migration, and we have favoured the hypothesis that basal cell proliferation is switched on by basal layer depletion; moreover, if the stimulus-responsive (?stem) cells at the centre of the EPU migrate (see Chapters 8 and 10), and if they are not already killed by the irradiation, then the possibility exists that the microcolony method may be measuring the radiosensitivity of a *more differentiated proliferative population* (Potten 1978*b*).

Figure 12.1 shows that, at lower dose levels, more macrocolonies than micro-colonies are scored, while at higher doses, more microcolonies are seen (Potten 1978*b*). This may indicate that the methods are not measuring the survival of the same population, but it is interesting to note that the macro- and microcolony methods in the intestine do not show similar discrepancies. Some 44 per cent of the microcolonies seen on the third day are adjacent to hair follicles, and it

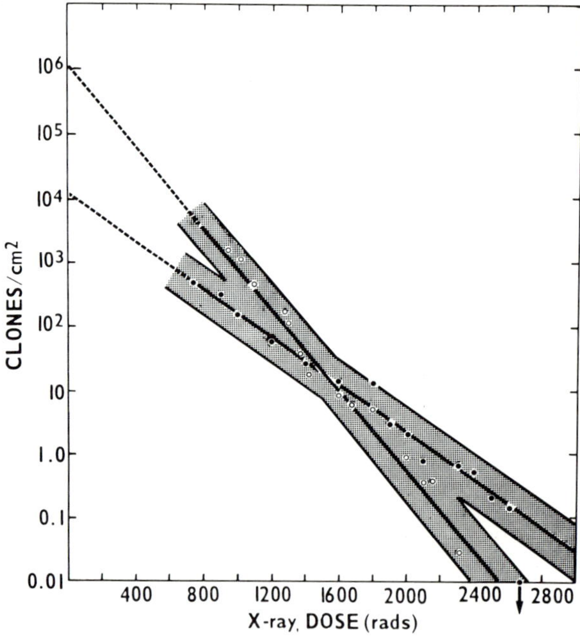


Fig. 12.1. The difference in clonogenic survival after graded doses of X-irradiation when measured by the macrocolony (○) or the microcolony method (●). (By courtesy of Dr C. S. Potten.).

seems distinctly likely that many are derived from *follicular stem cells*, and we have seen how Potten (1981) has proposed that such follicular stem cells may have a greater regenerative capacity (see Chapter 8, section 4). If the microcolony method measures follicular stem cells selectively, clones derived from clonogenic cells further down the follicle could take longer to form a macroscopically evident nodule, and the macrocolony method could thus be biased towards the interfollicular epidermis with its putative single clonogenic cell per EPU. It is then of course possible that these two populations have different radiosensitivities; Potten (1978*b*, 1981), on really no good grounds, has speculated that follicular stem cells are 'better' or 'the vital' squamous epithelial stem cells, and that the EPU stem cells are, in fact, early differentiated proliferative cells. On this hypothesis, the macrocolony method measures differentiated cell survival, with perhaps some stem cell contamination, and these 'differentiation-committed' clones may appear later; thus 'differentiation-committed' clonogenic cells are more radiosensitive than stem cells, although we should note that in the intestine the stem cells are radiosensitive compared with the committed cells (Cheng and Leblond 1974*c*), while in the bone marrow, Testa (1979) also reported that haemopoietic stem cells (CFU-S) have substantially lower $D_0$ values than committed granulocyte precursors (CFU-C). So, if this hypothesis obtains, epidermis is different from other renewing tissues; moreover, it is also contrary to the earlier hypothesis, advanced by Al-Barwari and Potten (1976), and referred to above, that the microcolony method measures differentiation-committed cells!

This discussion does indicate that, as in most fields of cell kinetics, the methodology is all important, and until it becomes clear what cells are measured by which technique, we cannot draw even tentative conclusions about regenerative populations.

Besides this disparity, Fig. 12.2 shows that there is some variability in the $D_Q$ value, and this is probably related to the irradiation conditions and the radiation quality; for example, if irradiated in three atmospheres of oxygen $D_Q$ is decreased (Withers 1967*b*), while the use of high LET irradiation, such as neutrons or helium ions (Denekamp *et al.* 1971), also leads to a marked decrease in $D_Q$. However it is interesting to note that the results with high LET irradiation do not give different $D_0$ values (Potten 1978*b*).

It will be noted that the skin is plucked before irradiation in most, if not all studies; we have seen how this increases the proliferative rate and decreases the cell cycle time in the interfollicular epidermis from a putative 110 hours to about 43 hours (Hegazy and Fowler 1973*a,b*; and see Table 10.2, p. 381). However, this manoeuvre apparently does not have any great effect on the survival data, although in Hegazy and Fowler's study the $D_0$ was decreased to 0.95 Gy, an effect which was probably unrelated to the fact that 28 MeV helium ions were used (see above). On the other hand, Potten (1978*b*) has expressed surprise that $D_0$ values as between the skin and the intestine are so similar, given the 7–10 times difference in the $T_C$ values, and the differing

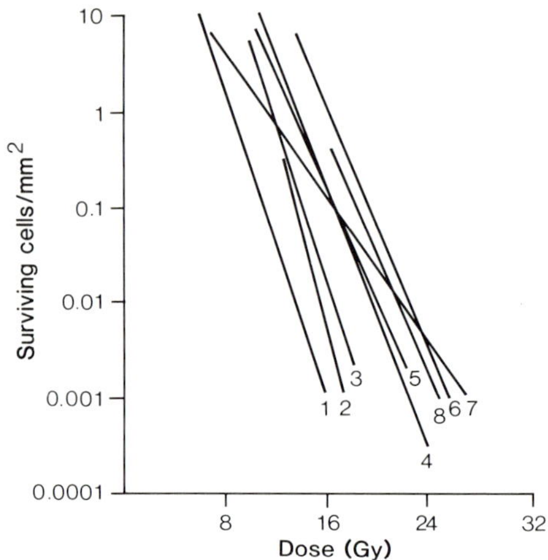



Fig. 12.2. Epidermal clonogenic cell survival curves. The average of the eight estimates for $D_0$ is 1.35 Gy.  1. 8 MeV neutrons at 16 hours after hair plucking (Denekamp *et al.* 1971). 2. 28 MeV helium ions (Leith *et al.* 1971).  3. 150 kV X-rays, 30 p.s.i. oxygen (Withers 1967*b*).  4. 29 kV X-rays at 1-12 hours post-plucking (Withers 1967*a*).  5. 150 kV X-rays (Withers 1967*b*). 6. 250 kV X-rays, 19 hours post-plucking (Emery *et al.* 1970). 7. 290 kV X-rays (Al-Barwari and Potten 1976).  8. 1 MeV electrons at 20-24 hours post-plucking (Denekamp *et al.* 1966). (Redrawn from Potten *et al.* (1983).)

radiosensitivities of the various phases. However, when comparing plucked epidermis with intestine, perhaps the $T_C$ values are not so different, particularly if the clonogenic intestinal cells are indeed the basal crypt columnar cells, which are known to be slowly cycling with respect to the remaining proliferative population.

## 12.3. Clonogenic cells: survival and numbers

Table 12.1 shows that most survival curves give extrapolation numbers of $<1.0 \times 10^7$ cells/cm$^2$, while basal cell density is about $1.5 \times 10^6$ cells/cm$^2$. We have already noted that mammalian radiation survival curves have a distinct shoulder, and even with the addition of the follicular stem cells, indicate that the basal cell cellularity is certainly more than the clonogenic cellularity; moreover, we have also discussed the investigation of the true origin of the survival curve by the determination of the recovery factor (see Chapter 8, section 4), and how Potten and Hendry (1973) used the data of Withers (1967*c*), Emery *et al.* (1970) and Denekamp *et al.* (1971) to estimate the recovery factor and equate it with the extrapolation number of the individual cell survival curve (see Fig. 8.18 and eqns 8.1 and 8.2); in all three cases (and note that different radiation quality was used in each), the origin of the cell survival curve was estimated at between $3 \times 10^4$ and $1.1 \times 10^5$, or about 10 per cent of all basal cells; if similar recovery

Table 12.1.   Radiation survival characteristics and extrapolation numbers for epidermal clonogenic cells in mice. (After Potten (1978) and Potten *et al.* (1983))

| Reference | Radiation quality* | Dose rate (Gy/min) | $D_0$ (Gy) | $N$† (cells/cm²) |
|---|---|---|---|---|
| Withers (1967*a*) | 29 kV X-rays | 7.7 | 1.35 | $1.7 \times 10^6$ |
| Withers (1967*b*) | 150 kV X-rays | 1.95 | 1.4 | $1.4 \times 10^6$ |
|  | 150 kV X-rays ($O_2$) | 1.95 | 1.07 | $1.0 \times 10^7$ |
| Emery *et al.* (1970) | 250 kV X-rays | 3.0 | 1.35 | $1.2 \times 10^7$ |
| Denekamp *et al.* (1971) | 8 MeV neutrons | – | 1.09 | $2.3 \times 10^5$ |
| Leith *et al.* (1971) | 28 MeV helium ions | 30.0 | 0.95 | $3.1 \times 10^7$ |
| Al-Barwari and Potten (1976)‡ | 290 kV X-rays ($O_2$) | 1.47 | 2.33 | $1.2 \times 10^4$ |

*Irradiation was carried out in an air atmosphere unless specified.   †$N$, extrapolation number.   ‡This measurement was made using the microcolony method – all other measurements used the macrocolony method (see text). Unirradiated basal cell density $\sim 1.5 \times 10^6$ cells/cm.  Overall $D_0 \sim 1.35$ Gy.

factors apply to the microcolony method, then the clonogenic cell number in the basal cells could in fact be very low. However, it is not yet clear what effects, if any, plucking has on these estimates, or, of course, how advisable it is to equate recovery potential in cells which have been irradiated with that of unirradiated cells. On the other hand, we have seen how such information has been used by Potten (1981) to support his concept of proliferative epidermal subpopulations (Chapter 8, section 4). The possible role of follicular stem cells in survival studies is also discussed in Chapter 8, section 4.

## 12.4.  Radiation-induced responses in squamous epithelium

### (i)  *Cell death*

Potten (1978*b*) has drawn attention to an interesting anomaly; from Fig. 8.18 (p. 320) it is clear that 12 Gy, given in a single dose, leads to a cell survival of about 100 cells/cm². With a normal value of about $1.5 \times 10^6$ basal cells/cm², this means that at least 99.9 per cent of the basal cells have been killed. If cell death had occurred on this scale, then many dead cells would be seen in epidermal sheet preparations or in sections, and recognized as frankly pyknotic nuclei (Brown 1970; Fowler and Denekamp 1976) or as apoptotic bodies (Kerr *et al.* 1972; Olson and Everett 1975). However, even given the problems of recognizing and counting pyknotic/apoptotic nuclei in epidermis, counts made by Potten (1978*b*) indicate that very low numbers of dead cells are seen in epidermis after irradiation; similarly the pyknotic index of Hegazy and Fowler (1973*b*) peaked at less than 10 per cent at 4–5 days after a dose of 10 Gy to mouse epidermis, a dose claimed to produce 90 per cent clonogenic cell kill, while Brown (1970) also noted very small degenerate cell indices after 10 Gy given to the hamster cheek pouch. Moreover, the counts do not appear to be dose-related, and there is no correlation with labelling or mitotic activity (Hegazy and Fowler 1973*b*; Fowler and Denekamp 1976; Al-Barwari and Potten 1976), so that mitotic death would not appear to be selectively the mode of cell death after irradiation. In this respect, however, the fraction of degenerate cells appeared to vary diurnally in irradiated cheek pouch (Brown 1970), in a manner indistinguishable from the diurnal $I_M$ curve; the fraction of degenerate cells did not rise above control values until after the period of mitotic inhibition; both these observations were interpreted as indicating that cell death in irradiation damaged cells is associated with mitosis.

There are of course several reasons why the yield of dead cells should not correspond with the known survival statistics; we really have no experimental justification for relating clonogenic survival to cell killing in this situation; it is conceivable that most cells are reproductively sterilized but not killed, and migrate after irradiation, either giving rise to no progeny at all, or to only a few non-clonogenic cells after a limited number of divisions; basal cell depletion would occur, and limited cell survival would be accomplished without cell death. Assuming that these apoptotic bodies observed by Potten (1978*b*) are not in fact the nuclear and cytoplasmic material which is thought to be extruded from cells after irradiation (Barnes 1983), cells must be dying; the fact that the apoptotic yield is low could be explained by a rapid removal process, since little is known about the kinetics of apoptosis. If these explanations are not tenable, then the views of Potten (1978*b*) are interesting: Potten has suggested that this cell death discrepancy could be explained by a subpopulation of cells, possibly less than 10 per cent (?stem cells, ?stimulus-responsive cells in $G_2$), whose response to irradiation is cell death; if these cells are in $G_2$, then this is of course known to be a very radiosensitive phase of the cell cycle. We have seen that there is a pronounced circadian variation in the proportion of cells in $G_2$ (Clausen *et al.* 1979, 1981), and cells in $G_2$ do reach quite high numbers at some time points in the day (see Fig. 9.8). Potten (1978*b*) considered that the death statistics were consistent with about 10 per cent of the population dying; now it may be purely fortuitous that 10 per cent is the proposed clonogenic fraction in mouse inter-follicular epidermis, or, of course, they could be identical populations. However, if this is the case, why is the observable death process non-dose-dependent? This could be caused by this subpopulation having a low irradiation threshold for cell death, as has been suggested for basal crypt columnar cells in the small intestine (Potten 1977), or perhaps only large changes in the fraction of dead cells in such a small subpopulation can be detected (Potten 1978*b*). In this context, Al-Barwari and Potten (1979), comparing ordinal scaled skin reactions in several skin sites in mice which differed widely in their cell kinetics (cf. tail and dorsal epidermis), found that X-rays produced later reactions (12 days after treatment) in the more rapidly proliferating tail than in the more slowly dividing dorsum, and concluded that the appearance of a skin reaction was unrelated to any cell kinetic parameter; again they attempted to explain the data on the basis of cell death occurring only in the small clonogenic subpopulation.

An alternative way of looking at the problem is to count the number of surviving cells, and obtain an estimate of the basal cell density. This has been done in sections by De Rey and Klein Szanto (1972) and Etoh *et al.* (1973, 1977) and in epidermal sheet preparations (Hamilton and Potten 1972) by Al-Barwari and Potten (1976). We have seen (Chapter 10) that plucking produces a small decrease in basal cell density, with a return to normal values at day 3–4, followed by the now familiar overshoot; 12 Gy, which produces 99.9 per cent clonogenic cell kill (Fig. 12.2), provoked basal cell depletion which reaches a minimum of 32–40 per cent loss between days 3 and 5 (De Rey and Klein Szanto 1972; Potten 1978*b*; and see Fig. 12.3). Similarly, a dose of 10 Gy, which sterilized

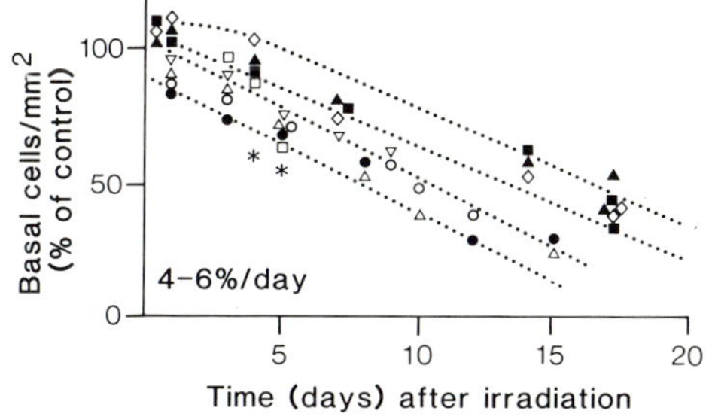


Fig. 12.3.   The density of mouse epidermal basal cells plotted as cells/mm² against time (days) after X-irradiation. The density falls steadily at a rate of about 5 per cent per day, which is quite independent of dose (■ 17, ◇ 23, ▼ 27 Gy, pig skin; ○ 22, △ 30, ● 50, ▽ 30 Gy, guinea pig skin; □ 80, * 160 Gy, rat tail skin). (Redrawn from Potten *et al.* (1983).)

90 per cent of basal cells, only induced a 30–40 per cent basal cell depopulation at 10 days after irradiation in mouse epidermis (Fowler and Denekamp 1976), while 23 Gy to swine epidermis caused an approximately 30 per cent decrease in basal cell density (Archambeau 1972). In this respect, the results of De Rey and Klein Szanto (1972) indicate that, even after 160 Gy, there are still about 50 per cent of the basal cell population remaining in rat epidermis; and even after chronic neutron irradiation over 24 weeks, Page and Toto (1978) could not detect any decrease in the basal cell density, although $\beta$-irradiation appears associated with a larger basal cell loss (Etoh *et al.* 1977).

It is therefore evident that basal cell density does not correlate with the notional $D_0$ of 1.35 Gy, which is often quoted for basal cells. Potten (1978*b*) has argued strongly that there are several competing processes which are responsible for the basal cell density at any point in the radiation response: overall cell loss would be direct loss by a death process, by continued cell migration and differentiation, and inhibition of mitosis, while cells would be added by stem cell regeneration together with some limited division of differentiated cells and/or otherwise sterilized stem cells, and also by migration and division of cells from outside the irradiated area. If the regeneration processes predominate, there may in fact be early restoration of basal cell numbers (Potten 1978*b*).

We thus have a situation where, although there is direct evidence of limited clonogenic cell survival, direct observation of cell death and the physical presence of surviving basal cells indicate little *actual* cell loss. The position can be explained by a small subpopulation, responsible for clonogenic cell number (Potten 1978*b*); this subpopulation would be small in number, so that microscopic evidence of cell death would not be a major post-irradiation phenomenon, and nor would basal cell depletion.

### (ii) *Cell migration and proliferation after irradiation*

In discussing the normal mechanisms of basal cell migration, we noted that, even when mitotic activity was blocked by large doses of $\beta$-irradiation, there was continued migration (see Chapter 8, section 6); thus, even at doses up to 30 Gy, Etoh *et al.* (1973) reported basal cell migration at the apparently normal rate, although at higher doses the migration rate may be stimulated and account for the basal cell depletion (De Rey and Klein Szanto 1972). As noted before (Chapter 8, section 6) the mechanism of this migration is obscure: since mitosis is inhibited for up to four days (Etoh *et al.* 1973), many cells could be blocked in $G_2$. These will be larger than average, compared with cells with a uniform or exponential age distribution, and thus might account for at least some lateral pressure; whether or not this is the whole story is difficult to say (see Chapter 8, section 6, and Fowler and Denekamp 1976). It is also interesting to note that, even at doses up to 12 Gy, which leads to loss of $>90$ per cent of the clonogenic cells, there is little in the way of ultrastructural abnormalities in the surviving basal and suprabasal cells; at higher doses nuclear abnormalities do occur (Potten 1978*b*). Neither is gross damage a major factor after 10–15 Gy: doses above 17 Gy give visible desquamation which begins at 10–12 days after irradiation, peaking at 18–20 days.

The rate of *repopulation* can be studied by giving graded test doses of X-rays to the tissue after the original irradiation, and the size of the test dose needed to bring the subthreshold damage to a constant level can be used as a measure of repopulation, of the depleted epidermis, using either equal skin reactions for comparison (Denekamp *et al.* 1969; Fowler and Denekamp 1976), or an epidermal clone iso-effect (Withers 1967*c*; Emery *et al.* 1970). No additional irradiation is necessary to compensate for proliferation until about a week after 10 Gy and then 0.5 Gy/day are required for the next week, equivalent to a doubling time of about two days. In plucked epidermis, repopulation ensues in about two days, indicating a more rapid onset of proliferation. It is interesting to note that this is sooner than indicated by measurements of the proliferative indices, possibly indicating the difficulties in making kinetic observations on such perturbed populations.

Although doses of irradiation as large as 30 Gy depress proliferative activity for some days, in the dose range 9–12 Gy however, although proliferative indices are depressed, they are not reduced to zero, and rise steadily with time (Etoh *et al.* 1977; Fowler and Denekamp 1976; Hegazy and Fowler 1973*b*), and in fact only rise above normal levels at about 7–9 days after 10 Gy, accompanied by a halving of the cell cycle time, due mainly to a decrease in $t_{G_1}$ (Devik 1961*a*,*b*; Denekamp 1973; Yamaguchi and Tabachnik 1972); the increase occurs even later in swine epidermis, where indices are maximal at 25 days after 23 Gy, where $T_C$ may be as short as 16 hours (Archambeau 1972). These rises in proliferative indices occurred after the peak of overtly pyknotic cells; moreover the onset of this proliferative wave does appear to be dose-dependent — for unplucked skin, 5 Gy evoked a rise in $I_S$ at five days, increasing to 18 days and $> 28$ days after 15 and 20 Gy respectively. Similar observations were made for plucked skin, and in guinea pig skin after β-irradiation (Etoh *et al.* 1977).

Brown (Brown and Berry 1969; Brown 1970) also reported that 10 Gy of X-rays to the hamster cheek pouch decreased the cell cycle time as evidenced by both FLM measurements and continuous labelling methods, and, although there was a lengthening of $t_{G_2}$, no detectable change was discerned in $t_S$ (Song and Tabachnik 1969). Because lower doses (i.e. $<15$ Gy) produce no desquamation, loss of the more superficial cells would not appear to be the cause of this proliferative wave; moreover, even with the induction of desquamation with 17 Gy, the enhanced proliferation precedes the desquamative event by two or three days (Fowler and Denekamp 1976).

Iso-effect experiments after fractionated irradiation also indicate that compensatory proliferation is delayed for some time (Denekamp 1973; Fowler and Denekamp 1976); although four 13 Gy fractions, given daily, induced no repopulation, after nine fractions 0.6, and after 14 fractions 1.3 Gy/day were required to compensate for the induced proliferation; these figures correspond to a doubling time of 36–48 hours and 16–24 hours respectively, and this was also supported by continuous labelling data (see Fig. 12.4). However, the $I_S$ would again appear to be but a poor estimator of proliferative rate in this

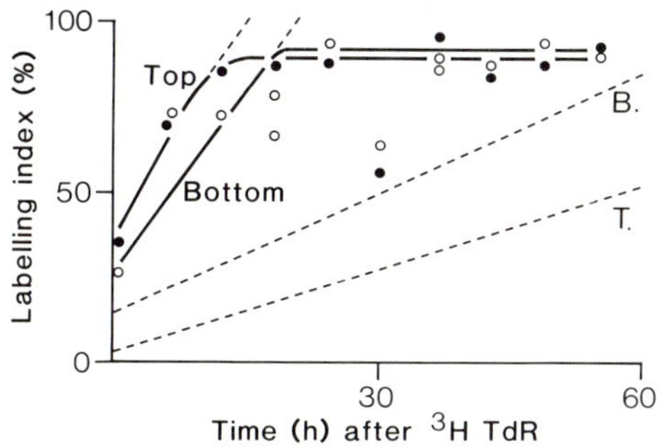


Fig. 12.4. Continuous [³H]-TdR labelling curves in the bottom (○) and top (●) of the mouse foot epidermis after 14 × 3 Gy fractions. Lines B and T relate to results in normal epidermis. (Redrawn from Denekamp *et al.* (1976).)

situation. Again there appeared to be a halving of the cell cycle time, and remembering the problems of measuring $I_P$ by the continuous labelling method, $I_P$ did appear to be reduced (Denekamp *et al.* 1976); in kinetic terms, this is a singular reaction to induced cell loss. But both iso-effect data, and continuous labelling results, indicated that proliferative rate was not increased until about 10 days after the commencement of the irradiation, suggesting a role for basal layer depletion as a trigger. Although this does correspond with the hypothesis relating basal cell density and the induction of cell proliferation after wounding, we should note that other authors, for example Brown (1970) have reported increases in $I_S$ as soon as 2–3 days after 10 Gy in the hamster cheek pouch.

The loss of prelabelled cells by migration in irradiated epidermis apparently occurs at the same rate as in unirradiated epidermis (Etoh *et al.* 1973), while Potten (1978*b*) found $I_S$ values which effectively remained the same as in unplucked epidermis after 12 Gy; even in epidermis receiving repeated doses of 3 Gy, continuous labelling studies show a similar rate of accumulation of labelled cells (Denekamp *et al.* 1976).

Thus it would appear that, even with doses as large as 12 Gy, which produce a 99.9 per cent reduction in clonogenic cell number, there are many cells which continue to divide and migrate in more or less the same manner. Again, the two-compartment model of Potten (cf. Potten 1976, 1978*b*; Potten *et al.* 1982*b*) would attempt to explain the phenomenon by clonogenic cell kill among the small subpopulation of stem cells, while the differentiating proliferative cells (i.e. cells in positions 2 to 6, Fig. 8.14, p. 313) are relatively resistant to irradiation, and continue to divide and migrate. This hypothesis would also relate the timing of the gross skin reaction to stem cell killing and not to general basal cell loss; if death of stem cells was the essential factor, then this might explain the time (5–7 days) taken for the reaction to occur, and also the problematical observation that the time of onset of the skin reaction is not dose-dependent, notwithstanding the dose-dependency of the onset of proliferative activity (Hegazy and Fowler 1973*b*).

Again, we might usefully ask after the cause of the elevated proliferative rate which ensues after irradiation, and also after possible signals. We have seen how proliferative rate does not increase until after some 7–9 days after single doses, and not until 9–14 days after multiple small doses; furthermore, the precedence of proliferation over desquamation, and of induced proliferation without desquamation, would indicate that basal cell loss, produced by continued migration of basal cells without reciprocal proliferation, is the trigger for proliferation. This is in accord with the concept of Fowler and Denekamp (1976); this of course equilibrates regenerative responses after mechanical injury, such as tape stripping or surgical wounds, with basal cell migration, and here both migration and the onset of proliferation are both so much more rapid (see Table 10.1, p. 374). This unifying concept, part of which also includes the induction of proliferation in the EPU as a result of local basal cell depletion caused by suprabasal migration of cells in position 5 in the EPU model (see Chapter 8, section 2), points to local basal cell loss as a general trigger for cell proliferation in the epidermis.


## 12.5. Ultraviolet irradiation

According to the hypothesis propounded by Fowler and Denekamp (1976), proliferative activity should follow basal layer depletion; this appears to hold for wound healing (Chapter 10), and for X-rays (section 12.4). However, after damage induced by ultraviolet irradiation (UVL), events are somewhat different. After an initial depression of DNA synthesis, there is a uniform sparse [$^3$H]-TdR labelling over viable epidermal nuclei, which probably reflects unscheduled or repair DNA synthesis, whereas scheduled DNA synthesis follows relatively rapidly after exposure—between 24 and 48 hours later, and much sooner than most reports with X-rays (Epstein *et al.* 1968; Cripps *et al.* 1972; Lowe and Breeding 1980). But it is evident that UVL has several actions besides the early inhibitory effect on DNA synthesis: Epstein *et al.* (1970) demonstrated an early depression of DNA and protein synthesis, and, as far as cell death is concerned, it is not generally appreciated that the early death of cells after UVL has long been recognized (Daniels *et al.* 1961), in the form of so-called 'sunburn cells', which are regarded as apoptotic bodies (Olson and Everett 1975) and these are readily apparent before 24 hours. These occur mainly in the Malpighian layer, and it is conceivable that cell death in this layer induces rapid basal cell migration, in the same way as that provoked by tape stripping (Chapter 10), with consequent basal layer depletion, and the induction of mitosis. This early cell death could thus account for the rapid onset of cell proliferation after UVL irradiation.


## 12.6. Conclusions

There is no doubt that we have many incongruous facts to explain here, prominent amongst which are the lack of correlation of clonogenic cell kill with measurements of pyknotic or apoptotic cells or surviving cell numbers, and continued proliferation and migration of surviving cells after most clonogenic cells are lost. However, the survival curve data (see section 12.2 and Chapter 8, section 4) indicate that only a small fraction of the basal cell population is clonogenic; if these cells are the epidermal stem cells, and feed a much larger compartment of dividing transit cells, this two-compartment model can rectify many of the apparent anomalies; if it is assumed that irradiation-induced clonogenic cell kill only occurs in this subpopulation, and that the majority of basal cells are unaffected by irradiation, then these latter cells would progress to division and migration.

It must be conceded that the evidence for this concept is, as yet, circumstantial, and moreover, the estimates of clonogenic cell number from the survival curve are, at best, very approximate. Furthermore, the role of $G_2$ cells, or of putative $G_0$ cells in the response has yet to be investigated.

The data also support the hypothesis that basal cell depletion is a necessary trigger for cell proliferation after epidermal damage, however caused.

# 13 Cell proliferation during carcinogenesis in squamous epithelium

There is little doubt that the proliferative activity of a tissue is an important variable in the carcinogenic process (Warwick 1971), and, in the classical experiments of Berenblum (1959), the critical role of the induction of epidermal hyperplasia in relation to tumour promotion or development has been emphasized. The long familiar multistage theory of carcinogenesis was formulated with skin in mind; *initiators*, such as 9,10-dimethyl-1,2-benzanthracene (DMBA), were seen as irritants, one role of which was to induce increased cell proliferation (Boutwell 1961, 1964), whilst *promoters*, such as the phorbol esters found in croton oil, converted and propagated the initiated cancer cell to its ultimate, invasive form.

Thus Pound (1968) demonstrated that preliminary treatment of mouse skin with croton oil induced increased cell proliferation in the epidermis, and if this was followed by application of chemicals such as urethane, benzpyrene or dimethylbenzanthracene, then increased tumour yields were obtained. In similar vein, Frei and Ritchie (1964) have claimed that painting the epidermis with DMBA, a complete carcinogen, induces more tumours at 2200 h, when the diurnal peak of DNA synthesis occurs, than painting at 1000 h, the nadir of proliferative activity.

Convinced of the critical role of cell kinetic studies during carcinogenesis, many authors have studied the effect of carcinogens on cell proliferation, and proliferative patterns during carcinogenesis itself, although kinetic studies on squamous or basal cell carcinoma, in both animals and man, are rather few. This chapter examines these studies, in various squamous epithelia, in a search for some common thread underlying cell proliferation changes in carcinogenesis.

## 13.1. The acute action of carcinogens on cell proliferation in squamous epithelium: the cell kinetics of cancer initiation

Much of the stimulus to the earlier work in this field came from the pioneering studies of Iversen and Evensen in Oslo (Evensen 1961, 1963; Iversen and Evensen 1962; Iversen and Bjerknes 1963; Iversen 1964; Elgjo and Skjaeggstad 1965). The model they used was the hairless mouse epidermis treated with a single initiating application of the complete carcinogen, 3-methylcholanthrene, dissolved

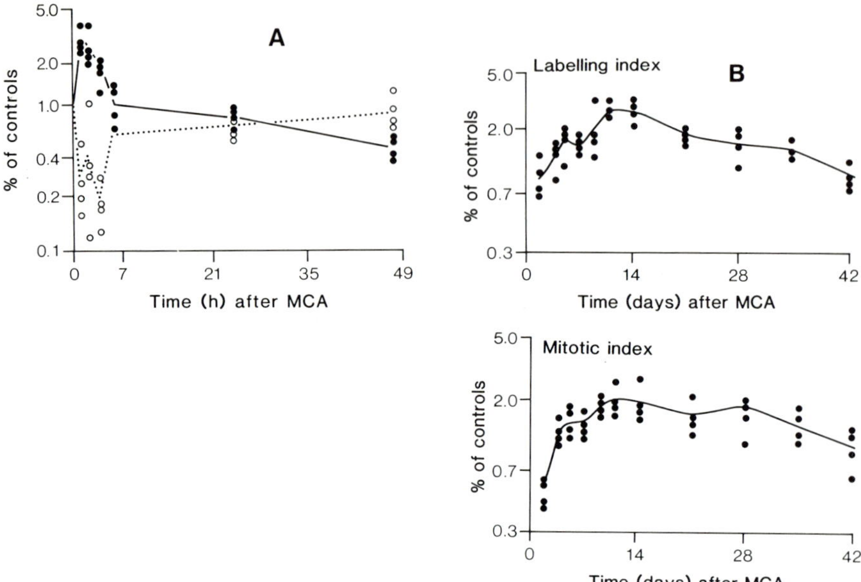



Fig. 13.1.   The effect of treatment with methylcholanthrene (1 per cent) on proliferative indices in the hairless mouse epidermis. (A) Changes in the relative labelling (o---o) and mitotic (●——●) indices during the first 48 hours after painting. (B) Changes in the relative labelling and mitotic index over the next 40 days. (Redrawn from Iversen and Evensen (1962).)

in benzene (Fig. 13.1). Within two hours of the application, the mitotic index rose to a peak, but the flash labelling index rapidly fell over the same period; measurement of the rates of entry into mitosis and DNA synthesis, using colcemid metaphase arrest and a double-labelling procedure with [$^3$H]-TdR, respectively, indicated that mitosis itself was blocked, with a marked lengthening of the mitotic duration and increase in the mitotic index, while the flux into DNA synthesis had apparently ceased, and that this was responsible for the reduction in the $I_S$. Both computer simulation studies and direct observation (see also Skjaeggstad 1964) indicated that there was a considerable degree of cell loss over the next 72 hours, occurring from both the basal and suprabasal layers, and, following this, there was a sustained rise in both $I_S$ and $I_M$, starting at about four days after the application and lasting as long as 30 days, which led to a prominent epidermal hyperplasia; the cell damage and death were interpreted as direct carcinogen-related events, while the increased cell production, which led to the hyperplasia, was regarded as a secondary phenomenon, probably due to a protracted regenerative response to the damage, further interpreted as a proliferative response to a negative feedback loop.

Elgjo (1968a,b) continued and extended these studies using the metaphase arrest technique: when the effects of single applications of *cantharidin*, a non-carcinogenic irritant, were compared to those of 3-methylcholanthrene, although similar initial changes were evident, the proliferative response was curtailed at three days, while the methylcholanthrene-painted epidermis was still hyper-proliferative seven days after application; although no attempt was made to assess the relative amount of cell damage and loss, Elgjo concluded that such sustained cell proliferation could not be regarded as a mere simple regenerative response. Similarly, when repeated applications of the two agents were made over a period of two weeks, cantharidin induced a hyperplasia with an elevated $r_M$, whereas methylcholanthrene, while increasing the proliferative rate, also induced a two- to threefold lengthening of $t_M$; thus, although apparently acting, at least initially, in the same manner as a hyperplasiogenic substance, the carcino-gen does appear to excite specific changes, possibly in the form of an exaggerated early proliferative response, and later, a large increase in the duration of mitosis.

Perhaps the most detailed study of the early events in initiation was reported by Clausen (1978), applying autoradiographic, stathmokinetic, and flow cyto-metric methods to methylcholanthrene-treated mouse epidermis. The early response was similar to that found by Iversen, with an immediate depression of the $S$ phase fraction, and subsequent recovery, with apparent overshoot at about 16 hours after application. However, there was a marked discrepancy between the $S$ phase fraction, as indicated by autoradiography with $[^3H]$-TdR, and by flow cytometric measurements in the early period (see Fig. 13.2) which was associated with very low mean interphase grain counts. This was interpreted in terms of a very slow rate of DNA synthesis, practically amounting to a block in those cells in the $S$ phase; this would leave the same number with an $S$ phase DNA content, but, as measured autoradiographically, the $S$ phase fraction would fall. Thus there is a marked increase in the duration of $S$ in this early period; cells flash labelled with $[^3H]$-TdR during the first peak only entered mitosis after the second peak (see Fig. 13.2), as evidenced by a large proportion of labelled mitoses that could be arrested with colcemid at this time; this would indicate a $t_S + t_{G_2}$ in excess of 23 hours, which is not really in accord with the proposed early increase in proliferative rate. Moreover, the correlation between the $S$ phase fractions measured by the two methods during the initial depression of DNA synthesis would indicate that this was due to a genuine decrease in the influx into $S$, as occurs after tape stripping; interestingly enough, there was also a transient early accumulation of cells in $G_2$, again in line with initial events after stripping. Although inspection of the $I_S$ curve would suggest an early stimulation of DNA synthesis and cell proliferation, as seen, for example, after cantharidin treatment or tape stripping (Clausen and Lindmo 1976), in fact progress through the $S$ phase is slowed, indicating a depression in proliferative rate. These sophisticated studies show that the cell kinetic reaction of epidermis to a single application of carcinogen is very different from non-specific damage.

How important are these several phases in initiation? We could regard the augmented cell proliferation that occurs after cytotoxic injury with carcinogens as a mainly non-specific response subsequent to injury and death, and possibly

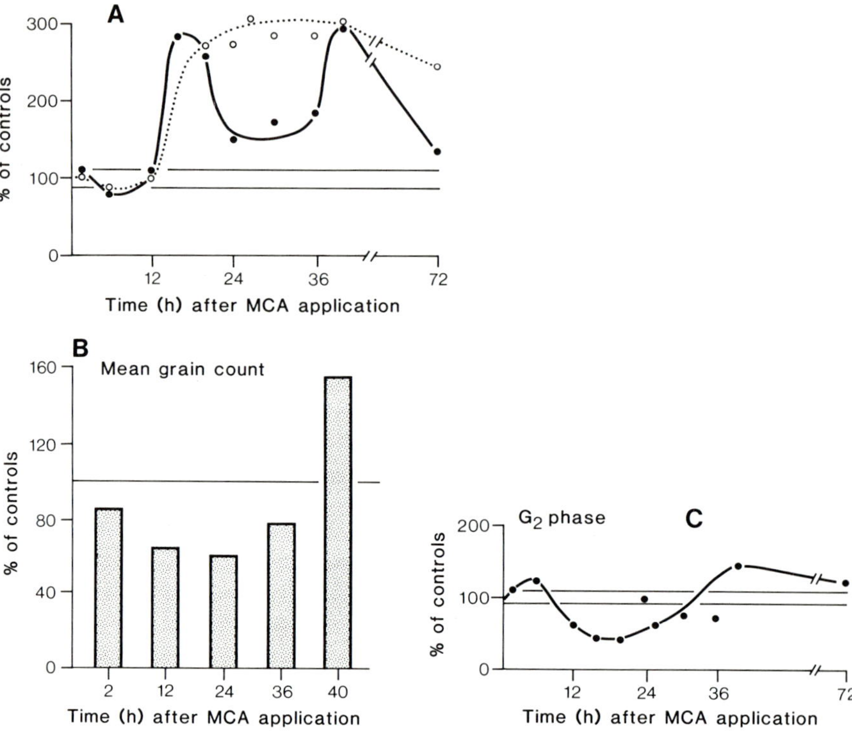

Fig. 13.2. Changes in (A) the proportion of cells with $S$ phase DNA content (o- - -o) and in $I_S$ (●——●) in isolated mouse epidermal cells after 20-methylcholanthrene application. The discrepancy in the two values between 20 and 40 hours indicates a block in DNA synthesis. (B) The mean interphase grain count after MCA treatment. (C) The fraction of cells in $G_2$. (Redrawn after Clausen (1978).)

independent of the carcinogen-induced inhibition of DNA synthesis; the latter is regularly observed after a variety of carcinogens in many tissues (Clausen 1978) and of course carcinogens do combine covalently with DNA, and indeed, correlations between carcinogenicity and covalent binding of hydrocarbons to DNA are reported (Brooks 1975). The delay in transit through the $S$ phase found by Clausen (1978) may be related to carcinogen–DNA interaction, with subsequent repair of the induced damage. In this respect, autoradiographic studies with ³H-labelled carcinogens (e.g. methylcholanthrene, DMBA) have shown that local application is followed by peak uptake in the basal cells at 2–4 hours later, and that the carcinogen is bound to nuclear and cytoplasmic proteins (Oehlert 1973); loss of labelling is detectable in the basal layer at two days, and after 3–4 days basal cells contain no label. Furthermore, there appears to be good correlation between the length of inhibition of DNA synthesis and the persistence of the carcinogen in the basal layer.

What is the biological significance of the early inhibition of DNA synthesis? Hennings and Boutwell (1969) could not relate the inhibition of DNA synthesis, produced by a small series of initiators, to the subsequent yield of papillomas induced in mouse epidermis; on the other hand, Slaga *et al.* (1974) found that a single initiating dose of DMBA depressed DNA synthesis for 24 hours without a subsequent increase, whereas DMBA, given in sufficient dosage to result in tumour formation itself, resulted in a protracted inhibition of DNA synthesis, followed by prominent rises above control levels. In similar vein, the dose of eight carcinogenic N-nitroso compounds (powerful carcinogens for the rat oesophageal epithelium) which causes 50 per cent early inhibition of DNA synthesis, was positively correlated with the lowest dose that would produce oesophageal tumours, and also with the percentage incidence at that dose (Mirvish *et al.* 1978).

The fact remains that increased proliferative rate is of critical importance during the initiation stage: if mouse skin is pretreated with 0.5 per cent croton oil to induce a wave of DNA synthesis 18–20 hours later, at which time the skin is treated with N-methyl-N-nitro-N-nitrosoguanidine, followed by croton oil promotion, a three- to four-fold increase in the subsequent tumour yield occurs (Hennings and Boutwell 1969); similar results have been obtained by Pound (1968; see above), and indicate an important role for induction of increased cell proliferation in initiation.

We can therefore summarize the kinetic events following the application of an initiator of carcinogenesis, and indeed of many complete carcinogens; (a) an early block in mitosis giving a prolonged mitotic duration; (b) a prolonged inhibition of DNA synthesis, with prolonged transit times for cells in the $S$ phase. This may be associated with covalent carcinogen:DNA binding; (c) cells which can repair the induced damage do so, others die, and this cell death process can be prominent, and is proportional to dosage; possibly as a response to cell death, there is a sustained proliferative response leading to hyperplasia. There is some evidence that the mitotic duration is prolonged in this later phase (Elgjo 1968*b*), and indeed morphologically abnormal mitoses do appear at this stage (Oehlert 1973). It is difficult to say how specific these responses are for carcinogens; certainly transient decreases in the $G_2$ and $S$ phase fractions do occur with non-specific stimuli (cf. tape stripping, Chapter 10), but the prolonged, apparently dose-responsive (Mirvish *et al.* 1978) inhibition of transit through the $S$ phase would appear to be specific; later, the sustained hyperplasia could be explained by a cell loss process and recovery, although the evidence that cells are experiencing difficulties with the mitotic process at this stage could indicate early genetic damage to the basal cells.


## 13.2. Cell proliferation and tumour-promoting action in squamous epithelia

It has long been known that croton oil is a promoter for mouse skin (Berenblum

1959), but a significant advance occurred when Van Duuren (1969) identified the active components of croton oil as certain phorbol esters; the most effective ester is 12-O-tetradecanoylphorbol-13-acetate (TPA), which promotes skin tumours in mice at doses as low as 1 nanomole per application, and is also active *in vitro* in nanomolar concentrations.

Again, there is good evidence that cell proliferation is induced by the action of promoters in epidermis, and notably by croton oil and phorbol esters. Indeed, Frei and Stephens (1968) concluded that an index of the efficacy of an agent as a tumour promoter, evoking a malignant transformation from a dormant tumour cell, is its ability to provoke a prominent hyperplasia; after a short period of depression, lasting about 12 hours, topical croton oil, in a dosage which was effective as a tumour promoter, stimulated [3H]-TdR incorporation into DNA in mouse epidermis, reaching a maximum at 24 hours, and ending at five days after application (Hennings and Boutwell 1970; Frankfurt and Raitcheva 1972; Fig. 13.3A). However, similar results were also evident with cantharidin, a weak

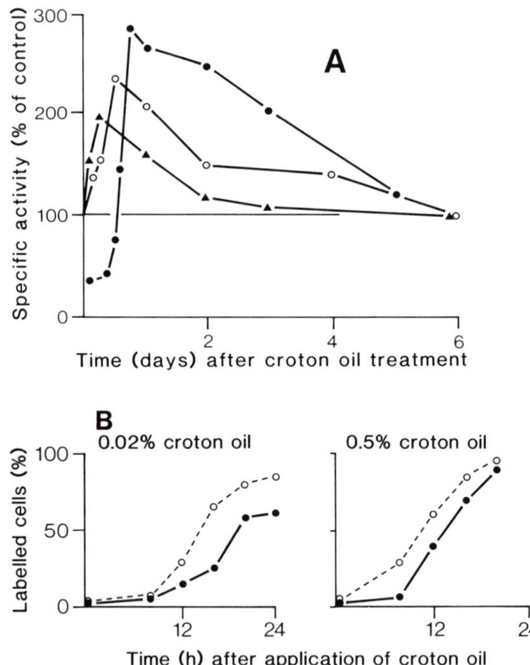



Fig. 13.3. (A) The effect of a single application of 0.5 per cent croton oil on the specific activity of mouse skin DNA (●) after [3H]-TdR injection, RNA (▲) after [3H]-cytidine, and protein (○) after [3H]-leucine injection. (Redrawn from Hennings and Boutwell (1970).) (B) Continuous labelling curves for normal (●——●) and DMBA-initiated mouse epidermis (○---○) after application of various concentrations of croton oil. (Redrawn from Frankfurt and Raitcheva (1973).)

tumour promoter, which also stimulates cell proliferation in mouse epidermis, while induction of cell proliferation by a local surgical incision in an initiated area produced some tumours along the line of the cut.

Thus the relationship between cell proliferation and promoting action cannot be simple; all hyperplasiogenic agents are not promoters (Berenblum 1954), and some authors have argued strongly for a dissociation of cell proliferation from promoting action. Low doses of TPA induce epidermal cell proliferation and thickening, but do not have a promoting action (Raick 1973). Moreover, Raick (1974*a,b*) showed that ethylphenylpropiolate and turpentine, both agents with minimal or no promoting activity, provoked a sustained proliferative response in DMBA-initiated epidermis; furthermore, a small, non-promoting dose of TPA induced a considerable proliferative response in the absence of promoting action (see above), and application of ethylphenylpropiolate or turpentine after a promoting dose of TPA *inhibited* the promoting action, the converse of what would be expected if a summation of the proliferative responses were responsible for the promoting action. This work would indicate an epiphenomenological role for cell proliferation in promotion.

Raick (1974*a*) has described groups of small dark atypical cells, similar to those seen in papillomas, which appear some 48 hours after TPA application to DMBA-initiated epidermis; these were considered to be precursor neoplastic cells which give rise to the developed papillomas; non-promoting hyperplasiogenic agents do not produce these cells, but TPA application to normal epidermis does, although TPA has little activity as a complete carcinogen; wounding, a weak promoter, also gives rise to cells resembling the putative neoplastic precursors. Hence definitive evidence of the specific neoplastic progenitor role of these cells is presently lacking.

Hennings and Boutwell (1969) distinguished between 'initiated' and 'dormant' tumour cells; if initiated cells were all dormant tumour cells, then any non-specific stimulus to proliferate would cause tumours, whereas most hyperplasiogenic agents are only weak promoters. Dormant tumour cells have a selective growth advantage when stimulated to divide, which is not possessed by initiated cells, and there was some evidence that promotion with croton oil was associated with the synthesis of a specific species of RNA (Hennings and Boutwell 1969). While on the subject of 'growth advantage', we could complain that there does not appear to be much difference between the cell cycle time of cells stimulated by active promoters (croton oil, Tween 60) in normal and in DMBA-initiated epidermis; FLM measurements indicated a median $T_C$ of 16 hours in each case (i.e. similar to some values obtained in epidermis regenerating after wounding — see Chapter 10), although there was some evidence that one effect of promoters in initiated epidermis was to produce a more powerful proliferative stimulus, as evidenced by a greater production of suprabasal cells (Frankfurt and Raitcheva 1972). However, it does appear that promoters induce DNA synthesis *faster* in initiated epidermis than in untreated epidermis; continuous labelling studies by

Frankfurt and Raitcheva (1973) showed that 0.02 per cent croton oil induced cell proliferation at eight hours in DMBA-initiated epidermis, but took 16 hours in normal skin (Fig. 13.3B). These results indicate that the main response of epidermis to croton oil is a shortening of $t_{G_1}$, with increased flux into $S$, and subsequent rises in $I_S$ and $I_M$. However, it appears that a preliminary increase in $I_M$, occurring before the main rise produced by increased numbers of cells in $S$ reaching $M$, is due to a stimulation of blocked/resting cells in $G_2$ (but see Clausen and Thorud 1980, and Chapter 10 for an alternative explanation of this phenomenon, which has also been described in wounded epidermis).

Studies with the *active* components of croton oil, i.e. the phorbol esters themselves, also indicate an effect on cell proliferation: treatment with TPA stimulates epidermal DNA synthesis (Paul and Hecker 1969; Raick *et al*. 1972; Kreig *et al*. 1974; Rohrbach *et al*. 1968; Balmain *et al*. 1977), while Baird *et al*. (1971) showed that the potent promoters TPA and phorbol-12,13-didecanoate, after a brief *inhibition* of DNA synthesis which lasted about six hours, evoked a powerful stimulation of $[^3\text{H}]$-TdR incorporation and $I_S$ values, which lasted for two to four days (see Fig. 13.4), events which are largely mimicked by the

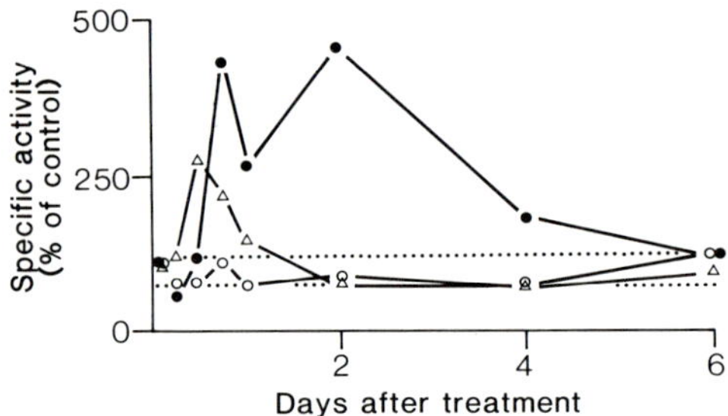


Fig. 13.4. The effect of a single application of various phorbol esters on the specific activity of DNA in mouse epidermis. The dotted lines indicate the average SD value for the controls; ● tetradecanoyl phorbol acetate, △ phorbol dibenzoate, and ○ phorbol acetate. (Redrawn from Baird *et al*. (1971).)

action of TPA on keratocytes maintained in culture (Yuspa *et al*. 1976). There is an associated increase in epidermal mass, and in the number of nucleated cell layers, entirely due to an increase in the number of suprabasal cells; basal cell number is decreased by 50 per cent at 24 hours after application, remains low for two days, and reaches normal levels over the next 10 days (Argyris 1980). Subsequent applications of TPA stimulate DNA synthesis, but do not produce the brief early inhibition evident at the first application (Raick *et al*. 1972). On the other hand, phorbol-12,13-dibenzoate, a weak promoter, produced changes which lasted for only 24 hours, while the very weak agents phorbol and phorbol-12,13-diacetate showed no stimulation of DNA synthesis. It would appear that promoting activity occurs in direct relationship to the ability to stimulate DNA synthesis; however, it is again interesting that weak, but evident promoters are active without stimulating DNA synthesis, and thus the relationship is again not simple, and may indicate some selective change independent of cell proliferation.

But Slaga *et al.* (1976) claimed a strong correlation between the ability of a series of phorbol esters to evoke hyperplasia, and their promoting ability. However, hyperplasiogenic agents not related to the phorbol series showed no such correlation, and this particular study relied only upon counts of nucleated layers as an index of cell proliferation, so that minor degrees of stimulation of cell division could, in fact, have been missed. Moreover, Frei (1977) was concerned that most workers, in investigating the relationship between hyperplasia and promoting action, have compared the hyperplasiogenic effect of a *single application*, with the promoting action of *repeated applications*. When the data of Raick (Raick *et al.* 1972; Raick 1974*a, b*) are replotted, once a certain level of hyperplasia is reached, as indicated by four cell layers, measurements of the degree of hyperplasia and of promotion fall on a straight line, and the relationship does hold.

How can we then regard the cell proliferation which occurs after TPA application? Argyris (1980) has reported that histologically evident damage occurs in the first two days after treatment, i.e. when basal cell number is falling, and this damage is also evident in EM studies (Raick *et al.* 1972); consequently, we could be looking at a mere regenerative response. However, can we distinguish the response of epidermis to promoters from the response to non-specific hyperplasiogenic agents? Hennings and Boutwell (1969) have proposed that a specific species of RNA is associated with croton oil action, while O'Brien *et al.* (1975) suggested that the induction of ornithine decarboxylase (ODC), an enzyme involved in polyamine synthesis, is a critical component of a tumour phenotype in epidermal cells, produced by phorbol esters; a gene controlling ODC activity is deleted by the initiator, and after promotion, a high level of ODC becomes evident; indeed, four hours after topical TPA there is a large increase in ODC levels, before DNA synthesis occurs (Marks *et al.* 1979; Lowe 1980). However, there is now good evidence that polyamine synthesis is associated with, and precedes, a number of proliferative responses, including that seen in psoriatic epidermis, so that this may be a non-specific event related to the induction of cell division. However, acetic acid, which actively stimulates epidermal proliferation, has no effect on epidermal levels of ODC, but does stimulate S-adenosylmethionine decarboxylase (AMDC), another biosynthetic enzyme (Slaga *et al.* 1975*a*), and the hyperplasiogenic, non-irritant and practically non-promoting 4-O-methyl ether of TPA does not affect polyamine metabolism (Marks *et al.* 1979). Moreover, retinoic acid not only inhibited the TPA-induced stimulation of ODC in mouse epidermis, but also substantially reduced the number of papillomas per mouse (Verma and Boutwell 1977; see also Sporn and Newton 1979). The relationship of this action to the induction of cell proliferation is not clear, but AMDC synthesis was not affected. In addition, there is good correlation between the ability of phorbol esters to induce ODC and DNA synthesis in epidermal cells *in vitro*, and their ability to promote epidermal carcinogenesis *in vivo* (Yuspa *et al.* 1977). However, removal of the horny layer induces a considerable increase in ODC levels after 4–8 hours, without promoting effect, and Marks *et al.* (1979) concluded that there was no clear correlation between cell proliferation and promotion on the one hand, and ODC activation on the other; mice maintained on a vitamin $B_6$ (pyridoxine) deficient diet, and therefore unable to sustain an ODC response to TPA, nevertheless still showed increases in cell proliferation similar to those seen in control mice (Murray 1978). Unfortunately, the effect on promotion of a pyridoxine deficient diet was not investigated.

Whether or not TPA acts via the putative cyclic nucleotide mechanism (see Chapter 14, section 2) is open to debate; Belmann and Trott (1972) and Grimm and Marks (1974) reported decreases in epidermal levels of cyclic AMP at times from 6 to 24 hours after application, evidently due to increased synthesis of cAMP phosphodiesterase with consequent cAMP breakdown (Marks *et al.* 1978); this was not promoter-specific, since removal of the horny layer induces the same effect. There was a small initial rise in cAMP levels, and, after 24 hours, a sustained increase above control levels was evident; the results were consistent with an induction of cell proliferation by a fall in the concentration of cAMP, but Grimm and Marks (1974) did suggest that cAMP could be involved rather in the expression of epidermal function which follows the stimulation of cell proliferation. Further evidence that cAMP levels are critical in promotion comes from the observation that theophylline, a cAMP phosphodiesterase inhibitor (which thus maintains cAMP levels) inhibits the promoting effect of phorbol ester (Belmann and Trott 1974).

These results could not be confirmed by Mufson *et al.* (1977), although in this study TPA did appear to diminish the accumulation of cAMP in response to the $\beta$-agonists isoproterenol and adrenaline (Grimm and Marks 1974): since the $\beta$-adrenergic receptor of adenyl cyclase is membrane-bound, TPA may be acting directly on the cell membrane. Mufson concluded that changes in the intracellular cAMP levels were probably not related to the tumour-promoting action of TPA.

A further clue to the importance (or otherwise) of cell proliferation in promotion is the fact that steroid anti-inflammatory agents, which are powerful inhibitors of the phorbol ester-induced proliferative wave (Viaje *et al.* 1977), are also potent inhibitors of its promoting action (Belmann and Trott 1972); non-steroid anti-inflammatory agents such as indomethacin did not inhibit TPA-induced cell proliferation, and were only very weak inhibitors of promotion. In the steroid group, there was a generally good correlation between the inhibition of promotion and of DNA synthesis (Viaje *et al.* 1977); additionally, polyinosinic-polycytidylic acid, which effectively prevents 3-methylcholanthrene-induced cell proliferation, also reduces tumour yield in mouse epidermis (Elgjo and Degré 1975). These experiments again point to the importance of cell proliferation in the promotion event.

The evidence is thus conflicting; we are left with the impression that (a) most, if not all, promoters induce cell proliferation; (b) many hyperplasiogenic agents are weak promoters, or not promoters at all; (c) there is some evidence that the induction of cell proliferation and promoting activity can be dissociated, although this is controversial; conversely, inhibition of cell proliferation can prevent promotion; (d) a search for a specific marker of promoter-induced proliferation, which discretely separates it from a non-specific proliferative stimulus, has not yet been successful; (e) we cannot, at this stage, exclude the possibility that the cell proliferation evoked by promotion is not a simple regenerative response induced by cell damage.

Promoters stimulate cell proliferation in a large number of basal cells, not all of which give rise to tumours; presumably initiated or dormant tumour cells are among those which are stimulated to divide, but, because we cannot identify them prior to promotion, their proliferative behaviour is swamped in the general wave and is obscured; thus we cannot yet answer the question of which constitutes the *target cell*, either for initiation, or the dormant tumour cell for promotion. Indeed, we know very little about which epidermal cells are affected in carcinogenesis in terms of the stem cell proliferation/differentiation sequence which is now unfolding (see Chapter 8). There is some evidence that promoters are not effective stimulators of cell proliferation in neonatal epidermis (Bertsch and Marks 1974), interpreted as a lack of effectiveness in a tissue where basal cells are 'pluripotential' (but see Chapter 2, section 9); but we have seen that few proliferative responses can be induced in neonatal epidermis (see Chapter 10). The relationship of cell proliferation to tumour promotion remains intriguing, but enigmatic.

## 13.3. Changes in cell proliferation during carcinogenesis

While we have not been very successful in elucidating the *role* of cell proliferation in carcinogenesis, there have been several salient studies on the changes in cell kinetic parameters which occur in squamous epithelium during the process of carcinogenesis. A favourite model for such studies has been the hamster cheek pouch treated topically with the complete carcinogen 9,10-dimethyl-1,2-benz-anthracene (DMBA) for several weeks, by which time gross tumours are evident. Thus Reiskin and Mendelsohn (1964) found marked changes in the tumours compared with the normal epithelium (see Table 13.1); $t_S$ and $t_{G_2}$ were FLM-derived, and $T_C$ was calculated by a stage-duration method, and calculations were made for the intervening hyperplastic epithelium, assuming that $t_S$ was the same as that found in the normal epithelium. Between the normal epithelium and the tumours $t_S$ was shortened from 10.2 to 6.2 hours, and, if the assumptions upon which the calculation are based (viz. $I_P = 1$, lack of diurnal variation in $I_S$, rectangular age distribution for the normal epithelium) are valid (see Chapter

Table 13.1. Kinetic parameters in normal, hyperplastic, and neoplastic squamous epithelia during carcinogenesis

| | $I_S$ (%) | $I_M$ (%) | $k_B$ (cells/%/h) |
|---|---|---|---|
| **(A) Normal epithelium** | | | |
| *(i) Hamster cheek pouch* | | | |
| Reiskin and Mendelsohn (1964) | 7.2 | 1.7 | – |
| Reiskin and Berry (1968) | 6.0 | 1.1 | – |
| Thilagarathram and Main (1972) | 6.1 | 1.1 | – |
| Izquierdo (1977) | – | – | – |
| Nagamine (1978) | 4.0 | – | – |
| *(ii) Mouse epidermis* | | | |
| Fukada *et al.*. (1978) | – | – | – |
| Chopra and Forbes (1974) | 6.7 | – | – |
| **(B) Hyperplastic epithelium** | | | |
| *(i) Hamster cheek pouch* | | | |
| Reiskin and Mendelsohn (1964) | 24.6 | 5.3 | – |
| Thilagarathram and Main (1972) | 11.0 | 2.0 | $1.3^3$ |
| Nagamine (1978) | 24.2 | – | – |
| *(ii) Mouse epidermis* | | | |
| Fukada *et al.* (1978) | – | – | – |
| Chopra and Forbes (1974) | 10.2 | – | – |
| **(C) 'Preneoplastic epithelium'** | | | |
| *(i) Hamster cheek pouch* | | | |
| Reiskin and Berry (1968) | 15.7 | 1.8 | – |
| Thilagarathram and Main (1972) | – | – | – |
| **(D) Neoplastic epithelium (tumour)** | | | |
| *(i) Hamster cheek pouch* | | | |
| Reiskin and Mendelsohn (1964) | 29.5 | 7.0 | – |
| Reiskin and Berry (1968) | 24.9 | 1.6 | – |
| Brown and Berry (1969) | – | – | – |
| Thilagarathram and Main (1972) | 21.7 | 2.04 | $2.6^3$ |
| Izquierdo (1977) | 12.6-20.4 | 1.1-2.6 | – |
| Nagamine (1978) | 33.1 | – | – |
| *(ii) Mouse epidermis* | | | |
| Fukada *et al.* (1978) | – | – | – |
| Chopra and Forbes (1974) | 19 | – | – |
| *(iii) Mouse squamous forestomach* | | | |
| Frankfurt (1966) | 29-40 | – | – |

All cell cycle phase durations are FLM derived except those marked: (1) continuous labelling method; (2) double labelling method; (3) metaphase arrest method; (4) indicates those measurements which are $t_2$, i.e. ($t_{G_2} + \frac{1}{2}t_M$) rather than $t_{G_2}$; (5) stage-duration calculation; (6) direct measurement.

| $I_P$ | $T_C$ (h) | $t_{G_1}$ (h) | $t_S$ (h) | $t_{G_2}$ (h) | $t_M$ (h) | $t_D$ (d) | $\phi$ (%) |
|---|---|---|---|---|---|---|---|
| — | 142[5] | 128[5] | 10.2 | 1.6 | 2.6 | — | — |
| 1.0[1] | 155[5] | 142[5] | 9.3 | 1.1 | 2.1 | — | — |
| — | 164[5] | 150[5] | 10.0 | 1.9 | 1.8[3] | — | — |
| — | — | — | 9.0 | 2.5 | — | — | — |
| 0.73 | 92[6] | 82[6] | 6.9[6] | 3.2[4] | — | — | — |
| 0.34[1] | — | — | — | — | — | — | — |
| — | 95[2,5] | — | 6.2[2] | — | — | — | — |
| — | 42[5] | 29[5] | 10.2 | 2.2 | 0.6 | — | — |
| — | 91[5] | 81[5] | 8.33 | — | 1.6[3] | — | — |
| 3.0 | 61[6] | 51[6] | 6.6 | 3.0[4] | — | — | — |
| 1.0[1] | 15–39[1] | — | 6–9[1] | — | — | — | — |
| — | 43[2,5] | — | 4.5[2] | — | — | — | — |
| — | 61[5] | 47[5] | 10.4 | 2.0 | 1.2 | — | — |
| — | 23[6] | 11.9[6] | 8.2 | 1.7 | 0.9[3] | — | — |
| — | 17.6[6] | 8.6[6] | 6.2 | 2.2 | 0.5 | — | — |
| 0.6 | 22[6] | 11[6] | 6.5 | 2 | 0.56 | 4.7 | 93 |
| — | 10.7[6] | 2.8[6] | 5.9 | 1.6 | 0.4 | — | — |
| — | 15[6] | 7.1[6] | 6.2 | 1.4 | 0.7[3] | — | — |
| — | — | — | 6 | 2.5 | — | — | — |
| 1.7 | 17.2[6] | 8[6] | 6.4 | 2.2[4] | — | 9–10 | — |
| 1.0[1] | 14–58[1] | — | 7–17[1] | — | — | — | — |
| — | 16[2,5] | — | 3.0[2] | — | — | — | — |
| — | 8–12 | | 3.6–5 | 2.2–2.5 | — | — | — |

6, section 2), then there was a shortening of the cell cycle time from 142 hours in the control, through 42 hours in the hyperplastic epithelium, to 17.6 hours in the tumours. This study can of course be criticized on the basis of its numerous assumptions, but the results were extended by Reiskin and Berry (1968), and Brown and Berry (1969), who directly measured $t_S$ in preneoplastic epithelium (which presumably showed atypical rather than mere hyperplastic changes) at 10.4 hours with the FLM method, and who were also able to measure $T_C$ directly in the tumours, also by the FLM method; the doubling time of the tumours varied from 4.7 to 12.3 days depending on the strain, and although Reiskin and Berry calculated $I_P$ as low as 4-7 per cent from the relationship $\ln(1 + I_P)/T_C = \ln2/t_D$, more orthodox calculations by the usual method of $I_{S(\text{expt.})}/I_{S(\text{theor})}$ gives values from 60 to 75 per cent according to the strain of hamster (see Table 13.1).

Nagamine (1978) has reported a peculiar phenomenon in the hyperplastic and neoplastic epithelium of the DMBA-treated hamster cheek pouch (see Table 13.1): calculation of $I_P$, apparently from the relationship $I_{S(\text{expt.})}/I_{S(\text{theor})}$, gave values in excess of unity. Nagamine explained this apparent anomaly by postulating that carcinogen-affected, and indeed the actual tumour cells, enter DNA synthesis, thus contributing to the $I_{S(\text{expt.})}$ but fail to divide, and are therefore not picked up on the plateau value of the FLM curve, or in the calculation of $I_{S(\text{theor})}$. However, a more likely explanation is that the $T_C$ results generated by the Takahashi *et al.* (1971) method employed to analyse the FLM curve (see Chapter 3, section 4) are too long, certainly as evidenced by visual inspection of the FLM curve for the tumours, at least.

Thilagarathram and Main (1972), also in the hamster cheek pouch, clearly distinguished between *hyperplastic epithelium*, which merely showed increased numbers of cell layers, and *preneoplastic epithelium*, which showed atypical changes progressing as far as carcinoma *in situ*, but without invasion. Table 13.1 shows that this study suffers from the usual drawback of absence of direct measurement of $T_C$ in the normal epithelium, but note that Nagamine (1978) did perform protracted FLM studies in normal, hyperplastic and neoplastic epithelium for up to 80 hours after [$^3$H]-TdR injection, and analysed them by the Takahashi method (Takahashi *et al.* 1971)—see Table 13.1. In *all this data*, there is evidence of a progressive increase in proliferative rate, with decreases in $t_S$, $t_M$, and probably in $t_{G_1}$, associated with increases in $I_M$ and in $k_B$. But Fukada *et al.* (1978), using a continuous labelling method, considered that $t_S$ and $T_C$ were prolonged in papillomas and in carcinomas compared to the intervening hyperplastic epithelium (Table 13.1). However, we should note that Fukada *et al.* extrapolated their continuous labelling curves to obtain both $t_S$ and $T_C$; we have already given our reasons for regarding this as an ineffective method of analysis (Chapter 3, section 5). Thus the bulk of the evidence, from studies in the cheek pouch epithelium, indicates that the evolution of squamous carcinoma, from normal epithelium through the stages of hyperplasia and of atypia/carcinoma *in situ*, is associated with a prominent elevation of the proliferative rate.

It is diverting to speculate on the mechanism of these changes: Izquierdo (1977) considered that the increase in the cell production rate in carcinogenesis was produced by two independent kinetic changes—an enlargement of the proliferative compartment and a decrease in the cell cycle time. Because we are unclear about the value (and nature) of the growth fraction in squamous epithelium, it is difficult to know how much this has changed, and its temporal nature *vis à vis* the decrease in $T_C$ and its component phases. However, the data of Nagamine (1978), although compromised by evidently inaccurate $T_C$ estimates, do indicate that the growth fraction in the basal layer is increased. We must remember (Chapter 7, section 4) to distinguish between a relative increase in $I_P$ in the basal layer, and an absolute increase in the number of proliferating cells, mediated by an expansion of the proliferative zone; simple inspection of the preneoplastic and neoplastic epithelium shows that an absolute increase is present, but it is conjectural whether relative changes are evident (see Table 13.1). In this context, Fukada *et al.* (1978) maintained that, although the relative growth fraction in the basal layer was increased in the hyperplastic epithelium, the proliferating cells were confined to the basal layer; with the occurrence of atypia, of course, multiple layers of proliferating cells are manifest. But in the human skin disease *actinic keratosis*, which is a dysplastic lesion showing atypia, and capable of progressing to invasive squamous carcinoma, not only does the involved epidermis show high $I_S$ values (17 per cent), but the hyperplastic perilesional epidermis shows a prominent elevation in $I_S$ (11.2 per cent), with flash-labelled cells high in the epidermis, indicating an absolute expansion in the size of the proliferative compartment (Pearse and Marks 1977). Moreover, in UV-induced carcinogenesis, the hyperplastic stage is marked by the large numbers of suprabasal labelled cells (Chopra and Forbes 1974).

Izquierdo (1977) interpreted the increase in labelling and mitotic indices as evidence of an enlarged proliferative pool, but of course these changes could be caused by an increased rate of transit through the cell cycle without invoking $I_P$ changes. However, it is highly likely that increases in $I_P$ in the basal layer do occur, and Izquierdo (1977), on the basis of a $G_0$ model of kinetic organization in the basal layer (see Chapter 7, section 4), attributed this $I_P$ change to the increased numbers of cells which are triggered to re-enter the cell cycle; Fukada *et al.* (1978) held that some 66 per cent of normal basal cells were in $G_0$, which were claimed, on the basis of continuous labelling studies, to leave $G_0$ and re-enter the cell cycle, leading to an $I_P$ of unity. Such conclusions are, however, essentially non-robust, because of the intrinsic difficulties in the continuous labelling method. The continuous labelling studies of Chopra and Forbes (1974), carried out on UV-treated mouse epidermis, indicated an $I_P$ of unity at all stages of carcinogenesis, including the normal epidermis. We have already discussed the pitfalls of such studies (see Chapter 7, section 4), but Chopra and Forbes did conclude that the proliferative change was mediated via changes in $T_C$ only, although an absolute increase in proliferative cell number did occur. We have argued (see Chapter 8, section 5) that there is little experimental evidence for a classical $G_0$ compartment in squamous epithelia, and it appears more likely that any $I_P$ increase is due to the ejection of post-mitotic maturing cells from the basal layer, their place being taken by proliferating (?initiated, ?dormant) tumour cells, and the basal layer becomes merely the first layer of a multilayered proliferative system.

There is probably also a marked decrease in the cell cycle time, and the now actively dividing proliferative zone contributes more cells to the tissue; the normal regulatory mechanisms of differentiation and cell loss cannot maintain the steady state, and the system increases in size (hyperplasia); sooner or later, autonomy takes over, proliferative hyperactivity is perpetuated, and we progress through carcinoma *in situ* to invasive cancer, as the cells acquire invasive properties.

The cytokinetic alternatives in carcinogenesis as seen by Izquierdo (1977), are summarized in Fig. 13.5; the system can be perturbed by one of three

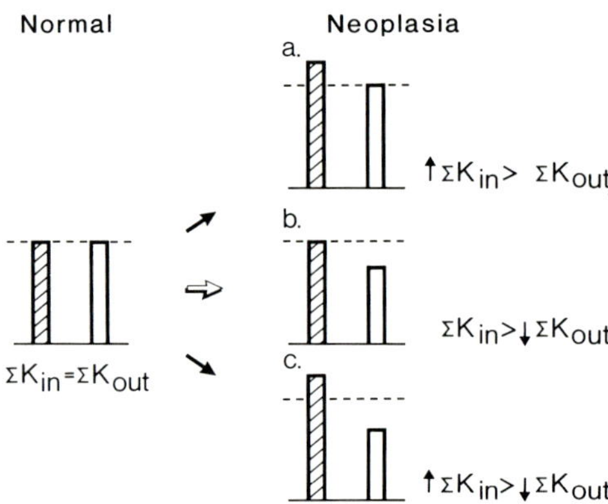


Fig. 13.5. The three mechanisms envisaged by Izquierdo for population enlargement during carcinogenesis. The equilibrium is disturbed respectively by (a) an increased cell production rate, (b) a reduced cell loss rate, and (c) by a simultaneous change. (Redrawn from Izquierdo (1977).)

mechanisms: (b), the second alternative, can be dismissed, because the data is uniform about the increase in proliferative rate (Table 13.1). However, since we cannot, as yet, accurately measure surface cell loss, we cannot distinguish between schemes (a) and (c). Note that we can only speculate on the mechanism of the increased cell number; the mechanisms of the development of anaplasia and invasive characteristics are not a cell kinetic problem.

We have seen that the data strongly support the conclusion that carcinogenesis is associated with a marked increase in the rate of cell proliferation; up to and including the stages of preneoplasia and carcinoma *in situ* the change would appear to involve all cells in the preceding neoplastic field, as evidenced by the finding of high proliferative indices in the perilesional epidermis in actinic keratosis (Pearse and Marks 1977).

Medline and Farber (1981) have argued that the selective proliferation that follows promotion can occur by at least three different mechanisms: (i) *selective inhibition*, whereby the initiated cells respond to a stimulus for cell proliferation while surrounding uninitiated cells do not respond; this is a favoured model in diethylnitrosamine-induced hepatic carcinogenesis (see Chapter 25, section 8); (ii) *selective stimulation*, in which initiated cells respond repeatedly to a mitogen, whereas uninitiated cells respond once or twice, as seen in the liver treated by α-hexachlorocyclohexane; or (iii) *selective recovery*, where both initiated and surrounding cells respond to the promoting agent, but the initiated cells do not return to the normal, non-hyperplastic state. On this hypothesis, it would appear that carcinogenesis in squamous epithelium is a modified process of selective recovery; however, the measurements made in the epidermis between neoplasms (see Table 13.1), and in the perilesional epidermis of actinic keratosis, would indicate that the uninitiated epidermis *remains* hyperplastic. Why do some areas become persistent and lack reversibility? Could this be related to the intensity of the selective stimulation of cell proliferation? Why do cells then acquire atypical characteristics, and invasive potential?

However, according to the monoclonal theory of the origin of tumours, each papilloma/carcinoma arises from transformation of a single cell; this single cell then grows through the epidermis, displacing and probably destroying its non-neoplastic neighbours, until an area of carcinoma *in situ* has developed; invasion would then occur. This pathological sequence of events is now well established, for such squamous cell neoplasms as carcinoma of the cervix, while in the epidermis the progression from carcinoma *in situ* (actinic keratosis, Bowen's disease) to invasive squamous carcinoma is now widely appreciated. Only rarely does carcinoma of the cervix become invasive without a preceding *in situ* phase, as in the so-called 'spray carcinoma'. The evolution of squamous carcinoma is thus thought to traverse a dysplasia/carcinoma *in situ* spectrum before becoming invasive. Is this initially a field change, or a monoclonal proliferation? Despite some objections, most evidence, for *in situ* proliferations at least, support a monoclonal origin. In the mouse, using allophenic animals chimaeric for glucose phosphate isomerase, Iannaconne *et al.* (1978) have shown that epidermal tumours produced by a variety of carcinogens are probably monoclonal. In human *in situ* squamous carcinoma of the cervix, arising in patients who are heterozygous for the polymorphic varieties of glucose-6-phosphate dehydrogenase, nearly every sample analysed had an enzyme phenotype indicating a clonal origin (Park and Jones 1968; Smith *et al.* 1971); karyotypic analysis also supports a clonal origin for these proliferations (Spriggs *et al.* 1971). While several examples of invasive cervical carcinoma appeared monoclonal, many tumours studied have shown a double enzyme phenotype, although most tumours studied by Fialkow (1976) were monoclonal, and normal cell contamination was suggested to account for the disparate results. In *epidermal* squamous carcinoma *in situ*, or Bowen's disease, a single enzyme phenotype has been found (Smith *et al.* 1971). Inherent in this argument is the proposal that the transformed cell has some advantage over its neighbours, and this is commonly assumed to be a *proliferative advantage*.

In a stimulating and thought-provoking paper, Williams and Bjerknes (1972) proposed a stochastic model for abnormal clone spread through the basal layer of a squamous epithelium; assuming Marques-Periera type migration kinetics (see Chapter 8, section 6), they simulated a single transformed cell which divided $k$ times as fast as its neighbours. Thus $k$ ($>1$) is the *carcinogenic advantage*. If, with this advantage, the transformed cell divides, one neighbour will be ejected (?at random) and two transformed cells will lie on the basement membrane; division of these cells will result in more basal layer transformed cells, and ejection of normal ones, although of course a transformed cell next to a normal cell in mitosis could well be ejected in its turn. This sequence will result in an abnormal clone occupying the basal layer. Inside this basal layer clone, if a transformed cell divides, a transformed cell migrates and in turn is replaced by a transformed cell. The situation can only change around the periphery, with transformed cells exerting a 'thrust' of $k-1$ against normal neighbours. If $N$ is the number of transformed cells, and $n$ the number of transformed peripheral cells,

$$\mathrm{d}N/\mathrm{d}t = (k-1)n.$$

Unexpectedly, $n$ is not proportional to $N^{0.5}$, but to $N^{0.55}$, because the dimension applied to the margin is 1.1, probably because of the inherent 'crinkliness' of the margin (see Fig. 13.6, which shows the results of some of Williams and Bjerknes's simulations, for $N = 100, 400, 800$, and 1600 cells, for several different values of $k$). If $k$ is very large (sequence 1), the transformed cell always pushes out the normal cell and regression does not occur. Enclaves of normal cells appear, but because of the proliferative advantage, the converse effect cannot occur. If $k = 2$, then, with a probability calculated at $1/k$, some 50 per cent of the abnormal clones will regress; enclaves of normal cells are larger, and take longer to fill in, and, notwithstanding the proliferative advantage, islands of transformed cells appear. If $k = 1.1$, more clones will regress, and the picture looks more infiltrating (see Fig. 13.6). The mean time to any total $N$ (say 50 000 cells for a 2 mm lesion), is $\bar{t} = 1.25N^{0.45}/(k-1)$; with $k = 2$, the induction period would be 1.3 years, rather long for experimental systems.

There are several points to make here: (i) in the studies outlined in Table 13.1, there would appear to be a field change, with all cells sharing in the increased proliferative rate; nevertheless, if we compare the mean of the $T_C$ values for hyperplastic epithelium, 60 hours, with the mean for the tumours, 22 hours, then $k$ will be about 3, which would decrease the induction period to more realistic levels. If $k$ does equal 3, then there would be no need to assume the several hundred initial hits postulated by Williams and Bjerknes (1972) to reduce

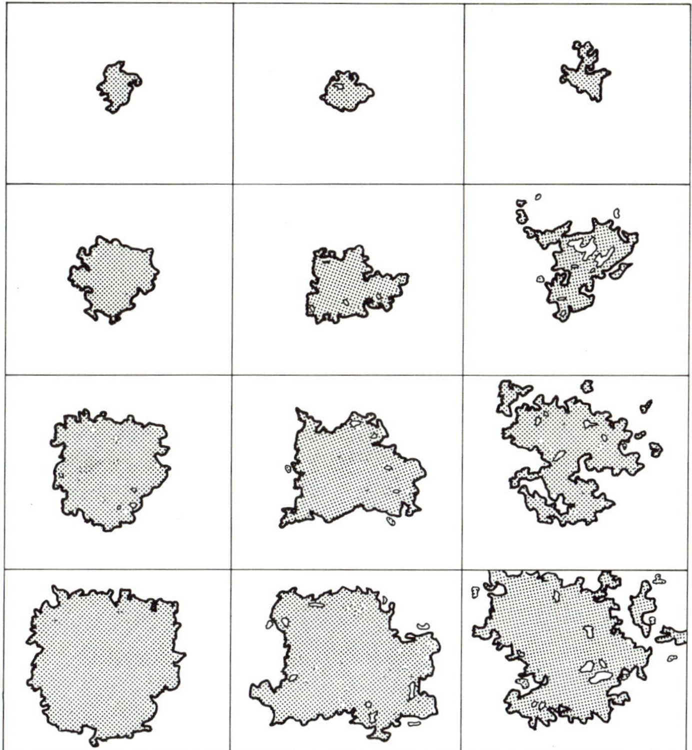

Fig. 13.6. Configuration of normal and transformed cells in the basal layer of a stratified squamous epithelium during clonal growth of an *in situ* carcinoma. The abnormal cells, shown by shading are considered to divide $k$ times as fast as the normal cells. Cells divide in the plane of the basal layer, and on division, one of the nearest six neighbouring cells, at random, is pushed suprabasally. The situations are shown for 100, 400, 800, and 1600 abnormal cells where in the left column $k = \infty$, in the middle $k = 2$ and in the right hand column, $k = 1.1$. (Redrawn from Williams and Bjerknes (1972).)

the latent period to a reasonable value; (ii) the model assumes that the junctions between transformed and normal cells are broken with the same ease, and that a normal neighbour is replaced by an abnormal daughter cell with probability $k/(k + 1)$. Although we have no good reason for disbelieving that transformed cells have the same adherence/migration characteristics, Williams and Bjerknes point out that if increased basement membrane adherence is the carcinogenic advantage, then the result would be effectively the same. In the hamster cheek pouch, if $k = 3$, there is no need to postulate adherence changes; (iii) the model, thus far, ignores the fact that stratification of transformed cells does occur, until the whole thickness of the epithelium is replaced with malignant cells; these retain the capacity to divide, and presumably could spread in suprabasal layers. Nevertheless, this is a fascinating link between theory and data.

This leads to a final question—what are the target cells? We can only speculate. If there is a stem cell hierarchical system, as envisaged in the EPU argument (see Chapter 8, section 2), then common sense would insist that stem cells are the target, since only the DNA of these cells remains to be affected over the prolonged (in relationship to $T_C$ values and transit times) period of carcinogenesis. On the other hand, carcinogen treatment of structured epidermis leads to early hyperplasia and loss of ordered structure; what happens to putative stem cells here? Perhaps the answer may lie in the study of stem cell kinetics in thick, non-stacked epidermis, and in carcinogen-treated epidermis.

## 13.4. Cell proliferation in squamous epithelial tumours

It is germane to complete this discussion of carcinogenesis in squamous epithelium with a look at the proliferation profile of the end product, the established tumour. Where primary tumours in experimental animals are concerned, Table 13.1 shows that, whatever the site, the squamous carcinomas are a more or less uniform group of rapidly proliferating tumours with short cell cycle times, and, where the data are complete (cf. Reiskin and Berry 1968), have high growth fractions and high cell loss factors. The difference between the normal counterpart and the malignant tissue is self-evident.

Table 13.2 (pp. 442–3) shows a selection of the more reliable kinetic parameters which have been published for a variety of human tumours which are of direct squamous epithelial origin, or for tumours which are associated with squamous epithelium (cf. basal cell carcinoma, malignant melanoma). There are naturally many problems encountered in the collection of data in these human tumours and in their interpretation, exemplified in the results tabulated for squamous carcinomas in Table 13.2. There are very few direct measurements of the cell cycle time, for example, and most estimates are stage-duration calculations, and the resulting values are therefore better described as *potential doubling times*. Because so few of the FLM curves are without second peaks, calculation of the growth fraction has usually been done by the Mendelsohn (1963) method of counting the FLM at some time when the labelled mitoses: total mitoses ratio has hopefully stabilized, although often there is no proof of this. However, Bresciani *et al.* (1974) were able to analyse seven cases of head and neck squamous cancer by computer analysis of FLM curves, most of which had a well-defined second peak, and obtained values for $T_C$ ranging from 52 to 88 hours. The duration of DNA synthesis in most studies has been relatively short compared with normal epidermis (Weinstein and Frost 1969), although some tumours do have very long $t_S$ values. Where the estimate of $I_P$ is meaningful, values are very variable, but the lack of reasonable information about the volume doubling time makes $\phi$ values rather rare. Bresciani *et al.* (1974) were only able to observe diameter changes in their cases of head and neck cancer for a short period (for obvious ethical reasons), but $t_D$ was variable, from 13 to 68 days, and, because of the limited time of observation, these readings are probably suspect. Steel (1977) culled many values from the older literature, and obtained a value of 56.7 days (95 per cent confidence limits, 42.9 and 75.1 days) for a group of 29 pulmonary metastases from head and neck carcinomas, which value is of course difficult to compare with primary tumours. However, the data of Bresciani *et al.* do indicate that most tumours are in a high cell loss state. Because they obtained quite complete data from seven patients, Bresciani *et al.* were able to compare kinetic parameters one with another, and it was interesting to see that the cell cycle time remained constant over a considerable range of volume doubling times, whereas $I_P$ was seen to fall as $t_D$ increased, which of course are well recognized phenomena for transplantable tumours growing in experimental animals (Steel 1977). However, there was no change in $\phi$ with increasing volume doubling time, whereas, in most experimental solid tumours, there is a distinct tendency for $\phi$ to vary directly with $t_D$.

It is difficult to relate basal cell carcinoma to squamous cell cancer, since there is no overall agreement about its histogenesis. In many instances, basal cell carcinomas can be seen actually linked (? arising from) the epidermis, and, while the most common form is disposed in solid sheets of cells, there is a capacity to differentiate into structures resembling skin adnexae; basal cell carcinomas are usually regarded as the least differentiated of the organoid adnexal tumours (Pinkus and Mehregan 1981). There have been no really reliable direct measurements of the cell cycle time in these tumours, but values for $t_{PD}$ appear to vary from 90 to 200 hours (Table 13.2). It is interesting that most estimates of $t_S$, derived by both the FLM and the double-labelling methods, give long values; this may be a characteristic of these tumours. Heenen *et al.* (1975) showed that, in *seborrheic verruca*, or, to give it its older name, *basal cell papilloma*, a benign lesion composed of basaloid cells, $t_S$ was nine hours; in the *premalignant fibroepithelial tumour (of Pinkus)*, which commonly gives rise to an invasive basal cell carcinoma, $t_S$ was lengthened to 19 hours, similar to that seen in the unequivocal basal cell carcinoma. However, in lesions of the *naevoid basalioma syndrome*, where basal cell naevi occur, $t_S$ was 14 hours (see Table 13.2). These are not really naevi at all, but are true basal cell carcinomas *ab initio*; this finding is not in keeping with the general trend for basal cell carcinomas to have long $S$ phase durations, and warrants further investigation.

Both Weinstein and Frost (1969) and Frindel *et al.* (1968) remarked on the evident disparity between the 'cell cycle time' and the actual doubling time of the lesion; notwithstanding the smallish growth fractions indicated by the Mendelsohn (1963) method in the paper of Frindel *et al.* (1968), these are notoriously slowly growing neoplasms; indeed, as long ago as 1953 Willis remarked on the enigmatically high incidence of mitoses in basal cell carcinoma. Kerr and Searle (1972) have suggested that there is in fact a high rate of cell loss in basal cell carcinoma, as evidenced by the large numbers of apoptotic bodies present in the lesions. We cannot dimensionalize the cell loss rate as we have no $t_D$ measurements, but it would appear to be high.

Table 13.2. Kinetic parameters in squamous cell carcinoma, basal cell carcinoma, and in malignant melanoma in humans

|  | $I_S$ (%) | $I_M$ (%) | $k_B$ (cells/%/h) |
|---|---|---|---|
| **(A)  Squamous cell carcinoma** | | | |
| Frindel *et al.* (1968) | | | |
| Squamous carcinoma | 10 | 1.4 | – |
| Spindle cell carcinoma | 7 | 1.4 | – |
| Bennington (1969) | | | |
| Squamous carcinoma of cervix | 0.4 | 0.13 | – |
| Meyer and Donaldson (1969) | | | |
| Squamous carcinomas of oral cavity | – | – | – |
| Bresciani *et al.* (1974) | | | |
| Squamous carcinomas of head and neck (7 cases) | 11–36 | – | 63–183 |
| Steel (1977) | | | |
| Pulmonary metastases of squamous carcinoma (from literature) | – | – | – |
| Friedland and Weinstein (1977) | | | |
| Primary cutaneous squamous carcinoma (12 cases) | 24.9 | – | – |
| Chavandra *et al.* (1979) | | | |
| Squamous carcinomas | 12–30 | – | – |
| Sakuma (1980) | | | |
| Primary squamous carcinomas of oral cavity (8 cases) | 19–20 | – | – |
| Meck *et al.* (1981) | | | |
| Vulvar squamous carcinoma xenografted in mice | 5–20 | 1.2–3.1 | – |
| **(B)  Basal cell lesions** | | | |
| Heenen *et al.* (1973, 1975) | | | |
| Seborrheic verruca | 1.4 | – | – |
| 'Basal cell naevus' | 8.2 | – | – |
| Premalignant fibroepithelioma | 8.4 | – | – |
| Basal cell carcinoma | 9.2 | – | – |
| Frindel *et al.* (1968) | | | |
| Basal cell carcinoma | 5–9 | 0.7 | – |
| Weinstein and Frost (1969, 1970) | | | |
| Basal cell carcinoma (9 cases) | 9.2 | – | – |
| **(C)  Malignant melanoma** | | | |
| Shirakawa *et al.* (1970) | | | |
| Secondary skin deposits | 3.8–6.2 | – | – |
| Young and DeVita (1970) | | | |
| Secondary skin deposits | 17–35 | 1.1–1.4 | – |
| Steel (1977) | | | |
| Pulmonary metastatic deposits | – | – | – |
| Newburger and Weinstein (1980) | 6.7[10]–18.7[11] | – | – |
| Primary melanomas | | | |

(1) FLM derived; (2) Continuous labelling; (3) Grain count halving; (4) Mendelsohn (1963) method;   (5) *in vitro* double labelling;   (6) direct measurement;   (7) $t_2$ rather than $t_{G_2}$;

| $I_P$ | $T_C$ (h) | $t_{G_1}$ (h) | $t_S$ (h) | $t_{G_2}$ (h) | $t_M$ (h) | $t_D$ (d) | $t_{PD}$ (d) | $\phi$ (%) |
|---|---|---|---|---|---|---|---|---|
| 0.4[4] | 25[3] | – | 11 | 4.5[7] | – | – | – | – |
| 0.82 | 30[6]<br>15.5[6,8] | – | 20 | – | – | – | – | – |
| 0.53 | 15.5[6,8] | | 11.8 | 1.3 | – | – | – | – |
| – | – | – | – | – | – | – | 6–9[14] | – |
| 0.31–0.84 | 52–88[1,6,12] | | 18–34 | 5–14 | – | 13–68 | – | 78–93 |
| – | – | – | – | – | – | 56.73<br>(42.9–75.1)[15] | | – |
| – | 50.2[13] | 28.8 | 12.5 | 6.9 | 2.0 | – | – | – |
| – | – | – | – | – | – | – | – | – |
| – | 22–55[5,13] | – | 4–14 | – | – | – | – | – |
| 0.2–0.8[1,2] | 43[6] | – | 8.6 | 8.9 | – | – | 5–30 | – |
| – | – | – | 9.2 | – | – | – | 26 | – |
| – | – | – | 14.3 | – | – | – | 7 | – |
| – | – | – | 18.4 | – | – | – | 9 | – |
| – | 207[5,13] | – | 19.2[5] | – | – | – | – | – |
| 0.29[4] | 97[13] | 66 | 19.0 | 12[7] | – | – | – | – |
| 0.39[4] | 93[13]<br>36[6] | | | | | | | |
| 1.0 | 217[13] | 188[13] | 20 | 7 | 1.5 | – | – | – |
| 0.25–0.40[1,2] | 71[1]<br>7–9[3] d | 45 | 25 | 7.5 | – | – | – | 76[9] |
| 0.56–0.58 | 47 | – | 22–24 | 4–8 | – | – | – | – |
| – | – | – | – | – | – | 53.47 | – | – |
| 0.7[9] | 65[13,10]<br>178[13,11] | | 18.1 | 8.3 | – | – | – | – |

(8) questionable;  (9) assumed;  (10) low subgroup;  (11) high subgroup;  (12) computer fit;
(13) stage-duration calculation;  (14) metaphase arrest;  (15) 95 per cent confidence limits.

There have been very few accurate measurements of cell kinetic parameters in primary malignant melanoma, and, although purportedly a rapidly growing neoplasm, Steel (1977) calculated a mean $t_D$ of 53 days, similar to squamous carcinoma (see Table 13.2), for published values for secondary melanoma deposits growing in the lung. Shirakawa *et al.* (1970) produced several reasonable FLM curves for metastatic cutaneous melanoma, with $T_C$ values of about 70 hours, although grain count halving experiments suggested a cell cycle time of 7-9 days, possibly indicating that some cells were decycling during the experiment; $\phi$ was high at 0.76, while $I_P$ estimates have been variable. Interestingly enough, most values for $t_S$ have been long, in excess of 20 hours in some reports. Newburger and Weinstein (1980) considered that there were two groups of melanomas, those with high $I_S$ values, and those with low ones. Differential cell kinetics were not carried out, however, so the kinetic cause of this disparity cannot be guessed.

## 13.5. Conclusions

It is clear that we know little about the relationship of cell proliferation to carcinogenesis: we are reasonably sure that cell proliferation is, in some way, involved in both initiation and promotion in epidermis, but whether this is an epiphenomenon, a regenerative response to cytotoxic cell death or a mandatory step which 'fixes' in some way the initiating or promoting action of the agent (Medline and Farber 1981) is not yet clear. However, squamous epithelium does show, with a few anomalies, rather a neat progression in proliferative rate as carcinogenesis proceeds, and experimentally-produced squamous carcinomas are uniformly rapidly proliferating neoplasms, but, because of the paucity of good data (both in normal epidermis—see Table 7.4 (p. 280), and tumours—see Table 13.2), we cannot say for certain whether a similar relationship exists in humans, although some human squamous cell carcinomas are rapidly proliferating. We know little or nothing about which cells are carcinogenic targets.

# 14 Growth control in squamous epithelium

We have already considered the general aspects of growth control in Chapter 2, section 10; there is a good deal of evidence that negative feedback systems are operative in epithelial tissues, but as yet there is no general agreement as to whether the information elicited by the control mechanism is inhibitory or stimulatory. In squamous epithelium it is true to say that the accent has been firmly on inhibitors—the chalone hypothesis was first proposed for epidermis, and latterly there has been the contentious issue of the inhibitory effect of cyclic AMP on epidermal cell proliferation. However, there are also factors which are held to have a stimulatory action: epidermal growth factor and cyclic-GMP are among these.

There is little doubt, from our previous discussions, that growth control in epithelia does involve the transfer of information between cells: current thought maintains that this information is chemical in nature, and it is the purpose of this chapter to look closely (and critically) at the various candidates which have been proposed for this role in squamous epithelium

## 14.1. The epidermal chalones

Since much of the earlier history of the chalones in fact encompasses the epidermal chalone, we have already encountered the basic propositions made by Bullough (1962), which form the basis of the chalone hypothesis of growth control. For many years this concept has provoked perhaps more heat than light, with not a little in the way of emotional fervour, both *pro* and *con*, with on the one hand, some protagonists for whom the hypothesis had become a 'belief system' (Rijke 1980), and still others who reach for their revolvers (Goering, attrib.) every time the word chalone was mentioned. While we have stated our opinion that much of the chalone work is founded upon suspect methodology, we are committed neither one way nor the other, and we hope, therefore capable of an objective viewpoint.

We have agreed that the rate of cell loss, the rate of cell production and of cell differentiation must, of necessity, be exactly co-ordinated to maintain a steady state condition; it is probable that stability can only really be maintained by some sort of *negative feedback mechanism* (Weiss and Kavanau 1957; and see Chapter 2, section 10), and such inhibiting information can be produced (a) either by the proliferating cells, and inhibit the rate of differentiation, or (b) by the differentiating cell mass to limit cell proliferation. This second concept has found most favour in the epidermis.

### (i) *Evidence for negative feedback in the epidermis*

We might start with the proposal that some sort of feedback is operative, as indicated by the ability of the epidermis to reach a new steady state after, say, a hyperplastic perturbative stimulus, as occurs for example after chronic friction (MacKenzie 1974*a,b*); moreover, the curtailment of proliferative activity after a wound (cf. tape stripping) indicates a control mechanism which recognizes when cell number is again at normal levels. The other salient point to make is that the proliferative response is intrinsically *local*, and is confined to the environs of the damaged tissue.

The main reason for the preference for inhibiting information in the epidermis were the early experiments of Bullough and Laurence (1960*a,b*); using the thin mouse ear, wounding of one side of which might reasonably be expected to allow diffusion of any proliferative control substance across to the other side, and wounds of various sizes, Bullough and Laurence showed that the pattern of response on the side opposite to the cut distinctly favoured the loss of a local concentration of an inhibitor, rather than the release of a stimulating wound hormone from the wound margins (Abercrombie 1957)—see Fig. 14.1. That this result was not due to increased blood flow or other factors which accompany wound healing (cf. leucocyte emigration, plasma protein exudation, etc.) was indicated by the observation that selective damage to the underlying connective tissue, which would produce all the epiphenomena of wounding, failed to produce

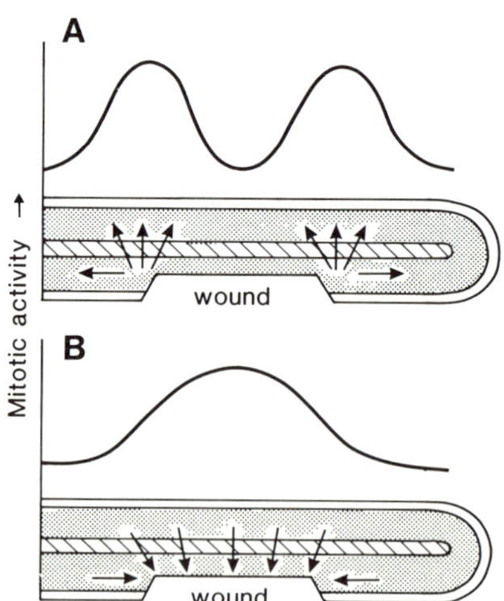


Fig. 14.1. Mitotic activity expected in the uninjured mouse ear epidermis after a 3 mm² wound in the opposite epidermis and superficial dermis, assuming (A) that an epidermal 'wound stimulator' hormone is produced, and (B) that the concentration of an epidermal chalone is reduced in the wounded region. (Redrawn from Bullough and Laurence (1960*a*).)

a proliferative response in the overlying epidermis (Bullough and Laurence 1960*a*). Although these experiments apparently strongly support the concept of a lack of inhibitor, theoretically they could be explained by the production of a localized epidermal stimulator from the dermis (Allen and Smith 1979).

However, Bullough and Laurence's results were supported by the findings of Finegold (1965), who showed that the proliferative response in hamster ear epidermis could be prevented by skin grafting the wounded area, while Iversen *et al.* (1974*a*), using the equally thin web membrane of the African fruit bat, were able to confirm, not only the results of Bullough and Laurence, but also Finegold's observations concerning the suppression of the proliferative response by grafting. In this context, Marks (1976*b*) has drawn attention to some interesting experiments of Mizuno and Fujii (1969); when an impermeable barrier such as cellophane is inserted between two opposing wound margins, the encroaching epidermis grows underneath the barrier to form a pocket; if, however, the barrier is made of porous material, the epidermis remains separated by the implant, indicating that some inhibitory information was penetrating the filter, suppressing proliferation on each side.

We have already seen that the amount of skin regenerated after abrasion describes a series of damped oscillations about the control value with time after wounding (Argyris 1976, 1977), characteristic of a negative feedback loop (see Chapter 2, section 10), and we might note that Iversen and Bjerknes (1963) were able to model the cell loss and cell proliferation evoked by carcinogen treatment of mouse epidermis with a computer model incorporating a negative feedback term.

These experimental observations all support the presence of negative feedback in epidermis, and we believe that this evidence is comparatively strong; it is when we begin to discuss the chemical nature of the inhibitory information that the sparks and the speculations begin to fly.

### (ii) *The nature of the epidermal chalone*

Since the active principle in the epidermis must be a freely diffusible substance, it seemed reasonable to assume that it was water soluble, and aqueous extracts of epidermis (Bullough and Laurence 1964*b*) or of whole skin (Iversen *et al.* (1968) were tested for inhibitory activity by means of the metaphase arrest method (see below); unfortunately the action proved to be on the $G_2$–$M$ flux only, thus placing its physiological role in jeopardy, since Bullough (1965) had declared that proliferation control is firmly the province of the $G_1$–$S$ boundary; in order to overcome this defect, Bullough (1965) speculated that chalones were able to slow down the whole cell cycle; this, however, was not to be, since this '$G_2$ chalone' has been shown to be devoid of any action on DNA synthesis (Baden and Sviokla 1968; Thornley and Laurence 1975).

### (iii) *The $G_2$ chalone*

This was then freely extractable in aqueous solvents (Baden and Sviokla 1968;

Frankfurt 1971; Marrs and Vorhees 1971*b*; Chopra *et al*. 1972; Wiley *et al*. 1973).

(a) *Assay systems*

As far as *assay systems* go, because the putative action of the chalone is solely on the $G_2$–*M* flux, the only available method was the metaphase arrest method, either upon epidermal cells *in vivo* (Iversen *et al*. 1968; Hondius Boldingh and Laurence 1968; Baden and Sviokla 1968; Elgjo 1969*a,b*) or on epidermal or squamous carcinoma cells growing *in vitro* (Bullough and Laurence 1961*a*; Marrs and Vorhees 1971*a,b*; Chopra *et al*. 1972; Isaakson-Forsén *et al*. 1977). We have already discussed, in our general critique of the chalone concept, the problems involved with injecting large amounts of foreign protein into animals (i.e. pig epidermal chalone into mice), the choice of appropriate controls, and the problems involved in studying cell kinetics in tissue explants (see Chapter 7). There is also the consideration that, apart from the detailed work of Iversen *et al*. (1968), the application of the metaphase arrest method usually leaves a good deal to be desired (see Chapter 3, section 3), particularly the lack of any confidence limits for the $r_M$ in the publications of Bullough and Laurence (Bullough and Laurence 1964*a*; Bullough *et al*. 1968*b*), and the fact that Elgjo (1969*a*) found that metaphase collection with colcemid, after the injection of epidermal extracts, could be non-linear.

Bullough and Laurence (1964*c*) considered that chalone action was only vouchsafed if, after the effect had worn off, the inhibitory activity was restored by washing in a high concentration adrenaline solution; this unstable 'adrenaline: chalone complex' then restored inhibition. However, no evidence was forthcoming that a complex was actually formed, and Ascheim (1968*a,b*) wondered about the stability of the unstable adrenaline in culture conditions, and the possible primary inhibitory action of its breakdown products. Despite the apparent confirmation of this effect by Iversen (1968), Marrs and Vorhees (1971*a,b*), and numerous other authors (cf. Chopra *et al*. 1972; Frankfurt 1971), little further light has been cast on this inter-relationship, and all attempts to isolate a chalone:adrenaline complex have failed (Laurence 1973; and see section 14.2). Furthermore, increasing doses of chalone given *in vivo* lead quickly to a plateau effect at 50 per cent inhibition for mouse ear epidermis and 20 per cent for tongue, but combined with adrenaline a 75 per cent inhibition could be obtained (Laurence and Randers Hansen 1971, 1972). We would wonder about the advisability of injecting quantities of adrenaline, which after all has profound systemic effects, as a means of studying its effects on cell proliferation (Ascheim 1968*a*). Moreover, adrenalectomized animals still respond to the $G_2$ chalone, impuning a direct effect of adrenaline upon chalone (Elgjo 1972).

Laurence (Laurence *et al*. 1972; Laurence 1973) proposed that chalone activity is modulated by an antagonist, or *anti-chalone*, and adrenaline (and hydrocortisone) blocks the effectiveness of this antagonist; whilst this remains entirely speculative, we might note that epidermal growth factor (EGF), a potential 'antichalone' (Marks 1976*b*) is under $\beta$-adrenergic control (Scheving *et al.* 1979); alternatively the antichalone may be synonymous with the nebulous 'mesenchymal factor' of Bullough (1975*a,b*), a hypothetical chemical messenger synthesized by connective tissue cells in the dermis, which acts as a non-tissue specific antichalone, and is responsible for the well-known ability of connective tissue to modulate epithelial growth and function (Bullough 1975*a,b*). In this respect, Bullough and Mitrani (1976), as a consequence of a wholly theoretically based study, proposed that a 'gradient' of chalone effectiveness was present through the epidermis, created largely by the opposing action of the mitotic stimulant or antichalone (the mesenchymal factor), emanating from the dermis; thus the chalone influence is weakest in the basal layer, and strongest in the suprabasal cells, which are post-mitotic; an important action of the chalone, in high concentrations, was to maintain cells in the post-mitotic state, and to control the rate of keratin synthesis and the speed of aging (Bullough 1972, 1975*a*).

In this context, Elgjo (Elgjo 1974*b*; Elgjo and Edgehill 1973) reported that the $G_2$ chalone could diffuse into the dermis, and that its inhibitory activity was inhibited by serum; the ability of connective tissue (and serum) to stimulate epidermal cell growth was then seen as a consequence of its ability to bind and/or inactivate the $G_2$ chalone. Also on the topic of putative antichalones, Frankfurt (1971) prepared rat skin extracts which *stimulated* migration of basal cells in $G_2$; some 12 hours after the extract was given, suprabasal cells with grain counts approaching the original flash interphase grain count were apparent above the basal layer. Since basal cells do not normally migrate in $G_2$ as a normal phenomenon, the significance of this finding is obscure. A further candidate antichalone might lie in the work of Hell (1970), who found a stimulator of DNA synthesis in guinea pig epidermis, but who did have some dermal contamination in her preparations, which could therefore conceivably be the source of the factor, which was dialysable and of low molecular weight.

Here we are faced with interpretation in the absence of specific chemical information; are these the same factors which Melbye and Karasek (1973) isolated from dermis, which stimulated epidermal cell proliferation *in vitro*? Marks (1976*b*) has pointed out that if these putative substances do act by neutralizing chalones, then exogenous chalone, in sufficient amounts, should be able to completely inhibit epidermal mitosis; despite the claim of Hall (1969*a,b*) to have achieved a dose-dependent total inhibition of DNA synthesis with epidermal extracts *in vitro*, most workers have found this to be impossible (Laurence 1979). Additionally, we have no quantitative information on the effects of chalones on differentiation (Laurence and Thornley 1976*b*) nor are we sure that a mesenchymal factor exists. Until more information is available, we must reserve judgement on this speculative interpretation, and beware of fanciful theoretical models founded upon little in the way of experimental data, in this already highly contentious field.

### (b) *Chemical nature of the $G_2$ chalone*

Attempts to purify this chalone have been largely unsuccessful; mouse $G_2$ chalone was considered to be a large molecule, non-dialysable and heat-labile (Bullough *et al.* 1967; Elgjo 1969*b*), but most work has been done with the 71–82 per cent ethanol precipitate of commercial pig rind extract (Hondius Boldingh and Laurence 1968) which contained a large amount of proline and hydroxyproline, with about 15 per cent carbohydrate, and was considered to be either a glycoprotein or a mixture of proteins and carbohydrates, with a molecular weight of $2.5-4 \times 10^4$ daltons. Isaakson-Forsén *et al.* (1977) reported a partial purification of mouse $G_2$ chalone by ammonium sulphate precipitation, affinity chromatography and gel filtration, with a 300-fold increase in biological activity, concluding that the material was a glycoprotein with a molecular weight of 20 000 daltons; even so, there were multiple bands on SDS polyacrylamide gel electrophoresis, and the material was still likely to be very heterogeneous.

However, Marks (1976*b*) considered that any conclusion that the $G_2$ chalone was a glycoprotein to be premature; Marks (1973*a*) showed that the lyophilized ethanolic precipitate is, in fact, highly aggregated, and speculated that the activity could well be confined to a small molecule, bound possibly to collagen (cf. the high hydroxyproline content); in this context, Iversen (1969) did find, in his own investigations, that most activity was associated with a fraction of apparent molecular weight of only 3000 daltons, although the existence of such small molecular weight inhibitors introduces its own problems in interpretation (see Chapter 2, section 10). The use of ethanolic precipitation has also been questioned (Marks 1976*b*), since it appears to give maximum activity in different fractions depending on the investigation (cf. Hondius Boldingh and Laurence 1968; Chopra *et al.* 1972; Marrs and Vorhees 1971*a,b*), which also may indicate the presence of aggregation; moreover, Marks (1976*b*) also doubts that the strongly stainable bands on polyacrylamide gel represent the chalone, as claimed by Chopra *et al.* (1972), since any chalone would be present in very low concentrations; again protein aggregation is probably the answer.

We must conclude that we do not know what the $G_2$ chalone, or more properly, the $G_2$–$M$ inhibitory action associated with epidermal extracts, is. But most investigators appear clear about where it occurs and what it does. It is practically ubiquitous in the epidermis of most species (Bullough *et al.* 1967; Marks (1976*b*), and in other squamous epithelia such as that of the oesophagus (Laurence 1973), palate and gingiva (Laurence and Randers Hansen 1972) and also in squamous carcinomas from various sites (Bullough and Laurence 1968*b,d*; Bullough and Deol 1971): its action is to inhibit $G_2$–$M$ flux (Bullough and Laurence 1964*c*; Elgjo 1969*a*). An action in late $S$ may also be possible (Clausen *et al.* 1979); we should perhaps note that not all investigators have been satisfied with their response, since Schilling *et al.* (1972) could only effect a *delay* in the wave of proliferation which follows explantation of human prepucial cells, with their chalone preparation.

(c) *Specificity*

As far as specificity goes, extracts of other tissues do not apparently affect $G_2$-$M$ flux in epidermis (Bullough and Laurence 1964c; Elgjo and Hennings 1971). The epidermal chalone itself does not affect cell division in non-related tissues such as rectal crypts (Bullough and Laurence 1964c; Chopra *et al.* 1972; Nome 1975b), but the original pig skin extract did affect $G_2$-$M$ flux in other squamous epithelia such as the oesophagus (Bullough and Laurence 1964c; Frankfurt 1971; Nome 1975b) and in sebaceous glands, although not hair follicle epithelium however, which is also thought to be a derivative of the primordial epithelial germ cell (Lever and Lever 1979) or the adult adnexal matrix cell (Pinkus and Mehregan 1981); because of this rather wide action, Bullough and Laurence (1964c) did wonder about the tissue specificity of the extract; in this respect, oral epithelial extracts *did* affect tongue epithelium more than epidermal extracts, and vice versa (Laurence and Randers Hansen 1972; Nome 1975b), so the possibility, not only of a chalone specific for squamous epithelia, but also of structural differences between squamous epithelia-specific chalones, exists (Marks 1976b). Furthermore, on additional purification, the sebaceous gland chalone could be separated from the epidermal $G_2$ chalone (Bullough and Laurence 1970); there are, apparently, distinctive sweat gland (Bullough and Deol 1972), melanocyte (Bullough and Laurence 1968c; and see Chapter 5, section 2), and hair follicle (Bullough 1967) chalones, none of which has yet been anything like fully characterized.

More recent investigations have not exactly cleared the water: Nome (1975b), working with crude extracts, claimed that $G_2$ chalone acted only on keratinizing squamous epithelia, but Laurence (1980) maintains that the non-keratinizing corneal epithelium does respond to purified $G_2$ chalone, although the endo-dermally derived oesophagus does not; indeed, perhaps a more realistic way of looking at the problem of tissue specificity is to consider that the cell line-specific nature of chalone action is related to the embryological origin of the tissue in question, rather than to its morphological appearance (Laurence and Thornley 1976a,b; Thornley and Laurence 1976). Although tongue epithelium responds less actively to epidermal extracts than to extracts from the palate (Laurence and Randers-Hansen 1972b), it is singular that $G_2$ chalone activity (i.e. activity against *epidermis*) is evident in oesophageal, palatal, and gingival extracts (see also Nome 1975b). The embryologically-specific hypothesis would appear to founder on this observation, and recently, Thornley (see Allen 1978) has discovered that his epidermal chalone ($?G_2$) also inhibits lung, trachea, and bladder epithelial cells.

(d) *The physiological role of the $G_2$ epidermal chalone*

There is, in fact, very little that we can say for certain concerning the physio-logical role of this inhibitory principle (since the *physiological role* of any chalone is yet to be demonstrated); perhaps we should say 'the potential physiological role'. Well, its phase-specific site in $G_2$ precludes any major role in cell proliferation control, but it will be recalled that Gelfant (Gelfant and Candelas 1972; Gelfant 1976, 1977) has advocated the presence of a minor cohort of $G_2$ blocked cells in the epidermis, and it is of course conceivable that the $G_2$ chalone could be the cause of this block; Potten (1974) has suspected, on admittedly non-quantitative evidence, that the 'stimulus-responsive cells' in the centre of the EPU are possibly in the $G_2$ phase (see Chapter 8, sections 2 and 4), perhaps another target cell? However, we do have reservations about the evidence for $G_2$ cells in epidermis (see Chapter 8, section 4), and anyway, if this is its only role, then but a very small proportion of proliferative epidermal cells are $G_2$ chalone-sensitive (Marks 1976$b$).

Just as its physiological role is open to doubt, its mechanism of action remains entirely speculative, as the mechanism of action of any substance whose nature is disputed must be; its rapidity of action is also a singular feature (Iversen 1969); nevertheless, Bullough (1975$a,b$) considered that the $G_2$ chalone was a repressor of the 'mitosis operon'; Gelfant and Candelas (1972) have suggested that it acts as a post-transcriptional repressor for an mRNA coding for some protein essential for $G_2$–$M$ flux, but Elgjo (Elgjo and Edgehill 1973; Elgjo 1975) maintains that the $G_2$ chalone may be primarily involved in differentiation, and only inhibits $G_2$–$M$ flux as a secondary phenomenon.

At this point, at least to those viewing the field from without, the chalone hypothesis seemed to be foundering, particularly when Baden and Sviolka (1968) failed to demonstrate any short-term effects of epidermal extracts on epidermal DNA synthesis, but then, in 1969, Hall reported that extracts of hamster cheek pouch inhibited DNA synthesis in pouch epithelium *in vitro*, and the studies of Hennings *et al.* (1969) revealed that mouse skin extract had a reversible inhibitory action on [$^3$H]-TdR incorporation in mouse epidermis *in vivo*, which reached a maximum at 9–12 hours after injection. Elgjo *et al.* (1971) and Elgjo and Hennings (1971) claimed that DNA synthesis in mouse epidermis and in hamster squamous carcinoma was inhibited by extracts made from skin treated with actinomycin D (the inhibitor of DNA-directed RNA synthesis), which evidently destroyed the $G_2$ chalone; Frankfurt (1971) also noted that a series of injections of an 80 per cent ethanolic extract of rat skin gave a depression in the flash labelling index in both mouse epidermis and forestomach.

These findings were confirmed and extended by Marks (1971, 1973$a$), who showed that, unlike the $G_2$ chalone, this '$G_1$ chalone' was resistant to heat, proteolytic enzymes and denaturing agents, and its elution profile on Sephadex indicated an apparent molecular weight of $10^5$ daltons, whereas Hondius Boldingh and Laurence (1968) considered that the $G_2$ chalone behaved as a molecule of $2$–$4 \times 10^4$ daltons molecular weight.

Elgjo *et al.* (1971, 1972) advanced the concept that basal cells and suprabasal differentiating cells contain different chalones; this resulted from work involving the preparation of extracts from basal cells and differentiating suprabasal cells separated by differential trypsinization (with minimal contamination, 13 and 27 per cent respectively); most of the $G_2$ activity was found in the basal layer, while the $G_1$ chalone appeared restricted to the suprabasal layers; in accord with the original Weiss and Kavanau (1957) negative feedback hypothesis, the $G_1$ chalone should therefore be the appropriate inhibitory signal; the cell cycle phase specificity of this material seemed much more appropriate for a real-life control factor.

### (iv) *The $G_1$ chalone*

This $G_1$ chalone is isolated in much the same manner, from skin and epidermal homogenates, as is the $G_2$ chalone.

### (a) *$G_1$ chalone assay*

The assay method is usually by measuring the incorporation of $[^3H]$-TdR into DNA in epidermal cells, usually estimated by scintillation counting. Maurer (Maurer and Laerum 1976, Maurer 1981) has a good deal to say regarding the use of such methods in chalone research, much of it highly deprecatory (see Chapter 3, section 2), and it is only comparatively recently that investigators have taken this advice to heart (Thornley *et al.* 1977; Laurence *et al.* 1979). There is also the salutary experiences of Duell *et al.* (1975), who observed that epidermal extracts inhibited $[^3H]$-TdR incorporation into fibroblast cultures without a corresponding inhibition of cell number increase; this action on $[^3H]$-TdR incorporation was found to be due to interference with both TdR transport, and phosphorylation to di- and triphosphate nucleotides; how many investigators have really looked at this?

Nevertheless, this has been the chosen method, but even if we accept the appropriateness of the assay (which we personally do not), both *in vivo* and *in vitro* assay methods are fraught with difficulty; *in vivo*, unless hairless mice are used, the telogen phase of the hair cycle is mandatory, since DNA synthesis in follicular cells can overwhelm the comparatively slow rates in interfollicular epidermis (Marks 1971, 1973*a*)—an artefact which could be effectively obviated by using the labelling index; moreover, it is fruitless to carry out such measurements without taking cognizance of the circadian rhythm (see Chapter 9). Laurence *et al.* (1979) have especially drawn attention to this aspect: because of the extreme variation in rates of DNA synthesis in epidermis, it is mandatory to first examine in some detail the circadian variation in $I_S$ or $[^3H]$-TdR uptake, and also to ensure that the putative chalone is given at the appropriate time point in the circadian cycle (see below). In this salient paper, which appears to denote a new critical phase in chalone research, Laurence *et al.* emphasize the importance of appropriate tissue controls; there is little point in comparing the action of the chalone on the target tissue with a control tissue where the kinetics of DNA synthesis are completely different, particularly with respect to the duration of DNA synthesis.

Even so, Marks (1976*b*) regards the *in vivo* assay as the only test which can be considered reasonably reliable, notwithstanding the problem of the administration of foreign protein. In short-term organ culture, epidermal DNA synthesis is unstable after explantation (Marks *et al.* 1971*a*; Dahl and Shuster 1971; Wiley *et al.* 1973; Young *et al.* 1975; Gradwohl 1978*a,b*) and unsuitable for use as a test system, and, presumably, if flux into $S$ is compromised, one wonders about the suitability of *in vitro* systems for assessment of $G_2$-$M$ flux inhibition also (Bullough and Laurence 1964*a,c*), especially if such explants are composed of newborn epidermis (Marrs and Vorhees 1971*a,b*), whose chalone-reactivity characteristics might differ from adult tissues (Bertsch and Marks 1974). However, we should note the experiments of Gradwohl (1978*a,b*), who tested the effects of epidermal extracts, prepared according to the Bullough and Laurence (1964*c*) procedure, on organ cultured mouse epidermis and on several cell lines in culture; while DNA synthesis and mitosis were both depressed in epidermis, the effect was non-cell line specific, and importantly, dye exclusion tests (which would tend to underestimate the damage), indicated a highly cytotoxic effect, even with fractionated extracts. Now the chalone protagonists would maintain that this is, by definition, not a chalone effect, but how many investigators test for cytotoxicity? As for explanted epidermis, Marks (1976*b*) considers that such explants may resemble wounded epidermis, which, it should be remembered, may not respond to $G_1$ chalone in the same manner as does intact tissue (Bertsch and Marks 1979*a*). Longer-term cultures of epidermal cells have been used as test systems in assays (Delescluse *et al.* 1974; Wiley *et al.* 1973) and have been criticized on the basis of the susceptibility of DNA synthesis to non-specific external effects (Marks 1976*b*). Marks would thus discount the reports that pig epidermal extracts inhibit DNA synthesis in both guinea pig epidermal and dermal cells (and also bladder cells) in prolonged culture (Delescluse *et al.* 1974), and also that the 80–90 per cent ethanolic fraction of pig skin contains a stimulant of epidermal DNA synthesis, in addition to an inhibitor in the 70–80 per cent fraction (Wiley *et al.* 1973).

### (b) *Chemical nature of the $G_1$ chalone*
The purification of the $G_1$ chalone has been rather more successful than that of the $G_2$ chalone; starting with the original pig skin extract referred to above (Hondius Boldingh and Laurence 1968), Marks (1971, 1973*a*) found most $G_1$ inhibition in the 55–72 per cent ethanolic fraction; the material was undialysable and resistant to proteolytic enzyme digestion and heat, with a molecular weight in excess of $3 \times 10^5$ daltons. Marks (1974, 1975) showed that this apparently high molecular weight was due to aggregation, possibly due to the use of distilled water in the extraction procedure (Allen and Smith 1979), and disaggregation showed an apparent molecular weight of only $2 \times 10^4$ daltons. Precipitation with cetylpyridinium chloride (which complexes with polyanionic compounds such as nucleic acids and mucopolysaccharides) and phenol extraction

gave a 2000-fold purification over the ethanolic fraction, and Marks (1976*b*) considered that the material represented a mucopolysaccharide-containing glycoprotein, although we cannot exclude a firm aggregation with a mucopolysaccharide. A dose as small as 0.05–0.1 $\mu$g of this preparation given to an adult mouse is sufficient for 50–70 per cent inhibition of DNA synthesis, and Marks (1976*b*) claimed that this was therefore in the range of a hormone; a similar $G_1$ to $S$ inhibitor of higher molecular weight was isolated from mouse epidermis by Marks (1976*b*), but its relationship to the low molecular weight form is not yet clear. More recently, Quirk *et al.* (1981) have partially purified a low molecular weight ($< 10\,000$) substance which inhibited [$^3$H]-TdR incorporation into mouse epidermis in organ culture; alcohol precipitation, ethyl acetate extraction and silicic acid chromatography produced a 15-fold increase in inhibitory action.

(c) *Action and specificity of the $G_1$ chalone*
The action of the $G_1$ chalone is to depress [$^3$H]-TdR incorporation into DNA at some time after injection. The time of action does vary with individual reports: for example, Elgjo *et al.* (1972) reported a delay period as long as 12 hours after injection, while Marks (1971) found a delay period of only five hours; maximum depression was obtained at 10–20 hours after injection. Grain count analysis (Yamaguchi *et al.* 1974) appears to also indicate that the $G_1$ chalone action includes a depression of DNA synthesis itself, rather than just an inhibition of flux into $S$: this smacks of a toxic action, rather than an effect of genuine potential physiological importance.

Nome (1975*b*) has claimed that such action is tissue-specific, and that the chalone acts only upon squamous epithelium, while Karienemi (1976) reports that serial injection of the 70–80 per cent ethanolic fraction of pig skin (Hondius Boldingh and Laurence 1968) for 48 hours not only selectively depresses $I_S$ in mouse ear epidermis, but also of human psoriatic keratocytes cultured in diffusion chambers in the same mice; however, no action upon dorsal or sole epidermis, or on other non-target tissues (cf. marrow, intestine) was noted. Extracts of other organs (e.g. liver, lung, kidney) do not apparently affect DNA synthesis in squamous epithelia (Marks 1971, 1973*a*). Taylor (1978) has pointed to the considerable difficulties which arise in interpretation of these and similar studies; unless detailed time course studies are done, effects on other tissues may be missed; in Taylor's experience, epidermal chalone effects on liver and kidney, for example, were evident only after 9–10 hours after injection of the preparation, later than the action on epidermis, so great care and considerable effort are needed to exclude effects on other tissues; conclusions regarding specificity may be premature. *In vitro*, Gradwohl (1978*a,b*) and Rothberg and Arp (1973, 1975) have produced evidence of epidermal chalone action against P-815 mastocytoma cells, and connective tissue cells, actions which were ascribed by Marks (1976*b*) to vagaries in the *in vitro* method, and possibly also the fact that Rothberg and

Arp (1973, 1975) used newborn mouse epidermis as the extracted tissue, and embryo chick epidermis as the target tissue; embryonic and neonatal epidermis has special properties *vis à vis* chalones, and should be used with discrimination (Bertsch and Marks 1974).

The action of the $G_1$ chalone is apparently fully reversible (Elgjo 1974*a*), and it does not appear to affect neonatal mouse epidermis (Bertsch and Marks 1974), an action which Marks (1976*b*) regarded as evidence of its non-toxic mechanism of action.

As to the actual mode of action, many workers agree that there is a genuine decrease in the $G_1$–$S$ flux of treated epidermis; this conclusion is based mainly on studies of the flash [³H]-TdR labelling index, and most studies have detected no change in the mean interphase grain count (Frankfurt 1971; Elgjo, see Marks 1976*b*). However, we have already noted that Yamaguchi *et al.* (1974) did find a reduction in grain count in regenerating mouse ear epidermis at one hour after injection of guinea pig epidermal extract; inhibition of DNA synthesis itself could not really be considered a potential control mechanism, and this effect can probably be ascribed to the heterogeneous nature of the extract (?containing TdR which would affect pool sizes).

#### (d) *Physiological role of the $G_1$ chalone*
In effect, the $G_1$ chalone is produced by the differentiating cells, inhibits the proliferating cells at an appropriate cell cycle stage, and is active at very low concentration; it may be tissue specific and appears non-toxic. How it is stored or secreted, and its physiological role, is still obscure; however, Marks (1973*a*, 1976*b*) has speculated that, since Odland bodies (see Chapter 4) contain mucopolysaccharide (Ohashi *et al.* 1973), and are apparently discharged by keratocytes into the intercellular epidermal space before keratinization is complete, this may represent chalone secretion; alternatively, since mucopolysaccharide also accumulates in desmosomes, if this represents chalone, then it may act on the cell membrane to inhibit migration and/or cell proliferation; since migration rather than proliferation is the initial response to epidermal injury (see Chapter 10), it is conceivable that the primary event may be an action on migration rather than proliferation (Marks 1976*b*).

#### (v) *Epidermal chalones and cancer*

If the chalone hypothesis is substantiated, there are evident implications for carcinogenesis: is the level of chalone *production* itself, or the chalone *responsiveness* reduced in epithelium undergoing neoplastic change (Bullough 1965, 1972)? Topical application of phorbol esters to mouse epidermis (see Chapter 13, section 2) produces a refractory state to $G_1$ chalone in the basal cells, even though there were normal levels of $G_1$ chalone in the treated epidermis (Kreig *et al.* 1974; Marks 1976*b*). That this response is not likely to be specific is indicated by the refractory response of the basal cells to $G_1$ chalone which occurs in the epidermis after stripping (Bertsch *et al.* 1976) and also the refractory state described by Bertsch and Marks (1974) in neonatal epidermis.

We have seen that the application of a carcinogen to epidermis is associated with a rapid onset of cell proliferation, a state which, on the results of Kreig *et al.* (1974), would be explained on the basis of refractoriness to the $G_1$ chalone; however, an alternative viewpoint was given by Rohrbach (Rohrbach 1972; Rohrbach *et al.* 1972; Rohrbach and Laerum 1974) who claimed that the $G_1$ chalone content of methylcholanthrene-treated mouse epidermis was inversely proportional to its proliferative activity; thus, when proliferative indices were maximal, chalone activity was absent from the epidermis. Two conflicting claims which, on the chalone hypothesis, would both fit the kinetic observations.

Established epidermal tumours appear to contain the epidermal $G_2$ chalone (Bullough and Laurence 1968*b*; Elgjo and Hennings 1971; Bullough and Deol 1971), although Bertsch and Marks (1979*b*) could find no $G_1$ chalone activity in extracts of the Hewitt mouse squamous carcinoma; as judged by the usual criteria, squamous carcinoma cells are responsive to $G_2$ chalone both *in vivo* and *in vitro*, and detailed studies by Bertsch and Marks (1979*b*) also showed that carcinoma cells responded to a $G_2$ inhibitor contained in epidermal extracts; mouse epidermal $G_2$ chalone also apparently inhibits mitotic activity in human pulmonary squamous carcinoma cells maintained *in vitro*, but adenocarcinomas from the same site did not appear to respond (Korsgaard *et al.* 1978).

During the course of production of tumours of epidermal origin, the *normal*, unaffected epidermis apparently shows selective inhibition of DNA synthesis, possibly indicating excess chalone production by the tumour; thus Karienemi and Rytomaa (1976) claimed that squamous epithelial tissues showed selective inhibition of DNA synthesis during growth of the Hewitt carcinoma, whereas lymphoid and haemopoietic cell lines did not; psoriatic epidermis in diffusion chambers also apparently showed decreased $I_S$ values in the same animals. There was, however, no attempt made to compare the results in animals with tumours of non-epidermal derivation, and the end point used was a curious mixture of scintillation counting and labelling indices without grain counts, with no attempt being made to correct for pool size changes; it is difficult to know what to make of such a study.

Squamous carcinomas are also apparently responsive to the $G_1$ chalone: thus Elgjo and Hennings (1971) and Bertsch and Marks (1979*b*) agreed that crude epidermal extracts depress labelling indices in the hamster Czernozemski and mouse Hewitt carcinomas respectively. When highly purified $G_1$ chalone was used, on the other hand, no such inhibition was apparent (Bertsch and Marks 1979*b*), an effect ascribed to the toxic or TdR pool diluent effects of the crude extract. It would thus appear that, in point of fact, tumours are not responsive to the $G_1$ chalone but the $G_2$ chalone retains its effect, and, from these results, Bertsch and Marks (1979*b*) opined that, whereas the $G_1$ chalone may control proliferation of committed epidermal stem cells, $G_2$ chalone may act on more primitive cell types, including the 'dedifferentiated' cells encountered in squamous carcinomas, and, if the possibility of cancer treatment with chalones does exist, then it lies with the $G_2$ chalone.

(vi) *What are the chalone target cells?*

We have seen that the $G_2$ chalone is supposedly produced by the basal cells, and the $G_1$ chalone by the differentiating cells, but on which cell population do these chalones act? By definition they must act upon proliferating cells; Bertsch and Marks (1979*b*) propose that the $G_1$ chalone acts on 'committed stem cells', while the $G_2$ chalone affects more 'primitive' cells (?pluripotent cells). Is there a primitive cell in the epidermis? Laurence (1979, 1980) argues against the existence of such a pluripotential cell in adult epidermis, although in fetal and in neonatal epidermis such cells were considered a possibility, and also complains that this stem cell terminology is both unconventional and confusing; if our previous discussions on the likelihood of epidermal stem cells (which must be *committed* in the sense that they are embarked on the course of production of exclusively epidermal cells) being a minor subpopulation have any credibility, then it is surprising that any kinetic effect of the chalone on any such small subpopulation was discernible. Since the action of chalones is so rapid, it must be some sort of direct action upon proliferating cells at large, rather than a remote action upon a stem cell subpopulation. In terms of the EPU hypothesis, could the $G_2$ chalone maintain Potten's stimulus-responsive cells in $G_2$? Are these cells committed stem cells, or differentiating proliferative cells, or primitive cells? In this respect Prunieras and Delescluse (1979) have claimed that, in culture, $G_1$ chalone blocks 85 per cent of basal cells in $G_1$, while $G_2$ chalone is only active upon 5-10 per cent of basal cells; could these be Gelfant's $G_2$ population? Possibly. However, it is clear that, in order to produce the level of DNA synthesis inhibition claimed for it, the $G_1$ chalone must affect differentiating proliferative cells, i.e. the cells in cell positions 3 to 4 in the EPU model (see Fig. 8.14, p. 313), in which case, what is the role of the $G_2$ chalone? Is it more relevant to differentiation, with its proliferative role only secondary, as Elgjo and Cromarty (1977) propose? It is plain that little attention has been given to these important questions.

Do chalones act equally upon epidermal proliferating populations in all states? We have seen that neonatal epidermis does not respond to the $G_1$ chalone (Bertsch and Marks 1974), even though neonatal epidermis does contain chalones. Further investigation (Bertsch and Marks 1979*b*) indicates that early, non-keratinizing fetal epidermis (see Chapter 5, section 1) does not contain $G_1$ chalone activity at nine days, but when keratinization ensues, $G_1$ chalone activity appears; chalone levels then apparently remain constant in the epidermis until death from old age (Leith 1978), and the cause of the increase in proliferative activity with age noted by Leith (and see Chapter 7, section 9) may therefore be due to a reactivation of the refractoriness which occurs at birth. In this respect the putative action of chalones on regenerating epidermis is interesting; Bullough (1969) reported a reduced effect of the 72-81 per cent ethanolic fraction on wounded epidermis compared with normal epidermis, while Bertsch *et al.* (1976) found that, although the elevated proliferative rate in massaged epidermis was successfully inhibited by the 72-81 per cent fraction (containing both $G_1$ and $G_2$ activity), epidermis regenerating after stripping did not respond for some 48 hours after wounding; it may be that proliferating cells in regenerating and neonatal epidermis share a degree of 'primitiveness' (or modulation, Pinkus and Mehregan 1981), which disappears with time and presumably differentiation; if this is so, we could suggest that the differentiation of surface chalone receptors, and the time required for their synthesis, could be the mechanism underlying these crude speculations; the neonatal epidermis contains proliferating cells without receptors, which however, appear during neonatal growth—wounding evokes the appearance of these cells again. Definitive evidence must await the identification of chalone:receptor complexes, but before accepting the experimental basis of this speculation, we should note that Elgjo and Cromarty (1977), while conceding that the epidermal response to chalones was very variable, were able to elicit inhibitory responses at most days studied over a period of 1-45 days after birth.

A related problem is the site of chalone synthesis in the cells' life cycle; can keratinizing cells synthesize and release chalone, according to the conventional idea (see above, and Marks 1976*b*)? In this respect, how is the chalone concentration maintained? By some putative but as yet unidentified mesenchymal stimulative factor? These questions must be answered.

## (vi) *Conclusions*

So how can we assess the current position? The major problem at the moment, of course, is the fact that until the chemical composition of the effector molecule is established, its receptor site recognized, and the sequence of biochemical events by which the effector molecule exerts its effect are known, the only way of recognizing the regulator is its behaviour in a bioassay, which, of course, must involve the study of some parameter of cell proliferation; having done all this, the effects must be tissue specific (Laurence *et al.* 1979; Laurence 1980).

We have already observed that the major reservation of most cell kineticists to the chalone concept is the methodology of the assay systems, and the choice of controls, particularly the control tissues. However, Laurence *et al.* (1979) have made a serious attempt to get to grips with some of these problems, in a methodological analysis of the $G_1$ chalone action: they remarked that (i) unless target and control tissues share the same cell cycle phase durations, and especially $t_S$, there will be grave problems in comparing a $G_1$ to $S$ inhibition sequence; in this respect, mouse sebaceous glands shared much the same median $t_S$ value (18.8 hours) with mouse epidermis (18.4 hours), and so makes a reasonable control; in this study, the Marks (1975) chalone had no effect on sebaceous glands, but inhibited DNA synthesis in the epidermis; however, the oesophagus and tongue showed short $t_S$ values of only 10 and 11 hours respectively, and the timing of any inhibition could well be quite different in these sets of tissues; (ii) the circadian rhythm is critical; there is little point in giving an inhibitor of DNA synthesis before a natural decrease in $I_S$, and similarly it is fruitless to compare target and control tissues where the circadian rhythmicity may differ; epidermis and sebaceous gland rhythms in fact correlate very well, but this has been *experimentally* verified; (iii) Laurence *et al.* (1979) have made a start on the numerous criticisms of the [³H]-TdR incorporation method by showing that the chalone preparation did not affect [³H]-TdR availability, as judged by measurements of the radioactivity of the acid-soluble washings, which elementary step appears to have been overlooked by previous investigators.

We also wonder about the advisability of using epidermal *extracts* as sources of putative chalone; the use of a tissue soup as a starting point is scarcely physiological, and naturally evokes the question of whether the active substance is actively exported by the cell; 'the fact that chalones can be *extracted* from tissues does not automatically guarantee that their physiological function is to control cell proliferation' (Argyris 1977). Surely a more reasonable approach would be the study of eluates of intact cells, which has been used, with some success in the bone marrow, by Lord and his colleagues (1979). In this respect, the preparation of eluates of a solid tissue presents its own problems, but surely this is no reason to revert to crude methodology.

Furthermore, we are surprised that, by and large, the only detailed studies on the effects of chalones on cells has been on the proliferative aspects—few authors have even looked at the action of their extract on cell viability, let alone oxidative metabolism; we all assume that, if chalones act, they act on 'restriction points' or 'mitosis operons' etc. There is no evidence that they do: there could as well be a (reversible) effect on general cell metabolism. It is necessary to look.

It is difficult to reach any conclusion in so nebulous and fractious a field: the non-aligned reader must make up his own mind; in one respect it seems puerile to make such a song and dance about the minutiae of the biological functions of preparations whose purity is highly questionable, while, on the other hand, the biological attractiveness of the proposal is obvious. So who is to blame for this present invidious position? The biochemists who cannot purify the active constituent in epidermal extracts? Possibly; but we would also apportion some of the blame to the cell kineticists who have, in many ways, allowed chalone research to become renowned for sloppy, inaccurate, and uninterpretable assay systems, which must, of necessity, be kinetically based.

Unless the whole field is one of the multilateral analysis of an artefact, then endogenous growth inhibitors do exist. There are some, of course, for whom the argument is not one of evidence, but one of belief, while others (cf. Pardee, see Allen 1978) consider that chalones, as defined by Bullough, do not exist. The problem now is one of (a) biochemical isolation, and (b) the design of assay systems which do mean something. There is room for vast progress in each.

## 14.2. Cyclic nucleotides and cell proliferation in squamous epithelium

There is little doubt that cyclic nucleotides are involved in the regulation of cell proliferation, at least *in vitro*. However, since 1970 there has been a minor information explosion, originating with the suggestion by Vorhees (1970) that abnormal cyclic AMP (cAMP) metabolism might account for the proliferative defect in psoriasis; this was exacerbated when Vorhees *et al.* (1972*a,b*) claimed that levels of cyclic AMP were decreased in lesional psoriatic epidermis, and when Vorhees *et al.*, in the following year, reported that cyclic GMP levels were elevated in the same tissue, a major breakthrough was apparent. Since then this field has abounded with claim, counter-claim and controversy, and much of the confusion has been caused by the original defects in the techniques of measuring cyclic nucleotides and their associated enzymes, and more recent results (Vorhees 1978; Marcello and Vorhees 1980) have seriously modified our ideas of this subject.

Consequently, we should look briefly at the technology available, which will facilitate the ensuing discussion; a further point to remember, particularly in association with cAMP and the control of cell proliferation, is that correlation does not prove causality (see, for example, Vorhees 1976).

### (i) *The measurement of cyclic nucleotides in epidermis*

Normal human epidermis contains very small quantities of cyclic AMP; usually about 0.2-0.4 pmol/$\mu$g DNA; after removal of interfering substances by column chromatography (Marcello *et al.* 1979), a protein binding assay (Gilman 1970), or a radioimmunoassay (Steiner *et al.* 1969) will detect this amount. Cyclic GMP (cGMP) levels in human epidermis range from 5 to 7 fmol/$\mu$g DNA; the low cGMP content of small epidermal samples means that extensive purification is mandatory (Marcello *et al.* 1979); after acetylation, cGMP levels are measured by radioimmunoassay (Harper and Booker 1975).

There has been considerable discussion concerning the choice of a denominator for the measurement of cyclic nucleotides (Vorhees *et al.* 1975*a*), whether to use unit amounts of DNA, protein or wet weight, and, while the best data base was not readily apparent, it may be advisable to use all three for comparative purposes.

A further problem in the measurement of cyclic nucleotides (or any other biochemical in epidermis for that matter), is the achievement of good separation of the epidermis from the dermis in the fresh material; this is usually carried out, in the human epidermis, with a keratotome, and there is difficulty in setting the keratotome at the right level to include the germinative layer but to exclude the dermis; obviously, good histological control is mandatory.

### (ii) *Cyclic nucleotides and cell proliferation*

The relationship between the control of cell proliferation *in vitro* and cyclic nucleotides has been largely established by the work of Pastan and his group (see, for example, Pastan 1971), from which the general association of low intracellular levels of cAMP with elevated proliferative rates has evolved; when cAMP levels rise, then cell proliferation is again generally restricted. Adductive evidence is available in that dibutyryl cAMP, and papaverine (a drug which inhibits the enzyme phosphodiesterase, which in turn breaks down cAMP, thus elevating intracellular cAMP levels, see Fig. 14.2A), inhibits neuroblastoma cell proliferation *in vitro*, together with theophylline, isoproterenol, and prostaglandin $E_2$,

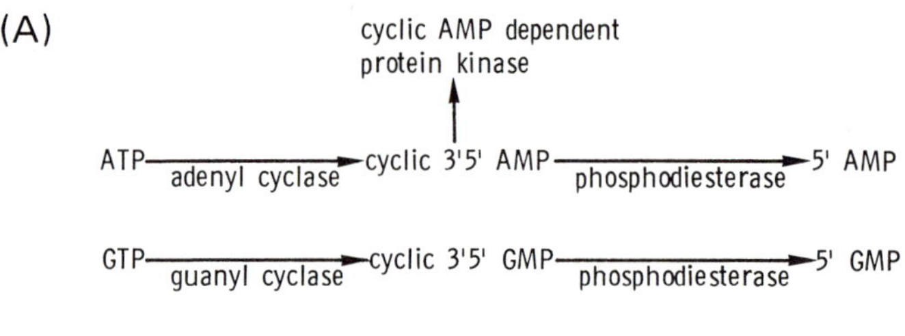


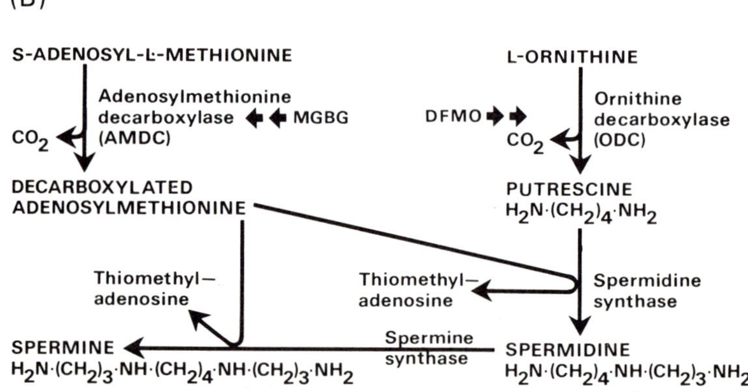


Fig. 14.2.   (A) Pathways of cyclic nucleotide metabolism.   (B) The metabolic pathways involving polyamine synthesis.

which all elevate cAMP levels and also decrease [3H]-TdR incorporation in cultured guinea pig epidermal cells (Delescluse *et al.* 1976). In adult human keratocytes maintained in culture, elevation of cAMP levels causes inhibition of proliferation (Flaxman and Harper 1975; Chopra 1977); again, we have an association between cAMP levels and cell proliferation rate.

The mechanism of this inhibition is in dispute: Willingham *et al.* (1972) and Froebach and Rachmeler (1974) consider that cAMP does actually inhibit cells in the $G_1$ phase of the cell cycle, although there is some evidence that an action in the $G_2$ phase may also operate (Willingham *et al.* 1972; Marks and Rebien 1972; Vorhees *et al.* 1972*a,b*). Unfortunately, like most findings in this field, these conclusions are disputed, with Vorhees *et al.* (1976) unable to show that dibutyryl cAMP could inhibit [³H]-TdR incorporation *in vitro*, while there is some evidence that cAMP can stimulate cell proliferation in some cell types (Vorhees *et al.* 1973*a,b*), mainly in newborn epidermal cells *in vitro*. Marcelo (1979) has claimed that both dibutyryl and 8-bromo cAMP, and cholera toxin (which elevates intracellular levels of cAMP) all *stimulate* cell proliferation in primary cultures of *neonatal* mouse keratocytes, as evidenced by measurement of the incorporation of [³H]-TdR into DNA. It is interesting to note that Marcelo's [³H]-TdR incorporation values do not correlate well with her labelling indices (see Chapter 3, section 2). Agents which increase intracellular cAMP levels also stimulate neonatal keratocyte *colony* growth *in vitro* (Green 1978). Thus *newborn* and *adult* cell responses to cAMP may differ, as may also be the case with the epidermal chalone (Bertsch and Marks 1974). To further complicate the picture, studies by Marcelo and Duell (1979), using *adult* keratocytes grown on a collagen gel, indicate that cAMP can both stimulate and inhibit keratocyte growth, while Delescluse *et al.* (1976) have claimed that increased levels of cAMP promotes differentiation in cultured guinea pig keratocytes; MSH, by promoting elevated intracellular cAMP levels, causes cessation of growth and increases melanin synthesis in cultured Cloudman S-91 melanoma cells (Pawelek 1976). Possibly to account for some of this conflicting data, Vorhees (1976) has suggested that the net effect of a change in the intracellular levels of cyclic AMP may depend on whether a cell is proliferating or resting at the time (by whatever mechanism?).

There is evidence that elevated *cGMP* levels are associated with *increased cell proliferation*. For example, Hadden *et al.* (1972) found that PHA-induced lymphocyte proliferation is accompanied by increased intracellular levels of cGMP, while the phorbol esters, which stimulate proliferation, also elevate cGMP levels in both 3T3 cells and in epidermal cells *in vitro* (Estensen *et al.* 1974; Vorhees *et al.* 1973*b*). But Marcelo (1979) could not relate cGMP levels to the increased proliferative rate induced by cAMP in newborn mouse epidermal cells *in vitro*.

Also based on work carried out *in vitro*, it appears that an inverse relationship may exist between levels of cAMP and cGMP; Goldberg *et al.* (1974) detected elevated cGMP levels in exponentially growing cells, decreasing as the plateau phase ensued, whereas Otten *et al.* (1971) found the converse, i.e. that cAMP levels increased with the onset of the plateau phase in cultured fibroblasts; the seal appeared set when Vorhees *et al.* (1973*b*) claimed that the rapidly dividing epidermis in psoriasis was associated with increased levels of cGMP and decreased levels of cAMP (but see below); Goldberg *et al.* (1974) referred to this as the *dualistic action* of cyclic nucleotides. Kram and Tomkins (1973) considered that such dualistic action may account for all pleiotypic responses.

### (iii) *Cyclic nucleotide metabolism in psoriasis*

#### (a) *Factors affecting cyclic nucleotide levels in epidermis*

Adrenaline, which possesses both $\alpha$- and $\beta$-agonist action, is a potent inhibitor of cell division, and it is interesting to note that Bronstad *et al.* (1971) found an associated accumulation of cAMP in adrenaline-treated hamster skin *in vitro*, while isoproterenol, a pure $\beta$-agonist, also elevates cAMP levels in cultured rat epidermis, and also in mouse epidermis *in vivo* (Verma *et al.* 1976), a rise which is prevented by the $\beta$-antagonist propanolol (Duell *et al.* 1971; Vorhees *et al.* 1976); moreover, while the action of adrenaline was also largely prevented by propanolol, phentolamine, an $\alpha$-antagonist had no effect on the adrenaline-induced rise in cAMP (Vorhees *et al.* 1976), and, from these results, it does appear as if stimulation of $\beta$ receptors produces elevated cAMP levels (Harper *et al.* 1974; Yoshikawa *et al.* 1975*b,c*); in addition, salbutamol, a $\beta$-agonist, also elevates cAMP levels *in vitro* (Vorhees *et al.* 1976), and Gaylarde *et al.* (1978) have reported that topical application of propanolol to guinea pig skin gives rise to a psoriasiform change 'similar to psoriasis' in man. It also appears that histamine, which inhibits epidermal cell proliferation *in vitro* (Harper and Flaxman 1975), will also elevate cAMP levels in the epidermis (Iizuka *et al.* 1977; Yoshikawa *et al.* 1980), but neither adrenaline nor papaverine affect epidermal cGMP levels (Marcelo and Vorhees 1980).

Relating this topic to a later section (see section 14.3), EGF treatment is reported to increase cAMP and adenyl cyclase levels in epidermis (Frati *et al.* 1977), but Aoyagi *et al.* (1980) have recently claimed that EGF had no effect on cAMP or adenyl cyclase levels, but stimulated cGMP levels in cultured pig skin; they concluded that, since the effect took several hours to manifest, these changes were not related to a potential action on cell proliferation.

Do cyclic nucleotides correlate with any *physiological* variation in cell proliferation in epidermis? Well, levels of these metabolites do vary diurnally in epidermis, and there has been speculation that this variation is correlated with changes in proliferative indices (Marks and Grimm 1972; Garte and Belman 1980; and see Chapter 9, section 5); from a purely practical viewpoint, it would be well to remember this variation in quantitative work *in vivo*.

#### (b) *Enzymes which affect cyclic nucleotide levels in epidermis*

The conversion of ATP to cyclic AMP is catalysed by adenylate cyclase, which, in most cells, is found in the cell membrane (Vorhees *et al.* 1976), whereas cAMP is hydrolyzed to $5'$ AMP by cyclic nucleotide phosphodiesterase, an enzyme which appears to be present in multiple forms in the epidermis (Murray and Rogers 1978; Rusin *et al.* 1978; and see Fig. 14.2). Similarly, cGMP levels are the net result of the activity of guanylate cyclase and cGMP phosphodiesterase (see Fig. 14.2; Marks 1973*b*). Adenylate cyclases from newborn rat epidermis and mouse epidermis are stimulated by adrenaline and isoproterenol *in vitro*, which stimulation appears to be inhibited by propanolol, but not by phentolamine (Duell *et al.* 1971; Marks and Rebien 1972). However, guanylate cyclase activity in mouse epidermis is unchanged by the effect of adrenaline and isoproterenol (Marks 1973*b*).

The $\beta$-agonist stimulation of cAMP levels thus appears to be largely due to an action on the membrane-bound adenyl cyclase, particularly since theophylline, an inhibitor of epidermal cAMP phosphodiesterase (Vorhees *et al.* 1974*a,b*), only causes a minimal elevation of cAMP levels (Vorhees 1976); but theophylline does appear to inhibit cell proliferation in the $G_2$ phase (Marks and Rebien 1972), an action which may thus be cAMP-independent.

The group of protein kinases which activates phosphorylase kinase (an enzyme which in turn degrades glycogen to glucose-1-phosphate units available for glycolysis) and also inhibits glycogen synthetase, are also cAMP dependent; these enzymes are present in mouse skin (Mier and Cotton 1967; Mier and Sartorius 1972), and it therefore appears that cAMP has important functions in controlling epidermal glycogen metabolism.

### (c) *Cyclic nucleotide levels in proliferative skin disease*

The evidence suggests that decreased cAMP levels are associated with increased rates of epidermal cell proliferation, so a natural channel for such studies was the investigation of cyclic nucleotide activity in psoriatic epidermis. That such studies have at least potential relevance in epidermis is indicated by the immunocytochemical studies of Verma *et al.* (1976), which showed that cAMP is found predominantly in the germinative, proliferating cells of the epidermis. We have already mentioned the problems involved in the measurement of cyclic nucleotides in epidermis, and we are probably justified in saying that our concepts have varied with the available technology (see Table 14.1).

Vorhees *et al.* (1972*a,b*), using the protein binding method on snap frozen material, and expressing the results per unit amount of DNA, protein and wet weight, claimed that there was a significant *decrease* in cAMP levels in psoriatic lesional epidermis compared with non-lesional epidermis, a decrease which ranged from 15 to 56 per cent; Vorhees *et al.* also considered that glycogen levels were increased in psoriatic epidermis, and naturally connected the two. But Hopsu-Havu *et al.* (1973) and Harkonen *et al.* (1974) were unable to confirm this finding, and instead reported that cAMP levels were *elevated* in lesional epidermis compared with normal epidermis, using wet weight as the denominator; however, later work by the Vorhees group (Marcelo *et al.* 1979), using radioimmunoassay, indicated that lesional epidermis contained *more* cAMP than uninvolved epidermis, although both tissues contained much less cAMP than normal epidermis. Moreover, Hammar (1973) and Yoshikawa *et al.* (1975*a*) found little difference in cAMP levels between lesional and non-lesional epidermis,

Table 14.1. Values for cyclic nucleotide levels in normal, involved, and uninvolved epidermis in psoriasis (after Marcelo and Vorhees 1980)

| Reference | Psoriatic Involved | Uninvolved | Normal |
|---|---|---|---|
| | | pmol cAMP/$\mu$g DNA | |
| Vorhees *et al.* (1972*a*) | 0.40  ± 0.10 | 0.70  ± 0.10 | 0.50  ± 0.10 |
| Vorhees *et al.* (1973*b*) | 0.72  ± 0.11 | 1.08  ± 0.09 | 0.86  ± 0.08 |
| Yoshikawa *et al.* (1975*a*) | 0.196 ± 0.025 | 0.186 ± 0.034 | – |
| Marcelo *et al.* (1979) | 0.198 ± 0.04 | 0.148 ± 0.02 | 0.367 ± 0.05 |
| | | pmol cAMP/mg wet weight | |
| Harkonen *et al.* (1974) | 2.10  ± 0.35 | 1.62  ± 0.36 | 0.94  ± 0.17 |
| Wadskov *et al.* (1979) | 1.44  ± 0.35 | 2.66  ± 0.51 | 1.31  ± 0.10 |
| | | fmol cGMP/$\mu$g DNA | |
| Vorhees *et al.* (1973*b*) | 11.83 ± 3.4 | 6.1 ± 1.0 | – |
| Marcelo *et al.* (1979) | 11.4  ± 2.2 | 4.9 ± 0.7 | 6.5 ± 1.08 |

although in the latter study the normal epidermal levels were not given; both these groups used unit amounts of DNA as the denominator; in this context, Saihan *et al.* (1980), employing the radioimmunoassay method, also failed to show any difference between cAMP levels in the several different epidermal states.

Wadskov *et al.* (1979), using a competitive protein-binding assay, have reconfirmed the low levels of cAMP in lesional epidermis, and both they and Marcelo and Vorhees (1980) have attempted to interpret these various findings by speculating that the cAMP levels may be related to the chronological stage of the lesion in its development, i.e. whether it is evolving, steady state or regressing. We have seen how cAMP may have both inhibitory and stimulatory actions on cell proliferation, possibly depending on the cell proliferative status at the time: this, however, remains hypothetical.

Given that some relationship exists, Vorhees *et al.* (1976) considered that possible reasons for a *low* level of cAMP in psoriatic epidermis were: (i) decreased adenyl cyclase activity; (ii) increased cAMP phosphodiesterase activity; or (iii) excessive leakage of cAMP from the damaged areas. Vorhees *et al.* (1973*a,b*, 1976), while not measuring adenyl cyclase activity directly, but using an indirect method, which measures the net result of adenylate cyclase and phosphodiesterase activity, report no evidence of a rise in adenylate cyclase activity, but propounds either a *decrease* in cyclase activity or a stimulation of phosphodiesterase activity; Hsai *et al.* (1972) and Wright *et al.* (1973) consider that adenylate cyclase activity is defective in psoriatic epidermis. As to the possible leakage of cAMP, although this can occur, for example from human fibroblasts in culture, Vorhees *et al.* (1976) could find no evidence for leakage from *epidermal* cells in culture, although it is difficult to comment on the significance of these findings, at least for the *in vivo* situation.

Possibly more edifying is the observation that greatly increased levels of cGMP are reported in lesional psoriatic epidermis compared with uninvolved tissue (and normal epidermis) (Vorhees *et al.* 1973*b*; Vorhees *et al.* 1976; Marcelo *et al.* 1979; Cantieri *et al.* 1980). Since it appears that cGMP can stimulate the breakdown of cAMP *in vitro* (Goldberger *et al.* 1974), this could account for any possible decrease in cAMP levels in psoriatic epidermis, acting via phosphodiesterase activity. Vorhees (1976) has speculated that the dualistic concept, relating cell proliferation changes to variation in the cAMP:cGMP ratio (see above), results in the increased epidermopoiesis which occurs in psoriasis; they have evolved a hypothetical scheme whereby effector molecules such as hormones and chalones react with receptors at the cell membrane level and effect changes in the cAMP:cGMP ratio, resulting in the glycogen accumulation and increased epidermopoiesis of psoriatic epidermis. But the significance of elevated cGMP levels in lesional epidermis is open to doubt, since increased proliferative activity is present in non-lesional epidermis (Goodwin and Fry 1977; Marks 1978), which does not show any change in cGMP levels (Marcelo *et al.* 1979).

Interestingly enough, a double-blind trial of a topical papaverine cream (which would presumably elevate intracellular cAMP levels), tested against vehicle alone, in 45 patients with psoriasis, indicated a small but significant effect, in favour of papaverine (which, however, was less effective than the powerful fluorinated glucocorticoids) (Stawiski *et al.* 1975).

There is evidence, obtained largely from the *in vitro* work of Halprin and his group, that cAMP levels in psoriatic epidermis show decreased responsiveness to stimulation with $PGE_1$ (see below) and $\beta$-adrenergic agents (Hsai *et al.* 1972; Iizuka *et al.* 1978; Yoshikawa *et al.* 1975*c*, 1980). In addition, treatment with blockers such as propanolol induce psoriasiform reactions in guinea pig skin (Gaylarde *et al.* 1978), and Wiley and Weinstein (1977) report that propanolol selectively induces proliferative abnormalities in the non-lesional psoriatic epidermis compared to normal skin. This evidence then, does support the hypothesis of Marcelo and Vorhees (1980) that some $\beta$-adrenergic-cAMP abnormality occurs in psoriatic epidermis.

It is difficult to accommodate the careful studies of Taylor *et al.* (1980), on cultured pig epidermis, in the general picture: cAMP and agents which elevate intracellular cAMP levels (histamine and $\beta$-agonists) had *no* effect on DNA synthesis; furthermore, the cAMP phosphodiesterase inhibitor theophylline reversibly inhibited DNA synthesis by a process independent of cAMP (also see above).

What do we make of all this? First, that there does not appear to be any constant relationship between cyclic nucleotide levels and epidermal proliferative state, and any hypothesis that cAMP levels are correlated with the evolutionary lesional state must be supported by a detailed clinical and histokinetic analysis; the dualistic hypothesis is more persuasive, but any global approach must account for the proliferative defect known to exist in non-lesional epidermis; there may be a defect associated with the $\beta$-adrenergic–cAMP mechanism in psoriasis, but any direct relationship with epidermal cell proliferation is equivocal; it would appear that this field is ripe for some rather widespread collaboration in an attempt to make some sense of the many contentious issues, and to test the hypothesis of a relationship between cyclic nucleotide metabolism and epidermal cell proliferation in psoriasis; in particular, and yet again, we must plead for more thought to be given to the cell proliferation assays which are used; there is little point in carrying out the experiments if, at the end of the day, cell kineticists cannot accept the cell proliferation methodology used (in this field, almost exclusively $[^3H]$-TdR incorporation; see Chapter 3, section 2).

### (iv) *Prostaglandins and cyclic nucleotides in epidermis*

Certain prostaglandins, particularly those of the E series, appear to stimulate adenyl cyclase activity and thus elevate cAMP levels (Kuehl 1973), while topically applied $PGE_2$ was apparently effective in the treatment of the hyperproliferative state which accompanies essential fatty acid deficiency in rats (Ziboh and Hsai 1972); direct observation of the action of prostaglandins reveals that $PGE_1$ and $PGE_2$ increase cAMP levels in mammalian epidermis (Vorhees *et al.* 1973*a*; Adachi *et al.* 1975; Azo *et al.* 1975). Since $PGE_2$ and $PGE_{2\alpha}$ levels are said to be decreased in psoriatic epidermis, Hammerstrom *et al.* (1975) have claimed that there might be a relative decrease in PGE biosynthesis in lesional psoriatic epidermis (even though actual levels of $PGE_2$ and $PGE_{2\alpha}$ were slightly increased); these might be additional reasons why cAMP levels are said to be decreased in psoriatic epidermis (see above). However, to support this concept, Azo *et al.* (1975) have reported that $PGE_1$, given intradermally into psoriatic epidermis, leads to a decrease in $I_S$ and $I_M$, associated, in some cases, with histological and clinical amelioration, but other similar work has not been so successful; thus Eaglestein and Weinstein (1975) found that $PGE_2$ causes large increases in $I_S$ 48 hours after intradermal injection in human epidermis, while $PGE_1$ induced a sustained stimulation of the labelling index in guinea pig epidermis; $PGE_{2\alpha}$ had no apparent effect, so the story is evidently not straightforward. On the other hand, $PGF_2$ appears to stimulate cGMP levels, and there is early evidence that increased formation of $PGF_2$ from $PGE_2$ could be demonstrated in psoriatic epidermis, which could account for the observed increase of cGMP levels in psoriasis.

In conclusion, although Furstenburger and Marks (1980) consider that a stimulation of PGE biosynthesis is an early and obligatory event in the induction of cell proliferation in mouse epidermis by phorbol ester (see Chapter 13, section 3), and, while conceding that there may be some relationship between cell proliferation in the epidermis and prostaglandins, we might agree with Raab (1980) that it is too early to comment on the exact nature of the relationship; as to the effect *apropos* cAMP: cGMP and its role in cell proliferation, until that itself is clarified, it is difficult to comment further.

### (v) *The action of adrenaline*

It has been known for some time that adrenaline has an action on the incidence of mitosis in mouse epidermis, but actual results have differed as between different groups of investigators; most reports (e.g. Bullough and Laurence 1966*b*) suggest a depressive effect on mitotic activity which is evident within three hours of injection, although Evensen (1964) has found an *increase* in the mitotic rate at one hour after adrenaline. It is also reported that adrenaline and other $\beta$-agonists increase cAMP formation (Bronstad *et al.* 1971), which, in certain circumstances, would be consistent with an inhibitory action on cell proliferation (see Chapter 14, section 2).

Most earlier work had used the metaphase arrest method, indifferently applied (see Chapter 3, section 3; cf. Bullough and Laurence 1966*b*), or mere [$^{3}$H]-TdR incorporation to detect a proliferative effect, but a recent study by Clausen *et al.* (1982), using a combination of flow cytometric, stathmokinetic and short-term FLM methods indicates that adrenaline administration *in vivo* induces a partially synchronized cohort of mouse epidermal cells to leave the $S$ phase rapidly, and to traverse $G_2$ at an increased rate. Some of these cells rapidly reach metaphase, while others are delayed in prophase. Following the release of the prophase/ metaphase block, there was a significant increase in the $S$ phase index, indicating a delay in the $S$ phase transit rate, which results in a reduction in the mitotic rate at 2–3 hours after adrenaline administration.

These studies therefore explain both the early increase in the mitotic rate described by Evensen (1964) and the later reduction in mitotic rate seen by most workers at about three hours after adrenaline. It will be remembered that Clausen *et al.* (1979) have claimed that the large circadian swings in $I_S$ and $I_M$ are caused primarily by changes in the rate of efflux from the $S$ phase, which would indicate a possible role for adrenaline in the establishment of circadian rhythms (see Chapter 9); however, operationally it is difficult to divorce direct effects of the agent from changes in blood flow due to its pressor action.

### (vi) *Chalones and cyclic nucleotides*

We have already discussed the action of adrenaline, which appears to be an inhibitor of cell proliferation, and conceivably acts as a component part of the chalone:adrenaline complex of Bullough and Laurence (1964*c*). Adrenaline probably acts via the cell membrane to stimulate the surface receptor for adenylate cyclase and to elevate cAMP levels, which, as we have seen, in certain instances can inhibit cell proliferation, although this relationship is by no means clear. Indeed, Iversen (1969) has speculated that the chalone–adrenaline complex, rather than adrenaline alone, might activate the cAMP system in the epidermis. This is an attractive hypothesis, but there has been a consistent failure to demonstrate an effect of the $G_2$ chalone either on cAMP levels, or on the $\beta$-adrenergic stimulation of adenylate cyclase (Vorhees and Duell 1971; Marks and Grimm

1972; Marks and Rebiel 1972; Marks 1973*b*; Marks and Raab 1974), and thus the postulated relationship between $G_2$ chalone and cAMP is more complex.

However, Elgjo (1975) reports that the $\beta$-blocker propanolol abolished the inhibitory activity of the $G_2$ chalone on mitotic activity in mouse epidermis, but caffeine, an inhibitor of nucleotide phosphodiesterase, restored the depressive mitotic action; thus a critical intracellular level of cAMP may be necessary for $G_2$ chalone to act. It is possible that cAMP may be the limiting factor in defining the level of mitotic depression achievable with chalones (Marks 1976*b*). But there is no suggestion of a correlation between the potentially more important $G_1$ chalone and cyclic nucleotides.

### (vii) *Ornithine decarboxylase and polyamines*

There is little doubt that ornithine decarboxylase (ODC) activity, and the levels of its biosynthetic product, putrescine, are elevated in proliferating squamous epithelium. The interrelationships of ODC and the polyamines are shown in Fig. 14.2B. The increased proliferative rate associated with phorbol ester treatment (O'Brien *et al.* 1975), with ultraviolet irradiation (Verma *et al.* 1979), hair plucking (Morrison and Goldsmith 1978), and psoriasis (Russell *et al.* 1978) are all associated with high ODC levels. It is thought that ODC is induced specifically during the late $G_1$ period (Lowe 1980), and is thus a candidate for a compound with specific controlling action; an alternative view is that ODC induction is an epiphenomenon of induced DNA synthesis in squamous epithelium; while experience with ODC inhibitors might suggest that ODC induction and putrescine synthesis is a necessary event in the induction of DNA synthesis *in vitro*, it is not yet clear whether the same conclusion is justified *in vivo*.

### (viii) *Conclusions*

Again, what are we to make of this conglomerate of conflicting information concerning the cyclic nucleotides and the control of their concentration, tenuously linked, as they are, to cell proliferation by a mish-mash of highly inappropriate assays? It is very hard to obtain any unified picture; there does appear to be a relationship between increased intracellular levels of cAMP and the inhibition of cell proliferation in work on cell lines *in vitro*, but even here the correlation is not absolute (see above); in epidermal tissues, even *in vitro*, there is no general agreement, even given that we can measure the levels of cyclic nucleotides adequately in the more important *in vivo* conditions. We are left in some wonder at the fluctuations in the published values for cAMP levels in psoriatic epidermis, and await with interest a definitive study showing that the proliferative activity of each psoriatic plaque studied, at whatever degree of evolution, is correlated unequivocally with cAMP levels, cAMP:cGMP ratios, or whatever. Then we would expect a demonstration that perturbations of the cyclic nucleotide levels are associated with changes in the proliferative rate in the direction propounded; all this, of course, using appropriate proliferative assays.

We would appear to be a long way from showing that proliferative reactions in stratified squamous epithelium are mediated through the cyclic nucleotide system.


## 14.3. Epidermal growth factor (EGF)

EGF was first detected by the ability of extracts of mouse submaxillary gland to induce precocious opening of the eyelids and early eruption of incisors in neonatal mice (Cohen 1960); the active factor was shown to be a polypeptide (Cohen 1962).

Mouse EGF is now prepared using its selective and reversible absorption to polyacrylamide columns, from crude homogenates of mouse submaxillary glands (Savage and Cohen 1973); it is a heat-stable, non-dialysable single polypeptide chain of 53 amino-acid residues, of molecular weight between 6000 and 7000. Human EGF has been prepared from human urine (Cohen and Carpenter 1975), and demonstrates biological properties similar to those described for mouse EGF, and has a similar molecular weight; interestingly, $\beta$-urogastrone, a gastric inhibitory hormone isolated from human urine, has remarkable amino acid homologies with EGF (Gregory 1975). There are several derivatives of EGF, including high molecular weight forms. The submaxillary gland is the most fruitful source of EGF in the mouse, where it is stored in the secretion granules of the granular convoluted tubule cells (see Chapter 27), while in the human, EGF has also been detected in Brunner's glands in the duodenum (Elder *et al.* 1978).

There is little doubt that EGF is mitogenic for a large number of *in vitro* cell lines (see Carpenter 1978), both epithelial and non-epithelial; these include human keratocytes (Rheinwald and Green 1977) and human cervical epithelial cells (Stanley and Parkinson 1979). The addition of EGF to quiescent cells leads to activation and entry of cells into $S$ about 12 hours later in most cell lines (Carpenter and Cohen 1976); specific saturable receptors for EGF have now been demonstrated using [125]I-labelled mouse EGF in a wide variety of cultured cells, and have been isolated in an active, solubilized form (Carpenter 1978). In organ culture EGF elevates $I_S$ and increases thymidine incorporation into DNA in chick embryo skin (Cohen 1965), fetal scalp skin (Cohen and Savage 1974) and bovine corneal epithelium (Gospadarowicz *et al.* 1980). In epidermis, ODC activity, RNA and protein synthesis, and eventually keratinization are also stimulated (Hoober and Cohen 1967).

There is no doubt that EGF is a mitogen for squamous epithelial cells *in vitro*; its action *in vivo*, and in particular, the part played by EGF in the maintenance of normal epidermal homeostasis, is more problematical. The original effect on eye opening in newborn animals was ascribed to an enhancement of epidermal growth and keratinization (Cohen and Elliott 1963); there was an associated increase in the thickness of the epithelium of the mouth and oesophagus. Studies of the effect of EGF on epidermal cell proliferation *in vivo* have been few:

Birnbaum *et al.* (1976) showed a dose-responsive stimulation of colchicine-arrested mitoses in mouse ear epidermis explanted *in vitro*, after several days EGF treatment *in vivo*. However, there is also evidence of stimulation of net epidermal growth from measurements of DNA and RNA content, and of stimulated keratinization from an increase in disulphide group content and a reduction in free sulphahydryl groups (Frati *et al.* 1977), although it is not clear whether this is a direct stimulation or an action via increasing basal cell production (Carpenter 1981). The enhancing effect of EGF on methylcholanthrene-induced carcinogenesis (Rose *et al.* 1976) is probably related to its mitogenic action (Carpenter 1981).

It thus appears that EGF can stimulate epidermal cell production *in vivo*; however, the biological function of EGF *vis à vis* its action *in vivo* has yet to be determined; it is apparently without action on human epidermal cell proliferation when applied topically to epidermal wounds (Greaves 1980). Does EGF have a role in the control of epidermal cell growth? Its occurrence in gastrointestinal-associated glands might suggest a primary action on gut epithelia (see Chapter 23).

# References


Aarnes, E., Thorud, E., and Clausen, O. P. (1981). Model analysis of circadian rhythms in mouse epidermal cell proliferation, *J. theoret. Biol.* **88**, 355–70.

Abel, E. A. and Farber, E. M. (1980). Photochemotherapy. In *Recent advances in dermatology* (ed. A. J. Rook and J. Savin) pp. 259–316. Churchill Livingstone, Edinburgh.

Abercrombie, M. (1946). Estimation of nuclear population from microtome sections. *Anat. Rec.* **94**, 239–47.

—— (1957). Localised formation of new tissue in an adult mammal. *Symp. Soc. exp. Biol.* **11**, 235–54.

Adachi, K., Yoshikawa, K., Halprin, K., and Levine, V. (1975). Prostaglandin and cyclic AMP in epidermis. *Br. J. Derm.* **92**, 381–8.

Agduhr, E. (1941). A contribution to the techniques of determining the number of nerve cells per unit volume of tissue. *Anat. Rec.* **80**, 191–202.

Aherne, W. A. (1967). Methods of counting discrete tissue components in microscopical sections. *J. R. microscop. Soc.* **87**, 493–508.

—— and Buck, P. (1971). The potential cell doubling time in neuroblastoma and nephroblastoma. *Br. J. Cancer* **25**, 691–6.

—— and Camplejohn, R. S. (1972). On correcting the error due to metaphase degeneration in stathmokinetic studies. *Expl. Cell Res.* **74**, 496–502.

—— —— and Wright, N. A. (1977a). *An introduction to cell population kinetics.* Edward Arnold, London.

—— —— Al-Wiswazy, M., Ford, D., and Kellerer, A. M. (1977b). Assessment of inherent fluctuations of mitotic and labelling indices of human tumours. *Br. J. Cancer* **36**, 577–91.

—— and Dunnill, M. (1982). *Morphometry.* Edward Arnold, London.

Alavaikko, M. (1971). Selective separation of mouse epidermal cells and cytophotometric estimation of their nuclear DNA. *Acta path. microbiol. scand.* Suppl. A **226**, 1–76.

Al-Barwari, S. E. and Potten, C. S. (1976). Regeneration and dose response characteristics of irradiated mouse dorsal epidermal cells. *Int. J. radiat. Biol.* **30**, 201–16.

—— —— (1979). A cell kinetic model to explain the time of appearance of skin reactions after X-rays or UV light irradiation. *Cell Tissue Kinet.* **12**, 281–9.

Al-Dewachi, H. S., Appleton, D. R., Watson, A. J., and Wright, N. A. (1979). Variation in the cell cycle time in the crypts of Lieberkuhn in the mouse. *Virchows Arch. Cell Path.* **31**, 37–44.

—— Wright, N. A., Appleton, D. R., and Watson, A. J. (1974). The cell cycle time in the rat jejunal crypt. *Cell Tissue Kinet.* **7**, 587–94.

—— —— —— —— (1975a). The effect of starvation and refeeding on cell population kinetics in the rat small bowel mucosa. *J. Anat.* **119**, 105–16.

—— —— —— —— (1975b). Cell population kinetics in the mouse jejunal crypt. *Virchows Arch. Cell. Path.* **18**, 225–42.

—— —— —— —— (1976). Studies on the mechanism of diurnal variation of proliferation in the small bowel mucosa of the rat. *Cell Tissue Kinet.* **10**, 459–67.

— — — — (1977). The effects of a single injection of hydroxyurea on cell population kinetics in the small bowel mucosa of the rat. *Cell Tissue Kinet.* **10**, 203–13.

— — — — (1980). The effect of a single injection of cytosine arabinoside on cell population kinetics in the mouse jejunal crypt. *Virchows Arch. Cell Path.* **34**, 299–309.

Alison, M. R., Appleton, D. R., Morley, A. R., and Wright, N. A. (1974). Cell population growth in the castrate mouse prostate complex: experimental verification of computer simulation. *Cell Tissue Kinet.* **7**, 425–31.

— McHanwell, S., and Wright, N. A. (1979). Failure of induced functional activity and cell deficit to restore preoperative cell number in paired accessory sex glands following unilateral ablation in castrated male mice. *Anat. Rec.* **193**, 903–12.

— Wright, N. A., Morley, A. R., and Appleton, D. R. (1976). Cell proliferation in the prostate complex of the castrated mouse. *J. Microsc.* **106**, 221–37.

Allegra, F. and DePanfilis, G. (1974). An in vivo method of studying the kinetics of cell proliferation in normal human epidermis. *Acta dematovener., Stockh.* **54**, 87–96.

Allen. J. C. (1978). Damn the dogma. *Nature, Lond.* **226**, 669.

— and Smith, C. J. (1979). Chalones: a reappraisal. *Biochem. Soc. Trans.* **7**, 584–92.

— — Curry, M. C., and Gaugas, J. M. (1977). Identification of a thymic inhibitor ('chalone') of lymphocyte transformation as a spermine complex. *Nature, Lond.* **267**, 623–5.

Allen, T. D. and Potten, C. S. (1974*a*). Fine structural identification and organisation of the epidermal proliferative unit. *J. cell Sci.* **15**, 291–319.

— — (1974*b*). Desmosomal form, fate and function in mammalian epidermis. *J. ultrastruct. Res.* **51**, 94–105.

— — (1976*a*). Ultrastructural site variations in mouse epidermal organisation. *J. cell Sci.* **21**, 341–59.

— — (1976*b*). Significance of cell shape in tissue architecture. *Nature, Lond.* **264**, 545–7.

Al-Mukhtar, M. Y. T., Polak, J., Bloom, S. R., and Wright, N. A. (1982*a*). The search for appropriate measurements of proliferative and morphological status in studies of intestinal adaptation. In *Mechanisms of intestinal adaptation* (ed. J. W. L. Robinson, H. R. Dowling, and E. O. Reicken) pp. 3–28. MTP Press, Lancaster.

— Sagor, G. R., Ghatei, M. A., Polak, J. M., Koopmans, H. S., Bloom, S. R., and Wright, N. A. (1982*b*). The relationship between endogenous gastrointestinal hormones and cell proliferation in models of intestinal adaptation. In *Mechanisms of intestinal adaptation* (ed. J. W. L. Robinson, H. R. Dowling, and E. O. Riecken) pp. 243–53. MTP Press, Lancaster.

Alper, T. (1979). *Cellular radiobiology*. Cambridge University Press.

Altmann, G. G. (1971). Influence of bile and pancreatic secretions on the size of the intestinal villi in the rat. *Am. J. Anat.* **132**, 167–78.

— and Enesco, M. (1967). Cell number as a measure of distribution and renewal of epithelial cells in the small intestine of growing and adult rats. *Am. J. Anat.* **121**, 319–36.

Andersson, P. (1973). Effect of an aqueous extract of small intestinal mucosa on the $G_2$ phase of jejunal proliferative cells in mice. *Virchows Arch. Cell Path.* **13**, 233–46.

Aoyagi, T., Adachi, K., Halprin, K., and Levine, V. (1980). The effects of epidermal growth factor on the cyclic nucleotide system in pig epidermis. *J. invest. Derm.* **74**, 238–41.

Appleton, D. R., Morley, A. R., and Wright, N. A. (1973). Cell proliferation in the castrate mouse seminal vesicle in response to testosterone propionate. II Theoretical considerations. *Cell Tissue Kinet.* **6**, 247–58.

—— and Sunter, J. P. (1979). Estimating the proportion of proliferating cells in a population. *Virchows Arch. Cell Path.* **32**, 69–73.

—— Wright, N. A., and Dyson, P. (1977). The age distribution of cells in stratified squamous epithelium. *J. theoret. Biol.* **65**, 769–79.

Archambeau, J. O. (1972). Epidermal cell population changes produced by exposure to 2300 R. *Radiology* **103**, 191–6.

Argyris, T. S. (1968). Growth induced by damage. *Adv. Morphol.* **7**, 1–43.

—— (1976). Kinetics of epidermal production during epidermal regeneration following abrasion in mice. *Am. J. Path.* **83**, 329–40.

—— (1977). Kinetics of regression of epidermal hyperplasia in the skin of mice following abrasion. *Am. J. Path.* **88**, 575–83.

—— (1979*a*). Epidermal growth and ribosomal RNA accumulation in regenerating mouse epidermis following abrasion. *J. invest. Derm.* **70**, 267–71.

—— (1979*b*). Ribosome accumulation and the regulation of epidermal hyperplastic growth. *Life Sci.* **24**, 1137–47.

—— (1980). Epidermal growth following a single application of 12-O-tetradeca-noyl-phorbol-13 acetate in mice. *Am. J. Path.* **98**, 639–48.

Arndt, R. A. and Freidberg, I. M. (1971). Recovery of $^3$HTdR after intradermal injection. *Archs Derm.* **103**, 304–5.

Aroesty, J., Lincoln, T., Morrison, P., and Carter, G. (1977). Cell kinetic models of transient states: a preliminary investigation of lymphocyte stimulation. In *Growth kinetics and biochemical regulation of normal and malignant cells* (ed. B. Drewinko and K. H. Humphrey) pp. 435–53. Williams and Wilkins, Baltimore.

Ascheim, E. (1968*a*). Experimental approach to the renewal of the skin surface. *Nature, Lond.* **220**, 1242–3.

—— (1968*b*). Epidermal homeostasis – a numerical model. Kinetics of epidermal cells. *Experientia* **24**, 94–8.

Aso, K., Orenberg, K., and Farber, E. M. (1975). Reduced epidermal cyclic AMP accumulation following prostaglandin stimulation: its possible role in the pathophysiology of psoriasis. *J. invest. Derm.* **65**, 375–8.

—— and Rabinowitz, I. (1977). Cyclic AMP and differentiation of epidermal cells in culture: biphasic concentration change during growth. In *Biochemistry of cutaneous differentiation* (ed. M. Seijii and I. A. Bernstein) pp. 309–18. University Park Press, Baltimore.

Baden, A. P. and Sviokla, S. (1968). The effect of chalone on epidermal DNA synthesis. *Expl Cell Res.* **50**, 644.

Baird, W. M., Sedgwick, J. A., and Boutwell, R. K. (1971). Effects of phorbol and four diesters of phorbol on the incorporation of tritiated precursors into DNA, RNA and protein in mouse epidermis. *Cancer Res.* **31**, 1434–9.

Baisch, H., Beck, H. P., Christensen, I. J., Hartmann, N. R., Fried, J., Dean, P. N., Gray, J. W., Jett, J. H., Johnston, D. A., White, R. A., Nicolini, C., Zeitz, S., and Watson, J. V. (1982). A comparison of mathematical methods for the analysis of DNA histograms obtained by flow cytometry. *Cell Tissue Kinet.* **15**, 235–50.

Baker, H. and Blair, C. (1963). Cell replacement in the human stratum corneum in old age. *Br. J. Derm.* **80**, 367–74.

— and Kligman, A. (1967). Technique for estimating turnover times of human stratum corneum. *Archs Derm.* **95**, 405–11.

Baker, T. G. (1972). Oogenesis and ovarian development. In *Reproductive biology* (ed. H. Balin and S. Glasser) pp. 398–437. Excerpta Medica, Amsterdam.

Balmain, A., Alonso, A., and Fischer, J. (1977). Histone phosphorylation and synthesis of DNA and RNA during phases of proliferation and differentiation induced in mouse epidermis by the tumour promoter 12-O-tetradecanoyl-phorbol-13 acetate. *Cancer Res.* **37**, 1548–55.

Banoczy, J. (1971). Keratinisation in explanted oral mucosal flaps of rabbits. *Archs oral Biol.* **16**, 1289–94.

Barfod, J. (1980). Cell production rates estimated by the use of vincristine and flow cytometry. *Cell Tissue Kinet.* **13**, 9–19.

Barlow, P. W. (1984). Double labelling techniques using two doses of tritiated thymidine in root meristems (in preparation).

— and MacDonald, P. D. M. (1973). An analysis of the mitotic cell cycle in the root meristem of *Zea mays*. *Proc. R. Soc. B* **183**, 385–98.

Barlow, S. D. and Ord, M. G. (1974). Thymidine transport in phytohaemag-glutinin stimulated pig lymphocytes. *Biochem. J.* **148**, 595–602.

Barnes, P. (1983). Studies on the regeneration of intestinal epithelium after irradiation. D. Phil. thesis, University of Oxford.

Barrett, J. C. (1966). A mathematical model of the mitotic cycle and its application to the interpretation of percentage labelled mitosis data. *J. natl Cancer Inst.* **37**, 443–8.

— (1970). Optimised parameters for the mitotic cycle. *Cell Tissue Kinet.* **3**, 349–56.

Baserga, R. (1962). A study of nucleic acid synthesis in ascites tumor cells by two emulsion autoradiography. *J. cell Biol.* **12**, 633–77.

— (1976). *Multiplication and division of mammalian cells*. Marcel Dekker, New York.

— (1978). Resting cells and the $G_1$ phase of the cell cycle. *J. cell Physiol.* **95**, 377–82.

— Henegar, G. C., Kisieleski, W. E., and Lisco, H. (1962). Uptake of tritiated thymidine by human tumours in vivo. *Lab. Invest.* **11**, 360–4.

— and Malamud, D. (1969). *Autoradiography, techniques and applications*. Hoeber, New York.

— and Nemeroff, K. (1962). Factors which affect efficiency of autographs with tritiated thymidine. *Stain Technol.* **37**, 21–32.

— Tyler, S. A., and Kisieleski, W. E. (1963). The kinetics of growth of the Ehrlich tumour. *Archs Path.* **76**, 9–13.

Basset, F. and Tariaf, J. (1965). Identification par la microscopie electronique du particles des nature problement virale dans la liaisons granulomateuses d'une histiocytose 'X' pulmonaire. *C. r. hebd Séanc. Acad. Sci., Paris* **261**, 3701.

Bauer, F. W. and de Grood, R. M. (1975). Impulse cytophotometry in psoriasis. *Br. J. Derm.* **93**, 225–8.

— — (1976). Improved technique for epidermal cell cycle analysis. *Br. J. Derm.* **95**, 565–6.

— Crombag, N. H. C. M. N., Boezemann, J. B. M., and de Grood, R. M. (1981). Flow cytometry as a tool for the study of cell kinetics in skin. 2. Cell kinetic data for psoriasis. *Br. J. Derm.* **104**, 271–6.

—— —— de Grood, R. M., and de Jough, C. J. (1980). Flow cytometry as a tool for the study of cell kinetics in epidermis. *Br. J. Derm.* **102**, 629–39.

Baxter, D. L. and Stoughton, R. B. (1970). Mitotic index of psoriatic lesions treated with anthralin, glucocorticoid and occlusion only. *J. invest. Derm.* **54**, 410–12.

Beagrie, G. S. (1963). An autoradiographic study of the gingival epithelium of mice and monkeys with tritiated thymidine. *Dent. Pract.* **14**, 18.

—— and Skougaard, M. R. (1962). Observations on the life cycle of the gingival epithelium cells of mice as revealed by autoradiography. *Acta odont. scand.* **20**, 15–31.

Beavo, J. A., Harman, J. G., and Sutherland, E. W. (1971). Stimulation of adenosine $3',5'$ monophosphate hydrolysis by guanosine $3',5'$ monophosphate. *J. biol. Chem.* **246**, 3841–6.

Beck, H.-P. (1982). Radiotoxicity of incorporated [$^3$H]-thymidine. Consequences for the interpretation of FLM data of irradiated populations. *Cell Tissue Kinet.* **15**, 469–80.

Becker, A. J., McCulloch, E. A., Simnovitch, C., and Till, J. E. (1965). The effect of differing demands for blood cell production on DNA synthesis by haemopoietic colony forming cells of mice. *Blood* **26**, 296–308.

Becker, H., Stanners, C. P., and Kudlow, J. E. (1971). Control of macromolecular synthesis in proliferating and resting Syrian hamster cells in monolayer culture. II Ribosome complement in resting and early $G_1$ cells. *J. cell Physiol.* **77**, 43–50.

Becker, S. W., Fitzpatrick, T. B., and Montgomery, H. (1952). Human melanogenesis; cytology and histology of pigment cells. *Archs Derm.* **65**, 511.

Begg, A. C. (1977). Cell loss from several types of solid murine tumour: a comparison of $^{125}$Iododeoxyuridine and tritiated thymidine methods. *Cell Tissue Kinet.* **10**, 409–27.

Belman, S. and Trott, W. (1972). The inhibition of croton oil-promoted mouse tumourigenesis by steroid hormones. *Cancer Res.* **32**, 450–4.

—— —— (1974). Phorbol 12-myristate acetate effect on cyclic adenosine $3',5'$ monophosphate levels in mouse skin and inhibition of phorbol myristate acetate tumourigenesis by theophylline. *Cancer Res.* **34**, 3446–55.

Benassi, M., Paoluzi, R., and Bresciani, F. (1973). Computer subtraction of background in autoradiography with tritiated thymidine. *Cell Tissue Kinet.* **6**, 81–5.

Bennington, J. L. (1969). Cellular kinetics of invasive squamous carcinoma of the human cervix. *Cancer Res.* **29**, 1082–8.

Bentley-Phillips, C. B., Pauli-Jorgensen, H., and Marks, R. (1977). The effects of prostaglandins $E_1$ and $F_2$ alpha on epidermal growth. *Archs. Derm. Res.* **257**, 223–37.

Berenblum, I. (1954). Speculative review: probable nature of promoting action and its significance in understanding the mechanism of carcinogenesis. *Cancer Res.* **14**, 471–7.

—— (1959). Mechanisms of carcinogenesis: the role of carcinogenic metabolites. *Acta un. int. Cancr* **15**, 22–30.

Bergstrasser, P. R. and Chapman, S. L. (1980). Maturation of normal human epidermis without an ordered structure. *Br. J. Derm.* **102**, 641–8.

—— Pariser, R. J., and Taylor, J. R. (1978). Counting and sizing of epidermal cells in normal skin. *J. invest. Derm.* **70**, 280–4.

—— and Taylor, J. R. (1977). Epidermal turnover times—a new examination. *Br. J. Derm.* **96**, 503–9.

Bertalanffy, F. D. (1960). Mitotic rates and renewal times for the digestive tract epithelia of the rat. *Acta anat.* **40**, 130–48.

— (1964). Tritiated thymidine versus colchicine technique in the study of cell population cytodynamics. *Lab. Invest.* **13**, 871–83.

— Pussy, V., and Abbott, M. O. (1965). Mitotic rates of rat epidermis during growth, maturity, senility and regeneration. *Archs Derm.* **92**, 91–108.

Bertalanffy, L. von (1960). Principles and theory of growth. In *Fundamental aspects of normal and malignant growth* (ed. W. W. Nowinsky). Elsevier, Amsterdam.

Bertsch, S., Csontos, K., Schweizer, J., and Marks, F. (1976). Effect of mechanical stimulation on cell proliferation in mouse epidermis and on growth regulation by endogenous factors. *Cell Tissue Kinet.* **9**, 445–57.

— and Marks, F. (1974). Lack of an effect of tumour promoting phorbol esters and of epidermal $G_1$ chalone on DNA synthesis in the epidermis of new born mice. *Cancer Res.* **34**, 3283–8.

— — (1978). Removal of the horny layer does not stimulate cell proliferation in newborn mouse epidermis. *Cell Tissue Kinet.* **11**, 651–8.

— — (1979a). Appearance of growth promoting inhibitor ($G_1$ chalone) during ontogenetic development of rat and chick epidermis in vivo and in vitro. *Experientia* **35**, 897–8.

— — (1979b). Effects of epidermal chalone and epidermal growth factor on a transplantable epidermal carcinoma (Hewitt) of the mouse in vivo. *Cancer Res.* **39**, 239–43.

Billingham, R. E. and Medewar, P. B. (1953). A study of the branched cells of the mammalian epidermis, with special reference to the fate of their division products. *Phil. Trans. R. Soc. B* **237**, 151–75.

— and Silvers, W. K. (1960). The melanocytes of animals. *Q. Rev. Biol.* **35**, 1–40.

Birnbaum, J. E. (1976). Cyclic AMP phosphodiesterase and epidermal mitosis. *J. invest. Derm.* **67**, 235–9.

— Sapp, T. M., and Moore, J. B. (1976). Effects of reserpine, epidermal growth factor and cyclic nucleotides on epidermal mitosis. *J. invest. Derm.* **66**, 313–18.

Bischoff, R. (1964). Inhibition of mitosis by homologous tissue extracts. *J. cell Biol.* **24**, 10A.

Bishop, G. H. (1945). Regeneration after experimental removal of the skin in man. *Am. J. Anat.* **76**, 153–81.

Bishop, S. C. and Cox, A. J. (1974). Autoradiographic study of the DNA synthetic activity in human skin cultured *in vitro*: effects of stripping and the inhibition by glucosamine. *J. invest. Derm.* **62**, 74–9.

Bjerknes, M. and Cheng, H. (1981a). The stem cell zone of the mouse small intestinal epithelium. I. Evidence from Paneth cells in the adult mouse. *Am. J. Anat.* **160**, 51–64.

— — (1981b). The stem cell zone of the small intestinal epithelium. II. Evidence from Paneth cells in the newborn mouse. *Am. J. Anat.* **160**, 65–76.

— — (1981c). The stem cell zone of the small intestinal epithelium. III. Evidence from columnar, enteroendocrine and mucous cells in the adult mouse. *Am. J. Anat.* **166**, 77–92.

— — (1981d). The stem cell zone of the small intestinal epithelium. IV. Effects of resecting 30% of the small intestine. *Am. J. Anat.* **160**, 93–104.

Bjerknes, R. and Iversen, O. H. (1974). 'Antichalone', a theoretical treatment of the role of antichalone in the growth control system. *Acta path. microbiol.*

*scand. A.* Suppl. **248**, 33–42.

Blenkinsopp, W. K. (1967). Effect of tritiated thymidine on cell proliferation. *J. cell Sci.* **2**, 305–8.

— (1968*a*). Cell proliferation in stratified squamous epithelium in mice. *Expl cell Res.* **50**, 265–76.

— (1968*b*). Duration of availability of tritiated thymidine following intra-peritoneal injection. *J. cell Sci.* **3**, 91–5.

— (1969). Cell proliferation in the epithelium of the oesophagus, trachea and ureter in mice. *J. cell Sci.* **5**, 393–403.

— and Gilbert, C. W. (1969). Comparison of multiple injections with con-tinuous infusion of tritiated thymidine, and estimation of the cell cycle time. *J. cell Sci.* **5**, 575–82.

Block, P., Seiter, I., and Oehlert, W. (1963). Autoradiographic studies of the initial cellular response to injury. *Expl cell Res.* **30**, 311–21.

Bok, S. T. and Kip, M. J. (1939). The size of the body and size of number of nerve cells in tissue sections. *Acta neurol. morph.* **3**, 1–33.

Born, W. (1969). Die epidermale DNS synthese wahrend des anflaus einer experimentallen akanthose beim Meirschweinchen. *Arch. klin. exp. Derm.* **236**, 53–60.

— (1977). Proliferation kinetic model for the understanding of the epidermal cell cycle in psoriasis. *Fortschr. Med.* **95**, 1395–402.

— and Kalkoff, K. W. (1969). Zur DNS synthesese der psoriatischen Epider-miszelle. *Arch. klin. exp. Derm.* **236**, 43–52.

Boutwell, R. K. (1961). Further dissection of the process of skin tumour for-mation in mice. *Proc. Am. Ass. Cancer Res.* **3**, 211.

— (1964). Some biological aspects of skin carcinogenesis. *Prog. exp. Tumour Res.* **4**, 207–50.

Bowden, G. T. and Boutwell, R. K. (1974). Studies on the role of stimulated epidermal DNA synthesis in the initiation of skin tumours in mice by N methyl-N′ nitro N-nitrosoguanidine. *Cancer Res.* **34**, 1552–63.

Bradley, M. O. and Sharkley, N. A. (1978). Mutagenicity of thymidine to cul-tured Chinese hamster cells. *Nature, Lond.* **274**, 607–9.

Bransome, E. D. (1968). Regulation of adrenal growth. Differences in the effects of ACTH in normal and dexamethasone-suppressed guinea pigs. *Endocrinology* **83**, 956–64.

Branson, R. G. (1968). Orthogonal arrays of microtubules in flattening cells of the epidermis. *Anat. Rec.* **160**, 109–22.

Braun Falco, O., Burg, G., and Schoefinius, H. A. (1971). Uber der Wirking von Dithranol (Cignolin) bei Psoriasis vulgaris; Cyto- und histochemische Unter-suchungen. *Arch. Derm. Forsch.* **241**, 217–37.

Bray, G. and Brent, T. (1972). Deoxyribonucleotide 5′ triphosphate pool fluctuations during the mammalian cell cycle. *Biochim. biophys. Acta* **269**, 184–9.

Breathnach, A. S. (1963). A new concept of the relation between the Langerhans cells and the melanocyte. *J. invest. Derm.* **40**, 279–81.

— (1965). The cell of Langerhans. *Int. Rev. Cytol.* **18**, 1–28.

— (1969). Melanocytes in early regenerated human epidermis. *J. invest. Derm.* **35**, 245–51.

— (1971). Embryology of human skin: a review of ultrastructural studies. *J. invest. Derm.* **57**, 133–43.

— (1975). Aspects of epidermal ultrastructure. *J. invest. Derm.* **65**, 2–15.

**480     References**

— (1980). Branched cells in the epidermis — an overview. *J. invest. Derm.* **75**, 6–11.
— and Robins, J. (1969). Ultrastructural features of the epidermis of a 14 mm (6 weeks) human embryo. *Br. J. Derm.* **81**, 504–16.
— — (1970). Ultrastructural observations on Merkel cells in human foetal skin. *J. Anat.* **106**, 411–22.
— Silbers, W. K., Smith, J., and Heynor, S. (1965). Langerhans cells in mouse skin experimentally deprived of its neural crest component. *J. invest. Derm.* **50**, 147–60.
— and Wyllie, L. M. A. (1966). The problem of the Langerhans cell. *Adv. Biol. Skin* **8**, 97–113.
Bresciani, F. (1965). A comparison of the cell generative cycle in normal, hyperplastic and neoplastic mammary gland. *Expl Cell Res.* **38**, 13–32.
— (1968). Cell proliferation in cancer. *Eur. J. Cancer* **4**, 343–66.
— Poaluzo, R., Benassi, M., Nerri, C., Casale, C., and Ziparo, E. (1974). Cell kinetics and growth of squamous cell carcinoma in man. *Cancer Res.* **34**, 2405–15.
Britton, N. F., Wright, N. A., and Murray, J. D. (1982). A mathematical model for cell population kinetics in the intestine. *J. theoret. Biol.* **98**, 531–42.
Bronstad, G. O., Elgjo, K., and Oye, I. (1971). Adrenaline increases cyclic $3',5'$ AMP formation in hamster epidermis. *Nature, New Biol.* **233**, 78.
Brookes, P. (1975). Covalent interaction of carcinogens with DNA. *Life Sci.* **16**, 331.
Brooks, R. F. (1976). Regulation of the fibroblast cell cycle by serum. *Nature, Lond.* **260**, 248–50.
— Bennett, D. C., and Smith, J. A. (1980). Mammalian cell cycles need two random transitions. *Cell* **19**, 493–504.
Brophy, D. and Lobitz, W. C. (1959). Injury and reinjury to the human epidermis II. Epidermal basal cell response. *J. invest. Derm.* **32**, 495–9.
Brown, J. M. (1968). Long $G_1$ or $G_0$ state; a method of resolving the dilemma for the cell cycle of an *in vivo* population. *Expl Cell Res.* **52**, 565–70.
— (1970). The effect of acute X-irradiation on the cell proliferation kinetics of induced carcinomas and their normal counterpart. *Radiat. Res.* **43**, 627–42.
— and Berry, R. J. (1968). The relationship between diurnal variation of the number of cells in mitosis and the number of cells synthesising DNA in the epithelium of the hamster cheek pouch. *Cell Tissue Kinet.* **1**, 23–31.
— — (1969). Effect of X-irradiation on the cell population kinetics in a model tumour and normal tissue system; implications for treatment of human malignancies. *Br. J. Radiol.* **42**, 372–7.
— and Oliver, R. (1968). A new method for estimating the cell cycle time in epithelial tissues of long duration. *Cell Tissue Kinet.* **1**, 11–21.
Bruce, W. R. (1976). In discussion. In *Stem cells of renewing cell populations* (ed. A. B. Cairnie, P. K. Lala, and D. G. Osmond) p. 138. Academic Press, New York.
Bryant, B. J. (1963). Reutilisation of lymphocyte DNA by cells of intestinal crypts and regenerating liver. *J. cell Biol.* **18**, 515–23.
— (1966). The incorporation of tritium from thymidine into proteins of the mouse. *J. cell. Biol.* **29**, 29–34.
Budtz, P. E. (1982). Time dependent effects of removal of the pars distalis of the pituitary gland on toad epidermal cell and tissue kinetic parameters. *Cell Tissue Kinet.* **15**, 507–16.
Bullough, W. S. (1949). Age and mitotic activity in the male mouse *Mus mus-*

*culus* L. *J. exp. Biol.* **26**, 261–86.

— (1950). Mitotic activity in tissues of dead mice, and tissues kept in physiological salt solutions. *Expl Cell Res.* **1**, 410–20.

— (1962). Control of mitotic activity in adult mammalian tissues. *Biol. Rev.* **37**, 307–42.

— (1965). Mitotic and functional homeostasis: a speculative review. *Cancer Res.* **25**, 1683–727.

— (1967). *The evolution of differentiation.* Academic Press, New York.

— (1969). Epithelial repair. In *Repair and regeneration* (ed. J. E. Dunphy and W. van Winkle) pp. 15–39. McGraw Hill, New York.

— (1972). The control of epidermal thickness. *Br. J. Derm.* **87**, 187–99; 347–56.

— (1975*a*). Chalone control mechanisms. *Life Sci.* **16**, 323–30.

— (1975*b*). Mitotic control in adult mammalian tissues. *Biol. Rev. Camb. phil. Soc.* **50**, 99–127.

— and Deol, J. N. R. (1971). The pattern of tumour growth. *Symp. Soc. exp. Biol.* **25**, 255–75.

— — (1972). Chalone control of mitotic activity in eccrine sweat glands. *Br. J. Derm.* **86**, 586–92.

— — (1975). Dermoepidermal adhesion and its effect on epidermal structure in the mouse. *Br. J. Derm.* **93**, 417–24.

— and Ebling, F. J. (1952). Cell replacement in the epidermis and sebaceous glands in the mouse. *J. Anat.* **86**, 29–36.

— Hewett, C. L., and Laurence, E. B. (1960). The epidermal chalone; a preliminary attempt at isolation. *Expl Cell Res.* **36**, 192.

— Homan, J. D. H., Laurence, E. B., and Overbeck, G. A. (1968). Prevention by nandrolone decanoate of dexamethasone-induced mitotic inhibition in mouse ear epidermis *in vitro*. *J. Endocr.* **41**, 453–4.

— and Laurence, E. B. (1957). Mitotic activity in healing wounds: a technique for the study of small epidermal wounds. *Br. J. exp. Path.* **38**, 273–83.

— — (1960*a*). The control of mitotic activity in the mouse. *Proc. R. Soc.* **151**, 517–36.

— — (1960*b*). Control of mitotic activity in mouse skin. *Expl Cell Res.* **21**, 394–405.

— — (1961*a*). The study of mammalian epidermal mitosis *in vitro*. A critical analysis of technique. *Expl Cell Res.* **24**, 289–96.

— — (1961*b*). Stress and adrenalin in relation to the diurnal cycle of epidermal mitotic activity in adult male mice. *Proc. R. Soc. B* **154**, 540–9.

— — (1964*a*). Duration of epidermal mitosis *in vitro*. *Expl Cell Res.* **35**, 629–35.

— — (1964*b*). Production of epidermal cells. *Symp. zool. Soc.* **12**, 1–23.

— — (1964*c*). Mitotic control by internal secretion; the role of the chalone: adrenalin complex. *Expl Cell Res.* **33**, 176–94.

— — (1966*a*). The diurnal cycle in epidermal mitotic duration and its relation to chalone and adrenalin. *Expl Cell Res.* **43**, 343–52.

— — (1966*b*). Accelerating and decelerating actions of adrenalin on epidermal mitotic activity. *Nature, Lond.* **210**, 715.

— — (1968*a*). The role of glucocorticoid hormones in the control of epidermal mitosis. *Cell Tissue Kinet.* **1**, 5–10.

— — (1968*b*). Epidermal chalone and mitotic control in the VX2 epidermal tumour. *Nature, Lond.* **220**, 134–5.

— — (1968*c*). Control of mitosis in mouse and hamster melanomata by means of the melanocyte chalone. *Eur. J. Cancer* **4**, 607–15.

— — (1968*d*). Control of mitosis in rabbit VX2 epidermal tumours by means of the epidermal chalone. *Eur. J. Cancer* **4**, 587–94.

— — (1968*e*). Melanocyte chalone and mitotic control in melanomata. *Nature, Lond.* **220**, 137–8.

— — (1970). Chalone control of mitotic activity in sebaceous glands. *Cell Tissue Kinet.* **3**, 291–7.

— — Iversen, O. H., and Elgjo, K. (1967). The vertebrate epidermal chalone. *Nature, Lond.* **214**, 578.

— and Mitrani, E. (1976). An analysis of the epidermal chalone control mechanism. In *Chalones* (ed. J. Houck) pp. 7–36. North Holland, Amsterdam.

— — (1978). The significance of vertical mitosis in epidermis. *Br. J. Derm.* **99**, 603–70.

— and Stolze, J. U. R. (1978). A new form of cellular arrangement in guinea pig epidermis. *Br. J. Derm.* **99**, 519–28.

Burholt, D. R., Schultze, B., and Maurer, W. (1976). Mode of growth of the jejunal crypt cells of the rat; an autoradiographic study using double labelling with $^3$H- and $^{14}$C-thymidine in lower and upper parts of the crypts. *Cell Tissue Kinet.* **9**, 107–17.

Burns, E. R. (1981). A critique of the practice of comparing control data obtained at a single time point to experimental data obtained at multiple timepoints. *Cell Tissue Kinet.* **14**, 219–24.

— and Scheving, I. E. (1973). Isoproterenol induced phase shifts in circadian rhythm of mitoses in murine corneal epithelium. *J. cell Biol.* **56**, 605–8.

— — (1975). Circadian influence on the wave form of the frequency of labelled mitoses in mouse corneal epithelium. *Cell Tissue Kinet.* **8**, 61–6.

— — Fawcett, D. F., Gibbs, W. M., and Galatzan, R. E. (1975). Circadian influence on the cell kinetic data obtained from the frequency of labelled mitoses method. *Anat. Rec.* **181**, 321–2.

— — Fawcett, D. F., Gibbs, W. M., and Galatzan, R. E. (1976). Circadian influence on the frequency of labelled mitoses method in the stratified squamous epithelium of the mouse oesophagus and tongue. *Anat. Rec.* **184**, 265–73.

Burns, F. J. and Tannock, I. F. (1970). On the existence of a $G_0$ phase in the cell cycle. *Cell Tissue Kinet.* **3**, 321–34.

— and Vanderlaan, M. (1977). Split dose recovery for radiation induced tumours in rat skin. *Int. J. radiat. Biol.* **32**, 135–44.

Cairnie, A. B. (1976). Homeostasis in the small intestine. In *Stem cells of renewing cell populations* (ed. A. B. Cairnie, P. K. Lala, and D. G. Osmond) pp. 67–78. Academic Press, New York.

— Lamerton, L. F., and Steel, G. G. (1965*a*). Cell proliferation studies in the intestinal epithelium of the rat. I. Determination of the kinetic parameters. *Expl Cell Res.* **39**, 528–38.

— — — (1965*b*). Cell proliferation studies in the epithelium of the rat. II. Theoretical aspects. *Expl Cell Res.* **39**, 539–53.

Cairns, J. (1975). Mutation, selection and the natural history of cancer. *Nature, Lond.* **225**, 197–200.

Cameron, I. L. (1966). Cell proliferation, migration and synchronisation in the epithelium of the mouse tongue. *J. exp. Zool.* **163**, 271–84.

— (1970). Cell renewal in the organs and tissues of the non-growing adult mouse. *Texas Rep. Biol. Med.* **28**, 203–48.

— (1972*a*). Cell proliferation and renewal in ageing mice. *J. Gerontol.* **27**, 162–72.

— (1972*b*). Minimum number of cell doublings in an epithelial cell population during the life span of the mouse. *J. Gerontol.* **27**, 157–61.

— Gosbe, D. G., and Pilgrim, C. (1965). The spatial distribution of dividing and synthesising cells in mouse epithelium. *J. cell. comp. Physiol.* **66**, 431.

—– and Greulich, R. C. (1963). Evidence for an essentially constant duration of DNA synthesis in renewing epithelia of the adult mouse. *J. cell Biol.* **18**, 34–40.

Campbell, R. D. and Campbell, J. H. (1971). Origin and continuity of desmosomes. In *Origin and continuity of cell organelles* (ed. J. Reinert and H. Ursprung) pp. 251–93. Springer, Berlin.

Camplejohn, R. S., Bone, G., and Aherne, W. A. (1973). Cell proliferation in rectal carcinoma and rectal mucosa. A stathmokinetic study. *Eur. J. Cancer* **9**, 577–81.

— Gelfant, S., and Chalker, D. (1981). An attempt to use vincristine and colcemid to measure proliferative rates in normal human epidermis *in vivo*. *Br. J. Derm.* **104**, 243–7.

— — — (1984). Circadian and seasonal variation in labelling indices and mitotic indices in human epidermis. *Cell Tissue Kinet.* in press.

— Schultze, B., and Maurer, W. (1980). An *in vivo* double labelling study of the subsequent fate of cells arrested in metaphase by vincristine in the JB-1 mouse ascites tumour. *Cell Tissue Kinet.* **13**, 239–52.

Cantieri, J. S., Graff, G., and Goldberg, N. D. (1980). Cyclic GMP metabolism in psoriasis: activation of soluble epidermal guanylate cyclase by arachidonic acid and by hydroxy 5,8,10,14 eicosatetraenonic acid. *J. invest. Derm.* **74**, 234–7.

Carpenter, G. (1978). The regulation of cell proliferation; advances in the biology and mechanism of action of epidermal growth factor. *J. invest. Derm.* **71**, 283–7.

— (1981). Epidermal growth factor. In *Tissue growth factors* (ed. R. Baserga), pp. 63–98. Springer-Verlag, Heidelberg.

— and Cohen, S. (1976). Human epidermal growth factor and the proliferation of fibroblasts. *J. cell Physiol.* **88**, 227–37.

Castor, L. N. (1980). A $G_1$ rate model accounts for cell cycle kinetics attributed to transition probability. *Nature, Lond.* **287**, 857–9.

Cattaneo, S. M., Quastler, H., and Sherman, F. G. (1960). DNA synthesis in irradiated hair follicles of the mouse. *Radiat. Res.* **12**, 587–96.

Cayama, E., Tsuda, H., Sarma, D. S. R., and Farber, E. (1978). Initiation of chemical carcinogenesis requires cell proliferation. *Nature, Lond.* **275**, 60–2.

Chambers, R. and Renyi, G. S. (1925). The structure of cells in tissues as revealed by microdissection 1. The physical relationships of the cells in epithelia. *Am. J. Anat.* **35**, 385–99.

Chang, W. W. L. (1978). Histogenesis of symmetrical 1,2 dimethylhydrazine induced neoplasms of the colon in the mouse. *J. natl Cancer Inst.* **60**, 1415–18.

— (1980). Pathogenesis and biological behaviour of 1,2-dimethylhydrazine-induced colonic neoplasms in the mouse. In *Cell proliferation in the gastro-intestinal tract* (ed. D. R. Appleton, J. P. Sunter, and A. J. Watson) pp. 273–97. Pitman, London.

Chaudry, A. P., Halberg, F., and Bittner, J. J. (1956). Reduction of mitotic activity in pinna epidermis of mice given cortisol or 9α fluorocortisol. *Proc. Soc. exp. Biol. Med.* **94**, 602–4.

Chavandra, N., Richard, J. M., and Malaise, E. P. (1979). Labelling index of human squamous carcinomas. Comparisons of *in vivo* and *in vitro* labelling methods. *Cell Tissue Kinet.* **12**, 145–52.

Cheng, H. and Leblond, C. P. (1974). Origin, differentiation and renewal of the four main epithelial cell types in the mouse small intestine. V. Unitarian theory of the origin of the four epithelial cell types. *Am. J. Anat.* **141**, 537–62.

Chevalier, S. and Verly, W. G. (1975). Identification of the inhibitor of labelled thymidine incorporation into Hela cell DNA present in endometrial extract. *Eur. J. Cancer* **11**, 657–60.

Chiakulas, J. J. and Scheving, L. E. (1966). Circadian phase relationships of $^3$H-thymidine uptake, numbers of labelled nuclei, grain counts and cell division rate in larval epidermis. *Expl Cell Res.* **64**, 256–67.

Choie, D. D. and Richter, G. W. (1978). $G_2$ subpopulation in mouse liver induced into mitosis by lead acetate. *Cell Tissue Kinet.* **11**, 235–9.

Chopra, D. P. (1973). Regulation of mitosis in the embryonic kidney (*Xenopus laevis*) by kidney growth inhibitor. *Nat. Cancer Inst. Monogr.* **38**, 189–94.

—— (1977). Effects of theophylline and dibutyryl cyclic AMP on proliferation and keratinisation of human keratinocytes. *Br. J. Derm.* **96**, 255–62.

— and Flaxman, B. A. (1972). Human epidermal cell cycle *in vitro*. *Br. J. Derm.* **87**, 13.

— — (1974). Comparative proliferation kinetics of cells from normal human epidermis and benign epidermal hyperplasia (psoriasis) *in vitro*. *Cell Tissue Kinet.* **7**, 69–76.

— — (1975). The effect of vitamin A on growth and differentiation of human keratinocytes *in vitro*. *J. invest. Derm.* **64**, 19–22.

— and Forbes, P. D. (1974). Analysis of cell kinetics during ultraviolet light-induced epidermal carcinogenesis in hairless mice. *Cancer Res.* **34**, 454–7.

— Riley, J. Y., and Flaxman, B. A. (1972). Demonstration of a tissue specific inhibitor of mitosis of human epidermal cells *in vitro*. *J. invest. Derm.* **59**, 207.

Christensen, I., Hartmann, N. R., Keiding, N., Larsen, J. K., Noer, H., and Vindelov, L. (1978). Statistical analysis of DNA distributions from cell populations with partial synchrony. In *2nd International Symposium on Pulse Cytophotometry* (ed. D. Lutz) pp. 71–8. European Press, Brussels.

Christophers, E. (1970). Autoradiographic analysis of the post-mitotic cell transitions of the epidermis. *Arch. klin. exp. Derm.* **236**, 161–72.

—— (1971*a*). The architecture of the stratum corneum after wounding. *J. invest. Derm.* **57**, 241–5.

—— (1971*b*). Cellular architecture of the stratum corneum. *J. invest. Derm.* **56**, 165–9.

—— (1971*c*). The columnar structure of the epidermis; possible mechanism of differentiation. *Z. Zellforsch.* **114**, 441–50.

—— (1972*a*). Correlation between column formation, tissue size and new cell production in guinea pig epidermis. *Virchows Arch. Cell Path.* **10**, 286–92.

—— (1972*b*). Kinetic aspects of wound healing. In *Epidermal wound healing* (ed. H. I. Maibach and D. T. Rovee) pp. 53–70. Year Book, Chicago.

—— (1974). Growth stimulation of cultured postembryonic epidermal cells by vitamin A acid. *J. invest. Derm.* **63**, 450–5.

— and Braun Falco, O. (1967). Mechanisms of parakeratosis. *Br. J. Derm.* **82**, 268–75.

—— —— (1970*a*). The architecture of the stratum corneum after wounding. *J. invest. Derm.* **54**, 437–40.

—— —— (1970*b*). Psoriatic hyperplasia—some measurements. *Br. J. Derm.* **83**, 63–8.

—— and Kligman, A. M. (1964). Visualisation of the cell layers of the stratum corneum. *J. invest. Derm.* **42**, 407–9.

—— and Laurence, E. B. (1973). Regional variations in mouse skin; quantitation of epidermal components in two different body sites. *Virchows Arch. Cell. Path.* **12**, 212–23.

—— —— (1976). Kinetics and structural organisation of the epidermis. *Curr. Prob. Derm.* **6**, 87–106.

—— and Petzolt, V. (1969). Epidermal cell replacement: topographical variations in albino guinea pig skin. *Br. J. Derm.* **81**, 598–606.

—— and Schaumloffel, E. (1967). Zur DNA-syntheseit in der menschlichen epidermis. *Arch. klin. exp. Derm.* **728**, 57–68.

—— and Wolff, H. H. (1975). Effects of vitamin A acid in skin—*in vivo* and *in vitro* studies. *Acta Derm.* Suppl. **74**, 43–9.

—— —— and Laurence, E. B. (1974). The formation of epidermal cell columns. *J. invest. Derm.* **62**, 555–9.

Cipriani, C. (1976). Adenyl cyclase activity in rat hair cycle. *Archs Derm. Res.* **256**, 319–25.

Claesson, M. H. and Hartmann, N. R. (1976). Cytodynamics in the thymus of young mice: a quantitative study on the loss of thymic blast cells and non-proliferating small thymocytes. *Cell Tissue Kinet.* **9**, 273–91.

Clarke, R. M. (1968). A comparative analysis of methods of estimating the size of cell populations from microtome sections. *J. Ry. microsc. Soc.* **88**, 189–95.

—— (1970*a*). The duration of mitosis in the crypts of Lieberkuhn of the terminal ileum of the albino rat. *Cell Tissue Kinet.* **3**, 27–34.

—— (1970*b*). Mucosal architecture and epithelial cell production rate in the small intestine of the albino rat. *J. Anat.* **107**, 519–29.

—— (1971). A comparison of metaphase arresting agents and tritiated thymidine in measurement of the rate of entry into mitosis in the crypts of Lieberkuhn of the rat. *Cell Tissue Kinet.* **4**, 263–72.

—— (1973). Progress in measuring epithelial turnover in the villus of the small intestine. *Digestion* **8**, 161–75.

Clarkson, B., Ohkita, T., Kazno, O., and Fried, J. (1967). Studies of cellular proliferation in acute leukaemia. Estimation of growth rates of leukaemic and normal haemopoietic cells in two adults with acute leukaemia given single injections of tritiated thymidine. *J. clin. Invest.* **46**, 506–29.

Clausen, O. P. (1975). The effect of Bleomycin on rapidly proliferating epidermis. A comparative investigation using microflow fluorometry, $^{3}$HTdR incorporation and a stathmokinetic method. *Virchows Arch. Cell Path.* **19**, 337–48.

—— (1978). Regenerative proliferation of mouse epidermal cells following application of a carcinogenic skin irritant (MCA). *Virchows Arch. Cell Path.* **27**, 205–15.

—— (1979). Regenerative proliferation of mouse epidermal cells following application of a skin irritant (cantharidin). *Cell Tissue Kinet.* **12**, 135–44.

—— and Lindmo, T. (1976). Regenerative proliferation of mouse epidermal cells following adhesive tape stripping. *Cell Tissue Kinet.* **9**, 573–89.

—— —— Sandnes, K., and Thorud, E. (1976). Separation of mouse epidermal cells and differentiating cells for micro-flow fluorometric measurements. *Virchows Arch. Cell Path.* **20**, 261–75.

## 486    References

— and Thorud, E. (1980). Perturbation of cell cycle progression in mouse epidermis prior to the regenerative response. *J. invest. Derm.* **74**, 433–6.

— — and Aarnes, E. (1981). Evidence of rapid and slow progression of cells through $G_2$ phase in mouse epidermis: a comparison between phase durations measured by different methods. *Cell Tissue Kinet.* **14**, 354–71.

— — Bjerknes, R., and Elgjo, K. (1979). Circadian rhythms in mouse epidermal cell proliferation. Variations in compartment size, flux and phase duration. *Cell Tissue Kinet.* **12**, 314–37.

— — and Bolund, L. (1980). DNA synthesis in mouse epidermis; labelled and unlabelled basal cells in *S* after administration of tritiated thymidine. *Virchows Arch. Cell Path.* **34**, 1–11.

— — and Iversen, O. H. (1982). Adrenalin has differential effects on epidermal cell cycle progression in mice. *J. invest. Derm.* **78**, 472–6.

Cleaton-Jones, P. (1970). Mitotic activity in the oral epithelium of the albino rat. *J. dent. Res.* **55**, 859–63.

Cleaver, J. (1965). The relationship between the duration of the *S* phase and the fraction of cells which incorporate $^3$H-thymidine during exponential growth. *Expl Cell Res.* **39**, 697–700.

— (1967). *Thymidine metabolism and cell kinetics*. North Holland, Amsterdam.

Clemmesen, J. (1965). Statistical studies in the aetiology of malignant neoplasms II. *Acta path. microbiol. scand.* Suppl. 174.

— (1969). Statistical studies in the aetiology of malignant neoplasms III. *Acta path. microbiol. scand.* Suppl. 209.

— (1974). Statistical studies in the aetiology of malignant neoplasms. IV. *Acta path. microbiol. scand.* Suppl. 247.

Clow, D. J. and Gillette, E. L. (1970). Survival of type A spermatogonia following X-irradiation. *Radiat. Res.* **42**, 397–404.

Cohen, S. (1960). Purification of a nerve growth promoting protein from the mouse salivary gland and its neuro-cytotoxic antiserum. *Proc. natl. Acad. Sci. USA* **46**, 302–11.

— (1962). Isolation of a mouse submaxillary gland protein accelerating incisor eruption and eyelid opening in the new-born animal. *J. biol. Chem.* **237**, 1555–62.

— (1965). The stimulation of epidermal proliferation by a specific protein (EGF). *Devl. Biol.* **12**, 394–407.

— and Carpenter, G. (1975). Human epidermal growth factor; isolation and chemical and biological properties. *Proc. natl. Acad. Sci. USA* **72**, 1317–21.

— and Elliott, G. A. (1963). The stimulation of epidermal keratinisation by a protein isolated from the submaxillary gland of the mouse. *J. invest. Derm.* **40**, 1–5.

— and Savage, C. R. (1974). Recent studies in the chemistry and biology of epidermal growth factor. *Recent Prog. Horm. Res.* **30**, 551–74.

— and Taylor, J. M. (1974). Epidermal growth factor: chemical and biological characterisation. *Recent Prog. Horm. Res.* **30**, 533–50.

Cooper, E. H. and Wickramsinghe, S. N. (1969). Quantitative cytochemistry in the study of erythropoiesis. *Ser. Hematol.* **4**, 65–87.

Cooper, Z. K. and Franklin, A. C. (1940). Mitotic rhythm in the epidermis of the mouse. *Anat. Rec.* **78**, 1–7.

— and Schiff, P. (1938). Mitotic rhythm in human epidermis. *Proc. Soc. exp. Biol. Med.* **39**, 323–4.

— Thuringer, J. M., and Katzberg, A. A. (1951). Age relations in mitotic frequency and pattern of the rete in human epidermis. *J. Gerontol.* **6**, 73 (Suppl. 3).

Cowdry, E. V. (1950). *Textbook of histology.* Lea and Febiger, Philadelphia.

Cox, A. J. and Watson, I. V. (1971). Histological evidence on the nature of psoriasis. In *Psoriasis* (ed. E. M. Farber and A. J. Cox) pp. 151–9. Stanford University Press.

Cripps, D. J., Rumsey, C. A., and Carter, J. (1972). Effect of monochromatic irradiation on DNA synthesis with *in vivo* and *in vitro* autoradiography. *J. invest. Derm.* **58**, 312–14.

Croft, C. B. and Tarin, D. (1970). Ultrastructural studies of wound healing in mouse skin. *J. Anat.* **106**, 63–77.

Curnow, R. N. and Livingstone, D. J. (1984). Mitotic coincidence and the transition probability model of the cell cycle (in preparation).

Cutright, D. E. and Bauer, H. (1967). Cell renewal in the oral mucosa of rats. II. DNA conservation and utilisation during the renewal cycle in *in vivo* and in tissue explants. *Oral Surg. oral Med. oral Path.* **23**, 261–72.

Dahl, M. G. C. and Shuster, S. (1972). A rapid *in vitro* assay of epidermal chalone. *J. invest. Derm.* **58**, 253.

Daniels, F., Brophy, D., and Lobitz, C. L. (1961). Histochemical responses of human skin following ultraviolet irradiation. *J. invest. Derm.* **37**, 351–7.

Darzynkiewicz, Z., Andreef, M., Traganos, F., Sharples, T., and Melamed, M. R. (1978). Discrimination of cycling and non-cycling lymphocytes by BUdR suppressed acridine orange fluorescence in a flow cytometric system. *Expl cell Res.* **115**, 31–5.

— Evenson, D., Staiano-Coico, L., Sharples, T., and Melamed, M. R. (1979a). Relationship between RNA content and progression of lymphocytes through $S$ phase of cell cycle. *Proc. natl. Acad. Sci. USA* **76**, 358–62.

— Sharples, T., Staiano-Coico, L., and Melamed, M. R. (1980). Subcompartments of the $G_1$ phase of cell cycle detected by flow cytometry. *Proc. natl. Acad. Sci. USA* **77**, 6696–9.

— Staiano–Coico, L., and Melamed, M. R. (1981). Increased uptake of rhodamine-123 during lymphocyte stimulation. *Proc. natl. Acad. Sci. USA* **78**, 2383–7.

— and Traganos, F. (1982). RNA content and chromatin structure in cycling and noncycling cell populations studied by flow cytometry. In *Genetic expression in the cell cycle* (ed. G. M. Padilla and K. McCarty) pp. 103–28. Academic Press, New York.

— — Andreef, M., Sharples, T., and Melamed, M. R. (1979b). Different sensitivity of chromatin to acid denaturation in quiescent and cycling cells as revealed by flow cytometry. *J. Histochem. Cytochem.* **27**, 478–85.

Davidson, A., Wood, C. B., Traynor, O., Duffield, R. G., and Wright, N. A. (1981). Measurements of the viable tumour fraction in malignant colorectal tumours. *J. Path.* **134**, 381.

Davidson, P., Liu, S., and Karasek, M. (1979). Limitations in the use of [$^3$H]-thymidine incorporation into DNA as an indicator of epidermal keratinocyte proliferation *in vitro. Cell Tissue Kinet.* **12**, 605–14.

Davis, J. C. (1964). The effect of cortisol on mitosis in the skin of the mouse. *J. path. Bact.* **88**, 247.

Dean, P. N. and Jett, J. A. (1974). Mathematical analysis of DNA distributions derived from flow microfluorometry. *J. cell Biol.* **60**, 523–7.

De Bersaques, S. and Rothman, S. (1962). Mechanism of keratin formation. *Nature, Lond.* **193**, 147–9.

Delescluse, C. (1977). Studies on guinea pig skin cell cultures. VI. Growth kinetics of epidermal keratinocytes and dermal fibroblasts. *Acta Derm. Vener.* **57**, 469–75.

— (1978). Studies on guinea pig skin cell cultures VII. Statistical analysis of growth maturation. *Acta derm. vener.* **58**, 1–7.

— (1979). Microflowfluorimetric evaluation of homeostasis in cultured epidermal cells. *Cell Biol. Int. Rep.* **3**, 685–90.

— Fukuyama, K., and Epstein, W. L. (1976). Dibutyryl cyclic AMP-induced differentiation of epidermal cells in tissue culture. *J. invest. Derm.* **66**, 8–13.

— Regnier, M., and Prunieras, M. (1974). Studies on guinea pig skin cell cultures, II. Effect of pig skin extract on DNA synthesis. *Acta derm. vener.* **54**, 1–6.

Delforno, C., Holt, P. J. A., and Marks, R. (1978). Corticosteroid effect on epidermal cell size. *Br. J. Derm.* **98**, 619–23.

De Maertelaer, V. (1978). Cell loss from germinative cell populations. In *Biomathematics and cell kinetics* (ed. A. J. Valleron and P. D. M. MacDonald) pp. 288–96. Elsevier-North Holland, Amsterdam.

— and Galand, P. (1975). Some properties of a $G_0$-model of the cell cycle. *Cell Tissue Kinet.* **8**, 11–22.

— — (1977*a*). Cell kinetics: a model of stimulated populations. *J. theoret. Biol.* **69**, 429–45.

— — (1977*b*). Some properties of a $G_0$ model of the cell cycle. *Cell Tissue Kinet.* **10**, 35–42.

Dendy, P. P. and Cleaver, J. E. (1964). An investigation of (a) variation in rate of DNA synthesis during *S*-phase in mouse L cells; (b) effect of ultraviolet radiation on rate of DNA synthesis. *Int. J. radiat. Biol.* **8**, 301–15.

— — (1965). The relationship between the duration of the *S* phase and the fraction of cells which incorporate ³H-thymidine during exponential growth. *Expl Cell Res.* **39**, 597–600.

— and Smith, C. L. (1964). Effects on DNA synthesis of localised irradiation of cells in tissue culture with (i) a UV microbeam and (ii) a particle microbeam. *Proc. R. Soc. B* **160**, 328–44.

Denekamp, J. (1973). Change in the rate of repopulation during multifraction irradiation of mouse skin. *Br. J. Radiol.* **46**, 381–9.

— (1975). Change in the rate of proliferation in normal tissues after irradiation. In *Radiation research* (ed. O. F. Nygaard, J. I. Adler, and W. K. Sinclair) p. 810. Academic Press, New York.

— Ball, M. M., and Fowler, J. F. (1969). Recovery and depopulation in mouse skin as a function of time after irradiation. *Radiat. Res.* **37**, 361–70.

— Emery, E. W., and Field, S. B. (1971). Response of mouse epidermal cells to single and double doses of fast neutrons. *Radiat Res.* **45**, 80.

— Fowler, J. F., Kraft, J. K., Parnell, C. J., and Field, S. B. (1966). Recovery and repopulation in mouse skin after irradiation with cyclotron neutrons as compared with 250 kV X-rays or 15-meV electrons. *Radiat. Res.* **29**, 71–84.

— and Kallman, R. F. (1973). *In vivo* and *in vitro* labelling of animal tumours with tritiated thymidine. *Cell Tissue Kinet.* **6**, 217–27.

— and Michael, B. D. (1972). Preferential sensitisation of hypoxic cells to radiation *in vivo*. *Nature, New Biol.* **239**, 21–3.

— Stewart, F. A., and Douglas, B. G. (1976). Changes in the proliferation rate of mouse epidermis after irradiation: continuous labelling studies. *Cell Tissue Kinet.* **9**, 19–29.

De Rey, B. L. and Klein-Szanto, A. J. (1972). A histometric study of acute irradiation effects on rat tail epidermis. *Strahlentherapie* **143**, 699–704.

Dethlefsen, L. A. (1969). Reutilisation of $^{131}$I-IUdR as compared to $^3$H-thymidine in mouse duodenum and mammary tumour. *J. natn. Cancer Inst.* **44**, 827–40.

—— (1971). An evaluation of radioactive labelled IUDR as a tracer for measuring cell loss from solid tumours. *Cell Tissue Kinet.* **4**, 123–38.

Devik, F. (1955). Study of local roentgen reaction on skin of mice, with special reference to vascular effects. *Acta radiol.* Suppl. **119**, 1–72.

—— (1961*a*). Studies on the duration of DNA synthesis and mitosis in irradiated and regenerating epidermis cells in mice by means of tritium labelled thymidine. *Int. J. radiat. Biol.* **5**, 59–66.

—— (1961*b*). Studies on cell population kinetics of X-irradiated and shielded mouse epidermis by autoradiography after administration of tritiated thymidine. *Acta path. microbiol. scand.* Suppl. **148**, 35.

Dewey, D. L. (1973). The melanocyte chalone. *Natl Cancer Inst. Monogr.* **38**, 213–16.

—— (1978). The identification of cell culture inhibitor in a tumour extract. *Cancer Lett.* **4**, 77–84.

Dhawan, A. S. and Toto, P. D. (1965). Renewal of cell population in palate and tongue epithelia of mice. *J. dent. Res.* **44**, 989.

Dimassimo, C. (1963). M.Sc. Thesis, University of Rochester, NY. Quoted by Skongaard (1970).

Dodson, J. W. and Hay, E. D. (1971). Secretion of collagenous stroma by isolated epithelium grown *in vitro*. *Expl Cell Res.* **65**, 215–31.

Dombernowsky, P., Bichel, P., and Hartmann, N. R. (1974). Cytokinetic studies on the regenerative phase in the JB-1 ascites tumour. *Cell Tissue Kinet.* **7**, 47–60.

—— and Hartmann, N. R. (1972). Analysis of variations in the population kinetics with tumour age in the L1210 ascites tumour. *Cancer Res.* **32**, 2452–8.

Domingo, J., Serratosa, J., Vidal, C., and Ruis, E. (1978). Mitotic coincidence of chick embryo hepatocytes *in vivo* and the transition probability model of the cell cycle. *Nature, Lond.* **273**, 50–2.

Domon, M. (1980). A biological variability model of cell survival curves. *Radiat. Res.* **82**, 611–15.

Douglas, B. G., Henkelman, K. M., Lear, G. K. Y., Fowler, J., and Eaves, C. J. (1979). Practical and theoretical considerations on the use of the mouse foot system to derive epithelial stem cell survival parameters. *Radiat. Res.* **77**, 453–71.

Downes, R. M., Matoltsky, A., and Sweeney, T. M. (1967). Rate of turnover of the stratum corneum in hairless mice. *J. invest. Derm.* **49**, 400–6.

Drewinko, B., Roper, P. K., and Barlogie, B. (1979). Patterns of cell survival following treatment with antitumour agents *in vitro*. *Eur. J. Cancer* **15**, 93–9.

DuBrule, E. F. (1972). Fine structure of epidermal differentiation in the mouse. *J. exp. Zool.* **181**, 145–58.

Duell, E. A., Kelsey, W. A., and Vorhees, J. J. (1975). Epidermal chalone – past to present concepts. *J. invest. Derm.* **65**, 67–70.

—— Vorhees, J. J., Kelsey, W. A., and Hayes, E. (1971). Isoproterenol-sensitive adenyl cyclase in a particulate fraction of epidermis. *Archs Derm.* **104**, 601–9.

Duffill, M., Appleton, D. R., Shuster, S., Dyson, P., and Wright, N. A. (1977). The measurement of the cell cycle time in squamous epithelium using the metaphase arrest method with vincristine. *Br. J. Derm.* **96**, 493–502.

—— Wright, N. A., and Shuster, S. (1976). The cell proliferation kinetics of psoriasis examined by three different *in vivo* techniques. *Br. J. Derm.* **94**, 355–62.

Du Shane, G. I. P. (1936). Dopa reaction in amphibia. *Proc. Soc. exp. Biol. Med.* **33**, 892–5.

Duuren Van, B. L. (1969). Tumour promoting agents in two stage carcinogenesis. *Prog. exp. Tumour Res.* **11**, 31–68.

du Vivier, A., Bible, R., Mikiuriyn, R. K., and Stoughton, R. (1976). An animal model for screening drugs for anti-psoriatic properties using hydroxyapatite to isolate DNA rapidly from the epidermis. *Br. J. Derm.* **94**, 1–6.

—— and Stoughton, K. B. (1975). An animal model of psoriasis and its use for screening topical and systemic agents for antiproliferative properties in the skin. *J. invest. Derm.* **65**, 235–7.

Eaglestein, W. H. and Weinstein, G. D. (1975). Prostaglandins and DNA synthesis in human skin; possible relation to UV light effects. *J. invest. Derm.* **64**, 386–9.

—— —— and Frost, P. (1970). Fluorouracil; mechanism of action in human skin and actinic keratosis. 1. Effect on DNA synthesis *in vivo*. *Archs. Derm.* **101**, 132–9.

Ebbeson, S. O. E. and Tang, D. (1965). A method for estimating the number of cells in histological sections. *J. R. Microsc. Soc.* **84**, 449–64.

Edwards, J. L. and Klein, R. E. (1961). Cell renewal in adult mouse tissues. *Am. J. Path.* **38**, 437–53.

Eigstei, O. J. and Dustin, P. (1955). *Colchicine in agriculture, medicine, biology and chemistry*. Iowa State College Press, Ames, Iowa.

Elder, J. B., Williams, G., Lacey, E., and Gregory, H. (1978). Cellular localisation of human urogastrone/epidermal growth factor. *Nature, Lond.* **271**, 466–7.

Elgjo, K. (1968*a*). Epidermal cell population kinetics after repeated applications of some hyperplasia producing substances. *Eur. J. Cancer* **4**, 183–92.

—— (1968*b*). Epidermal cell population kinetics after a single application of some hyperplasia producing substances. *Eur. J. Cancer* **3**, 519–27.

—— (1968*c*). Growth kinetics of the mouse epidermis after a single application of cigarette smoke condensate. *Acta path. microbiol. scand.* **73**, 316–22.

—— (1969*a*). Epidermal cell proliferation during the first 24 h after injection of an aqueous skin extract (chalone). *Virchows Arch. Cell Path.* **4**, 119–25.

—— (1969*b*). The stability of the epidermal mitosis inhibiting factor (chalone) in water solution. *Acta path. microbiol. scand.* **76**, 31–4.

—— (1972). Chalone inhibition of cellular proliferation. *J. invest. Derm.* **59**, 81–3.

—— (1973). Epidermal chalone: cell cycle specificity of two epidermal growth inhibitors. *Natl. Cancer Inst. Monogr.* **38**, 71–6.

—— (1974*a*). Reversible inhibition of epidermal $G_1$ cells by repeated injections of aqueous skin extracts (chalone). *Virchows Arch. Cell Path.* **15**, 157–63.

—— (1974*b*). Evidence for the presence of epidermal $G_2$ inhibitors (epidermal chalone) in dermis. *Virchows Arch. Cell Path.* **16**, 243–7.

—— (1975). Epidermal chalone and cyclic AMP: an *in vivo* study. *J. invest. Derm.* **64**, 14–18.

—— and Cromarty, A. (1977). Growth kinetics in newborn mouse epidermis: response to epidermal chalone. *Virchows Arch. Cell Path.* **24**, 101–8.

—— and Degre, M. (1975). Polyinosinic-polycytidylic acid inhibition of cell proliferation in carcinogen treated epidermis and in carcinogen treated skin tumours in the mouse. *J. natn. Cancer Inst.* **54**, 219–21.

— and Devik, F. (1978). Growth regulation in X-irradiated mouse skin: the probable site of chalone. *Int. J. radiat. Biol.* **34**, 119–26.

— and Edgehill, W. (1973). Epidermal growth factor inhibitors (chalones) in dermis and serum. *Virchows Arch. Cell Path.* **13**, 14–23.

— and Hennings, H. (1971). Epidermal mitotic rate and DNA synthesis after injection of water extracts made from mouse skin treated with actinomycin D. Two or more growth regulating substances. *Virchows Arch. Cell Path.* **7**, 342.

— — Michael, D., and Yuspa, S. H. (1976). Natural synchrony of newborn mouse epidermal cells *in vitro*. *J. invest. Derm.* **66**, 292–6.

— Laerum, O. D., and Edgehill, W. (1971). Growth regulation in mouse epidermis. 1. $G_2$ inhibitors present in the basal cell layer. *Virchows Arch. Cell Path.* **8**, 277–87.

— — — (1972). Growth regulation in mouse epidermis II $G_1$ inhibitor present in the differentiating cells. *Virchows Arch. Cell Path.* **10**, 229–36.

— and Skjaeggstad, O. (1965). Cell population kinetics in the mouse epidermis after repeated applications of 20-methylcholanthrene. *Acta path. microbiol. scand.* **64**, 185–92.

Elkind, M. M., Hau, A., and Volz, K. W. (1963). Radiation response of mammalian cells grown in culture. IV. Dose dependence of division delay and post-irradiation growth of surviving and non-surviving Chinese hamster cells. *J. natl. Cancer Inst.* **30**, 705–21.

Ellis, J. N. J. (1980). Quantitative studies on differentiating cell populations exposed to radiation (using the fern *Osmunda regalis* and the murine intestine). Ph.D. thesis, University of Manchester.

Emery, E. W., Denekamp, J., and Ball, M. M. (1970). Survival of mouse skin epithelial cells following single and double doses of X-rays. *Radiat. Res.* **41**, 450–66.

Enesco, M. and Leblond, C. P. (1962). Increase in cell number as a factor in the growth of organs and tissues in the young male rat. *J. Embryol. exp. Morphol.* **10**, 530–62.

England, M. C. and Burke, G. W. (1968). Diurnal variation in deoxyribonucleic acid synthesis in selected epithelial tissues of male rats. *J. dent. Res.* **47**, 478–81.

— — (1969). Autoradiographic demonstration of DNA synthesis in selected epithelial tissues in male rats. *J. dent. Res.* **48**, 1219–23.

— — (1971). Diurnal variation in DNA synthesis in hamster oral epithelium. *J. dent. Res.* **50**, 976.

Engler, W. O., Ramjford, S. P., and Hiniker, J. J. (1965). Development of epithelial attachment and gingival sulcus in rhesus monkeys. *J. Peridontol.* **36**, 44–56.

Epifanova, O. I. and Terskikh, V. V. (1969). On the resting periods in the cell life cycle. *Cell Tissue Kinet.* **2**, 75–93.

Epstein, J. H. and Fukuyama, K. (1975). Effects of 8-methoxy psoralen-induced phototoxic effects on mammalian epidermal macromolecular synthesis in mice. *Photochem. Photobiol.* **21**, 325–30.

— — and Epstein, W. L. (1968). UVL-induced stimulation of DNA synthesis in hairless mouse epidermis. *J. invest. Derm.* **51**, 445–53.

— — and Fye, K. (1970). Effects of UV-irradiation on the mitotic cycle and DNA, RNA and protein synthesis in mammalian epidermis *in vivo*. *Photochem. Photobiol.* **12**, 57–65.

Epstein, W. and Maibach, H. (1965). Cell renewal in human epidermis. *Archs Derm.* **92**, 462–8.

— and Sullivan, D. J. (1963). Epidermal mitotic activity in wounded human skin. In *Advances in the biology of the skin* (ed. W. Montagna and R. E. Billingham). Pergamon Press, New York.

Estensen, R. D., Hadden, J. W., Hadden, E. M., Touraine, F., Touraine, J. L., Haddox, M. K., and Goldberg, N. D. (1974). Phorbol myristate acetate: effects of a tumour promoter on cyclic GMP in mouse fibroblasts and as a mitogen in human lymphocytes. *Cold Spring Harbour Symposia* **6**, 627–34 (ed. C. Clarkson and R. Baserga).

Etoh, H., Taguchi, Y. H., and Tabachnik, J. (1973). Movement of epidermal basal cells to the spinous layer in the absence of division pressure. *J. cell Biol.* **59**, 95A.

— — — (1977). Cytokinetics of regeneration in $\beta$-irradiated guinea pig epidermis. *Radiat. Res.* **71**, 109–18.

Evans, C. H. (1978). Is the A ($G_0$) state time independent? *J. theoret. Biol.* **79**, 259–62.

Evensen, A. (1961). Changes in the synthesis of deoxyribonucleic acid (DNA) and on mitotic count in epidermis of hairless mice after a single application of 1% 3-methylcholanthrene in benzene. *Acta path. microbiol. scand.* Suppl. **148**, 43–52.

— (1963). Kinetics of epidermal cell proliferation in experimental skin carcinogenesis. *Bull. Wld Hlth Org.* **28**, 513–15.

— (1964). The effect of adrenalin on the mitotic rate of the epidermis of hairless mice. *Acta path. microbiol. scand.* **61**, 55–69.

— (1965). Significance of mitotic duration in evaluating kinetics of cellular proliferation. *Nature, Lond.* **195**, 718–19.

— and Heldaas, O. (1964). The effect of adrenaline on the mitotic rate in the epidermis of hairless mice *in vitro*. *Acta path. microbiol. scand.* **62**, 24–8.

Fau, J., Schoenfeld, R. J., and Hunter, R. (1959). A study of the epidermal clear cells with special reference to their relationship to the cells of Langerhans. *J. invest. Derm.* **32**, 445–9.

Fausto, N., Uchiyama, T., and Van Lancker, J. L. (1964). Metabolic alterations after total body doses of X-radiation. *Archs Biochem. Biophys.* **106**, 447–54.

Fialkow, P. J. (1976). Clonal origin of human tumours. *Biochim. Biophys. Acta* **458**, 283–321.

— Martin, G. M., and Klein, G. (1972). Evidence for a clonal origin of head and neck tumours. *Int. J. Cancer* **9**, 133–42.

Field, S. B., Morris, C., and Denekamp. J. (1975). The response of mouse skin to fractionated X rays. *Eur. J. Cancer* **11**, 291–9.

Finegold, M. J. (1965). Control of cell multiplication in epidermis. *Proc. Soc. exp. Biol. Med.* **119**, 96–100.

Fisher, L. B. (1968*a*). Determination of the normal rate and duration of mitosis in human epidermis. *Br. J. Derm.* **80**, 24–33.

— (1968*b*). The diurnal mitotic rhythm in the human epidermis. *Br. J. Derm.* **80**, 75–80.

— (1972). The effect of occlusive and semipermeable dressings on the mitotic activity of normal and wounded epidermis. *Br. J. Derm.* **86**, 593–600.

— and Maibach, H. I. (1971*a*). The effect of corticosteroids on human epidermal mitotic activity. *Archs Derm.* **103**, 39–44.

— — (1971*b*). Topically active chemotherapy agents in psoriasis; antimitotic effects on human skin. In *Psoriasis* (ed. E. M. Farber and A. J. Cox) pp. 331–45. Stanford University Press.

—— (1972). The effect of occlusive and semipermeable dressings on the cell kinetics of normal and wounded epidermis. In *Epidermal wound healing* (ed. H. I. Maibach and D. Rovee) pp. 113–131. Year Book, Chicago.

—— (1975). The effect of anthralin and its derivatives on epidermal cell kinetics. *J. invest. Derm.* **64**, 339–41.

— and Wells, G. C. (1968). The mitotic rate and duration in lesions of psoriasis and ichthyosis. *Br. J. Derm.* **80**, 235–40.

Fitzgerald, P. H. and Brehaut, L. A. (1970). Depression of DNA synthesis and mitotic index by colchicine in cultured human lymphocytes. *Expl Cell Res.* **59**, 27–38.

Fitzgerald, P. J. (1970). Biochemical mechanisms involved in regeneration. *Fedn. Proc. Fedn Am. Socs exp. Biol.* **29**, 1429–32.

Fitzpatrick, T. B. and Breathnach, A. S. (1963). Das epidermale Melanin Einheit System. *Derm. Wschr.* **147**, 481.

Flaxman, B. A. and Chopra, D. P. (1972). Cell cycle of normal and psoriatic epidermis *in vitro. J. invest. Derm.* **59**, 102.

—— and Chiarella, S. (1977). Effects of methotrexate on proliferation of human keratocytes *in vitro. J. invest. Derm.* **68**, 66–9.

— and Harper, R. A. (1975). *In vitro* analysis of the control of keratinocyte proliferation in human epidermis by physiologic and pharmacologic agents. *J. invest. Derm.* **65**, 52–9.

Floderus, S. (1944). Untersuchungen uber den Bau der menschlichen Hypophyse mit besonderer Berucksichtigung der quantitativen microrphologischen Verhaltnisse. *Acta path. microbiol. scand.* **51**, Suppl. 59, 93.

Forsum, U., Klareskog, L., Tjernland, N. A., and Peterson, P. A. (1978). Significance of expression of HLA-DR antigen on epidermal Langerhans cells. *Acta derm.* **58**, 37–40.

Fowler, J. and Denekamp. J. (1976). Regulation of epidermal stem cells. In *Stem cells of renewing cell populations* (ed. A. B. Cairnie, P. K. Lala, and D. G. Osmond) pp. 117–34. Academic Press, New York.

—— and Delapeyre, C. (1974). Skin reactions in mice after multifraction X-irradiation. *Int. J. radiat. Biol.* **25**, 213–23.

Francis, A. J. and Marks, R. (1977*a*). The effects of antiprostaglandin agents on epidermal proliferation induced by dermal inflammation. *Br. J. Derm.* **97**, 395–400.

—— (1977*b*). Skin stretching and epidermopoiesis. *Br. J. exp. Path.* **58**, 35–9.

Frankfurt, O. S. (1966). Mitotic cell cycle and cell differentiation in squamous cell carcinoma. *Int. J. Cancer* **2**, 304–10.

— (1967). Cell proliferation and differentiation in the squamous epithelium of the forestomach of the mouse. *Expl Cell Res.* **46**, 603–6.

— (1968). Effect of hydrocortisone, adrenalin and actinomycin D on transition of cells to the DNA synthesis phase. *Expl Cell Res.* **52**, 222–32.

— (1971). Epidermal chalone-effect on cell cycle and development of hyperplasia. *Expl Cell Res.* **64**, 140–4.

— and Raitcheva, E. (1972). Effect of tumour promoters on cell kinetics in mouse epidermis. *J. natn. Cancer Inst.* **49**, 131–7.

—— (1973). Fast onset of DNA synthesis stimulated by tumor promoter in mouse epidermis at the initiation phase of carcinogenesis. *J. natn. Cancer Inst.* **51**, 1861–4.

Frati, L., D'Asarceito, M., Gulletta, E., Verna, A., and Conelli, I. (1977). The control of epidermis proliferation by epidermal growth factor (EGF). Relationship with cyclic nucleotides systems. *Pharmacol. Res. Commun.* **9**, 815–21.

Frei, E. W., Whang, J., Scoggins, R. B., Van Scott, E. J., Rall, D. P., and Ben, M. (1964). The stathmokinetic effect of vincristine. *Cancer Res.* **24**, 1918–22.

Frei, J. V. (1977). Comment on epidermal hyperplasia and tumour promotion. *J. natn. Cancer Inst.* **59**, 299.

— and Ritchie, A. C. (1964). Diurnal variation in the susceptibility of mouse epidermis to carcinogen and its relationship to DNA synthesis. *J. natn. Cancer Inst.* **32**, 1213–20.

— and Stephens, P. (1968). The correlation of promotion of tumour growth and induction of hyperplasia in epidermal two stage carcinogenesis. *Br. J. Cancer* **22**, 83–92.

— Waugh, W. O., and Ritchie, A. C. (1963). Mitosis: distribution in mouse ear epidermis. *Science, NY* **140**, 487.

Frelinger, J. G., Hood, L., Hill, S., and Frelinger, J. A. (1979). Mouse epidermal Ia molecules have a bone marrow origin. *Nature, Lond.* **22**, 324–6.

Frentz, G., Møller, U., and Larsen, J. K. (1980). DNA flow cytometry in human epidermis I. Methodological studies on normal skin. *J. invest. Derm.* **74**, 119–21.

Fried, J. (1969). Correction for threshold error in the determination of generation time by the grain count halving method. *J. theoret. Biol.* **24**, 108–19.

— (1976). Method for the quantitative evaluation of data from flow microfluometry. *Comp. Biomed. Res.* **9**, 263–9.

Friedland, A. M. and Weinstein, G. D. (1977). Cell proliferation in human cutaneous squamous cell carcinoma. *J. natn. Cancer Inst.* **59**, 3–6.

Friedman, M., Nervi, C., and Casale, C. (1973). Significance of growth rate, cell kinetics and histology in the irradiation and chemotherapy of squamous carcinoma of the mouth. *Cancer* **31**, 10–16.

Friedman-Kien, A. E., Morrill, P., Prose, H., and Liebhaber, H. (1966). Culture of adult human skin; *in vitro* growth and keratinisation of epidermal cells. *Nature, Lond.* **212**, 1583–4.

Frindel, E., Malaise, E., and Tubiana, M. (1968). Cell proliferation kinetics in 5 human solid tumours. *Cancer* **22**, 611–20.

Froelich, J. E. and Rachmeler, M. (1974). Inhibition of cell growth in $G_1$ phase by adenosine 3' 5' cyclic monophosphate. *J. cell Biol.* **60**, 249–57.

Frost, P. and Van Scott, E. (1966). Icthyosiform dermatoses. *Archs Derm.* **94**, 113–26.

— and Weinstein, G. D. (1969). Topical administration of vitamin A acid for icthyosiform dermatosis and psoriasis. *J. invest. Derm.* **19**, 1863.

— — and Van Scott, E. J. (1966). The ichthyosiform dermatoses II. Autoradiographic studies of epidermal proliferation. *J. invest. Derm.* **47**, 561–5.

Fry, L. and McMinn, R. M. H. (1966). Action of methotrexate on skin and intestinal epithelium in psoriasis. *Archs Derm.* **93**, 726–30.

— — (1968). The action of chemotherapeutic agents on psoriatic epidermis. *Br. J. Derm.* **80**, 373–83.

Fukuda, M., Okamura, L., Fujita, S., Bohm, N., Rohrbach, R., and Sandritter, W. (1978a). The different stem cell populations in the mouse epidermis and lingual epithelium. *Path. Res. Pract.* **163**, 205–27.

— — Rohrbach, R., Bohm, N., and Fujita, S. (1978b). Changes in cell population kinetics during epidermal carcinogenesis. *Cell Tissue Kinet.* **11**, 611–21.

Fukuyama, K. and Epstein, W. L. (1966). Epidermal keratinisation: localisation of isotopically labelled amino acids. *J. invest. Derm.* **47**, 551–9.

Fulker, M. J., Cooper, E. A., and Tanaka, T. (1971). Proliferation and ultrastructure of papillary transitional carcinoma of the human bladder. *Cancer* **27**, 71–7.

Furstenburger, G. and Marks, F. (1980). Early prostaglandin E synthesis is an obligatory event in the induction of cell proliferation in mouse epidermis *in vivo* by the phorbol ester TPA. *Biochem. Biophys. Commun.* **92**, 749.

Gahan, P. B. (ed.). (1972). *Autoradiography for biologists.* Academic Press, London.

Galand, P., Rodesch, F., Leroy, F., and Chretien, J. (1967*a*). Radiographic evaluation of the estrogen-dependent proliferative pool in the stem cell compartment of the mouse uterine and vaginal epithelia. *Expl Cell Res.* **48**, 595–604.

—— —— —— —— (1967*b*). Altered duration of DNA synthesis and cell cycle in non-target tissues of mice treated with oestrogens. *Nature, Lond.* **216**, 1211–12.

Galosi, H. (1980). Abnormal cell proliferation on the elbow in psoriatic and normal skin. *Archs Derm. Res.* **267**, 105–7.

Ganguli, P. C., Polak, J. M., Pearse, A. G. E., Elder, J. B., and Hegarty, M. (1974). Antral G cell hyperplasia in peptic ulcer disease; a new clinical entity. *Lancet* i, 583–6.

Garber, B. B. (1976). Control of epithelial development. *Curr. Probl. Derm.* **6**, 154–90.

Garcia-Arce, H. and Mizell, S. (1972). Mitotic activity in dorsal epidermis of *Rana pipiens. Comp. Biochem. Physiol. A* **42**, 501–9.

Garte, S. J. and Belman, S. (1980). Diurnal variation in cyclic nucleotide levels in normal and phorbol myristate acetate treated mouse epidermis. *J. invest. Derm.* **74**, 224–5.

Gasser, R. F., Scheving, L. E., and Pauly, J. E. (1972). Circadian rhythms in the mitotic activity of the basal epithelium and the uptake of $^3$H-thymidine by the tongue of the rat. *J. cell Physiol.* **80**, 437–42.

Gaulden, M. E. and Carlson, J. G. (1951). Cytological effects of colchicine on the grasshopper neuroblast in vitro with special reference to the origin of the spindle. *Expl Cell Res.* **2**, 416–23.

Gaylarde, P. M., Brock, A. P., and Sarkany, I. (1978). Psoriasiform changes in guinea pig skin from propanolol. *Clin. exp. Derm.* **3**, 157–60.

—— and Sarkany, I. (1975). Cell migration and DNA synthesis in organ culture of human skin. *Br. J. Derm.* **92**, 375–80.

Gelfant, S. (1959). A study of mitosis *in vitro. Expl Cell Res.* **16**, 527–36.

—— (1962). Initiation of mitosis in relation to the cell division cycle. *Expl Cell Res.* **26**, 395–403.

—— (1963*a*). A new theory on the mechanism of cell division. *Symp. Int. Soc. cell Biol.* **2**, 229–37.

—— (1963*b*). Inhibition of cell division: critical and experimental analysis. *Int. Rev. Cytol.* **14**, 1–39.

—— (1963*c*). Patterns of epidermal cell proliferation. I. Genetic behaviour of $G_1$ cell population. *Expl Cell Res.* **32**, 521–8.

—— (1966). Patterns of cell division: the demonstration of discrete cell populations. In *Methods in cell physiology* (ed. D. Prescott) Vol. 25, pp. 359–95. Academic Press, New York.

—— (1975). Temperature-induced cell proliferation in mouse ear epidermis *in vivo. Expl Cell Res.* **90**, 458–61.

— (1976). The cell cycle in psoriasis: a reappraisal. *Br. J. Derm.* **95**, 577–90.

— (1977). A new concept of tissue and tumour cell proliferation. *Cancer Res.* **37**, 3845–62.

— (1978*a*). Tissue and tumour cell proliferation. In *Cell reproduction* (ed. E. R. Dirksen, D. Prescott, and M. Fox) pp. 689–99. Academic Press, New York.

— (1978*b*). Understanding chemotherapy of psoriasis in terms of epidermal cell proliferation. *Cell Tissue Kinet.* **11**, 577–8.

— (1981). Cycling-non-cycling cell transitions in tissue aging, immunological surveillance, transformation and tissue growth. *Int. Rev. Cytol.* **70**, 1–26.

— and Candelas, G. C. (1972). Regulation of epidermal mitosis. *J. invest. Derm.* **59**, 7–11.

— Drewinko, B., Chalker, D., and Eisinger, G. (1983). Non-cycling–cycling cell transitions in psoriatic epidermis. In *Psoriasis: cell proliferation* (ed. N. A. Wright and R. S. Camplejohn). Churchill Livingstone, Edinburgh.

— and Smith, J. G. (1972). Aging: non-cycling cells an explanation. *Science, NY* **178**, 357–61.

Gerstein, W. (1971). Cell proliferation in human fetal epidermis. *J. invest. Derm.* **57**, 262–5.

Ghadially, F. N. and Green, H. M. (1954). The effect of cortisone on chemical carcinogenesis in mouse skin. *Br. J. Cancer* **8**, 291–5.

Giacometti, L. (1967). The healing of skin wounds in primates. I. The kinetics of cell proliferation. *J. invest. Derm.* **48**, 133–7.

— (1969). The healing of skin wounds in primates III. Behaviour of the cells of Langerhans. *J. invest. Derm.* **53**, 151–5.

— and Allegra, F. (1967). The effect of wounding upon the uptake of $^3$H-thymidine by guinea pig epidermal melanocytes. *Adv. Biol. Skin* **8**, 89–95.

— and Montagna, W. (1967). Langerhans cells: uptake of tritiated thymidine. *Science, NY* **157**, 439–40.

— — (1968). The healing of skin wounds in primates. II. The proliferation of epidermal melanocytes. *J. invest. Derm.* **50**, 273–5.

— — (1969). Healing of skin wounds in primates. In *Repair and regeneration* (ed. T. E. Dunphy and H. W. VanWinkle) pp. 47–56. McGraw Hill, New York.

— — Bell, M., and Hu, F. (1972). Epidermal melanocyte proliferation in skin wound healing after UV-irradiation in Rhesus monkeys. In *Pigmentation; its genesis and biologic control* (ed. V. Riley) pp. 451–61. Appleton Century Crofts, New York.

Gibbins, J. R. (1969). Migration of stratified squamous epithelium after injury *in vivo* and in organ culture. *Pathology* **1**, 225–38.

Gibbs, S. and Casserett, G. (1969). Influences of a circadian rhythm and mitotic delay from tritiated thymidine on cytokinetic studies in hamster cheek pouch epithelium. *Radiat. Res.* **40**, 588–600.

— — (1972). Spatial distribution of cells in mitotic and DNA synthesis phases of the cell cycle in hamster cheek pouch epithelium. *J. dent. Res.* **51**, 30–9.

Gilbert, C. W. (1972). The labelled mitoses curve and the estimation of the parameters of the cell cycle. *Cell Tissue Kinet.* **5**, 53–63.

— and Lajtha, L. G. (1965). The importance of cell population kinetics in determining response to irradiation of normal and malignant tissues. In *Cellular radiation biology*, p. 474. The University of Texas M. D. Anderson Hospital Tumour Institute.

Gilman, A. G. (1970). A protein binding assay for adenosine 3′, 5′ monophosphate. *Proc. natn. Acad. Sci. USA* **67**, 305–12.

Goerttler, K., Haag, D., and Tschargane, C. (1970). Differences with age in nucleic acid content of cell nuclei in human skin. *Naturwissenschaften* **57**, 249–50.

— Reuter, M., and Strabener, H. E. (1973). Morphologische Untersuchungen zur Proliferationskinetik der Mausehaut. *Z. Zellforsch. Mikroscop. Anat.* **142**, 131–46.

Goldberg, N. D., Haddox, M. K., Durham, E., Lopez, C., and Hadden, J. W. (1974). The Yin Yang hypothesis of biological control: opposing influences of cyclic AMP in the regulation of cell proliferation and other biological processes. In *Cold Spring Harbour Symposia* **6**, 609–25 (ed. B. Clarkson and R. Baserga).

— — Narble, D. K., and Hadden, J. W. (1973). The biological role of cyclic 3′, 5′ guanosine monophosphate. *Proc. 5th Int. Congr. Pharmacol.* **5**, pp. 146–69. Karger, Basel.

Goldschmidt, H. and Kligman, A. M. (1967). Exfoliative cytology of human horny layer. *Archs Derm.* **96**, 572–9.

Goldsmith, D. P. J. (1973). Changes in desquamation rate of jejunal epithelium in cats during fasting. *Digestion* **8**, 130–41.

Goldspink, D. F. and Goldberg, A. L. (1973). Problems in the use of (me-³H) thymidine for the measurement of DNA synthesis. *Biochim. biophys. Acta* **299**, 521–32.

Goodenough, D. A. and Revel, J. P. (1970). A fine structural analysis of intercellular junctions in the mouse liver. *J. cell Biol.* **45**, 272–90.

Goodwin, P. G. and Fry, L. (1974). Turnover times in psoriasis. *Proc. R. Soc. Med.* **67**, 598–600.

— — (1977). The cell cycle time in psoriasis. The uninvolved epidermis. *Clin. exp. Derm.* **2**, 259–64.

— Hamilton, S., and Fry, L. (1973). Comparison between DNA synthesis and mitosis in uninvolved and involved psoriatic epidermis and normal epidermis. *Br. J. Derm.* **89**, 619–28.

— — — (1974). The cell cycle time in psoriasis. *Br. J. Derm.* **90**, 517–25.

— — — (1975). On the cell cycle time in psoriasis. *Br. J. Derm.* **92**, 231–2.

Gospadarowicz, D., Vlodavsky, I., Savion, N., and Johnson, L. K. (1980). The effect of EGF on cell proliferation and gene expression. In *Control mechanisms in animal cells.* (ed. L. Jeminez de Asna, R. Levi-Montalini, R. Shields, and S. Iacobelli) pp. 61–83. Raven Press, New York.

Goss, R. J. (1964). *Adaptive growth.* Logos, Academic Press, London.

— (1965). Kinetics of compensatory growth. *Q. Rev. Biol.* **40**, 123–46.

— (1978). *The physiology of growth.* Academic Press, New York.

Gradwohl, P. R. (1978*a*). The proliferation of epidermal cells in mouse ear in organ culture. *Archs derm. Res.* **263**, 273–81.

— (1978*b*). Mouse epidermal extracts tested for cytostatic activity (chalones). *Archs derm. Res.* **263**, 283–95.

Graem, N. (1979). Mitotic activity and delay in fixation of tumour tissue. *Acta path. microbiol. scand.* **87A**, 375–8.

Gray, J. W. (1976). Cell cycle analysis of perturbed cell populations: computer simulation of sequential DNA distributions. *Cell Tissue Kinet.* **9**, 499–512.

Gray, L. H. and Scholes, M. E. (1951). Effect of ionising radiations on broad bean root. *Br. J. Radiol.* **24**, 348–52.

Greaves, M. W. (1980). Lack of specific effect of topically applied epidermal growth factor (EGF) on epidermal growth in man *in vivo*. *Clin. exp. Derm.* **5**, 101–3.

Green, H. (1979). Growth of cultured human epidermal cells into multiple epithelia suitable for grafting. *Proc. natn. Acad. Sci. USA* **76**, 5665–8.

Green, P. J. (1980). A 'random transition' in the cell cycle? *Nature, Lond.* **285**, 116–18.

Gregory, H. (1975). Isolation and structure of urogastrone and its relationship to epidermal growth factor. *Nature, Lond.* **257**, 325–7.

Greulich, R. C. (1964). Aspects of cell individuality in the renewal of stratified squamous epithelium. In *The epidermis* (ed. W. Montagna and W. C. Lobitz) pp. 117–33. Academic Press, New York.

— Cameron, I. L., and Thrasher, J. D. (1961). Stimulation of mitosis in adult mice by administration of thymidine. *Proc. natn. Acad. Sci. USA* **47**, 743–6.

Grimm, W. and Marks, F. (1974). Effect of tumour-promoting phorbol esters on the normal and the isoproterenol elevated level of adenosine $3', 5'$ cyclic monophosphate in mouse epidermis *in vivo*. *Cancer Res.* **34**, 3128–34.

Grove, G. L. (1979). Epidermal cell kinetics in psoriasis. *Int. J. Derm.* **18**, 111–22.

— (1980). Proliferative response to fluid injection in psoriatic and normal epidermis. *Br. J. Derm.* **102**, 281–3.

— Anderton, R. L., and Graham-Smith, J. (1976). Cytophometric studies of epidermal proliferation in psoriatic and normal skin. *J. invest. Derm.* **66**, 236–9.

— and Cristafalo, V. J. (1976). The 'transition probability model' and the regulation of proliferation of human diploid cell cultures during aging. *Cell Tissue Kinet.* **9**, 395–404.

— — (1978). Transition probability models and aging human diploid cell-cultures. *Cell. Biol. Int. Rep.* **2**, 185–8.

— and Graham-Smith, J. (1976). Examination of cell proliferation in normal and psoriatic epidermis by Feulgen DNA cytophotometry. In *Psoriasis* (ed. G. M. Farber and A. J. Cox) pp. 365–7. Stanford University Press.

Grube, D. D., Auerbach, H., and Brues, A. M. (1970). Diurnal variation in the labelling index of mouse epidermis: a double isotope autoradiographic demonstration of changing flow rates. *Cell Tissue Kinet.* **3**, 363–79.

Grunicke, H., Hirsch, F., Wolf, H., Bauer, U., and Keifer, G. (1975). Selective inhibition of thymidine transport at low doses of the alkylating agent tri-ethyleneiminobenzoquinone (Trenimon). *Expl Cell Res.* **90**, 346–53.

Gschnait, F. and Brenner, W. (1979). Kinetics of epidermal Langerhans cells. *J. invest. Derm.* **73**, 566–9.

Guidotti, D., Weiss, B., and Costa, E. (1972). Adenosine $3', 5'$-monophosphate concentrations and isoproterenol-induced synthesis of deoxyribonucleic acid in mouse parotid gland. *Molec. Pharmacol.* **8**, 521–30.

Guignet, M., Collyn-d'Hooghe, M., and Valleron, A. J. (1980). S, $G_2$ and M phase durations are relatively as variable as $G_1$ duration. *Cell Tissue Kinet.* **13**, 689–97.

— Klein, B., and Valleron, A. J. (1978). Diurnal variation and the analysis of percent labeled mitoses curves. In *Biomathematics and cell kinetics* (ed. A. J. Valleron and P. D. M. MacDonald) pp. 191–8. Elsevier–North Holland, Amsterdam.

Haag, D. (1975). Isolation of single cell nuclei from human epidermis for cyto-photometric DNA measurements. *Archs Derm. Res.* **253**, 301–10.

Hadden, J. W., Hadden, G. M., Haddox, M. K., and Goldberg, N. D. (1972). Guanosine 3′,5′ cyclic monophosphate: a possible intracellular mediator of mitogenic influences in lymphocytes. *Proc. natn. Acad. Sci. USA* **69**, 3024–7.

Haddow, A. (1935). Influence of certain polycyclic hydrocarbons on the growth of the Jensen rat sarcoma. *Nature, Lond.* **136**, 868–9.

— (1938). Cellular inhibition and the origin of cancer. *Acta Un. int. Canc.* **3**, 342–53.

Hagemann, R. F. and Lesher, S. (1971*a*). Intestinal crypt survival and total and percent levels of crypt proliferative cellularity following irradiation: age response and animal lethality. *Radiat. Res.* **47**, 159–67.

— — (1971*b*). Irradiation of the gastrointestinal tract; compensatory response of stomach, jejunum and colon. *Br. J. Radiol.* **44**, 599–602.

Halberg, F., Barnum. C. P. Silber, R. H., and Bitner, J. J. (1958). 24-hour rhythms at several levels of integration in mice on different lighting regimens. *Proc. Soc. exp. Biol. Med.* **97**, 897–900.

— Peterson, R. E., and Selber, R. H. (1959). Phase relations of 24 hour periodicities in blood corticosterone, mitoses in cortical adrenal parenchyma and total bodily activity. *Endocrinology* **64**, 222–30.

Hale, A. J., Cooper, E. H., and Milton, J. D. (1965). Studies of the incorporation of pyrimidines into DNA in single leukaemic and other proliferating leucocytes. *Br. J. Haematol.* **11**, 144–61.

Hall, E. J. (1975). Biological problems in the measurement of survival at low doses. In *Cell survival after low doses of irradiation* (ed. T. Alper) pp. 13–24. London Institute of Physics and Wiley, London.

Hall, R. G. (1969*a*). DNA synthesis in organ cultures of the hamster cheek pouch. Inhibition by homologous extract. *Expl Cell Res.* **58**, 429–31.

— (1969*b*). Regeneration-like responses in organ cultures of adult hamster cheek pouch tissue. *Expl Cell Res.* **55**, 323–9.

Hally, A. D. (1964). A counting method for measuring the volumes of tissue components in microscopical sections. *Q. J. microsc. Sci.* **105**, 503–14.

Halprin, K. M. (1972). Epidermal turnover times—a reexamination. *Br. J. Derm.* **86**, 14–29.

Hambrick, G. W. and Handwurger, R. L. (1969). The behaviour of psoriatic explants *in vitro*. *J. invest. Derm.* **52**, 126–35.

Hamilton, A, I. (1972). Keratoblast and keratocyte; not keratinocyte. *Nature, Lond.* **238**, 98.

— and Blackwood, H. J. J. (1974). Cell renewal of oral mucosa in the rat. *J. Anat.* **117**, 313–27.

Hamilton, E. and Dobbin, J. (1982). ³H-thymidine labels less than half the DNA synthesising cells in the mouse tumour carcinoma NT. *Cell Tissue Kinet.* **15**, 405–13.

— and Potten, C. S. (1972). Influences of hair plucking on the turnover time of the epidermal basal layer. *Cell Tissue Kinet.* **5**, 507–17.

Hammar, H. (1973). ATP and ADP levels and epidermal replacement in the normal skin and in some papulosquamous diseases of the skin. *Archs Derm. Vener.* **53**, 251–8.

Hammarstrom, S., Hamberg, M., Samuelsson, B., Duell, E. A., Stawiski, M., and Vorhees, J. J. (1975). Increased concentrations of non-esterified arachidonic acid, 12-hydroxy-5,8,10,14-eicotetranoic acid, prostaglandin E2 and prostaglandin F2α in epidermis of psoriasis. *Proc. natn. Acad. Sci. USA* **72**, 5130–4.

Hampton, J. C. (1968). Further evidence for the presence of a Paneth cell precursor in mouse intestine. *Cell Tissue Kinet.* **1**, 309–17.

Hanson, J. (1947). The histogenesis of the epidermis in the rat and mouse. *J. Anat.* **81**, 174–96.

Harkonen, M., Hopsu-Havu, V. K., and Raij, K. (1974). Cyclic adenosine monophosphate, adenyl cyclase and cyclic nucleotide phosphodiesterase in psoriatic epidermis, with a comment on urinary cAMP. *Acta derm. vener.* **54**, 13–21.

Harper, J. F. and Brooker, G. (1975). Femtomole sensitive radioimmunoassay for cyclic AMP and cyclic GMP after 2′ O acetylation by acetic anhydride in aqueous solution. *J. cyclic Nucleotide Res.* **1**, 207–18.

Harper, R. A. and Flaxman, B. A. (1974). Mitotic index of normal and psoriatic keratinocytes *in vitro* to compounds known to affect intracellular cAMP. *J. invest. Derm.* **62**, 384–7.

—— —— (1975). Effect of pharmacological agents on human keratinocyte mitosis *in vitro*. III. Inhibition by histamine and methylated analogues. *J. invest. Derm.* **65**, 400–3.

—— Rispler, J., and Urbaulk, W. R. (1978). DNA synthesis among involved and uninvolved psoriatic epidermal cells *in vitro*. *J. invest. Derm.* **70**, 254–6.

Harriss, E. B. and Hoelzer, D. (1971). DNA synthesis time of leukaemic cells as measured by the double labelling and percentage labelled mitoses methods. *Cell Tissue Kinet.* **4**, 433–41.

Hartmann, N. R., Gilbert, C. W., Jansson, B., MacDonald, P. D. M., Steel, G. G., and Valleron, A. J. (1975). A comparison of computer methods for the analysis of labelled mitoses curves. *Cell Tissue Kinet.* **8**, 119–24.

—— and Møller, U. (1978). A compartment theory in the cell kinetics including considerations on circadian variations. In *Biomathematics and cell kinetics* (ed. A. J. Valleron and P. M. MacDonald) pp. 223–51. Elsevier–North Holland, Amsterdam.

—— and Pedersen, T. (1970). Analysis of the kinetics of granulosa cell populations in the mouse ovary. *Cell Tissue Kinet.* **3**, 1–12.

Hartschuh, W. (1979). The Merkel cell—a member of the APUD cell system? *Archs derm. Res.* **265**, 115–22.

Hashimoto, K. (1972). The ultrastructure of the skin of human embryos. X. Merkel tactile cells in the finger and nail. *J. Anat.* **111**, 99–120.

—— and Tarnowski, W. M. (1968). Some aspects of the Langerhans cells. *Archs Derm.* **97**, 450–64.

Haustein, I. L. (1979). Epidermal DNA synthesis in organ culture explants. *Virchows Arch. Cell Path.* **31**, 181–91.

Hayflick, L. (1965). The limited *in vitro* lifetime of human diploid strains. *Expl Cell Res.* **37**, 614–36.

Hayes, R. L., Silberkweit, M., and Soni, N. N. (1964). Pattern of mitotic activity and cell densities in normal gingival epithelium of children. *J. dent. Res.* **43**, 217–23.

Heenen, M., Achten, G., and Galand, P. (1973). Autoradiographic analysis of cell kinetics in human normal epidermis and basal cell carcinoma. *Cancer Res.* **33**, 123–7.

—— DeMaertelaer, V., and Galand, P. (1979). Psoriasis and the cell cycle; computer simulation. *Acta derm. vener.* Suppl. **87**, 73–6.

—— and Galand, P. (1971). Cell population kinetics in human epidermis: *in vivo* autoradiographic study by double labelling methods. *J. invest. Derm.* **56**, 425–9.

— Lambert, J. C., Achten, G., and Galand, P. (1975). Kinetics of cell proliferation in benign and premalignant tumours of the human epidermis. *J. natn. Cancer Inst.* **54**, 825–7.

Hegazy, M. A. H. and Fowler, J. F. (1973*a*). Cell population kinetics of plucked and unplucked mouse skin. I. Unirradiated skin. *Cell Tissue Kinet.* **6**, 17–33.

— — (1973*b*). Cell population kinetics of plucked and unplucked mouse skin. II. Irradiated skin. *Cell Tissue Kinet.* **6**, 587–602.

Hell, E. A. (1963). The effect of injury on epidermal mitotic indices in guinea pigs. *Expl Cell Res.* **32**, 354–7.

— (1970). A stimulant to DNA synthesis in guinea pig epidermis. *Br. J. Derm.* **83**, 632–6.

— and Cox, D. (1963). Effects of colchicine and colcemid on synthesis of deoxyribonucleic acid in the skin of the guinea pig *in vitro*. *Nature, Lond.* **197**, 287–8.

— and Cruikshank, C. N. D. (1963). The effect of injury upon the uptake of $^3$H-thymidine by guinea pig epidermis. *Expl Cell Res.* **31**, 128–39.

— and Hodgson, C. (1966). The uptake of $^3$H-thymidine by epidermal cells in normal and psoriatic subjects. *Br. J. Derm.* **78**, 262–8.

— — (1978). Human epidermal stimulants, normal and psoriatic. *Clin. exp. Derm.* **3**, 269–78.

— — and Manna, V. (1979). Psoralen photochemotherapy of psoriasis. *Br. J. Derm.* **101**, 293–8.

— and Maibach, H. (1972). A comparison of *in vivo* and *in vitro* methods of identifying human epidermal cells in DNA synthesis. *Br. J. Derm.* **86**, 506–7.

Hellman, S. and Botnick, L. E. (1977). Stem cell depletion: an explanation of the late effects of cytotoxics. *Int. J. Rad. Oncol. Biol. Phys.* **2**, 181–4.

Hemingway, J. T. (1960). Withdrawal of inhibition of mitosis as a mechanism influencing normal, and pathological growth. *Nature, Lond.* **185**, 106–7.

Hendry, J. H., Edmundson, J. M., and Potten, C. S. (1980). The radiosensitivity of hair follicles in the mouse dorsum and tail. *Radiat. Res.* **84**, 87–96.

— and Potten, C. S. (1974). Cryptogenic and proliferative cells in intestinal epithelium. *Int. J. radiat. Biol.* **25**, 583–8.

Henkolow, J. S., Slaaf, D. W., and Lenn, J. C. (1972). Cell communication in the basal cells of the human epidermis. *Biophys. J.* **12**, 1266–84.

Hennings, H. and Boutwell, R. K. (1969). The inhibition of DNA synthesis by initiators of mouse skin tumorigenesis. *Cancer Res.* **29**, 510–14.

— — (1970). Studies in the mechanism of skin tumor promoters. *Cancer Res.* **30**, 312–20.

— and Devik, F. (1971). Comparison of cytotoxicity of hydroxyurea in normal and rapidly proliferating epidermis and small intestine in mice. *Cancer Res.* **31**, 277–82.

— and Elgjo, K. (1970). Epidermal regeneration after cellophane tape stripping of hairless mouse skin. *Cell Tissue Kinet.* **3**, 243–52.

— — (1971). Hydrocortisone: inhibition of DNA synthesis and mitotic rate after local application to mouse epidermis. *Virchows Arch. Cell Path.* **8**, 42–9.

— — and Iversen, O. H. (1969). Delayed inhibition of epidermal DNA synthesis after injection of an aqueous skin extract ('chalone'). *Virchows Arch. Cell Path.* **4**, 45–54.

Henry, J. L., Meyer, J., Weinmann, P., and Schour, I. (1952). Pattern of mitotic activity in oral epithelium of rabbits. *Archs Path.* **54**, 281–94.

Hermens, A. F. and Barendsen, G. W. (1969). Changes of cell proliferation characteristics in a rat rhabdomyosarcoma before and after irradiation. *Eur. J. Cancer* **5**, 173–89.

Hill, M. W. (1977). Histology of Langerhans cells in neonatal rat palatal mucosa. *Archs oral Biol.* **22**, 641–5.

Hilscher, W. and Maurer, W. (1962). Autoradiographische Bestimmung der Dauer des DNS-Verdopplung und ihres zeitlichen Verlaufs bei Spermatogonier durch Doppelmarkierung mit $^{14}$C and $^{3}$H thymidine. *Naturwissenschaften* **49**, 352.

Hoffman, J. G. (1949). Quantitative analysis of the growth of epidermis. *Archs Path.* **47**, 37–43.

— and Post, J. (1967). *In vivo* studies of DNA synthesis in human normal and tumour cells. *Cancer Res.* **27**, 898–902.

Hoffman-Fezer, G., Gotze, D., Rodt, H., and Thierfelder, S. (1978). Immuno-chemical localisation of xenogeneic antibodies against Iak antigens on B cells and reticular cells. *Immunogenetics* **6**, 367–77.

Holbrook, K. A. and Odland, G. F. (1974). Regional differences in the thickness (cell layers) of the human stratum corneum; an ultrastructural analysis. *J. invest. Derm.* **62**, 415–22.

— — (1975). The fine structure of developing human epidermis: light, scanning and transmission electron microscopy of the periderm. *J. invest. Derm.* **65**, 16–38.

Holt, P. J. and Marks, R. (1977). The epidermal response to a change in thyroid status. *J. invest. Derm.* **68**, 299–301.

Holtzer, H. (1978). Cell lineages, stem cells and the 'quantal cell cycle' concept. In *Stem cells and tissue homeostasis* (ed. C. S. Potten, B. I. Lord, and R. Schofield) pp. 1–27. Cambridge University Press.

— (1979). Comment and reply to Lajtha. *Differentiation* **14**, 33–4.

Hondius Boldingh, W. and Laurence, E. B. (1968). Extraction, purification and preliminary characterisation of the epidermal chalone: a tissue specific mitotic inhibitor obtained from vertebrate skin. *Eur. J. Biochem.* **5**, 191–8.

Hoober, J. K. and Cohen, S. (1967). Epidermal growth factor I. The stimulation of protein and ribonucleic acid synthesis in chick embryo epidermis. *Biochim. biophys. Acta* **138**, 347–56.

Hopper, J. L. and Brockwell, P. J. (1978). A stochastic model for cell populations with circadian rhythms. *Cell Tissue Kinet.* **11**, 205–17.

— Radley, J. M., and Hodgson, G. S. (1980). A random walk model for the analysis of the isoprenaline-stimulated response of the rat submaxillary gland. *Cell Tissue Kinet.* **13**, 519–33.

Hopsu-Havu, V. K., Harkonen, M., and Raij, K. (1973). cAMP, adenyl cyclase and cyclic nucleotide phosphodiesterase in psoriatic epidermis. *Eur. Soc. derm. Res. Abstr.* 4–5.

Hornstein, O. P. (1979). An autoradiographic analysis of the mode of proliferation in the buccal mucosa of rats incubated up to 5 hours. *Virchows Arch. Cell. Path.* **31**, 57–66.

Houck, J. C. and Dougherty, W. F. (1974). *Chalones: a tissue-specific approach to mitotic control*, p. 41. Medcom Press, New York.

— Irausquin, H., and Leikin, S. (1971). Lymphocyte DNA synthesis inhibition. *Science, NY* **173**, 1139–41.

Howard, A. and Dewey, D. L. (1961). Non-uniformity of labelling rate during DNA synthesis. *Expl Cell Res.* **24**, 623–8.

Hoyes, A. D. (1968). Electron microscopy of the surface layer (periderm) of human foetal skin. *J. Anat.* **103**, 321–33.

Hsai, S. L., Wright, R. L., Stephen, M. S., Mandy, H., and Halprin, K. M. (1972). Adenyl cyclase in normal and in psoriatic skin. *J. invest. Derm.* **59**, 109–13.

Hsu, T. H. and Hsu, J. M. (1972). Zinc deficiency and epithelial wound repair: an autoradiographic study of $^3$H-thymidine incorporation. *Proc. Soc. exp. Biol. Med.* **140**, 157–60.

Hume, W. J. and Potten, C. S. (1976). The ordered columnar structure of mouse filiform papillae. *J. cell Sci.* **22**, 149–60.

—— (1979). Advances in epithelial kinetics – an oral view. *J. oral Path.* **8**, 3–22.

—— (1980). Changes in proliferative activity as cells move along undulating basement membranes in stratified squamous epithelium. *Br. J. Derm.* **103**, 499–504.

—— (1982). A long-lived thymidine pool in epithelial stem cells. *Cell Tissue Kinet.* **15**, 49–58.

—— (1983). Proliferative units in stratified squamous epithelium. *Clin. exp. Derm.* **8**, 93–106.

Hunter, R., Pinkus, H., and Steele, C. H. (1956). Examination of the epidermis by the strip method. III. The number of keratin cells in the human epidermis. *J. invest. Derm.* **27**, 31–4.

Iannaconne, P. M., Gardiner, R. L., and Harris, H. (1978). The cellular origin of chemically induced tumours. *J. cell. Sci.* **29**, 249–69.

Iizuka, H., Adachi, K., Halprin, K. M., and Levine, V. (1977). Epidermal adenylate cyclase: stimulation of the histamine (H2) receptor by tolazoline. *J. invest. Derm.* **69**, 442–5.

———— (1978). Epinephrine activation of pig skin adenylate cyclase *in vivo* and subsequent refractoriness. *J. invest. Derm.* **70**, 119–22.

Inoué, S. (1952). The effect of colchicine on the microscopic and submicroscopic structure of the mitotic spindle. *Expl Cell Res.* Suppl. 2, 305–9.

Isaacson, P. (1981). Crypt cell carcinoma of the appendix (so-called adenocarcinoid tumour). *Am. J. Surg. Path.* **5**, 213–24.

Isaakson-Forsén, G., Burton, D. R., Korsgaard, R., Elgjo, K., and Iversen, O. H, (1977). Partial purification of the epidermal $G_2$ chalone. *Virchows Arch. Cell Path.* **26**, 97–104.

Itoh, S., Katsuwa, G., and Hirota, R. (1980). Conditioned circadian rhythm of plasma corticosterone in the rat induced by food restriction. *J. Physiol.* **30**, 365–75.

Iversen, O. H. (1961). The regulation of cell numbers in epidermis. A cybernetic point of view. *Acta path. microbiol. scand.* Suppl. **148**, 91–5.

— (1964). Discussion on cell destruction and population dynamics in experimental skin carcinogenesis in mice. *Prog. exp. Tumour Res.* **4**, 169–206.

— (1967). Kinetics of cellular proliferation and cell loss in human carcinoma. *Eur. J. Cancer* **3**, 389–99.

— (1968). Effect of epidermal chalone on human epidermal mitotic activity *in vitro*. *Nature, Lond.* **29**, 75–6.

— (1969). Chalones of the skin. In *Ciba Symposium on Homeostatic Regulators* (ed. T. Wolstenholme and J. Knight) pp. 29–40. Churchill Livingstone, Edinburgh.

— (1973). The chalones. *Acta path. microbiol. scand.* Sect. A. Suppl. 236, 71–6.

— (1978*a*). Epidermal chalones and squamous carcinomas. The growth inhibiting effects of aqueous epidermal extracts ($G_1$ and $G_2$ chalones) on the epidermis and on a transplantable squamous carcinoma in mice. *Virchows Arch. Cell Path.* **27**, 229–36.

— (1978*b*). The effect of the epidermal $G_2$ chalone on the mitotic duration in nude mouse epidermis and in a transplantable squamous cell carcinoma. *Virchows Arch. Cell Path.* **28**, 274–8.

— (1981). The chalones. In *Tissue growth factors* (ed. R. Baserga). Handbook of Experimental Pharmacology, Vol. 57. Springer, Berlin.

— Aarndahl, E., and Elgjo, K. (1968). The effect of an epidermis specific mitotic inhibitor (chalone) extracted from epidermal cells. *Acta path. microbiol. scand.* **64**, 506.

— Bhangoo, K. S., and Hansen, K. (1974*a*). Control of epidermal cell renewal in the bat web. *Virchows Arch. Cell Path.* **16**, 159–79.

— and Bjerknes, R. (1963). Kinetics of epidermal reaction to carcinogens. *Acta path. microbiol. scand.* Suppl. 165.

— — and Devik, F. (1968). Kinetics of cell renewal, cell migration and cell loss in the hairless mouse dorsal epidermis. *Cell Tissue Kinet.* **1**, 365–78.

— — Raknerud, N., and Sjaeggestad, O. (1971). The fate of the nucleus in hairless mouse epidermis. *Arch. Derm. Forsch.* **241**, 396–405.

— Clausen, O. P. F., Elgjo, K., Iversen, U. M., and Rohrbach, R. (1977). Effect of Bleomycin on the epidermal content of growth regulating substances (chalones). *Cell Tissue Kinet.* **10**, 71–9.

— — Iversen, U. M., and Rohrbach, R. (1976). Some effects of Bleomycin on proliferation, maturation time and protein synthesis of hairless mouse epidermis. *Cell Tissue Kinet.* **9**, 77–97.

— and Evensen, A. (1962). *Experimental carcinogenesis in mice.* Norwegian Universities Press.

— and Iversen, U. M. (1976). Is there a diurnal variation in the susceptibility of mouse skin to the tumourigenic action of methylcholanthrene. *Acta path. microbiol. scand. A* **84**, 406–15.

— Iversen, U., Zeigler, J. L., and Bluming, A. Z. (1974*b*). Cell kinetics in Burkitt's lymphoma. *Eur. J. Cancer* **10**, 155–63.

Izquierdo, J. N. (1977). Increased cell proliferation with persistence of circadian rhythms in hamster cheek pouch neoplasms. *Cell Tissue Kinet.* **10**, 313–22.

— and Gibbs, S. J. (1972). Circadian rhythms of DNA synthesis and mitotic activity in hamster cheek pouch epithelium. *Expl Cell Res.* **71**, 402–8.

— — (1974). Turnover of cell renewing populations undergoing circadian rhythms in cell proliferation. *Cell Tissue Kinet.* **7**, 99–111.

Jansen, L. H., Hojyo Tomoko, M. T., and Kligman, A. M. (1974). Improved fluorescent staining technique for estimating turnover of the human stratum corneum. *Br. J. Derm.* **90**, 9–12.

Jarrett, A. (1974). The epidermis. In *Physiology and pathophysiology of the skin.* Academic Press, New York.

Jellinghaus, W., Schultze, B., and Maurer, W. (1977). The effect of vincristine on mouse jejunal cells of differing cell age: double labelling experiments using [$^3$H] and [$^{14}$C] TdR. *Cell Tissue Kinet.* **10**, 147–62.

Johannesen, A. and Hammar, H. (1980). The rate of formation of corneocyte layers after stripping the non-involved psoriatic skin and its relation to parakeratosis and epidermal enzyme levels. *J. invest. Derm.* **74**, 226–9.

Johnson, F. R. and MacMinn, R. M. (1960). The cytology of wound healing of body surfaces in mammals. *Biol. Rev.* **35**, 364–412.

Johnson, H. A. and Cronkite, E. P. (1959). The effect of tritiated thymidine on mouse spermatogonia. *Radiat. Res.* **11**, 825–31.

Johnson, L. R. (1976). The trophic action of gastrointestinal hormones. *Gastroenterology* **70**, 278–88.

— and Guthrie, P. D. (1980). Stimulation of rat oxyntic gland mucosa growth by epidermal growth factor. *Am. J. Physiol.* **238**, G45–49.

Kahn, G., Weinstein, G. D., and Frost, P. (1968). Kinetics of human epidermal cell proliferation: diurnal variations. *J. invest. Derm.* **50**, 459–62.

Karasek, M. A. (1971). Effects of water-soluble antimetabolites on epithelial cell growth *in vitro*. In *Psoriasis* (ed. E. M. Farber and A. J. Cox) pp. 271–6. Stanford University Press, California.

Karatschai, M., Kinzel, V., Goerttler, K., and Suss, R. (1971). Geography of mitosis and cell division in the basal layer of the mouse epidermis. *Z. Krebsforsch.* **76**, 59–64.

Karienemi, A. L. (1976). Chalone-induced inhibition of DNA synthesis of human psoriatic epidermal cells cultured in diffusion chambers in mice. *Ann. clin. Res.* **8**, 340.

— and Rytomaa, T. (1976). Effect of the Hewitt keratinising epidermal carcinoma on cell proliferation in various organs of the host mouse and on human psoriatic skin cultured in diffusion chambers. *Br. J. Derm.* **94**, 515–23.

Katz, S. I., Tamaki, K., and Sachs, D. H. (1979). Epidermal Langerhans cells are derived from cells originating in bone marrow. *Nature, Lond.* **382**, 324–6.

Katzberg, A. A. (1952). Influences of age on rate of desquamation of human epidermis. *Anat. Rec.* **112**, 418.

Kefalides, N. A. (1970). Comparative biochemistry of mammalian basement membranes. In *Chemistry and molecular biology of the intercellular matrix* (ed. E. A. Balazs) p. 535. Academic Press, New York.

Kember, N. F. (1969). Recent experimental data on growth of cartilage and skin. *Br. J. Radiol.* **42**, 799.

— Quastler, H., and Wimber, D. R. (1962). Adaptation of the rat intestine to continuous irradiation. *Br. J. Radiol.* **35**, 290–7.

Kerr, J. F. R. and Searle, K. (1972). A mode of cell loss in malignant neoplasms. *J. Path.* **106**, xi.

— — (1973). Deletion of cells by apoptosis during castration-induced involution of the rat prostate. *Virchows Arch. Cell Path.* **13**, 87–102.

— Wyllie, A. H., and Currie, A. R. (1972). Apoptosis: a basic phenomenon with wide ranging implications in tissue kinetics. *Br. J. Cancer* **26**, 239–57.

Kiger, N., Florentin, I., and Mathe, G. (1972). Some effects of a partially purified lymphocyte inhibitory factor from calf thymus. *Transplantation* **14**, 448–54.

— — — (1973). A lymphocyte inhibiting factor (chalone) extracted from thymus: immunosuppressive effects. *Natn. Cancer Inst. Monogr.* **38**, 135–43.

Kierney, C. E. and Rothberg, S. (1969). *In vitro* DNA synthesis in chick embryonic skin; a sensitive indicator for testing stimulatory or inhibitory actions. *J. invest. Derm.* **53**, 351–5.

Kisieleski, W. E., Baserga, R., and Lisco, H. (1961). Tritiated thymidine and the study of tumours. *Atompraxis* **7**, 81–8.

Kit, S., Dubbs, D. R., and Frearson, D. M. (1965). Decline of thymidine kinase activity in stationary phase fibroblast cells. *J. biol. Chem.* **240**, 2565–8.

Kitano, Y. and Hu, F. (1970). Proliferation and differentiation of pigment cells *in vitro*. *J. invest. Derm.* **55**, 444–51.

— and Okada, N. (1980). Keratinisation of human keratinocytes in culture. The effect of cyclic AMP. In *Biochemistry of normal and abnormal epidermal differentiation* (ed. I. A. Bernstein and M. Seiji) pp. 27–38. Karger, Basel.

Klareskog, L., Tjernlund, U. M., Forsum, U., and Peterson, I. A. (1977). Epidermal Langerhans cells express Ia antigens. *Nature, Lond.* **268**, 248–50.

Klein, B. and Giuget, M. (1978). Relative importance of the phases of the cell cycle for explaining diurnal rhythms in cell proliferation in the tissues with a long $G_1$ duration. In *Biomathematics and cell kinetics* (ed. A. J. Valleron and P. D. M. MacDonald) pp. 199–210. Elsevier–North Holland, Amsterdam.

— and Valleron, A. J. (1977). A compartmental model for the study of diurnal rhythms in cell proliferation. *J. theoret. Biol.* **64**, 27–42.

Klein, H. and Giesel, H. (1947). Nachweis eines 24-Stundenrhythmus der Mitosen bei Ratte und Maus. *Klin. Wehrshr.* **24-5**, 662–3.

Klotz, U. and Stock, J. (1972). Influence of cyclic guanosine $3', 5'$ monophosphate in the enzymatic hydrolysis of adenosine $3', 5'$ monophosphate. *Archs Pharmacol.* **264**, 54–62.

Knowlton, N. P. and Widner, W. R. (1950). Use of X-rays to determine mitotic and intermitotic time of various tissues. *Cancer Res.* **10**, 59–63.

Kobayashi, K. (1976). Regional differences in mitotic activity due to injury in mouse skin. *Cell Tissue Res.* **175**, 319–24.

— (1977). Effects of whole body and partial body X-irradiation upon epidermal mitotic activity during wound healing in mouse skin. *Radiat. Res.* **69**, 413–29.

Koberg, E. (1963). The use of grain counts in the study of cell proliferation. In *Cell proliferation* (ed. L. F. Lamerton and R. J. M. Fry) p. 62. Blackwell, Oxford.

— and Maurer, W. (1962). Autoradiographische Untersuchung mit $^3$H-thymidine uber die daner des deoxyribonukleinsauresynthese und ihren zeitlichen Verlauf bei den Darmeepithelien und anderen Zelltypen der Maus. *Biochim. biophys. Acta* **61**, 229–42.

Koch, A. L. (1980). Does the variability of the cell cycle result from one or many chance events? *Nature, Lond.* **286**, 80–2.

Kohn, A. (1975). Differential effects of isoleucine deprivation on cell motility, membrane transport and DNA synthesis in Nil 8 hamster cells. *Expl Cell Res.* **94**, 143–5.

Kollmorgen, G. M. and Griffin, M. J. (1969). The effect of hydrocortisone on Hela cell growth. *Cell Tissue Kinet.* **2**, 111–22.

Korsgaard, R., Iversen, O. H., Isaaksson-Forsen, G., and Burton, D. R. (1978). Specificity of epidermal chalone extracts on epidermal tumour cells *in vitro*: a preliminary report. *Cell Tissue Kinet.* **11**, 441–3.

Koschel, K. W., Radley, J. M. N., and Hodgson, G. S. (1977). The analysis of cell kinetics using cohort fraction labelled mitosis (COFLM) curves. *J. theoret. Biol.* **64**, 355–61.

Kovacs, C. J., Evans, M. J., Johnke, R. M., Emma, D. A., and Scarantino, C. W. (1982). Enhanced progenitor production in the normal tissues of tumour bearing animals. *Cell Tissue Kinet.* **15**, 680.

Kram, R. and Tomkins, G. M. (1973). Pleiotypic control by cyclic AMP: interaction with cyclic GMP and possible role of microtubules. *Proc. natn. Acad. Sci. USA* **70**, 1659–63.

Krause, W. J. and Cutts, J. H. (1981). *Concise text of histology*, p. 32. Williams and Wilkins, Baltimore.

Krawczyk, W. S. (1971). A pattern of epidermal cell migration during wound healing. *J. cell Biol.* **49**, 247–63.

— (1972). Some ultrastructural aspects of epidermal repair in two model wound healing systems. In *Epidermal wound healing* (ed. H. Maibach and D. Rovee) pp. 123–31. Year Book, Chicago.

Kreig, L., Kohlmann, I., and Marles, F. (1974). Effects of tumour promoting phorbol esters and of acetic acid on mechanisms controlling DNA synthesis and mitosis (chalones). *Cancer Res.* **31**, 3135–46.

Kreyberg, L., Evensen, A., and Iversen, O. H. (1965). Influence of stress on the diurnal rhythm in the epidermis of hairless mouse. *Acta path. microbiol. scand.* **64**, 176–84.

Kurman, M. and Argyris, T. S. (1975). The proliferative response of epidermis of hairless mice to full thickness wounds. *Am. J. Path.* **79**, 301–10.

Laerum, C. D. and Boyum, A. (1969). The separation and cultivation of basal and differentiating cells from hairless mouse epidermis. *J. invest. Derm.* **54**, 279–87.

Lahtiharju, A., Rasanen, T., and Teir, H. (1964). Inhibition of DNA synthesis in various organs of the mouse following a single corticosteroid injection. *Growth* **28**, 221–3.

Lajtha, L. G. (1963). On the concept of the cell cycle. *J. cell comp. Physiol.* **62**, Suppl. 1, 143–5.

— (1967). Cytokinetics and regulation of progenitor cells. *J. cell Physiol.* **57**, Suppl. 1, 133–48.

— (1979). Stem cell concepts. *Differentiation* **14**, 23–34.

— Oliver, R., Berry, R. J., and Hell, E. (1960). Analysis of metabolic rates at the cellular level. *Nature, Lond.* **187**, 919–22.

— — and Ellis, F. (1954). Incorporation of $^{32}$P and adenine $^{14}$C into DNA by human bone marrow cells *in vitro*. *Br. J. Cancer* **8**, 367–79.

— — and Gurney, G. W. (1962). Kinetic model of a bone marrow stem cell population. *Br. J. Haematol.* **8**, 442–60.

Lala, P. K., Maloney, M. A., and Patt, H. M. (1965). Measurement of DNA synthesis time in myeloid–erythroid precursors. *Expl Cell Res.* **38**, 626–34.

Lamerton, L. F. and Fry, R. J. M. (1963). *Cell proliferation*. Blackwell, Oxford.

Lapachelle, J. M. and Gillman, T. (1969). Tritiated thymidine labelling of normal epidermal nuclei. *Br. J. Derm.* **81**, 603–7.

Laurence, E. B. (1973). Chalones—concepts and current researches. *J. natn. Cancer Inst.* **38**, 61–8.

— (1979). The place of chalones among the regulators of cell production in inflammation. In *Chemical messengers of the inflammatory process* (ed. J. C. Houck). Elsevier–North Holland, Amsterdam.

— (1980). The significance of chalones in epidermal growth. In *The skin of vertebrates* (ed. R. I. Spearman and P. A. Riley) pp. 139–50. Academic Press, New York.

— and Christophers, E. (1976). Selective action of hydrocortisone on post-mitotic epidermal cells *in vivo*. *J. invest. Derm.* **66**, 222–9.

— and Elgjo, K. (1971). Epidermal chalone and cell proliferation in a transplantable squamous carcinoma in hamsters. II. *In vitro* results. *Virchows Arch. Cell Path.* **7**, 8–15.

— and Randers-Hansen, E. (1971). An *in vivo* study of epidermal chalone and stress hormones on mitosis in tongue epithelium and ear epidermis of the mouse. *Virchows Arch. Cell Path.* **9**, 271–9.

—— (1972). Regional specificity of the epidermal chalone extracted from two different body sites. *Virchows Arch. Cell Path.* **11**, 34–42.

—— and Christophers, E. (1972). Systemic factors influencing epidermal mitosis. *Rev. Eur. Etnd. clin. Biol.* **17**, 133–9.

— Spargo, D. J., and Thornley, A. L. (1979). Cell proliferation kinetics of epidermis and sebaceous glands in relation to chalone action. *Cell Tissue Kinet.* **12**, 615–33.

— and Thornley, A. L. (1976*a*). Chalone tissue specificity and the embryonic derivation of organs. An appraisal of the problems. In *Chalones* (ed. J. C. Houck) p. 273. Elsevier–North Holland, Amsterdam.

— and Thornley, A. L. (1976*b*). The influence of epidermal chalone on cell proliferation. In *Stem cells of renewing cell populations* (ed. A. B. Cairnie, P. K. Lala, and D. G. Osmond) p. 103. Academic Press, New York.

Leblond, C. P. (1964). Classification of cell populations on the basis of their proliferative activity. *Natn. Cancer Inst. Monogr.* **14**, 119–49.

— (1975). Quoted in discussion. In *Stem cells of renewing cell populations* (ed. A. B. Cairnie, P. K. Lala, and D. G. Osmond) pp. 85–8. Academic Press, New York.

— Clermont, Y., and Nadler, N. J. (1967). The pattern of stem cell renewal in three epithelia (oesophagus, intestine and testis). *Can. Cancer Conf.* **7**, 3–30.

— Greulich, R. C., and Periera, J. P. M. (1964). Relationship of cell formation and cell migration in the renewal of stratified squamous epithelia. *Adv. Biol. Skin.* **v**, 39–53. Pergamon, London.

— Messier, B., and Kopriwa, B. (1959). Thymidine-$^3$H as a tool for the investigation of the renewal of cell populations. *Lab Invest.* **8**, 296–306.

— and Walker, B. E. (1956). Renewal of cell populations. *Physiol. Rev.* **36**, 255–76.

Leith, J. T. (1978). Comparison of the effects of epidermal chalone on young and aging mice. *Cell Tissue Kinet.* **11**, 433–40.

— Schilling, W. A., and Lyman, J. T. (1975). Comparison of skin responses of mice after single fractionated exposure to cyclotron accelerated helium ions and 230 kV X-irradiation. *Radiat. Res.* **62**, 195–215.

—— and Welch, G. P. (1971). Survival of mouse skin epithelial cells after heavy particle irradiation. *Int. J. radiat. Biol.* **19**, 603–9.

Lenn, G. C., Lowe, L. B., and Beerens, E. G. Y. (1974). The influence of skin temperature on dermal:epidermal adherence. *J. invest. Derm.* **62**, 42–5.

Leroy, F., Bogaert, C., and Van Hoeck, J. (1978). Stimulation of cell division in the endometrial epithelium of the rat by uterine distension. *J. Endocr.* **70**, 517–18.

Lever, W. F. and Schaumberg-Lever, G. (1979). *Histopathology of the skin.* Lippincott, Philadelphia.

Levine, M. (1972). The growth of adult human skin *in vitro*. *Br. J. Derm.* **86**, 481–90.

Lewis, F. T. (1926). The effect of cell division on the shape and size of hexagonal cells. *Anat. Rec.* **33**, 331–55.

— (1928). The shape of cork cells; a simple demonstration that they are tetrakaidecahedral. *Science, NY* **63**, 607–9.

Lewis, P. D. (1978). The application of cell turnover studies to neuropathology. In *Recent advances in neuropathology* (ed. W. T. Smith and J. B. Cavanagh) pp. 41–65. Churchill Livingstone, Edinburgh.

Lezzard, R., Wolff, K., and Winkelmann, R. K. (1966). Induced 'shedding' of epidermal Langerhans cells. *Nature, Lond.* **212**, 628-9.
— — — (1968). The disappearance and regeneration of Langerhans cells following epidermal injury. *J. invest. Derm.* **50**, 171-9.
Lin, L. M., Goepp, R. A., and Sewell, A. F. (1977). Diurnal variation in epithelial cell population kinetics of young mouse tongue. *J. dent. Res.* **56**, 425-36.
Lipkin, M. (1965). Cell replication in the gastrointestinal tract of man. *Gastroenterology* **48**, 616-24.
Liu, S. C. (1978). Isolation and growth of adult epidermal keratinocytes in cell culture. *J. invest. Derm.* **71**, 157-62.
— (1979). Growth characteristics of human epidermal keratinocytes from newborn foreskin in primary and serial culture. *In vitro* **15**, 813-22.
— and Karasek, M. (1978). Isolation and serial cultivation of rabbit skin epithelial cells. *J. invest. Derm.* **70**, 288-93.
Livingstone, R. B., Ambros, U., George, S. C., Freireich, E. T., and Hart, J. S. (1974). *In vitro* determination of thymidine-$^3$H labelling index in human solid tumours. *Cancer Res.* **34**, 1376-80.
Loe, H., Karring, T., and Hasa, K. (1972). The site of mitotic activity in rat and human oral epithelium. *Scand. J. dent. Res.* **86**, 111-19.
Lord, B. I. (1976). The assay of cell proliferation inhibitors. In *Chalones* (ed. J. C. Houck) pp. 76-151. Elsevier–North Holland, Amsterdam.
— (1979). Proliferation regulators in haemopoiesis. *Clinics Haematol.* **8**, 435-51.
Lowe, N. J. (1980). Epidermal ornithine decarboxylase, polyamines cell proliferation and tumour promotion. *Archs Derm.* **116**, 822-35.
— and Breeding, J. (1980). Antiinflammatory drug effects on UV light induced epidermal ornithine decarboxylase and DNA synthesis. *J. invest. Derm.* **74**, 418-20.
— Stoughton, R. B., McCullough, J. L., and Weinstein, G. D. (1981). Topical drug effects on normal and proliferating epidermal cell models. *Archs Derm.* **117**, 394-8.
McCullough, J. L. (1978). Cell proliferation kinetics of epidermis in the essential fatty acid deficient rat. *J. invest. Derm.* **70**, 318-20.
— and Weinstein, G. D. (1979). The role of cell kinetics in the chemotherapy of hyperproliferative skin diseases. *Pharmacol. Ther.* **4**, 601-15.
McDonald, C. L. and Los, L. L. (1973). Studies on the *in vitro* incorporation of $^3$H-thymidine, $^3$H-uridine, $^3$H-leucine into mammalian epidermal cells. *Dermatologica* **147**, 391-8.
MacDonald, D. G. (1971). Cell renewal in oral epithelium. In *Current concepts of the histology of oral mucosa* (ed. C. A. Squier and J. Meyer) p. 61. Thoma, Washington.
MacDonald, P. D. M. (1970). Statistical inferences from the fraction labelled mitoses curve. *Biometrika* **57**, 489-503.
— (1978). Age distributions in the general cell kinetic model. In *Mathematical models in cell kinetics* (ed. A.-J. Valleron and P. D. M. MacDonald) pp. 3-20. European Press, Medikon, Ghent.
McGuire, J. and Arnesen, S. (1972). Control of keratinocyte division *in vitro*. *J. invest. Derm.* **59**, 85-9.
McHugh, W. D. (1959). The development and structure of gingival epithelium. Ph.D. thesis, St. Andrews University.
— (1960). The development of the gingival epithelium in the monkey. *Dent. Pract.* **11**, 314.

# 510    References

— and Zander, H. A. (1965). Cell division in the peridontium of developing and erupted teeth. *Dent. Pract.* **15**, 451-7.

MacKenzie, I. C. (1969). Ordered structure of the stratum corneum of mammalian skin. *Nature, Lond.* **222**, 881.

— (1970). Relationship between mitosis and the ordered structure of the stratum corneum in mouse epidermis. *Nature, Lond.* **226**, 653.

— (1972). The ordered structure of mammalian skin. In *Epidermal wound healing* (ed. H. I. Maibach and D. Rovee) pp. 5-26. Year Book, New York.

— (1974*a*). The effect of frictional stimulation on mouse ear epidermis. I. Cell proliferation. *J. invest. Derm.* **62**, 80-5.

— (1974*b*). The effects of frictional stimulation on mouse ear epidermis. II. Histologic appearance and cell counts. *J. invest Derm.* **63**, 194-8.

— (1975*a*). Spatial distribution of mitosis in mouse epidermis. *Anat. Rec.* **181**, 705-10.

— (1975*b*). Labelling of murine epidermal Langerhans cells with $^3$H thymidine. *Am. J. Anat.* **144**, 125-36.

— (1975*c*). Ordered structure of the epidermis. *J. invest. Derm.* **65**, 45-51.

— and Linder, J. E. (1973). An examination of cellular organisation within the stratum corneum by a silver staining method. *J. invest. Derm.* **61**, 245-50.

— and Zimmerman, K. (1981). The development of ordered structure in neonate rat epidermis. *J. invest. Derm.* **77**, 272-82.

— — and Peterson, L. (1981). The pattern of cellular organisation of human epidermis. *J. invest. Derm.* **76**, 459-61.

Mackey, M. C. and Dormer, P. (1981). Enigmatic hemopoiesis. In *Biomathematics and cell kinetics* (ed. M. Rotenberg) pp. 87-103. Elsevier–North Holland, Amsterdam.

McMinn, R. M. H. and Taylor, M. (1966). The cytology of repair in experimental perforations of the tympanic membrane. *Br. J. Surg.* **53**, 222-32.

Major, I. R. (1970). Correlation of initial changes in the mouse epidermal cell population with 2 stage carcinogenesis. A quantitative study. *Br. J. Cancer* **24**, 149-63.

Mantell, P. A. (1972). The circadian rhythm of mitotic activity in the incisor pulp and the cornea of *Rattus norwegicus*. *Archs oral Biol.* **18**, 403-7.

Marcelo, C. L. (1979). Differential effects of cAMP and cGMP in *in vitro* epidermal growth. *Expl Cell Res.* **120**, 201-10.

— and Duell, E. A. (1979). Cyclic AMP stimulates and inhibits adult human epidermal cell growth. *J. invest. Derm.* **72**, 279.

— — Streviski, M. A., Anderson, T. F., and Vorhees, J. J. (1979). Cyclic nucleotide levels in psoriatic and normal keratomed epidermis. *J. invest. Derm.* **72**, 20-4.

— Kim, Y. G., Kaine, J. L., and Vorhees, J. J. (1978). Stratification, specialization and proliferation of keratinocyte cultures. *J. cell Biol.* **79**, 356-70.

— and Vorhees, J. J. (1980). Cyclic nucleotides and the control of psoriatic cell function. *Adv. cyclic nucleotide Res.* **12**, 129-37.

Marengo, N. P. (1964). Paraffin section thickness—a direct method of measurement. *Stain Technol.* **19**, 1-10.

Marks, F. (1971). Direct evidence of two tissue specific chalone like factors regulating mitosis and DNA synthesis in mouse epidermis. *Hoppe Seylers Z. Physiol. Chem.* **352**, 1273-4.

— (1972). Diurnal fluctuation and adrenergic elevation of cyclic AMP in mouse epidermis *in vivo*. *Nature, New Biol.* **240**, 178-9.

— (1973*a*). A tissue specific factor inhibiting DNA synthesis in mouse epidermis. *J. natn. Cancer Inst.* **38**, 79–90.

— (1973*b*). The second messenger system of mouse epidermis. III. Guanyl cyclase. *Biochim. biophys. Acta* **300**, 349–56.

— (1974). Attempts to purify and characterise a factor inhibiting epidermal DNA synthesis. *Hoppe Seylers Z. Physiol. Chem.* **355**, 1228–9.

— (1975). Isolation of an endogenous inhibitor of DNA synthesis ($G_1$ chalone) from pig skin. *Hoppe Seylers Z. Physiol. Chem.* **356**, 1989–92.

— (1976*a*). Epidermal growth control mechanism, hyperplasia and tumour induction in the skin. *Cancer Res.* **36**, 2636–44.

— (1976*b*). The epidermal chalones. In *Chalones* (ed. J. C. Houck) p. 173. North Holland, Amsterdam.

— (1978). Homeostatic regulators of epidermal cell production. *Bull. Cancer, Paris* **65**, 207–22.

— Bertsch, S., and Furstenberger, G. (1979). Ornithine decarboxylase activity, cell proliferation and tumour promotion in mouse epidermis *in vivo. Cancer Res.* **39**, 4183–8.

— — Grimm, W., and Schweizer, J. (1978). Hyperplastic transformation and tumour promotion in mouse epidermis: possible consequences of disturbances of endogenous mechanisms controlling proliferation and differentiation. In *Carcinogenesis* (ed. T. J. Slaga, A. Swik, and R. K. Boutwell), Vol. 2, pp. 97–116. Raven Press, New York.

— and Grimm, W. (1972). Diurnal fluctuation and $\beta$-adrenergic elevation of cyclic AMP in mouse epidermis *in vivo. Nature, Lond.* **240**, 178–9.

— and Raab, I. (1974). The second messenger system of mouse epidermis. IV. Cyclic AMP and cyclic GMP phosphodiesterase. *Biochim. biophys. Acta* **334**, 368–77.

— and Rebien, P. (1972). Cyclic $3',5'$ AMP and theophylline inhibit epidermal mitosis in $G_2$ phase. *Naturwissenschaften* **59**, 41–2.

Marks, R. (1972). The role of chalones in epidermal homeostasis. *Br. J. Derm.* **86**, 543–8.

— (1975). Is there a relationship between clinical morphology and epidermal cell kinetics? *Proc. R. Soc. Med.* **68**, 161–7.

— (1978). Epidermal activity in the involved and uninvolved skin of patients with psoriasis. *Br. J. Derm.* **98**, 399–404.

— and Boghel, B. (1973). Scanning electron microscopy of the normal epidermis and the epidermis *in vitro. Br. J. Derm.* **90**, 387–96.

— Fukui, K., and Halprin, K. (1971*a*). The application of an *in vitro* technique to the study of epidermal replication and metabolism. *Br. J. Derm.* **84**, 453–7.

— Halprin, K., and Fukui, K. (1971*b*). Topically applied triamcinolone and macromolecular synthesis by human epidermis. *J. invest. Derm.* **56**, 470–3.

— Nicholls, F., and Dykes, P. (1980). Quantitative aspects of keratinisation. In *The epidermis in disease* (ed. R. Marks and E. Christophers) pp. 79–90. MTP Press, Lancaster.

— Pongschaun, D., and Sayhan, T. (1973). A method for the assay of topical corticosteroids. *Br. J. Derm.* **88**, 69–74.

Marquardt, H. (1974). Cell cycle dependence of chemically-induced malignant transformation *in vitro. Cancer Res.* **34**, 1612–15.

Marques-Periera, J. P. and Leblond, C. P. (1965). Mitosis and differentiation in the stratified squamous epithelium of the rat oesophagus. *Am. J. Anat.* **117**, 73–91.

Marrable, A. W. (1962). The counting of cells and nuclei in microtome sections. *Q. J. microsc. Sci.* **103**, 331–47.

Marrs, J. M. and Voorhees, J. J. (1971*a*). A method for bioassay of an epidermal chalone-like inhibitor. *J. invest. Derm.* **56**, 174–81.

— — (1971*b*). Preliminary characterisation of an epidermal chalone-like inhibitor. *J. invest. Derm.* **56**, 353.

Martinez, I. R. (1972). Fine structural studies of migrating epithelial cells following incision wounds. In *Epidermal wound healing* (ed. H. I. Maibach and D. T. Rovee) pp. 323–43. Year Book, New York.

Marwah, A. S., Weinmann, J. P., and Meyer, J. (1960). Effect of chronic inflammation on the epithelial turnover of the human gingiva. *Archs Path.* **69**, 147–53.

Mary, J. Y. (1981). Reference data on normal human granulopoiesis. In *Biomathematics and cell kinetics* (ed. M. Rotenberg) pp. 65–86. Elsevier–North Holland, Amsterdam.

Mathur, M. and Deo, M. G. (1976). Kinetics of proliferation and differentiation in the hair follicle and epidermis in neonatally undernourished rats. *Am. J. Path.* **82**, 9–24.

Matoltsky, A. G. and Viziam, C. B. (1970). Further observations on epithelialisation of small wounds. *J. invest. Derm.* **55**, 20–5.

Maurer, H. R. (1981). Potential pitfalls of [$^3$H]-thymidine techniques to measure cell proliferation. *Cell Tissue Kinet.* **14**, 111–20.

— and Laerum, O. D. (1976). Granulocyte chalone testing: a critical review. In *Chalones* (ed. J. C. Houck) pp. 331–50. North Holland, Amsterdam.

— Weiss, G., and Laerum, O. D. (1976). Evaluation of a short term *in vitro* test for granulocyte chalone activity. *Virchows Arch. Cell Path.* **20**, 229–38.

Maurer, W. and Primbusch, E. (1964). Grosse de $\beta$-selbstabsorption bei der $^3$H-autoradiographie. *Expl Cell Res.* **33**, 8–18.

— Schultze, B., Schmeer, A. C., and Haack, V. (1976). Autoradiographic studies on the mode of growth in jejunal crypt cells of the mouse. *J. Microsc.* **96**, 181–9.

Meck, R. A., Ingram, M., Meck, J. M., McCullough, J. L., Uw, M. C., and Yanis, A. A. (1981). Establishment and cell cycle kinetics of a human squamous cell carcinoma in nude mice and *in vitro. Cancer Res.* **41**, 1076–85.

Medline, A. and Farber, E. (1981). The multistep theory of neoplasia. In *Recent advances in histopathology* (ed. P. P. Anthony and R. N. M. MacSween) pp. 19–34. Churchill Livingstone, Edinburgh.

Melbye, S. W. and Karasek, M. A. (1973). Some characteristics of a factor stimulating skin epithelial growth *in vitro. Expl Cell Res.* **79**, 279–86.

Mendelsohn, M. L. (1962). Chronic infusion of tritiated thymidine into mice. *Science, NY* **135**, 213.

— (1963). Cell proliferation and tumour growth. In *Cell proliferation* (ed. L. F. Lamerton and R. J. Fry). Blackwell, Oxford.

— (1965). The kinetics of tumour cell proliferation. In *Cellular radiation biology*, pp. 219–28. Williams and Wilkins, Baltimore.

— and Takahashi, M. (1971). A critical evaluation of the labelled mitoses method as applied to the analysis of tumours and other cycles. In *The cell cycle and cancer* (ed. R. Baserga) pp. 72–91. Marcel Dekker, New York.

Menton, D. N. (1976*a*). A minimum surface mechanism to account for the organization of cells into columns in the mammalian epidermis. *Am. J. Anat.* **145**, 1–22.

—— (1976*b*). A liquid film model of tetrakaidecahedral packing to account for the establishment of epidermal columns. *J. invest. Derm.* **66**, 283–91.

—— and Eisen, A. Z. (1971*a*). Structure and organisation of mammalian stratum corneum. *J. ultrastruct. Res.* **35**, 247–64.

—— —— (1971*b*). Structural organisation of the human stratum corneum in certain scaling disorders. *J. invest. Derm.* **51**, 296–307.

Messier, B. and Leblond. C. P. (1960). Cell proliferation and migration as revealed by autoradiography after injection of thymidine-$H^3$ into male rats and mice. *Am. J. Anat.* **106**, 247–85.

Meyer, J. S. and Donaldson, R. C. (1969). Growth kinetics of squamous cell carcinoma in man. *Archs Path.* **87**, 479–90.

—— Medak, J., and Weinmann, J. P. (1960). Mitotic activity and rates of growth in regions of oral epithelium differing in width. *Growth* **24**, 29–46.

Mier, P. D. and Colton, D. W. K. (1967). Enzymes of the glycolytic pathway in skin. III. Phosphorylase. *Br. J. Derm.* **79**, 164–9.

—— and Sutorius, A. H. M. (1972). The control of glycogen utilisation in skin: phosphorylase B kinase and phosphorylase A phosphatase. *Br. J. Derm.* **86**, 49–53.

Miller, R. C. (1972). Prolonged utilization of DNA precursor by rat lens epithelium after single application of tritiated thymidine in the eye. MSc. thesis, University of Iowa.

Milstein, H. and Cornell, R. (1973). Diurnal mitotic studies of psoriatic epidermis. *J. invest. Derm.* **61**, 180–3.

Minor, P. D. and Smith, J. A. (1974). Explanation of the degree of correlation of sibling generation times in mammalian cells. *Nature, Lond.* **248**, 241–2.

Mirvish, S. S., Chu, C., and Clayson, D. B. (1978). Inhibition of $^3H$ thymidine incorporation in DNA of rat oesophageal epithelium and related tissues by carcinogenic nitroso compounds. *Cancer Res.* **38**, 458–66.

Mishima, Y. and Miller-Milinska, A. (1961). Junctional and high level dendritic cells revealed by osmium iodide reaction in human and animal epidermis under conditions of hyperpigmentation and depigmentation. *J. invest. Derm.* **37**, 107–11.

—— and Pinkus, H. (1968). Electron microscopy of keratin layer stripped human epidermis. *J. invest. Derm.* **50**, 89–96.

Mitrani, E. (1978). Possible role of connective tissue in epidermal neoplasia. *Br. J. Derm.* **99**, 233–44.

Mitchison, J. M. (1971). *The biology of the cell cycle.* Cambridge University Press.

—— (1977). The timing of cell cycle events. In *Mitosis: facts and questions* (ed. M. Little). Springer, Berlin.

Mizuno, T. and Fujii, T. (1969). The relation between wound repair processes and early skin carcinogenesis. *Proc. Jap. Acad.* **45**, 925–30.

Moffat, G. H. and Pelc, S. R. (1966). Delay after plucking hairs between the appearance of $^3H$-thymidine in cells and its incorporation into DNA. *Expl Cell Res.* **42**, 460–9.

Mohr, U., Althoff, J., Kinzel, V., Suss, R., and Volm, M. (1968). Melanoma regression induced by chalone; a tumour inhibiting principle acting *in vivo*. *Nature, Lond.* **220**, 138.

—— Boldingh, W. H., and Althoff, J. (1972*a*). Identification of contaminating clostridium spores as the oncolytic agent in some chalone preparations. *Cancer Res.* **32**, 1117.

— — Emminger, A., and Behagel, H. A. (1972*b*). Oncolysis by a new strain of clostridium. *Cancer Res.* **32**, 1122–6.

Møller, U. (1978). Interaction of external agents with the circadian mitotic rhythm in the epithelium of the hamster cheek pouch. *J. Interdisc. Cycle Res.* **9**, 105–14.

— Hartmann, N., and Faber, M. (1979). Mitotic index, influx and mean transit time in the hamster cheek pouch epithelium; a partially synchronised system. *Cell Tissue Kinet.* **12**, 581–92.

— and Keiding, N. (1982). Circadian variations in influx and efflux of the *S* phase in a partially synchronised cell system. Double labelling with ³H thymidine in the epithelium of the cheek pouch. *Cell Tissue Kinet.* **15**, 341–56.

— — and Engel, F. (1982). Subclassification of cells in S phase in a partially synchronised system. Methodological problems concerning pulse labelling of epithelial cells in the hamster cheek pouch. *Cell Tissue Kinet.* **15**, 157–74.

— and Larsen, J. K. (1978). Circadian variations in the epithelial growth of the hamster cheek pouch. *Cell Tissue Kinet.* **11**, 405–13.

— — (1979). DNA flow cytometry of isolated keratinised epithelia. *Cell Tissue Kinet.* **12**, 203–11.

— — and Faber, M. (1974). The influence of injected tritiated thymidine on the mitotic circadian rhythm in the epithelium of the hamster cheek pouch. *Cell Tissue Kinet.* **7**, 231–9.

Montagna, W. and Parakkal, P. F. (1974). *The structure and function of skin.* Academic Press, New York.

Moore, J. V. (1979). Ablation of murine jejunal crypts by alkylating agents. *Br. J. Cancer* **39**, 175–81.

Morgan, W. W. and Mizell, S. (1971). Diurnal fluctuation in the DNA content and DNA synthesis in the dorsal epidermis of *Rana pipiens*. *Comp. Biochem. Physiol.* **38**, 591–602.

Morley, A. R. (1970). Studies in the cell kinetics of the mouse accessory sex glands with special reference to testosterone propionate. MD thesis. University of Newcastle upon Tyne.

— and Wright, N. A. (1972). Androgen-induced cell proliferation and differentiation in the seminal vesicle and coagulating gland of the castrated mouse. *J. Endocr.* **54**, 465–71.

— — and Appleton, D. R. (1973). Cell proliferation in the castrate mouse seminal vesicle in response to testosterone propionate. *Cell Tissue Kinet.* **6**, 239–46.

Morley, C. G. D. and Kingdon, H. (1972). Use of ³H thymidine for measurement of DNA synthesis in rat liver—a warning. *Analyt. Biochem.* **45**, 298.

Morris, V. B., Cowan, R., and Culpin, D. (1979). Variability of cell cycle times measured *in vivo* in embryonic duck retina by continuous labelling with BudR. *Nature, Lond.* **280**, 68–70.

Morris, W. T. (1967). *In vivo* studies in the optimum time for the action of colchicine on mouse lymphoid tissues. *Expl Cell Res.* **48**, 209–14.

Morrison, D. M. and Goldsmith, L. A. (1978). Ornithine decarboxylase in rat skin. *J. invest. Derm.* **70**, 309–13.

Mufson, R. A. (1978). Effects of tumour promoters on cyclic nucleotide metabolism in mouse skin and epidermis *in vivo*. In *Carcinogenesis* (ed. Y. Slaga, A. Swak, and R. K. Boutwell), Vol. 2, pp. 135–52. Raven Press, New York.

— Simsiman, R. C., and Boutwell, R. K. (1977). The effect of the phorbol ester tumour promoters on the basal and catecholamine stimulated levels of cyclic adenosine $3', 5'$ monophosphate in mouse skin epidermis *in vivo*. *Cancer Res.* **37**, 665-9.

Muggia, F. M. and De Vita, V. T. (1972). *In vitro* tumour cell kinetic studies: use of local thymidine injection followed by fine needle aspiration. *J. lab. clin. Med.* **80**, 297-301.

Murray, A. W. (1978). Induction of epidermal cell proliferation by a tumour promoter in vitamin B6 deficient mice. *Experientia* **34**, 691.

— and Rogers. A. (1978). Calcium dependent protein modulator of cyclic nucleotide phosphodiesterases from mouse epidermis. *Biochem. J.* **176**, 727-32.

Nafstad, P. H. J. and Baker, R. E. (1973). Comparative ultrastructural study of normal and grafted skin in the frog, *Rana pipiens*, with special reference to neuroepithelial connections. *Z. Zellforsch.* **139**, 451-62.

Nagamine, G. Y. (1978). Changes of the cell cycle during carcinogenesis of the hamster cheek pouch epithelium by 7, 12-dimethylbenz(a)anthracene. *Gann* **69**, 317-22.

Nelson, S. and Green, M. (1981). The random transition model of the cell cycle: a critical review. *Cancer Chemother. Pharmacol.* **6**, 11-18.

Neuburger, A. E. and Weinstein, G. D. (1980). Cell proliferation patterns in human malignant melanoma. *Cancer* **46**, 308-13.

Nias, A. H. W. and Fox, M. (1968). Minimal clone size for estimating the normal reproductive capacity of cultured cells. *Br. J. Radiol.* **41**, 468-74.

— Gilbert, C. W., Lajtha, L. G., and Lange, C. S. (1965). Clone size analysis in the study of cell growth following single or during continuous irradiation. *Int. J. radiat. Biol.* **9**, 275-90.

Nicolini, C. (1975). The discrete phases of the cell cycle: autoradiographic, physical and chemical evidence. *J. natn. Cancer Inst.* **55**, 821-6.

Noble, W. C., White, P. M., and Baker, H. (1975). Assay of therapeutic doses of methotrexate in body fluids of patients with psoriasis. *J. invest. Derm.* **64**, 69-76.

Nome, O. (1975*a*). Demecolcin and vinblastine sulphate as stathmokinetic agents for different tissues of the hairless mouse. *Path. Eur.* **10**, 221-8.

— (1975*b*). Tissue specificity of epidermal chalones. *Virchows Arch. Cell Path.* **19**, 1-25.

Nurse, P. (1980). Cell cycle control—both deterministic and probabilistic. *Nature, Lond.* **286**, 9-10.

O'Brien, T. G., Simsiman, R. E., and Boutwell, R. K. (1975). Induction of the polyamine biosynthetic enzymes in mouse epidermis and their specificity for tumour promotion. *Cancer Res.* **35**, 2426-33.

Odell, T. T., Jackson, C. W., and Reiter, R. S. (1968). Generation cycle of rat megakaryocytes. *Expl Cell Res.* **53**, 321-8.

Odaye, O. (1975). Effects of various induced local environmental conditions on histopathological studies in experimental *Dermatophilus congolensis* infection of the bovine skin. *Res. vet. Sci.* **19**, 245-52.

Odland, G. and Ross, R. (1968). Human wound repair. I. Epidermal regeneration. *J. cell Biol.* **39**, 135-51.

Oehlert, W. (1973). Cellular proliferation in carcinogenesis. *Cell Tissue Kinet.* **6**, 325-35.

Ohashi, M., Sawada, Y., and Makita, R. (1973). Odland bodies and intracellular substances. *Acts derm. vener.* Suppl. **73**, 47-54.

Okun, M. (1965). Histogenesis of melanocytes. *J. invest. Derm.* **44**, 285-99.

Olson, R. L. and Everett, M. A. (1975). Epidermal apoptosis: cell deletion by phagocytosis. *J. cutan. Path.* **2**, 53–7.

—— Nordgrist, R. E., and Everett, M. A. (1969). Mitosis in human epidermis. An ultrastructural study. *Dermatologica, Basel* **138**, 268–83.

Olsson, L. (1976). Effects of tritium labelled pyrimidine nucleosides in the mouse. I. Cytodynamic perturbations in normal circadian rhythms after a single injection of $^3$H thymidine. *Radiat. Res.* **68**, 258–71.

Olvey, S. K. and MacDonald, C. J. (1972). The *in vitro* cell cycle of adult mammalian epidermis; a preliminary study. *J. invest. Derm.* **58**, 175–9.

Oscarson, J. E. A., Vien, H. F., Williamson, R. C. N., Ross, J. S., and Malt, R. A. (1977). Compensatory postresectional hyperplasia and starvation atrophy in small bowel: dissociation from endogenous gastrin levels. *Gastroenterology* **72**, 890–5.

Osgood, E. C. (1957). A unifying concept of the etiology of the leukaemias, lymphomas and cancers. *J. natn. Cancer Inst.* **18**, 155–66.

Otten, J., Johnson, G. S., and Pastan, I. (1971). Cyclic AMP levels in fibroblasts: relationship to growth rate and contact inhibition of growth. *Biochem. Biophys. Res. Commun.* **44**, 1192–8.

—— —— —— (1972). Regulation of cell growth by cyclic adenosine 3′,5′ monophosphate. *J. biol. Chem.* **247**, 7082–7.

Page, D. G. and Toto, P. D. (1977). The effects of chronic neutron radiation on the cell density of mouse oral epithelium. *J. dent. Res.* **56**, 990–7.

Painter, R. B., Drew, R. M., and Giauque, B. G. (1960). Further studies on deoxyribonucleic acid metabolism in mammalian cell cultures. *Expl Cell Res.* **21**, 98–105.

Palade, G. E. and Farquhar, M. G. (1965). A special fibril of the dermis. *J. cell Biol.* **27**, 215–21.

Pallavicini, M. G. (1982). Solid tissue dispersal for cell kinetics analysis. *Cell Tissue Kinet.* **15**, 686.

Pardee, A. B. (1974). A restriction point for control of normal animal cell proliferation. *Proc. natn. Acad. Sci. USA* **7**, 1286–90.

—— (1978). Conditional mutants of the cell cycle. *J. cell Physiol.* **95**, 833–6.

—— Dubrow, R., Hamlin, J. L., and Kletzien, R. F. (1978). Animal cell cycle. *A. Rev. Biochem.* **47**, 715–50.

Park, I. J. and Jones, H. W. (1968). Glucose-6-phosphate dehydrogenase and the histogenesis of epidermoid carcinoma of the cervix. *Am. J. Obstet. Gynec.* **102**, 106–9.

Pastan, I. (1972). Cyclic AMP. *Scient. Am.* **227**, 97–105.

Paul, D. and Hecker, E. (1969). On the biochemical mechanism of tumourigenesis in mouse skin. II. Early effects on the biosynthesis of nucleic acids induced by initiating doses of DMBA and by promoting doses of phorbol 12-13 diester TPA. *Z. Krebsforsch.* **72**, 49–63.

Pawelek, J. M. (1976). Factors regulating growth and pigmentation of melanoma cells. *J. invest. Derm.* **66**, 201–9.

Pearse, A. D. and Marks, R. (1977). Actinic keratoses and the epidermis on which they arise. *Br. J. Derm.* **96**, 45–50.

Pechl, D. M. (1980). Clonal growth of human keratinocytes with small amounts of dialysed serum. *In Vitro* **16**, 526–40.

Pedersen, T. and Gelfant, S. (1970). $G_2$-population cells in mouse kidney and duodenum and their behaviour during the cell division cycle. *Expl Cell Res.* **59**, 32–6.

Pelc, S. R. (1971). Influence of metabolic DNA on determination of the cell cycle. *Cell Tissue Kinet.* **4**, 577–84.

Penneys, N. S., Fulton, J. E., Weinstein, G. D., and Frost, P. (1970). Location of proliferating cells in human epidermis. *Archs Derm.* **101**, 323–9.

Pernice, B. (1889). Sulla caricinesii delle cellule epitheliali e dell' endotello dei vasi della mucosa della stomacho e dell' intestino, nella studio della gastro-enterite sperimentale (nell' arrelenamento pereolchico). *Sicilia Med.* **1**, 265. (Quoted by Eigsti and Dustin (1955).)

Peto, R. (1977). Epidemiology, multistage models and short term mutagenicity tests. Cold Spring Harbor Conference on cell proliferation, Vol. 4, pp. 1403–28 (ed. H. H. Hiatt, J. D. Watson, and J. A. Winston).

Pilgrim, C. (1967). Autoradiographic investigations with $^3$H thymidine on the influence of diurnal variation on cell proliferation kinetics. In *The cellular aspects of biorrhythms* (ed. H. von Mayersbach) pp. 100–4. Springer, New York.

—— Erb, W., and Maurer, W. (1963). Diurnal fluctuation in the numbers of DNA synthesising nuclei in various mouse tissues. *Nature, Lond.* **199**, 863.

Pinkus, H. (1952). Examination of the epidermis by the strip method of removing horny layers. 1. Observations on the thickness of the horny layer, and on mitotic activity after stripping. *J. invest. Derm.* **16**, 383–6.

—— (1970). The direction of growth of human epidermis. *Br. J. Derm.* **83**, 556–64.

—— (1972). Morphokinetics of the epidermis. *Arch. Derm. Forsch.* **244**, 13–21.

—— and Hunter, R. R. (1966). The direction of the mitotic axis in human epidermis. *Archs Derm.* **94**, 351–4.

—— and Mehregan, A. H. H. (1981). *A guide to dermatohistopathology*. Appleton Century Crofts, New York.

Piscitelli, N. (1956). Etude quantitative de l'inhibition mitotique des cellular medullaires de rats normaux et surrenalectmisis par la method colchicinique. *Bull. Classe Sci. Acad. belg.* **5**, ser. 42, 201–22.

Plewig, G. and Braun Falco, O. (1975). Kinetics of epidermis and adnexa following vitamin A acid in the human. *Acta derm. vener.* Suppl. **74**, 87–98.

Ponder, B. A. J. (1983). Lectin histochemistry. In *Immunocytochemistry: practical applications in pathology and biology* (ed. J. M. Polak and S. Van Noorden). Wright, Bristol.

Porter, D. and Shuster, S. (1967). A new method for measuring replacement of epidermis and stratum corneum in human skin. *J. invest. Derm.* **49**, 251–6.

Post, J. and Hoffmann, J. (1968). Early and late effects of $^3$H TdR-labelled DNA on ileal cell replication *in vivo*. *Radiat. Res.* **34**, 570–81.

Potten, C. S. (1968). Radiation depigmentation of mouse hair; a study of follicular melanocyte populations. *Cell Tissue Kinet.* **1**, 239–49.

—— (1971*a*). Early and late incorporation of tritiated thymidine into skin cells and the presence of a long-lived $G_0$ specific precursor pool. *J. cell Biol.* **51**, 855–63.

—— (1971*b*). Tritiated thymidine incorporation into hair follicle matrix and epidermal basal cells after stimulation. *J. invest. Derm.* **56**, 311–14.

—— (1972). Some observations on the post-plucking depression in tritiated thymidine utilisation in mouse skin and some tentative cell kinetic determinations. *J. invest. Derm.* **58**, 180–4.

—— (1973). Further observations on the late labelling associated with stimulus responsive cells in skin. *Cell Tissue Kinet.* **6**, 553–62.

—— (1974). The epidermal proliferative unit: the possible role of the central basal cell. *Cell Tissue Kinet.* **7**, 77–86.

—— (1975*a*). Epidermal transit times. *Br. J. Derm.* **93**, 649–58.

—— (1975*b*). Epidermal cell production rates. *J. invest. Derm.* **65**, 488–506.

—— (1976). Identification of clonogenic cells in the epidermis and the structural arrangement of the epidermal proliferative unit. In *Stem cells of renewing cell populations* (ed. A. B. Cairnie, P. K. Lala, and D. G. Osmond) pp. 91–102. Academic Press, New York.

—— (1977). Extreme sensitivity of some intestinal crypt cells to X and $\gamma$ irradiation. *Nature, Lond.* **269**, 518–21.

—— (1978*a*). Epithelial proliferative subpopulations. In *Stem cells and tissue homeostasis* (ed. B. I. Lord, C. S. Potten, and R. J. Cole) pp. 335–58. Cambridge University Press.

—— (1978*b*). The cellular and tissue response of skin to single doses of ionising radiation. In *Current topics in radiation research* (ed. M. Ebert and A. Howard) pp. 1–59. North Holland, Amsterdam.

—— (1980). Keratopoiesis in normal epidermis. In *The epidermis and disease* (ed. R. Marks and E. Christophers) pp. 171–91. MTP, Lancaster.

—— (1981). Cell replacement in epidermis (keratopoiesis) via discrete units of proliferation. *Int. Rev. Cytol.* **69**, 271–318.

—— Al-Barwari, S. E., Hume, W. J., and Searle, J. (1977). Circadian rhythms of presumptive stem cells in three different epithelia of the mouse. *Cell Tissue Kinet.* **10**, 557–68.

—— —— and Searle, J. (1978). Differential radiation response amongst proliferating epithelial cells. *Cell Tissue Kinet.* **11**, 149–62.

—— and Allen, T. D. (1975*a*). Control of epidermal proliferative units (EPUs). A hypothesis based on the arrangement of neighbouring differentiated cells. *Differentiation* **3**, 161–5.

—— —— (1975*b*). The fine structure and cell kinetics of mouse epidermis after wounding. *J. cell Sci.* **17**, 413–37.

—— —— (1976). A model implicating the Langerhans cell in keratinocyte proliferation control. *Differentiation* **5**, 43–7.

—— Chwalinski, S., Swindell, R., and Palmer, M. (1982*a*). The spatial organisation of the hierarchical proliferative cells of the crypts of the small intestine into clusters of synchronised cells. *Cell Tissue Kinet.* **15**, 351–69.

—— and Croxson, M. D. (1971). Fixative soluble tritium pool size after tritiated thymidine injection into mice with skin stimulated by plucking. *Br. J. Derm.* **85**, 566–76.

—— and Hendry, J. H. (1973). Clonogenic cells and stem cells in epidermis. *Int. J. radiat. Biol.* **24**, 537–40.

—— —— (1975). Differential regeneration of intestinal proliferative cells and cryptogenic cells after irradiation. *Int. J. radiat. Biol.* **27**, 413–24.

—— —— and Al-Barwari, S. E. (1983). A cellular analysis of radiation injury in epidermis. In *Cytotoxic injury to tissues; effects on cell lineages* (ed. C. S. Potten and J. H. Hendry). Churchill Livingstone, Edinburgh.

—— Hume, W. J., Reid, P., and Cairns, J. (1978). The segregation of DNA in epithelial stem cells. *Cell* **15**, 899–906.

—— Jessup, B. A., and Croxson, M. B. (1971*a*). Incorporation of tritiated thymidine into hair and skin follicles. 1. Oscillatory changes through the hair growth cycle. *Cell Tissue Kinet.* **4**, 241–52.

—— —— —— (1971*b*). Incorporation of tritiated thymidine into the skin and hair follicles. II. Daily fluctuations in $^3$HTdR and $^3$H UdR levels. *Cell Tissue Kinet.* **4**, 413–21.

—— Kovacs, L., and Hamilton, E. (1974). Continuous labelling studies on mouse skin and intestine. *Cell Tissue Kinet.* **7**, 271–83.

—— and Major, D. (1980). Repeated injection (continuous labelling) experiments in mouse epidermis. *J. theoret. Biol.* **82**, 465–72.

—— and Schaer, J. C. (1971). Labelling of murine mastocytoma cells *in vitro* with plasma from tritiated thymidine labelled animals. *J. cell Biol.* **51**, 862–7.

—— Schofield, R., and Lajtha, L. G. (1979). A comparison of cell replacement in bone marrow, and three regions of surface epithelium. *Biochim. biophys. Acta* **560**, 281–99.

—— Wichmann, H. E., Loeffler, M., Dobek, K., and Major, D. (1982*b*). Evidence for discrete cell kinetic subpopulations in mouse epidermis based on mathematical analysis. *Cell Tissue Kinet.* **15**, 305–21.

Potter, K., Chaloner-Larsson, G., and Yamazaki, H. (1974). Abnormally high rate of cyclic AMP excretion from an *Escherichia coli* mutant deficient in cyclic AMP receptor protein. *Biochem. Biophys. Res. Commun.* **57**, 379–85.

Pound, A. W. (1968). Carcinogenesis and cell proliferation. *N. Z. med. J.* Suppl. **67**, 88–99.

Pozharisski, K. M., Klimashevski, K., and Gushchin, V. A. (1980). Study of the kinetics of epithelial cell populations in normal tissues of the rats intestines and in carcinogenesis. I. A comparison of enterocyte population kinetics in different segments of the intestine and colon. *Exp. Path., Jena* **18**, 387–406.

Prescott, D. M. (1968). Regulation of cell reproduction. *Cancer Res.* **38**, 1815–20.

—— (1976). *Reproduction of eukaryotic cells.* Academic Press, New York.

Price, D. (1963). Comparative aspects of development and structure in the prostate. In *Biology of the prostate and related tissues.* Natn. Cancer Inst. Monogr. **12**, 1–27.

Pritchett, C. J., Senior, P. V., Sunter, J. P., Watson, A. J., Appleton, D., and Watson, R. G. (1982). Human colorectal tumours in short term organ culture: a stathmokinetic study. *Cell Tissue Kinet.* **15**, 555–68.

Prunieras, M. (1969). Interaction between keratinocytes and dendritic cells. *J. invest. Derm.* **52**, 1–8.

—— (1979). Recent advances in epidermal cell cultures. *Archs Derm. Res.* **264**, 243–7.

—— and Delescluse, C. (1979). Physiological regulation of epidermal growth. *Acta derm. vener.* Suppl. **87**, 55–8.

Prutkin, L. (1971). The effect of actinomycin D on the incorporation of labelled vitamin A acid in normal and tumour epithelia. *J. invest. Derm.* **57**, 323–9.

—— and Bogart, B. (1970). The uptake of labelled vitamin A acid in keratoacanthoma; an electronmicroscopic radioautographic study. *J. invest. Derm.* **57**, 323–9.

Puck, T. T., Sanders, P., and Petersen, D. (1964). Life cycle analysis of mammalian cells. II. Cells from the Chinese hamster ovary grown in suspension culture. *Biophys. J.* **4**, 441–53.

—— and Steffen, J. (1963). Life cycle analysis of mammalian cells. *Biophys. J.* **3**, 379–97.

Pullman, H. (1978). Pathogenesis of psoriasis. *Fortschr. Med.* **96**, 1335–8.

—— Lennartz, K. J., and Steigleder, G. K. (1974). *In vitro* examination of cell proliferation in normal and psoriatic epidermis with special regard to diurnal variations. *Arch. Derm. Forsch.* **250**, 177–84.

—— —— —— (1975). Kinetics of cell proliferation in normal epidermis before and after application of 10% vitamin A acid solution. *Archs Derm. Res.* **253**, 71–6.

## 520    References

—— —— —— (1977). Disturbance of DNA synthesis in early psoriasis. *Archs Derm. Res.* **258**, 211–18.

—— Steigleder, G. K., and Lennartz, K. J. (1980). Effects of selective UV photo-therapy (SUP) and local PUVA treatment on DNA synthesis in guinea pig skin. *Archs Derm. Res.* **267**, 37–45.

Quastler, H. (1963). The analysis of cell population kinetics. In *Cell proliferation* (ed. L. F. Lamerton and R. J. M. Fry) pp. 18–34. Blackwell, Oxford.

—— and Sherman, F. G. (1959). Cell population kinetics in intestinal epithelium in the mouse. *Expl Cell Res.* **17**, 420–38.

Quevodo, W. C. and Fleischmann, R. P. (1980). Developmental biology of mammalian melanocytes. *J. invest. Derm.* **75**, 116–70.

—— Szabo, G., Virk, J., and Senesi, S. J. (1965). Melanocyte populations in UV irradiated human skin. *J. invest. Derm.* **45**, 295–8.

Quirk, A. V., Pain, R. H., Davison, M., and Shuster, S. (1981). Partial purifi-cation of a low molecular weight inhibitor of epidermal DNA synthesis. *Cell Tissue Kinet.* **15**, 31–40.

Raab, W. P. (1980). Cyclic nucleotides and prostaglandins in psoriasis. *Int. J. Pharmacol. Ther. Toxicol.* **18**, 212–24.

Rafferty, N. S. and Gfeller, E. (1970). $^3$H thymidine-mediated and delayed nuclear labelling in frog lens epithelia. Evidence for reutilization of DNA precursors. *Expl Cell Res.* **59**, 249–58.

Raick, A. N. (1973). Ultrastructural, histological and biochemical alterations produced by 12-O-tetradecanoyl-phorbol-13-acetate on mouse epidermis and their relevance to skin tumour promotion. *Cancer Res.* **33**, 269–86.

—— (1974*a*). Cell proliferation and promoting action in skin carcinogenesis. *Cancer Res.* **34**, 920–6.

—— (1974*b*). Cell differentiation and tumour promoting action in skin carcino-genesis. *Cancer Res.* **34**, 2915–25.

—— Thumm, K., and Chivers, B. R. (1972). Early effects of 12-O-tetradecanoyl-phorbol-13-acetate on the incorporation of tritiated precursor into DNA and the thickness of the interfollicular epidermis and other relations to tumour promotion in mouse skin. *Cancer Res.* **32**, 1562–8.

Rajewsky, M. F. (1965). *In vitro* studies of cell proliferation in tumours. *Eur. J. Cancer* **1**, 281–7.

Raknurud, N. (1974). The ultrastructure of the intrafollicular epidermis of the hairless hr/hr mouse. II. *Virchows Arch. Cell Path.* **17**, 113–35.

—— (1977). The ultrastructure of the interfollicular epidermis of the hairless mouse. V. The cytoplasm of the horny cells. *Virchows Arch. Cell Path.* **24**, 179–96.

—— Hevig, T., and Iversen, O. H. (1971). The ultrastructure of the interfollicular epidermis of the hairless hr/hr mouse. *Virchows Arch. Cell Path.* **8**, 206–24.

Ralfs, I., Dawber, R., Ryan, T., Duffill, M., and Wright, N. A. (1981*a*). The kinetics of metaphase arrest in human psoriatic epidermis; an examination of optimal experimental conditions for determining the birth rate. *Br. J. Derm.* **104**, 231–42.

—— —— —— and Wright, N. A. (1981*b*). Pityriasis rubra pilaris: epidermal cell kinetics. *Br. J. Derm.* **104**, 249–53.

Randers Hansen, E. (1967). Mitotic activity and duration in tongue and gingival epithelium of mice. Effect of chalone. *Odont. Tidskr.* **75**, 480–7.

Rasmussen, H. (1970). Cell communication, calcium ions and cyclic adenosine monophosphate. *Science, NY* **170**, 404–12.

Rausch, E., Kaiserking, E., and Goss, M. (1977). Langerhans cells and interdigitating reticulum cells in the thymus-dependent region in human dermatopathic lymphadenitis. *Virchows Arch. Cell Path.* **25**, 327–43.

Rawles, M. E. (1947). Origin of pigment cells from neural crest in mouse embryo. *Physiol. Zool.* **20**, 248–66.

Reams, W. M. and Tompkins, S. P. (1973). A developmental study of murine epidermal Langerhans cells. *Devl Biol.* **31**, 114–23.

Reaven, E. P. and Cox, A. J. (1968). Behaviour of adult human skin in organ culture; effects of cellophane tape stripping, oxygen tension, pH and serum. *J. invest. Derm.* **50**, 118–28.

Reeve, D. R. (1977). The mitotic response of the stratified squamous epithelium at the edge of large perforations of the tympanic membrane in guinea pigs. *J. Anat.* **124**, 731–40.

Refsum, S. B. and Berdal, P. (1967). Cell loss in malignant tumours in man. *Eur. J. Cancer* **3**, 235–6.

Reiskin, A. B. and Berry, R. J. (1968). Cell proliferation and carcinogenesis in the hamster cheek pouch. *Cancer Res.* **28**, 898–905.

— and Mendelsohn, M. L. (1964). A comparison of the cell cycle in induced carcinoma and their normal counterpart. *Cancer Res.* **24**, 1131–6.

Rheinwald, J. G. and Green, H. (1977). Epidermal growth factor and the multiplication of cultured human epidermal keratinocytes. *Nature, Lond.* **265**, 421–4.

Riches, A. C., Gore, D., and Docherty, J. (1976). The use of $^{125}$Iodo-2′deoxyuridine for monitoring DNA synthesis in organ culture. *J. Anat.* **121**, 323–9.

— McQueen, L., Carr, H. M. H., and Thomas, D. B. (1979). The kinetic properties of murine foetal thymus. *J. Anat.* **128**, 442.

Rijke, R. P. C. (1980). Some speculations on control mechanisms of cell proliferation in intestinal epithelium. In *Cell proliferation in the gastrointestinal tract* (ed. D. R. Appleton, J. P. Sunter, and A. J. Watson). pp. 57–65. Pitman Medical, London.

Riley, P. A. (1966). Esterase in epidermal dendritic cells of the mouse. *Br. J. Derm.* **78**, 388–97.

— (1967). A model of the relationship between melanocytes and Langerhans cells. Melanocyte population dynamics. *Br. J. Derm.* **79**, 52–8.

— (1974). Melanin and melanocytes. In *The physiology and pathophysiology of the skin* (ed. A. Jarrett) pp. 1131–45. Academic Press, New York.

Roberts, D. (1980). The determination of regional and age variations in the rate of desquamation; a comparison of four techniques. *J. invest. Derm.* **74**, 13–16.

Roberts, K. B., Florey, H. W., and Joklik, W. K. (1952). The influence of cortisone on cell division. *Q. J. exp. Physiol.* **37**, 239–57.

Robinson, J. R. W., Dowling, H. R., and Reicken, E. O. (1982). *Mechanisms of intestinal adaptation*. MTP Press, Lancaster.

Robison, G. A., Butcher, R. W., and Sutherland, E. W. (1968). Cyclic AMP. *A. Rev. Biochem.* **37**, 149–74.

Rogers, A. W. (1973). *Techniques of autoradiography*. Elsevier, Amsterdam.

Rohrbach, R., Elgjo, K., Iversen, O. H., and Sandritter, W. (1972). Effects of methylcholanthrene on epidermal growth regulators. 1. Variations in the M factor. *Beitr. Path.* **147**, 21–32.

— Hecker, O., and Sandritter, W. (1968). Cytophotometrische messengen des DNS-Gelhaltes der epidermis nach unspezifischen, cocarcinogen und carcinogen Reizen. *Z. Krebsforsch.* **70**, 211–22.

## 522    References

— Iversen, O. H., Elgjo, K., Reide, U. N., and Sandritter, W. (1976*a*). Effects of croton oil on epidermal growth regulators (chalones). *Beitr. Path.* **159**, 143–56.

— — — and Sandritter, W. (1976*b*). Effects of methylcholanthrene on epidermal growth regulation. II. Variation in the S factor. *Beitr. Path.* **158**, 145–58.

— — Reide, U. N., and Sandritter, W. (1976*c*). Effects of cellophane tape stripping of mouse skin on epidermal growth regulators (chalones). *Beitr. Path.* **160**, 175–86.

— and Laerum, O. D. (1974). Variations of mitosis inhibiting chalone activity in epidermis and dermis after carcinogen treatment. *Cell Tissue Kinet.* **7**, 251–7.

Rolshoven, E. (1951). Ueber der Reifungsteilungen bei der Spermatogonese mit einer Kritik der bisherigender Zellteihergen. *Vert. Anat. Ges. Jena* **49**, 189–97.

Rose, S. P., Stahn, R., and Passovoy, D. S. (1976). Epidermal growth factor enhancement of skin tumour induction in mice. *Experientia* **32**, 913–15.

Rothberg, S. and Arp, B. C. (1973). Epidermal chalone and inhibition of DNA synthesis. *Natn. Cancer Inst. Monogr.* **38**, 93–8.

— — (1975). Inhibition of epidermal and diurnal DNA synthesis by mouse and cow-snout epidermal extracts. *J. invest. Derm.* **64**, 245–9.

— Grounse, R. G., and Lee, J. L. (1961). Glycine $C^{14}$ incorporation into the proteins of the normal stratum corneum and the abnormal stratum corneum of psoriasis. *J. invest. Derm.* **37**, 497–504.

— and Ekel, T. M. (1967). Mitotic activity in the chick embryo epidermis. *Nature, Lond.* **216**, 1352.

— — (1971). Variation in DNA synthesis with age in chick embryo skin. *Nature, Lond.* **229**, 341–2.

Rovee, D. T., Kurowsky, C. A., and Labun, J. (1972*a*). Local wound environment and epidermal healing. Mitotic response. *Archs Derm.* **106**, 330–4.

— — — and Downes, A. M. (1972*b*). Effect of local wound environment on epidermal healing. In *Epidermal wound healing* (ed. H. I. Maibach and A. Rovee) pp. 159–81. Year Book, New York.

Rovera, G. and Baserga, R. (1973). Effect of nutritional changes on chromatin template activity and non-histone chromosomal protein synthesis in WI-38 and 3T6 cells. *Expl Cell Res.* **78**, 118–26.

Rowden, G. (1977). Immunoelectron microscopic studies of surface receptors and antigens of human Langerhans cells. *Br. J. Derm.* **97**, 593–608.

— (1980). Expression of Ia antigens on Langerhans cells in mice, guinea pigs and man. *J. invest. Derm.* **75**, 22–31.

Rowe, L. and Dixon, W. J. (1972). Clustering and control of mitotic activity in human epidermis. *J. invest. Derm.* **58**, 16–23.

— — (1975). Mitosis in human epidermis, clustering of sister cells and findings suggestive of a spreading mitotic stimulus. *Oncology* **31** 133–8.

— — and Forsythe, A. (1978). Mitoses in normal and psoriatic epidermis. *Br. J. Derm.* **98**, 293–9.

Rubini, I. R. (1966). *In vitro* DNA labelling of bone marrow and leukaemic blood leukocytes with tritiated thymidine. II. $^3$H-thymidine biochemistry *in vitro. J. lab. clin. Med.* **68**, 566–74.

Rusin, L. J., Duell, E. A., and Vorhees, J. J. (1978). Papaverine and Ro25 1724 inhibit cyclic nucleotide phosphodiesterase activity and increase cyclic AMP levels in psoriatic epidermis *in vitro. J. invest. Derm.* **71**, 154–6.

Russell, D. H., Combest, W. L., Duell, E. A., Stawiski, M. A., Anderson, T. F., and Vorhees, J. J. (1978). Glucocorticoid inhibits elevated polyamine biosynthesis in psoriasis. *J. invest. Derm.* **71**, 177–81.

Ryan, T. J. (1966). The direction of growth of epithelium. *Br. J. Derm.* **78**, 403.

— (1970). Direction of epithelial growth. *Br. J. Derm.* **83**, 701–2.

Saihan, E. M., Albano, J. A., and Burton, J. L. (1980). The effect of steroids and dithranol therapy on cyclic nucleotides in psoriatic epidermis. *Br. J. Derm.* **102**, 565–9.

Sakuma, J. (1980). Cell kinetics of human squamous cell carcinoma in the oral cavity. *Bull. Tokyo Med. Dent. Univ.* **27**, 43–54.

Sander, G. and Pardee, A. B. (1972). Transport changes in synchronously growing CHO and L cells. *J. cell Physiol.* **80**, 267–72.

Sarkany, P. and Caron, G.A. (1965). Phytohaemagglutinin-induced mitotic stimulation of epithelial cells in organ culture of adult human skin. *Br. J. Derm.* **77**, 439–42.

Sasai, Y., Kamamura, K., and Namba, K. (1979). The separation of basal and differentiating cells from human epidermis for DNA cytofluorometry. *Histochemistry* **63**, 265–72.

Sassier, P. and Bergeron, M. (1980). Existence of an endogenous inhibitor of DNA synthesis in rabbit small intestine specifically effective on cell proliferation in adult mouse intestine. *Cell Tissue Kinet.* **13**, 251–61.

Sato, F., Muramatsu, S., Tsuchihashi, S., Shiragai, A., Hiraoka, T., Inada, J., Kawashima, K., Matsuzawa, H., Nakamura, W., Trucco, E., and Sacher, G. A. (1972). Radiation effects on cell populations in the intestinal epithelium of mice and its theory. *Cell Tissue Kinet.* **5**, 227–35.

Sato, T. and Kawada, A. (1972). Uptake of tritiated thymidine by epidermal melanocytes of hairless mice during ultraviolet irradiation. *J. invest. Derm.* **58**, 71–3.

Sauerborn, R., Balmain, A., Goerrtler, K., and Stohr, M. (1978). On the existence of arrested $G_2$ cells in mouse epidermis. *Cell Tissue Kinet.* **11**, 291–300.

Savage, C. R. and Cohen, S. (1972). Epidermal growth factor and a new derivative: rapid isolation procedures and histological and chemical characerisation. *J. biol. Chem.* **207**, 7609–11.

— (1973). Proliferation of corneal epithelium induced by epidermal growth factor. *Expl eye Res.* **15**, 361–6.

Sawicki, W. and Rowinski, J. (1968). Kinetics of proliferation and migration of epithelial cells in the guinea pig colon. *Expl Cell Res.* **50**, 93–103.

Sawyer, R. H. (1972). Avian scale development. II. A study of cell proliferation. *J. exp. Zool.* **181**, 385–408.

Schaier, J. C., Ramscier, L., and Schindler, R. (1971). Studies on the division cycle of mammalian cells. IV. Incorporation of labelled precursors with DNA of synchronously dividing cells in culture. *Expl Cell Res.* **65**, 17–22.

Schell, H., Hornstein, O. P., and Schwartz, W. (1980). Human epidermal cell proliferation with regard to circadian variation of plasma cortisol. *Dermatologica* **161**, 12–21.

— Rosenberger, H., Hornsteen, O. P., and Wawra, E. (1977). Autoradiographic *in vitro* studies on diurnal variation in human epidermal cell proliferation. *Archs derm. Res.* **257**, 267–72.

Schellander, F. (1969). Reaktion von epidermis und subepidermalen bindegewebe auf Hornschichabrisse. *Arch. klin. exp. Derm.* **234**, 158–69.

— and Marks, R. (1973). Thymidine uptake by high level epidermal cells. A potential error in studies of keratinocyte proliferation. *Acta derm. vener.* **53**, 31–4.

— and Wolff (1967). Zur autoradiographischen markierung von Langerhans

Zellen mit $^3$H thymidine. *Arch. klin. exp. Derm.* **230**, 140–5.

Scheving, L. E. (1959). Mitotic activity in human epidermis. *Anat. Rec.* **135**, 7–10.

— and Chiakulas, J. J. (1965). Twenty four hours periodicity in the uptake of tritiated thymidine and its relation to mitotic rate in urodele larval epidermis. *Expl Cell Res.* **39**, 161–8.

— and Pauly, J. E. (1960). Daily mitotic fluctuations in the epidermis of the rat and their relation to spontaneous activity and rectal temperature. *Acta anat., Basel* **43**, 337–43.

— — (1967). Circadian phase relationship of thymidine-$^3$H uptake, labelled nuclei, grain counts and cell division rate in corneal epithelium of rats. *J. cell Biol.* **32**, 677–86.

— — (1973). Cellular mechanisms involving biorhythms with emphasis on those rhythms associated with the S and M stages of the cell cycle. *Int. J. chron. Biol.* **1**, 269–86.

— Yeh, L. C., Tsai, L. H., and Scheving, L. E. (1979). Circadian phase dependent stimulated effects of epidermal growth factor on DNA synthesis in tongue, oesophagus and stomach of adult male mouse. *Endocrinology* **105**, 1475–80.

Schiffer, L. M., Markoe, A. M., and Nelson, S. R. (1976). Estimation of tumour growth fraction in murine tumours by the primer-available DNA-dependent DNA polymerase assay. *Cancer Res.* **36**, 2415–18.

Schilling, W. A., Magilen, G., and Leith, J. T. (1972). Response of human primary foreskin cells to a cell extract obtained from skin. *Experientia* **28**, 161–2.

Schmid, G. H. and Hornstein, O. P. (1974). Autoradiographische *in vitro* Untersuchungen uber die Einfluss der Incubationdauer auf die DNA synthese psoriatischer Epidermiszellen von Stanzbiopsien im Glutamin-frieem Nahrmedium. *Virchows Arch. Cell Path.* **16**, 165–76.

— Schell, H., Hornstein, O. P., Meisel, H. H., and Zeissler, H. J. (1974). Untersuchungen uber der DNS-Synthesephase in der Wangenschleim haut der mannlichen Ratte methoden. *Virchows Arch. Cell Path.* **17**, 177–84.

Schofield, R. (1978). The relationship between the spleen colony forming cell and the haemopoietic stem cell. *Blood Cells* **4**, 7–25.

Schotz, W. E. and Zelen, M. (1971). Effect of length sampling bias on labelled mitotic index waves. *J. theoret. Biol.* **32**, 383–404.

Schultze, B., Haak, V., Schmeer, A. C., and Maurer, W. (1972). Autoradiographic investigation of cell kinetics of crypt epithelia in the jejunum of the mouse. *Cell Tissue Kinet.* **5**, 131–45.

— Kellerer, A. M., and Maurer, W. (1979). Transit times through the cycle phases of jejunal crypt cells of the mouse. *Cell Tissue Kinet.* **12**, 344–59.

— Maurer, W., and Hagenbusch, H. (1976). A two emulsion autoradiographic technique and the discrimination of the three different types of labelling after double labelling with $^3$H and $^{14}$C thymidine. *Cell Tissue Kinet.* **9**, 245–55.

— and Oehlert, W. (1960). Autoradiographic investigation of incorporation of $^3$H-thymidine into cells of the rat and mouse. *Science, NY* **131**, 737–8.

Schweizer, J. and Marks, F. (1978). Lack of relationship between Langerhans cells, epidermal cell proliferation and epidermal $G_1$ chalone. *Experientia* **34**, 1235–6.

Seijii, M., Nokano, A., Akiba, H., and Kato, T. (1974). Inhibition of DNA and protein synthesis in melanomata by a melanoma extract. *J. invest. Derm.* **62**, 11–19.

Sengel, P. (1976). *Morphogenesis of skin*. Cambridge University Press.

Shackney, S. E. (1975*a*). The radioautographic transfer function and its implications for radioautographic methodology. *J. natn. Cancer Inst.* **55**, 811–20.

—— (1975*b*). On the discreteness of the phases of the cell cycle. *J. natn. Cancer Inst.* **55**, 827–9.

—— Bunn, P. A., and Ford, S. S. (1976). The effect of colcemid on mouse bone marrow. *Cell Tissue Kinet.* **9**, 363–72.

—— Ford, S. S., and Wittig, A. B. (1973). The effect of counting threshold and emulsion exposure duration on the percent labelled mitoses curve and their implications for cell cycle analysis. *Cancer Res.* **33**, 2726–31.

Shand, J. A. and de Sousa, M. (1974). The simultaneous detection of $^3$H and $^{14}$C labelled cells by double layer autoradiography. *J. molec. Meth.* **6**, 141–9.

Sharav, Y. and Massler, M. (1967). Age changes in oral epithelia. *Expl Cell Res.* **47**, 132–8.

Sheehy, P. F., Fried, J., and Winn, R. (1974). Cell cycle changes in ovarian cancer after arabinosylcytosine. *Cancer* **33**, 28–37.

Sherman, B. S. (1961). The effect of vitamin A on epithelial mitosis *in vivo* and *in vitro*. *J. invest. Derm.* **37**, 469–80.

Sherman, F. G., Quastler, H., and Wimber, D. R. (1961). Cell population kinetics in the ear epidermis of mice. *Expl Cell Res.* **25**, 114–25.

Shields, R. (1977). Transition probability and the origin of variation in the cell cycle. *Nature, Lond.* **267**, 704–5.

—— (1978). Further evidence for a random transition in the cell cycle. *Nature, Lond.* **273**, 755–8.

—— (1979). Transition probability and the division pattern of W1-38 cells. *Cell Biol. Int. Rep.* **3**, 659–62.

—— and Smith, J. A. (1977). Cells regulate their proliferation through alteration in transition probability. *J. cell Physiol.* **91**, 345–56.

Shirakawa, S., Luce, J. K., Tannock, I., and Frei, E. (1970). Cell proliferation in human melanoma. *J. clin. Invest.* **49**, 1188–99.

Shorter, R. G., Moertel, C. G., Titus, J. L., and Reitemeier, R. J. (1964). Cell kinetics in the jejunum and rectum of man. *Am. J. Digest. Dis.* **9**, 760–8.

Shukla, R. C. (1969*a*). Regeneration of the basal melanocyte in the skin of the black guinea pig. *Acta Anat.* **74**, 599–614.

—— (1969*b*). The origin of Langerhans cells from the melanocyte. *Acta anat.* **73**, 372–88.

—— (1977). Regeneration of the basal melanocytes of the human skin. *Acta anat.* **97**, 1–3.

—— and Das, M. (1976). Morphological, histochemical, autoradiographic and topographical evidences of the origin of the Langerhans cell from the melanocyte in the skin of the guinea pig. *Acta anat.* **96**, 610–29.

—— Rastogi, M., and Sharma, R. N. (1973). Autoradiographic and Feulgen reaction study of the regenerating basal melanocyte of the white guinea pig. *Acta anat.* **85**, 37–40.

Silberberg, I. (1973). Apposition of mononuclear cells to Langerhans cells in contact allergic reactions. *Acta derm. vener.* **53**, 1–12.

Sillstrom, J. (1963). Grain count corrections in autoradiography. *Int. J. appl. Radiat. Isot.* **14**, 113–18.

Silvers, W. K. (1957). A histological and experimental approach to determine the relationship between gold impregnated dendritic cells and melanocytes. *Am. J. Anat.* **100**, 278–82.

Simnett, J. D. (1968). The measurement of mitotic incidence and radioautographic labelling index from tissue sections: some mathematical considerations. *J. R. microsc. Soc.* **88**, 371–82.

— (1972). Reduplication of DNA content in nuclei from organ cultures of mouse lung. *Neoplasma* **19**, 11–18.

— and Fisher, J. M. (1973*a*). The relation between cell division rate and $^3$HTdR incorporation in organ culture. *J. cell Physiol.* **81**, 171–80.

— — (1973*b*). An assay system for humoral growth factors. *Natn. Cancer Inst. Monogr.* **38**, 19–23.

— — and Heppleston, A. G. (1969). Tissue-specific inhibition of lung alveolar mitosis in organ culture. *Nature, Lond.* **223**, 944–6.

— and Heppleston, A. G. (1966). Cell renewal in the mouse lung. The influence of sex, strain and age. *Lab. Invest.* **15**, 1793–801.

— — (1968). Rates of deoxyribonucleic acid synthesis in organ cultures of lung and prostate as measured by tritiated thymidine autoradiography. *Lab. Invest.* **19**, 333–8.

Simpson-Herren, L., Sanford, A. H., Holmquist, J. P., Springer, T. A., and Lloyd, H. H. (1976). Ambiguity of the thymidine index. *Cancer Res.* **36**, 4705–9.

Skjaeggstad, O. (1964). Cell population kinetics of the mouse ear epidermis following application of 3 methylcholanthrene. *Acta path. microbiol. scand.* **73**, 156–69.

Skougaard, M. (1965*a*). Turnover of the gingival epithelium in marmosets. *Acta odont. scand.* **23**, 623–39.

— (1965*b*). Duration of DNA synthesis in the gingival epithelial cells of marmosets. *Acta odont. scand.* **23**, 616–22.

— (1970). Cell renewal with special reference to the gingival epithelium. *Adv. oral Biol.* **4**, 261–72.

Slaga, T. J., Boroden, G. T., and Boutwell, R. K. (1975*a*). Acetic acid: a potent stimulator of epidermal macromolecular synthesis and hyperplasia, but with weak tumour promoting ability. *J. natn. Cancer Inst.* **55**, 933–87.

— Scribner, J. D., Thompson, S., and Viaje, A. (1976). Epidermal cell proliferation and promoting ability of phorbol ester. *J. natn. Cancer Inst.* **57**, 1145–9.

— Thompson, S., and Shuckler, E. A. (1974). Macromolecular synthesis following a single application of polycyclic hydrocarbons used as inhibitors of mouse skin tumorigenesis. *Cancer Res.* **34**, 771–7.

— — — (1975*b*). Prolonged inhibition of mouse epidermal DNA synthesis by dexamethasone. *J. natn. Cancer Inst.* **54**, 931–6.

Smart, I. H. M. (1970*a*). Variation in the plane of cell cleavage during the process of stratification in the mouse epidermis. *Br. J. Derm.* **82**, 276–87.

— (1970*b*). Changes in location and orientation of mitotic figures in mouse oesophageal epithelium during the development of stratification. *J. Anat.* **106**, 15–21.

Smets, L. A. (1969). Discrepancies between precursor uptake and DNA synthesis in mammalian cells. *J. cell Physiol.* **74**, 63–71.

Smith, C. L. and Dendy, P. P. (1962). Relation between mitotic index, duration of mitosis, generation time and fraction of dividing cells in a cell population. *Nature, Lond.* **193**, 555–6.

Smith, J. A. (1977). An application of the theory of transition probability on 'ageing' W1-38 cells; similar behaviour of clonogenic cells from early and late passage culture. *Cell Biol. Int. Rep.* **1**, 283–91.

— and Martin, L. (1973). Do cells cycle? *Proc. natn. Acad. Sci. USA* **70**, 1263–7.

— — (1974). Regulation of cell proliferation. In *Cell cycle controls* (ed. G. M. Padilla, I. L. Cameron, and A. Zimmerman). Academic Press, New York.

Smith, J. W., Townsend, D. E., and Sparks, R. S. (1971). Genetic variants of glucose-6-phosphate dehydrogenase in the study of carcinoma of the cervix. *Cancer* **28**, 259–67.

Smith, K. R. (1970). The ultrastructure of the human Haarscheibe and Merkel cell. *J. invest. Derm.* **54**, 150–9.

Smith, R. S., Thomas, D. B., and Riches, A. C. (1974). Cell production in tumour isografts measured using vincristine and colcemid. *Cell Tissue Kinet.* **7**, 529–36.

Snell, R. S. (1962). The effect of X-ray irradiation on melanocytes in the skin. *J. invest. Derm.* **40**, 233–41.

— (1963). A study of melanocytes and melanin in a healing deep wound. *J. Anat.* **97**, 243–52.

— and Bischitz, P. G. (1963). The melanocytes and melanin in human abdominal wall skin; a survey made at different ages in both sexes and during pregnancy. *J. Anat.* **97**, 361–76.

Solem, J. H. (1966). Plasma corticosteroids in mice with special regard to sexual difference in adrenocortical responsiveness to exogenous corticotrophin. *Scand. J. clin. lab. Invest.* **18**, Suppl. 93, 1–36.

Sondergaard, J. (1979). DNA in psoriatic epidermis. *Acta derm. vener.* Suppl. **87**, 59–63.

Song, C. W. and Tabachnick, J. (1968). Epidermal cell population kinetics in albino guinea pig skin. *Nature, Lond.* **217**, 650.

— — (1969). Cell population kinetics in radiation induced hyperplastic epidermis of guinea pig skin. *Int. J. radiat. Biol.* **15**, 171–4.

Spearman, R. F. (1977). Ultrastructure of the contrasting types of keratinisation seen in the tail epidermis of the laboratory mouse *Mus musculus*. *Archs Derm. Res.* **258**, 33–40.

Sporn, M. B. and Newton, D. L. (1979). Chemoprevention of cancer with retinoids. *Fedn. Proc. Fedn Am. Socs exp. Biol.* **38**, 2528–39.

Spriggs, A. I., Sowie, C. E., and Cowdell, R. H. (1971). Chromosomes of precancerous lesions of the cervix uteri. New data and a review. *Cancer* **27**, 1239–54.

Stankler, L. (1969). Blood and tissue factors influencing the Koebner reaction in psoriasis. *Br. J. Derm.* **81**, 207–12.

Stanley, M. A. and Parkinson, E. K. (1979). Growth requirements of cervical epithelial cells in culture. *Int. J. Cancer* **15**, 407–17.

Stawiski, M. A., Powell, J. A., Lang, P., Schark, A., Duell, E. A., and Vorhees, J. J. (1975). Papaverine: its effects on cyclic AMP *in vitro* and in psoriasis *in vivo*. *J. invest. Derm.* **64**, 124–7.

Steel, G. G. (1968). Cell loss from experimental tumours. *Cell Tissue Kinet.* **1**, 193–207.

— (1972). The cell cycle in tumours: an examination of data gained by the technique of labelled mitoses. *Cell Tissue Kinet.* **5**, 87–100.

— (1977). *Growth kinetics of tumours*. Oxford University Press.

— Adams, K., and Barratt, J. C. (1966). Analysis of the cell population kinetics of transplanted tumours of widely differing growth rate. *Br. J. Cancer* **20**, 784–92.

— and Hanes, S. (1971). The technique of labelled mitoses; analysis by automatic curve-fitting. *Cell Tissue Kinet.* **4**, 93–105.

Stefan Zil, P. (1972). Vitamin A acid effects on epidermal mitotic activity, thickness and cellularity in the hairless mouse. *J. invest. Derm.* **59**, 228–32.

**528    References**

Steigleder, G. K. and Pullman, H. (1979). Cell proliferation in psoriasis. *Acta derm. vener.* Suppl. **87**, 64–6.

Stein, G. and Baserga, R. (1972). Nuclear proteins and the cell cycle. *Adv. Cancer Res.* **15**, 287–330.

Steiner, D. L., Kipnis, D. M., and Utriger, R. (1969). Radioimmunoassay for the measurement of adenosine 3′,5′ cyclic phosphate. *Proc. natn. Acad. Sci. USA* **63**, 367–73.

Stern, I. B. (1974). The uptake of ³HTdR by human foetal epidermis. *J. invest. Derm.* **63**, 268–72.

— Dayton, L., and Denny, J. (1971). The uptake of tritiated thymidine by the dorsal epidermis of foetal and newborn rats. *Anat. Rec.* **170**, 225–35.

Stevens Hooper, C. (1961). Use of colchicine for the measurement of mitotic rate in the intestinal epithelium. *Am. J. Anat.* **108**, 231–45.

Stevens, L., Al-Mukhtar, M., and Wright, N. A. (1981). Pentagastrin is not trophic to the small intestine. *Gut* **22**, A.567.

Stingl, G., Wolff-Schreiner, E. C., Pickler, W. J., Gschnait, F., Knapp, W., and Wolff, K. (1977). Epidermal Langerhans cells bear Fc and C3 receptors. *Nature, Lond.* **218**, 245–6.

Storey, W. F. and Leblond, C. P. (1951). Measurement of the rate of proliferation of epidermis and associated structures. *Ann. NY Acad. Sci.* **53**, 537–45.

Stoughton, R. B. (1971). Effect of hydroxyurea and cytarabine on rodent epidermal cells. *Archs Derm.* **104**, 161–5.

— (1978). Crude coal tar plus ultraviolet light suppresses DNA synthesis in epidermis. *Archs Derm.* **114**, 43–5.

Sullivan, D. J. and Epstein, W. L. (1963). Mitotic activity of wounded human epidermis. *J. invest. Derm.* **41**, 39–43.

Sunkara, P. S. and Rao, P. N. (1981). Role of polyamines in the regulation of the cell cycle in normal and transformed mammalian cells. In *The transformed cell* (ed. I. L. Cameron and T. B. Pool) pp. 267–92. Academic Press, New York.

Sunter, J. P. (1981). Cell proliferation studies on normal, carcinogen-damaged and neoplastic intestinal epithelia. MD thesis, University of Newcastle upon Tyne.

— Appleton, D. R., Wright, N. A., and Watson, A. J. (1978*a*). Kinetics of changes in the crypts of the jejunal mucosa of dimethylhydrazine-treated rats. *Br. J. Cancer* **37**, 662–72.

— — — — (1978*b*). Pathological features of the colonic tumours induced in rats by the administration of 1.2 dimethylhydrazine. *Virchows Arch. Cell Path.* **29**, 211–23.

Susi, F. R. (1968). Studies of cellular renewal and protein synthesis in mouse oral mucosa utilising ³H thymidine and ³H cystine. *J. invest. Derm.* **51**, 403–7.

Szabo, G. (1967). The regional anatomy of the human integument with special reference to the distribution of hair follicles, sweat glands and melanocytes. *Phil. Trans. R. Soc. B* **252**, 467–72.

Tachibona, T. (1978). The Merkel cell in the labial ridge epidermis of Anuran tadpole. 1. Fine structure, distribution and cytochemical studies. *Anat. Rec.* **191**, 487–514.

Taguchi, Y. H. and Tabnachnik, J. (1973). Evaluation of section thickness of skin in determining the epidermal labelling index with tritiated thymidine. *Arch. Derm. Forsch.* **247**, 207–10.

— — and Manaka, K. (1975). Effect of needle puncture and intradermal fluid injection on epidermal cell kinetics of albino guinea pig skin. *Arch. Derm. Forsch.* **255**, 83–92.

Takahashi, M. (1968). Theoretical basis for cell cycle analysis. II. Further studies on labelled mitoses wave method. *J. theoret. Biol.* **18**, 195–209.

— Hogg, G. D., and Mendelsohn, M. L. (1971). The automatic analysis of FLM curves. *Cell Tissue Kinet.* **4**, 505–18.

Tamaki. K. and Katz, I. (1980). Ontogeny of Langerhans cells. *J. invest. Derm.* **75**, 12–13.

— Stingl, G., and Katz, S. I. (1980). The origin of Langerhans cells. *J. invest. Derm.* **74**, 309–31.

Tannock, I. F. (1967). A comparison of the relative efficiencies of various metaphase arrest agents. *Expl Cell. Res.* **47**, 345–56.

— (1968). The relation between cell proliferation and the vascular system in a transplanted mouse mammary tumour. *Br. J. Cancer* **22**, 258–73.

— (1970). Population kinetics of carcinoma cells, capillary endothelial cells, and fibroblasts in a transplanted mouse mammary tumour. *Cancer Res.* **30**, 2470–6.

Tarbutt, R. G. and Blackett, N. M. (1968). Cell population kinetics of the recognisable erythroid cells in the rat. *Cell Tissue Kinet.* **1**, 65–73.

Tarnowski, W. M. and Hashimoto, K. (1967). Langerhans cell granules in histiocytosis X. *Archs Derm.* **96**, 298–306.

Taylor, E. W. (1965). The mechanism of colchicine inhibition of mitosis. I. Kinetics of inhibition and the binding of $^3$H-colchicine. *J. cell. Biol.* **25**, 145–60.

Taylor, J. F., Iversen, O. H., and Bjerknes, R. (1977). Growth kinetics of Kaposi sarcoma. *Br. J. Cancer* **35**, 470–9.

Taylor, J. H., Woods, P. S., and Hughes, W. L. (1957). The organisation and duplication of chromosomes using tritium labelled thymidine. *Proc. natn. Acad. Sci. USA* **43**, 122–8.

Taylor, J. R. and Halprin, K. M. (1977). Methotrexate uptake by psoriatic epidermis. *Br. J. Derm.* **96**, 167–70.

— — and Levine, V. (1980). Inhibitors of epidermal cell DNA synthesis in surviving pig skin *in vitro*. *J. invest. Derm.* **74**, 125–30.

Taylor, M. (1977). Epidermal growth control mechanisms. PhD thesis, University of Newcastle upon Tyne.

Terasima, T., Takabe, K., and Watanabe, S. (1976). Effect of bleomycin on cell survival and some implications for tumour therapy. *Prog. Biochem. Pharmacol.* **11**, 68–77.

Terz, J. J., Crutchet, H. P., and Lawrence, W. (1971). Analysis of the cell kinetics of human solid tumours. *Cancer* **28**, 1100–10.

Testa, N. G. (1979). Erythroid progenitor cells: their relevance for the study of haematological disease. *Clin. Haematol.* **8**, 311–33.

Thilagarathram, C. N. and Main, J. H. P. (1972). Changes in cell cycle characteristics in hamster cheek pouch epithelium during treatment with DMBA. *J. oral Path.* **1**, 89–102.

Thorbecke, G. J., Silberberg-Sinakin, I., and Flotte, F. J. (1980). Langerhans cells as macrophages in skin and lymphoid organs. *J. invest. Derm.* **75**, 32–43.

Thornley, A. L. (1976). Chalone regulation of cell proliferation. *S. Afr. J. med. Sci.* **41**, 239–59.

— (1979). Quoted by Allen, J. C. (1979).

— and Laurence, E. B. (1975). Chalone regulation of the epidermal cell cycle. *Experientia* **31**, 1024.

— — (1976). The specificity of epidermal chalone action: the results of *in vivo* experimentation with two purified skin extracts. *Devl Biol.* **51**, 10–22.

— Spargo, D., and Laurence, E. B. (1977). A simple method for measuring DNA synthesis in epidermis and sebaceous glands and its application in chalone assays. *Br. J. Derm.* **97**, 11–23.

Thorud, E. and Clausen, O. P. F. (1978). Epidermal cell kinetics in hairless mice after Bleomycin. 1. Perturbations after different single doses. *Cancer Treat. Rep.* **62**, 351–6.

— — (1980). Epidermal cell kinetics after Bleomycin. II. Perturbations after multiple doses. *Cancer Treat. Rep.* **64**, 1039–45.

— — Bjerknes, R., and Aarnes, E. (1980). The stathmokinetic method *in vivo*. *Cell Tissue Kinet.* **13**, 625–34.

Thrasher, J. D. (1971). Age and the cell cycle of the mouse oesophageal epithelium. *Exp. Geront.* **6**, 19–24.

Thuringer, J. M. (1924). Regulation of stratified squamous epithelium. *Anat. Rec.* **25**, 28–31.

— (1928). Studies in cell division in the human epidermis. *Anat. Rec.* **40**, 1–13.

— and Katzberg, A. A. (1951). Aging of the human epidermis as reflected in dimensional changes and mitotic ratios. *J. Geront.* **6**, 158 Suppl. 3.

Till, J. E. and McCulloch, E. A. (1961). A direct measurement of the radiation sensitivity of normal mouse bone marrow. *Radiat. Res.* **14**, 213–22.

Toto, P. and Dhawan, S. A. (1966). Generation cycle of oral epithelium in 400 day old mice. *J. dent. Res.* **45**, 949.

Trott, J. R. and Gorenstein, S. C. (1963). Mitotic rates in the oral and gingival epithelium of the rat. *Archs. oral Biol.* **8**, 425–32.

Trucco, E. (1965). A note on Puck and Steffens equations for the life cycle analysis of mammalian cells. *Biophys. J.* **5**, 743–6.

— and Brockwell, P. J. (1968). Percentage labelled mitoses curves in exponentially growing cell populations. *J. theoret. Biol.* **20**, 321–7.

Tsuji, T., Sugai, T., and Saito, R. (1969). Ultrastructure of three types of epidermal dendritic cells in hairless mice. *J. invest. Derm.* **53**, 332–40.

Tutton, P. J. M. (1973*a*). Variation in crypt cell cycle time and mitotic time in the small intestine of the rat. *Virchows Arch. Cell Path.* **13**, 68–75.

— (1973*b*). Control of epithelial cell proliferation in the small intestinal crypt. *Cell Tissue Kinet.* **6**, 211–18.

— and Barkla, D. H. (1976). Cell proliferation in the descending colon of dimethylhydrazine-treated rats and in dimethylhydrazine-induced adenocarcinomata. *Virchows Arch. Cell Path.* **21**, 147–60.

Tvermyr, E. M. F. (1969). Circadian rhythms in epidermal mitotic activity. Diurnal variations of the mitotic index, the mitotic rate and mitotic duration. *Virchows Arch. Cell Path.* **2**, 318–25.

— (1972). Circadian rhythms in hairless mouse epidermal DNA synthesis as measured by double labelling with $^3$H thymidine. *Virchows Arch. Cell Path.* **11**, 43–56.

Valleron, A. J. and Frindel, E. (1973). Computer simulation of growing cell populations. *Cell Tissue Kinet.* **6**, 69–79.

Van Scott, E. J. (1972). Tissue compartments of the skin lesion of psoriasis. *J. invest. Derm.* **59**, 4–6.

Vaughan, F. L. and Bernstein, I. A. (1971). Studies of proliferative capabilities in isolated epidermal basal and differentiated cells. *J. invest. Derm.* **56**, 454–66.

— Mitra, R. S., and Hornstein, I. A. (1976). Synthesis of DNA in isolated nuclei from differentiated mammalian epidermal cells. *J. invest. Derm.* **66**, 355–9.

— and Trinkaus, J. P. (1966). Movements of epithelial sheets *in vitro*. *J. cell Sci.* **1**, 407–13.

Verhagen, A. M. W. (1960). Growth curves and their functional form. *Aust. J. Stat.* **2**, 122–7.

Verma, A. K. and Boutwell, R. K. (1977). Vitamin A acid (retinoic acid), a potent inhibitor of 12-O-tetradecanoyl phorbol-13-acetate-induced ornithine decarboxylase activity in mouse epidermis. *Cancer Res.* **37**, 2196–201.

— Dixon, K. E., Froscio, R. P., and Murray, A. W. (1976). Localisation of adenosine $3', 5'$ monophosphate in mouse epidermis by immunofluorescence. *J. invest. Derm.* **66**, 239–41.

— Lowe, N. J., and Boutwell, R. K. (1979). Induction of mouse ornithine decarboxylase activity and DNA synthesis by ultraviolet light. *Cancer Res.* **39**, 1035–40.

Viaje, A., Slaga, T. J., Wigler, M., and Weinstein, I. B. (1977). Effects of anti-inflammatory agents on mouse skin tumour promotion, epidermal DNA synthesis, phorbol ester-induced cellular proliferation and production of plasminogen activator. *Cancer Res.* **37**, 1530–6.

Viziam, C. B., Matoltsky, G., and Meseon, H. (1964). Epithelialisation of small wounds. *J. invest. Derm.* **43**, 499–507.

Vorhees, J. J. (1970). Chalone regulation of the epidermal cell cycle and psoriasis. 29th Annual Meeting American Academy of Dermatology.

— (1974). The epidermis and cyclic AMP. *Br. J. Derm.* **90**, 223–7.

— (1976). Cyclic AMP and cyclic GMP in epidermal physiology and patho-physiology. *Curr. Probl. Derm.* **6**, 107–53.

— (1977). Pathophysiology of psoriasis. *A. Rev. Med.* **28**, 467–73.

— (1978). Regulation of epidermal proliferation and differentiation in psoriasis. *Int. J. Derm.* **5**, 241–55.

— Chambers, D. A., and Duell, E. A. (1976). Molecular mechanisms in proliferative skin disease. *J. invest. Derm.* **67**, 442–50.

— and Duell, E. A. (1971). Psoriasis as a possible defect of the adenyl cyclase–cyclic AMP cascade. *Archs Derm.* **104**, 352.

— — Bass, L. J., and Harrell, E. R. (1973*a*). Role of cyclic AMP in the control of epidermal cell growth and differentiation. *Natn. Cancer Inst. Monog.* **38**, 47–59.

— — — Powell, J. A., and Harrell, E. R. (1972*a*). The cyclic AMP system in normal and psoriatic epidermis. *J. invest. Derm.* **89**, 115–20.

— — — — — (1972*b*). Decreased cyclic AMP in the epidermis of lesions of psoriasis. *Archs Derm.* **105**, 695.

— — and Kelsey, W. H. (1972*c*). Dibutyryl cyclic AMP inhibition of epidermal cell division. *Archs Derm.* **105**, 384–92.

— Kelsey, W., Stawiski, M., Smith, E., Duell, E. A., Haddox, M., and Goldberg, N. (1973*b*). Increased cyclic GMP and decreased cyclic AMP levels in the rapidly proliferating epithelium of psoriasis. In *The role of cyclic nucleotides in carcinogenesis* (ed. J. Schultz and H. G. Gratzner) pp. 325–73. Academic Press, New York.

Wadskov, S., Kassis, V., and Sondergaard, J. (1979). Cyclic AMP and psoriasis once more. *Acta derm. vener.* **59**, 525–7.

## 532    References

Walter, J. F., Stoughton, R. B., and de Quoy, P. (1978). Suppression of epidermal proliferation by ultraviolet light, coal tar and anthralin. *Br. J. Derm.* **99**, 89–96.

— Vorhees, W., Kelsey, W. H., and Duell, E. A. (1973). Psoralen plus black light inhibits epidermal DNA synthesis. *Archs Derm.* **107**, 861–5.

Warakulasuriya, K. A. A. S., MacDonald, D. G., and Rennie, J. S. (1975). Cell kinetics in normal human buccal epithelium. *J. dent. Res.* **54**, Special Issue A. 123–39.

Warwick, G. P. (1971). Effect of the cell cycle on carcinogenesis. *Fedn Proc. Fedn Am. Socs exp. Biol.* **30**, 1760–5.

Watson, J. V. (1977). The application of age distribution theory in the analysis of cytophotometric DNA histogram data. *Cell Tissue Kinet.* **10**, 157–68.

Weibel, E. R. (1963). *Morphometry of the human lung.* Springer, Berlin.

Weinstein, G. (1971). Biochemical and pathophysiological rationale for methotrexate in psoriasis. *Ann. NY Acad. Sci.* **186**, 452–66.

— (1975). On the cell cycle of psoriasis. *Br. J. Derm.* **92**, 229.

— Copron, A., and McCullough, J. L. (1983). Proliferative defects in psoriasis. In *Psoriasis: cell proliferation* (ed. N. A. Wright and R. S. Camplejohn). Churchill Livingstone, Edinburgh.

— and Frost, P. (1967). Kinetics of epidermal cell proliferation in psoriasis. *Clin. Res.* **15**, 256–67.

— — (1968). Abnormal cell proliferation in psoriasis. *J. invest. Derm.* **50**, 254–9.

— — (1969). Cell proliferation kinetics in benign and malignant skin disease in humans. *Natn. Cancer Inst. Monogr.* **30**, 225–46.

— — (1970). Cell proliferation in human basal cell carcinoma. *Cancer Res.* **30**, 724–8.

— — (1971). Methotrexate for psoriasis. *Archs Derm.* **103**, 33.

— and McCullough, J. L. (1973). Cytokinetics in diseases of epidermal hyperplasia. *A. Rev. Med.* **24**, 345–52.

— — (1975). Effects of methotrexate esters on normal and psoriatic skin. *Archs Derm.* **111**, 471–5.

— — (1976). Cytokinetics and chemotherapy for psoriasis. *J. invest. Derm.* **67**, 26–30.

— and Van Scott, E. J. (1965). Autoradiographic analysis of turnover times of normal and psoriatic epidermis. *J. invest. Derm.* **45**, 257–62.

— and Velasco, J. (1972). Selective action of methotrexate on psoriatic epidermal cells. *J. invest. Derm.* **59**, 122–8.

— — and Frost, P. B. (1971). Biochemical and pharmacological relationship of psoriasis and methotrexate. In *Psoriasis* (ed. G. Farber) pp. 393–407. Stanford University Press.

Weinstein, W. M. (1979). Epithelial cell renewal of the small intestinal mucosa. *Med. Clins. N. Am.* **58**, 1375–86.

— Tytgat, G. N., Chen, M., and Band, P. R. (1973). *In vivo* studies of cell proliferation and kinetics in the human jejunal mucosa. *Gastroenterology* **64**, A137/820.

Weir, K. A., Fukuyama, K., and Epstein, W. L. (1971). Nuclear changes during keratinisation of normal human epidermis. *J. ultrastruct. Res.* **37**, 138–45.

Weirich, E. G. (1978). Effect of topical salicylic acid on animal epidermopoiesis. *Dermatologia* **156**, 89–96.

Weiss, P. (1961). The biological foundations of wound repair. *Harvey Lect.* **55**, 13-42.

— and Kavanau, J. L. (1957). A model of growth and growth control in mathematical terms. *J. gen. Physiol.* **41**, 1-47.

Wessells, N. K. (1964*a*). Tissue interactions and cytodifferentiation. *J. exp. Zool.* **157**, 139-52.

— (1964*b*). Substrate and nutrient effects upon epidermal basal cell orientation and proliferation. *Proc. natn. Acad. Sci. USA* **52**, 252-9.

Wheldon, T. E. (1983). Mathematical models in experimental and clinical oncology. In *Mathematical methods in medicine* (ed. D. Ingram and R. Bloch). Wiley, Chichester.

White, J. E. (1971). The effect of methotrexate on newborn rat skin. *J. invest. Derm.* **57**, 163-5.

Whiteley, H. J. and Horton, D. L. (1963). The effect of age on the mitotic activity of the ear epithelium in the CBA mouse. *J. Geront.* **18**, 335-9.

Whitfield, J. F., Boynton, A. L., MacManus, J. P., Rixon, R. H., Sikorska, H., Tsang, B., and Walker, P. R. (1980). The roles of calcium and cyclic AMP in cell proliferation. *Ann. NY Acad. Sci.* **339**, 216-40.

Whittal, P., Appleton, D. R., and Wright, N. A. (1975). The robustness of standard techniques in cell population kinetics. In *Mathematical models in cell kinetics* (ed. A. J. Valleron) pp. 49-53. European Scientific Press, Brussels.

Whitton, J. D. and Everall, J. D. (1973). The thickness of the epidermis. *Br. J. Derm.* **89**, 467-76.

Wiley, C. L., Williams, W. W., and McDonald, C. J. (1973). The effects of the epidermal chalone on DNA synthesis in mammalian epidermal cells. *J. invest. Derm.* **60**, 160-5.

Wiley, H. E. and Weinstein, G. D. (1979). Abnormal proliferation of uninvolved psoriatic epidermis. Differential induction by saline, propanolol and tape stripping *in vivo*. *J. invest. Derm.* **73**, 545-7.

Wilkinson, D. I. (1978). Effect of vitamin A on the growth of keratinocytes in culture. *Archs Derm. Res.* **263**, 75-81.

Williams, J. P. G. (1973). The embryospecific action of stathmokinetic agents. In *The cell cycle in development and differentiation* (ed. M. Balls and F. S. Billett) p. 331. Cambridge University Press.

Williams, M. G. and Hunter, R. (1957). Studies on epidermal regeneration by means of the strip method. *J. invest. Derm.* **29**, 407-12.

Williams, R. P. (1979). Effects of intestinally absorbed thymidine on tritiated thymidine utilisation. *Cell Tissue Kinet.* **12**, 405-10.

Williams, T. and Bjerknes, R. (1972). Stochastic model for abnormal clone spread through epithelial basal layers. *Nature, Lond.* **236**, 19-21.

Williamson, R. C. N., Bauer, F. L. R., Ross, J. J., and Malt, R. A. (1978*a*). Contribution of bile and pancreatic juice to cell proliferation in ileal mucosa. *Surgery* **83**, 570-6.

— Buchholtz, T. N., and Malt, R. A. (1978*b*). Humoral stimulation of cell proliferation in small bowel after transection and resection. *Gastroenterology* **75**, 249-54.

Willingham, M. C., Johnson, G. S., and Pastan, I. (1972). Control of DNA synthesis and mitosis in 3T3 cells by cyclic AMP. *Biochem. Biophys. Res. Commun.* **48**, 743-8.

Willis, R. A. (1953). *The pathology of tumours*. Butterworths, London.

Wimber, D. R. and Lamerton, L. F. (1963). Cell population studies in the intestine of continuously irradiated rats. *Radiat. Res.* **18**, 137–49.

— and Quastler, H. (1963). A $^{14}$C- and $^{3}$H-thymidine double labelling technique in the study of cell proliferation in *Tradescantia* root tips. *Expl Cell Res.* **30**, 8–16.

Winkelmann, K. K. (1977). The Merkel cell system and a comparison between it and the neurosecretory or APUD cell system. *J. invest. Derm.* **69**, 41–6.

Winkelmann, R. and Breathnach, A. S. (1973). The Merkel cell. *J. invest. Derm.* **60**, 2–15.

Winstanley, E. W. (1975). The rate of mitotic division in regenerating epithelium of the dog. *Res. vet. Sci.* **18**, 144–8.

Winter, G. D. (1962). Formation of the scab and the rate of epithelialisation of superficial wounds in the skin of the young domestic pig. *Nature, Lond.* **193**, 293–5.

— (1972). Epidermal regeneration studied in the domestic pig. In *Epidermal wound healing* (ed. H. Maibach and D. Rovee) pp. 71–112. Year Book, New York.

Withers, H. R. (1966). The dose-response relationship for epithelial cells of skin. *Radiology* **86**, 1110–11.

— (1967a). Recovery and repopulation *in vivo* by mouse skin epithelial cells during fractionated irradiation. *Radiat. Res.* **32**, 227–34.

— (1967b). The effect of oxygen and anaesthesia on radiosensitivity *in vivo* of epithelial cells of mouse skin. *Br. J. Radiol.* **40**, 335–44.

— (1967c). The dose-survival relationship for irradiation of epithelial cells of mouse skin. *Br. J. Radiol.* **40**, 187–95.

— and Elkind, M. M. (1969). Radiosensitivity and fractionation response of crypt cells in mouse jejunum. *Radiat. Res.* **38**, 598–613.

— — (1970). Microcolony survival assay for cells of mouse intestinal mucosa exposed to irradiation. *Int. J. radiat. Biol.* **17**, 261–7.

— Mason, K., and Reid, B. O. (1974). Response of mouse intestine to neutrons and gamma rays in relation to dose fractionation and division cycle. *Cancer* **34**, 39–47.

Wolff, K. (1967). Die Langerhans Zelle. Ergebrisse neuerer experimenteller Untersuchungen. *Arch. klin. exp. Derm.* **229**, 54–101.

— (1972). The Langerhans cell. *Curr. Prob. Derm.* **4**, 79–145.

— and Winkelmann, R. K. (1967a). Quantitative studies on the Langerhans cell population of guinea pig epidermis. *J. invest. Derm.* **48**, 504–13.

— — (1967b). The influence of Langerhans cell population and its hydrolytic enzymes in guinea pigs. *J. invest. Derm.* **48**, 531–9.

— — (1967c). Ultrastructural localisation of nucleoside triphosphatase in Langerhans cells. *J. invest. Derm.* **48**, 50–4.

Wolfsberg, M. F. (1964). Cell population kinetics in the epithelium of the forestomach of the mouse. *Expl Cell Res.* **35**, 119–31.

Wong, Y. C. and Buck, R. C. (1971). Langerhans cells in epidermoid metaplasia. *J. invest. Derm.* **56**, 10.

Wright, M. L. (1973). DNA synthesis during differentiation of tadpole shank epidermis. *J. exp. Zool.* **186**, 237–56.

Wright, N. A. (1971a). Cell proliferation in the prepubertal male rat adrenal cortex: an autoradiographic study. *J. Endocr.* **49**, 599–609.

— (1971b). Variation in tritiated thymidine uptake during DNA synthesis in the adrenal cortex. *Histochemie* **28**, 99–102.

— (1973). Studies in cell population kinetics in the small intestinal mucosa: an experimental and clinical study. MD thesis, University of Newcastle upon Tyne.

— (1976). A new look at epidermal cell kinetics in psoriasis and other dermatoses. *Clin. exp. Derm.* **1**, 275–8.

— (1977). Cell population kinetics in human epidermis. *Int. J. Derm.* **16**, 449–63;

— (1978). The cell population kinetics of repopulating cells in the intestine. In *Stem cells and tissue homeostasis* (ed. B. I. Lord, C. S. Potten, and R. J. Cole) pp. 335–58. Cambridge University Press.

— (1980*a*). The kinetics of human epidermal cell populations in health and disease. In *Recent advances in dermatology* (ed. A. Rook and J. Hunter) pp. 317–43. Churchill Livingstone, Edinburgh.

— (1980*b*). Cell proliferation in normal experimental animals: implications for proliferative responses. In *Cell proliferation in the gastrointestinal tract* (ed. D. R. Appleton, J. P. Sunter, and A. J. Watson) pp. 3–21. Pitman Medical, London.

— (1980*c*). A methodological approach to epidermopoiesis. In *Epidermis in disease* (ed. R. Marks and E. Christophers) pp. 139–53. MTP Press, Lancaster.

— (1981). The tissue kinetics of cell loss. In *Cell death in biology and pathology* (ed. I. D. Bowen and R. A. Lockshin) pp. 171–209. Chapman and Hall, London.

— Al-Dewachi, H. S., Appleton, D. R., and Watson, A. J. (1975*a*). Cell population kinetics in the rat jejunal crypt. *Cell Tissue Kinet.* **8**, 361–8.

— and Al-Nafussi, A. (1982). Kinetics of villous cell populations in the mouse small intestine. II. Negative feedback control after death of proliferative cells. *Cell Tissue Kinet.* **15**, 610–22.

— and Appleton, D. R. (1980). The metaphase arrest method—a critical review. *Cell Tissue Kinet.* **13**, 643–58.

— Britton, D. R., Bone, G., and Appleton, D. R. (1977). Cell proliferation in gastric carcinoma; a stathmokinetic study. *Cell Tissue Kinet.* **10**, 429–38.

— and Camplejohn, R. S. (1983). *Psoriasis: cell proliferation*. Churchill Livingstone, Edinburgh.

— and Irwin, M. (1982). The kinetics of villous cell populations in the mouse. I. The steady state requirement. *Cell Tissue Kinet.* **15**, 595–609.

— Morley, A. R., and Appleton, D. R. (1972*a*). The action of testosterone on cell proliferation and differentiation in the small bowel. *J. Endocr.* **52**, 161–75.

— — — (1972*b*). Variation in the duration of mitosis in the crypts of Lieberkuhn of the rat; a cytokinetic study using vincristine. *Cell Tissue Kinet.* **5**, 351–64.

— and Salter, D. (1979). Human epidermal stimulants, normal and psoriatic. *Clin. exp. Derm.* **4**, 545–7.

— and Voncina, D. (1977). Studies on the postnatal growth of the rat adrenal cortex. *J. Anat.* **123**, 147–56.

— — and Morley, A. R. (1973*a*). An attempt to demonstrate cell migration from the zona glomerulosa in the prepubertal male rat adrenal cortex. *J. Endocr.* **59**, 451–9.

— Watson, A., Morley, A., Appleton, D., and Marks, J. (1973*b*). Cell kinetics in flat avillous mucosa of the human small intestine. *Gut* **14**, 701–10.

— — — — — and Douglas, A. (1973*c*). The cell cycle time in the flat (avillous) mucosa of the human small intestine. *Gut* **14**, 603–6.

Wright, R. A., Mandy, S. H., Halprin, K. M., and Hsia, S. L. (1973). Defects and deficiency of adenyl cyclase in psoriatic skin. *Archs. Derm.* **107**, 47–53.

Wyllie, A. H. (1981). Cell death: a new classification separating apoptosis from necrosis. In *Cell death in biology and pathology* (ed. I. D. Bowen and R. A. Lockshin) pp. 9–34. Chapman and Hall, London.

Yamaguchi, Y., Hirobe, T., Kinyo, Y., and Manaka, K. (1974). Effect of chalone on the cell cycle in the epidermis during wound healing. *Expl Cell Res.* **89**, 245–54.

— and Tabachnik, J. (1972). Cell kinetics of epidermal repopulation and persistent hyperplasia in locally irradiated guinea pig skin. *Radiat. Res.* **50**, 158–80.

Yassin, T. M. and Toner, P. G. (1977). Langerhans cells in the human oesophagus. *J. Anat.* **122**, 435–45.

Yoshikawa, K., Adachi, K., Halprin, K. M., and Levine, V. (1975*a*). Is the cyclic AMP in psoriatic epidermis low? *Br. J. Derm.* **93**. 253–8.

— — — — (1975*b*). On the lack of response to catecholamine stimulation by the adenyl cyclase system in psoriatic lesions. *Br. J. Derm.* **92**, 619–34.

— — — — (1975*c*). The effects of catecholamine and related compounds on the adenyl cyclase system in the epidermis. *Br. J. Derm.* **93**, 29–36.

— Mori, N., Hadame, K., and Sakakibara, S. (1980). Differences in response of psoriatic epidermis in cyclic AMP accumulation against certain adenyl cyclase agonists. *Acta derm. vener.* **60**, 95–8.

Youdale, T. and MacManus, J. P. (1975). Failure of tritiated thymidine incorporation into DNA to reflect the autoradiographically demonstrable calcium-induced increase in thymic lymphoblast DNA synthesis. *J. cell Physiol.* **86**, 495–502.

Young, J. M., Lawrence, H. S., and Cordell, S. L. (1975). *In vitro* epidermal cell proliferation in rat skin plugs. *J. invest. Derm.* **64**, 23–9.

Young, R. C. and de Vita, V. T. (1970). Cell cycle characteristics of human solid tumours *in vivo*. *Cell Tissue Kinet.* **3**, 285–90.

Yuspa, S. H., Ben, T., Patterson, E., Michaek, K., Elgjo, K., and Hennings, H. (1976). Stimulated DNA synthesis in mouse epidermal cell cultures treated with 12-O-tetradecanoyl-phorbol-13-acetate. *Cancer Res.* **36**, 4062–8.

— Elgjo, K., and Morse, M. A. (1977). Retinyl acetate modulation of cell growth kinetics and carcinogen–cellular interaction in mouse epidermal cell cultures. *Chem. Biol. interact.* **16**, 251–64.

— Slaga, T. J., Colburn, N., Kelsey, W., Lichti, U., Ben, T., Pattern, E., and Hunng, H. (1976). Phorbol esters stimulate DNA synthesis and ornithine decarboxylase activity in mouse epidermal cell cultures. *Nature, Lond.* **262**, 402–4.

Zelickson, A. S. and Mottaz, J. H. (1968). Epidermal dendritic cells: a quantitative study. *Archs Derm.* **98**, 652–9.

— — and Weiss, L. W. (1975). Effects of topical fluorouracil on normal skin. *Archs Derm.* **111**, 1301–6.

Ziboh, V. A. and Hsia, S. L. (1972). Effects of prostaglandins E2 on rat skin: inhibition of sterol ester biosynthesis and clearing of scaly lesions in essential fatty acid deficiency. *J. lipid Res.* **13**, 458–67.

# Index



*Volume 2 begins on page 539*

*Volume 2 begins on page 539*

*Volume 2 begins on page 539*

*Volume 2 begins on page 539*

*Volume 2 begins on page 539*

*Volume 2 begins on page 539*

*Volume 2 begins on page 539*

*Volume 2 begins on page 539*

*Volume 2 begins on page 539*

*Volume 2 begins on page 539*

# Combined index for volumes 1 and 2



*Volume 2 begins on page 539*

*Volume 2 begins on page 539*

# Combined index for volumes 1 and 2



*Volume 2 begins on page 539*

*Volume 2 begins on page 539*